Why Do You Need This New Edition?

The world of technical communication is constantly evolving. I you're wondering why you should buy this new edition of *Technical Communication Today*, here are a few great reasons:

1. **A new chapter on Using Web 2.0 in the Technical Workplace** discusses how job seekers can use social networking sites to make and advance professional connections as well as how employees can use and contribute to a workplace's official social networking outlets (Ch. 23).
2. **hapters on E-mail, Letters, and Memos have been combined** to better reflect how e-mail has become more common in ie workplace and how letters and memos are increasingly delivered within e-mail messages (Ch. 5).
3. **verage of doing research, evaluating sources, and writing with sources has been expanded** into two chapters to better lp you manage and work effectively with information (Chs. 14, 15).

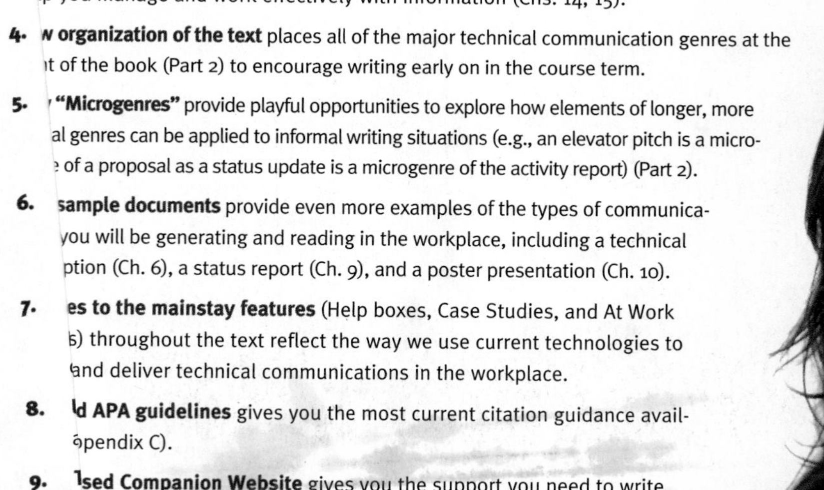

4. **w organization of the text** places all of the major technical communication genres at the it of the book (Part 2) to encourage writing early on in the course term.
5. **"Microgenres"** provide playful opportunities to explore how elements of longer, more al genres can be applied to informal writing situations (e.g., an elevator pitch is a micro- of a proposal as a status update is a microgenre of the activity report) (Part 2).
6. **sample documents** provide even more examples of the types of communica- you will be generating and reading in the workplace, including a technical ption (Ch. 6), a status report (Ch. 9), and a poster presentation (Ch. 10).
7. **es to the mainstay features** (Help boxes, Case Studies, and At Work s) throughout the text reflect the way we use current technologies to and deliver technical communications in the workplace.
8. **d APA guidelines** gives you the most current citation guidance avail- ppendix C).
9. **sed Companion Website** gives you the support you need to write ge of workplace situations and audiences and includes annotated ocuments, case studies, chapter quizzes, and flashcards.
10. **Th edition is available as a CourseSmart** ed you can subscribe to *Technical Ccation Today* at CourseSmart.com. The sies all of the book's content in a format thes you to search, bookmark, save n print.

RSON

Technical Communication Today

FOURTH EDITION

Richard Johnson-Sheehan
Purdue University

PEARSON

Boston Columbus Indianapolis New York San Francisco Upper Saddle River
Amsterdam Cape Town Dubai London Madrid Milan Munich Paris Montreal Toronto
Delhi Mexico City São Paulo Sydney Hong Kong Seoul Singapore Taipei Tokyo

To Tracey, Emily, and Collin

Senior Acquisitions Editor: Lauren A. Finn
Senior Development Editor: Anne Brunell Ehrenworth
Senior Supplements Editor: Donna Campion
Senior Media Producer: Stefanie Liebman
Executive Marketing Manager: Joyce Nilsen
Production Manager: Eric Jorgensen
Project Coordination, Text Design, and Electronic Page Makeup: Nesbitt Graphics, Inc.
Cover Design Manager: John Callahan
Cover Designer: Kay Petronio
Cover images (clockwise from top): © Monkey Business Images/Shutterstock, © Blazej Lyjak/Shutterstock, © Maksym Dykha/Fotolia, © Niki Crucillo/Shutterstock, © iconista/Shutterstock
Photo Researcher: Connie Gardner
Senior Manufacturing Buyer: Roy L. Pickering, Jr.
Printer and Binder: RR Donnelley & Sons Company / Crawfordsville
Cover Printer: Lehigh-Phoenix Color / Hagerstown

Credits and acknowledgments borrowed from other sources and reproduced, with permission, in this textbook appear on pp. C-1–C-2.

Library of Congress Cataloging-in-Publication Data
Johnson-Sheehan, Richard.
Technical communication today / Richard Johnson-Sheehan. -- 4th ed.
p. cm.
Includes bibliographical references and index.
ISBN-13: 978-0-205-17119-4
ISBN-10: 0-205-17119-2
1. Communication of technical information--Data processing. I. Title.
T10.5.J65 2011
601.4--dc23

2011021797

1 2 3 4 5 6 7 8 9 10—DOC—14 13 12 11

www.pearsonhighered.com

ISBN-13: 978-0-205-17119-4
ISBN-10: 0-205-17119-2

Contents

Preface *xx*

What's New in the Fourth Edition? *xx*
Guiding Themes *xxi*
Computers as Thinking Tools *xxi*
Genres as Pathways for Interpretation and Expression *xxii*
Visual-Spatial Reading, Thinking, and Composing *xxii*
The International, Cross-Cultural Workplace *xxiii*
The Activity of Technical Communication *xxiii*
Supplements to the Book *xxiii*
Acknowledgments *xxiv*

Part 1: Elements of Technical Communication

CHAPTER 1

Communicating in the Technical Workplace *1*

Developing a Workplace Writing Process *2*
Genres and the Technical Writing Process *4*
Stage 1: Planning and Researching *5*
Stage 2: Organizing and Drafting *7*
Stage 3: Improving the Style *9*
Stage 4: Designing *9*
Stage 5: Revising and Editing *9*
What Is Technical Communication? *10*
Technical Communication Is Interactive and Adaptable *11*
Technical Communication Is Reader Centered *12*
Technical Communication Relies on Teamwork *12*
Technical Communication Is Visual *13*
Technical Communication Has Ethical, Legal, and Political Dimensions *13*
Technical Communication Is International and Cross-Cultural *15*
How Important Is Technical Communication? *15*
Chapter Review *16*
Exercises and Projects *17*

CHAPTER 2 Readers and Contexts of Use *19*

Profiling Your Readers *20*
- Identifying Your Readers *22*
- Profiling Your Readers' Needs, Values, and Attitudes *23*

Profiling Contexts of Use *26*
- Identifying the Context of Use *26*

Using Profiles to Your Own Advantage *27*

International and Cross-Cultural Communication *28*
- Differences in Content *32*
- Differences in Organization *33*
- Differences in Style *34*
- Differences in Design *35*
- Listen and Learn: The Key to International and Cross-Cultural Communication *36*

Chapter Review *38*

Exercises and Projects *38*

Case Study: Installing a Medical Waste Incinerator *40*

CHAPTER 3 Working in Teams *44*

The Stages of Teaming *45*

Forming: Strategic Planning *45*
- Step 1: Define the Project Mission and Objectives *46*
- Step 2: Identify Project Outcomes *47*
- Step 3: Define Team Member Responsibilities *47*
- Step 4: Create a Project Calendar *48*
- Step 5: Write Out a Work Plan *48*
- Step 6: Agree on How Conflicts Will Be Resolved *49*

Storming: Managing Conflict *54*
- Running Effective Meetings *54*
- Mediating Conflicts *56*
- Firing a Team Member *58*

Norming: Determining Team Roles *58*
- Revising Objectives and Outcomes *58*
- **Help: Virtual Teaming** *59*
- Identifying Team Roles *61*
- Using Groupware to Facilitate Work *62*

Performing: Improving Quality *63*

The Keys to Teaming *63*

Chapter Review *65*

Exercises and Projects 65
Case Study: Not a Sunny Day 67

CHAPTER 4 Ethics in the Technical Workplace 68
What Are Ethics? 69
Where Do Ethics Come From? 72
Personal Ethics 73
Social Ethics 73
Conservation Ethics 76
Resolving Ethical Dilemmas 77
Help: Stopping Cyberbullying and Computer Harassment 78
Confronting an Ethical Dilemma 80
Resolving an Ethical Dilemma 81
When You Disagree with the Company 82
Ethics in the Technical Workplace 85
Copyright Law 85
Trademarks 85
Patents 86
Privacy 87
Information Sharing 87
Proprietary Information 87
Libel and Slander 88
Fraud 88
Copyright Law in Technical Communication 88
Asking Permission 89
Copyrighting Your Work 90
Plagiarism 90
Chapter Review 90
Exercises and Projects 91
Case Study: This Company Is Bugging Me 92

Part 2: Genres of Technical Communication

CHAPTER 5 Letters, Memos, and E-Mail 93
Features of Letters, Memos, and E-Mails 94
Planning and Researching 100
Determining the Rhetorical Situation 100

Organizing and Drafting 102
Introduction with a Purpose and a Main Point 102
Body That Provides Need-to-Know Information 105
Conclusion That Restates the Main Point 106
Types of Letters, Memos, and E-Mails 106
Inquiries 106
Responses 107
Transmittal Letters and Memos 107
Claims or Complaints 111
Adjustments 111
Refusals 114
Using Style and Design 114
Strategies for Developing an Appropriate Style 116
Designing and Formatting Letters and Memos 118
Formatting Letters 118
Formatting Envelopes 120
Formatting Memos 121
Using E-Mail Internationally 121
Microgenre: Texting at Work 124
Chapter Review 126
Exercises and Projects 126
Case Study: The Nastygram 130
CHAPTER 6 Technical Descriptions and Specifications 132
Planning and Researching 133
Planning 133
Addressing ISO 9000/14000 Issues 139
Researching 140
Partitioning the Subject 140
Organizing and Drafting 145
Specific and Precise Title 145
Introduction with an Overall Description 145
Description by Features, Functions, or Stages 146
Description by Senses, Similes, Analogies, and Metaphors 147
Conclusion 149
Help: Using Digital Photography in Descriptions 149
Using Style and Design 151
Plain, Simple Style 151
Page Layout That Fits 151
Graphics That Illustrate 151

Microgenre: Technical Definitions 155
Chapter Review 159
Exercises and Projects 159
Case Study: In the Vapor 163

CHAPTER 7 **Instructions and Documentation** 165

Planning and Researching 166
- Planning 168
- Researching 177

Planning for Cross-Cultural Readers and Contexts 179
- Verbal Considerations 180
- Design Considerations 180

Organizing and Drafting 181
- Specific and Precise Title 181
- Introduction 181
- List of Parts, Tools, and Conditions Required 182
- Sequentially Ordered Steps 183
- Safety Information 189
- Conclusion That Signals Completion of Task 192

Using Style and Design 192
- **Help: On-Screen Documentation** 194
- Plain Style with a Touch of Emotion 195
- Functional, Attractive Page Layout 196
- Graphics That Reinforce Written Text 197
- User-Testing Your Documentation 199

Microgenre: Emergency Instructions 199
Chapter Review 201
Exercises and Projects 202
Case Study: The Flame 204

CHAPTER 8 **Proposals** 205

Planning and Researching 206
- Planning 206
- Researching 212

Organizing and Drafting 213
- Writing the Introduction 214
- Describing the Current Situation 214
- Describing the Project Plan 218
- Describing Qualifications 219
- Concluding with Costs and Benefits 226

Using Style and Design 226
A Balance of Plain and Persuasive Styles 229
An Attractive, Functional Design 229
Microgenre: The Elevator Pitch 232
Chapter Review 234
Exercises and Projects 235
Case Study: The Mole 245

CHAPTER 9 Activity Reports 246
Types of Activity Reports 247
Progress Reports 247
Briefings and White Papers 247
Incident Reports 250
Laboratory Reports 250
Planning and Researching 254
Analyzing the Rhetorical Situation 258
Organizing and Drafting 260
Writing the Introduction 260
Writing the Body 260
Writing the Conclusion 261
Using Style and Design 263
Using a Plain Style 263
Using Design and Graphics 263
Microgenre: The Status Report 264
Chapter Review 266
Exercises and Projects 266
Case Study: Bad Chemistry 268

CHAPTER 10 Analytical Reports 269
Types of Analytical Reports 270
Planning and Researching 272
Planning 272
Researching 277
Organizing and Drafting 282
Writing the Introduction 282
Describing Your Methodology 284
Summarizing the Results of the Study 284
Discussing Your Results 285

Stating Your Overall Conclusions and Recommendations 285
Help: Using Google Docs to Collaborate with International Teams 298
Drafting Front Matter and Back Matter 299
Developing Front Matter 299
Developing Back Matter 303
Using Style and Design 303
Using Plain Style in a Persuasive Way 303
A Straightforward Design 304
Microgenre: The Poster Presentation 309
Chapter Review 310
Exercises and Projects 311
Case Study: The X-File 313

CHAPTER 11 Starting Your Career 314
Setting Goals, Making a Plan 315
Setting Goals 315
Using a Variety of Job-Seeking Paths 315
Preparing a Résumé 320
Types of Résumés 320
Chronological Résumé 320
Functional Résumé 329
Designing the Résumé 329
Writing Effective Application Letters 331
Content and Organization 331
Help: Designing a Scannable/Searchable Résumé 333
Style 335
Revising and Proofreading the Résumé and Letter 339
Creating a Professional Portfolio 339
Collecting Materials 340
Organizing Your Portfolio 341
Assembling the Portfolio in a Binder 341
Creating an Electronic Portfolio 342
Interviewing Strategies 343
Preparing for the Interview 343
At the Interview 344
Writing Thank You Letters and/or E-Mails 345
Microgenre: The Bio 347
Chapter Review 348

Exercises and Projects 349
Case Study: The Lie 351

Part 3: Planning and Doing Research

CHAPTER 12 **Strategic Planning, Being Creative** ***352***

Using Strategic Planning 353
Step 1: Set Your Objectives 353
Step 2: Create a List of Tasks (or Task List) 354
Step 3: Set a Timeline 355
Help: Planning with Online Calendars 356
Generating New Ideas 357
Tips for Being More Creative 357
Inventing Ideas 358
Chapter Review 363
Exercises and Projects 363
Case Study: Getting Back to Crazy 365

CHAPTER 13 **Persuading Others** ***366***

Persuading with Reasoning 367
Reasoning with Logic 370
Reasoning with Examples and Evidence 371
Persuading with Values 372
Help: Persuading Readers Online 373
Appealing to Common Goals and Ideals 374
Framing Issues from the Readers' Perspective 377
Persuasion in High-Context Cultures 378
Chapter Review 382
Exercises and Projects 382
Case Study: Leapfrogging with Wireless 384

CHAPTER 14 **Researching and Research Methods** ***385***

Beginning Your Research 387
Defining Your Research Subject 388
Narrowing Your Research Subject 389
Formulating a Research Question or Hypothesis 389

Developing a Research Methodology 391
Mapping Out a Methodology 391
Describing Your Methodology 391
Using and Revising Your Methodology 392
Triangulating Materials 393
Using Electronic Sources 394
Using Print Sources 396
Using Empirical Sources 398
Chapter Review 404
Exercises and Projects 404
Case Study: The Life of a Dilemma 406

CHAPTER 15 Using Sources and Managing Information 407
Taking Useful Notes 408
Managing Information 409
Careful Note Taking 409
Documenting Sources 413
Help: Using a Citation Manager 416
Appraising Your Information 417
Is the Source Reliable? 417
How Biased Is the Source? 418
Am I Biased? 418
Is the Source Up to Date? 419
Can the Information Be Verified? 420
Avoiding Plagiarism 420
Chapter Review 421
Exercises and Projects 422
Case Study: The Patchwriter 423

Part 4: Drafting, Designing, and Revising

CHAPTER 16 Organizing and Drafting 424
Basic Organization for Any Document 425
Using Genres to Organize Information 425
Outlining the Document 428

Organizing and Drafting the Introduction 430
Six Opening Moves in an Introduction 431
Drafting with the Six Moves 432
Organizing and Drafting the Body 433
Carving the Body into Sections 433
Patterns of Arrangement 437
Organizing and Drafting the Conclusion 446
Five Closing Moves in a Conclusion 446
Organizing Cross-Cultural Documents 448
Indirect Approach Introductions 450
Indirect Approach Conclusions 450
Chapter Review 451
Exercises and Projects 451
Case Study: The Bad News 453

CHAPTER 17 Using Plain and Persuasive Style 454
What Is Style? 455
Writing Plain Sentences 455
Basic Parts of a Sentence 456
Eight Guidelines for Plain Sentences 456
Creating Plain Sentences with a Computer 460
Help: Intercultural Style and Translation Software 462
Writing Plain Paragraphs 464
The Elements of a Paragraph 464
Using the Four Types of Sentences in a Paragraph 466
Aligning Sentence Subjects in a Paragraph 467
The Given/New Method 469
When Is It Appropriate to Use Passive Voice? 470
Persuasive Style 472
Elevate the Tone 472
Use Similes and Analogies 473
Use Metaphors 474
Change the Pace 475
Balancing Plain and Persuasive Style 477
Chapter Review 478
Exercises and Projects 478
Case Study: Going Over the Top 480

CHAPTER 18

Designing Documents and Interfaces 481

Five Principles of Design 482
Design Principle 1: Balance 482
Weighting a Page or Screen 483
Using Grids to Balance a Page Layout 486
Using Other Balance Techniques 490
Design Principle 2: Alignment 492
Design Principle 3: Grouping 493
Using Headings 495
Using Borders and Rules 498
Design Principle 4: Consistency 500
Choosing Typefaces 500
Labeling Graphics 500
Creating Sequential and Nonsequential Lists 502
Inserting Headers and Footers 504
Design Principle 5: Contrast 505
Adding Shading and Background Color 505
Highlighting Text 506
Using Font Size and Line Length 507
Cross-Cultural Design 508
Using the Principles of Design 511
Analyze Your Readers and the Document's Context of Use 511
Use Thumbnails to Sketch Out the Design 511
Design the Document 512
Revise and Edit the Design 513
A Primer on Binding and Paper 513
Binding 513
Selecting the Paper 515
Chapter Review 516
Exercises and Projects 516
Case Study: Scorpions Invade 518

CHAPTER 19

Creating and Using Graphics 522

Guidelines for Using Graphics 523
Guideline One: A Graphic Should Tell a Simple Story 524
Guideline Two: A Graphic Should Reinforce the Written Text, Not Replace It 525
Guideline Three: A Graphic Should Be Ethical 525
Guideline Four: A Graphic Should Be Labeled and Placed Properly 527

Displaying Data with Graphs, Tables, and Charts 528
Line Graphs 529
Bar Charts 530
Tables 531
Pie Charts 534
Flowcharts 535
Gantt Charts 536
Using Pictures, Drawings, and Screen Shots 536
Photographs 537
Inserting Photographs and Other Images 538
Illustrations 539
Screen Shots 541
Using Cross-Cultural Symbols 542
Using Video and Audio 545
Video 545
Audio, Podcasting, and Music 546
Chapter Review 547
Exercises and Projects 547
Case Study: Looking Guilty 549

CHAPTER 20 Revising and Editing for Usability 551
Levels of Edit 552
Revising: Level 1 Editing 552
Substantive Editing: Level 2 Editing 554
Copyediting: Level 3 Editing 556
Proofreading: Level 4 Editing 558
Grammar 558
Punctuation 559
Spelling and Typos 560
Word Usage 561
Using Copyediting Symbols 562
Lost in Translation: Cross-Cultural Editing 563
Documenting Cycling and Usability Testing 566
Document Cycling 566
Usability Testing 566
Chapter Review 570
Exercises and Projects 571
Case Study: Wrong Version 572

Part 5: Connecting with Clients

CHAPTER 21 Preparing and Giving Presentations 573
Planning and Researching Your Presentation 574
Defining the Rhetorical Situation 576
Allotting Your Time 578
Choosing the Right Presentation Technology 579
Organizing the Content of Your Presentation 582
Building the Presentation 582
The Introduction: Tell Them What You're Going to Tell Them 583
Help: Giving Presentations with your iPod, MP3, or Mobile Phone 586
The Body: Tell Them 589
The Conclusion: Tell Them What You Told Them 591
Preparing to Answer Questions 593
Choosing Your Presentation Style 594
Creating Visuals 596
Designing Visual Aids 596
Using Graphics 598
Slides to Avoid 598
Delivering the Presentation 599
Body Language 599
Voice, Rhythm, and Tone 601
Using Your Notes 601
Rehearsing 602
Evaluating Your Performance 602
Working Cross-Culturally with Translators 604
Chapter Review 608
Exercises and Projects 608
Case Study: The Coward 610

CHAPTER 22 Designing Websites 611
Basic Features of a Website 612
Planning and Researching a Website 615
Subject 615
Purpose 616
Readers 616
Context of Use 617
Websites for International and Cross-Cultural Readers 618

Organizing and Drafting a Website 620
Organizing the Website 620
Creating Levels in the Website 621
Drafting the Home Page 621
Drafting Node Pages 623
Drafting Basic Pages 623
Drafting Navigational Pages 624
A Warning About Copyright and Plagiarism 625
Using Style in a Website 626
Designing the Website 626
Designing the Interface 626
Adding Images 628
Uploading, Testing, and Maintaining Your Website 630
Uploading the Website 630
Testing the Website 630
Maintaining the Website 631
Chapter Review 632
Exercises and Projects 632
Case Study: Slowed to a Crawl 634

CHAPTER 23 **Using Social Networking Tools (Web 2.0)** 635
Social Networking in the Workplace 636
Using a Personal Social Networking Site 636
Using a Corporate Social Networking Site 637
Blogs and Microblogs 638
Internet Videos and Podcasts 640
Wikis 642
Help: Using Skype for Collaboration 642
Chapter Review 644
Exercises and Projects 644
Case Study: My Boss Might Not "Like" This 646

Appendix A: **Grammar and Punctuation Guide** A-1
The Top Ten Grammar Mistakes A-1
Comma Splice A-1
Run-On Sentence A-2
Fragment A-3
Dangling Modifier A-3

Subject-Verb Disagreement *A-4*
Pronoun-Antecedent Disagreement *A-5*
Faulty Parallelism *A-5*
Pronoun Case Error (*I* and *Me*, *We* and *Us*) *A-6*
Shifted Tense *A-7*
Vague Pronoun *A-7*

Punctuation Refresher *A-8*
Period, Exclamation Point, Question Mark *A-9*
Commas *A-9*
Semicolon and Colon *A-11*
Apostrophe *A-13*
Quotation Marks *A-14*
Dashes and Hyphens *A-16*
Parentheses and Brackets *A-17*
Ellipses *A-18*

Appendix B: English as a Second Language Guide *A-19*

Using Articles Properly *A-19*
Putting Adjectives and Adverbs in the Correct Order *A-20*
Using Verb Tenses Appropriately *A-21*

Appendix C: Documentation Guide *A-24*

APA Documentation Style *A-25*
APA In-Text Citations *A-25*
The References List for APA Style *A-27*
Creating the APA References List *A-30*

CSE Documentation Style (Citation-Sequence) *A-31*
The References List for CSE Citation-Sequence Style *A-31*
Creating the CSE References List (Citation-Sequence Style) *A-34*

MLA Documentation Style *A-35*
MLA In-Text Citations *A-35*
The Works Cited List for MLA Style *A-37*
Creating the MLA Works Cited List *A-39*

References *R-1*
Credits *C-1*
Index *I-1*
Sample Documents *Inside Back Cover*

Preface

People use their computers to help them research, compose, design, revise, and deliver technical documents and presentations. By making computers central to the writing process and exploring how we use them to join the ongoing conversation around us, *Technical Communication Today* helps students and professionals take full advantage of these important workplace tools.

New media and communication technologies are dramatically altering technical fields at an astounding rate. People are working more efficiently, more globally, and more visually. These changes are exciting, and they will continue to accelerate in the technical workplace. The fourth edition of *Technical Communication Today* continues to help writers master these changing communication tools that are critical to success in technical fields.

Today, as the technical workplace has expanded, almost all professionals find themselves needing to communicate technical information. To meet this need, this book addresses a broad range of people, including those who need to communicate in business, computer science, the natural sciences, the social sciences, public relations, medicine, law, and engineering.

What's New in the Fourth Edition?

The fourth edition of *Technical Communication Today* provides students with up-to-date information.

- **A new chapter on using Web 2.0 in the technical workplace** discusses how job seekers can use social networking sites to make and advance professional connections as well as how employees can use and contribute to a workplace's official social networking outlets (Ch. 23).
- **Genre chapters have been moved to the front of the text** to get students writing earlier in the semester (Part 2).
- **New "Microgenres"** demonstrate how elements of broad genres can be applied to narrower rhetorical situations (e.g., a status update is a microgenre of the activity report; an elevator pitch is a microgenre of a proposal). Each chapter in Part 2 features a microgenre that includes a description, example, and writing activity to provide students with informal opportunities to play with and stretch genre conventions.
- **Chapters on e-mail, letters, and memos have been combined** to better reflect how e-mail has become more common in the workplace and how letters and memos are increasingly delivered via e-mail messages (Ch. 5).
- **Coverage of doing research, evaluating sources, and writing with sources has been expanded** into two chapters to better help students manage and work effectively with information (Chs. 14, 15).
- **New chapters on using strategic planning (Ch. 12) and developing a persuasion strategy (Ch. 13)** pull apart the third edition's Chapter 6 to treat each topic discretely and provide expanded instruction.

- **New sample documents** provide even more examples of the types of communications students will encounter in the workplace, including a technical description, a status report, and a poster presentation (Chs. 6, 9, 10).
- **Four new case studies** prepare students for real workplace situations by presenting ethical challenges for reflection and rich class discussion.
    - Deciding whether to work with a prestigious company that conducts controversial research (Ch. 15)
    - Presenting distorted information and dealing with the ramifications (Ch. 19)
    - Handling a situation in which an error-filled document is sent to a potential client (Ch. 20)
    - Determining whether or not to "friend" a manager (Ch. 23)
- **New Help boxes** on persuading online (Ch. 13), using a citation manager (Ch. 15), and using Skype for collaboration (Ch. 23) provide students with the help they need to succeed in today's wired workplace.
- **The chapter on technical definitions has been telescoped** into the chapter on technical descriptions and specifications (Ch. 6) as most definitions of technical terms exist not as stand-alone texts but within precise documents such as technical descriptions, patents, specifications, experiments, and field observations.
- **New APA guidelines** are reflected in Appendix C to give students the most current citation guidance available.

Guiding Themes

In times of accelerated change, we must quickly adapt to new communication tools and strategies, while retaining proven approaches to writing and speaking. In this book, I have incorporated the newest technology in workplace communication. But the basics have not been forgotten. You will also find that the book is grounded in a solid core of rhetorical principles that have been around for at least two and a half millennia. In fact, these core principles hold up surprisingly well in this Information Age and are perhaps even more relevant as we return to a more visual and oral culture.

My intent was to develop a book that teaches students the core principles of rhetoric, while showing them how to use computers in a rapidly evolving information-based society.

Computers as Thinking Tools

The foremost theme of this book is that computers are integral and indispensable in technical communication. This premise may seem obvious to many readers; yet the majority of technical communication textbooks still do not successfully integrate computers into their discussions of workplace communication. These textbooks often limit computers to their word-processing abilities. They do not adequately show students how to fully use their computers to succeed in a networked technical workplace.

This book reconceptualizes the computer as a thinking tool in the technical workplace and in student learning. We need to recognize that students use their computers as thinking tools from beginning to end, inventing their ideas and composing text at the same time. In this book, the writing process has been redefined with the computer as a communication medium. As a result, the writing process described here is far more in line with the kinds of computer-centered activities in the technical workplace.

Genres as Pathways for Interpretation and Expression

This book follows a genre-based approach to writing and speaking in technical workplaces. Genres are relatively stable patterns that help people accomplish their goals in a variety of common rhetorical situations. Genres are not formulas or recipes to be followed mechanically. Instead, they offer flexible approaches that allow people to bring order to the evolving reality around them.

Genres can be used to interpret rhetorical situations, helping people in technical workplaces make decisions about what kinds of information they need to generate or collect. They can help individuals and collaborative teams plan projects and develop rhetorical strategies for responding appropriately to complex situations. They can then be used to guide invention, organization, style, and design.

A genre-based approach to technical communication provides students with a "genre set" that is applicable to a variety of technical communication situations. While practicing these genres, students will also learn how to adapt genres and cross genres in ways that help them respond appropriately to situations that are unique or new to them.

Visual-Spatial Reading, Thinking, and Composing

This book also reflects an ongoing evolution in technical communication from literal-linear texts toward visual-spatial documents and presentations. We now see documents as "spaces" where information is stored and flows. Visual-spatial reading, thinking, and composing involve interacting with text in three dimensions.

This book addresses this evolution toward visual-spatial thinking in four ways:

- First, this book shows writers and speakers how to use visual-spatial techniques to research, invent, draft, design, and edit their work.
- Second, it teaches students how to write and speak visually, while designing highly navigable documents and presentations.
- Third, the book shows how to compose visual-spatial documents like hypertexts, websites, and multimedia presentations. Writing in these environments is becoming increasingly important as companies move their communications and documentation online.
- Finally, it practices what it preaches by presenting information in a visual-spatial way that will be more accessible to today's students. Clearly, students learn differently now than they did even a couple of decades ago. This book reflects their ability to think visually and spatially.

This visual-spatial turn is an important intellectual shift in our culture—one that we do not fully understand at the moment. We do know, however, that communicating visually and spatially involves more than adding headings and charts to documents or using PowerPoint to enhance oral presentations. Instead, we must recognize that the advent of the computer, which is a visual-spatial medium, is revolutionizing how we conceptualize the world and how we communicate. Increasingly, people are thinking visually and spatially in addition to literally and linearly. This book incorporates this important change.

The International, Cross-Cultural Workplace

This edition of *Technical Communication Today* includes expanded coverage of international and cross-cultural issues. I have met with hundreds of technical communication instructors and have learned that they want even more coverage of the globalized, cross-cultural workplace.

International and cross-cultural issues are integrated into the main discussion rather than shunted off into special sidebars, because issues of globalization are no longer separable from technical communication. Today, we always need to think globally, because computers greatly expand our reach into the world.

The Activity of Technical Communication

In this computer-centered age, people learn by doing, not by passively listening or reading. This book continues to stress the activity of technical communication—producing effective documents and presentations. Each chapter follows a process approach that mirrors how professionals communicate in the technical workplace. Meanwhile, the book shows students how to pay close attention to the evolving workplace contexts in which communication happens.

Perhaps this theme comes about because of my experiences with students and my observations of people using books like this one. As someone who has consulted and taught technical communication for nearly two decades, I realize that today's students rarely read their textbooks. Instead, they raid their textbooks for the specific information they need to complete a task. They use their textbooks like they use websites. They ask questions of the text and then look for the answers.

Supplements to the Book

Accompanying this book are important tools that instructors and students will find especially helpful.

Instructor's Manual

The *Instructor's Manual* offers teaching strategies for each chapter while also providing prompts for class discussion and strategies for improving student writing and presentations. The *Instructor's Manual* is also available online and offers additional materials for downloading, such as slides. The online version offers additional ideas for assignments and projects.

Companion Website

The Companion Website (http://www.pearsonhighered.com/johnsonweb4) includes a wealth of materials and is designed to be used side by side with the book. Students are able to look at more samples, download worksheets, utilize flashcards, take practice quizzes, and learn from the excellent communication-related websites available on the Internet. The "For Instructors Only" area has materials and teaching strategies to maximize the use of *Technical Communication Today*.

MyTechCommLab

Instructors who package MyTechCommLab with *Technical Communication Today*, Fourth Edition, provide their students not only with the full text of *Technical Communication Today* in electronic format but also with a comprehensive resource that offers the very best multimedia support for technical writing in one integrated, easy-to-use site. Organized into five topical categories, MyTechCommLab includes tutorials, activities, case studies, interactive model documents, and gradeable quizzes, in addition to such unique features as MySearchLab to help students with the online research process, the full text of the *Longman Online Handbook*, and full texts of additional technical communication resources in PDF format on the Student Bookshelf. MyTechCommLab is available packaged with *Technical Communication Today* at no additional cost or for purchase at www.mytechcommlab.com.

CourseSmart

Students can subscribe to *Technical Communication Today*, Fourth Edition, as a CourseSmart eText (at CourseSmart.com). The site includes all of the book's content in a format that enables students to search the text, bookmark passages, save their own notes, and print reading assignments that incorporate lecture notes.

Resources for Technical Communication, Second Edition

(ISBN 0-321-45081-7)

A complement to Pearson's introduction to technical communication textbooks, *Resources for Technical Communication,* Second Edition, offers a wide range of sample documents for use in technical communication courses, including letters, proposals, reports, memos, e-mails, résumés, abstracts, instructions, descriptions, and slide presentations.

Acknowledgments

The fourth edition of *Technical Communication Today* has given me the opportunity to work with many people at Pearson and at colleges around the country. I wish to thank the following individuals for their insight and support:

Teresa Aggen, Pikes Peak Community College; Sherrie L. Amido, California Polytechnic State University—San Luis Obispo; James Baker, Texas A&M University; Lauri M. Baker, University of Florida; Russell Barrett, Blinn College; Eric Bateman, San Juan College; Norman Douglas Bradley, University of California—Santa Barbara; Lee Brasseur, Illinois State University; Stuart Brown, New Mexico State University; Ellie Bunting, Edison College; Maria J. Cahill, Edison State College; Tracy L. Dalton, Missouri State University; Roger Friedman, Kansas State University; Timothy D. Giles, Georgia Southern University; Jeffrey Jablonski, University of Nevada—Las Vegas; Rebecca Jackson, Texas State University; Leslie Janac, Blinn College—Bryan Campus; Miles A. Kimball, Texas Tech University; Christy L. Kinnion, Wake Technical Community College; Barry Lawler, Oregon State University; Barbara L'Eplattenier, University of Arkansas—Little Rock; Anna Maheshwari, Schoolcraft College; Barry Maid, Arizona State University; Jodie Marion, Mt. Hood Community College; Steve Marsden, Stephen F. Austin State University; Mary S. McCauley, Wake

Technical Community College; Kenneth Mitchell, Southeastern Louisiana University; Jacqueline S. Palmer, Texas A&M University; Andrea M. Penner, San Juan College; Cindy Raisor, Texas A&M University; Sherry Rankins-Robertson, Arizona State University; Carlos Salinas, The University of Texas at El Paso; Teryl Sands, Arizona State University; Jennifer Sheppard, New Mexico State University; Nancy Small, Texas A&M University; Krista Soria, University of Alaska Anchorage; Karina Stokes, University of Houston—Downtown; Christine Strebeck, Louisiana Tech University; Valerie Thomas, University of New Mexico; Christopher Toth, Iowa State University; Jack Trotter, Trident Technical College; Greg Wilson, Iowa State University; Alan Zemel, Drexel University.

Editors Lauren Finn and Anne Brunell Ehrenworth were essential in the revision of this book and I thank them for their ideas. Thanks also to my colleagues, Professors Scott Sanders, Charles Paine, and David Blakesley. Finally, thanks to Jeremy Cushman and Enrique Reynoso, Jr., for their assistance.

Most important, I would like to thank my wife, Tracey, and my children, Emily and Collin, for their continued support.

Richard Johnson-Sheehan
Purdue University

CHAPTER

1

Communicating in the Technical Workplace

Developing a Workplace Writing Process 2

Genres and the Technical Writing Process 4

What Is Technical Communication? 10

How Important Is Technical Communication 15

Chapter Review 16

Exercises and Projects 17

In this chapter, you will learn:

- How to develop a writing process that is suitable for the technical workplace.
- How genres are used in technical workplaces to develop documents.
- How to use your computer to overcome writer's block.
- To define technical communication as a process of managing information in ways that allow people to take action.
- The importance of communication in today's technical workplace.
- The importance of effective written and spoken communication to your career.

When college graduates begin their technical and scientific careers, they are often surprised by the amount of writing and speaking required in their new jobs. Of course, they knew technical communication would be important, but they never realized it would be so crucial to their success.

Effective communication is the cornerstone of the technical workplace, whether you are an engineer, scientist, doctor, nurse, psychologist, social worker, anthropologist, architect, technical writer, or any other professional in a technical field. People who are able to write and speak effectively tend to succeed. People who cannot communicate well often find themselves wondering why they didn't get the job or why they were passed over for promotions.

Developing a Workplace Writing Process

One of the major differences between workplace writing and college writing is the pace at which you need to work. Computers have greatly increased the speed of the technical workplace, and they allow people to work around the clock. So, you need to work smarter, not harder.

To be successful, you need to develop a writing process that helps you consistently produce high-quality documents, presentations, and multimedia materials. In this book, you will be learning a *genre-based approach* to the technical writing process. Genres are relatively stable patterns that reflect the activities and practices of the workplace. A genre shapes a document's content, organization, style, and

Computers Are the Central Nervous System of the Workplace

Your ability to communicate with others through computer networks will be critical to your career.

design, as well as the medium in which it is delivered. Genres also help you anticipate the needs of your readers and the situations in which they will use your documents and presentations.

For example, *technical specifications* follow a different genre than *analytical reports* (Figure 1.1). Specifications and analytical reports are written for different kinds of readers for different workplace situations. They include different kinds of information and follow their own organizational patterns. The style and design of these two genres are distinctly different. Yet, someone working in a technical workplace would need to know how to use both of these genres.

Genres do much more than help you organize your ideas. They help you interpret complex workplace situations and make sense of what is happening around you. For example, if you know you need to write an analytical report, the genre will help you figure out what kind of information you need to collect, how that information should be arranged, and how it should be presented. Your readers, meanwhile, will interpret your ideas through the genre. If you call something a "report," they will have specific expectations about the content, organization, style, design, and medium of the document.

Genres are not formulas or recipes to be followed mechanically. Instead, genres reflect the activities and practices of scientific and technical workplaces. Each genre should be adapted to fit the readers and the situations in which the document will be used.

Two Different Genres

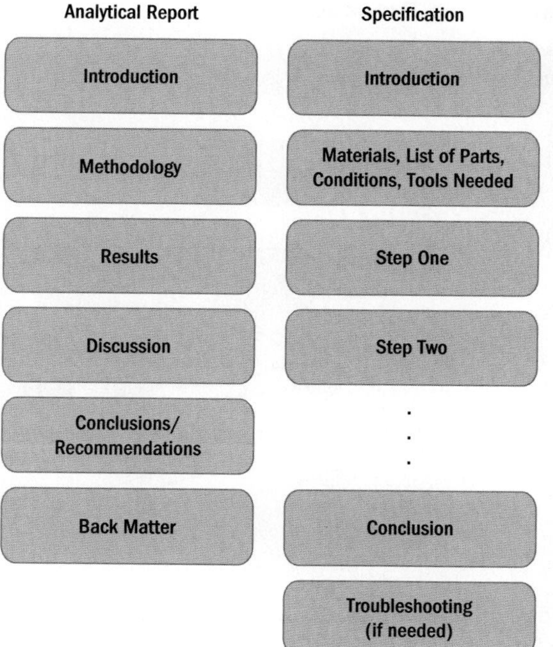

Figure 1.1: Each genre has its own content, organization, style, and design. Here are the outlines of two distinctly different genres set side by side.

GO TO THE NET To find websites that discuss various genres, go to **www.pearsonhighered.com/johnsonweb4/1.1**

Genres and the Technical Writing Process

Over time, you will develop your own writing process for the technical workplace. For now, though, you might find it helpful to view technical writing as a *process* that includes the stages shown in Figure 1.2:

- **Planning and researching**—planning the project, using research to collect information, and developing your own ideas.
- **Organizing and drafting**—deciding how to arrange your information and then turning those ideas into sentences, paragraphs, and sections.
- **Improving the style**—writing clearly and persuasively for your readers.
- **Designing the document**—developing an appropriate page design that improves the usability and attractiveness of your document.
- **Revising and editing**—improving the quality of your work by re-visioning, rewriting, and proofreading your writing.

As you write your document or develop your presentation, you will find yourself working back and forth among these stages, as shown in Figure 1.2. While drafting, for example, you may discover that you need to do more research on your topic. While editing, you may decide that you need to draft an additional section for the document. Overall, these stages will lead you from the beginning of a project to the end.

Meanwhile, the genre you are using will guide each stage in your writing process. The genre helps you make decisions about the content of the document, as well as the organization, style, design, and medium that would be best for readers.

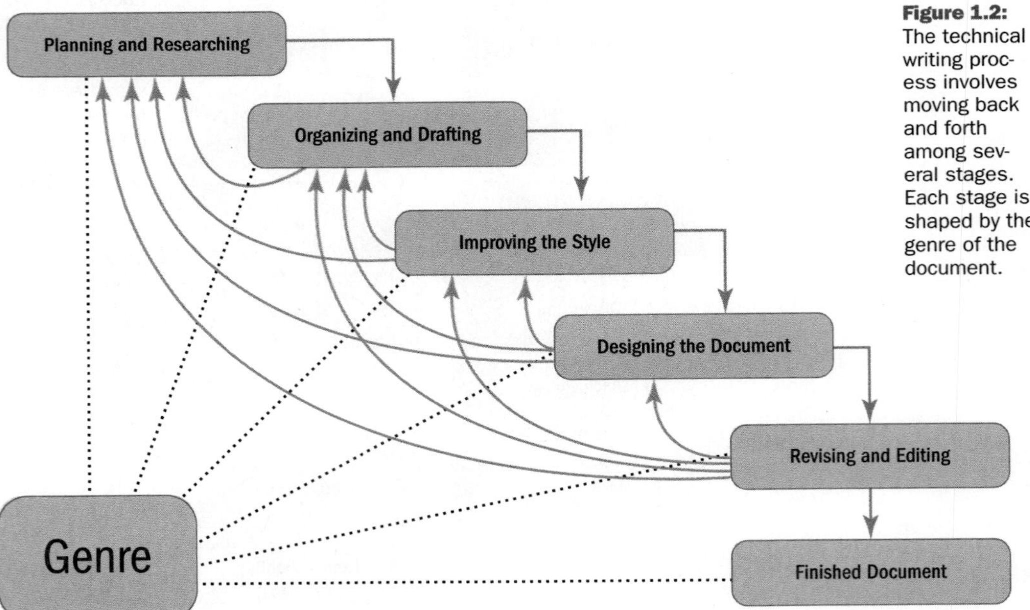

Figure 1.2: The technical writing process involves moving back and forth among several stages. Each stage is shaped by the genre of the document.

GO TO THE NET For more discussions of the writing process, go to **www.pearsonhighered.com/johnsonweb4/1.2**

Stage 1: Planning and Researching

When planning and researching, you should spend some time doing three activities:

> **Define the rhetorical situation**—Identify your document's subject, purpose, readers, and context of use.
>
> **State your purpose**—Sharpen your purpose into a one-sentence statement that will guide your research and drafting of the document.
>
> **Research your subject**—Use electronic, print, and empirical sources to collect information on your subject.

DEFINING THE RHETORICAL SITUATION A good first step is to define the *rhetorical situation* that will shape the content, organization, style, and design of your document. Understanding the rhetorical situation means gaining a firm grasp of your document's subject, purpose, readers, and context of use (Figure 1.3).

To define the rhetorical situation, start out by asking the *Five-W and How Questions:* who, what, why, where, when, and how.

- *Who* are my readers, and who else is involved with the project?
- *What* do the readers want and need, and what do I want and need?
- *Why* do the readers need the information in this document?
- *Where* do they need the information, and *where* will they use it?
- *When* will the information be used, and *when* is it needed?
- *How* should I achieve my purpose and goals?

The Five-W and How Questions will give you an overall sense of your document's rhetorical situation.

Defining the Rhetorical Situation

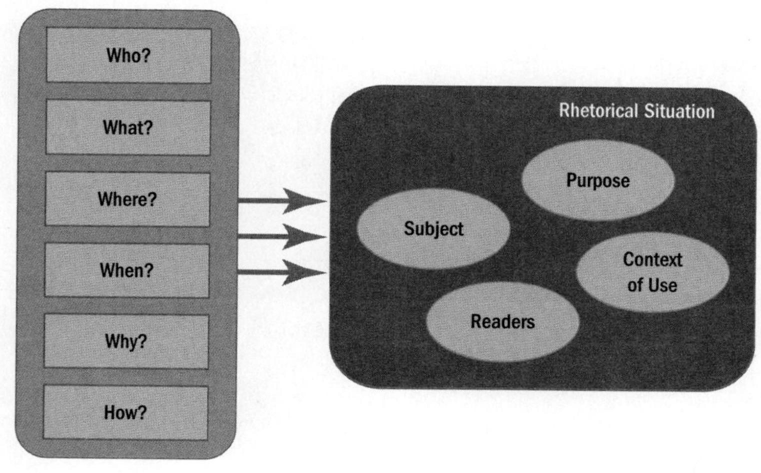

Figure 1.3: The Five-W and How Questions can help you determine the rhetorical situation for your technical document or presentation.

GO TO THE NET To find worksheets to help you analyze the rhetorical situation, go to **www.pearsonhighered.com/johnsonweb4/1.3**

Now, spend some time taking notes on the following four elements of the rhetorical situation:

> **Link**
> To learn about adapting texts to readers and contexts, go to Chapter 2, page 26.

Subject—What is the document about? What is it *not* about? What kinds of information will my readers need to make a decision or complete a task? What is the scope of the project?

Purpose—What does this document need to achieve or prove? Why do my readers need this document and what do they need to know?

Readers—Who are the readers of this document? What are their specific needs and interests? What are they looking for in this document?

Context of use—Where and when will this document be used? What physical, economic, political, and ethical constraints will shape this text?

Defining the rhetorical situation may seem like an added step that will keep you from writing. Actually, knowing your document's rhetorical situation will save you time and effort, because you will avoid dead ends, unnecessary revision, and writer's block.

DEFINING YOUR PURPOSE Among the four elements of the rhetorical situation, your document's purpose is probably the most important. It is what you want to do—and what you want the document to achieve.

Your purpose statement is like a compass for the document. Once you have clearly defined your purpose for yourself and your readers, you can use that purpose statement to guide your decisions about the content, organization, style, and design of your document.

When defining your purpose, try to express exactly what you want your document to achieve. Sometimes it helps to find an appropriate action verb and then build your purpose statement around it. Here are some useful action verbs that you might use:

Informative Documents	Persuasive Documents
to inform	to persuade
to describe	to convince
to define	to influence
to review	to recommend
to notify	to change
to instruct	to advocate
to advise	to urge
to announce	to defend
to explain	to justify
to demonstrate	to support

GO TO THE NET For more help writing a purpose statement, go to **www.pearsonhighered.com/johnsonweb4/1.4**

Once you have chosen an action verb, try to state your purpose in one sentence. It might help to finish the phrase "The purpose of my document is to . . . ."

> The purpose of my report is to review the successes and failures of wolf re-introduction programs in the western United States.

> The purpose of my proposal is to recommend significant changes to flood control strategies in the Ohio River Valley.

Hammering your purpose statement down into one sentence is hard work but worth the effort. Your one-sentence purpose statement will focus your writing, saving you time. Chapter 12 on strategic planning provides some helpful ideas for figuring out your purpose statement, especially with larger, more complex projects.

RESEARCHING YOUR SUBJECT Solid research is your next step. You need to gather information from a variety of sources, including the Internet, print documents, and empirical methods (e.g., experiments, surveys, observations, interviews). Chapters 14 and 15 will help you do effective research and evaluate your sources.

Computers have significantly changed the way we do research in technical workplaces. Before computers, finding enough information was usually a writer's main challenge. Today, there is almost too much information available on any given subject. So, it is important that you learn how to *manage* the information you collect, sorting through all the texts, scraps, junk, and distortions to uncover what you need. Your documents should give your readers only the information they require to make a decision or take action. Leave out anything else.

Stage 2: Organizing and Drafting

Organizing and drafting is usually the hardest part of the writing process. While organizing and drafting, you are essentially doing two things at the same time:

> **Organizing the content**—Using common genres to shape your ideas into documents that will be familiar to readers.
>
> **Drafting the content**—Generating the content of your document by including facts, data, reasoning, and examples.

Here's where the concept of genres is especially helpful. If you understand the genre, you will understand how to organize the information you've collected in a way that achieves your purpose. For example, the document in Figure 1.4 is easily recognizable as a *set of instructions* because it is following the genre.

Chapters 5 through 11 will teach you how to use the most common genres in technical workplaces. In most situations, you will already know which genre you need because your supervisor or instructor will ask you to write a "specification," "report," or "proposal." But if you are uncertain which genre suits your needs, pay attention to your document's purpose. Then, find the genre that best suits the purpose you are trying to achieve.

For helpful tools for doing research online, go to **www.pearsonhighered.com/johnsonweb4/1.5**

Sample of Genre: Instructions

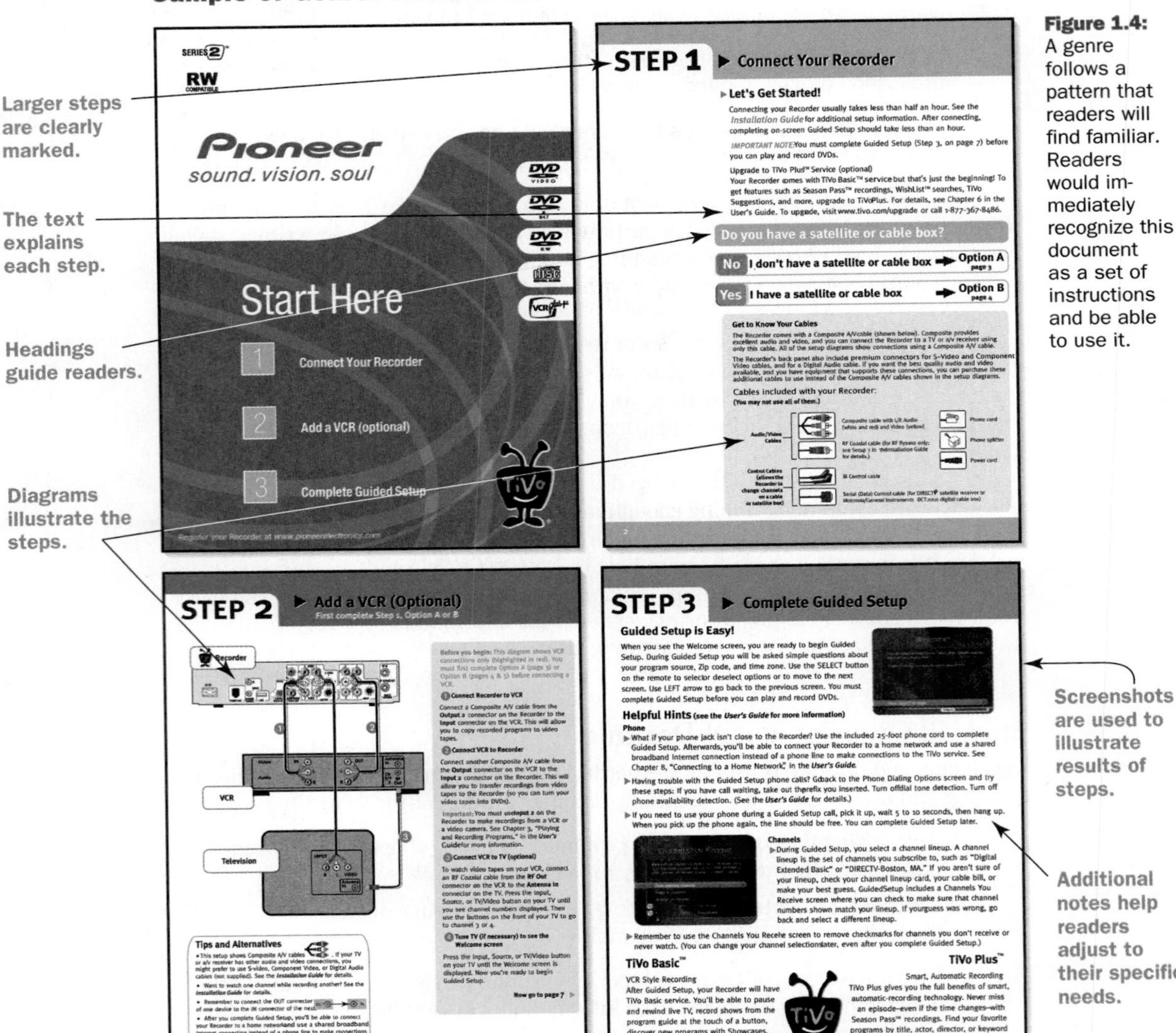

Figure 1.4: A genre follows a pattern that readers will find familiar. Readers would immediately recognize this document as a set of instructions and be able to use it.

Source: TiVo.

GO TO THE NET Want to improve your writing style? Go to **www.pearsonhighered.com/johnsonweb4/1.6**

Stage 3: Improving the Style

All documents have a style, effective or not. Good style is a choice you can and should make. In Chapter 17, you will learn about two kinds of style that are widely used in technical documents: plain style and persuasive style.

> **Plain style**—This style stresses clarity and accuracy. By paying attention to your sentences and paragraphs, you can make your ideas clearer and easier to understand.
>
> **Persuasive style**—You can use persuasive style to motivate readers by appealing to their values and emotions. You can use similes and analogies to add a visual quality to your work. You can use metaphors to change your readers' perspective on issues. Meanwhile, you can use tone and pace to add energy and color to your work.

Most workplace texts are written in the plain style, but technical documents sometimes need the extra energy and vision provided by the persuasive style. Your goal should always be to make information as clear and concrete as possible. When persuasion is needed, you will want to energize your readers with persuasive style techniques.

Stage 4: Designing

Designing a document only takes minutes with a computer, and you can create graphics with a few clicks of a button. So, design is not only possible—your readers will *expect* your technical documents to be well designed.

As you think about the design of your document, keep this saying in mind: *Readers are "raiders" for information.* Your readers want the important parts highlighted for them. They prefer documents that use effective graphics and layout to make the information more accessible, interesting, and attractive (Figure 1.5).

Chapter 18 will show you how to design workplace documents. Chapter 19 will show you how to create and place graphics in your documents. As you draft and revise your document, look for places where you can use visual design to help readers locate the information they need. Look for places where graphics might support or reinforce the written text. The design of your document should make it both attractive and easy to read.

Stage 5: Revising and Editing

When you have finished drafting and designing the document, you are only a little over halfway finished. In technical communication, it is crucial to leave plenty of time for revising, editing, and proofreading. Clarity and accuracy are essential if your readers are going to understand what you are trying to tell them.

In Chapter 20, you will learn about four levels of revising and editing:

> **Level 1: Revising**—Re-examine your subject and purpose while thinking again about the information your readers need to know.
>
> **Level 2: Substantive editing**—Look closely at the content, organization, and design of the document to make sure your readers can find the information they need.

If you want more information on document design, go to **www.pearsonhighered.com/johnsonweb4/1.7**

Document Design Is Very Important

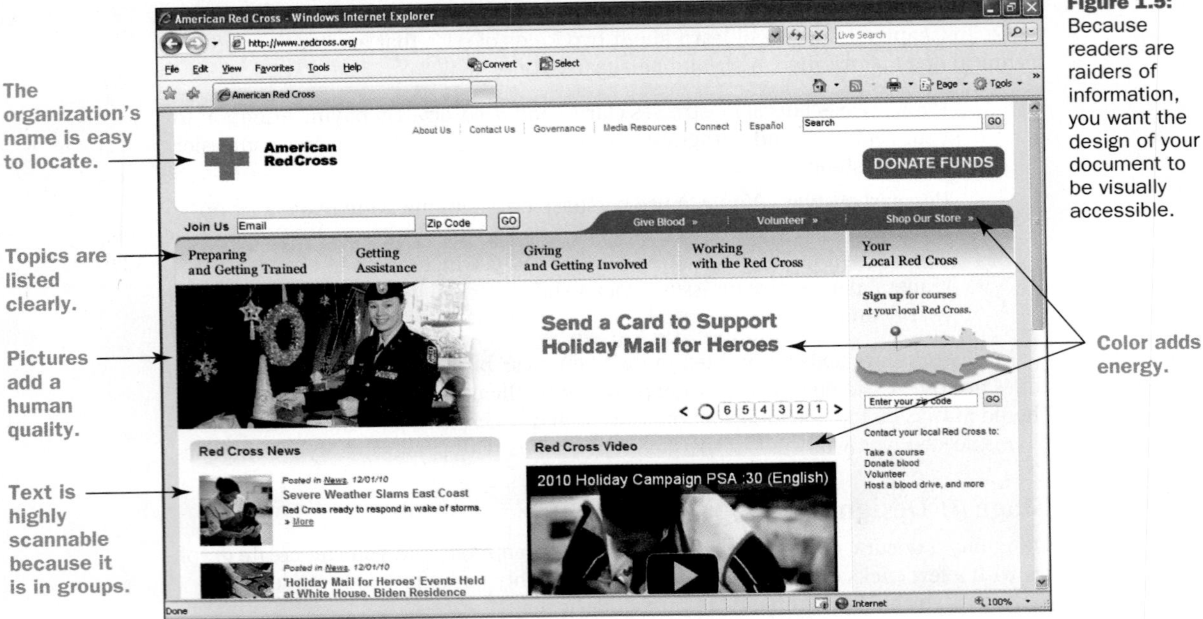

Figure 1.5: Because readers are raiders of information, you want the design of your document to be visually accessible.

Source: American Red Cross, http://www.redcross.org.

> **Level 3: Copyediting**—Pay close attention to the document's sentences, paragraphs, and graphics to make sure they are clear, accurate, and efficient.
>
> **Level 4: Proofreading**—Carefully proofread your document to eliminate grammar problems, typos, spelling errors, and usage mistakes. In workplace documents, errors are a signal of low-quality work.

Revising and editing is a crucial step in the technical workplace, where clarity and accuracy are essential. Your supervisors will ask you to do much more revising and editing than your college professors.

What Is Technical Communication?

Let's step back for a moment to look at the big picture. This chapter hasn't given you a definition of technical communication yet—on purpose. That's because you first needed to understand that technical communication is a *process*. Here is the definition of technical communication that will be used throughout this book:

> **Technical communication is a process of managing technical information in ways that allow people to take action.**

GO TO THE NET

For other definitions of technical communication, go to **www.pearsonhighered.com/johnsonweb4/1.8**

The Qualities of Technical Communication

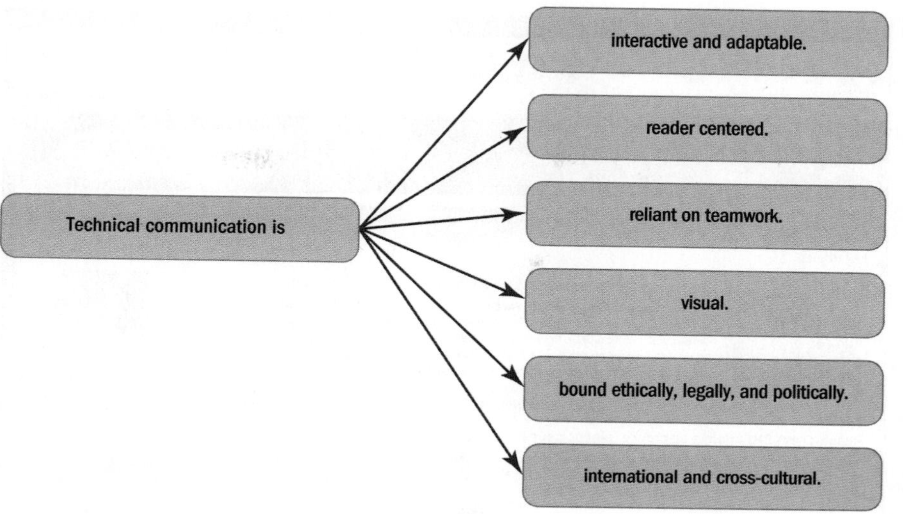

Figure 1.6: Technical communication puts much more emphasis on managing information and taking action than most other forms of writing.

> **AT A GLANCE**
>
> **Qualities of Technical Communication**
>
> Technical communication is:
> - interactive and adaptable.
> - reader centered.
> - produced in teams.
> - visual.
> - influenced by ethics, laws, and politics.
> - international and cross-cultural.

The key words in this definition are *process, manage,* and *action*. In this book, you will learn the *process* of technical communication so you can *manage* large amounts of information in ways that allow you to take *action*. Technical communication involves learning how to manage the flow of information so you can get things done (Figure 1.6).

Technical Communication Is Interactive and Adaptable

One of the most significant changes brought about by computers is the amount of *interactivity* and *collaboration* among people in the technical workplace. In the computer-networked workplace, people are constantly communicating with each other and sharing their ideas.

As a result of this interactivity, it is possible for you to quickly adapt documents and presentations to fit the specific needs of many different kinds of readers and situations. Websites are an especially interactive form of technical communication (Figure 1.7). Using a website, people can find the information that is most helpful to them. And, if they cannot find the information they are looking for on the website, they can send an e-mail to get the answers they need.

Similarly, paper-based documents can also be adapted to the changing needs of readers. Before computers, it was difficult to adjust and revise paper-based documents. Once they were printed, documents were hard to change. Today, with computers, you can easily update documents to reflect changes in your company's products and services. Or, you can quickly revise documents to address unexpected changes in the workplace.

Sample Webpage

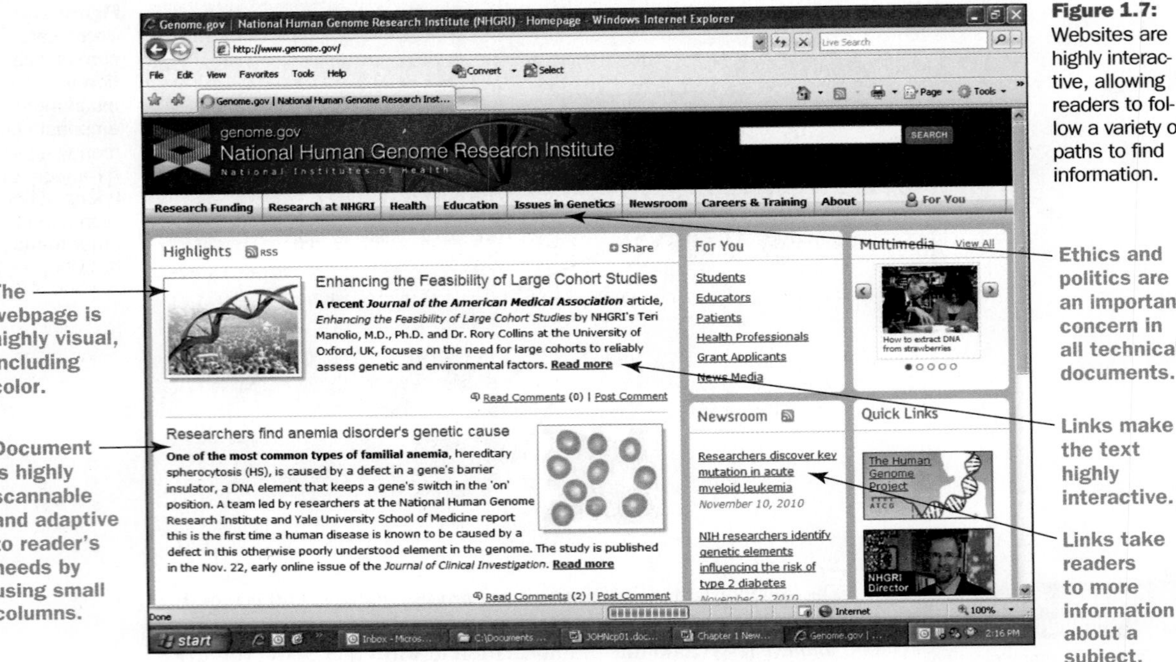

The webpage is highly visual, including color.
Document is highly scannable and adaptive to reader's needs by using small columns.
Ethics and politics are an important concern in all technical documents.
Links make the text highly interactive.
Links take readers to more information about a subject.
National Human Genome Research Institute
Research Funding
Research at NHGRI
Health
Education
Issues in Genetics
Newsroom
Careers & Training
About
For You
Highlights
Enhancing the Feasibility of Large Cohort Studies
Researchers find anemia disorder's genetic cause
Multimedia
Newsroom
Quick Links

Figure 1.7: Websites are highly interactive, allowing readers to follow a variety of paths to find information.

Source: National Human Genome Research Institute, http://www.genome.gov.

Technical Communication Is Reader Centered

In technical communication, readers play a much more significant role than they do in other kinds of writing. When writing a typical college essay, you are trying to express *your* ideas and opinions. Technical communication turns this situation around. It concentrates on what the readers "need to know" to take action, not only what you, as the writer, want to tell them.

Because it is reader centered, effective technical communication tends to be highly pragmatic. Technical communication needs to be efficient, easy to understand, accessible, action oriented, and adaptable.

Link

For more information on working in teams, see Chapter 3, page 45.

Technical Communication Relies on Teamwork

Technical workplaces are highly collaborative, meaning you will likely work with a team of specialists on almost every project. Writing and presenting with a team are crucial skills in any technical workplace.

Computers have only heightened the team orientation of the technical workplace. Today, because documents can be shared through e-mail or the Internet,

GO TO THE NET

To find other resources on technical communication, go to **www.pearsonhighered.com/johnsonweb4/1.9**

it is common for many people to be working on a document at the same time. In some cases, your team might be adjusting and updating documents on an ongoing basis.

Link

For more information on visual design, see Chapter 18, page 482.

Technical Communication Is Visual

By making texts highly visual, you can help readers quickly locate the information they need. Visual cues, like headings, lists, diagrams, and margin comments, are common in technical documents (Figure 1.8). Graphics also play an important role in technical communication. By using charts, graphs, drawings, and pictures, you can clarify and strengthen your arguments in any technical document. Today's readers quickly grow impatient with large blocks of text. They prefer graphics that reinforce the text and help them quickly gain access to important information.

Link

To learn about using graphics in documents, turn to Chapter 19, page 523.

Technical Communication Has Ethical, Legal, and Political Dimensions

In the increasingly complex technical workplace, issues involving ethics, laws, and politics are always present. Ethical and legal standards can be violated if you aren't careful. Moreover, computers have created new micro- and macropolitical challenges that need to be negotiated in the workplace. To communicate effectively in the technical workplace, you need to be aware of the ethical, legal, and political issues that shape your writing and speaking.

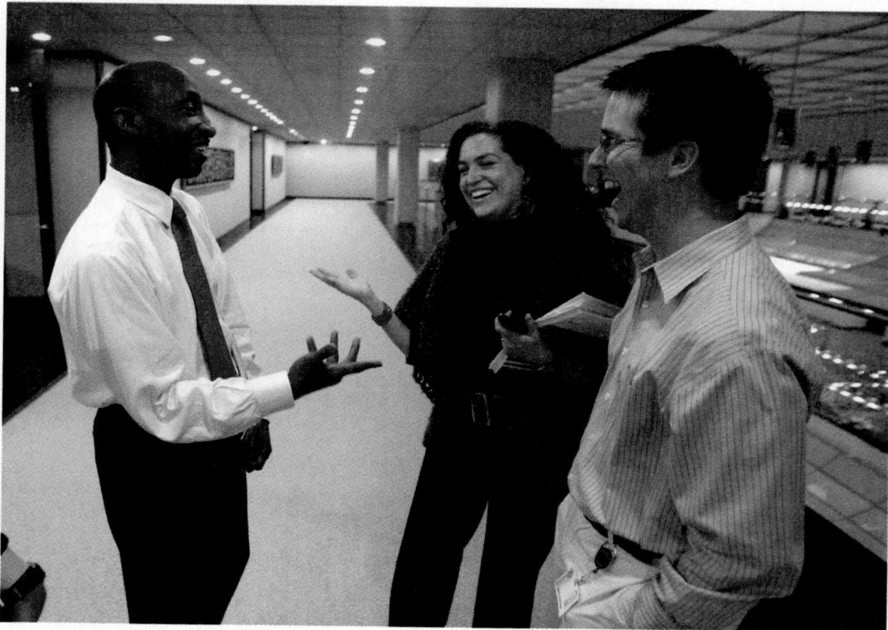

Working with a team can be fun and rewarding. Teams take advantage of the strengths and knowledge of different people to succeed.

GO TO THE NET To learn more about writing in teams, go to **www.pearsonhighered.com/johnsonweb4/1.10**

The Importance of Visual Design

Visuals add color and emotion.

An easy-to-read title identifies the document's subject.

Headings make the text highly scannable.

The two-column format makes the text easy to scan.

Ecosystems

NOS Releases Two National Progress Reports on Reef Conservation

This year, NOS released two major progress reports on coral reef research, monitoring and management. *The State of Coral Reef Ecosystems of the United States and Pacific Freely Associated States: 2005* established the first quantitative baseline of the conditions of shallow water coral reef ecosystems in the U.S., the Republic of Palau, the Republic of the Marshall Islands, and the Federated States of Micronesia. More than 160 scientists and resource managers contributed to the report, which documents the geographic extent of reef ecosystems and the status of water quality, benthic habitats, associated biological communities and key threats to coral ecosystem health. The second report, *Implementation of the National Coral Reef Action Strategy: Report on U.S. Coral Reef Agency Activities from 2002 to 2003*, highlights the activities of NOAA and the U.S. Coral Reef Task Force under each of the 13 national conservation goals defined by the 2002 U.S. National Coral Reef Action Strategy. The report indicates that collective research and management actions are moving in the right direction, citing examples like the creation of 14 new coral reef protected areas and the creation of Local Action Strategies for conservation.

8

Tortugas Ecological Reserve Show Signs of Species Abundance

Four years after the establishment of the Tortugas Ecological Reserve, NOS scientists are studying how the ecosystem is changing as a result of reserve status. This year, scientists conducted 253 dives to collect data and fish samples, and found that certain fish species are increasingly abundant.

New Tide and Water Quality Monitoring Station Includes Multiple Features

In August, NOS installed a tide and water quality monitoring station at the Wells National Estuarine Research Reserve (NERR), in Wells, Maine. The station combines the capabilities of the National Water Level Observation Network (NWLON) and the System-wide Monitoring Network. The station, which is the first of its kind installed at a NERR, includes primary and backup water level sensors, a suite of meteorological sensors, and a water quality sensor that measures several parameters. The NWLON technology allows Wells NERR staff to access water level, weather, and water quality data all from the same platform at the same time. Products generated from these data will benefit both short-term (such as habitat restoration) and long-term (such as sea level trends) applications, as well as research and education objectives.

Restoration Efforts at Blackwater National Wildlife Refuge

NOS and NOAA Fisheries are working with the U.S. Geological Survey (USGS), U.S. Army Corps of Engineers, U.S. Fish and Wildlife Service, National Aquarium in Baltimore, and others to restore 8,000 acres of wetlands at the Blackwater National Wildlife Refuge in eastern Maryland. Under common observing and data management principles of the Integrated Ocean Observing System, the partners are collecting water level data so that NOAA can process and conduct analyses of the data to apply to the restoration project. The Refuge also hosted a workshop on the importance of geodetic control for tidal analysis and applications. After the workshop, a global positioning system survey was conducted to connect NOAA's and USGS's water level stations, and USGS's surface elevation tables to the same geodetic network.

NOAA'S NATIONAL OCEAN SERVICE: ACCOMPLISHMENTS 2005

Figure 1.8: Visual design is an essential part of technical communication.

Source: *National Ocean Service, 2005.*

GO TO THE NET

For more on the importance of visual design in technical communication, go to **www.pearsonhighered.com/johnsonweb4/1.11**

In the global market, the ability to communicate is the key to success.

As management structures become flatter—meaning there are fewer layers of management—employees are being asked to take on more decision-making responsibilities than ever. In most corporations, fewer checks and balances exist, meaning that all employees need to be able to sort out the ethical, legal, and political aspects of a decision for themselves.

Link
Ethical, legal, and political issues are discussed in Chapter 4, starting on page 69.

Technical Communication Is International and Cross-Cultural

Computers have also increased the international nature of the technical workplace. Today, it is common for professionals to regularly communicate with people around the world. Almost all companies and institutions compete in a global marketplace. Many have offices, communication hubs, and manufacturing sites in Europe, Asia, Africa, Australia, and South America. The growth of international and cross-cultural trade means you will find yourself working with people who speak other languages and have other customs. They will also hold different expectations about how technical documents and presentations should work.

Link
To learn about communicating internationally and cross-culturally, go to Chapter 2, page 28.

How Important Is Technical Communication?

At this point, you're probably still wondering how important technical communication will be to your career. Surveys regularly show that oral and written communication skills are among the most important in the technical workplace. A survey of Silicon Valley recruiters found that "employers were not fully satisfied with the business communication skills (writing, speaking, interpersonal) of their newly hired college graduates" (Stevens, 2005, p. 5). This survey found that 40 percent of employers wanted new hires to have better speaking skills, and 25 percent wanted better writing skills.

These findings are in line with conclusions from other surveys. When members of the American Institute of Aeronautics and Astronautics (AIAA) were asked to evaluate their educational preparation for their jobs, they ranked the following areas as the ones needing improvement in engineering education.

Areas of Needed Improvement in Education for Engineers

1. Oral communication
2. Visualization in three dimensions
3. Technical writing
4. Understanding the processes of fabrication and assembly
5. Using CAD, CAM, and solid modeling
6. Estimating solutions to complex problems without using computer models
7. Sketching and drawing

The membership of the AIAA is made up mostly of engineers, so it is interesting that two out of three of the top skills they listed stress the importance of technical communication.

Corporations spend billions each year to improve the writing skills of their employees, according to the 2004 report "A Ticket to Work . . . or a Ticket Out," from the National Commission on Writing. Poor writing skills are the "kiss of death," according to the report, because 51 percent of companies say they "frequently or almost always take writing into consideration when hiring salaried employees" (p. 9).

Fortunately, you can learn how to write and speak effectively in the technical workplace. The ability to communicate effectively is not something people are born with. With guidance and practice, anyone can learn to write and speak well. Right now, you have a golden opportunity to develop these important technical communication skills. They will help you land the job you want, and they will help you succeed.

If you are reading this book, you are probably in a class on technical communication or are looking to improve your skills in the technical workplace. This book will give you the tools you need for success.

CHAPTER REVIEW

- By consciously developing a writing process, you will learn how to write more efficiently. In other words, you will "work smarter, not harder."
- A useful workplace writing process includes the following stages: planning and researching, organizing and drafting, improving the style, designing, and revising and editing.
- Technical writing genres are helpful for organizing information into patterns that your readers will expect.
- Computers, the Internet, and instant forms of communication have had an enormous impact on communication in the technical workplace.

GO TO THE NET — To see the AIAA survey, go to **www.pearsonhighered.com/johnsonweb4/1.13**

- Technical communication is defined as a process of managing technical information in ways that allow people to take action.
- Technical communication is a blend of actions, words, and images. Readers expect technical documents to use writing, visuals, and design to communicate effectively.
- Technical communication is interactive, adaptable, reader centered, and often produced in teams.
- Technical communication has ethical, political, international, and cross-cultural dimensions that must be considered.
- Effective written and spoken communication will be vital to your career.

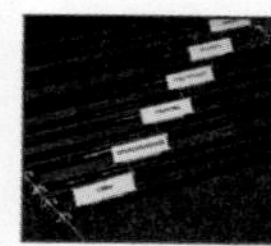

EXERCISES AND PROJECTS

Individual or Team Projects

1. Locate a document that is used in a technical workplace through a search engine like Google.com, Altavista.com, or Yahoo.com. To find documents, type in keywords like "report," "proposal," "instructions," and "presentation." Links to sample documents are also available at www.pearsonhighered.com/johnsonweb4/1.10.
   What characteristics make the document you found a form of technical communication? Develop a two-minute presentation for your class in which you highlight these characteristics of the document. Compare and contrast the document with academic essays you have written for your other classes.

2. Using a search engine on the Internet, locate a professional who works in your chosen field. Write an e-mail asking that person what kinds of documents or presentations he or she needs to produce. Ask how much time he or she devotes to communication on the job. Ask whether he or she has some advice about how to gain and improve the communication skills that you will need in your career. Write a memo to your instructor in which you summarize your findings.

3. Using the information in this chapter, write a memo to your instructor in which you compare and contrast the kinds of writing you have done for classes in the past (e.g., essays, short answer, short stories) with the kinds of writing you expect to do in your career. Then, tell your instructor how this class would best help you prepare for your career in a technical workplace.

Collaborative Project: Writing a Course Mission Statement

As you begin this semester, it is a good idea for your class to develop a common understanding of the course objectives and outcomes. Companies develop mission statements to help focus their efforts and keep their employees striving toward common ends. Corporate mission statements are typically general and nonspecific, but they set an agenda or tone for how the company will do business internally and with its clients.

Your task in this assignment is to work with a group to develop a "Course Mission Statement" in which you lay out your expectations for the course, your instructor, and yourselves. To write the mission statement, follow these steps:

1. Go to www.pearsonhighered.com/johnsonweb4/1.10 to find links to sample mission statements. Or, you can use an Internet search engine to find your own examples of mission statements. Just type "mission statement" in Lycos.com, Webcrawler.com, or Google.com.

2. In class, with your group, identify the common characteristics of these mission statements. Pay special attention to their content, organization, and style. Make note of their common features.

3. With your group, write your own course mission statement. Be sure to include goals you would like the course to meet. You might also want to develop an "ethics statement" that talks about your approach to ethical issues associated with assignments, course readings, and attendance.

4. Compare your group's course mission statement with other groups' mission statements. Note places where your statement is similar to and different from their statements.

When your course mission statement is complete, it should provide a one-paragraph description of what you are trying to achieve in your class.

For additional technical writing resources, including interactive sample documents, document design tutorials and guidelines, and more, go to **www.mytechcommlab.com.**

CHAPTER

2

Readers and Contexts of Use

Profiling Your Readers *20*

Profiling Contexts of Use *26*

Using Profiles to Your Own Advantage *27*

International and Cross-Cultural Communication *28*

Chapter Review *38*

Exercises and Projects *38*

Case Study: Installing a Medical Waste Incinerator *40*

In this chapter, you will learn:

- How to develop a comprehensive profile of a document's readers.
- How to use the computer as a reader analysis tool.
- How to sort your readers into primary, secondary, tertiary, or gatekeeper audiences.
- Techniques for identifying readers' needs, values, and attitudes about you and your document.
- How to analyze the physical, economic, political, and ethical contexts of use that influence how readers will interpret your text.
- How to anticipate the needs of international and cross-cultural readers.

Knowing your readers is vital to effective technical communication. Your readers will have their own needs, values, and attitudes about what you are saying. Meanwhile, more than ever, readers don't have time to slog through information they don't need. So, you should find out exactly what your readers need to know and how they want that information presented.

Another concern is the ever-increasing importance of international communication through electronic networks. In technical fields, you *will* find yourself regularly communicating with people who speak other languages, have different customs, and hold different expectations. Computers have broken down many of the geographical barriers that once separated people and cultures. It is now common to communicate with people around the world on a daily basis.

Profiling Your Readers

In technical communication, documents are designed to suit the needs of specific types of readers. For this reason, early in the writing process, you should profile the types of people who might be interested in your document.

Reader profiles are sketches of your readers' tendencies, abilities, experiences, needs, values, and attitudes. To build a profile, begin by asking yourself the Five-W and How Questions about your readers (Figure 2.1).

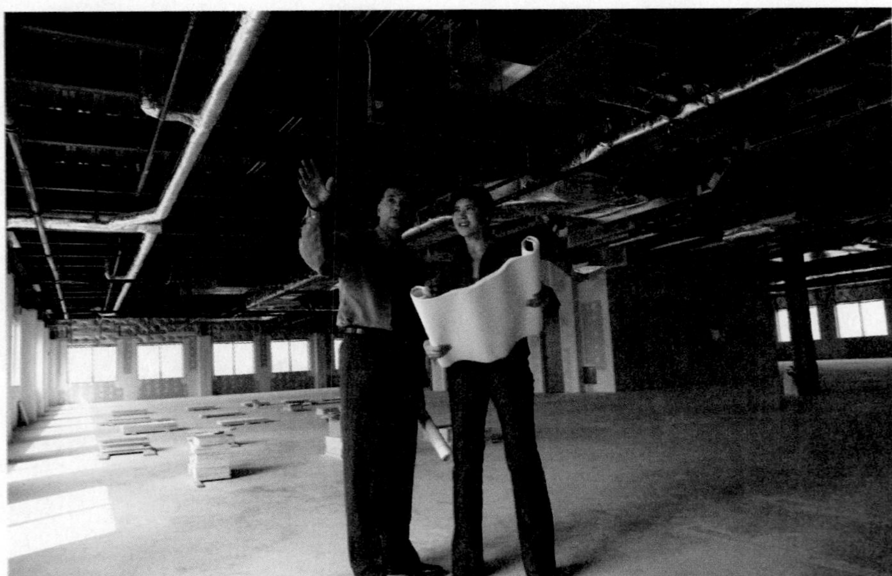

Your readers only want the information they need to make a decision or take action. As the writer, it is your job to find out what they need and how they want the information presented.

For websites that offer other ways to use the Five-W and How Questions, go to **www.pearsonhighered.com/johnsonweb4/2.1**

Developing a Reader Profile

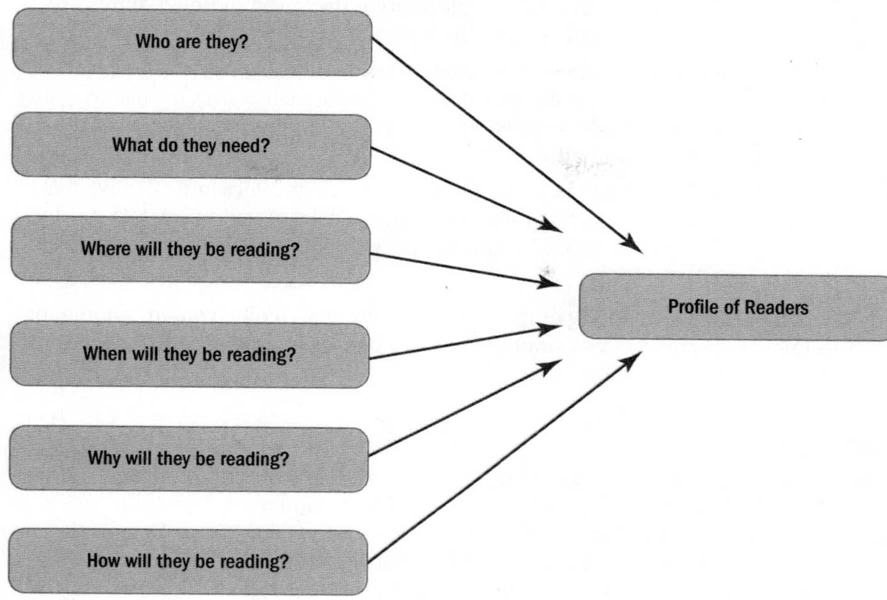

Figure 2.1: To develop a profile of your readers, use the Five-W and How Questions to look at them from a variety of perspectives.

AT A GLANCE

About Your Readers

- Readers are "raiders" for information.
- Readers are wholly responsible for interpreting your text.
- Readers want only "need-to-know" information.
- Readers prefer concise texts.
- Readers prefer documents with graphics and effective page design.

Who might read this document?
What information do they need?
Where will they read the document?
When will they read the document?
Why will they be reading it?
How will they be reading it?

As you answer these questions, keep in mind the following guidelines about your readers and how they prefer to read.

Guideline One: Readers are "raiders" for information—People don't read technical documents for pleasure. Instead, most readers are *raiding* your document for the information they need to make decisions or take action.

Guideline Two: Readers are wholly responsible for interpreting your text—You won't be available to explain what your document means, so your readers need to be able to easily figure out on their own what you are telling them.

GO TO THE NET What else do readers want? For answers, go to **www.pearsonhighered.com/johnsonweb4/2.2**

Guideline Three: Readers want only "need-to-know" information—Readers want you to give them only the information they need, nothing more. Any additional material only makes the information they want harder to find.

Guideline Four: Readers prefer concise texts—The shorter, the better. Usually, the longer the document is, the less likely it is that people are going to read it. Your readers prefer documents that get to the point and highlight the important information.

Guideline Five: Readers prefer documents with graphics and effective page design—We live in a visual culture. Large blocks of text intimidate most readers. So include graphics and use page design to make your document more readable.

Think about how you are reading this book. More than likely, you are raiding for need-to-know information. You want the book to be concise and visually interesting. Your readers want these things, too.

Identifying Your Readers

You should always begin by identifying the readers of your document. Figure 2.2 shows a Writer-Centered Analysis Chart that will help you locate the various kinds of people who might look over your text (Mathes & Stevenson, 1976). You, as the writer, are in the center ring. Each ring in the chart identifies your readers from most important (primary readers) to least important (tertiary readers).

Writer-Centered Analysis Chart

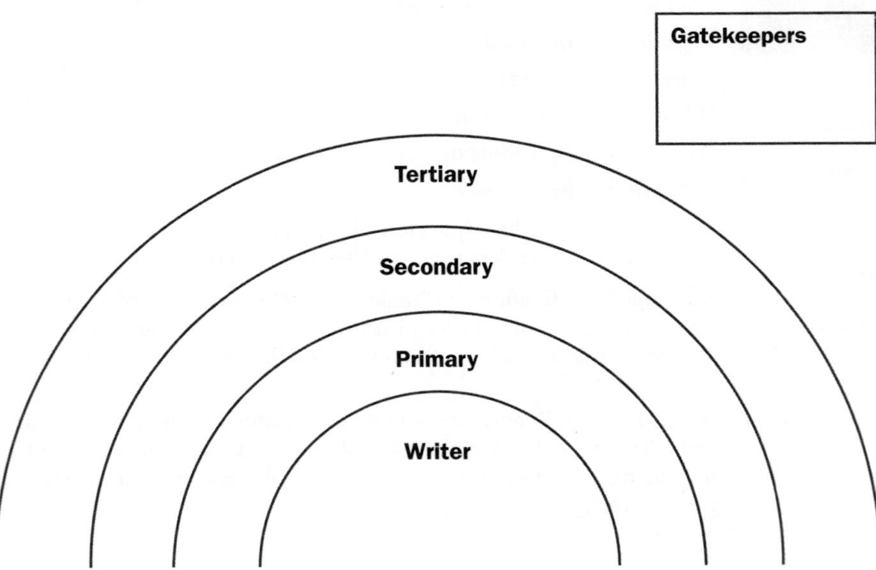

Figure 2.2: A Writer-Centered Analysis Chart starts with you in the center and identifies the various people who may be interested in your document.

GO TO THE NET For a downloadable version of the Writer-Centered Analysis Chart, go to **www.pearsonhighered.com/johnsonweb4/2.3**

To use the Writer-Centered Chart, begin filling in the names and titles of the primary, secondary, tertiary, and gatekeeper readers who will or might look over your work.

PRIMARY READERS (ACTION TAKERS) The primary readers are the people to whom your document is addressed. They are usually *action takers* because the information you are providing will allow them to do something or make a decision. Usually, your document will have only one or two primary readers or types of primary readers.

SECONDARY READERS (ADVISORS) The secondary readers are people who *advise* the primary readers. Usually, they are experts in the field, or they have special knowledge that the primary readers require to make a decision. They might be engineers, technicians, lawyers, scientists, doctors, accountants, and others to whom the primary readers will turn for advice.

AT A GLANCE

Types of Readers

- Primary readers: Action takers
- Secondary readers: Advisors
- Tertiary readers: Evaluators
- Gatekeepers: Supervisors

TERTIARY READERS (EVALUATORS) The tertiary readers include others who may have an interest in your document's information. They are often *evaluators* of you, your team, or your company. These readers might be local news reporters, lawyers, auditors, historians, politicians, community activists, environmentalists, or perhaps your company's competitors. Even if you don't expect your document to ever fall into these readers' hands, you should keep them in mind to avoid writing anything that could put you or your company at risk. Figure 2.3, for example, shows a memo in which the tertiary readers were not kept in mind.

GATEKEEPERS (SUPERVISORS) The gatekeepers are people who will need to look over your document before it is sent to the primary readers. Your most common gatekeeper is your immediate supervisor. In some cases, though, your company's lawyers, accountants, and others may need to sign off on the document before it is sent out.

Each of these four types of readers will look for different kinds of information. The primary readers are the most important, so their needs come first. Nevertheless, a well-written document also anticipates the needs of the secondary, tertiary, and gatekeeper readers.

Profiling Your Readers' Needs, Values, and Attitudes

Now that you have identified the readers of your document, you should develop profiles that describe their needs, values, and attitudes. Don't assume that your readers have the same needs, values, and attitudes as you do. Readers often have very different characteristics and expectations than the writers of a document.

As you begin considering your readers, think about some of the following issues:

- Readers' familiarity with the subject
- Readers' professional experience
- Readers' educational level
- Readers' reading and comprehension level
- Readers' skill level

A Memo That Does Not Consider Tertiary Readers

Figure 2.3: This memo was leaked to the press by someone at the U.S. Department of Homeland Security (DHS). In it, the DHS Deputy Director discusses the results of a survey of government employees. The survey's results reveal the incredibly low opinion that DHS's employees have of their department and its management. The author tries to spin the results while conceding the obvious. This memo caused some public embarrassment when it slipped out.

From: [DHS] Broadcast
Sent: Tuesday, January 30, 2007
Subject: MESSAGE FROM [DHS] DEPUTY SECRETARY (Michael) JACKSON: DHS FHC SURVEY RESULTS

Importance: High

January 30, 2007

Primary readers are identified here. → MEMORANDUM FOR ALL DHS EMPLOYEES

FROM: MICHAEL P. JACKSON

SUBJECT: Federal Human Capital Survey Results

The purpose and main point are stated up front.

The Office of Personnel Management (OPM) surveyed federal employees last summer about various measures of job satisfaction and agency performance, and the results will be released today. Over 10,400 DHS employees responded and, candidly, what you said shows that DHS is not where any of us wants to be.

The survey results will be posted on the OPM website (www.opm.gov) and our own DHS intranet, and we encourage you to review them in detail. In brief, of 36 peer federal agencies surveyed, DHS ranks as follows:

- 36th on the job satisfaction index
- 35th on the leadership and knowledge management index
- 36th on the results-oriented performance culture index
- 33rd on the talent management index

These results deliver a clear and jolting message from managers and line employees alike. On whole, it is not significantly changed since OPM's 2004 employee survey. Secretary Chertoff and I discussed these results with concern.

Initial details indicate that we get low marks in basic supervision, management and leadership. Some examples are:

This list summarizes the current situation.

- Promotion and pay increase based on merit
- Dealing with poor performance
- Rewarding creativity and innovation
- Leadership generating high levels of motivation in the workforce
- Recognition for doing a good job
- Lack of satisfaction with various component policies and procedures
- Lack of information about what is going on with the organization

I am writing to assure you that, starting at the top, the leadership team across DHS is committed to address the underlying reasons for DHS employee dissatisfaction and suggestions for improvement.

Standing up this new and vital Department is clearly not a walk in the park, but our employees bring a passion for this mission, great professionalism and outstanding performance every single day. DHS employees have shouldered the weight of long hours, complex integration assignments, multiple reorganizations, and no small amount of criticism. In some cases you've had to wait too long for tools you need to suceed.

These are not excuses to rationalize where we stand, rather an acknowledgement on my part of how much our team is doing. And there are good news items in the survey for DHS. As chief operating officer of DHS, I commit to improve results. We will need your help.

Several months ago, the Secretary asked the Homeland Security Advisory Council to study and suggest a strategy for creating a stronger common culture. This month, drawing on the experience of top executives in the private sector, the Council has delivered a set of recommendations for promoting a culture of excellence in DHS.

The author looks to the future.

In the days ahead, our Under Secretary for Management, Paul Schneider, will join the Secretary and me in evaluating carefully the details of the OPM survey and the HSAC report. Our first steps will be to analyze thoroughly the survey data, including specific attention to those government organizations that are recognized for their high performance in these areas, and determine the specific steps to improvement. This process wil include the leadership team in each operating component and every headquarters unit to discuss details of the survey with our workforce. We will do so with a sense of urgency and seriousness.

The memo returns to its main point.

Strengthening core management is one of the Secretary's highest priorities and the key elements are effective communications and proper recognition of our workforce. You deserve nothing less. We will build on some good work that has already been done to chart a path forward on these issues. We will then go where you point us, to improve job satisfaction for the DHS team.

Along the way, I will continue to ask for your help and guidance. Thanks in advance for that assistance, and thanks for what you are doing each day for DHS.

Source: Slate Magazine, *http://www.slate.com/id2158997.*

Want to see other controversial documents? Go to **www.pearsonhighered.com/johnsonweb4/2.4**

Reader Analysis Chart

Readers	Needs	Values	Attitudes
Primary			
Secondary			
Tertiary			
Gatekeepers			

Figure 2.4: To better understand your readers, fill in this Reader Analysis Chart with notes about their characteristics.

With these reader characteristics in mind, you can begin viewing your document from their perspective. To help you deepen this perspective, a Reader Analysis Chart, like the one shown in Figure 2.4, can help you identify your readers' needs, values, and attitudes toward your text.

To use the Reader Analysis Chart, fill in what you know about your readers' needs, values, and attitudes.

AT A GLANCE

Determining How Readers Make Decisions

- *Needs*—information the readers need to take action or make a decision
- *Values*—issues, goals, or beliefs that the readers feel are important
- *Attitudes*—the readers' emotional response to you, your project, or your company

NEEDS What information do your primary readers require to make a decision or take action? What do the secondary readers need if they are going to make positive recommendations to the primary readers? What are the tertiary and gatekeeper readers looking for in your document?

VALUES What do your readers value most? Do they value efficiency and consistency? Do they value accuracy? Is profit a key concern? How much do they value environmental or social concerns?

ATTITUDES What are your readers' attitudes toward you, your company, and the subject of your document? Will your readers be excited, upset, wary, positive, hopeful, careful, concerned, skeptical, or heartened by what you are telling them?

GO TO THE NET

For a downloadable version of the Reader Analysis Chart, go to **www.pearsonhighered.com/johnsonweb4/2.5**

As you fill in the Reader Analysis Chart, you will be making strategic guesses about your readers. Put a question mark (?) in spaces where you aren't sure about your readers' needs, values, or attitudes. These question marks highlight where you need to do more research on your readers.

To find out more, you might interview people who are Subject Matter Experts (SMEs) at your company or who hire themselves out as consultants. These experts may be able to give you insights into your readers' likely characteristics.

Above all, your goal is to view the situation from your readers' perspective. Your profile will help you anticipate how your readers act, react, and make decisions.

Profiling Contexts of Use

The places where people will read your document can strongly influence how they interpret what you say. So, you should also build a profile of the *contexts of use* in which they will read or use your document.

Identifying the Context of Use

Perhaps the most obvious concern is the *physical context* in which the document will be used. Will your readers be in their office or at a meeting? Will they be on the factory floor, trying to repair a robotic arm? Or are they in the emergency room, trying to save someone's life? Each of these physical contexts will alter the way your readers interpret your document.

But context of use goes beyond your readers' physical context. Your readers may also be influenced by the economic, ethical, and political issues that shape how they see the world. To help you sort out these various contexts, you can use a Context Analysis Chart like the one shown in Figure 2.5.

To use the Context Analysis Chart, fill in what you know about the physical, economic, political, and ethical issues that might influence the primary readers, their company, and their industry.

AT A GLANCE

Context of Use

- *Physical context*—the places where the readers will use your document
- *Economic context*—the money-related issues that will restrict the kinds of actions possible
- *Political context*—the micropolitical and macropolitical trends that will guide your readers
- *Ethical context*—the personal, social, and environmental issues that shape the readers' responses

PHYSICAL CONTEXT Where will your readers use your document? How do these various places affect how they will read your document? How should you write and design the document to fit these places?

ECONOMIC CONTEXT What are the economic issues that will influence your readers' decisions? What are the costs and benefits of your ideas? How would accepting your ideas change the financial situation of your readers, their company, or their industry?

POLITICAL CONTEXT What are the political forces influencing you and your readers? On a micropolitical level, how will your ideas affect your readers' relationships with you, their supervisors, or their colleagues? On a macropolitical level, how will political trends at the local, state, federal, and international levels shape how your readers interpret your ideas?

Context Analysis Chart

	Physical Context	Economic Context	Political Context	Ethical Context
Primary Readers				
Readers' Company				
Readers' Industry				

Figure 2.5: Each reader is influenced by physical, economic, political, and ethical concerns. A Context Analysis Chart anticipates these concerns for the primary readers, their company, and their industry.

ETHICAL CONTEXT How will your ideas affect the rights, values, and well-being of others? Does your document involve any social or environmental issues that might be of concern to your readers? Will any laws or rules be bent or broken if your readers do what you want?

Put a question mark (?) in spaces where you don't have specific information about your readers' physical, economic, political, and ethical contexts. You can then turn to the Internet for answers, or you can interview Subject Matter Experts who may have the answers you need.

Link

For more help with identifying ethical issues, see Chapter 4, page 69.

Using Profiles to Your Own Advantage

You are now ready to use your Reader Analysis and Context Analysis charts to strengthen your writing and make it more informed and persuasive. In your charts, circle or highlight the most important terms, concepts, and phrases. The items you circle are the *tensions* that you will need to address as you collect information and draft the document.

As you draft your document, your analysis of readers and contexts of use will help you

- make strategic decisions about what information to include in your document.
- organize your document to highlight the information that is most important to your readers.

For a downloadable version of the Context Analysis Chart, go to **www.pearsonhighered.com/johnsonweb4/2.6**

- develop a persuasive style that will appeal to your readers.
- design the document for the places it will be used.

Figures 2.6 and 2.7 show documents from the same website about the same topic, West Nile virus, that are written for two different types of readers. The first document, Figure 2.6, is written for the general public. Notice how it uses content, organization, style, and design to appeal to this audience.

The second document, Figure 2.7, is written for medical personnel. Notice how the content is far more complex, and the style is less personal in the document for medical personnel. Effective reader analysis and context analysis allowed the author of these documents (probably the same person) to effectively present the same information to two very different kinds of readers.

International and Cross-Cultural Communication

Computers have greatly blurred geographical and political boundaries. Whether you are developing software documentation or describing a heart transplant procedure, your documents will be read and used by people from different cultures. The use of computers, especially the Internet, has only heightened the necessity of working and communicating with people from different cultures (Hoft, 1995; Reynolds & Valentine, 2004). It's all very exciting—and very challenging.

International and cross-cultural issues will affect the content, organization, style, and design of your document.

Your documents will likely be used by people around the world.

Want more help with reader analysis? Go to **www.pearsonhighered.com/johnsonweb4/2.7**

Document Written for the General Public

Style is action oriented.

Images reinforce written text.

Headings help readers scan document.

Common examples are used to illustrate points.

Fight the Bite!

Avoid Mosquito Bites to Avoid Infection

Human illness from West Nile virus is rare, even in areas where the virus has been reported. The chance that any one person is going to become ill from a mosquito bite is low.

Three Ways to Reduce Your West Nile Virus Risk

1. Avoid Mosquito Bites
2. Mosquito-Proof Your Home
3. Help Your Community

Avoid Mosquito Bites!

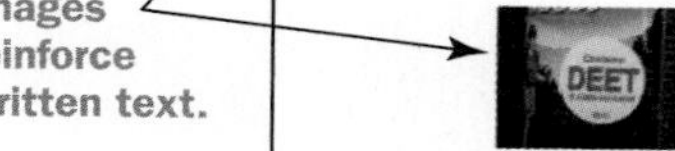

Apply Insect Repellent Containing DEET (Look for: *N,N-diethyl-meta-toluamide*) to exposed skin when you go outdoors. Even a short time sitting outdoors can be long enough to get a mosquito bite. For details on when and how to apply repellent, see CDC's Insect Repellent Use and Safety page: http://www.cdc.gov/ncidod/dvbid/westnile/qa/insect_repellent.htm .

Clothing Can Help Reduce Mosquito Bites. When possible, wear long-sleeves, long pants and socks when outdoors. Mosquitoes may bite through thin clothing, so spraying clothes with repellent containing permethrin or DEET will give extra protection. Don't apply repellents containing permethrin directly to skin. Do not spray repellent containing DEET on the skin under your clothing.

Be Aware of Peak Mosquito Hours. The hours from dusk to dawn are peak mosquito biting times. Consider avoiding outdoor activities during these times -- or take extra care to use repellent and protective clothing during evening and early morning.

Mosquito-Proof Your Home

Drain Standing Water. Mosquitoes lay their eggs in standing water. Limit the number of places around your home for mosquitoes to breed by getting rid of items that hold water. Need examples? Learn more on the Prevention of West Nile Virus question and answer page: http://www.cdc.gov/ncidod/dvbid/westnile/qa/prevention.htm

Install or Repair Screens. Some mosquitoes like to come indoors. Keep them outside by having well-fitting screens on both windows and doors. Offer to help neighbors whose screens might be in bad shape.

Help Your Community

Clean Up. Mosquito breeding sites can be everywhere. Neighborhood clean up days can be organized by civic or youth organizations to pick up containers from vacant lots, parks and to encourage people to keep thier yards free of standing water. Mosquitoes don't care about fences, so breeding sites anywhere in the neighborhood are important.

Source: Centers for Disease Control and Prevention (CDC), http://www.cdc.gov/ncidod/dvbid/westnile/index.htm.

Figure 2.6: This document on West Nile virus was written for the general public. It is action oriented and not very technical. The images also help reinforce the message, and the layout makes it highly scannable.

Figure 2.6: *(continued)*

Report Dead Birds to Local Authorities. Dead birds may be a sign that West Nile virus is circulating between birds and the mosquitoes in an area. Over 110 species of birds are known to have been infected with West Nile virus, though not all infected birds will die.

By reporting dead birds to state and local health departments, the public plays an important role in monitoring West Nile virus. Because state and local agencies have different policies for collecting and testing birds check the Links to State and Local Government Sites page to find information about reporting dead birds in your area: http://www.cdc.gov/ncidod/dvbid/westnile/city_states.htm. This page contains more information about reporting dead birds and dealing with bird carcasses: http://www.cdc.gov/ncidod/dvbid/westnile/qa/wnv_birds.htm

Mosquito Control Programs. Check with local health authorities to see if there is an organized mosquito control program in your area. If no program exists, work with your local government officials to establish a program. The American Mosquito Control Association (www.mosquito.org) can provide advice, and their book Organization for Mosquito Control is a useful reference. More questions about mosquito control? A source for information about pesticides and repellents is the National Pesticide Information Center: http://npic.orst.edu/ , which also operates a toll-free information line: 1-800-858-7378 (check their Web site for hours).

Find out more about local prevention efforts. Find state and local West Nile virus information and contacts on the Links to State and Local Government Sites page.

Links provide more information.

Document Written for Experts

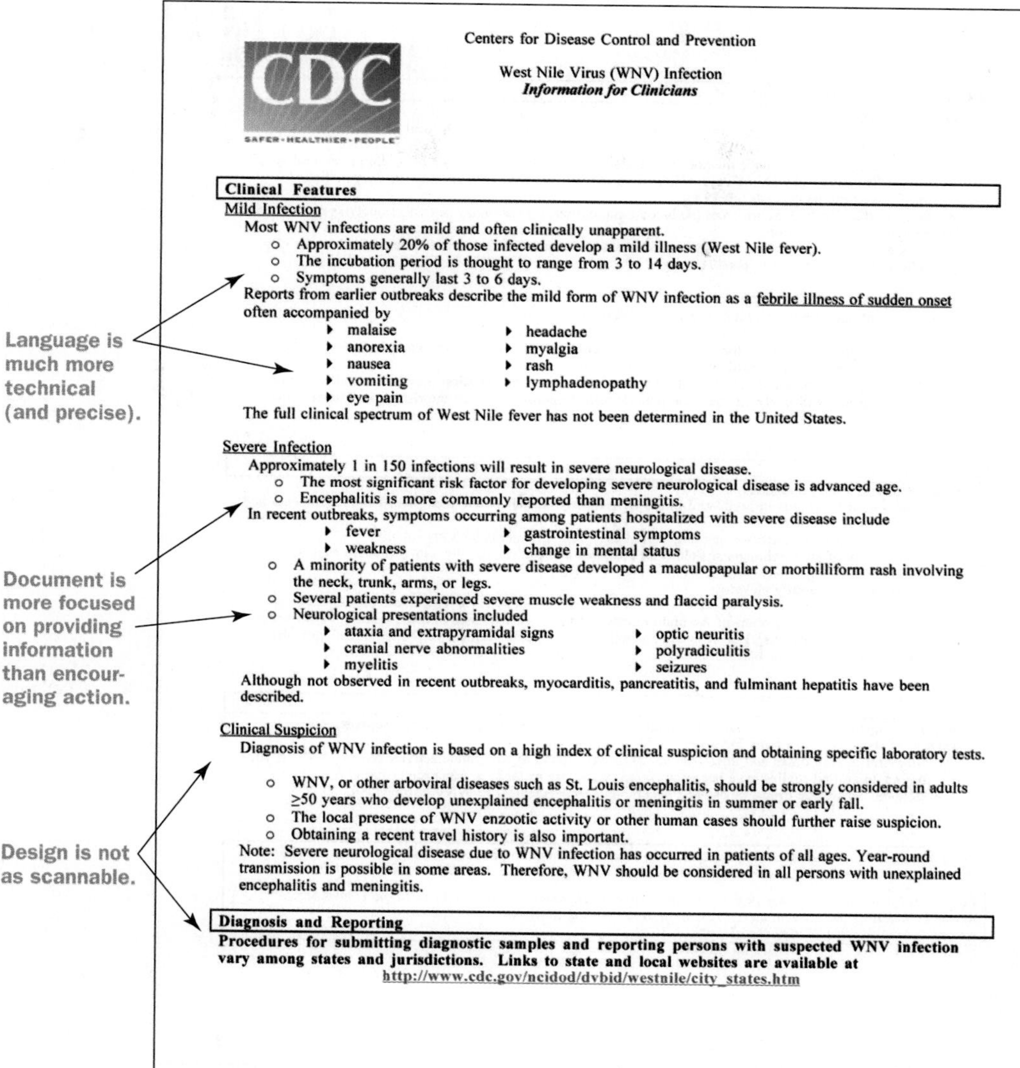

Centers for Disease Control and Prevention

CDC
SAFER·HEALTHIER·PEOPLE™

West Nile Virus (WNV) Infection
Information for Clinicians

Clinical Features

Mild Infection

Most WNV infections are mild and often clinically unapparent.

- Approximately 20% of those infected develop a mild illness (West Nile fever).
- The incubation period is thought to range from 3 to 14 days.
- Symptoms generally last 3 to 6 days.

Reports from earlier outbreaks describe the mild form of WNV infection as a febrile illness of sudden onset often accompanied by

- malaise
- anorexia
- nausea
- vomiting
- eye pain
- headache
- myalgia
- rash
- lymphadenopathy

The full clinical spectrum of West Nile fever has not been determined in the United States.

Severe Infection

Approximately 1 in 150 infections will result in severe neurological disease.

- The most significant risk factor for developing severe neurological disease is advanced age.
- Encephalitis is more commonly reported than meningitis.

In recent outbreaks, symptoms occurring among patients hospitalized with severe disease include

- fever
- weakness
- gastrointestinal symptoms
- change in mental status

- A minority of patients with severe disease developed a maculopapular or morbilliform rash involving the neck, trunk, arms, or legs.
- Several patients experienced severe muscle weakness and flaccid paralysis.
- Neurological presentations included
  - ataxia and extrapyramidal signs
  - cranial nerve abnormalities
  - myelitis
  - optic neuritis
  - polyradiculitis
  - seizures

Although not observed in recent outbreaks, myocarditis, pancreatitis, and fulminant hepatitis have been described.

Clinical Suspicion

Diagnosis of WNV infection is based on a high index of clinical suspicion and obtaining specific laboratory tests.

- WNV, or other arboviral diseases such as St. Louis encephalitis, should be strongly considered in adults ≥50 years who develop unexplained encephalitis or meningitis in summer or early fall.
- The local presence of WNV enzootic activity or other human cases should further raise suspicion.
- Obtaining a recent travel history is also important.

Note: Severe neurological disease due to WNV infection has occurred in patients of all ages. Year-round transmission is possible in some areas. Therefore, WNV should be considered in all persons with unexplained encephalitis and meningitis.

Diagnosis and Reporting

Procedures for submitting diagnostic samples and reporting persons with suspected WNV infection vary among states and jurisdictions. Links to state and local websites are available at http://www.cdc.gov/ncidod/dvbid/westnile/city_states.htm

Language is much more technical (and precise).
Document is more focused on providing information than encouraging action.
Design is not as scannable.

Figure 2.7: This document was written for clinicians and other medical personnel. It is far more technical in style and design than the document written for the general public. Nevertheless, it contains much of the same information.

Source: Centers for Disease Control and Prevention (CDC), http://www.cdc.gov/ncidod/dvbid/westnile/index.htm.

Diagnosis and Reporting – *continued*

Diagnostic Testing

WNV testing for patients with encephalitis or meningitis can be obtained through local or state health departments.

- The most efficient diagnostic method is detection of IgM antibody to WNV in serum or cerebral spinal fluid (CSF) collected within 8 days of illness onset using the IgM antibody capture enzyme-linked immunosorbent assay (MAC-ELISA).
- Since IgM antibody does not cross the blood-brain barrier, IgM antibody in CSF strongly suggests central nervous system infection.
- Patients who have been recently vaccinated against or recently infected with related flaviviruses (e.g., yellow fever, Japanese encephalitis, dengue) may have positive WNV MAC-ELISA results.

Reporting Suspected WNV Infection

Refer to local and state health department reporting requirements: www.cdc.gov/ncidod/dvbid/westnile/city_states.htm

- WNV encephalitis is on the list of designated nationally notifiable arboviral encephalitides.
- Aseptic meningitis is reportable in some jurisdictions.

The timely identification of persons with acute WNV or other arboviral infection may have significant public health implications and will likely augment the public health response to reduce the risk of additional human infections.

Laboratory Findings

Among patients in recent outbreaks

- Total leukocyte counts in peripheral blood were mostly normal or elevated, with lymphocytopenia and anemia also occurring.
- Hyponatremia was sometimes present, particularly among patients with encephalitis.
- Examination of the cerebrospinal fluid (CSF) showed pleocytosis, usually with a predominance of lymphocytes.
- Protein was universally elevated.
- Glucose was normal.
- Computed tomographic scans of the brain mostly did not show evidence of acute disease, but in about one-third of patients, magnetic resonance imaging showed enhancement of the leptomeninges, the periventricular areas, or both.

Treatment

Treatment is supportive, often involving hospitalization, intravenous fluids, respiratory support, and prevention of secondary infections for patients with severe disease.

- Ribavirin in high doses and interferon alpha-2b were found to have some activity against WNV in vitro, but no controlled studies have been completed on the use of these or other medications, including steroids, antiseizure drugs, or osmotic agents, in the management of WNV encephalitis.

For additional clinical information, please refer to Petersen LR and Marfin AA, "West Nile Virus: A Primer for the Clinician[Review]," Annals of Internal Medicine (August 6) 2002: 137:173-9.

For clinical and laboratory case definitions, see "Epidemic/Epizootic West Nile Virus in the United States: Revised Guidelines for Surveillance, Prevention, and Control, 2001," at www.cdc.gov/ncidod/dvbid/westnile/surv&control.htm

Links provide additional information.

Summaries are used to present important information.

Places where more information can be found are identified.

Figure 2.7: *(continued)*

Differences in Content

Cultures have different expectations about content in technical documentation:

- In China, the content of your documents and presentations should be fact based, and you should focus on long-term benefits for your readers and you, not short-term gains. In business, the Chinese tend to trust relationships above all, so they look for facts in documents and they do not like overt attempts to persuade.

- In Mexico, South America, and many African countries, family and personal backgrounds are of great importance. It is common for family-related issues to be mentioned in public relations, advertising, and documentation. Business relationships and meetings often start with exchanges about families and personal interests.
- In the Middle East, Arabs often put a premium on negotiation and bargaining, especially when it comes to the price of a service or product. As a result, it is crucial that all the details in documents are spelled out exactly before the two sides try to work out a deal. In most cases, though, the first offer you make will rarely be considered the final offer.
- In Asian countries, the reputation of the writer or company is essential for establishing the credibility of the information (Haneda & Shima, 1983). Interpersonal relationships and prior experiences can sometimes even trump empirical evidence in Asia.
- Also in Asia, contextual cues can be more important than content. In other words, *how* someone says something may be more important than *what* he or she is saying. For example, when Japanese people speak or write in their own language, they rarely use the word *no*. Instead, they rely on contextual cues to signal the refusal. As a result, when Japanese is translated into English, these "high-context" linguistic strategies are often misunderstood (Chaney & Martin, 2004). Similarly, in Indonesia, the phrase "Yes, but" actually means "no" when someone is speaking.
- In India, business is often conducted in English because the nation has over a dozen major languages and hundreds of minor languages. So, don't be surprised when your Indian partners are very fluent in English and expect you to show a high level of fluency, especially if you are a native English speaker.
- In several African countries like Tunisia and Morocco, business tends to be conducted in French, even though the official language of the country is Arabic.

Link

To learn more about working in high-context cultures, go to Chapter 13, page 378.

Differences in Organization

The organization of a document often needs to be altered to suit an international audience. Organizational structures that Americans perceive to be "logical" or "common sense" can seem confusing and even rude in some cultures.

- In Arab cultures, documents and meetings often start out with statements of appreciation and attempts to build common bonds among people. The American tendency to "get to the point" can be seen as rude.
- Also in Arabic cultures, documents rely on repetition to make their points. To North Americans, this repetition might seem like the document is moving one step back for every two steps forward. To Arabs, American documents often seem incomplete because they lack this repetition.
- Asians often prefer to start out with contextual information about nonbusiness issues. For example, it is common for Japanese writers to start out letters by

Want to learn more about international readers? Go to **www.pearsonhighered.com/johnsonweb4/2.8**

Computers, especially networked computers, have increased opportunities to work across cultures.

saying something about the weather. To some Asians, American documents seem abrupt, because Americans tend to bluntly highlight goals and objectives up front.

- In India, the term *thank you* is considered a form of payment and is used more formally than in American culture. So, if someone has done you a favor, you should accept the favor with appreciation but without saying thank you. It's fine to say thank you when doing other kinds of business.

Differences in Style

Beyond difficulties with translation, style is usually an important difference among cultures:

- In China, overt attempts at persuasion are often seen as rude and undesirable. Instead, documents and meetings should be used to build relationships and present factual information. Strong relationships lead to good business, not the other way around.
- Arabic style may seem overly ornamental to North Americans, making Arabic documents and presentations seem colorful to non-Arabs. On the other hand, the American reliance on "plain language" can offend the sensibilities of Arabs, who prefer a more ornate style in formal documents.

GO TO THE NET For information sources on other countries, go to **www.pearsonhighered.com/johnsonweb4/2.9**

- In Mexico and much of South America, an informal style often suggests a lack of respect for the project, the product, or the readers. Mexicans especially value formality in business settings, so the use of first names and contractions in business prose can be offensive.
- In sub-Saharan Africa, readers prefer a document's tone to stress a win–win situation. Your tone, therefore, should imply that both sides will benefit from the arrangement.
- Some Native Americans prefer the sense that everyone had input on the document. Therefore, a direct writing or presentation style will meet resistance because it will seem to represent the opinion of only one person.
- In North America, women are more direct than women in other parts of the world, including Europe. This directness often works to their advantage in other countries, because they are viewed as confident and forward thinking. However, as writers and speakers, women should not be too surprised when people of other cultures resist their directness.

Differences in Design

Even the design of documents is important when you are working with international and cross-cultural readers:

- Arabic and some Chinese scripts are read from right to left, unlike English, which is read from left to right (Figure 2.8). As a result, Arabic and some Chinese readers tend to scan pages and images differently than Americans or Europeans do.
- Some icons that show hand gestures—like the OK sign, a pointing finger, or a peace sign with the back of the hand facing outward—can be highly offensive in some cultures. Imagine a document in which a hand with the middle finger extended is used to point at things. You get the picture.
- In many South American and Asian cultures, the use of the right hand is preferred when handing items (e.g., business cards, documents, products) to people. Therefore, pictures or drawings in documents should show people using their right hands to interact.

Different Ways of Scanning a Page

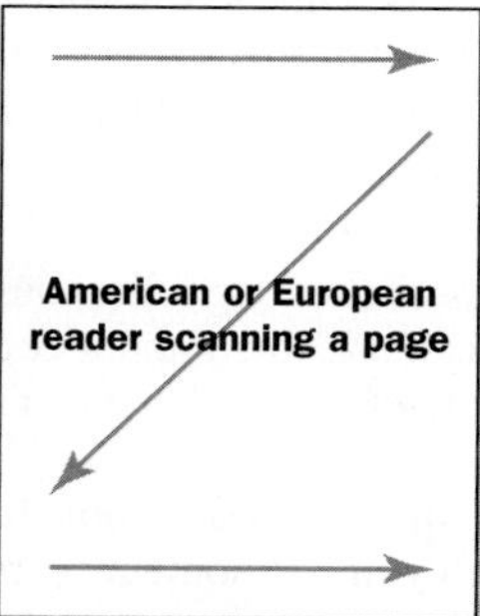

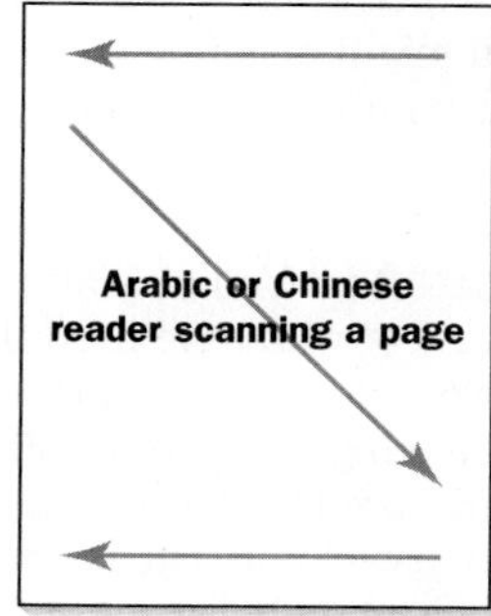

Figure 2.8: Readers from other cultures may scan the page differently. The design needs to take their preferences into account.

For more about using graphics in other cultures, go to **www.pearsonhighered.com/johnsonweb4/2.10**

- In some Asian cultures, a white flower or white dress can symbolize death. As a result, a photograph of white flowers or white dresses can signal a funeral or mourning.
- Europeans find that American texts include too many graphics and use too much white space. Americans, meanwhile, often find that the small margins in European texts make the documents look crowded and cramped.
- Graphs and charts that seem to have obvious meanings to Americans can be baffling and confusing to readers from other cultures. If your international document includes graphs and charts, you should seek out someone from the readers' culture to help you determine whether your visuals will be understood.
- When giving presentations in some Native American cultures, hand gestures should be limited and eye contact should be minimized. Ironically, this advice is exactly the opposite of what most public speaking coaches suggest for non-Native audiences.

Listen and Learn: The Key to International and Cross-Cultural Communication

With all these differences in content, organization, style, and design, how can you possibly write for international or cross-cultural readers? Here are four helpful strategies:

LISTEN CAREFULLY Careful listening is a valued skill in all cultures, and you will learn a great amount by simply paying attention to what your readers expect the document to include and how it should look.

BE POLITE Politeness in one culture tends to translate well into other cultures. For example, words like *please* and *thank you* are universally seen as polite. Smiles and a friendly tone are almost always welcome. There are subtle differences in how these words and gestures are used in other cultures, but your readers will understand that you are trying to be polite.

RESEARCH THE TARGET CULTURE Use the Internet to do some research into your readers' cultural expectations for technical documents. On the Internet or at your workplace, you might also find some model texts from the readers' culture. Use them to help guide your decisions about content, organization, style, and design.

TALK TO YOUR COLLEAGUES You may also seek out co-workers or colleagues who are from the target culture or who have lived there. You can ask them about conventions that might make your document or presentation more effective. They can also help you avoid doing anything awkward or offensive.

Overall, when you are communicating with international readers or people from different cultures, be observant and listen to what they tell you. Do some research into their expectations, and be ready to learn from your mistakes.

Want to learn more about Native American cultures? Go to **www.pearsonhighered.com/johnsonweb4/2.11**

AT WORK

Carol Leininger, Ph.D.

GLOBAL PROJECT LEADER, F. HOFFMAN-LA ROCHE, SWISS HEADQUARTERS

F. Hoffman-La Roche is a pharmaceutical company that works closely with a partner company, Chugai Pharmaceuticals, in Tokyo, Japan (8 hours time difference). In 2009, La-Roche integrated with their California partner, Genentech (9 hours time difference).

What are some strategies for communicating with people from another culture?

Physical distance is the biggest hurdle in global communication, affecting how people work together even more than language or culture does. Effects of distance are felt through the difficulty in meeting face-to-face and, perhaps even worse, in not sharing similar time zones. When people are meeting face to face, they work out issues through interacting, even when they are not fluent in the same languages. Awkward moments may arise, which is to be expected in real-time interactions in any language

Time zone differences necessitate constant calculation and recalibration, which gets easier with practice yet never goes away. The time zone differences (distance) represent a logistics challenge in teleconferencing, videoconferencing, and even e-mail and texting. For example, a teleconference starting five minutes late may be more upsetting to those at the end of the local working day than to someone just starting their workday; the value of a "wasted" five minutes may be higher at 6 P.M. than at 9 A.M. So working across distance, even in the same language, can still be a communication challenge.

Solid *preparation* is the key to working across distances and languages. Communication needs to be structured and simple. Your readers may be reading your text in their third or fourth language, so make their job easier by writing as simply and clearly as you can.

All the rules for good technical communication in the United States apply to international communication—only more so. What helps second-language or non-U.S. English speakers?

- State your objectives and purpose clearly.
- Use language consistently (i.e., use the same terms for the same things). While this may seem wooden, it will increase clarity.
- Do not even attempt humor until a relationship has been well-established.
- Rank issues by importance.
- Handle only one message per e-mail.
- Use headings and subheadings that convey a specific meaning.
- Minimize use of adjectives and adverbs.
- Minimize prepositional phrases.
- Highlight actions, deadlines, and dates.
- Spell everything correctly (always check the spelling).

Always be as polite as you can by your own cultural standards (e.g., formal language, politeness markers like "please" and "thank you," use of full names in greetings and salutations). Even if your cultural view of what is "polite" is different from that of your audience, your intention to be polite will be recognized by international readers as courtesy and civility.

GO TO THE NET

To find websites that discuss politeness strategies, go to **www.pearsonhighered.com/johnsonweb4/2.12**

CHAPTER REVIEW

- Early in the writing process, you should begin developing a profile of the types of people who may be interested in your document.
- You can use your computer as a reader analysis tool to better tailor your document's content, organization, style, and design to the needs of your readers.
- Your readers will include *primary readers* (action takers), *secondary readers* (advisors), *tertiary readers* (evaluators), and *gatekeepers* (supervisors).
- In your documents and presentations, you should anticipate various readers' needs, values, and attitudes.
- You should anticipate the document's contexts of use, which include the physical, economic, political, and ethical factors that may influence a reader's ideas.
- The emergence of the Internet has heightened the importance of international and cross-cultural communication. You need to adjust the content, organization, style, and design of your text to be sensitive to cross-cultural needs.

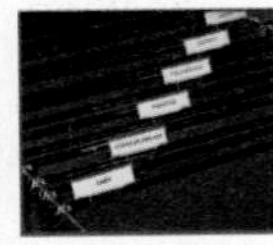

EXERCISES AND PROJECTS

Individual or Team Projects

1. Choose two websites that are designed for very different types of readers. Write a memo to your instructor in which you compare and contrast the websites, showing how they approach their readers differently. How do they use content, organization, style, and design to meet the needs, values, and attitudes of their readers?

   Some pairs of websites you might consider include websites for cars (chevrolet.com versus www.miniusa.com), magazines (time.com versus outsidemag.com), or computers (dell.com versus apple.com). Look for websites for products that are similar but pursue different kinds of customers.

2. Consider the advertisement in Figure 2.9 and "reverse-engineer" its reader analysis. Using a Writer-Centered Analysis Chart and a Reader Analysis Chart, identify the primary, secondary, and tertiary readers of the text. Then, make guesses about the needs, values, and attitudes of these readers.

   Write a report to your instructor in which you use your charts to discuss the readers of this document. Then, show how the document anticipates these readers' needs.

3. Choose a country or culture that interests you. Then, find two texts written by people from that country or culture. Write a memo to your instructor in which you discuss any similarities or differences between the texts and your own expectations for texts. Pay close attention to differences in content, organization, style, and design of these texts.

Source: Microsoft Game Studios, www.microsoft-careers.com/go/343-Industries-Jobs/190537.

Figure 2.9: This advertisement is actually part of a help wanted ad from Microsoft. Who are the intended readers of this advertisement? What are their needs, values, and attitudes?

Collaborative Project

With a group of people from your class, create a website that explores the needs, values, and attitudes of people from a different country or culture. The website does not need to be complex. On the Internet, identify various websites that offer information on that country or culture. Then, organize those websites by content and create links to them. Specifically, pay attention to the ways in which this country's physical, economic, political, and ethical contexts shape the way its people live their lives.

When you are finished with the website, give a presentation to your class in which your group discusses how this country or culture differs from your own. Answer the following question: If you were going to offer a product or service to the people of this country or culture, what considerations would you need to keep in mind? If you needed to write a proposal or a set of instructions to people from this country or culture, how might you need to adjust it to fit their unique qualities?

For additional technical writing resources, including interactive sample documents, document design tutorials and guidelines, and more, go to **www.mytechcommlab.com.**

CASE STUDY

Installing a Medical Waste Incinerator

Duane Jackson knew this decision was going to be difficult. As the assistant city engineer for Dover City, he was frequently asked to study construction proposals sent to the city council. He would then write a report with a recommendation. So, when the proposal for constructing a medical waste incinerator crossed his desk, he knew there was going to be trouble.

Overall, the proposal from Valley Medical, Inc., looked solid. The incinerator would be within 3 miles of the two major hospitals and a biotech research facility. And, it would bring about 30 good jobs to the Blue Park neighborhood, an economically depressed part of town.

The problem was that people in Blue Park were going to be skeptical. Duane grew up in a neighborhood like Blue Park, primarily African-American and lower middle class. He knew that hazardous industries often put their operations in these kinds of neighborhoods because the locals did not have the financial resources or political clout to fight them. In the past, companies had taken advantage of these neighborhoods' political weaknesses, leaving the areas polluted and unhealthy.

Powerful interests were weighing in on this issue. Dover City's mayor wanted the incinerator badly because she wanted the economic boost the new business would provide. Certainly, the executives at the hospitals and research laboratory were enthusiastic, because a nearby incinerator would help them cut costs. The city councilor who represented Blue Park wanted the jobs, but not at the expense of his constituents' health. Environmental groups, health advocates, and neighborhood associations were cautious about the incinerator.

Analyzing the Readers

After a few weeks of intense study, Duane's research convinced him that the incinerator was not a health hazard to the people of Blue Park. Similar incinerators built by Valley Medical had spotless records. Emissions would be minimal because advanced "scrubbers" would remove almost all the particles left over after incineration. The scrubbers were very advanced, almost completely removing any harmful pollutants such as dioxin and mercury.

Also, the company had a good plan for ensuring that medical waste would not sit around in trucks or containers, waiting to be burned. The waste would be immediately incinerated on arrival.

GO TO THE NET

For more information about waste incinerators, go to **www.pearsonhighered.com/johnsonweb4/2.13**

Duane's Writer-Centered Analysis Chart

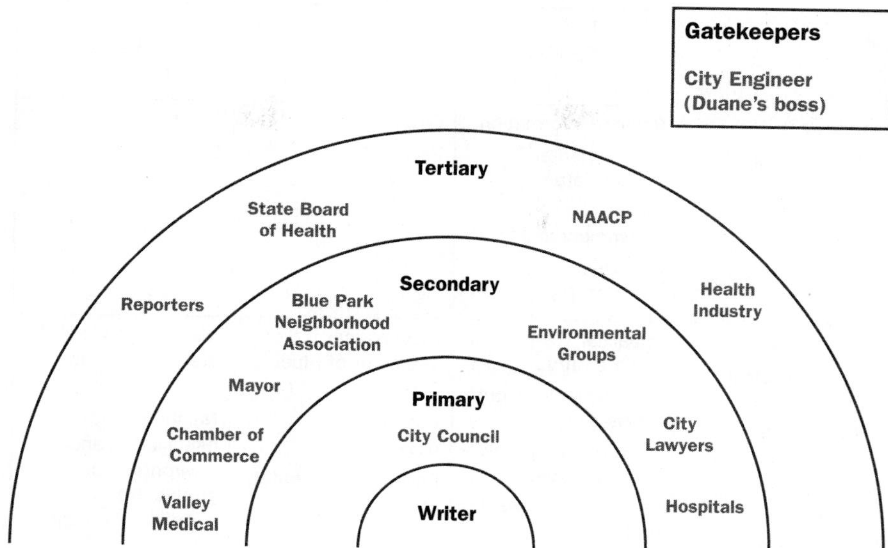

Figure 2.10: Duane's Writer-Centered Analysis Chart showed how many readers his report would have.

Duane decided to write a report to the city council that recommended the incinerator be built. That decision was the easy part. Now he needed to write a report that would convince the skeptics.

After identifying the subject and purpose of the report, Duane decided to do a thorough analysis of his readers and the report's contexts of use. He began with a Writer-Centered Analysis Chart (Figure 2.10). He then used a Reader Analysis Chart to identify the various readers' needs, values, and attitudes (Figure 2.11). Finally, Duane filled out a Context Analysis Chart to identify the physical, economic, political, and ethical issues involved (Figure 2.12).

(continued)

Duane's Reader Analysis Chart

	Needs	Values	Attitudes
Primary • City Council	Reliable information Environmental impact data Clear recommendation Impartial commentary	Citizen safety Economic development Fairness	Optimistic Cautious
Secondary • Valley Medical • Chamber of Commerce • Mayor • Neighborhood Association • Environmental Groups • City Lawyers • Hospitals	Impartial commentary Specific facts about emissions Cultural and social considerations Valley Medical wants profits	Maintaining character of Blue Park Economic development Environmental safety Safe disposal of waste	Mayor and hospitals are positive and hopeful Neighborhood association and environmental groups are skeptical, perhaps resistant Valley Medical hopeful
Tertiary • Reporters • State Board of Health • NAACP • Health Industry	Impartial decision Reassurance that race or poverty are not factors Basic facts about incinerator	Fairness and honesty Lack of bias Protection of people with little political power	Skeptical Open-minded
Gatekeepers • City Engineer	Reliable information Clear decision based on reliable data	Minimal trouble with mayor and council Low profile; stay out of politics Honesty	Impartial to project Concerned that report may cause tensions

Figure 2.11: Duane filled in a Reader Analysis Chart, noting everything he knew about the various readers of his report. He noticed that most readers wanted impartial and reliable information. Some readers were positive; others were skeptical.

Duane's Context Analysis Chart

	Physical Context	Economic Context	Political Context	Ethical Context
Primary Readers • City Council	Initially at their office Later, in city council meeting	Looking to improve city economics	Re-election is always an issue Mayor wants it BP city councilor would like it	Exploiting poor neighborhood? Racial issues? Environmental issues
Readers' Company • Dover City	City Hall Engineers' office City website	The jobs are a financial plus. Home values in neighborhood decline? Good for hospitals	Voters are wary Neighborhood and environmental groups draw attention Racial politics?	City liable if mistake? Environmental impact Infringement of people's rights? Legal issues
Readers' Industry • City government	Reports on Internet about project	Job creation Economic growth	Pressure on mayor Shows Dover City is positive about this kind of business	Public relations Don't want to seem exploitative

Figure 2.12: Duane's Context Analysis Chart revealed some interesting tensions between the economics and politics of the decision on the incinerator.

What Should Duane Do Now?

On the facts alone, Duane was convinced that the incinerator would be well placed in the Blue Park neighborhood. But his reader analysis charts showed that facts alone were not going to win over his primary audience, the members of the city council. They would have numerous other economic, political, and ethical issues to consider besides the facts. In many ways, these factors were more important than the empirical evidence for the incinerator.

One important thing Duane noticed was that the city council members, although they were the "action takers," were heavily influenced by the secondary readers. These secondary readers, the "advisors," would play a large role in the council's decision.

How can Duane adjust the content, organization, style, and design to better write his report? If you were Duane's readers, what kinds of information would you expect in this kind of report?

CHAPTER

3

Working in Teams

The Stages of Teaming *45*

Forming: Strategic Planning *45*

Storming: Managing Conflict *54*

Norming: Determining Team Roles *58*

Help: Virtual Teaming *59*

Performing: Improving Quality *63*

The Keys to Teaming *63*

Chapter Review *65*

Exercises and Projects *65*

Case Study: Not a Sunny Day *67*

In this chapter, you will learn:

- Why working in teams is essential in technical workplaces.
- The four stages of teaming in the workplace: forming, storming, norming, and performing.
- How to use strategic planning to form a team and begin a project.
- Strategies for managing team conflict in the storming stage.
- How to define team roles in the norming stage to improve productivity.
- How to improve performance with Total Quality Management (TQM) strategies.
- How to work as part of a "virtual" team.

Working in teams happens every day in the technical workplace. In fact, when managers are surveyed about the abilities they look for in new employees, they often put "works well with a team" near the top of their list. The ability to collaborate with others is an essential skill if you are going to succeed in today's networked workplace.

Computers have only increased the ability and necessity to work in teams. Communication tools like e-mail, instant messaging, chat rooms, and websites allow people to work together electronically. Telecommuting, teleworking, and "virtual offices" are becoming more common, and people are increasingly finding themselves working outside the traditional office setting. Now, more than ever, it is essential that you learn how to use computers to help you work effectively with a team of others.

The Stages of Teaming

It would be nice if people worked well together from the start. But in reality, team members often need time to set goals and adjust to each other's working styles and abilities. In 1965, Bruce Tuckman introduced a model of how teams learn to work together (Figure 3.1). He pointed out that teams go through four stages.

These stages are not rigid. Instead, a team tends to move back and forth among the stages as the project evolves and moves forward.

Forming: Strategic Planning

Forming is an important part of the team-building process. When a team is first created, the members are usually excited and optimistic about the new project. They are often a little anxious, because each person is uncertain about the others' expectations.

Tuckman's Four Stages of Teaming

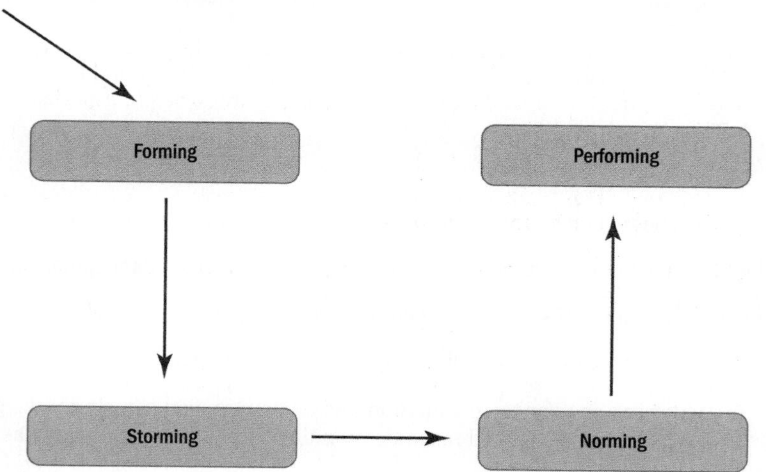

Figure 3.1: A team will typically go through four stages: forming, storming, norming, and performing. Give your team time to properly evolve as a unit.

GO TO THE NET

To learn more about Tuckman's teaming theories, go to **www.pearsonhighered.com/johnsonweb4/3.1**

So, in the forming stage, members should spend time getting to know each other and assessing each other's strengths and abilities.

When forming a new team, strategic planning is the key to effective teamwork. By working through the following steps at the beginning of a project, you will give your team time to form properly, often saving yourselves time and frustration as the project moves forward.

Step 1: Define the Project Mission and Objectives

Link

For more help defining the rhetorical situation, turn to Chapter 1, page 5.

Don't just rush in. Take some time to first determine what your team is being asked to accomplish. A good way to start forming is to define the rhetorical situation in which your team is working:

Subject—What are we being asked to develop? What are the boundaries of our project? What are we *not* being asked to do?

Purpose (mission statement)—What is the mission of the project? Why are we being asked to do this? What are the end results (deliverables) that we are being asked to produce?

Readers—Who are our clients? What are their needs, values, and attitudes? Who will be evaluating our work?

Context—What are the physical, economic, political, and ethical factors that will influence this project? How should we adjust to them?

Your statement of the purpose is your *mission statement* for the project. You and your team members should first agree on this mission statement before you do anything else. Your mission statement (purpose) is your primary objective. Next, list two to five secondary objectives that you intend to reach along the way.

Our Mission:

The purpose of this Staph Infection Task Force is to determine the level of staph infection vulnerability at St. Thomas Medical Center and to develop strategies for limiting our patients' exposure to staph, especially MRSA.

Secondary Objectives:

- Gain a better understanding of current research on staph and its treatments.
- Raise awareness of staph infections among our medical staff and patients.
- Assess the level of staph risk, especially MRSA, here at the hospital.
- Develop methods for controlling staph on hospital surfaces.

By first agreeing on your team's mission and objectives, you can clarify the goals of the project. You can also avoid misunderstandings about what the team was formed to accomplish.

For websites that describe other planning strategies, go to **www.pearsonhighered.com/johnsonweb4/3.2**

Step 2: Identify Project Outcomes

Outcomes are the tangible results of your project. Your project outcomes describe the measurable results of the team's efforts. To identify the outcomes of your project, simply convert your objectives into measurable results (Figure 3.2). Then, specify the *deliverables* that the project will produce. Deliverables are the products or services that you will deliver to the client during the project and after it is completed.

This process of turning objectives into outcomes and deliverables can take some time. But in the end, the process will save you time, because everyone in your team will have a clear understanding of the expected results and the products that will be created.

Step 3: Define Team Member Responsibilities

Not everyone is good at everything. So, once you have defined your objectives, ask each member of the team to talk about his or her abilities and previous experiences. Ask team members to identify ways they could contribute to the project. Discuss any time limitations or potential conflicts with other projects.

Defining the Project Objectives and Outcomes

Objectives	Outcomes
Gain a better understanding of current research on staph and its treatments	Outcome: Collection of current data and research literature on staph infections. Interview other hospitals about successful control methods. **Deliverable: A report on the findings, due May 25**
Raise awareness of staph infections among our medical staff and patients	Outcome: More awareness of the problem here at the hospital. People taking precautions against staph. **Deliverable: Pamphlets and posters raising awareness and describing staph control procedures**
Assess the level of staph risk, especially MRSA, here at the hospital	Outcome: Experiments that quantify the staph risk at the hospital **Deliverable: Report to the administration about the extent of the problem, due June 2**
Develop methods for controlling staph on hospital surfaces	Outcome: Develop strategies for preventing and containing staph infections. **Deliverable: Create a contingency plan that offers concrete steps for controlling staph. Make training modules to educate staff about the problem.**

Figure 3.2: By transforming your objectives into outcomes and deliverables, you can show how abstract goals become measurable results.

GO TO THE NET To see other sample mission statements, go to **www.pearsonhighered.com/johnsonweb4/3.3**

If your team's project involves writing a document, you should identify each team member's responsibilities while dividing up the writing task. For example, here are four jobs you might consider:

> **Coordinator**—The coordinator is responsible for maintaining the project schedule and running the meetings. The coordinator is not the "boss." Rather, he or she is a facilitator who helps keep the project on track.
>
> **Researchers**—One or two people in the group should be assigned to collect information. They are responsible for doing Internet searches, digging up materials in the library, and coordinating the team's empirical research.
>
> **Editor**—The editor is responsible for the organization and style of the document. He or she identifies places where the document is missing content or where information needs to be reorganized to achieve the project's purpose.
>
> **Designer**—The designer is responsible for laying out the document, collecting images, and making tables, graphs, and charts.

Notice that there is no "writer" among these roles. Instead, everyone in the group is responsible for writing some part of the document.

Step 4: Create a Project Calendar

Project calendars are essential for meeting deadlines. Numerous project management software packages like Microsoft Project, ArrantSoft, and Artemis Project Management help teams lay out calendars for completing projects. These programs are helpful for setting deadlines and specifying when interrelated parts of the project need to be completed (Figure 3.3).

You don't need project management software for smaller projects. A reliable time management technique is to use *backward planning* to determine when you need to accomplish specific tasks and meet smaller and final deadlines.

To do backward planning, start out by putting the project's deadline on a calendar. Then, work backward from that deadline, writing down the dates when specific project tasks need to be completed (Figure 3.4).

The advantage of a project calendar is that it keeps the team on task. The calendar shows the milestones for the project, so everyone knows how the rest of the team is progressing. In other words, everyone on the team knows when his or her part of the project needs to be completed.

Step 5: Write Out a Work Plan

A work plan is a description of how the project will be completed (Figure 3.5). Some work plans are rather simple, perhaps using an outline to describe how the project will go from start to finish. Other work plans, like the one in Figure 3.5, are very detailed and thorough.

A work plan will do the following:

- Identify the mission and objectives of the project.
- Lay out a step-by-step plan for achieving the mission and objectives.

For more on sharing team responsibilities, go to **www.pearsonhighered.com/johnsonweb4/3.4**

Project Planning Software

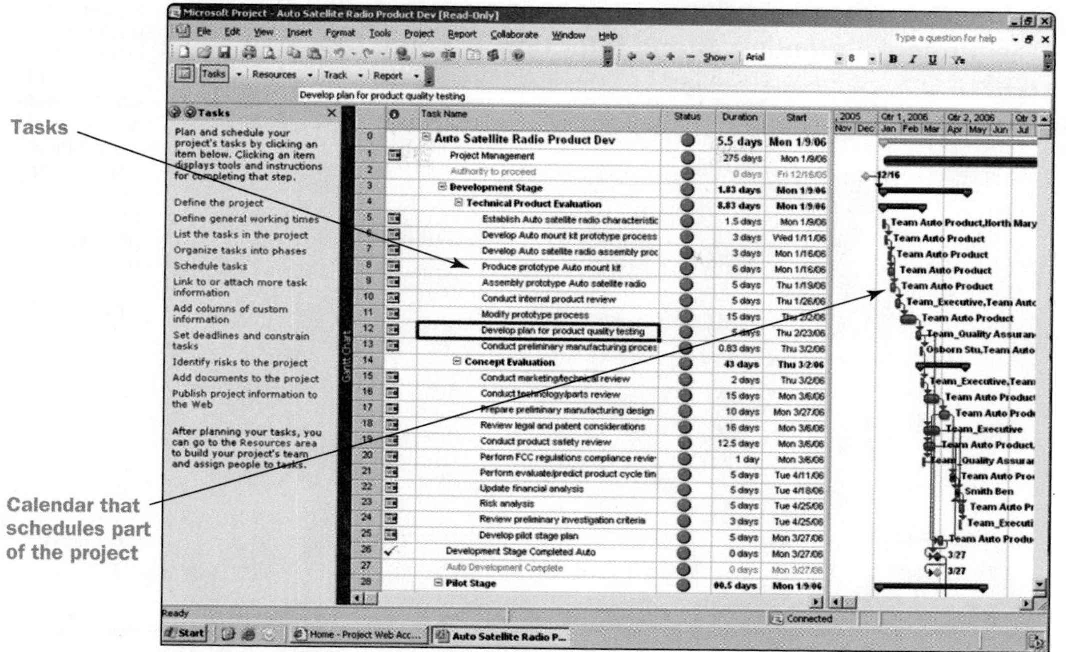

Figure 3.3: Project planning software can be helpful when setting a calendar for the team. In this screen, the calendar is represented visually to show how tasks relate to each other over time.

Source: Microsoft Inc., http://www.office.microsoft.com/en-us/project/HA101656381033.aspx.

- Establish a project calendar.
- Estimate a project budget if needed.
- Summarize the results/deliverables of the project.

A work plan is helpful for both small and large projects because team members need to see the project in writing. Otherwise, they will walk away from meetings with very different ideas about what needs to be accomplished.

By writing up a work plan, your team specifies how the project will be completed and who is responsible for which parts of the project. That way, team members can review the work plan if they are uncertain about (1) what tasks are being completed, (2) when the tasks will be finished, and (3) who is responsible for completing them.

Link

For help writing work plans as proposals, turn to Chapter 8, page 219.

Step 6: Agree on How Conflicts Will Be Resolved

Finally, your team should talk about how it will handle conflicts. Conflict is a natural, even healthy, part of a team project. However, the worst time to figure out how your team will handle conflicts is when you are in the middle of one.

Instead, in advance, talk with your team about how conflicts should be handled.

GO TO THE NET

For links to project software sites, go to **www.pearsonhighered.com/johnsonweb4/3.5**

Backward Planning Calendar

Project Calendar: Staph Infection Training Modules				
Monday	**Tuesday**	**Wednesday**	**Thursday**	**Friday**
4 Staff Meeting	**5** ←	**6** ←	**7** ←	**8** Complete Collection of Data
11 ←	**12** ←	**13** Report on Findings Due	**14** ←	**15** ←
18 ←	**19** Brochures, Pamphlets Printed	**20** ←	**21** ←	**22** Proofread Training Materials
25 ←	**26** ←	**27** Training Modules Completed	**28** ←	**29** Deadline: Training Day

Figure 3.4: Backward planning is a process of working backward from the deadline. Starting with the deadline, chart out when each task needs to be completed.

Here is the deadline. Work backward from this date.

Will the team take votes on important topics? Will the majority rule?

Will the team rely on the judgment of the team coordinator or the supervisor?

Does the team need to reach full consensus on decisions?

Can any team member call a team meeting to discuss conflicts?

Should agreements be written down for future reference?

AT A GLANCE

Six Steps for Strategic Planning

- Define the project mission and objectives.
- Identify project outcomes.
- Define team member responsibilities.
- Create a project calendar.
- Write out a work plan.
- Agree on how conflicts will be resolved.

You should not shy away from conflict in your team, because conflict is a natural part of the teaming process. Constructive conflict often leads to more creativity and closer bonds among team members. But destructive conflict can lead to dysfunctional working relationships, frustration, and lower morale. To foster constructive conflict and minimize destructive conflict, you should spend some time during the forming phase talking about the ways conflicts will be resolved.

A Work Plan

Figure 3.5: A work plan specifies the who what, where, when, why, and how of the project.

St. Thomas Medical Center

Date: April 28, 2011
To: Staph Infection Task Force (M. Franks, C. Little, J. Archuleta, L. Drew, V. Yi, J. Matthews, J. McManus)
From: Alice Falsworthy, Infections Specialist
Re: Work Plan for Combating Staph Infection

Last week, we met to discuss how we should handle the increase in staph infections here at the St. Thomas Medical Center. We defined our mission, defined major tasks, and developed a project calendar. The purpose of this memo is to summarize those decisions and lay out a work plan that we will follow.

Purpose of the work plan

I cannot overstate the importance of this project. Staph infections are becoming an increasing problem at hospitals around the country. Of particular concern are antibiotic-resistant bacteria called methicillin-resistant *Staphylococcus aureus* (MRSA), which can kill patients with otherwise routine injuries. It is essential that we do everything in our power to control MRSA and other forms of staph.

Please post this work plan in your office to keep you on task and on schedule.

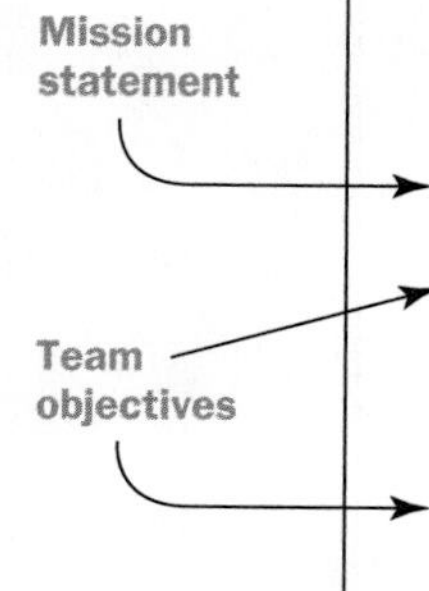

Project Mission

Our Mission: The purpose of this Staph Infection Task Force is to determine the level of staph infection vulnerability at St. Thomas Medical Center and to develop strategies for limiting our patients' exposure to staph, especially MRSA.

Secondary Objectives:

- Gain a better understanding of current research on staph and its treatments.
- Raise awareness of staph infections among our medical staff and patients.
- Assess the level of staph risk, especially MRSA, here at the hospital.
- Develop methods for controlling staph on hospital surfaces.

Figure 3.5: *(continued)*

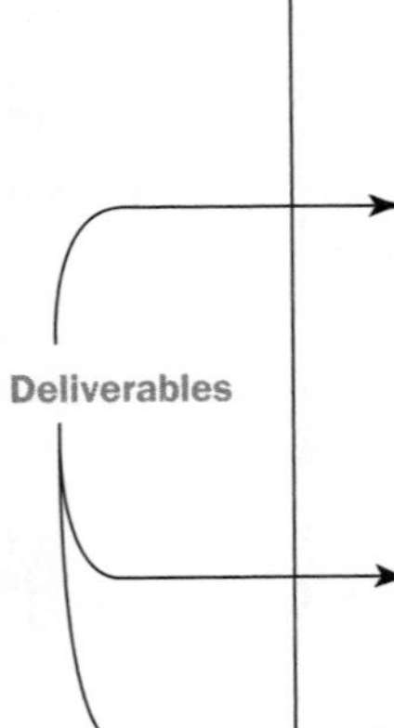

Project Plan
To achieve our mission and meet the above objectives, we developed the following five-action plan:

Action One: Collect Current Research on Staph Infections and Their Treatments
Mary Franks and Charles Little will collect and synthesize current research on staph infections, the available treatments, and control methods. They will run Internet searches, study the journals, attend workshops, and interview experts. They will also contact other hospitals to collect information on successful control methods. They will write a report on their findings and present it to the group on May 25 at our monthly meeting.

Action Two: Create Pamphlets and Other Literature That Help Medical Personnel Limit Staph Infections
Juliet Archuleta will begin developing a series of pamphlets and white papers that stress the importance of staph infections and offer strategies for combating them. These documents will be aimed at doctors and nurses to help them understand the importance of using antibiotics responsibly. Juliet will begin working on these documents immediately, but she will need the information collected by Mary and Charles to complete her part of the project. The documents will be completed by July 13, so we can have them printed to hand out at training modules.

Action Three: Conduct Experiments to Determine Level of Staph Risk at St. Thomas Medical Center
Lisa Drew and Valerie Yi will collect samples around the medical center to determine the risk of staph infection here at St. Thomas. The Center's Administration has asked us to develop a measurable knowledge of the problem at the Center. Lisa and Valerie will write a report to the Administration in which they discuss their findings. The report will be completed and delivered to the task force by June 2.

Action Four: Develop Contingency Plans for Handling Staph Instances
John Matthews and Joe McManus will develop contingency plans for handling instances of staph infections at the Center. These plans will offer

Figure 3.5: (continued)

concrete steps that the staph task force can take to limit exposure to staph bacteria. These plans will be based on the research collected by Mary and Charles. They will be completed by July 13.

Deliverables

Action Five: Develop Training Modules
When our research is complete, all of the members of this task force will develop training modules to raise awareness of staph infections, offer prevention strategies, and provide information on proper use of antibiotics. Different modules will be developed for doctors, nurses, and custodial staff. Alice Falsworthy and Charles Little will coordinate the development of these training modules. The training modules will be ready by August 29.

Project Calendar
Here is a chart that illustrates the project calendar and its deadlines:

Timeline for achieving goals

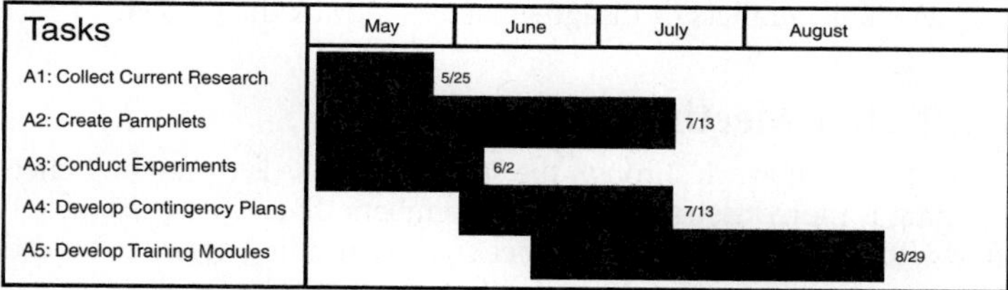

Conclusion
If anyone on the task force would like to change the plan, we will call a meeting to discuss the proposed changes. If you wish to call a meeting, please contact me at ext. 8712, or e-mail me at Alice_Falsworthy@stthomasmc.com.

Storming: Managing Conflict

Not long after the forming stage, a team will typically go through a storming phase. When the actual work begins, some tension will usually surface among team members. At this point, team members will need to negotiate, adapt, and compromise to achieve the team's mission.

During the storming stage, team members may:

- resist suggestions for improvement from other members.
- have doubts about the work plan's ability to succeed.
- compete for resources or recognition.
- resent that others are not listening to their ideas.
- want to change the team's objectives.
- raise issues of ethics or politics that need to be addressed.
- believe they are doing more than their share of the work.

Storming is rarely pleasant, but it is a natural part of the teaming process. When storming, teams realize that even the best work plans are never perfect and that people don't always work the same way or have the same expectations. The important thing is to not let small conflicts or disagreements sidetrack the project.

Running Effective Meetings

One way to constructively work through the storming phase is to conduct effective meetings. Nothing is more frustrating to team members than having to waste their time and effort sitting in an unproductive meeting. By running organized meetings, your team can maintain the structure needed to keep people on track.

CHOOSE A MEETING FACILITATOR In the workplace, usually a manager or supervisor runs the meeting, so he or she is responsible for setting the time and agenda. An interesting workplace trend, though, is to rotate the facilitator role among team members. That way, everyone has a chance to run the meeting, allowing everyone on the team to take on leadership roles. In classroom situations, your team should rotate the facilitator role to maintain a more democratic approach.

SET AN AGENDA An agenda is a list of topics to be discussed at the meeting (Figure 3.6). The meeting coordinator should send out the meeting agenda at least a couple of days before the meeting. That way, everyone will know what issues will be discussed and decided on. Begin each meeting by first making sure everyone agrees to the agenda. Then, during the meeting, use the agenda to avoid going off track into nonagenda topics.

START AND END MEETINGS PROMPTLY If team members are not present, start the meeting anyway. Waiting for latecomers can be frustrating, so you should insist that people arrive on time. If people know the meeting will start on time, they will be there on time. Likewise, end meetings on time. Meetings that drag on endlessly can be equally frustrating.

Want to learn more about resolving conflict in the workplace? Go to **www.pearsonhighered.com/johnsonweb4/3.6**

An Agenda

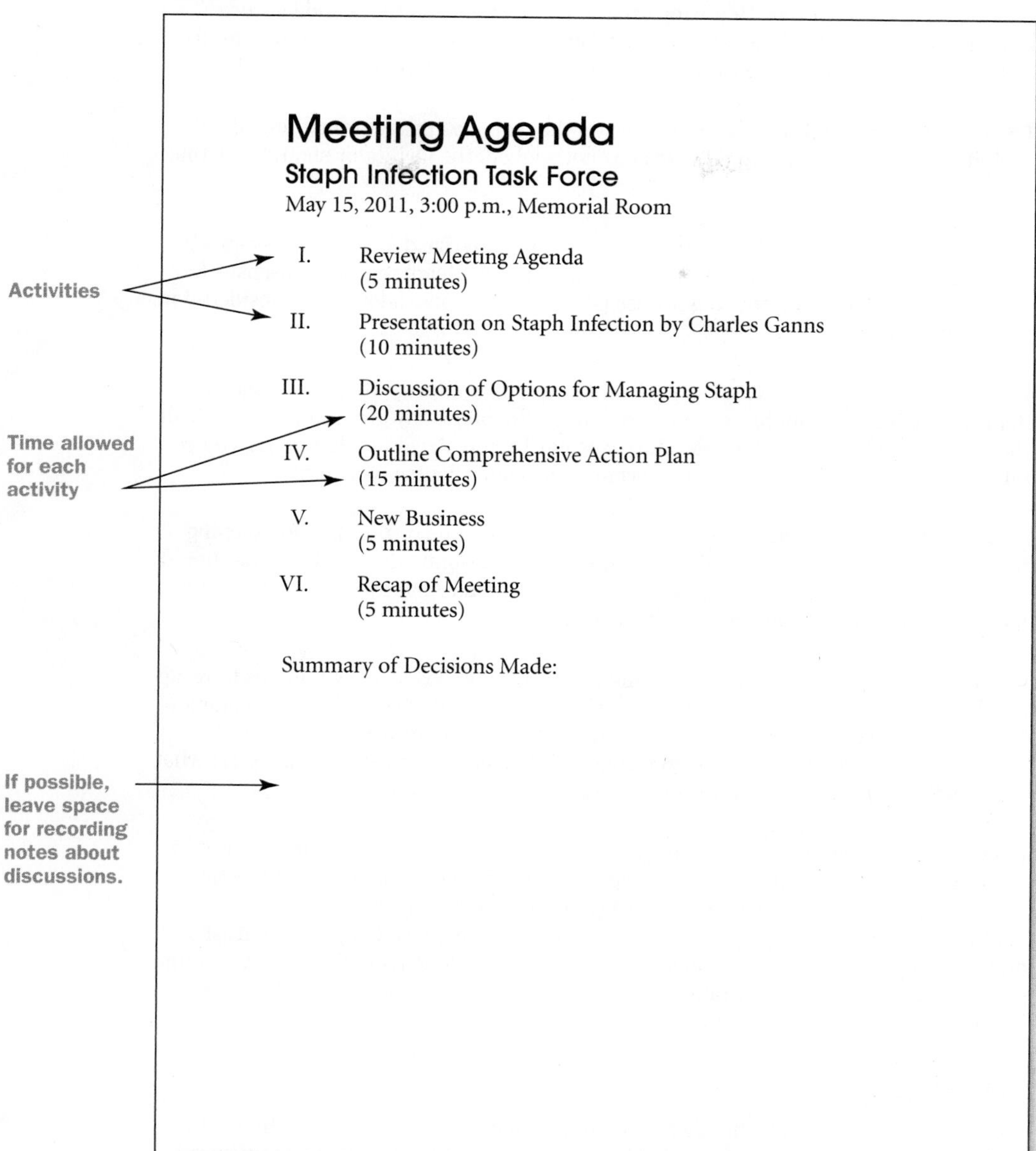

Meeting Agenda
Staph Infection Task Force
May 15, 2011, 3:00 p.m., Memorial Room
I. Review Meeting Agenda
(5 minutes)
II. Presentation on Staph Infection by Charles Ganns
(10 minutes)
III. Discussion of Options for Managing Staph
(20 minutes)
IV. Outline Comprehensive Action Plan
(15 minutes)
V. New Business
(5 minutes)
VI. Recap of Meeting
(5 minutes)
Summary of Decisions Made:
Activities
Time allowed for each activity
If possible, leave space for recording notes about discussions.

Figure 3.6: A simple agenda is a helpful tool for keeping the meeting on track.

GO TO THE NET Want to know more about running effective meetings? Go to **www.pearsonhighered.com/johnsonweb4/3.7**

ADDRESS EACH AGENDA ITEM SEPARATELY Discuss each agenda item before moving on to the next one. Bouncing around among items on the agenda ensures only that the meeting will be inefficient. If someone wants to move ahead to a future agenda item, first make sure the current item of discussion has been addressed.

ENCOURAGE PARTICIPATION Everyone on the team should say something about each item. If one of the team members has not spoken, the facilitator should give that person an opportunity to speak.

ALLOW DISSENT At meetings, it is fine to disagree. Active debate about issues will help everyone consider the issues involved. In fact, if the team reaches consensus too quickly on an issue, someone might raise possible objections, allowing a consideration of alternatives.

REACH CONSENSUS AND MOVE ON People can talk endlessly about each agenda item, even after the team has reached consensus. So, allow any group member to "call the question" when he or she feels consensus has been reached. At that point, you can take an informal or formal vote to determine the team's course of action.

RECORD DECISIONS During meetings, someone should be responsible for keeping the minutes. The minutes record the team's decisions. Minimally, all decisions should be written down. After the meeting, the facilitator should send these notes or the full minutes to the team members, usually via e-mail.

RECAP EACH AGENDA ITEM At the end of the meeting, leave a few minutes to recap the decisions made by the team and to clarify who is doing what. It is not uncommon for teams to have a "great meeting" that still leaves people unsure of what was decided and who is doing what. So, go through each agenda item, summarizing (1) what action will be taken and (2) who is responsible for taking that action.

LOOK AHEAD Discuss when the team will meet again and the expectations for that meeting. If necessary, clarify what should be accomplished before the next meeting. Also, decide who will be responsible for facilitating the next meeting.

Typically, storming becomes most evident during meetings. People grow frustrated and even angry as the team struggles to accomplish its objectives. That's why running effective meetings is so important. Creating a predictable structure for the meeting can lower the level of frustration, allowing you to get work done.

Mediating Conflicts

Smaller conflicts should be handled using the conflict resolution methods that your team discussed when it was forming. Make sure everyone is encouraged to express his or her views openly. Make sure everyone is heard. Then, use the conflict resolution methods (vote, decision of the team leader, appeal to supervisor, reach full consensus) to decide which way to go.

The Steps of Mediation

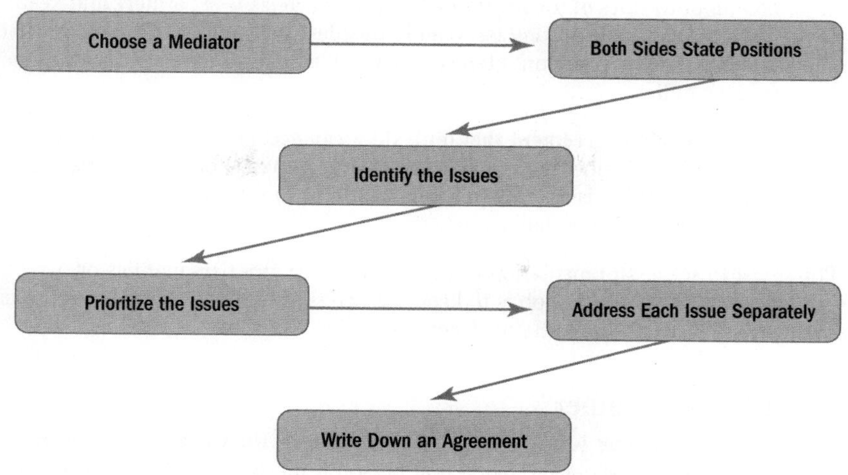

Figure 3.7: When more formal mediation is needed, you can follow these steps toward a resolution of the problem.

There will be situations, however, when personalities or ideas clash in ways that cannot be easily resolved. At these times, you may want to use mediation techniques to help your team move forward (Figure 3.7).

1. **Choose a mediator**—A mediator is like a referee. He or she does not take sides. Instead, it is the mediator's job to keep both sides of an argument talking about the issue. When the dialogue becomes unfriendly or goes off track, the mediator brings both sides back to the issues. In formal mediation, the mediator is someone who is not part of the group and has no connections to either side. In informal mediation, one of the group members who hasn't chosen sides can often serve as the referee.

2. **Ask both sides to state their positions**—Often, conflicts arise simply because each side has not clearly stated its position. Once each side has had a chance to explain its ideas clearly, the conflict should seem more resolvable.

3. **Identify the issues**—Both sides should discuss and identify the "issues" they disagree about. Often, disputes hinge on a small number of issues. Once these issues are identified, it becomes easier to talk about them.

4. **Prioritize the issues from most to least important**—Some issues are more important than others. By prioritizing the issues, both sides can usually find places where they already agree. Or, in some cases, a top priority to one side is not a foremost concern to the other side. By prioritizing the issues, both sides often begin to see room for negotiation.

GO TO THE NET

To learn more about mediation, go to **www.pearsonhighered.com/johnsonweb4/3.8**

Link

Ethical issues are often a source of conflict. To learn more about ethics, go to Chapter 4, page 69.

5. **Address each issue separately, trying to find a middle ground that is acceptable to both sides of the dispute**—Focus on each issue separately and keep looking for middle ground between both sides. When both sides realize that they have many common interests, they will usually come up with solutions to the conflict.

6. **Write down an agreement that both sides can accept**—As the mediation continues, both sides usually find themselves agreeing on ways to resolve some or all of the issues. At this point, it helps to write down any compromises that both sides can accept.

The secret to successful mediation is a focus on issues, not personalities or perceived wrongs. The mediator's job is to keep the two sides discussing the issues, steering the discussion away from other distractions.

Firing a Team Member

Sometimes a team member is not doing his or her share of the work. When this happens, the team might consider removing that person from the project. The best way to handle these situations is to first mediate the problem. The members of the team should meet to talk with this person about their expectations, giving the person a chance to explain the situation.

After hearing this person's side of the story, the team might decide to give him or her a second chance. At that point, a work contract should be written that specifies exactly what this person needs to do for the project.

If the team still wants to let the person go, the supervisor (perhaps your instructor) should be asked about removing the person from the team. The supervisor should be present when the team tells the problematic team member that he or she is being removed.

Norming: Determining Team Roles

The storming period can be frustrating, but soon afterward, your team should enter the *norming* stage. In this stage, members of your team will begin to accept their responsibilities and their roles in the project. A sense of team unity will develop as people begin to trust each other. Criticism will become increasingly constructive as team members strive to achieve the project's mission and objectives.

Revising Objectives and Outcomes

The storming stage often reveals the flaws in the work plan. So, when norming, you might find it helpful to revisit and refine the team's original decisions about objectives and outcomes. The team may also want to revise the project schedule and reallocate the workload.

You don't need to completely rewrite the work plan from scratch. You should stay with your original work plan in most cases. The plan probably just needs to be revised and refined, not completely redone.

Need to fire a team member? Go to **www.pearsonhighered.com/johnsonweb4/3.9**

HELP

Virtual Teaming

Increasingly, technical workplaces are turning to virtual teaming to put the right people on any given project. Virtual teaming allows people to work on projects collaboratively through electronic networks, using e-mail, instant messaging, and phones to stay in contact. Electronic networks, called *intranets*, are often used to share information and documents among team members.

Virtual teams, just like teams working together in an office, schedule regular meetings, share ideas and documents, and work toward achieving specific goals. The main differences between on-site teams and virtual teams are how people communicate and where they are located. Software, such as Virtual Office, MS Outlook, Lotus Notes, and CyberMatrix Office, help people stay in touch and collaborate on projects.

Trends in the technical workplace support virtual teaming. Today, more people are "telecommuting" or "teleworking" from home or remote sites. Also, the global economy sometimes results in people on the same project working thousands of miles away from each other. Meanwhile, wireless technologies allow people to work just about anywhere. Chances are, you will find yourself working with a virtual team in the near future—if you aren't already.

Interestingly, virtual teaming does not change traditional teaming strategies—it only makes them more necessary. A virtual team will go through the forming, storming, norming, and performing stages, just like an on-site team. If good planning, communication, and conflict resolution are important with on-site teaming, they are even more important in virtual teams. After all, communicating with your virtual team is a little more difficult, because you cannot physically visit each other.

Here are some strategies for managing a successful virtual team:

Develop a work plan and stick to it—Members of virtual teams do not bump into each other in the hallway or the break room. So, they need a clear work plan to keep everyone moving together toward the final goal. Your team's work plan should (1) define the mission, (2) state objectives and measurable outcomes, (3) spell out each stage and task in the project, (4) specify who is responsible for each task, and (5) lay out a project calendar.

Communicate regularly—In virtual teams, the old saying "out of sight, out of mind" now becomes "out of communication, out of mind." Each member of the virtual team should agree to communicate with the others regularly (e.g., two times a day, two times a week). Your team can use e-mail, phones, instant messaging, or chat rooms to contact each other. You and your team members should constantly keep each other up to date on your progress. And if someone does not communicate for a day or two, the team leader should track him or her down and urge the team member to resume communications.

Hold teleconferences and videoconferences—There are many ways to hold real-time virtual meetings with team members. Your team members can teleconference over the phone, or you can use a chat room or instant messaging to exchange ideas. Increasingly, broadband technology is allowing people to hold videoconferences in which people meet and see each other with Internet

(continued)

GO TO THE NET

To find websites that discuss virtual teaming, go to **www.pearsonhighered.com/johnsonweb4/3.10**

Teleconferencing

A webcam projects images to the other participants in the virtual meeting.
The monitor can be used to show people, documents, or presentations.

Figure A: Teleconferencing allows team members to hold meetings virtually.

cameras (webcams) through computer screens (Figure A). Like on-site meetings, virtual meetings should be preplanned and follow an agenda. The only significant difference between on-site meetings and virtual meetings is that people are not in the same room.

Build trust and respect—One of the shortcomings of virtual teaming is the lack of nonverbal cues (smiles, shrugs, scowls) that help people to avoid misunderstandings. As a result, people in virtual teams can feel insulted or disrespected much more easily than people in on-site teams. So it is doubly important that team members learn how to build trust with others and show respect. Trust is built by communicating effectively, meeting deadlines, and doing high-quality work. Respect is fostered by giving compliments and using "please" and "thank you" in messages. When conflicts do arise (and they will), focus on issues and problem solving, not personalities or perceived slights.

Keep regular hours—Time management is always important, even if you are working in a virtual team. During regular office hours, team members should be confident that they can contact each other. You and your team members should be ready to answer the phone, use instant messaging, or answer e-mail as though you were all working together in a typical office.

More than likely, virtual offices, teleworking, and virtual teams will be a common part of the workplace in the near future. Like on-site teaming, virtual teaming requires you to learn how to work effectively with others.

Identifying Team Roles

When planning the project, your team divided up the work, giving each person specific responsibilities. As the team begins norming, though, you will notice that team members tend to take on *team roles* that reflect their personalities, capabilities, and interests.

A management specialist, Meredith Belbin (1981), developed a description of nine team roles that people generally follow. He also organized these nine roles into three categories: people-oriented roles, action-oriented roles, and cerebral roles.

PEOPLE-ORIENTED ROLES The people in these roles are responsible for managing the activities of the team members:

> The **coordinator** sets the agenda and keeps track of the team's objectives; asks broader questions and occasionally summarizes the team's decisions; keeps an eye on the project calendar and coordinates the work of various team members.
>
> The **resource investigator** goes out to find information, bringing new ideas and strategies into the discussion; looks outside the team for ways to improve the project.
>
> The **team worker** focuses on getting the work done; may not be fully invested in planning the project but will do his or her part of it.

ACTION-ORIENTED ROLES The people in these roles are responsible for getting things done:

> The **shaper** focuses on team tasks while looking for patterns in team discussions; emphasizes completing the project.
>
> The **implementor** stresses the "how to" nature of the project and is eager to develop methods for turning abstract objectives and plans into real actions.
>
> The **completer/finisher** stresses attention to details and the overall quality of the project; is concerned about meeting deadlines and maintaining a schedule.

CEREBRAL ROLES The people in cerebral roles are responsible for planning, creating, and providing expertise in a project:

> The **monitor/evaluator** keeps the team on task by critiquing poor decisions or pointing out any flaws in reasoning; tends to focus on achieving outcomes.
>
> The **plant** thinks creatively, often providing original suggestions and innovative solutions to problems; tends to stress the big picture over smaller details.
>
> The **specialist** contributes special skills and knowledge to the team; masters a specific topic or area of research, adding depth to the team's discussions.

As the team begins norming, you and the other members of the team might take some time to identify the roles each of you is playing. By identifying team roles, you can take advantage of each member's natural strengths and interests.

Not all the roles will be filled, especially in a smaller team. Instead, each team member might take on two or three roles, depending on the project. Roles may change and evolve as the project moves forward.

In other words, let the "completer/finisher" in the team worry about the deadlines and quality issues. Let the "team workers" concentrate on achieving specific tasks. Encourage

To learn more about team role theory, go to **www.pearsonhighered.com/johnsonweb4/3.11**

AT A GLANCE

Norming

- Revise objectives and outcomes.
- Identify team roles.
- Use groupware to facilitate work.

the “coordinators” and “shapers” to keep an eye on the overall objectives and mission of the team.

Using Groupware to Facilitate Work

When working in a team, you might need to use *groupware*, a kind of software that helps move information and documents around. Groupware allows team members to communicate and work collaboratively through a local area network (LAN) or an intranet.

Perhaps the most common use of groupware is sharing documents and sending messages. The two most popular groupware packages are IBM Lotus Notes® (Figure 3.8) and Microsoft Outlook®. These software packages and others like them support the following kinds of activities:

Scheduling and calendaring—One of the more powerful features of groupware is the ability to schedule meetings and keep a common calendar for the team. With this feature, team members can regularly check the calendar to stay on task.

Discussion lists and instant messaging—Groupware offers easy access to discussion lists (usually via e-mail) and instant messaging. With these tools, team members can post notes to each other on a discussion list or have real-time discussions through instant messaging.

Using Groupware

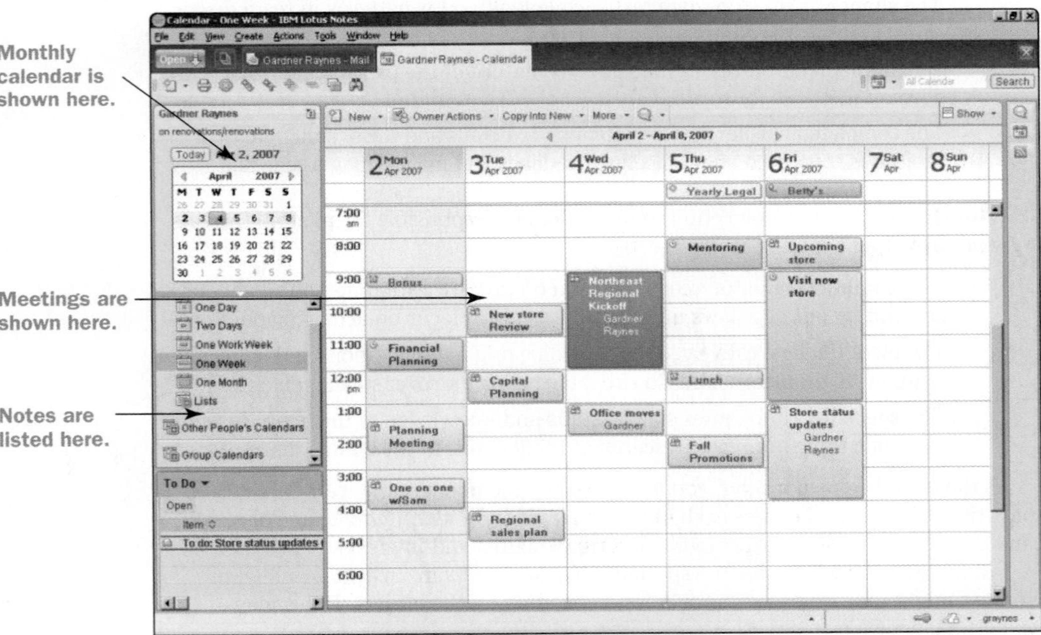

Figure 3.8: With groupware like Lotus Notes, shown here, members of a group can keep a common calendar, post notes, and send e-mails to each other. In this screen, the user can perform a variety of functions using the menu at the top.

Source: Reprint courtesy of International Business Machines Corporation, copyright 2011 © International Business Machines Corporation.

GO TO THE NET — To learn more about groupware, go to **www.pearsonhighered.com/johnsonweb4/3.12**

Document posting and commenting—Files can be posted to a common site, allowing team members to view documents, comment on them electronically, or download them.

Using groupware effectively takes some practice. Eventually, the team will begin using the groupware as a meeting place and a posting board. It will become an integral part of the project.

Performing: Improving Quality

Link

For more ideas about improving quality through document cycling, see Chapter 20, page 566.

Your team is *performing* when members are comfortable with the project and their roles in it. Team members will recognize the other members' talents and weaknesses. They also begin to anticipate each other's needs and capabilities.

When your team is performing, you can start looking for ways to improve the quality of your work. One of the gurus of quality was W. Edwards Deming, who developed many of the principles behind Total Quality Management (TQM) and Continuous Quality Improvement (CQI), which are widely used in technical workplaces. Deming argued that teams should put an emphasis on improving the *process* rather than simply exhorting people to improve the product (Deming, 2000).

How can you improve quality in your team? While performing, a helpful technique is to develop *quality feedback loops* in which your team regularly compares outcomes to the project objectives.

An effective quality feedback loop should include methods to collect feedback on the outcomes of the project. To collect feedback, perhaps "focus groups" of customers might be consulted. Perhaps the product might be thoroughly user-tested. Perhaps supervisors or outside experts might be called in to offer suggestions for improvement. The type of feedback you need depends on the kind of product your team is being asked to produce.

Teams also need to regularly review the performance of their own members. Figure 3.9 shows a "Team Performance Review" form that is similar to ones found in the workplace. Your team and instructor can use this form to assess the performance of your team and look for places to improve. A performing team will have its ups and downs. There may even be times when the team regresses into the norming or even the storming stages. Eventually, though, the performing team usually regroups and puts the focus back on quality.

The Keys to Teaming

The keys to good teaming are good planning and effective communication. The planning strategies discussed in this chapter might seem like extra work, especially when time is limited and your team is eager to start working on the project. But good planning will save your team time in the long run. Each person needs a clear understanding of the mission and the steps in the project. Then, you need to keep the communication lines open. Plan to communicate regularly by phone, e-mail, and instant messaging.

Telecommuting, or teleworking, is becoming much more common in today's technical workplace. Sometimes your team members will be working a few days per week at home. Or, they may be working while they are on the road. Some telecommuters work almost exclusively from home, going to the office only when absolutely necessary. Good planning and effective communication are the keys to success in these virtual workplace environments.

GO TO THE NET

Are you interested in TQM or CQI? Go to **www.pearsonhighered.com/johnsonweb4/3.13**

Team Performance Review

Date: ______________

Name: ____

Name of Project: ____

The purpose of this performance review is to evaluate the participation of you and your team members on the project we just completed. Your feedback on this form will be taken into account when determining each member's "participation" part of the grade for this assignment.

Describe your role on the team and list your top four contributions to the project.

My role:

My contributions to the project:

1.

2.

3.

4.

Describe the role and list the contributions of your team members.

Team Member	a.	b.	c.
Role on team			
Contributions			

Rate the participation (1–10 with 10 being the highest) of your team members.

Team Member	a.	b.	c.
Attended team meetings			
Did her/his share of the project			
Contributed good ideas			
Respected the ideas of others			
Handled conflict and stress			
Communicated with team			
Overall Participation Rating (1–10)			

On the back of this sheet, write detailed comments about how well your team worked together. What were its strengths? What could have been improved? Which team members were essential to the project and why were they so important? Which team members should have done more?

Figure 3.9: Performance reviews are an important way to improve a team's effectiveness.

CHAPTER REVIEW

- Teams have several advantages. They can concentrate strengths, foster creativity, share the workload, and improve morale. Disadvantages include conflict with other members and disproportionate workloads.
- Tuckman's four stages of teaming are forming, storming, norming, and performing.
- In the forming stage, use strategic planning to define the mission, set objectives, define roles, and establish a project schedule.
- In the storming stage, use conflict management techniques to handle emerging disagreements, tension and anxiety, leadership challenges, and frustration.
- In the norming stage, revise the work plan to form consensus, refine the team's objectives and outcomes, and solidify team roles.
- In the performing stage, pay attention to improving the process in ways that improve the quality of the work.
- Virtual teaming, or working together from a distance, requires good planning and effective communication.
- W. Edwards Deming developed many of the principles behind Total Quality Management (TQM) and Continuous Quality Improvement (CQI), which are common in technical workplaces today.

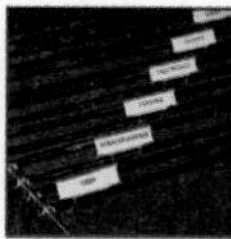

EXERCISES AND PROJECTS

Individual or Team Projects

1. If you have a job now, write a report to your instructor in which you talk about how your co-workers fluctuate among Tuckman's four stages. What are some of the indications that the team is forming, storming, norming, and performing? How does the team tend to react during each of these stages? Does your team aid in forming strategic planning? How do you mediate conflicts during storming? How is norming achieved at your workplace? What does performing look like?

2. Imagine that your class is a workplace. What are the objectives and outcomes of this course? How are conflicts resolved in the classroom? Do members of your class take on various team roles in the classroom? How could you and your instructor create quality feedback loops to improve your learning experience? In a class, discuss how teaming strategies might be helpful in improving how the class is managed.

3. On the Internet, research the theories and writings of Tuckman, Belbin, or Deming. Write a report to your instructor in which you summarize the principles of one of their theories. Then, discuss the ways in which you might apply this theory to your own life and work.

Collaborative Project

With a team, try to write a report in one hour on a topic of interest to all of you. For example, you might write about a problem on campus or at your workplace. While you are writing the report, pay attention to how your group forms, norms, storms, and performs. Pay attention to how the group plans and divides up the responsibilities. Then, as the project moves forward, pay attention to the ways conflict is resolved. Finally, identify the different roles that group members tend to play as the group develops norms for the project.

When your one-hour report is finished, talk among yourselves about how the group project went. Did the group form and plan properly? Did you handle conflict well? Did the group members take on identifiable roles? Do you think you ever reached the performing stage? If you were going to do the project over, how might your group do things differently?

For additional technical writing resources, including interactive sample documents, document design tutorials and guidelines, and more, go to **www.mytechcommlab.com.**

CASE STUDY

Not a Sunny Day

Veronica Norton liked working on teams. She liked interacting with others, and she liked how working with others allowed her to accomplish projects that were too big for one person. Veronica also had a strong interest in building solar motors, like the ones that went into solar-powered cars. She hoped to find a job in the automotive industry building solar vehicles.

So, she decided to join her university's team in the American Solar Challenge race competition. She and a group of ten other students were going to build a solar car to race across the country against cars from other universities.

The first meeting of the team went great. Everyone was excited, and they found out there were funds to support the project. Several engineering departments were contributing money, and a local aerospace engineering firm was matching university funds dollar for dollar. The firm was also letting the team use its wind tunnel to help streamline the car. There were other local sponsors who wanted to participate.

The team rushed right into the project. They sketched out a design for the car, ordered supplies, and began welding together a frame for the car. Veronica began designing a solar motor that would be long lasting and have plenty of power. She noticed that the overall plan for the car was not well thought out, but she went along with it anyway. At least they were making progress.

One of the team members, George Franks, began emerging as the leader. He was finishing up his engineering degree, so he had only a couple of classes left to take. He had plenty of time to work on the solar car.

Unfortunately, he didn't like to follow plans. Instead, he just liked to tinker, putting things together as he saw fit. It wasn't long before the team started running into problems. George's tinkering approach was creating a car that would be rather heavy. Also, each time Veronica visited the shop, the dimensions of the car had changed, so the motor she was building needed to be completely redesigned.

Perhaps the worst problem was that they were running out of money. George's tinkering meant lots of wasted materials. As a result, they had almost used up their entire budget. Sally, who was in charge of the finances for the project, told everyone they were going to be out of money in a month. Everyone was getting anxious.

Finally, at one of the monthly meetings, things fell apart. People were yelling at each other. After being blamed for messing up the project, George stormed out of the meeting. Things looked pretty hopeless. Everyone left the meeting unsure of what to do.

Veronica still wanted to complete the project, though. If you were her, how might you handle this situation? What should she do to get the project going again?

GO TO THE NET

For more information on the American Solar Challenge, go to **www.pearsonhighered.com/johnsonweb4/3.14**

CHAPTER

4

Ethics in the Technical Workplace

What Are Ethics? 69

Where Do Ethics Come From? 72

Resolving Ethical Dilemmas 77

Help: Stopping Cyberbullying and Computer Harassment 78

Ethics in the Technical Workplace 85

Copyright Law in Technical Communication 88

Chapter Review 90

Exercises and Projects 91

Case Study: This Company Is Bugging Me 92

In this chapter, you will learn:

- A working definition and understanding of ethics.
- To identify three ethical systems: personal, social, and conservation.
- To consider social ethics as issues of rights, justice, utility, and care.
- Strategies for resolving ethical conflicts in technical workplaces.
- How to balance the many issues involved in an ethical dilemma.
- How copyright law affects technical communication.
- About the new ethical challenges that face the computer-centered workplace.

In the technical workplace, you will regularly encounter ethical dilemmas. In these situations, you need to be able to identify what is at stake and make an informed decision. Ethical behavior is more than a matter of personal virtue—it is good business.

What Are Ethics?

For some people, ethics are about issues of morality. For others, ethics are a matter of law. Actually, ethics bring together many different ideas about appropriate behavior in a society.

Ethics are systems of moral, social, or cultural values that govern the conduct of an individual or community. For many people, acting ethically simply means "doing the right thing," a phrase that actually sums up ethics quite well. The hard part, of course, is figuring out the right thing to do. Ethical choices, after all, are not always straightforward.

Every decision you make has an ethical dimension, whether it is apparent or not. In most workplace situations, the ethical choice is apparent, so you do not pause to consider whether you are acting ethically. Occasionally, though, you will be presented with an *ethical dilemma* that needs more consideration. An ethical dilemma offers a choice among two or more unsatisfactory courses of action. At these decision points, it is helpful to ponder the ethics of each path so you can make the best choice.

AT A GLANCE

Definitions of *Ethics* and *Ethical Dilemma*

- Ethics—systems of moral, social, or cultural values that govern the conduct of an individual or community
- Ethical dilemma—a choice among two or more unsatisfactory courses of action

In technical workplaces, resolving ethical dilemmas will be a part of your job. Resources, time, and reputations are at stake, so you will feel pressure to overpromise, underdeliver, bend the rules, cook the numbers, or exaggerate results. Technical fields are also highly competitive, so people sometimes stretch a little further than they should. Ethical dilemmas can force us into situations where all choices seem unsatisfactory.

Why do some people behave unethically? People rarely set out to do something unethical. Rather, they find themselves facing a tough decision in which moving forward means taking risks or treating others unfairly. In these situations, they may be tempted to act unethically due to a fear of failure, a desire to survive, pressure from others, or just a series of bad decisions. Small lies lead to bigger lies until the whole house of cards collapses on them.

Keep in mind, though, that ethics are not always about deception or fraud. A famous decision involving Albert Einstein offers an interesting example. Figure 4.1 shows a letter that Einstein wrote to President Franklin Roosevelt encouraging research into the development of the atom bomb. Throughout the rest of his life, Einstein, who was an ardent pacifist, was troubled by this letter. Five months before his death, he stated:

> I made one great mistake in my life . . . when I signed the letter to President Roosevelt recommending that atom bombs be made; but there was some justification—the danger that the Germans would make them. (Clark, 1971, p. 752)

Want to see some other definitions of ethics? Go to **www.pearsonhighered.com/johnsonweb4/4.1**

Einstein's Letter to Roosevelt About the Atom Bomb

Albert Einstein
Old Grove Rd.
Nassau Point
Peconic, Long Island

August 2nd 1939

F.D. Roosevelt
President of the United States
White House
Washington, D.C.

Sir:

Some recent work by E.Fermi and L. Szilard, which has been communicated to me in manuscript, leads me to expect that the element uranium may be turned into a new and important source of energy in the immediate future. Certain aspects of the situation which has arisen seem to call for watchfulness and, if necessary, quick action on the part of the Administration. I believe therefore that it is my duty to bring to your attention the following facts and recommendations:

In the course of the last four months it has been made probable - through the work of Joliot in France as well as Fermi and Szilard in America - that it may become possible to set up a nuclear chain reaction in a large mass of uranium, by which vast amounts of power and large quantities of new radium-like elements would be generated. Now it appears almost certain that this could be achieved in the immediate future.

This new phenomenon would also lead to the construction of bombs, and it is conceivable - though much less certain - that extremely powerful bombs of a new type may thus be constructed. A single bomb of this type, carried by boat and exploded in a port, might very well destroy the whole port together with some of the surrounding territory. However, such bombs might very well prove to be too heavy for transportation by air.

Here is Einstein's main point.

Einstein expresses the imminent problem.

He points out the possible threat of these new kinds of weapons.

Figure 4.1: In 1939, Einstein wrote this letter to President Franklin Roosevelt. The atom bomb would have been built without Einstein's letter, but his prodding jump-started the U.S. nuclear program.

Source: Argonne National Laboratory, http://www.anl.gov/OPA/frontiers96arch/aetofdr.html.

Figure 4.1: *(continued)*

-2-

The United States has only very poor ores of uranium in moderate quantities. There is some good ore in Canada and the former Czechoslovakia, while the most important source of uranium is Belgian Congo.

In view of the situation you may think it desirable to have more permanent contact maintained between the Administration and the group of physicists working on chain reactions in America. One possible way of achieving this might be for you to entrust with this task a person who has your confidence and who could perhaps serve in an inofficial capacity. His task might comprise the following:

a) to approach Government Departments, keep them informed of the further development, and put forward recommendations for Government action, giving particular attention to the problem of securing a supply of uranium ore for the United States;

b) to speed up the experimental work, which is at present being carried on within the limits of the budgets of University laboratories, by providing funds, if such funds be required, through his contacts with y private persons who are willing to make contributions for this cause, and perhaps also by obtaining the co-operation of industrial laboratories which have the necessary equipment.

I understand that Germany has actually stopped the sale of uranium from the Czechoslovakian mines which she has taken over. That she should have taken such early action might perhaps be understood on the ground that the son of the German Under-Secretary of State, von Weizsäcker, is attached to the Kaiser-Wilhelm-Institut in Berlin where some of the American work on uranium is now being repeated.

Yours very truly,

A. Einstein

(Albert Einstein)

Einstein offers a potential solution.

He points out that the Nazis may already be working on nuclear technology, potentially a bomb.

Einstein with Robert Oppenheimer

Einstein meets with Robert Oppenheimer, the leader of the U.S. efforts to develop an atom bomb. Later, Einstein regretted his involvement, though minimal, with its development.

In this quote, you see the ethical dilemma weighing on Einstein. He deeply regretted the atom bomb's development and use on Japan. However, he also recognized that his letter may have alerted Roosevelt to a very real danger. Historians have pointed out that Einstein's letter may have helped prevent the Nazis from creating an atom bomb. Ethical dilemmas put people in these kinds of quandaries.

Where Do Ethics Come From?

How can you identify ethical issues and make appropriate choices? To begin, consider where values come from:

Personal ethics—Values derived from family, culture, and faith

Social ethics—Values derived from constitutional, legal, utilitarian, and caring sources

Conservation ethics—Values that protect and preserve the ecosystem in which we live

These ethical systems intertwine, and sometimes they even conflict with each other (Figure 4.2).

GO TO THE NET To learn more about Einstein, go to **www.pearsonhighered.com/johnsonweb4/4.2**

Intertwined Ethical Systems

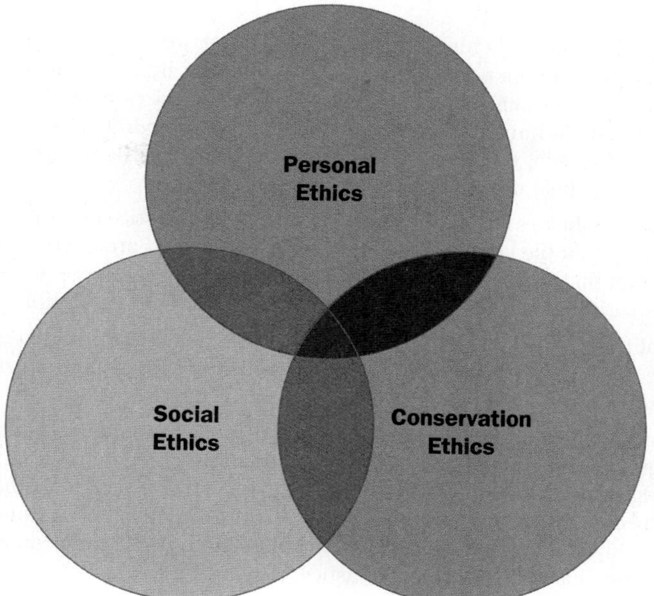

Figure 4.2: Ethics come from a variety of sources that overlap. Your personal sense of ethics guides the majority of your daily decisions. Social and conservation ethics play significant roles in the technical workplace.

Personal Ethics

By this point in your life, you have developed a good sense of right and wrong. More than likely, your personal ethics derive from your family, your culture, and your faith. Your family, especially your parents, taught you some principles to live by. Meanwhile, your culture, including the people in your neighborhood or even the people you watch on television, has shaped how you make decisions. And for many people, faith gives them specific principles about how they should live their lives.

In the technical workplace, a strong sense of personal values is essential, because these values offer a reliable touchstone for ethical behavior. A good exercise is making a list of values that you hold dear. Perhaps some of these values are honesty, integrity, respect, candor, loyalty, politeness, thoughtfulness, cautiousness, thriftiness, and caring. By articulating these values and following them, you will likely find yourself acting ethically in almost all situations.

Social Ethics

In technical workplaces, the most difficult ethical dilemmas are usually found in the social realm. Social ethics require you to think more globally about the consequences of your or your company's actions.

GO TO THE NET For websites that discuss sources of personal ethics, go to **www.pearsonhighered.com/johnsonweb4/4.3**

Ethics scholar Manuel Velasquez (2002) offers a helpful four-part categorization of social ethical situations:

Rights—Rights are fundamental freedoms that are innate to humans or granted by a nation to its citizens. *Human rights,* like those mentioned in the *U.S. Declaration of Independence* (life, liberty, and the pursuit of happiness), are innate to humans and cannot be taken away. *Constitutional rights* (freedom of speech, right to bear arms, protection against double jeopardy) are the rights held in common by citizens of a nation.

Justice—Justice involves fairness among equals. Justice takes its most obvious form in the laws that govern a society. Our laws are a formalized ethical system that is designed to ensure that people are treated equally and fairly. Similarly, *corporate policies* are the rules that ensure fairness within a company.

Utility—Utility suggests that the interests of the majority should outweigh the interests of the few. Of paramount importance to utilitarianism is *the greatest good for the greatest number of people.*

Care—Care suggests that tolerance and compassion take precedence over rigid, absolute rules. Ethics of care suggest that each situation should be judged on its own, putting heightened attention on concern for the welfare of people and preserving relationships. It also recognizes that some relationships, like those involving friends and family, will often lead to ethical choices that transcend rights, justice, and utility.

Legal issues, usually involving rights and justice, are especially important in technical communication, because the temptation to break the law to gain a competitive edge can be great. Legal issues of copyright law, patent law, liability, privacy, and fraud (all of which are discussed later in this chapter) are crucial concerns that affect how individuals and companies conduct themselves. You should be aware of the laws that apply to your discipline.

When facing an ethical dilemma or a controversy involving ethics, you should first identify which of these four ethical categories applies to your situation. Ethical issues that involve human or constitutional rights usually have the most gravity (Figure 4.3). Issues involving care are still important, but they have the least gravity. In other words, if an ethical decision involves human or constitutional rights, it will take on much more importance than a decision that involves issues of justice, utility, or care.

By sorting ethical dilemmas into these four categories, you can often decide which course of action is best. For example, consider the following case study:

Link

For more information on working with international readers, see Chapter 2, page 28.

Case Study: Your company makes a popular action figure toy with many tiny accessories. Countless children enjoy the toy, especially all those tiny boots, hats, backpacks, weapons, and so on. However, a few children have choked on some of the small pieces that go with the toy. How would the four levels of ethics govern how to handle this situation?

Answer: In this case, a human right (life) has more weight than utility (thousands of children versus a few). So, the ethical choice would be to alter the toy or stop selling it.

GO TO THE NET

More information on Manuel Velasquez's ethical system is available at **www.pearsonhighered.com/johnsonweb4/4.4**

Four Categories of Social Ethics

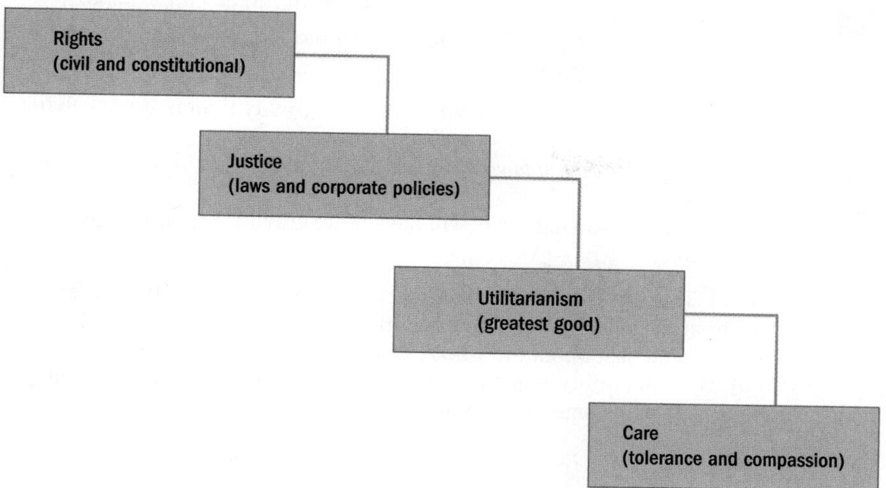

Figure 4.3: Social ethics can be ranked. Concerns about rights usually have more gravity than justice issues, and so on.

Social ethical issues are rarely this clear cut, though. After all, rights, justice, utility, and care are open to interpretation and debate. A union organizer, for example, may see a company's resistance to unionizing as a violation of the "right to free assembly" provided by the U.S. Constitution. The company, on the other hand, may point to federal laws that allow it to curb union activities. This kind of debate over rights and justice happens all the time.

Another problem is that people miscategorize their ethical issue.

> **Case Study:** Your town's city council has decided to implement a no-smoking policy that includes all public property and restaurants. Many smokers now find it impossible to have their smoke break around public buildings and in restaurants. They argue that their "right to smoke" is being violated. How might you resolve this ethical issue?
>
> **Answer:** Actually, there is no such thing as a right to smoke. Smoking is a legal issue (a matter of justice) and a utility issue (the health interests of the nonsmoking majority versus the interests of a smoking minority). So, if the city chooses to ban smoking on public property or in restaurants, it can do so legally as long as it applies the law fairly to all. It is not violating anyone's human or constitutional rights.
>
> Of course, defenders of smoking may point out that restricting smoking may hurt businesses like bars and restaurants (a utility argument). Advocates of nonsmoking places might counter that secondhand smoke may cause cancer in patrons and employees of these establishments (also a utility argument). The city council may take these arguments into consideration before passing a new law.

GO TO THE NET

Want to see the U.S. Constitution? Go to **www.pearsonhighered.com/johnsonweb4/4.5**

AT A GLANCE

Four Categories of Social Ethics

- Rights—Civil rights and constitutional rights
- Justice—Laws and corporate policies
- Utility—Greatest good (majority rules)
- Care—Tolerance and compassion for others

When making decisions about social ethical issues, it is important to first decide which ethical categories fit the ethical dilemma you are pondering. Then, decide which set of ethics has more significance, or gravity. In most cases:

- issues involving *rights* will have more gravity than issues involving *laws*.
- issues involving *laws* will have more gravity than issues involving *utility*.
- issues involving *utility* will have more gravity than issues involving *care*.

There are, of course, exceptions. In some cases, utility may be used to argue against laws that are antiquated or unfair. For example, it was once legal to smoke just about anywhere, including the workplace (and the college classroom). By using utility arguments, opponents of smoking have successfully changed those laws. Today, smoking is ever more restricted in public.

Conservation Ethics

Increasingly, issues involving our ecosystems are becoming sources of ethical dilemmas. With issues such as global warming, nuclear waste storage, toxic waste disposal, and overpopulation, we must move beyond the idea that conservation is a personal virtue. We are now forced to realize that human health and survival are closely tied to the health and survival of the entire ecosystem in which we live. Conservation ethics involve issues of water conservation, chemical and nuclear production and waste, management of insects and weeds in agriculture, mining, energy production and use, land use, pollution, and other environmental issues.

One of America's prominent naturalists, Aldo Leopold, suggested that humans need to develop a *land ethic*. He argued:

> All ethics so far evolved rest upon a single premise: that the individual is a member of a community of interdependent parts. . . . The land ethic simply enlarges the boundaries of the community to include soils, waters, plants, and animals, or collectively: the land. (1996, p. 239)

In other words, your considerations of ethics should go beyond the impact on humans and their communities. The health and welfare of the ecosystem around you should also be carefully considered.

People who work in technical fields need to be especially aware of conservation ethics, because we handle so many tools and products that can damage the ecosystem. Without careful concern for use and disposal of materials and wastes, we can do great harm to the environment.

Ultimately, conservation ethics are about *sustainability*. Can humans interact with their ecology in ways that are sustainable in the long term? Conservation ethics recognize that resources must be used. They simply ask that people use resources in sustainable ways. They ask us to pay attention to the effects of our decisions on the air, water, soil, plants, and animals on this planet.

Conservation ethics are becoming increasingly important. The twenty-first century has been characterized as the "Green Century," because humans have reached a point

For more information on conservation ethics, go to **www.pearsonhighered.com/johnsonweb4/4.6**

Aldo Leopold and the Land Ethic

Aldo Leopold, a naturalist and conservationist, developed the concept of a "land ethic," which defines a sustainable relationship between humans and nature.

where we can no longer ignore the ecological damage caused by our decisions. For example, within this century, estimates suggest that human-caused climate change will raise global temperatures between 2 and 10 degrees. Such a rise would radically alter our ecosystem.

Moreover, as emerging markets such as China and India continue to grow, the world economy will need to learn how to use its limited resources in ways that are fair and conscientious.

Resolving Ethical Dilemmas

No doubt, you will be faced with numerous ethical dilemmas during your career. There is no formula or mechanism you can use to come up with the right answer. Rather, ethical dilemmas usually force us to choose among uncomfortable alternatives.

Doing the right thing can mean putting your reputation and your career on the line. It might mean putting the interests of people above profits. It also might mean putting the long-term interests of the environment above short-term solutions to waste disposal and use of resources.

HELP

Stopping Cyberbullying and Computer Harassment

Have you ever been cyberbullied or harassed via computer? Cyberbullying is the use of a computer to harm or threaten others psychologically, economically, or in a way that damages their careers or personal lives. Computer harassment involves using a computer to disturb or threaten others because of their race, gender, sexual orientation, disability, ancestry, religious affiliation, or other inherent characteristics. People also use computers to sexually harass others by making unwanted sexual advances and making demeaning or humiliating sexual comments.

To combat cyberbullying and computer harassment, anti-harassment policies are being developed by schools, workplaces, video games, Internet service providers, and social networking sites. Figure A, for example, shows the World of Warcraft®'s policies for dealing with harassment.

In some cases, people don't even realize they are bullying or harassing others. When e-mailing or texting others, they use aggressive language such as "I'll have you fired for this," or "I'm going to kill myself," or "Tomorrow, I think I'll bring my assault rifle to the meeting." Or, they might think it's funny to e-mail dirty jokes or pornographic cartoons or images to their co-workers.

Figure A: World of Warcraft's Anti-Harassment Policies

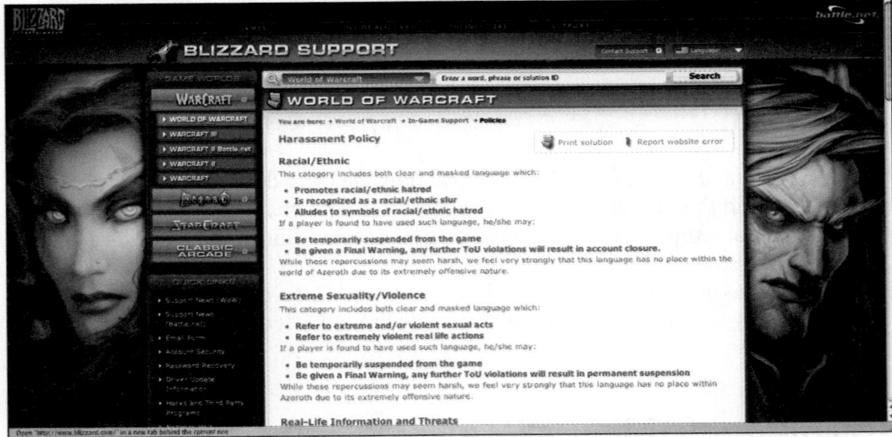

Bullying and harassment are not new in school or the workplace, but computers add some important new elements that make these activities especially harmful. Bullies and harassers can mask their identities or pretend to be someone else. Video games often allow people to do things they would not do in real life. Meanwhile, people can spread rumors and harmful information quickly through texting, games, Listservs, and social networking.

Source: Blizzard Entertainment, http://us.blizzard.com/support/article.xml?articleId=20226&categoryId=2415&parentCategoryId=2318&pageNumber=1

For more information on cyberbullying, go to **www.pearsonhighered.com/johnsonweb4/4.7**

Unfortunately, these kinds of behaviors are not uncommon on campuses and in today's workplace. So what should you do if someone is bullying or harassing you electronically? Here are some tips for preventing and stopping cyberbullying and harassment:

Prevention

NEVER GIVE OUT PERSONAL INFORMATION ONLINE—This includes social networking sites. You might think only your friends can access your Facebook or MySpace page, but it is not that difficult for someone else to gain access to that information through one of your "friends." Your e-mails and text messages can also be easily forwarded, so any personal information may fall into the wrong hands.

DON'T PUT COMPROMISING CONTENT ONLINE—Compromising pictures, video, or statements have a way of being copied, leaked, and forwarded. If you have done or written something that would look bad to your classmates, parents, instructors, or a future employer, you should not put it online or send it through your phone. Once it's out there, it can be saved and used against you.

REFUSE TO PASS ALONG MESSAGES FROM CYBERBULLIES AND HARASSERS—Instead, tell the cyberbully or harasser to stop sending the messages and report the incident to your instructors, the Dean of Students, or your supervisor at work. Even if you are just making a victim aware that harmful things are being said about him or her, your forwarding of the message to the victim or others is really doing work for the bully or harasser.

KNOW WHO YOU ARE TALKING TO—Cyberbullies, harassers, and predators often try to build relationships with their victims over time. Then, once a "friendship" is established, they try to exploit their victim's trust. As soon as your "friend" does or says something strange or offensive or makes an inappropriate request, you should end the relationship.

Stopping It

TELL THE PERSON TO STOP—The person who is bullying or harassing you may not realize he or she is being intimidating or offensive. Firmly tell that person that his or her messages are not welcome and that they should cease.

BLOCK MESSAGES—Social networking sites, video games, Internet service providers, and mobile phones almost always give you the ability to block messages from people you don't want to hear from.

SAVE AND PRINT MESSAGES—By saving and printing messages, you are collecting evidence that can be used against the cyberbully or harasser. Even if you don't want to take action now or you don't know who is doing it, you should keep any messages in case the problem escalates or the person is identified.

(continued)

FILE A COMPLAINT—Most networking and game sites have a procedure for filing a complaint that will warn, suspend, or ban the person who is bullying or harassing you. If you are being bullied or harassed in college, you can file a complaint with your university's Dean of Students office or the Equal Opportunity Office at your campus. At a workplace, you should contact the Human Resources office at your company.

Networking sites and video games are going to challenge ethics in new ways. These games can be problematic because players earn virtual money that can be converted into real money. So, we are already seeing instances of cybermuggings, fraud, and even virtual rape. Take steps to avoid being a victim.

Confronting an Ethical Dilemma

When faced with an ethical dilemma, start considering it from all three ethical perspectives: personal, social, and conservation (Figure 4.4).

> **Personal ethics**—How does my upbringing in a family, culture, and faith guide my decision? How can I do unto others as I would have them do unto me?
>
> **Social ethics**—What rights or laws are involved in my decision? What is best for the majority? How can I demonstrate caring by being tolerant and compassionate?
>
> **Conservation ethics**—How will my decision affect the ecosystem? Will my choice be ecologically sustainable in the long term?

With most ethical dilemmas, you will find that ethical stances conflict. To resolve the dilemma, it helps to first locate the "ethical tension"—the point where two or more ethical stances are incompatible. For example:

- An emergency room doctor who treats gunshot victims believes gun ownership should be highly restricted. Constitutional law, however, makes gun ownership a right. *Here, rights are in tension with utility.*
- Someone offers you a draft of your competitor's proposal for an important project. With that information, your company would almost certainly win the contract. However, your industry's code of ethics regarding proprietary information forbids you from looking at it. *Here, justice is in tension with utility.*
- Your company owns the rights to the timber in a forest, but an endangered species of eagle lives there and its habitat, by law, should be protected. *Here, justice, rights, and conservation are in tension.*
- A legal loophole allows your company to pump tons of pollution into the air even though this pollution harms the health of the residents in a small town a few miles downwind. *Here, personal and conservation ethics are in tension with justice.*

Do you want to try solving some other ethical dilemmas? Go to **www.pearsonhighered.com/johnsonweb4/4.8**

Balancing the Different Issues in an Ethical Dilemma

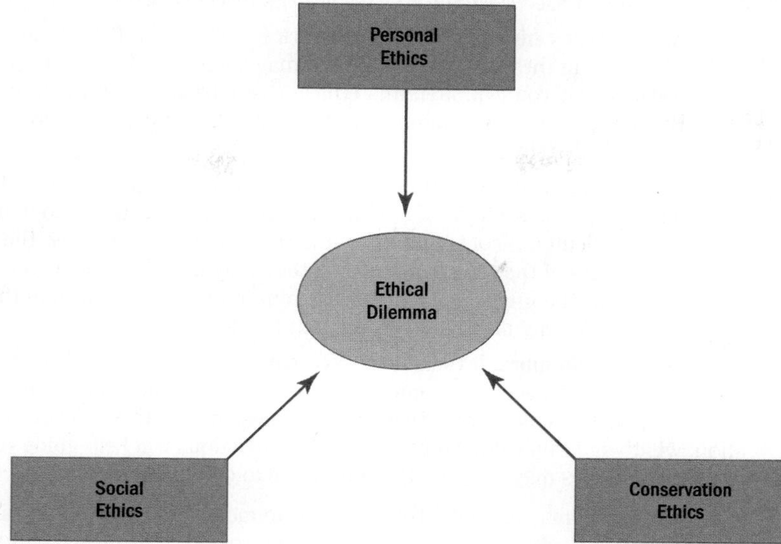

Figure 4.4: Resolving an ethical dilemma requires you to consider it from various ethical perspectives.

Link

For more information on liability in technical documents, see Chapter 7, page 177.

Product liability is a place where these kinds of conflicts become especially important. If your company produces a product that harms people unintentionally, the company may still be found liable for damages. It is not enough to simply cover your documentation with warnings. Even if warnings are provided, your company might be found negligent by the courts and ordered to pay damages.

Resolving an Ethical Dilemma

When faced with an ethical dilemma, you can use the following five questions to help you resolve it. These questions are a variation of the ones developed by Professor Sam Dragga in an article on ethics in technical communication (1996).

Do any laws or rules govern my decision?—In many cases, laws at the federal, state, and local levels will specify the appropriate action in an ethical case. You can look to your company's legal counsel for guidance in these matters. Otherwise, companies often have written rules or procedures that address ethical situations.

Do any corporate or professional codes of ethics offer guidance?—Most companies and professional organizations have published codes of ethics. They are usually rather abstract, but they can help frame ethical situations so you

AT A GLANCE

Resolving Ethical Dilemmas

- Do any laws or rules govern my decision?
- Do any corporate or professional codes of ethics offer guidance?
- Are there any historical records to learn from?
- What do my colleagues think?
- What would moral leaders do?

can make clearer decisions. Figure 4.5 shows the code of ethics for the Institute of Electrical and Electronics Engineers (IEEE).

Are there any historical records to learn from?—Look for similar situations in the past. Your company may keep records of past decisions, or you can often find ethical cases discussed on the Internet. By noting successes or failures in the past, you can make a more informed decision.

What do my colleagues think?—Your co-workers, especially people who have been around for a while, may have some insight into handling difficult ethical situations. First, they can help you assess the seriousness of the situation. Second, they may be able to help you determine the impact on others. At a minimum, talking through the ethical dilemma may help you sort out the facts.

What would moral leaders do?—You can look for guidance from moral leaders that you respect. These might include spiritual leaders, civil rights advocates, business pioneers, or even your friends and relatives. In your situation, what would they do? Sometimes their convictions will help guide your own. Their stories may give you the confidence to do what is right.

Facing an ethical dilemma, you will probably need to make a judgment call. In the end, you want to make an informed decision. If you fully consider the personal, social, and conservation perspectives, you will likely make a good decision.

When You Disagree with the Company

Ethical conflicts between you and your company need to be handled carefully. If you suspect your company or your supervisors are acting unethically, there are a few paths you can take:

Persuasion through costs and benefits—After you have collected the facts, take some time to discuss the issue with your supervisors in terms of costs and benefits. Usually, unethical practices are costly in the long term. Show them that the ethical choice will be beneficial over time.

Link

For more on persuasion strategies, see Chapter 13, page 367.

Seek legal advice—Your company likely has an attorney who can offer legal counsel on some issues. You may visit legal counsel to sort out the laws involved in the situation. If your company does not have legal counsel or you don't feel comfortable using it, you may need to look outside the company for legal help.

Mediation—Companies often offer access to mediators who can facilitate meetings between you and others. Mediators will not offer judgments on your ethical case, but they can help you and others identify the issues at stake and work toward solutions.

Link

For more information on resolving conflicts, see Chapter 4, page 81.

Memos to file—In some cases, you will be overruled by your supervisors. If you believe a mistake is being made, you may decide to write a *memo to file* in which you express your concerns. In the memo, write down all the facts and your concerns. Then, present the memo to your supervisors and keep a copy for yourself. These memos are usually filed for future reference to show your doubts.

To find websites discussing moral leaders, go to **www.pearsonhighered.com/johnsonweb4/4.9**

Sample Code of Ethics

IEEE CODE OF ETHICS

WE, THE MEMBERS OF THE IEEE, in recognition of the importance of our technologies in affecting the quality of life throughout the world and in accepting a personal obligation to our profession, its members and the communities we serve, do hereby commit ourselves to the highest ethical and professional conduct and agree:

1. to accept responsibility in making decisions consistent with the safety, health and welfare of the public, and to disclose promptly factors that might endanger the public or the environment;
2. to avoid real or perceived conflicts of interest whenever possible, and to disclose them to affected parties when they do exist;
3. to be honest and realistic in stating claims or estimates based on available data;
4. to reject bribery in all its forms;
5. to improve the understanding of technology, its appropriate application, and potential consequences;
6. to maintain and improve our technical competence and to undertake technological tasks for others only if qualified by training or experience, or after full disclosure of pertinent limitations;
7. to seek, accept, and offer honest criticism of technical work, to acknowledge and correct errors, and to credit properly the contributions of others;
8. to treat fairly all persons regardless of such factors as race, religion, gender, disability, age, or national origin;
9. to avoid injuring others, their property, reputation, or employment by false or malicious action;
10. to assist colleagues and co-workers in their professional development and to support them in following this code of ethics.

Approved by the IEEE Board of Directors | February 2006

Source: Institute of Electrical and Electronics Engineers. © 2006 IEEE. Reprinted with permission of the IEEE.

Figure 4.5: Like the IEEE, just about every established field has a code of ethics you can turn to for guidance.

Want to see other codes of ethics? Go to **www.pearsonhighered.com/johnsonweb4/4.10**

Whistle-blowing—In serious cases, especially where people's lives are at stake, you may even choose to be a whistle-blower. Whistle-blowing usually involves going to legal authorities, regulatory agencies, or the news media. Being a whistle-blower is a serious decision. It will affect your career and your company. Federal laws exist that protect whistle-blowers, but there is always a personal price to be paid.

Ethical situations should be carefully considered, but they should not be ignored. When faced with an ethical dilemma, it is tempting to walk away from it or pretend it isn't there. In any ethical situation, you should take some kind of action. Inaction on your part is both ethically wrong and might leave you or your company vulnerable to liability lawsuits. At a minimum, taking action will allow you to live with your conscience.

Websites exist that can help you make your decision by considering ethical case studies. For example, the Online Ethics Center at the National Academy of Engineering offers many case studies that are discussed by ethics experts (Figure 4.6). Perhaps one of these cases is similar to the one you face, and you can use the wisdom of these experts to make the ethical decision.

Online Ethics

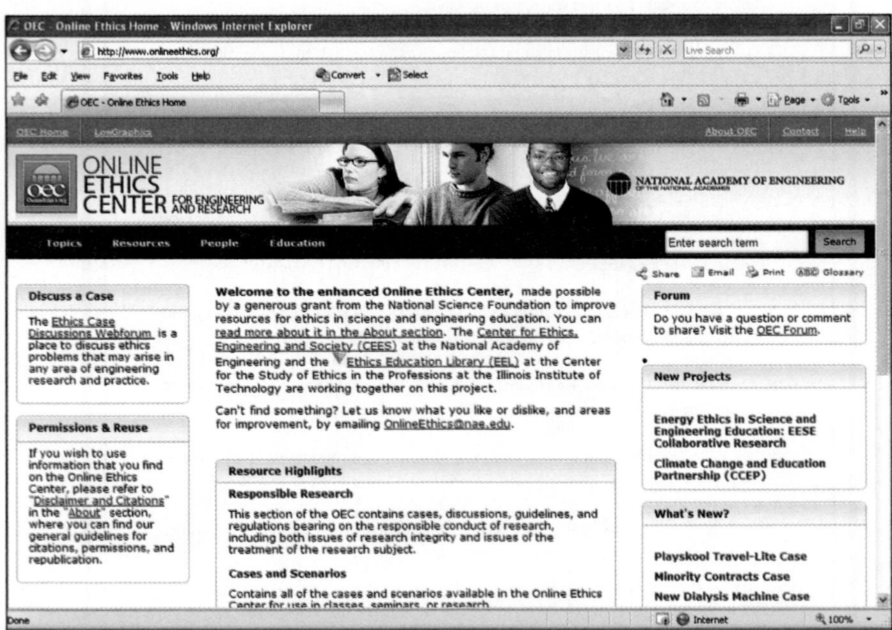

Figure 4.6: The Online Ethics Center at the National Academy of Engineering is a great place to learn about ethics in scientific and technical disciplines.

Source: The Online Ethics Center at the National Academy of Engineering, http://www.onlineethics.org.

GO TO THE NET

For more strategies for handling unethical situations at work, go to **www.pearsonhighered.com/johnsonweb4/4.11**
Does whistle-blowing really work? Go to **www.pearsonhighered.com/johnsonweb4/4.12**

Ethics in the Technical Workplace

Some legal and ethics scholars have speculated that the Information Age requires a new sense of ethics, or at least an updating of commonly held ethics. These scholars may be right. After all, our ethical systems, especially those involving forms of communication, are based on the printing press as the prominent technology. Laws and guidelines about copyright, plagiarism, privacy, information sharing, and proprietary information are all based on the idea that information is "owned" and shared on paper.

The fluid, shareable, changeable nature of electronic files and text brings many of these laws and guidelines into discussion. For example, consider the following case study:

> **Case Study:** You and your co-workers are pulling together a training package by collecting information off the Internet. You find numerous sources of information on the websites of consultants and college professors. Most of it is well written, so you cut and paste some of the text directly into your materials. You also find some great pictures and drawings on the Internet to add to your presentation. At what point does your cutting and pasting of text become a violation of copyright law? How can you avoid any copyright problems?

In the past, these kinds of questions were easier to resolve, because text was almost exclusively paper-based. Printed text is rather static, so determining who "owns" something is a bit easier. Today, the flexibility and speed of electronic media make these questions much more complicated.

At this time, many of our laws governing the use of information and text are evolving to suit new situations.

Copyright Law

Today, copyright law is being strained by the electronic sharing of information, images, and music. In legal and illegal forms, copies of books, songs, and software are all available on the Internet. According to the law, these materials are owned by the people who wrote or produced them; however, how can these materials be protected when they can be shared with a few clicks of a mouse? You can find out more about copyright law at the website of the U.S. Copyright Office (www.copyright.gov). Later in this chapter, U.S. copyright law is discussed in depth.

Trademarks

People or companies can claim a symbol, word, or phrase as their property by trademarking it. Usually, a trademark is signaled with a ™ symbol. For example, the Internet search engine Google™ is a trademarked name. The trademark signals that the company is claiming this word for its use in the area of Internet search engines.

To gain further protection, a company might register its trademark with the U.S. Patent and Trademark Office, which allows the company to use the symbol ® after the logo, word, or phrase. Once the item is registered, the trademark owner has exclusive rights to use that symbol, word, or phrase. For example, IBM's familiar blue symbol is its registered trademark, and it has the exclusive right to use it.

Want to learn more about wired ethics? Go to **www.pearsonhighered.com/johnsonweb4/4.13**
For more information on trademark and patent laws, go to
www.pearsonhighered.com/johnsonweb4/4.14

AT WORK

Caroline Whitbeck, Ph.D.

DIRECTOR OF THE ONLINE ETHICS CENTER FOR ENGINEERING AND SCIENCE, NATIONAL ACADEMY FOR ENGINEERING

The Online Ethics Center for Engineering and Science in Cleveland, Ohio, is an academic division that researches issues involving technology.

Why should technical professionals learn about ethics?

The practice of a profession, such as the profession of engineering, is characterized by two elements: (1) the practice directly influences one or more major aspects of human well-being and (2) it requires mastery of a complex body of knowledge and specialized skills. To become a professional in a field requires both formal education and practical experience.

The responsibility to achieve certain ends characterizes the core of professional ethics. For example, engineers have a responsibility for the public safety, and research investigators have a responsibility for the integrity of research. Achieving ends requires judgment in the application of professional knowledge.

Following moral rules, such as "Do not offer or accept bribes," important as they are, does not demand as much exercise of judgment as is required to fulfill responsibilities. Because the judgment needed to fulfill responsibilities requires professional knowledge, those without professional knowledge cannot judge whether a professional is making responsible judgments, that is, behaving both competently and with due concern. This is the reason why professions establish standards of responsible practice for their practitioners.

Of course, individual practitioners and sometimes the professions themselves may prove untrustworthy, but when they do, everyone loses. When professionals prove irresponsible, they may be monitored more closely. Monitoring may work to see if the rules are being followed, but it is not a ready check on the trustworthiness of professional judgments, precisely because they are judgments.

If society loses trust in a profession, people avoid relying on the service of members of that profession.

There are exceptions, though. The First Amendment of the U.S. Constitution, which protects free speech, has allowed trademarked items to be parodied or critiqued without permission of the trademark's owner.

Patents

Inventors of machines, processes, products, and other items can protect their inventions by patenting them. Obtaining a patent is very difficult because the mechanism being patented must be demonstrably unique. But once something is patented, the inventor is protected against others' use of his or her ideas to create new products.

To access the Online Ethics Center and similar sites, go to **www.pearsonhighered.com/johnsonweb4/4.15**

Privacy

Whether you realize it or not, your movements are regularly monitored through electronic networks. Surveillance cameras are increasingly prevalent in our society, and your online movements are tracked. Websites will send or ask for *cookies* that identify your computer. These cookies can be used to build a profile of you. Meanwhile, at a workplace, your e-mail and phone conversations can be monitored by your supervisors. Privacy laws are only now being established to cover these issues.

Information Sharing

Through electronic networks, companies can build databases with information about their customers and employees that can be shared with other companies. Information sharing, especially involving medical information, is an important issue that will probably be resolved in the courts.

Proprietary Information

As an employee of a company, you have access to proprietary information that you are expected not to share with others outside the company. In government-related work, you may even have a *security clearance* that determines what kinds of information you have access to. When you leave that company, you cannot take copies of documents, databases, or software with you. In some cases, you may even be asked to sign papers that prevent you from sharing your previous employer's secrets with your new employer.

Privacy in an Electronic World

Electronic networks offer many more opportunities for surveillance.

Libel and Slander

You or your company can be sued for printing falsehoods (libel) or speaking untruths (slander) that damage the reputation or livelihood of another person or company. With the broadcasting capabilities of the Internet, libel and slander have much greater reach. For example, a website that libels another person or company may be a target for legal retaliation. If you use e-mail to libel others, these messages could be used against you.

Fraud

The Internet is also opening whole new avenues for fraud. Con artists are finding new victims with classic fraud schemes. Websites, especially, can sometimes give the appearance of legitimacy to a fraudulent operation. Meanwhile, con artists use e-mail to find victims. (Have you received any e-mails from the widows of wealthy Nigerian dictators recently?)

Copyright Law in Technical Communication

An interesting flash point today is copyright law. A copyright gives someone an exclusive legal right to reproduce, publish, or sell his or her literary, musical, or artistic works. Copyright law in the United States was established by Article 1 of the U.S. Constitution. The U.S. law that governs copyright protection is called "Title 17" of the U.S. Code. You can find this code explained on the U.S. Copyright Office website at www.copyright.gov (Figure 4.7).

The U.S. Copyright Office Website

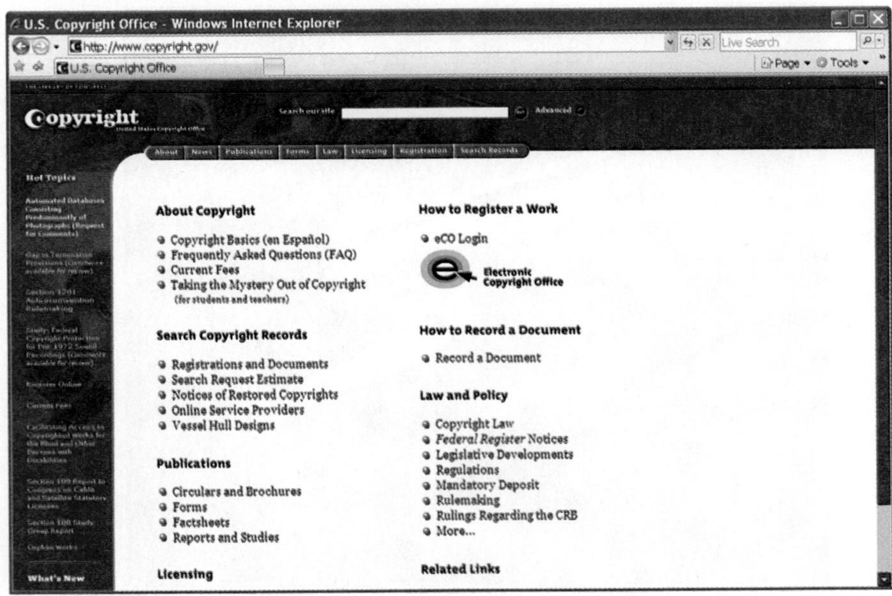

Figure 4.7: You can visit the U.S. Copyright Office website to learn more about copyright law or to protect your own work.

Source: United States Copyright Office, http://www.copyright.gov.

GO TO THE NET

Want to learn more about Internet libel, slander, and fraud? Go to **www.pearsonhighered.com/johnsonweb4/4.16**

Essentially, a copyright means that a creative work is someone's property. If others would like to duplicate that work, they need to ask permission and possibly pay the owner. Authors, musicians, and artists often sign over their copyrights to publishers, who pay them royalties for the right to duplicate their work.

New electronic media, however, have complicated copyright law. For example,

- When you purchase something, like a music CD, you have the right to duplicate it for your own personal use. What happens if you decide to copy a song off a CD and put it on your website for downloading? You might claim that you put the song on your website for your personal use, but now anyone else can download the song for free. Are you violating copyright law?
- According to Title 17, section 107, you can reproduce the work of others "for purposes such as criticism, comment, news reporting, teaching (including multiple copies for classroom use), scholarship, or research." This is referred to as "fair use." So, is it illegal to scan whole chapters of books for "teaching purposes" and put them on a CD for fellow students or co-workers?
- New technology like webcasting (using digital cameras to broadcast over the Internet) allows people to produce creative works. If you decided to webcast your and your roommates' dorm room antics each evening, would you be protected by copyright law?
- Blogs, or web logs, have become a popular way to broadcast news and opinions. Are these materials copyrighted?

The answer to these questions is "yes," but the laws are still being worked out. It is illegal to allow others to download songs off your website. It would be illegal to scan large parts of a book, even if you claimed they were being used for educational purposes. Meanwhile, you can protect webcasting and blogs through the copyright laws.

The problem is the ease of duplication. Before computers, copyrights were easier to protect because expensive equipment like printing presses, sound studios, and heavy cameras were required to copy someone else's work. Today, anyone can easily duplicate the works of others with a scanner, CD/DVD recorder, or digital video recorder.

Ultimately, violating copyright is like stealing someone else's property. The fact that it is easier to steal today does not make it acceptable. Nevertheless, a few scholars have argued that copyright law is antiquated and that this kind of electronic sharing is how people will use text and music in the future.

Asking Permission

To avoid legal problems, it is best to follow copyright law as it is currently written. You need to ask permission if you would like to duplicate or take something from someone else's work. You can ask permission by writing a letter or e-mail to the publisher of the materials. Publishers can almost always be found on the Internet. On their websites, they will often include a procedure for obtaining permissions. Tell them exactly what you want to use and how it will be used.

In some cases, especially when you are a student, your use may fall under the "fair use clause" of the Copyright Act. Fair use allows people to copy works for purposes of "criticism, comment, news reporting, teaching (including multiple copies for classroom use), scholarship, or research" (17 U.S. Code, sec. 107). If your use of the materials falls under these guidelines, you may have a *limited* right to use the materials without asking permission.

For example, fair use would likely allow you to use a song legally downloaded from the Internet as background music in a presentation for your class. However, it does not allow you to distribute that song freely to your friends, even if you claim you are doing so for educational purposes.

Copyrighting Your Work

What if you write a novel, take a picture, produce a movie, or create a song? How do you copyright it? The good news is that you already have. In the United States, a work is copyrighted as soon as it exists in written form. If you want, you can add the copyright symbol "©" to your work to signal that it is copyrighted. The copyright symbol, however, is no longer necessary to protect a work.

If you want to formally protect your work from copyright infringement (i.e., so you can sue someone who uses your work without your permission), you should register your copyright with the U.S. Copyright Office. This step is not necessary to protect your work, but it makes settling who owns the material much easier.

Plagiarism

One type of copyright infringement is plagiarism. In Chapter 15, plagiarism is discussed in depth, but the subject is worth briefly mentioning here. Plagiarism is the use of someone else's text or ideas as your own without giving credit. Plagiarism is a violation of copyright law, but it is also a form of academic dishonesty that can have consequences for your education and your career.

For example, cutting and pasting words and images off the Internet and "patchwriting" them into your documents is a form of plagiarism, unless those materials are properly cited. To avoid questions of plagiarism, make sure you cite your sources properly and, when needed, ask permission to use someone else's work.

CHAPTER REVIEW

- Ethics are systems of moral, social, or cultural values that govern the conduct of an individual or community.
- Ethical dilemmas force us to choose among uncomfortable alternatives.
- When you are faced with an ethical dilemma, consider it from all three ethical perspectives: personal, social, and conservation.
- You can turn to sources like laws, professional codes of ethics, historical records, your colleagues, or moral leaders to help you make ethical choices.
- When you disagree with the company, use persuasion first to discuss costs and benefits. You may turn to legal avenues if persuasion doesn't work.
- Ethical guidelines are evolving to suit the new abilities of computers.
- Copyright law and plagiarism are two rapidly evolving areas of ethics in this computer-centered world.
- Privacy and information sharing are also becoming hot topics, because computer networks facilitate the collection of so much information.

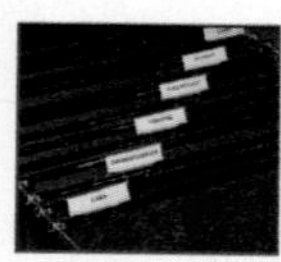

EXERCISES AND PROJECTS

Individual or Team Projects

1. Describe a real or fictional situation that involves a communication-related ethical dilemma. As you describe the situation, try to bring personal, social, and conservation ethics into conflict. At the end of the situation, leave the readers with a difficult question to answer.

   In a memo to your instructor, identify the ethical issues at stake in the situation you described and offer a solution to the ethical dilemma. Then, give your description to someone else in your class. He or she should write a memo to you and your instructor discussing the ethical issue at stake and offering a solution to the problem. Compare your original solution to your classmate's solution.

2. Find examples of advertising that seem to stretch ethics by making unreasonable claims. Choose one of these examples and write a short report to your instructor in which you discuss why you find the advertisement unethical. Use the terminology from this chapter to show how the advertisement challenges your sense of personal ethics, social ethics, or conservation ethics.

3. After researching conservation ethics on the Internet (see www.pearsonhighered.com/johnsonweb4/4.6), develop a Conservation Code of Ethics for your campus. In your code, you might discuss issues of recycling, water usage, chemical usage, testing on animals, or release of pollution. If you were an administrator at your university, how might you go about putting your code of ethics into action?

Collaborative Project

The case study at the end of this chapter discusses a difficult case in which a company might be doing something unethical. Read and discuss this case with a group of others. Sort out the ethical issues involved by paying attention to the personal, social, and conservation factors that shape the ethical dilemma.

If you were Hanna, how would you react to this ethical dilemma? How would you use writing to turn your reaction into action? Would you write a memo to your supervisor? Would you contact a government agency? Would a memo to file be appropriate? Would you blow the whistle?

Whichever path you choose, write a letter or memo to a specific reader (e.g., your supervisor, corporate management, the human resources office, a newspaper journalist) that urges the reader to take action. Summarize the situation for the reader of your document, and then suggest an appropriate course of action. Support your recommendation by highlighting the ethical issues involved and discussing the ramifications of inaction.

For additional technical writing resources, including interactive sample documents, document design tutorials and guidelines, and more, go to **www.mytechcommlab.com.**

CASE STUDY

This Company Is Bugging Me

Last week, Hanna Simpson's employer, Ventron United, sent out an identical memo and e-mail to all 50 employees. The memo/e-mail announced that all employees would be required to have a Radio Frequency Identification (RFID) chip inserted into their left or right hand. The RFID would allow employees to open secure doors, use equipment, and log onto computers throughout the company's campus in Portland, Oregon.

The announcement explained that the company would be working on some projects that required high security. So, management wanted to ensure that only employees who were supposed to enter secure areas or use secure computers would have access. Also, the company experienced a break-in a couple of months ago. The thieves took a few laptop computers, but the rumor was that the thieves were really looking for information on Ventron United's new products. They had used an employee's access card, which was reported missing the next day.

Hanna's two closest co-workers, Jim Peters and Georgia Miller, had very different reactions to the memo. Jim thought it was a great idea. He already had a personal RFID implanted in one of his hands. "It's about the size of a grain of rice. I don't notice it at all. But now I don't need to carry keys anymore. It opens the doors to my house, and I can turn on my car with just a button. I'm looking forward to the day when I can pay for my groceries with the wave of my hand."

Georgia was less enthusiastic, even paranoid. She confided to Hanna, "I think they're just trying to track our movements around here. Besides, my body is my personal space. These RFIDs are an invasion of my privacy."

"But would you quit your job over this?" Hanna asked Georgia.

"I hope it doesn't come to that, but yes I would," Georgia replied. "I'm not some dog that needs a chip implanted in me to keep track of me."

Other employees were worried about the health implications of the chips. Some just felt uncomfortable about the idea of a chip being injected into their hands. Some were considering a lawsuit if the company didn't back down on the issue.

Do some research on the Internet about RFIDs. What ethical issues are in conflict here?

To learn more about the ethics of RFIDs, go to **www.pearsonhighered.com/johnsonweb4/4.17**

CHAPTER

5

Letters, Memos, and E-Mail

Features of Letters, Memos, and E-Mails *94*

Planning and Researching *100*

Organizing and Drafting *102*

Types of Letters, Memos, and E-Mails *106*

Using Style and Design *114*

Using E-Mail Internationally *121*

Microgenre: Texting at Work *124*

Chapter Review *126*

Exercises and Projects *126*

Case Study: The Nastygram *130*

In this chapter, you will learn:

- The role of correspondence in the technical workplace.
- The basic features of letters, memos, and e-mail.
- How to plan, organize, and draft letters, memos, and e-mails.
- Common patterns for letters, memos, and e-mails.
- How to choose an appropriate style for correspondence.
- How to design and format letters and memos.
- How to revise, review, and proofread letters, memos, and e-mails.

Writing letters, memos, and e-mail will be a regular part of your job, but it can also be a drain on your time. The key is to learn how to write these documents quickly and efficiently, within the natural flow of your workday.

Letters and Memos Usually Convey Formal Messages

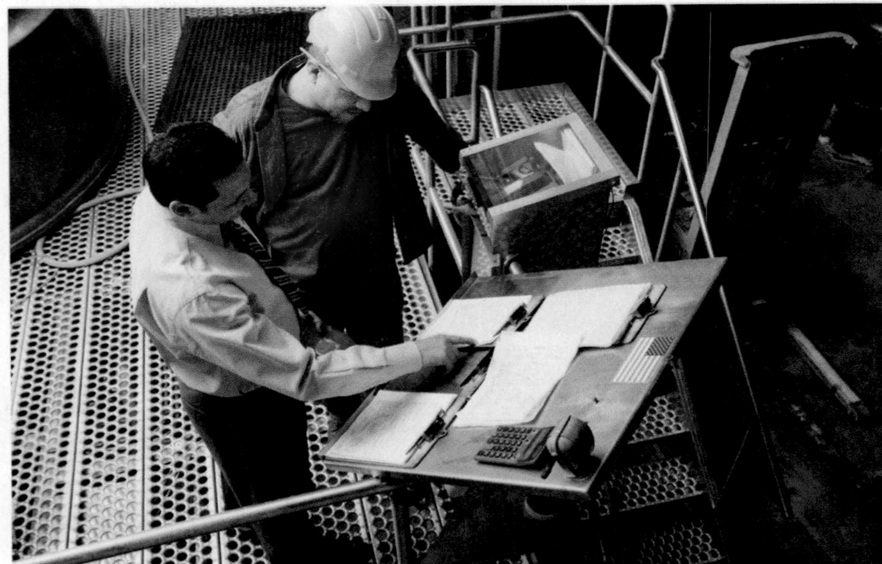

Letters and memos are generally used to convey formal messages, while e-mail is used for less formal messages.

Features of Letters, Memos, and E-Mails

Letters, memos, and e-mails are similar in most ways. These documents look different because they follow different *formats*. But in the end, they tend to use the same kinds of content, organization, style, and design to get their message across. So, how are they different?

Letters are written to people *outside* the company or organization. Primarily, letters are used in formal situations in which an employee is acting as a representative of the company. Letters can be used to make requests or inquiries, accept or refuse claims, communicate important information, record agreements, and apply for jobs.

Memos are written to people *inside* the company or organization. They usually contain meeting agendas, policies, internal reports, and short proposals. When a message is too important or proprietary for e-mail, most people will

To learn about the history of e-mail, go to **www.pearsonhighered.com/johnsonweb4/5.1**
To learn about the history of letter writing, go to **www.pearsonhighered.com/johnsonweb4/5.2**

Letters, Memos, and E-mails

This model shows a typical organizational pattern for a letter, memo, or e-mail. You should alter this pattern to fit your unique writing situation.

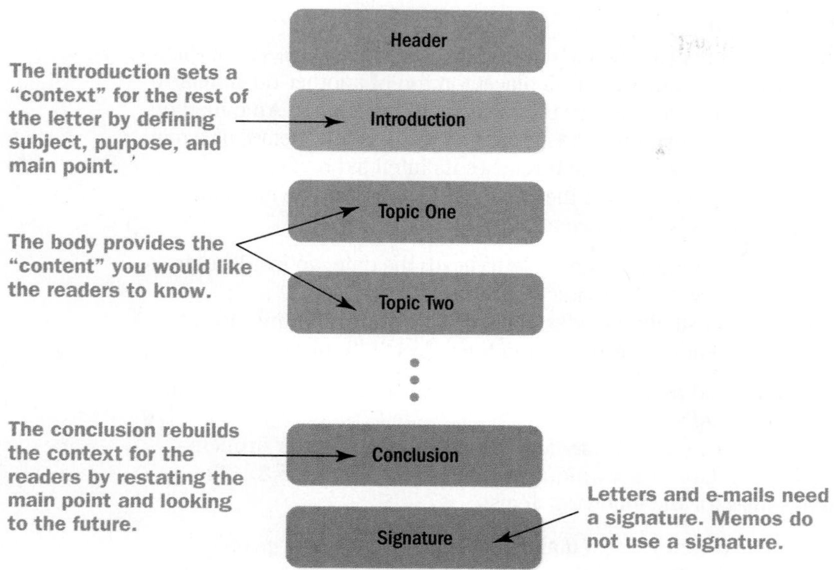

Header
The introduction sets a "context" for the rest of the letter by defining subject, purpose, and main point.
Introduction
Topic One
The body provides the "content" you would like the readers to know.
Topic Two
The conclusion rebuilds the context for the readers by restating the main point and looking to the future.
Conclusion
Signature
Letters and e-mails need a signature. Memos do not use a signature.

Basic Features of Letters, Memos, and E-mails

A letter, memo, or e-mail will generally have the following features:

- Header
- Salutation (for letters and some e-mails but not memos)
- Introduction that states a clear main point
- Body paragraphs with need-to-know information
- Conclusion that restates the main point
- Signature (for letters and e-mails but not memos)

send a memo instead. Memos are still more reliable than e-mails for information that should not be broadly released.

E-mails can be written to people *inside* or *outside* the company or organization. E-mail is used in situations that once called for less formal memos, letters, or phone calls. Increasingly, e-mail is being used for formal communication, too.

Letters, memos, and e-mails are also regularly used as *transmittal documents.* In the workplace, a letter or memo of "transmittal" is placed on top of another document to explain the document's purpose and clearly state who should receive it. An e-mail of transmittal is often used to send documents as attachments. A letter, memo, or e-mail of transmittal helps make sure your document reaches its intended readers.

Besides the inside the company/outside the company distinction, the main difference between letters and memos is their formatting:

- The format for a letter usually includes: a letterhead, the date, an inside address, a greeting, and a closing with the writer's signature.
- The format for a memo usually includes: a header, the date, and lines for the addressee ("To:"), the sender ("From:"), and the subject ("Subject:").

Figure 5.1 shows a letter and a memo with basically the same content so you can see the differences in formatting.

The format of an e-mail includes a *header* and *body*, similar to a letter or memo, but these formatting features have a few unique characteristics (Figure 5.2). The header of an e-mail has lines for the following items:

To line—This line contains the e-mail address of the person or people to whom you are sending the e-mail.

Cc and Bcc lines—The cc and bcc lines are used to copy the message to people who are not the primary readers, including your supervisors or others who might be interested in your conversation. The cc line shows your message's recipient that others are receiving copies of the message, too. The bcc line ("blind cc") allows you to copy your messages to others without anyone else knowing.

Subject line—The subject line signals the topic of the e-mail with a concise phrase.

Attachments line—This line signals whether there are any additional files, pictures, or programs "attached" to the e-mail message.

The *message area* is where you will type your message to your readers. Like other written documents, your message should have an introduction, a body, and a conclusion. You should also be as brief as possible, because most people won't read long e-mail messages.

Introduction—The *introduction* should (1) define the subject, (2) state your purpose, and (3) state your main point. If you want the reader to do something, you should mention that "action item" here in the introduction, because most recipients will not read your whole message.

To learn about how e-mail has evolved, go to **www.pearsonhighered.com/johnsonweb4/5.3**
For helpful advice about using e-mail, go to **www.pearsonhighered.com/johnsonweb4/5.4**

Letter and Memo

(a)

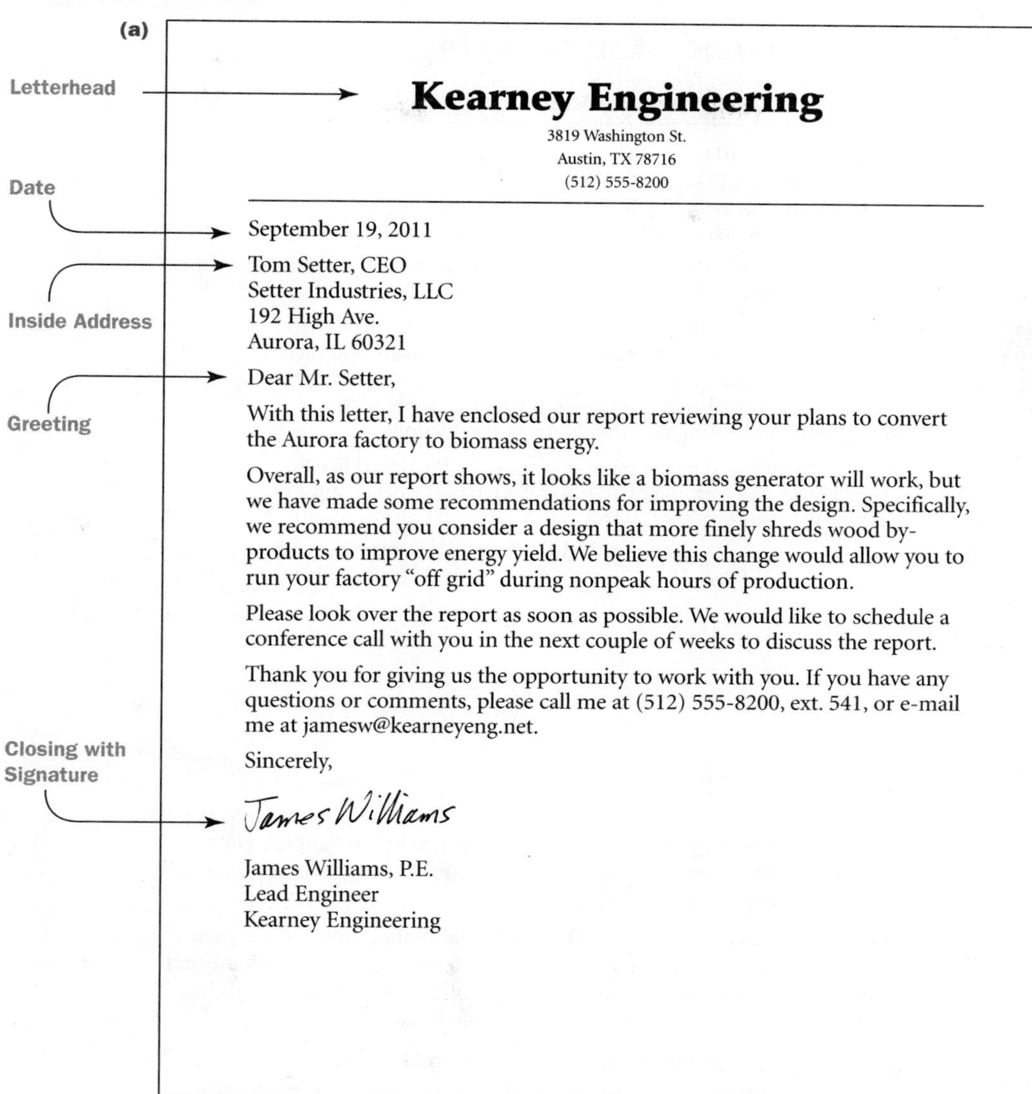

Kearney Engineering

3819 Washington St.
Austin, TX 78716
(512) 555-8200

September 19, 2011

Tom Setter, CEO
Setter Industries, LLC
192 High Ave.
Aurora, IL 60321

Dear Mr. Setter,

With this letter, I have enclosed our report reviewing your plans to convert the Aurora factory to biomass energy.

Overall, as our report shows, it looks like a biomass generator will work, but we have made some recommendations for improving the design. Specifically, we recommend you consider a design that more finely shreds wood by-products to improve energy yield. We believe this change would allow you to run your factory "off grid" during nonpeak hours of production.

Please look over the report as soon as possible. We would like to schedule a conference call with you in the next couple of weeks to discuss the report.

Thank you for giving us the opportunity to work with you. If you have any questions or comments, please call me at (512) 555-8200, ext. 541, or e-mail me at jamesw@kearneyeng.net.

Sincerely,

James Williams

James Williams, P.E.
Lead Engineer
Kearney Engineering

Figure 5.1: Letters (a) and memos (b) are basically the same, except in their formatting. The main differences are that letters are written to readers outside the company, whereas memos are written to readers inside the company.

To see other sample letters and memos, go to **www.pearsonhighered.com/johnsonweb4/5.5**

Figure 5.1: *(continued)*

(b)

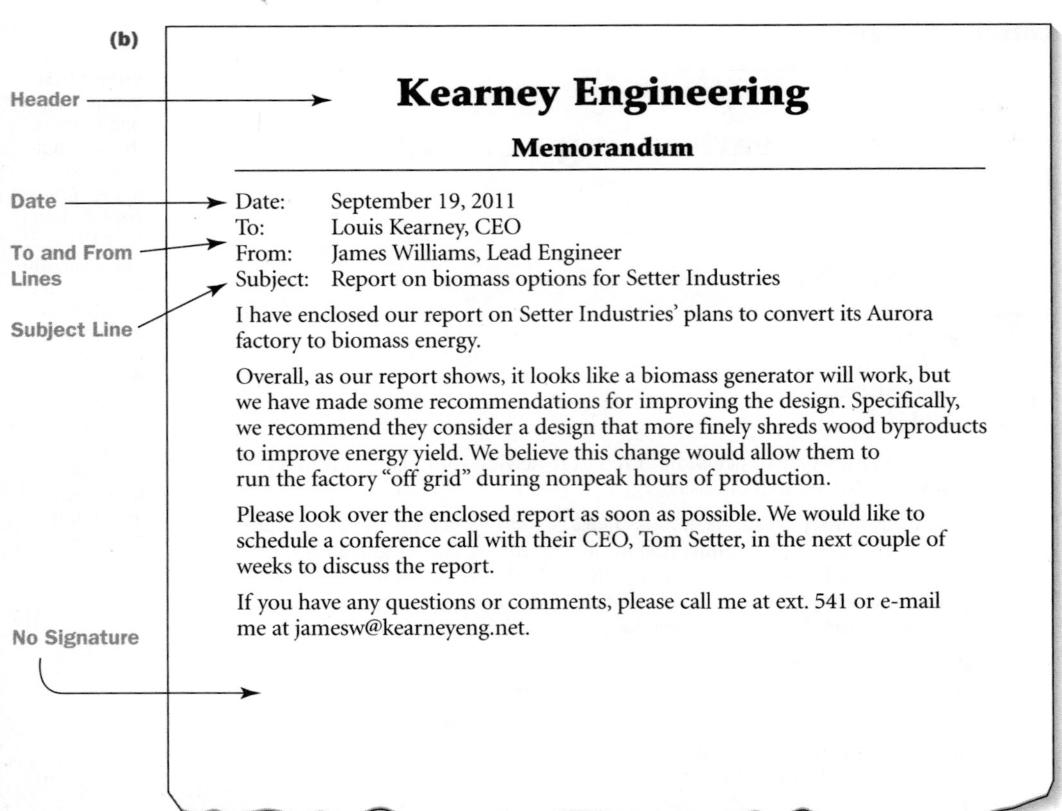
Header
Date
To and From Lines
Subject Line
No Signature
Kearney Engineering
Memorandum
Date: September 19, 2011
To: Louis Kearney, CEO
From: James Williams, Lead Engineer
Subject: Report on biomass options for Setter Industries
I have enclosed our report on Setter Industries' plans to convert its Aurora factory to biomass energy.
Overall, as our report shows, it looks like a biomass generator will work, but we have made some recommendations for improving the design. Specifically, we recommend they consider a design that more finely shreds wood byproducts to improve energy yield. We believe this change would allow them to run the factory "off grid" during nonpeak hours of production.
Please look over the enclosed report as soon as possible. We would like to schedule a conference call with their CEO, Tom Setter, in the next couple of weeks to discuss the report.
If you have any questions or comments, please call me at ext. 541 or e-mail me at jamesw@kearneyeng.net.

Body—The *body* should provide the information needed to support your e-mail's main point or achieve your e-mail's purpose. Strip your comments down to only need-to-know information.

Conclusion—The *conclusion* should (1) restate the main point and (2) look to the future. Here is not the place to tell your reader something important that you didn't mention earlier in the e-mail or ask him or her to do something, because most recipients will not read your whole e-mail.

The message area of an e-mail might also include these other kinds of text:

Reply text—When you reply to a message, most e-mail programs allow you to include parts of the original message in your message. These parts are often identified with arrows (>) running down the left margin (Figure 5.2).

Links—You can also include direct links to websites (Figure 5.2). Most e-mail software will automatically recognize a webpage address like http://www.predatorconservation.org and make it a live link in the e-mail's message area.

Features of an E-Mail

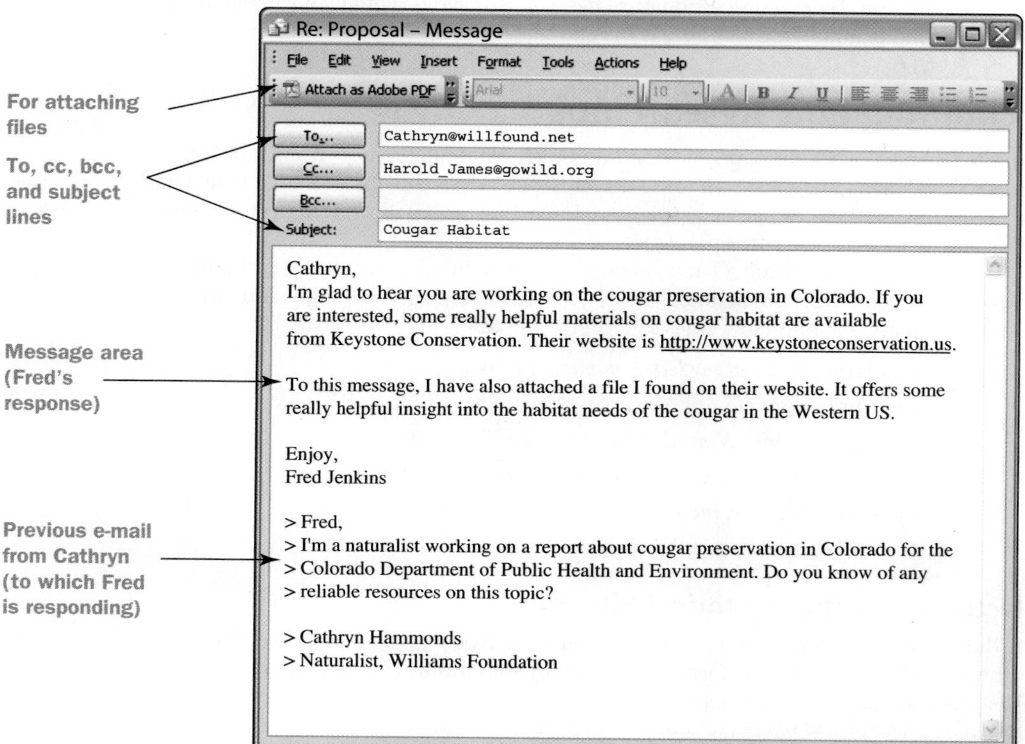

Figure 5.2: A basic message has several parts. In this message, the sender, Fred, is replying to Cathryn and copying (cc'ing) the message to Harold.

Attachments—Attachments are files, pictures, or programs that your readers can download to their own computers. If you would like to add an attachment to your e-mail message, click on the button that says "Attach Document" or "Attachment" in your e-mail software program. Most programs will then open a window that allows you to find and select the file you want to attach. If you attach a file to your e-mail message, you should notify readers in the message area that a file is attached. Otherwise, they may not notice it. You might write something like, "To this e-mail, I have attached a file that. . . ."

Signature—E-mail programs usually let you create a *signature file* that automatically puts a *signature* at the end of your messages.

Frank Randall, Marketing Assistant
Genflex Microsystems
612–555–9876
frandall@genflexmicro.net

For examples of corporate e-mail policies, go to **www.pearsonhighered.com/johnsonweb4/5.6**

Signature files allow you to personalize your message and add contact information. By creating a signature file, you can avoid typing your name, title, phone number, and so on at the end of each message you write.

Planning and Researching

In the technical workplace, people don't do much planning before writing a letter, memo, or e-mail. They just write it. This approach works fine for everyday letters and memos, but, as the message becomes more important, you will want to prepare more thoroughly to write the correspondence.

When you begin writing a letter, memo, or e-mail on your computer, first consider how your readers will use the information you are providing. You might start by answering the Five-W and How Questions:

Who is the reader of my letter, memo, or e-mail?

Why am I writing to this person?

What is my point? What do I want my reader to do?

Where will the letter, memo, or e-mail be read?

When will the letter, memo, or e-mail be used?

How will the reader use this document now and in the future?

Link

For more information about distinguishing between need-to-know and want-to-tell information, go to Chapter 15, page 409.

Determining the Rhetorical Situation

If the message is informal or routine, you might be ready to start typing right now. However, if your message is formal or especially important, you should explore the rhetorical situation in more depth.

SUBJECT Pay attention to what your readers need to know to take action. Letters and memos should be as concise as possible, so include only need-to-know information. Strip away any want-to-tell information that will distract your readers from your main point.

PURPOSE The purpose of your letter, memo, or e-mail should be immediately obvious to your readers. You should include a purpose statement in the first paragraph, perhaps even in the first sentence. Some key words for the purpose might include the following:

to inform	*to apologize*
to explain	*to discuss*
to complain	*to clarify*
to congratulate	*to notify*
to answer	*to advise*
to confirm	*to announce*
to respond	*to invite*

Want to find a cool signature file? Go to **www.pearsonhighered.com/johnsonweb4/5.7**

Your purpose statement might sound like one of the following:

> We are writing to inform you that we have accepted your proposal to build the Washington Street overpass.
>
> I would like to congratulate the Materials Team for successfully patenting the fusion polymer blending process.
>
> This memo explains and clarifies the revised manufacturing schedule for the remainder of this year.

Your purpose should be obvious to your readers as soon as they read the first paragraph. They should not have to guess why you are writing to them.

Link

For more guidance on defining a document's purpose, see Chapter 1, page 6.

READERS Letters, memos, and e-mails can be written to individuals or to whole groups of people. Since these documents are often shared or filed, you need to anticipate all possible readers who might want a copy of your document.

Primary readers (action takers) are the people who will take action after they read your message. Your letter, memo, or e-mail needs to be absolutely clear about what you want these readers to do. It should also be tailored to their individual motives, values, and attitudes about the subject.

Secondary readers (advisors) are the people to whom your primary readers will turn if they need advice. They may be experts in the area, support staff, supervisors, or colleagues. You should anticipate these readers' concerns, but your focus should still be on the primary readers' needs.

Tertiary readers (evaluators) are any other people who may have an interest in what you are saying. These readers may be more important than you expect. Letters, memos, and e-mails have a strange way of turning up in unexpected places. For example:

- That "confidential" memo you wrote to your research team might end up in your competitor's hands.
- The local newspaper might get a hold of a letter you sent to your company's clients explaining a problem with an important new product.
- A potentially embarrassing private e-mail to a co-worker might end up copied to the bottom of an e-mail that was sent to your supervisor.

Before sending any correspondence, think carefully about how the document would look if it were made public. Anticipate how it might be used against you or your company.

Gatekeeper readers (supervisors), such as your supervisor or legal counsel, may want to look over an especially important correspondence before it is sent out. You should always keep in mind that you are representing your company in your letters, memos, and e-mails. Your supervisor or the corporate lawyer may want to ensure that you are communicating appropriately with clients.

Link

For more information on analyzing readers, turn to Chapter 2, page 20.

Link

For strategies to help identify contextual issues, go to Chapter 2, page 26.

CONTEXT OF USE Imagine all the different places your letter, memo, or e-mail may be used. Where will readers use this document now and in the future? Where will the document be kept (if at all) after it is read? Be sure to consider the physical, economic, political, and ethical factors that will influence how your readers will interpret and respond to your message. Put yourself in their place, imagining their concerns as they are reading your document.

Organizing and Drafting

Some messages require more time and care than others, but in most cases, you should be able to generate letters, memos, and e-mails within the natural flow of your workday.

How can you write these documents efficiently? Keep in mind that the introductions and conclusions of these texts tend to make some predictable moves. If you memorize these moves, you can spend more time concentrating on what you need to say in the body of your document.

Introduction with a Purpose and a Main Point

In the introduction, you should make at least three moves: (a) identify the *subject,* (b) state the *purpose,* and (c) state the *main point* (Figure 5.3). Depending on your message, you might also make two additional moves: (d) offer some *background information* and (e) stress the *importance of the subject.*

SUBJECT Your *subject* should be stated or signaled in the first or second sentence of the introduction. Simply tell your readers what you are writing about. *Do not assume* that they already know what you are writing about.

> Recently, the Watson Project has been a source of much concern for our company.
>
> This memo discusses the equipment thefts that have occurred in our office over the last few months.

PURPOSE Your *purpose* for writing should also be stated almost immediately in the first paragraph, preferably in the first or second sentence.

> Now that we have reached Stage Two of the Oakbrook Project, I would like to refine the responsibilities of each team member.
>
> The purpose of this letter is to inform you about our new transportation policies for low-level nuclear waste sent to the WIPP Storage Facility in New Mexico.

MAIN POINT All letters, memos, and e-mails should have a *main point* that you want your readers to grasp or remember. In many cases, "the point" is something you want

Need help writing an introduction? Go to **www.pearsonhighered.com/johnsonweb4/5.8**

Morris Blue Industries

Date: November 18, 2011
To: Hanna Marietta, Chief Executive Officer
From: Jason Santos, Corporate Health Officer
Subject: Bird Flu Contingency Plan

Last week, the Executive Board inquired about our company's contingency plans if a bird flu pandemic occurs. As the Board mentioned, the exposure of our overseas manufacturing operations, especially in the Asian Pacific region, puts our company at special risk. At this point, we have no approved contingency plan, but my team strongly believes we need to create one as soon as possible. In this memo, I will highlight important issues to consider, and my team requests a meeting with you to discuss developing a plan for Board approval.

Despite the media hype, a bird flu pandemic is not imminent. A remote possibility exists that the H5N1 avian influenza virus could mutate into a form that can be transmitted among humans. To this point, though, only a small number of bird flu infections have occurred in humans. In these cases, birds have infected humans through close contact. The World Health Organization (WHO) reported in April 2011 that only 318 confirmed deaths had occurred worldwide, almost all in Asia. Human-to-human transmissions of bird flu are extremely rare.

Nevertheless, the risk of a pandemic is real and the WHO recommends the immediate development of contingency plans. We recommend the following actions right now:

A. Develop a decision tree that outlines how our company will respond to a pandemic.
B. Design an alert system that notifies managers how to identify bird flu symptoms, when to be watchful, when to send employees home, and how to evacuate them.

The subject is identified in the first sentence.

Background information is offered to remind the reader about the subject.

The body provides need-to-know information.

Figure 5.3: This memo shows the basic parts of a correspondence. The introduction sets a context, the body provides information, and the conclusion restates the main point.

The main point and purpose are clearly stated up front.

This paragraph uses facts to inform the readers.

This list makes important details easy to find.

To read about some letter and memo blunders, go to **www.pearsonhighered.com/johnsonweb4/5.9**

Figure 5.3: *(continued)*

C. Strengthen ties with local health authorities and law enforcement near our factories to speed the flow of information to local managers.
D. Create a training package for managers to educate them about bird flu and our company's response to a pandemic.

The Executive Board should also consider (a) whether we want to procure stocks of antiviral drugs like Tamiflu and Relenza, (b) whether our sick leave policies need to be adjusted to handle a pandemic, and (c) how our medical insurance would cover prevention and recovery for employees. These issues will require legal counsel from each country in which we have employees.

Thank you for contacting me about this matter. We believe a contingency plan should be developed as soon as possible. To get things rolling, we would like to schedule an appointment with you to go over these issues in more depth. You or your assistant can reach me at ext. 2205 or e-mail me at tjackson@morrisblue.com.

A "thank you" signals the conclusion of the memo.

The conclusion restates the main point and action item, while looking to the future.

Contact information is provided.

your readers to do (an "action item") when they are finished reading. In other words, state the *action* you want readers to take.

> We request the hiring of three new physician's assistants to help us with the recent increases in emergency room patients.
>
> Put bluntly, our subcontractors must meet ISO-9001 quality standards. It is our job to make sure that they comply.

It may seem odd to state your main point up front. Wouldn't it be better to lead up to the point, perhaps putting it in the conclusion? No. Most of your readers will only scan your message. By putting your main point (the action item) up front, you will ensure that they do not miss it.

BACKGROUND INFORMATION Writers often like to start their letters, memos, and e-mails with a statement that gives some background information or makes a personal connection to readers.

> Our staff meeting on June 28 was very productive, and I hope we all came away with a better understanding of the project. In this memo. . . .
>
> When you and I met at the NEPSCORE Convention last October, our company was not ready to provide specifics about our new ceramic circuit boards. Now we are ready. . . .

IMPORTANCE OF THE SUBJECT In some cases, you might also want your introduction to stress the importance of the subject.

> This seems like a great opportunity to expand our network into the Indianapolis market. We may not see this opportunity again.

> If we don't start looking for a new facility now, we may find ourselves struggling to keep up with the demand for our products.

Introductions should be as concise as possible. At a minimum, the introduction should tell readers your subject, purpose, and main point. Background information and statements about the importance of the subject should be used where needed.

Link
For more information on writing introductions, see Chapter 16, page 430.

Body That Provides Need-to-Know Information

The body is where you will provide your readers with the information they need to make a decision or take action. As shown in Figure 5.3, the body is the largest part of the memo or letter, and it will take up one or more paragraphs.

As you begin drafting the body of your text, divide your subject into the two to five major topics you need to discuss with your readers. Each of these major topics will likely receive a paragraph or two of discussion.

If you are struggling to develop the content, you can use mapping to put your ideas on the screen or a piece of paper (Figure 5.4). Start out by putting the purpose statement in the center of the screen or at the top of a piece of paper. Then, branch out into two to five major topics. You can use mapping to identify any supporting information that will be needed for those topics.

While drafting, keep looking back at your purpose statement in the introduction. Ask yourself, "What information do I need to provide to achieve this purpose?" Then,

Link
For more information on using logical mapping, go to Chapter 14, page 338.

Using Mapping to Generate Content

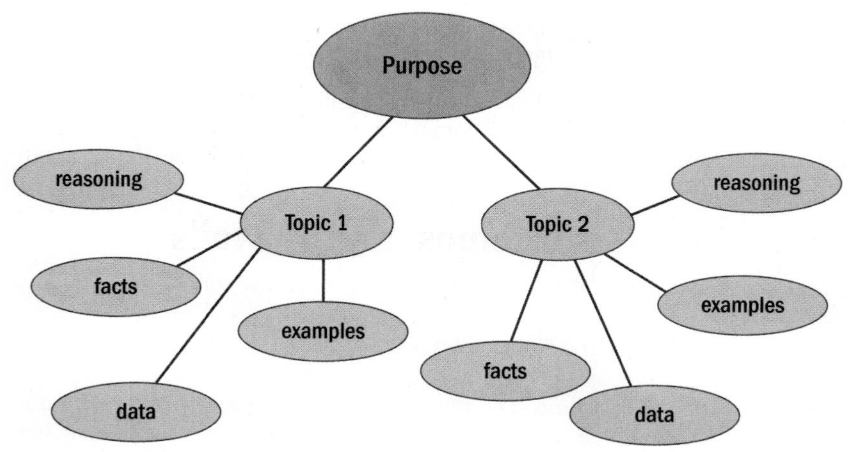

Figure 5.4: Using your purpose as a guide, identify the topics you will need to cover in your correspondence.

include any facts, examples, data, and reasoning that will help support your argument.

AT A GLANCE

Elements of a Letter, Memo, or E-mail

- Header
- Introduction—subject, purpose, main point, background information, importance of the subject
- Body—discussion topics, usually with one paragraph per topic
- Conclusion—thank you, main point (restated), and a look to the future

Conclusion That Restates the Main Point

The conclusion of your letter, memo, or e-mail should be short and to the point. Nothing essential should appear in the conclusion that has not already been stated in the introduction or body.

Conclusions in these documents tend to make three moves: *thank the readers, restate your main point,* and *look to the future.*

THANK THE READERS Tell them that you appreciate their attention to your message. By thanking them at the end, you leave them with a positive impression as you conclude.

> Thank you for your time and attention to this important matter.
>
> We appreciate your company's efforts on this project, and we look forward to working with you over the next year.

RESTATE YOUR MAIN POINT Remind your readers of the action you would like them to take.

> Time is short, so we will need your final report in our office by Friday, September 15, at 5:00.
>
> Please discuss this proposal right away with your team so that we can make any final adjustments before the submission deadline.

LOOK TO THE FUTURE Try to end your correspondence by looking forward in some way.

> When this project is completed, we will have taken the first revolutionary step toward changing our approach to manufacturing.
>
> If you have questions or comments, please call me at 555–1291 or e-mail me at sue.franklin@justintimecorp.com.

Link

To learn more about writing conclusions, go to Chapter 16, page 443.

Your conclusion should run about one to three sentences. If you find yourself writing a conclusion that is more than a small paragraph, you probably need to trim the added information or move some of it into the body of the letter, memo, or e-mail.

Types of Letters, Memos, and E-Mails

In the technical workplace, letters, memos, and e-mails are used for a variety of purposes.

Inquiries

The purpose of an *inquiry* is to gather information, especially answers to questions about important or sensitive subjects. In these situations, you could use e-mail, but a printed document is sometimes preferable because the recipients will view it as a formal request.

Here are some guidelines to follow when writing a letter, memo, or e-mail of inquiry:

- Clearly identify your subject and purpose.
- State your questions clearly and concisely.
- Limit your questions to five or fewer.
- If possible, offer something in return.
- Thank readers in advance for their response.
- Provide contact information (address, e-mail address, or phone number).

Figure 5.5 shows a typical letter of inquiry. Notice how the author of the letter is specific about the kinds of information she wants.

Responses

A response is written to answer an inquiry. A response should answer each of the inquirer's questions in specific detail. The amount of detail you provide will depend on the kinds of questions asked. In some situations, you may need to offer a lengthy explanation. In other situations, a simple answer or referral to the corporate website or enclosed product literature will be sufficient.

Here are some guidelines to follow when writing a response:

- Thank the writer for the inquiry.
- Clearly state the subject and purpose of the letter, memo, or e-mail.
- Answer any questions point by point.
- Offer more information, if available.
- Provide contact information (address, e-mail address, or phone number).

Figure 5.6 shows an example of a response letter. Pay attention to the author's point-by-point response to the questions in the original letter of inquiry (Figure 5.5).

Transmittal Letters and Memos

When sending documents or materials through the mail, you should include a letter or memo of transmittal. Also called "cover letters" or "cover memos," the purpose of these documents is to explain the reason the enclosed materials are being sent. For example, if you were sending a proposal to the vice president of your company, you would likely add a memo of transmittal like the one shown in Figure 5.7. Earlier in this chapter, the documents in Figure 5.1 also showed a letter and a memo of transmittal.

A transmittal letter or memo should do the following:

- Identify the materials enclosed.
- State the reason the materials are being sent.
- Summarize the information being sent.
- Clearly state any action requested or required of readers.
- Provide contact information.

For more examples of inquiry and response letters, go to **www.pearsonhighered.com/johnsonweb4/5.10**

Letter of Inquiry

State the subject and purpose of the letter.

State questions clearly and concisely.

Offer something in return.

Thank the readers.

Provide contact information.

Arctic Information Associates
2315 BROADWAY, FARGO, ND 58102

February 23, 2011

Customer Service
Durable Computers
1923 Hanson Street
Orono, Maine 04467

Dear Customer Service:

My research team is planning a scientific expedition to the northern Alaskan tundra to study the migration habits of caribou. We are looking for a rugged laptop that will stand up to the unavoidable abuse that will occur during our trip. Please send us detailed information on your Yeti rugged laptop. We need answers to the following questions:

- How waterproof is the laptop?
- How far can the laptop fall before serious damage will occur?
- How well does the laptop hold up to vibration?
- Does the laptop interface easily with GPS systems?
- Can we receive a discount on a purchase of 20 computers?

Upon return from our expedition, we would be willing to share stories about how your laptops held up in the Alaskan tundra.

Thank you for addressing our questions. Please respond to these inquiries and send us any other information you might have on the Yeti rugged laptop. Information can be sent to me at Arctic Information Associates, 2315 Broadway, Fargo, ND 58102. I can also be contacted at 701-555-2312 or salvorman@arcticia.com.

Sincerely,

S Vorman

Sally Vorman, Ph.D.
Arctic Specialist

Figure 5.5: A letter of inquiry needs to be clear about the information it is seeking. In this letter, notice how the writer has listed her questions in an unmistakable way.

Response Letter

Thank the readers for their inquiry.

State the subject and purpose of the letter.

Answer the questions point by point.

Offer more information, if available.

Provide contact information.

Durable Computers

1923 Hanson Street, Orono, Maine 04467

March 7, 2011

Sally Vorman, Arctic Specialist
Arctic Information Associates
2315 Broadway, Fargo, ND 58102

Dear Dr. Vorman:

Thank you for your inquiries regarding our Yeti Rugged Laptop. This computer is one of our most durable products, and it is particularly suited to the kinds of arctic climates you will be experiencing.

Here are the answers to your questions:

Waterproofing: The Yeti stands up well to rain and other kinds of moisture. It can be submersed briefly (a few seconds), but it cannot be left underwater for a sustained amount of time.

Damage Protection: The Yeti can be dropped from 20 feet onto concrete without significant damage. Its magnesium alloy casing provides maximum protection.

Vibration: The Yeti meets the tough US-MIL 810E standards, which pay close attention to vibration, especially across rough terrain in a vehicle.

GPS Compatibility: The Yeti is compatible with all GPS systems we are aware of.

Discounts: We offer a discount of 10% for orders of 10 or more Yetis.

I am also enclosing some of our promotional literature on the Yeti, including the technical specifications. In these materials you will find the results of our endurance testing on the laptop.

We would very much like to hear about your trip and your experiences with the Yeti. If you have any more questions or would like to place an order, please call me at 293-555-3422. Or, e-mail me at garys@duracomps.net.

Sincerely,

Gary Smothers

Gary Smothers
Design Engineer
Durable Computers

Figure 5.6: A response letter should answer the inquirer's questions point by point and offer additional information, if available.

Memo of Transmittal

Rockford Services

MEMORANDUM

Date: May 8, 2011
To: Brenda Young, VP of Services
From: Valerie Ansel, Outreach Coordinator
cc: Hank Billups, Pat Roberts
Re: Outreach to Homeless Youth

Identify the enclosed materials.

State the reason materials are being sent.

Enclosed is the Proposal for the Rockford Homeless Youth Initiative, which you requested at the Board Meeting on February 16. We need you to look it over before we write the final version.

Summarize the enclosed materials.

The proposal describes a broad-based program in which Rockford Services will proactively reach out to the homeless youth in our city. In the past, we have generally waited for these youths to find their way to our shelter on the west side of town. We always knew, though, that many youths are reluctant or unable to come to the shelter, especially the ones who are mentally ill or addicted to drugs. The program described in this proposal offers a way to reach out to these youths in a nonthreatening way, providing them a gateway to services or treatment.

State the action item clearly.

Please look over this proposal. We welcome any suggestions for improvement you might have. We plan to submit the final version of this proposal to the Board on May 24th at the monthly meeting.

Provide contact information.

Thank you for your help. You can contact me by phone at 555-1242, or you can e-mail me at valansel@rockfordservices.org.

Enclosed: Proposal for the Rockford Homeless Youth Initiative

Figure 5.7: A transmittal memo should be concise. Make sure any action items are clearly stated.

You should keep your comments brief in transmittal letters or memos. After all, readers are mostly interested in the enclosed materials, not your transmittal letter or memo.

Why should you include a letter or memo of transmittal in the first place? There are a few good reasons:

- If a document, such as a report, shows up in your readers' mail without a transmittal letter or memo, they may not understand why it is being sent to them and what they should do with it.
- Transmittal letters and memos give you an opportunity to make a personal connection with the readers.
- They also give you an opportunity to set a specific tone for readers, motivating them to respond positively to the document or materials you have enclosed.

An effective transmittal letter or memo welcomes your readers to the materials you have sent.

Claims or Complaints

In the technical workplace, products break and errors happen. In these situations, you may need to write a claim, also called a complaint. The purpose of a claim is to explain a problem and ask for amends. Here are some guidelines to follow when writing a claim:

- State the subject and purpose clearly and concisely.
- Explain the problem in detail.
- Describe how the problem inconvenienced you.
- State what you would like the receiver to do to address the problem.
- Thank your reader for his or her response to your request.
- Provide contact information.

Figure 5.8 shows a claim letter with these features.

A claim should always be professional in tone. Angry letters, memos, and e-mails might give you a temporary sense of satisfaction, but they are less likely to achieve your purpose—to have the problem fixed. If possible, you want to avoid putting readers on the defensive, because they may choose to ignore you or halfheartedly try to remedy the situation.

Adjustments

If you receive a claim or complaint, you may need to respond with an *adjustment* letter, memo, or e-mail. The purpose of an adjustment is to respond to the issue described by the client, customer, or co-worker. These documents, though, need to do more than simply respond to the problem. They should also try to rebuild a potentially damaged relationship with the reader.

Here are some guidelines to follow when writing an adjustment:

- Express regret for the problem *without directly taking blame.*
- State clearly what you are going to do about the problem.
- Tell your reader when he or she should expect results.
- Show appreciation for his or her continued business with your company.
- Provide contact information.

Figure 5.9 shows an adjustment letter with these features.

For more samples of claim and adjustment letters, go to **www.pearsonhighered.com/johnsonweb4/5.11**

Claim Letter

Figure 5.8: A claim letter should explain the problem in a professional tone and describe the remedy being sought.

Outwest Engineering

2931 Mission Drive, Provo, UT 84601 (801) 555-6650

June 15, 2011

Customer Service
Optima Camera Manufacturers, Inc.
Chicago, IL 60018

Dear Customer Service:

We are requesting the repair or replacement of a damaged ClearCam Digital Camcorder (#289PTDi), which we bought directly from Optima Camera Manufacturers in May 2011.

Here is what happened. On June 12, we were making a promotional film about one of our new products for our website. As we were making adjustments to the lighting on the set, the camcorder was bumped and it fell ten feet to the floor. Afterward, it would not work, forcing us to cancel the filming, causing us a few days' delay.

We paid a significant amount of money for this camcorder because your advertising claims it is "highly durable." So, we were surprised and disappointed when the camcorder could not survive a routine fall.

Please repair or replace the enclosed camcorder as soon as possible. I have provided a copy of the receipt for your records.

Thank you for your prompt response to this situation. If you have any questions, please call me at 801-555-6650, ext. 139.

Sincerely,

Paul Williams

Paul Williams
Senior Product Engineer

State the subject and purpose of the letter.

Explain the problem in detail.

Describe how the problem inconvenienced you.

State what the reader should do.

Thank the reader for the anticipated response.

Provide contact information.

Adjustment Letter

Figure 5.9: An adjustment letter should express regret for the problem and offer a remedy.

OCM

Optima Camera Manufacturers, Inc.
Chicago, IL 60018 312-555-9120

July 1, 2011

Paul Williams, Senior Product Engineer
Outwest Engineering Services
2931 Mission Drive
Provo, UT 84601

Dear Mr. Williams,

We are sorry that the ClearCam Digital Camcorder did not meet your expectations for durability. At Optima, we take great pride in offering high-quality, durable cameras that our customers can rely on. We will make the repairs you requested.

After inspecting your camera, our service department estimates the repair will take two weeks. When the camera is repaired, we will return it to you by overnight freight. The repair will be made at no cost to you.

We appreciate your purchase of a ClearCam Digital Camcorder, and we are eager to restore your trust in our products.

Thank you for your letter. If you have any questions, please contact me at 312-555-9128.

Sincerely,

Ginger Faust

Ginger Faust
Customer Service Technician

Express regret for the problem.

State what will be done.

Tell when results should be expected.

Show appreciation to the customer.

Provide contact information.

Why shouldn't you take direct blame? Several factors might be involved when something goes wrong. So, it is fine to acknowledge that something unfortunate happened. For example, you can say, "We are sorry to hear about your injury when using the Zip-2000 soldering tool." But it is something quite different to say, "We accept full responsibility for the injuries caused by our Zip-2000 soldering tool." This kind of statement could make your company unnecessarily liable for damages.

Ethically, your company may need to accept full responsibility for an accident. In these situations, legal counsel should be involved with the writing of the letter.

Refusals

Refusals, also called "bad news" letters, memos, or e-mails, always need to be carefully written. In these documents, you are telling the readers something they don't want to hear (i.e., "no"). Yet, if possible, you want to maintain a professional or business relationship with these customers or clients.

When writing a refusal, show your readers how you logically came to your decision. In most cases, you will not want to start out immediately with the bad news (e.g., "We have finished interviewing candidates and have decided not to hire you"). However, you also do not want to make readers wait too long for the bad news.

Here are some guidelines for writing a refusal:

- State your subject.
- Summarize your understanding of the facts.
- Deliver the bad news, explaining your reasoning.
- Offer any alternatives, if they are available.
- Express a desire to retain the relationship.
- Provide contact information.

Keep any apologizing to a minimum, if you feel you must apologize at all. Some readers will see your apology as an opening to negotiate or complain further. An effective refusal logically explains the reasons for the turndown, leaving your reader satisfied with your response—if a bit disappointed. Figure 5.10 shows a sample refusal letter with these features.

Using Style and Design

The style and design of a letter, memo, or e-mail can make a big difference. One thing to keep in mind is this: *All letters, memos, and e-mails are personal.* They make a one-to-one connection with readers. Even if you are writing a memo to the whole company or sending out a form letter to your company's customers, you are still making a personal, one-to-one connection with each of those readers. People will read your correspondence as a message written to them individually. So, you should take some time to ensure that the style and design of your correspondence fits the rhetorical situation.

To see samples of other refusal letters, go to **www.pearsonhighered.com/johnsonweb4/5.12**

Refusal Letter

State the subject.

Summarize what happened.

Deliver the bad news, explaining your reasoning.

Offer alternatives.

Provide contact information.

Express a desire to retain the relationship.

O C M

Optima Camera Manufacturers, Inc.
Chicago, IL 60018 312-555-9120

July 1, 2011

Paul Williams, Senior Product Engineer
Outwest Engineering Services
2931 Mission Drive
Provo, UT 84601

Dear Mr. Williams,

We are sorry that the ClearCam Digital Camcorder did not meet your expectations for durability. At Optima, we take great pride in offering high-quality, durable cameras that our customers can rely on.

According to the letter you sent us, the camcorder experienced a fall and stopped working. After inspecting your camcorder, we have determined that we will need to charge for the repair. According to the warranty, repairs can only be made at no cost when problems are due to manufacturer error. A camcorder that experienced a fall like the one you described is not covered under the warranty.

We sent your camcorder to the service department for a repair estimate. After inspecting your camera, they estimate the repair will take two weeks at a cost of $156.00. When it is repaired, we will return it to you by overnight freight.

If you would like us to repair the camcorder, please send a check or money order for $156.00. If you do not want us to repair the camcorder, please call me at 312-555-9128. Upon hearing from you, we will send the camcorder back to you immediately.

Again, we are sorry for the damage to your camcorder. We appreciate your purchase of a ClearCam Digital Camcorder, and we are eager to retain your business.

Sincerely,

Ginger Faust

Ginger Faust
Customer Service Technician

Enclosed: Warranty Information

Figure 5.10: A refusal letter should deliver the bad news politely and offer alternatives if available. You should strive to maintain the relationship with the person whose request is being refused.

Strategies for Developing an Appropriate Style

Since letters, memos, and e-mails are personal documents, their style needs to be suited to their readers and contexts of use. Here are some strategies for projecting the appropriate style:

- Use the "you" style.
- Create an appropriate tone.
- Avoid bureaucratic phrasing.

USE THE "YOU" STYLE When you are conveying neutral or positive information, you should use the word *you* to address your readers. The "you" style puts the emphasis on the readers rather than on you, the author.

> Well done. Your part of the project went very smoothly, saving us time and money.
>
> We would like to update your team on the status of the Howards Pharmaceutical case.
>
> You are to be congratulated for winning the Baldrige Award for high-quality manufacturing.

In most cases, negative information should not use the "you" style, because readers will tend to react with more hostility than you expect.

> **Offensive:** Your lack of oversight and supervision on the assembly line led to the recent work stoppage.
>
> **Improved:** Increased oversight and supervision will help us avoid work stoppages in the future.
>
> **Offensive:** At our last meeting, your ideas for new products were not fully thought through. In the future, you should come more prepared.
>
> **Improved:** Any ideas for new products should be thoroughly considered before they are presented. In the future, we would like to see presenters more prepared.

Link

For more advice about choosing an appropriate style, see Chapter 17, page 455.

Don't worry about whether your readers will notice that you are criticizing them. Even without the "you" style, they will figure out that you are conveying negative information or criticisms. By avoiding "you" in these negative situations, though, you will create a constructive tone and avoid an overly defensive reaction from your readers.

CREATE A TONE Think about the image you want to project. Put yourself into character as you compose your message. Are you satisfied, hopeful, professional, pleased, enthusiastic, or annoyed? Write your message with that tone in mind.

Mapping is an especially good way to project a specific tone in your correspondence (Figure 5.11). For example, perhaps you want to argue that you are an "expert." Put the word "expert" in the middle of the screen or a piece of paper. Then, write down words associated with this word.

GO TO THE NET

Want more advice about avoiding bureaucratic language? Go to **www.pearsonhighered.com/johnsonweb4/5.13**

Mapping a Tone

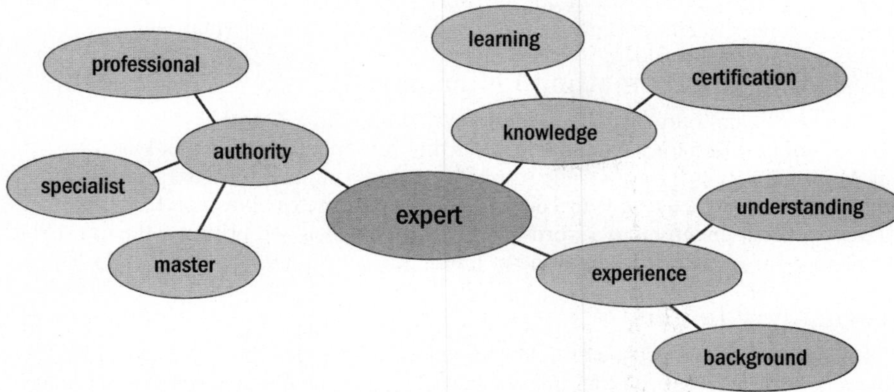

Figure 5.11: To set a tone, you can use mapping to develop words that are associated with the tone you want to create.

Once you have mapped out the tone you want to create, you can then weave these words into your message. If the words are used strategically, your readers will subconsciously sense the tone you are trying to create.

AVOID BUREAUCRATIC PHRASING When writing correspondence, especially a formal letter, some people feel a strange urge to use phrasing that sounds bureaucratic:

Bureaucratic: Pursuant to your request, please find the enclosed materials.

Nonbureaucratic: We have included the materials you requested.

Do you notice how the bureaucratic phrasing in the first sentence only makes the message harder to understand? It doesn't add any information. Moreover, this phrasing depersonalizes the letter, undermining the one-to-one relationship between writer and reader.

Here are a few other bureaucratic phrases and ways they can be avoided:

Bureaucratic Phrase	Nonbureaucratic Phrase
Per your request	As you requested
In lieu of	Instead of
Attached, please find	I have attached
Enclosed, please find	I have enclosed
Contingent upon receipt	When we receive
In accordance with your wishes	As you requested
In observance with	According to
Please be aware that	We believe
It has come to our attention	We know
Pursuant to	In response to
Prior to receipt of	Before receiving

GO TO THE NET To download letter and memo templates, go to **www.pearsonhighered.com/johnsonweb4/5.14**

A simple guideline is not to use words and phrases that you would not use in everyday speech. If you would not use words like *lieu, contingent,* or *pursuant* in a conversation, you should not use them in a letter, memo, or e-mail.

Designing and Formatting Letters and Memos

Letters and memos are usually rather plain in design. In the workplace, they typically follow standardized formats and templates that prescribe how they will look.

Most companies have premade word-processing templates for letters and memos that you can download on your computer. These templates allow you to type your letter or memo directly into a word-processing file. When you print out the document, the letterhead or memo header appears at the top.

Formatting Letters

As stated earlier in this chapter, letter formats typically include some predictable features: a header (letterhead), an inside address, a greeting, the message, and a closing with a signature (Figure 5.12).

LETTERHEAD Companies typically have letterhead available as a premade word-processor template or as stationery. Letterhead includes the company name and address. If letterhead is not available, you should enter your return address followed by the date. Do not include your name in the return address.

> 1054 Kellogg Avenue, Apt. 12
> Hinsdale, Illinois 60521
> December 19, 2010

The return address is best set along the left margin of the letter.

INSIDE ADDRESS The address of the person to whom you are sending the letter (called the *inside address*) should appear two lines below the date or return address.

> George Falls, District Manager
> Optechnical Instruments
> 875 Industrial Avenue, Suite 5
> Starkville, New York 10034

The inside address should be the same as the address that will appear on the letter's envelope.

GREETING Include a greeting two lines below the inside address. It is common to use the word "Dear," followed by the name of the person to whom you are sending the letter. A comma or colon can follow the name, although in business correspondence a colon is preferred.

If you do not know the name of the person to whom you are sending the letter, choose a gender-neutral title like "Human Resources Director," "Production Manager," or "Head Engineer." A generic greeting like "To Whom It May Concern" is inappropriate because it is too impersonal. With a little thought, you can usually come up with a neutral title that better targets the reader of your letter.

Formatting a Letter

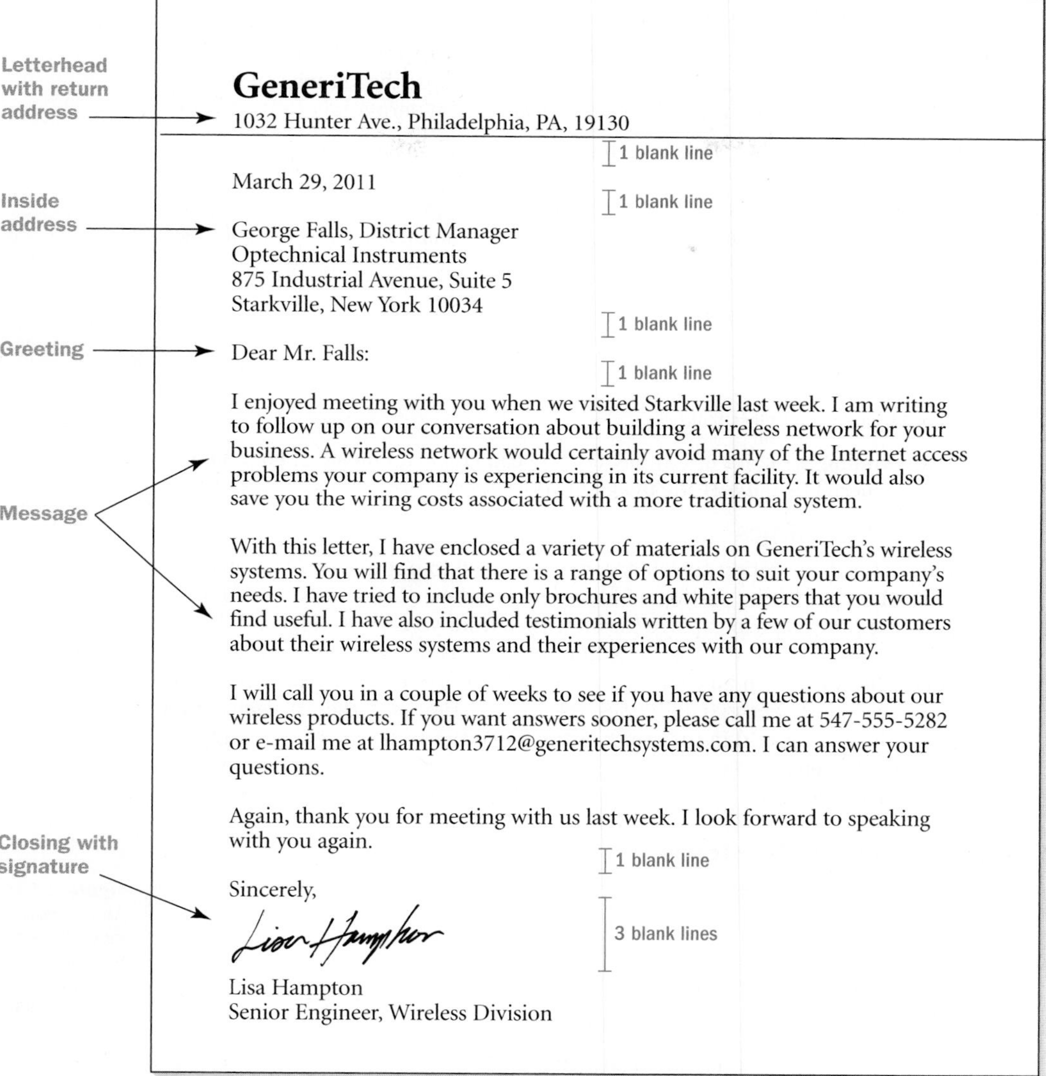

GeneriTech
1032 Hunter Ave., Philadelphia, PA, 19130

March 29, 2011

George Falls, District Manager
Optechnical Instruments
875 Industrial Avenue, Suite 5
Starkville, New York 10034

Dear Mr. Falls:

I enjoyed meeting with you when we visited Starkville last week. I am writing to follow up on our conversation about building a wireless network for your business. A wireless network would certainly avoid many of the Internet access problems your company is experiencing in its current facility. It would also save you the wiring costs associated with a more traditional system.

With this letter, I have enclosed a variety of materials on GeneriTech's wireless systems. You will find that there is a range of options to suit your company's needs. I have tried to include only brochures and white papers that you would find useful. I have also included testimonials written by a few of our customers about their wireless systems and their experiences with our company.

I will call you in a couple of weeks to see if you have any questions about our wireless products. If you want answers sooner, please call me at 547-555-5282 or e-mail me at lhampton3712@generitechsystems.com. I can answer your questions.

Again, thank you for meeting with us last week. I look forward to speaking with you again.

Sincerely,

Lisa Hampton
Senior Engineer, Wireless Division

Figure 5.12: The format of a letter has predictable features, like the letterhead, inside address, greeting, message, and a closing with a signature.

Want more help with formatting your letter or memo? Go to **www.pearsonhighered.com/johnsonweb4/5.15**

Also, remember that it is no longer appropriate to use gender-biased terms like "Dear Sirs" or "Dear Gentlemen." You will offend at least half the receivers of your letters with these kinds of gendered titles.

MESSAGE The message should begin two lines below the greeting. Today, most letters are set in *block format,* meaning the message is set against the left margin with no indentation. In block format, a space appears between each paragraph.

CLOSING WITH SIGNATURE Two lines below the message, you should include a closing with a signature underneath. In most cases, the word "Sincerely," followed by a comma, is preferred. Your signature should appear next, with your name and title typed beneath it. To save room for your signature, you should leave three blank lines between the closing and your typed name.

Sincerely,

Lisa Hampton
Senior Engineer, Wireless Division

If you are sending the letter electronically, you can create an image of your signature with a scanner. Then, insert the image in your letter.

Formatting Envelopes

Once you have finished writing your letter, you will need to put it in an envelope. Fortunately, with computers, putting addresses on envelopes is not difficult. Your word-processing program can capture the addresses from your letter (Figure 5.13). Then, with the Envelopes and Labels function (or equivalent), you can have the word processor put the address on an envelope or label. Most printers can print envelopes.

An envelope should have two addresses, the *return address* and the *recipient address.* The return address is printed in the upper left-hand corner of the envelope, a couple of lines from the top edge of the envelope. The recipient address is printed in the center of the envelope, about halfway down from the top edge of the envelope.

Formatting for an Envelope

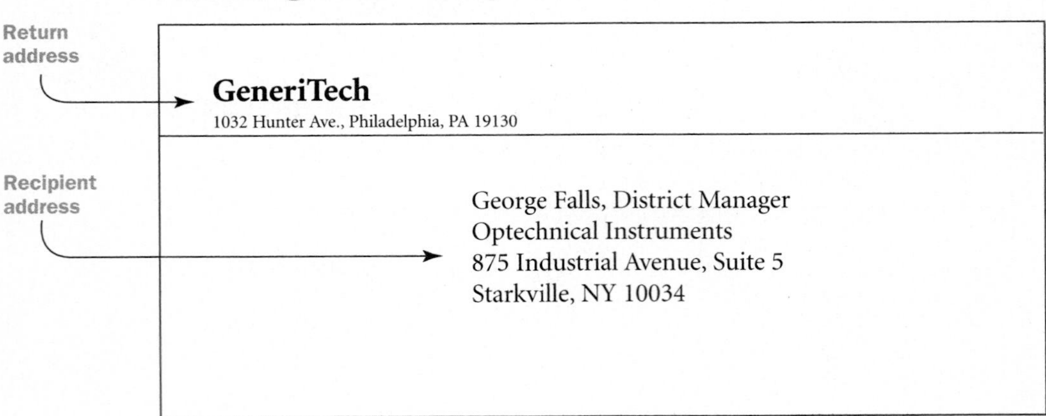

Figure 5.13: An envelope includes a return address and a recipient address.

If your company has premade envelopes with the return address already printed on them, printing an envelope will be easier. You will only need to add the recipient address.

Formatting Memos

Memos are easier to format than letters because they include only a header and message.

HEADER Most companies have stationery available that follows a standard memo format (Figure 5.14). If memo stationery is not available, you can make your own by typing the following list:

Date:
To:
cc:
From:
Subject:

The "Subject" line should offer a descriptive and specific phrase that describes the content of the memo. Most readers will look at the subject line first to determine if they want to read the memo. If it is too generic (e.g., "Project" or "FYI"), they may not read the memo. Instead, give them a more specific phrase like "Update on the TruFit Project" or "Accidental Spill on 2/2/11."

The "cc" line (optional) includes the names of any people who will receive copies of the memo. Often, copies of memos are automatically sent to supervisors to keep them informed.

If possible, sign your initials next to your name on the "From" line. Since memos are not signed, these initials serve as your signature on the document.

MESSAGE Memos do not include a "Dear" line or any other kind of greeting. They just start out with the message. The block style (all lines set against the left margin and spaces between paragraphs) is preferred, though some writers indent the first line of each paragraph.

Longer memos should include headings to help readers identify the structure of the text. In some cases, you might choose to include graphics to support the written text.

It is important to remember that memos do *not* include a closing or signature. When your conclusion is complete, the memo is complete. No closing or signature is needed.

Link

For more ideas about designing documents, go to Chapter 18, page 482.

Using E-Mail Internationally

The speed of e-mail makes it an ideal way to communicate and build relationships with international clients and co-workers. In many cases, e-mail has replaced both phone calls and letters, because it has the immediacy of the phone while giving readers the time to translate and consider the message being sent. When working internationally, you will discover that many clients and co-workers prefer to conduct business via e-mail.

Sample Memo

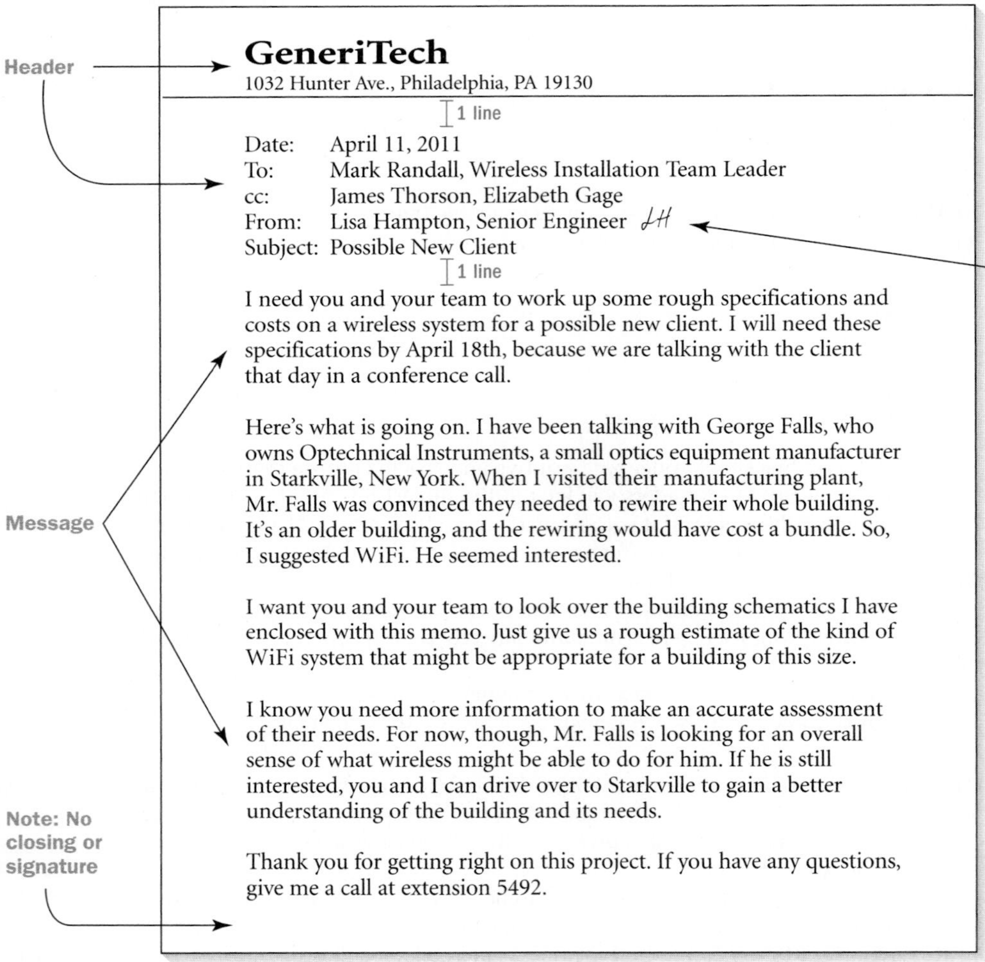

GeneriTech

1032 Hunter Ave., Philadelphia, PA 19130

Date: April 11, 2011
To: Mark Randall, Wireless Installation Team Leader
cc: James Thorson, Elizabeth Gage
From: Lisa Hampton, Senior Engineer LH
Subject: Possible New Client

I need you and your team to work up some rough specifications and costs on a wireless system for a possible new client. I will need these specifications by April 18th, because we are talking with the client that day in a conference call.

Here's what is going on. I have been talking with George Falls, who owns Optechnical Instruments, a small optics equipment manufacturer in Starkville, New York. When I visited their manufacturing plant, Mr. Falls was convinced they needed to rewire their whole building. It's an older building, and the rewiring would have cost a bundle. So, I suggested WiFi. He seemed interested.

I want you and your team to look over the building schematics I have enclosed with this memo. Just give us a rough estimate of the kind of WiFi system that might be appropriate for a building of this size.

I know you need more information to make an accurate assessment of their needs. For now, though, Mr. Falls is looking for an overall sense of what wireless might be able to do for him. If he is still interested, you and I can drive over to Starkville to gain a better understanding of the building and its needs.

Thank you for getting right on this project. If you have any questions, give me a call at extension 5492.

Figure 5.14: The formatting of a memo uses a head that differs from a letterhead. Notice that there is no closing or signature.

A challenge with using e-mail internationally is that North Americans tend to view e-mail as an "informal" or even "intimate" medium for communication. As a result, they regularly stumble over social norms and conventions of other cultures. Too quickly, Americans try to become too friendly and too informal. Also, in e-mail, Americans can be sloppy with grammar, spelling, and word usage, causing significant problems for non-English speakers who are trying to translate.

Want to learn more about e-mail netiquette? Go to **www.pearsonhighered.com/johnsonweb4/5.16**

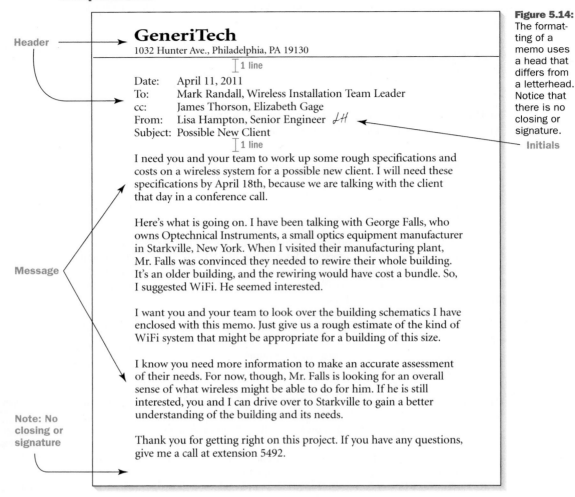

Here are some tips for using e-mail internationally:

Allow time to form a relationship—Introduce yourself by name and title, and provide some background information about your company and yourself. Tell the readers where you are writing from and where you are in relation to a major city. Don't rush into making a request, because doing so will often come across as pushy or rude.

Use titles and last names—Titles are often much more important in other cultures than in the United States. Minimally, you should use titles like Mr., Ms., or Dr. (Mrs. can be risky). If you know the proper titles from the readers' culture, such as Madame, Herr, Signora, then you should use them. Eventually, your international clients or co-workers may want to move to a first-name relationship, but it's often a good idea to let them make that first move.

Focus on the facts—In your message, concentrate on factual issues regarding the who, what, where, when, how, and why. Cut out other nonessential information because it will cloud your overall message.

Talk about the weather—If you want to personalize your message, talking about the weather is a safe topic. People are often curious about the weather in other parts of the world. It's a safe, universal topic that will allow you to get beyond just factual information. (*Hint:* Convert all temperatures to Celsius and measurements into metric.)

Use attachments only when needed—In some parts of the world, e-mail systems cannot handle large attachments. Inbox quotas may be too small or baud rates may be too slow. A good approach is to send an initial e-mail that asks if an attachment would be welcome. Then, if the reader tells you it will work, you can send it in a follow-up e-mail.

Use plain text—You should assume that your readers can only receive plain text. So, turn off any special characters (smart quotes, dashes, emoticons, etc.) because they will often come out as gibberish at the other end. Also, assume that any embedded website addresses and e-mail addresses won't be shown as links. You should spell these addresses out in full, so readers can cut and paste them for their own use.

Limit or avoid photographs and graphics—Photographs, background images, and other graphics don't always transfer properly when they are sent internationally, and they can require a great amount of memory. Plus, photographs can mean something unexpected to readers from other cultures.

Avoid clichés at the closing—Commonly used closings like "Do not hesitate to contact me," "If there's a problem, just holler" or "Don't be afraid to call," do not translate well into other languages and may be confusing to the readers.

Avoid humor—Attempts to be funny or tell jokes can backfire. Humor usually relies on cultural knowledge, so its meaning can be misinterpreted by readers. In some cases, the humor might be seen as insulting.

Create a simple signature file with your contact information—International e-mails are often printed out, which can cause the sender's e-mail address and other contact information to be separated from the message. A concise signature file that appears at the bottom of each e-mail should include your name, title, e-mail address, postal address, phone number, and corporate website.

Use simple grammar and proofread carefully—Simple sentences are easier for human and machine translators to interpret. Complex grammar or grammatical errors greatly increase problems with translation.

MICROGENRE

Texting at Work

Workplace texting has become a popular way to communicate with co-workers on projects. Not long ago, supervisors were concerned that employees texting at work were really just wasting time. They suspected that texting was primarily being used to socialize with friends rather than work. However, as the generation comfortable with texting has entered the workplace—and become supervisors themselves—employers are more comfortable with texting as a workplace tool. In fact, sometimes texting is the best way to communicate at the office.

Here are some tips for effective texting at work:

Write longer text messages. Your messages should offer useful information and/or make a specific request. A typical workplace text will run one or two full sentences. When texting your friends, a ping is fine (e.g., "wassup"), but these kinds of brief messages shouldn't be sent to co-workers because they can be disruptive.

Spell out most words and punctuate. Text messages are usually limited to 160 characters, so abbreviating is common. However, in a workplace text message, you should spell out most words and punctuate properly, so your co-workers can understand what you are saying. It's better to send longer texts that are spelled out than a garble of abbreviations that people won't understand.

E-mail or call when it's important. Many of your co-workers won't use texting at work, so if something is important, you'd better send an e-mail or call them.

Make sure you're doing work. It's easy to get caught up texting with a co-worker, but remember, you're on company time. If you're chatting about something other than work, it's probably time to end the conversation.

Don't text during meetings. Supervisors will often react negatively if people are looking down at their phones during a meeting or presentation, even if the texting is work-related. Put the phone in your pocket or leave it at your desk if you can't stop checking it for messages.

Don't use texting to flirt at work. There is a fine line between playful text messages and sexual harassment in the workplace. Messages can be saved and used against you.

Remember: Texts sent with company phones are not private. If your company issues you a mobile phone, do not use it for private texting. Your company can access those messages, which has led to people being fired for misuse of company property.

Write

Try using texting to communicate on your next project. As you work with your team in class, try following the above texting guidelines instead of talking. When you are finished for the day, print out your conversation. What are the pros and cons of texting while working on a project?

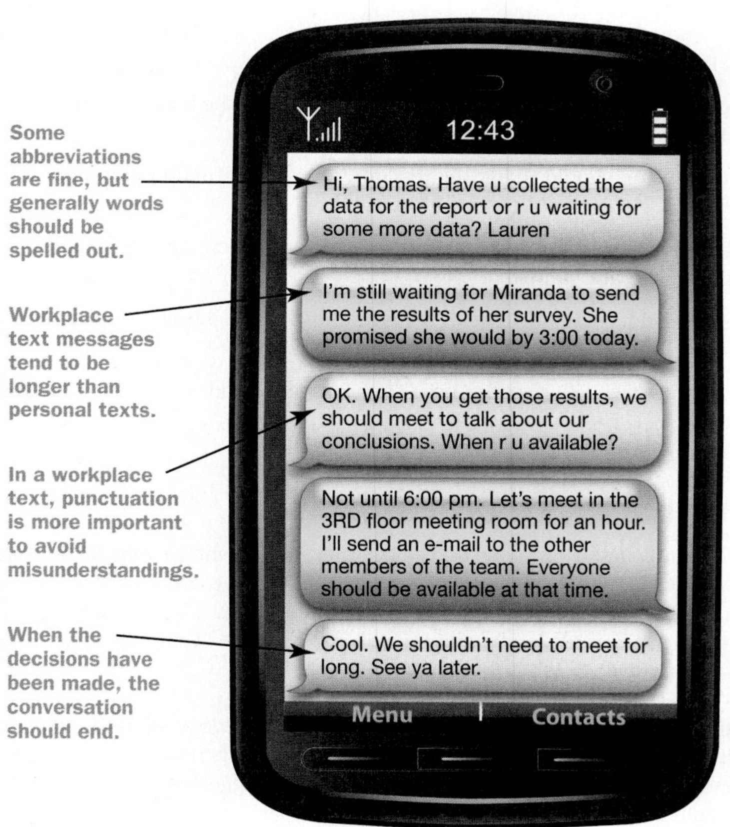

12:43
Hi, Thomas. Have u collected the data for the report or r u waiting for some more data? Lauren
I'm still waiting for Miranda to send me the results of her survey. She promised she would by 3:00 today.
OK. When you get those results, we should meet to talk about our conclusions. When r u available?
Not until 6:00 pm. Let's meet in the 3RD floor meeting room for an hour. I'll send an e-mail to the other members of the team. Everyone should be available at that time.
Cool. We shouldn't need to meet for long. See ya later.
Menu
Contacts
Some abbreviations are fine, but generally words should be spelled out.
Workplace text messages tend to be longer than personal texts.
In a workplace text, punctuation is more important to avoid misunderstandings.
When the decisions have been made, the conversation should end.

Texting at work is less cryptic and more formal than personal texting.

CHAPTER REVIEW

- Writing letters, memos, and e-mails can be a drain on your time. You should learn how to write them quickly and efficiently.
- Letters, memos, and e-mails are essentially the same kind of document, called a *correspondence.* They use different formats, but they tend to achieve the same purposes. Letters are used for messages that go outside the company. Memos are for internal messages.
- Letters, memos, and e-mails are always personal. They are intended to be a one-to-one communication to a reader, even if they are sent to more than one person.
- Letters, memos, and e-mails share the same basic features: header, introduction, informative body, and conclusion. They differ mostly in *format,* not content or purpose.
- The introduction should identify the subject, purpose, and point of the correspondence. If you want readers to take action, put that action up front.
- The body of the correspondence should give readers the information they need to take action or make a decision.
- The conclusion should thank the readers, restate your main point, and look to the future.
- To develop an appropriate style, use the "you" style, create a deliberate tone, and avoid bureaucratic phrasing.
- Use standard formats for letters, memos, and e-mails. These formats will make the nature of your message easier to recognize.
- Be sure to revise your work as necessary and proofread it when you are finished.

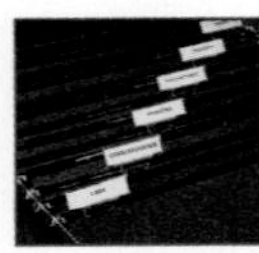

EXERCISES AND PROJECTS

Individual or Team Projects

1. Find a sample letter or memo on the Internet. In a memo to your instructor, discuss why you believe the letter is effective or ineffective. Discuss how the content, organization, style, and design are effective/ineffective. Then, make suggestions for improvement. Sample letters and memos are available at www.pearsonhighered.com/johnsonweb4/5.5.

2. Think of something that bothers you about your college campus. To a named authority, write a letter in which you complain about this problem and discuss how it has inconvenienced you in some way. Offer some suggestions about how the problem might be remedied. Be sure to be tactful.

3. Imagine that you are the college administrator who received the complaint letter in Exercise 3. Write a response letter to the complainant in which you offer a reasonable response to the writer's concerns. If the writer is asking for a change that requires a great amount of money, you may need to write a letter that refuses the request.

Collaborative Project

With a group, choose three significantly different cultures that interest you. Then, research these cultures' different conventions, traditions, and expectations concerning letters, memos, and e-mails. You will find that correspondence conventions in countries such as Japan or Saudi Arabia are very different from those in the United States. The Japanese often find American correspondence to be blunt and rude. Arabs often find American correspondence to be bland (and rude, too).

Write a brief report, in memo form, to your class in which you compare and contrast these three different cultures' expectations for correspondence. In your memo, discuss some of the problems that occur when a person is not aware of correspondence conventions in other countries. Then, offer some solutions that might help the others in your class become better intercultural communicators.

Present your findings to the class.

For additional technical writing resources, including interactive sample documents, document design tutorials and guidelines, and more, go to **www.mytechcommlab.com.**

Revision Challenge

The memo shown in Figure A needs to be revised before it is sent to its primary readers. Using the concepts and strategies discussed in this chapter, analyze the weaknesses of this document. Then, identify some ways it could be improved through revision.

- What information in the memo goes beyond what readers need to know?
- How can the memo be reorganized to highlight its purpose and main point?
- What is the "action item" in the memo, and where should it appear?
- How can the style of the memo be improved to make the text easier to understand?
- How might design be used to improve the readers' understanding?

An electronic version of this memo is available at www.pearsonhighered.com/johnsonweb4/5.17. There, you can also download the file for revision.

GO TO THE NET

To download an electronic version of this revision challenge, go to **www.pearsonhighered.com/johnsonweb4/5.17**

Figure A: This memo needs some revision. How could it be improved?

ChemConcepts, LLC

Memorandum

Date: November 15, 2010
To: Laboratory Supervisors
cc: George Castillo, VP of Research and Development
From: Vicki Hampton, Safety Task Force
Re: FYI

It is the policy of the ChemConcepts to ensure the safety of its employees at all times. We are obligated to adhere to the policies of the State of Illinois Fire and Life Safety Codes as adopted by the Illinois State Fire Marshal's Office (ISFMO). The intent of these policies is to foster safe practices and work habits throughout companies in Illinois, thus reducing the risk of fire and the severity of fire if one should occur. The importance of chemical safety at our company does not need to be stated. Last year, we had four incidents of accidental chemical combustion in our laboratories. We needed to send three employees to the hospital due to the accidental combustion of chemicals stored or used in our laboratories. The injuries were minor and these employees have recovered; but without clear policies it is only a matter of time before a major accident occurs. If such an accident happens, we want to feel assured that all precautions were taken to avoid it, and that its effects were minimized through proper procedures to handle the situation.

In the laboratories of ChemConcepts, our employees work with various chemical compounds that cause fire or explosions if mishandled. For example, when stored near reducing materials, oxidizing agents such as peroxides, hydroperoxides and peroxyesters can react at ambient temperatures. These unstable oxidizing agents may initiate or promote combustion in materials around them. Of special concern are organic peroxides, the most hazardous chemicals handled in our laboratories. These

Figure A: *(continued)*

compounds form extremely dangerous peroxides that can be highly combustible. We need to have clear policies that describe how these kinds of chemicals should be stored and handled. We need policies regarding other chemicals, too. The problem in the past is that we have not had a consistent, comprehensive safety policy for storing and handling chemicals in our laboratories. The reasons for the lack of such a comprehensive policy are not clear. In the past, laboratories have been asked to develop their own policies, but our review of laboratory safety procedures shows that only four of our nine laboratories have written safety policies that specifically address chemicals. It is clear that we need a consistent safety policy that governs storage and handling of chemicals at all of our laboratories.

So, at a meeting on November 3, it was decided that ChemConcepts needs a consistent policy regarding the handling of chemical compounds, especially ones that are flammable or prone to combustion. Such a policy would describe in depth how chemicals should be stored and handled in the company's laboratories. It should also describe procedures for handling any hazardous spills, fires, or other emergencies due to chemicals. We are calling a mandatory meeting for November 28 from 1:00–5:00 in which issues of chemical safety will be discussed. The meeting will be attended by the various safety officers in the company, as well as George Castillo, VP of Research and Development. Before the meeting, please develop a draft policy for chemical safety for your laboratory. Make fifteen copies of your draft policy for distribution to others in the meeting. We will go over the policies from each laboratory, looking for consistencies. Then, merging these policies, we will draft a comprehensive policy that will be applicable throughout the corporation.

CASE STUDY

The Nastygram

Shannon Phillips is the testing laboratory supervisor at the Rosewood Medical Center. It is her laboratory's job to test samples, including blood samples, drawn from patients. After the samples are tested, one of her technicians returns a report on the sample to the doctor.

Last week, Shannon's laboratory was in the process of transferring over to a new facility. It was a complex move because the samples needed to be delivered on rolling carts to the new laboratory. There was no other way to make the transfer.

Then, the accident happened. One of the technicians who was pushing a cart of blood samples left it alone for a moment. The cart rolled down an incline, into the lobby, and turned over. Samples were strewn all over the floor.

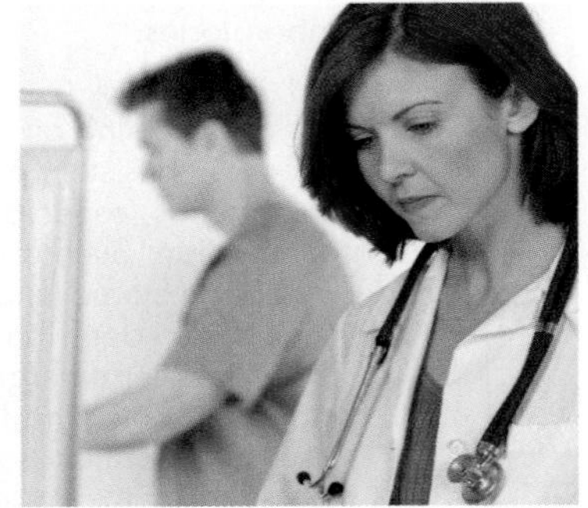

The hazardous materials team cleaned up the mess, but the samples were all destroyed. In an e-mail, Shannon notified the doctors who had ordered the tests that they would need to draw new samples from their patients. With her apologies, she tried her best to explain the situation and its remedy.

Most doctors were not pleased, but they understood the situation. However, Dr. Alice Keenan, the director of the AIDS Center at the hospital, was quite angry about the accident. She wrote the memo shown in Figure B to Shannon.

Needless to say, Shannon was quite upset about the memo. She had done everything she could to avoid the accident, and now a doctor was going to ask the board of directors to fire her. At a minimum, Shannon might need to fire her technician, who had never made a mistake up to this point. She was also quite angry about the tone of the letter. It seemed so unprofessional.

If you were Shannon, how might you respond to this situation?

Rosewood Medical Center

712 Hospital Drive, Omaha, Nebraska 68183

Memorandum

Date: February 13, 2011
To: Shannon Phillips, Laboratory Supervisor
From: Dr. Alice Keenan, MD, Director of AIDS Center
cc: George Jones, CEO Rosewood Medical Center
Subject: Your Idiotic Blunder

I cannot believe you morons in the lab ruined the blood samples we sent from our patients. Do you realize how difficult it is to draw these samples in the first place? Our patients are already in great discomfort and in a weakened state. So, asking them for a second blood draw within only a few days is a big problem.

Not to mention the hazardous situation created by this idiotic blunder. Do you realize that someone in the lobby could have come into contact with these samples and perhaps contracted HIV? Are you too obtuse to realize what kind of horrible consequences your mistake might have caused? You would have made us liable for huge damages if we were sued (which we would be).

Someone down there should be fired over this mistake! In my opinion, Shannon, you should be the first to go. I'm very concerned about your leadership abilities if you could allow something like this to happen. If I don't hear about a firing or your resignation in the next week, I'm going to bring a formal complaint to the Board of Directors at their next meeting.

Are you people all idiots down there? I've never heard of something this stupid happening at this hospital.

Figure B: Sometimes it is difficult to respond professionally to a nasty memo or letter like this one.

CHAPTER

6

Technical Descriptions and Specifications

Planning and Researching *133*

Partitioning the Subject *140*

Organizing and Drafting *145*

Help: Using Digital Photography in Descriptions *149*

Using Style and Design *151*

Microgenre: Technical Definitions *155*

Chapter Review *159*

Exercises and Projects *159*

Case Study: In the Vapor *163*

In this chapter, you will learn:

- How descriptions and specifications are used in technical workplaces.
- Common features of descriptions and specifications.
- How to determine the rhetorical situation for a description.
- Strategies for partitioning objects, places, or processes into major and minor parts.
- Techniques for organizing and drafting descriptions and specifications.
- How to use plain style to make descriptions and specifications understandable.
- How to use page layout and graphics to highlight and illustrate important concepts.

Technical descriptions are detailed explanations of objects, places, or processes. There are several types of technical descriptions, written for various purposes in the technical workplace:

Technical description—Manufacturers use technical descriptions to describe their products for patents, quality control, and sales.

Patents—An application for a patent requires a detailed technical description of an invention.

Specifications (often referred to as "specs")—Engineers write specifications to describe a product in great detail, providing exact information about features, dimensions, power requirements, and other qualities.

Field notes—Naturalists, anthropologists, sociologists, and others use field notes to help them accurately describe people, animals, and places.

Observations—Scientists and medical personnel need to accurately describe what they observe, so they can measure changes in their patients' conditions.

Descriptions appear in almost every technical document, including experimental reports, user's manuals, reference materials, proposals, marketing literature, magazine articles, and conference presentations. Specifications, meanwhile, are used to establish a standard and an exact set of requirements for a product or service. Often, when a product or service does not meet these standards, it is referred to as "out of spec."

Figure 6.1 shows a technical description of the Canadarm, a robotic arm that is used in space. The exactness of the description allows a variety of readers to better understand the function and abilities of this important tool.

Planning and Researching

During the planning and researching phase, you should identify what kinds of information your readers need to know, how they will use that information, and the contexts in which they will use it.

Planning

As you begin planning your technical description, it is important that you first have a good understanding of the situation in which your description will be used. Start by considering the Five-W and How Questions:

Who might need this description?

Why is this description needed?

What details and facts should the description include?

Where will the description be used?

When will the description be used?

How will this description be used?

To see examples of technical descriptions, go to **www.pearsonhighered.com/johnsonweb4/6.1**

Technical Descriptions and Specifications

This model shows a typical organizational pattern for a technical description. You should alter this pattern to fit the unique features of the object, place, or process that you are describing.

Basic Features of Technical Descriptions

A technical description can be part of a larger document, or it can stand alone as a separate document. A stand-alone technical description will generally have the following features:

- Specific and precise title
- Introduction with an overall description
- Description of features, functions, or stages of a process
- Use of senses, similes, analogies, and metaphors
- Graphics
- Conclusion that shows the thing, place, or process in action

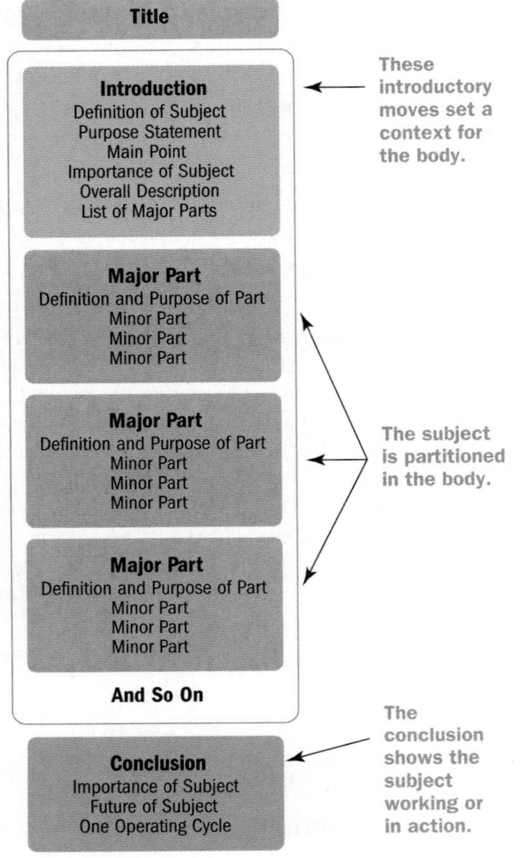

GO TO THE NET

Need help defining the rhetorical situation? Go to **www.pearsonhighered.com/johnsonweb4/6.2**

An Extended Technical Description

Technical Description: Canadarm

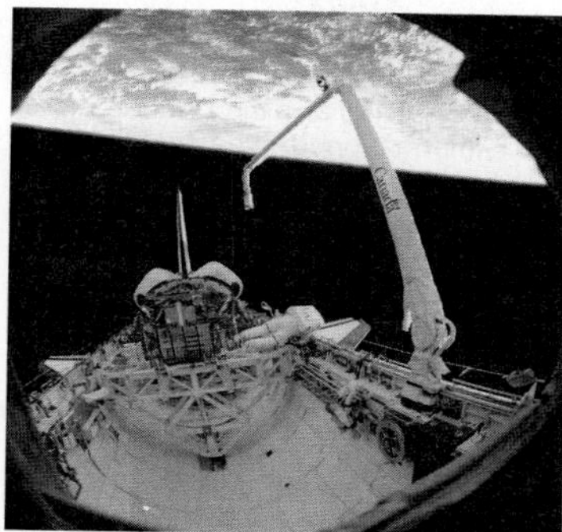

Since its maiden voyage aboard U.S. Space Shuttle Columbia in 1981, the Shuttle Remote Manipulator System (SRMS), known as Canadarm, has demonstrated its reliability, usefulness, and versatility and has provided strong, yet precise and delicate handling of its payloads.

Canadarm was designed, developed and built by MDA, under contract to the National Research Council of Canada. The first arm was Canada's contribution to NASA's Space Shuttle Program. Subsequently, NASA ordered four additional units which have resulted in over $900 million in export sales for Canada.

The Shuttle Remote Manipulator System consists of a shoulder, elbow and wrist joint separated by an upper and lower arm boom. The shoulder joint has two degrees of freedom, the elbow joint has one degree of freedom, and the wrist joint has up to three degrees of freedom.

JOINTS

Each subassembly component of the SRMS i.e. the shoulder, elbow or wrist is made up of a basic element called a joint one-degree-of-freedom or JOD. The JOD's are simply motor driven gearboxes that allow the basic structure of the arm to articulate much like the human arm. There are two JOD's in the shoulder joint which allow the whole arm to pitch (up and down motion) and yaw (side to side motion). One in the elbow joint to allow the lower arm to pitch and three in the wrist joint to allow the tip of the arm to pitch, yaw and roll (rotating motion). SRMS is much more articulate than even the human arm and can therefore accomplish very complex manoeuvres. The JOD motors are equipped with their own brakes and joint motor speed control. Each JOD also incorporates a device called an encoder, which accurately measures joint angles. Thus each joint is capable of moving independently at different speeds and in different directions with respect to any or all the other JOD's.

BOOMS

Linking the shoulder, elbow and wrist joints are the upper and lower arm booms. These booms are constructed of graphite-epoxy. The upper arm boom is approximately 16 ft. long by 13.0 in. in diameter comprising of 16 plies of graphite-epoxy (each ply is .005" thick) for a total weight of just under 50 lbs. The lower arm boom is approximately 19 ft. long by 13.0 in. in diameter comprising of 11 plies of

Figure 6.1: Here is a technical description of the Canadarm, a robotic arm that is used in space.

Introduction identifies subject and discusses its importance.

The subject is partitioned into its major parts.

Major feature

Minor features

Source: MDA. NASA. Images courtesy of nasaimages.org, http://nasaimages.org.

(continued)

Figure 6.1: *(continued)*

graphite-epoxy for a total weight of just over 50 lbs. Each boom is protected with a Kevlar bumper (the same material used in bulletproof vests) to preclude the possibility of dents or scratches on the carbon composite.

Major feature →

WIRING HARNESS

Just as the arm booms linked the shoulder, elbow and wrist joints mechanically, the wiring harness (electrical cabling) accomplishes the same thing only electrically. The wiring harness provides electrical power to all the joints and the End Effector (mechanical hand) as well as data and feed back information from each of the joints. This link continues from the SRMS in the payload bay and continues on into cabin of the space shuttle where astronauts control the actions of the arm remotely.

END EFFECTOR

The End Effector or mechanical hand of the SRMS allows the arm to capture stationary or free flying payloads by providing a large capture envelope (a cylinder 8 in. in diameter by 4 in. deep) and a mechanism/structure capable of soft docking and rigidizing. This action is accomplished by a two stage mechanism in the End Effector which closes three cables (like a snare) around a grapple probe (knobbed pin) bolted onto the payload and then draws it into the device until close contact is established and a load of approximately 1100 lbs. is imparted to the grapple probe. The forces developed by the End Effector on the payload through the grapple probe will allow for manoeuvring of the payload without separation from the remainder of the SRMS to the positional accuracy's previously stated.

CLOSED CIRCUIT TELEVISIONS (CCTV)

The SRMS has two CCTVs, one at the elbow joint and one at the wrist joint. The CCTV units are used to aid the astronauts in the positioning of the arm for payload capture/retrieval or payload by capture/deployment.

Details add realism to the text.

SRMS CONTROL SYSTEM

The space shuttles general-purpose computer (GPC) controls the movement of the SRMS. The hand controllers used by the astronauts tell the computer what the astronauts would like the arm to do. Built in software examines what the astronauts commanded inputs are and calculates which joints to move, what direction to move them in, how fast to move them and what angle to move to. As the computer issues the commands to each of the joints it also looks at what is happening to each joint every 80 milliseconds. Any changes inputted by the astronauts to the initial trajectory commanded are re-examined and recalculated by the GPC and updated commands are then sent out to each of the joints. The SRMS control system is continuously monitoring its "health" every 80 milliseconds and should a failure occur the GPC will automatically apply the brakes to all joints and notify the astronaut of a failure condition. The control system also provides a continuous display of joint rates and speeds, which are displayed on monitors located on the flight deck in the orbiter. As with any control system, the GPC can be over-rided and the astronaut can operate the joints individually from the flight deck.

Figure 6.1: *(continued)*

THERMAL PROTECTION SYSTEM

The SRMS is covered over its entire length with a multi-layer insulation thermal blanket system, which provides passive thermal control. This material consists of alternate layers of godized Kapton, Dacron scrim cloth and a Beta cloth outer covering. In extreme cold conditions, thermostatically controlled electric heaters (resistance elements) attached to critical mechanical and electronic hardware can be powered on to maintain a stable operating temperature.

Clearly labeled diagram

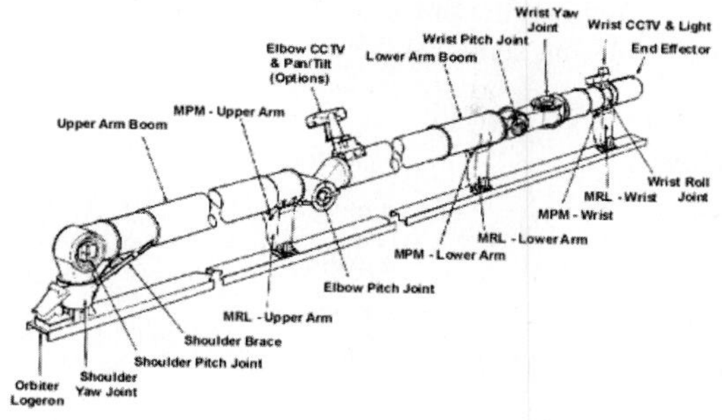

Technical Details

A table that summarizes the technical specifications

Length	15.2m (50ft.)
Diameter	38cm (15 in.)
Weight on Earth	410Kg (905 lbs.)
Speed of Movement	- unloaded 60 cm/sec. (2 ft./sec.) - loaded 6 cm/sec. (2.4 in./sec.)
Upper & Lower Arm Boom	Carbon Composite Material
Wrist Joint	Three degrees of movement (pitch +/- 120°, yaw +/- 120°, roll +/- 447°)
Elbow Joint	One degree of movement (pitch +2° to - 160°)
Shoulder Joint	Two degrees of movement (pitch +145° to -2, yaw +/- 180°)
Translational Hand Controller	Right, up, down forward, and backward movement of the arm
Rotational Hand Controller	Controls the pitch, roll, and yaw of the arm

Now, define the subject, purpose, readers, and context of your description.

Link
For more help defining your subject, see Chapter 1, page 5.

SUBJECT What exactly is your subject? What are its major features? What are its boundaries?

PURPOSE What should your technical description achieve? Do the readers want exact detail, or do they want an overall familiarity with the subject?

In one sentence, write down the purpose of your description. Here are some verbs that might help you write that sentence:

to describe	*to clarify*
to illustrate	*to reveal*
to show	*to explain*
to depict	*to portray*
to characterize	
to represent	

Link
For more information on defining a document's purpose, see Chapter 1, page 6.

Your purpose statement might say something like the following:

> The purpose of this description is to show how a fuel cell generates power.
>
> In this description, I will explain the basic features of the International Space Station.

If your purpose statement goes beyond one sentence, you likely need to be more specific about what you are trying to achieve.

READERS Technical descriptions tend to be written for readers who are unfamiliar with the subject. So, you will also need to adjust the detail and complexity of your description to suit their specific interests and needs.

Primary readers (action takers) are readers who most need to understand your description. What exactly do they need to know?

Secondary readers (advisors) will likely be experts who understand your subject, such as engineers, technicians, or scientists. How much technical detail and accuracy will these readers need to advise the primary readers?

Link
For more information on analyzing readers, see Chapter 2, page 20.

Tertiary readers (evaluators) could include just about anyone who has an interest in the product, place, or process you are describing, including reporters, lawyers, auditors, or concerned citizens. What kinds of information should be given to them and what should be held back?

Gatekeeper readers (supervisors) within your company will want to review your materials for exactness and correctness.

Worksheets are available to help you define readers and contexts of use. Go to **www.pearsonhighered.com/johnsonweb4/6.3**

CONTEXT OF USE Imagine your primary readers using your technical description. Where are they likely to read and use the document? What economic, ethical, and political factors will influence how they interpret the text?

Link

For more ideas about analyzing the context of use, see Chapter 2, page 26.

Addressing ISO 9000/ISO 14000 Issues

One important issue involving context of use is whether your technical description needs to conform to ISO 9000 or ISO 14000 standards. These voluntary standards are accepted internationally and managed by the International Organization for Standardization (ISO). ISO 9000 standards involve quality management systems, while ISO 14000 standards involve environmental management systems. Many high-tech companies, especially ones working for the U.S. government, follow these quality management and environmental management standards. Figure 6.2 shows an introduction to these standards drawn from the ISO website (www.iso.org).

The ISO standards cannot be discussed in sufficient depth here, but you should be aware that they exist. If your company follows ISO standards, any descriptions you write will need to reflect and conform to these standards.

ISO 9000 and ISO 14000

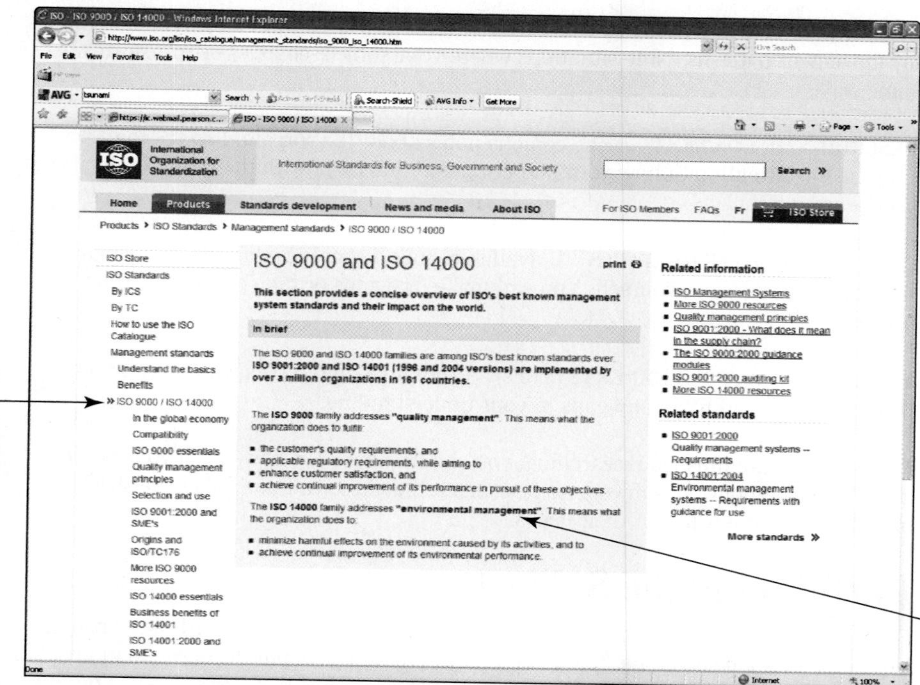

Certification is hard to achieve, but necessary in many fields.

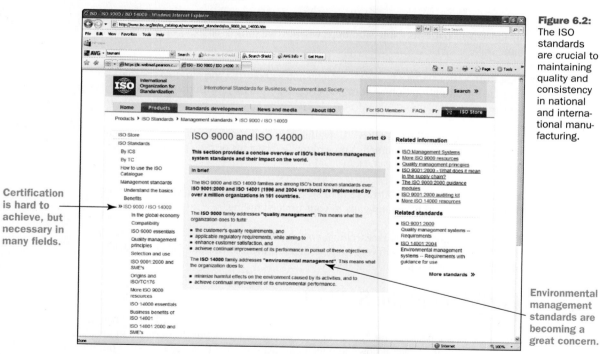

Environmental management standards are becoming a great concern.

Figure 6.2: The ISO standards are crucial to maintaining quality and consistency in national and international manufacturing.

Source: International Organization for Standardization, http://www.iso.org/iso/iso_catalogue/management_standards/iso_9000_iso_14000.htm.

To learn more about ISO guidelines, go to **www.pearsonhighered.com/johnsonweb4/6.4**

Link

For more information on doing research, go to Chapter 14, page 387.

Researching

In most cases, doing research for a technical description is primarily experiential. In other words, you will likely need to personally observe the object, thing, or process you are describing. Here are some strategies that are especially applicable to writing descriptions and specifications.

DO BACKGROUND RESEARCH Before making direct observations, you should know as much as possible about your subject. On the Internet, use search engines to find as much information as you can. Then, collect print sources, like books, documents, and other literature.

USE YOUR SENSES While making direct observations, use all of your available senses. As much as possible, take notes about how something looks, sounds, smells, feels, and tastes. Pay special attention to colors and textures, because they will add depth and vividness to your description.

TAKE MEASUREMENTS When possible, measure qualities like height, width, depth, and weight. Exact measurements are especially important if you are writing a specification that will set standards for a product or device.

DESCRIBE MOTION AND CHANGE Pay attention to your subject's movements. Look for movement patterns. Note situations where your subject changes or transforms in some way.

DESCRIBE THE CONTEXT Take copious notes about the surroundings of the subject. Pay attention to how your subject acts or interacts with the things and people around it.

CREATE OR LOCATE GRAPHICS If available, collect graphics that illustrate your subject, or create them yourself. You can make drawings or take pictures of your subject.

ASK SUBJECT MATTER EXPERTS (SMEs) If possible, find SMEs who can answer your questions and fill in any gaps in your understanding.

When you are finished researching, you should figure out how much your readers already know about the subject and how much they need to know. You can then prioritize your notes to suit their needs.

Partitioning the Subject

Once you are familiar with the subject, you can start describing it in words and images. Your first question should be, "How can I partition it?" Or, to put the question more simply, "How can I break it down into its major features, functions, or stages?" Your answer to this question will determine your *partitioning strategy* for describing your subject.

Want to learn how to closely observe things? Go to **www.pearsonhighered.com/johnsonweb4/6.5**

Chris Peterson

CAD TECHNICIAN, LARON, INC., PHOENIX, AZ

Laron, Inc., is an engineering company that specializes in heavy equipment.

How does computer-aided drafting (CAD) help write descriptions?

"A picture is worth a thousand words." It just doesn't get any better than that. A drawing can show a potential client your expertise in his or her particular needs. With the newer 3D CAD programs, such as Autodesk's® Inventor® or SolidWorks®, you can produce awesome pictures. For video presentations, you can use CAD to walk your clients through anything, from their new home or new plant, to a tour through a new machine. When new video games come out, like PS3® or Xbox®, what I hear kids (including big kids) say is "the graphics are awesome." CAD is a way to make your descriptions "awesome."

In order to impress your clients, you must master the CAD programs as well as your product. You can have the greatest talent to produce a "mousetrap," but if you lack the skills to present it, you will look like all the others, or worse! You may also be the greatest illustrator, but lack of understanding of the product you are promoting will stick out like a sore thumb.

Another place CAD is very helpful is for "internal clients" such as machinists, welders, carpenters, and pipe fitters. Here again you must have a good knowledge of your discipline as well as the CAD program. There is no substitute for knowledge of what you are trying to convey to the client.

AT WORK

By features—You might partition the subject by separately describing its parts or features. For example, a description of a computer might describe it part by part by first partitioning it into a monitor, keyboard, external hard drives, and a central processing unit (CPU).

By functions—You might partition the subject by noting how its different parts function. A description of the International Space Station, for example, might partition it function by function into research, power generation, infrastructure, habitation, and docking sections.

By stages of a process—You could partition the subject chronologically by showing how it is assembled or how it works. A description of Hodgkin's disease, for example, might walk readers step-by-step through detection, diagnosis, staging, and remission stages.

At this point, logical mapping can help you break down (partition) your subject into major and minor parts (Figure 6.3).

For model descriptions that use each partitioning strategy, go to **www.pearsonhighered.com/johnsonweb4/6.6**

Partitioning with Logical Mapping

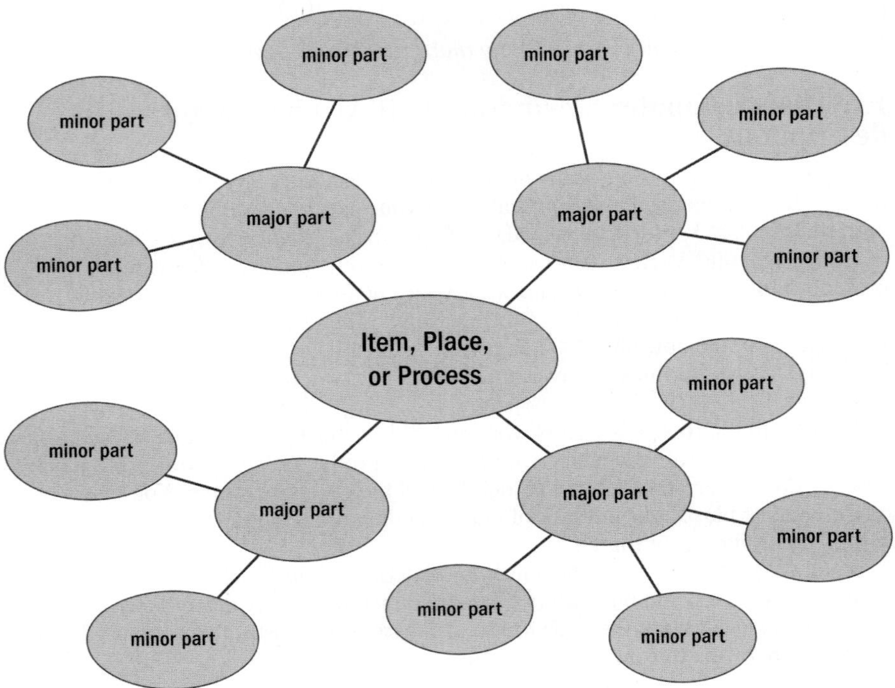

Figure 6.3: Partitioning means dividing the whole into its major and minor parts.

Link

For more help using logical mapping, see Chapter 14, page 388.

To use logical mapping to help you describe something, follow these steps:

1. Put the name of your subject in the middle of your screen or a sheet of paper.
2. Write down the two to five major parts in the space around it.
3. Circle each major part.
4. Partition each major part into two to five minor parts.

For example, in Figure 6.4, NASA's description of the Mars Exploration Rover partitions the subject into

Core structure

Suspension system

Navigation system

Instrument deployment device

Batteries

Radioscope heater

Computer

In this example, notice how the description carves the subject into its major parts. Then, each of these major parts is described in detail by paying attention to its minor parts.

Mars Exploration Rover

At the heart of each Mars Exploration Rover spacecraft is its rover. This is the mobile geological laboratory that will study the landing site and travel to examine selected rocks up close.

The Mars Exploration Rovers differ in many ways from their only predecessor, Mars Pathfinder's Sojourner Rover. Sojourner was about 65 centimeters (2 feet) long and weighed 10 kilograms (22 pounds). Each Mars Exploration Rover is 1.6 meter (5.2 feet) long and weighs 174 kilograms (384 pounds). Sojourner traveled a total distance equal to the length of about one football field during its 12 weeks of activity on Mars. Each Mars Exploration Rover is expected to travel six to 10 times that distance during its three-month prime mission. Pathfinder's lander, not Sojourner, housed that mission's main telecommunications, camera and computer functions. The Mars Exploration Rovers carry equipment for those functions onboard and do not interact with their landers any further once they roll off.

On each Mars Exploration Rover, the core structure is made of composite honeycomb material insulated with a high-tech material called aerogel. This core body, called the warm electronics box, is topped with a triangular surface called the rover equipment deck. The deck is populated with three antennas, a camera mast and a panel of solar cells. Additional solar panels are connected by hinges to the edges of the triangle. The solar panels fold up to fit inside the lander for the trip to Mars, and deploy to form a total area of 1.3 square meters (14 square feet) of three-layer photovoltaic cells. Each layer is of different materials: gallium indium phosphorus, gallium arsenide and germanium. The array can produce nearly 900 watt-hours of energy per martian day, or sol. However, by the end of the 90-sol mission, the energy generating capability is reduced to about 600 watt-hours per sol because of accumulating dust and the change in season. The solar array repeatedly recharges two lithium-ion batteries inside the warm electronics box.

Doing sport utility vehicles one better, each rover is equipped with six-wheel drive. A rocker-bogie suspension system, which bends at its joints rather than using any springs, allows rolling over rocks bigger than the wheel diameter of 26 centimeters (10 inches). The distribution of mass on the vehicle is arranged so that the center of mass is near the pivot point of the rocker-bogie system. That enables the rover to tolerate a tilt of up to 45 degrees in any direction without overturning, although onboard comput-

Figure 6.4: This description of the Mars Exploration Rover shows how a subject can be partitioned into major and minor parts.

Definition of subject

Overall description of subject

Major part: Core Structure

Minor parts

Major part: Suspension System

Source: NASA, http://www.jpl.nasa.gov/news/press_kits/merlandings.pdf.

(continued)

To see some interesting descriptions from NASA, go to **www.pearsonhighered.com/johnsonweb4/6.7**

Figure 6.4:
(continued)

Minor parts →

ers are programmed to prevent tilts of more than 30 degrees. Independent steering of the front and rear wheels allows the rover to turn in place or drive in gradual arcs.

Major part: Navigation System →

Minor parts →

The rover has navigation software and hazard-avoiding capabilities it can use to make its own way toward a destination identified to it in a daily set of commands. It can move at up to 5 centimeters (2 inches) per second on flat hard ground, but under automated control with hazard avoidance, it travels at an average speed about one-fifth of that. Two stereo pairs of hazard-identification cameras are mounted below the deck, one pair at the front of the rover and the other at the rear. Besides supporting automated navigation, the one on the front also provides imaging of what the rover's arm is doing. Two other stereo camera pairs sit high on a mast rising from the deck: the panoramic camera included as one of the science instruments, and a wider-angle, lower-resolution navigation camera pair. The mast also doubles as a periscope for another one of the science instruments, the miniature thermal emission spectrometer.

Major part: Instrument Deployment Device →

The rest of the science instruments are at the end of an arm, called the "instrument deployment device," which tucks under the front of the rover while the vehicle is traveling. The arm extends forward when the rover is in position to examine a particular rock or patch of soil.

Major part: Batteries →

Minor parts →

Batteries and other components that are not designed to survive cold martian nights reside in the warm electronics box. Nighttime temperatures may fall as low as minus 105°C (minus 157°F). The batteries need to be kept above minus 20°C (minus 4°F) for when they are supplying power, and above 0°C (32°F) when being recharged. Heat inside the warm electronics box comes from a combination of electrical heaters, eight radioisotope heater units and heat given off by electronics components.

Major part: Radioisotope Heater →

Minor parts →

Each radioisotope heater unit produces about one watt of heat and contains about 2.7 grams (0.1 ounce) of plutonium dioxide as a pellet about the size and shape of the eraser on the end of a standard pencil. Each pellet is encapsulated in a metal cladding of platinum-rhodium alloy and surrounded by multiple layers of carbon-graphite composite material, making the complete unit about the size and shape of a C-cell battery. This design of multiple protective layers has been tested extensively, and the heater units are expected to contain their plutonium dioxide under a wide range of launch and orbital-reentry accident conditions. Other spacecraft, including Mars Pathfinder's Sojourner rover, have used radioisotope heater units to keep electronic systems warm and working.

Major part: Computer →

Minor parts →

The computer in each Mars Exploration Rover runs with a 32-bit Rad 6000 microprocessor, a radiation-hardened version of the PowerPC chip used in some models of Macintosh computers, operating at a speed of 20 million instructions per second. Onboard memory includes 128 megabytes of random access memory, augmented by 256 megabytes of flash memory and smaller amounts of other non-volatile memory, which allows the system to retain data even without power.

Organizing and Drafting

With your subject partitioned into major and minor features, you are ready to start organizing and drafting your description. You can describe your subject in many ways, but it is best to choose an organizational pattern that demonstrates an obvious logic that readers will immediately recognize. The Quick Start at the beginning of this chapter shows a basic model to help you get started.

Specific and Precise Title

The title of your technical description should clearly identify the purpose of the document. For example,

> Description of the Mars Exploration Rover
>
> Specifications for the XC-9000 Microprocessor
>
> How Does a Fuel Cell Work?
>
> Lung Cancer: Profile of a Killer

Your title should clearly distinguish your document as a technical description.

Introduction with an Overall Description

Typically, the introduction will set a framework or context by including some or all of the following features:

DEFINITION OF SUBJECT A sentence definition of your subject includes three parts: the *term*, the *class* in which the subject belongs, and the *characteristics* that distinguish the subject from the other members in its class.

> The International Space Station is a multinational research facility that will house six state-of-the-art laboratories in orbit.
>
> Hodgkin's disease is a type of cancer that starts in the lymph nodes and other organs that are the body's system for making blood and protecting against germs.

The definition of the subject should appear early in the introduction, preferably in the first sentence.

Link

For more information on writing definitions, see the Microgenre on page 155.

PURPOSE STATEMENT Directly or indirectly state that you are describing something.

> This description of the International Space Station will explain its major features, highlighting its research capabilities.
>
> In this article, we will try to demystify Hodgkin's disease, so you can better understand its diagnosis and treatment.

For examples of good introductions, go to **www.pearsonhighered.com/johnsonweb4/6.8**

MAIN POINT Give your readers an overall claim that your description will support or prove.

> Building the International Space Station is an incredible engineering feat that will challenge the best scientists and engineers in many different nations.
>
> To fight Hodgkin's disease, you first need to understand it.

IMPORTANCE OF THE SUBJECT For readers who are unfamiliar with your subject, you might want to include a sentence or paragraph that stresses its importance.

> The ISS, when completed, will provide scientists and other researchers an excellent platform from which to study space.
>
> Hodgkin's disease is one of the most acute forms of cancer, and it needs to be aggressively treated.

OVERALL DESCRIPTION OF THE SUBJECT Descriptions sometimes offer an overall look at the item being described.

> From a distance, the International Space Station looks like a large collection of white tubes with two rectangular solar panels jutting out like ears from its side.
>
> Hodgkin's disease spreads through the lymphatic vessels to other lymph nodes. It enlarges the lymphatic tissue, often putting pressure on vital organs and other important parts of the body.

This overall description will help your readers visualize how the parts fit together as they read further.

LIST OF THE MAJOR FEATURES, FUNCTIONS, OR STAGES In many descriptions, especially longer descriptions, the introduction will list the major features, functions, or stages of the subject.

> The International Space Station includes five main features: modules, nodes, trusses, solar power arrays, and thermal radiators.
>
> Once Hodgkin's has been detected, doctors will usually (1) determine the stage of the cancer, (2) offer treatment options, and (3) make a plan for remission.

Link

For more information on writing introductions, see Chapter 16, page 430.

You can then use this list of features, functions, or stages to organize the body of your description.

Description by Features, Functions, or Stages

The body of your description will be devoted to describing the subject's features, functions, or stages.

Address each major part separately, defining it and describing it in detail. Within your description of each major part, identify and describe the minor parts.

Definition of major part → **Modules** are pressurized cylinders of habitable space on board the Station. They may contain research facilities, living quarters, and any vehicle operational systems and equipment the astronauts may need to access. ← Minor parts

If necessary, each of these minor parts could then be described separately. In fact, you could extend your description endlessly, teasing out the smaller and smaller features of the subject.

Figure 6.5 shows a description of a subject by "stages in a process." In this description of a fuel cell, the author walks readers through the energy generation process, showing them step by step how the fuel cell works.

Description by Senses, Similes, Analogies, and Metaphors

The key to a successful technical description is the use of vivid detail to bring your subject to life—to make it seem real. To add this level of detail, you might consider using some of the following techniques:

DESCRIPTION THROUGH SENSES Consider each of your five senses separately, asking yourself, "How does it look?" "How does it sound?" "How does it smell?" "How does it feel?" and "How does it taste?"

> A visit to a Japanese car manufacturing plant can be an overwhelming experience. Workers in blue jumpsuits seem to be in constant motion. Cars of every color—green, yellow, red—are moving down the assembly line with workers hopping in and out. The smell of welding is in the air, and you can hear the whining hum of robots at work somewhere else in the plant.

SIMILES A simile describes something by comparing it to something familiar to the readers ("A is like B").

> The mixed-waste landfill at Sandia Labs is like a football field with tons of toxic chemical and nuclear waste buried underneath it.

Similes are especially helpful for nonexpert readers, because they make the unfamiliar seem familiar.

ANALOGIES Analogies are like similes, but they work on two parallel levels ("A is to B as C is to D").

> Circuits on a semiconductor wafer are like tiny interconnected roads crisscrossing a city's downtown.

Link

For more information on using similes, analogies, and metaphors, see Chapter 17, page 473.

METAPHORS Metaphors are used to present an image of the subject by equating two different things ("A is B"). For example, consider these two common metaphors:

> The heart is a pump: it has valves and chambers, and it pushes fluids through a circulation system of pipes called arteries and veins.

> Ants live in colonies: a colony will have a queen, soldiers, workers, and slaves.

For more advice on using the senses, go to **www.pearsonhighered.com/johnsonweb4/6.9**

A Technical Description: Stages in a Process

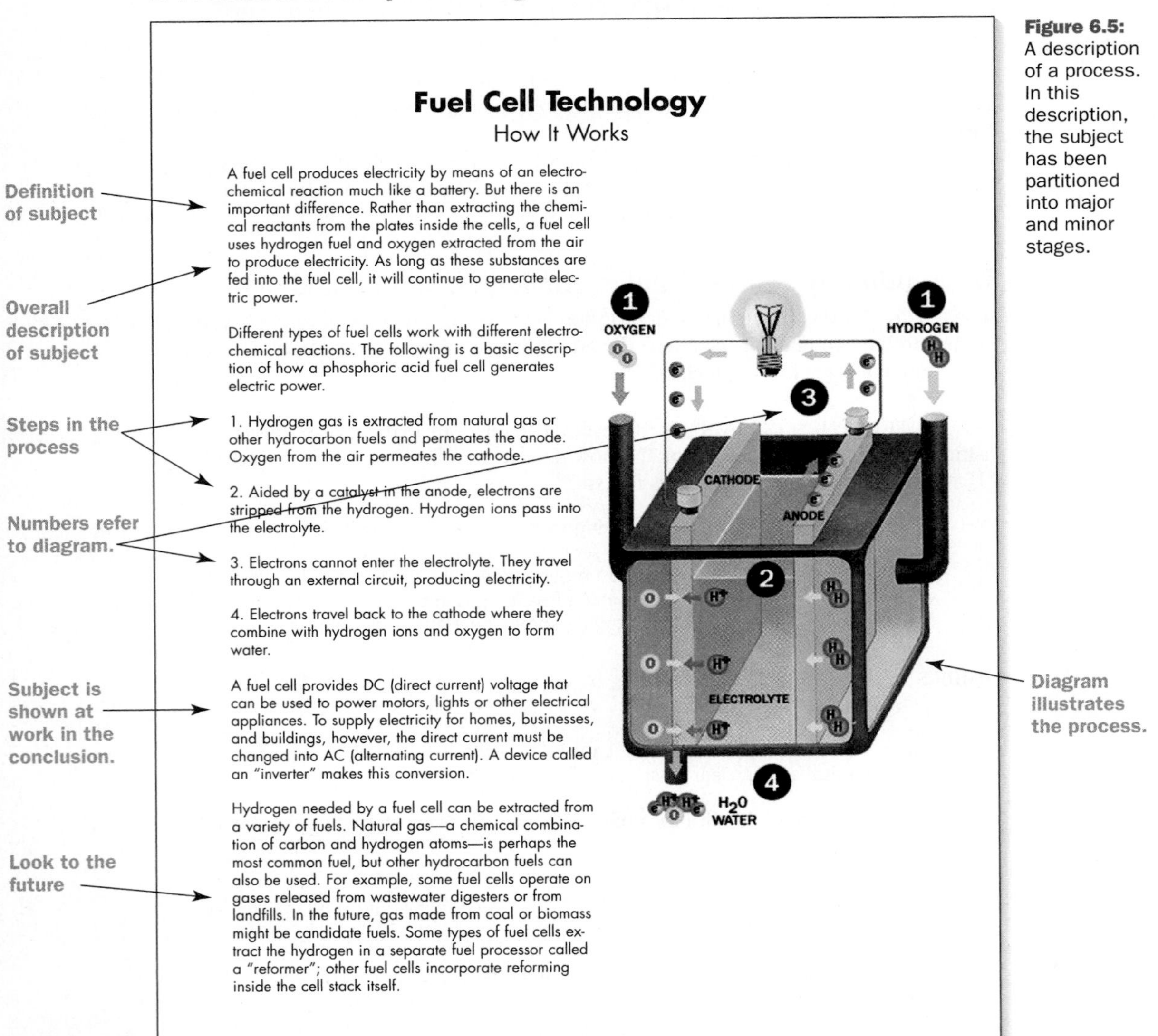

Fuel Cell Technology

How It Works

A fuel cell produces electricity by means of an electrochemical reaction much like a battery. But there is an important difference. Rather than extracting the chemical reactants from the plates inside the cells, a fuel cell uses hydrogen fuel and oxygen extracted from the air to produce electricity. As long as these substances are fed into the fuel cell, it will continue to generate electric power.

Different types of fuel cells work with different electrochemical reactions. The following is a basic description of how a phosphoric acid fuel cell generates electric power.

1. Hydrogen gas is extracted from natural gas or other hydrocarbon fuels and permeates the anode. Oxygen from the air permeates the cathode.

2. Aided by a catalyst in the anode, electrons are stripped from the hydrogen. Hydrogen ions pass into the electrolyte.

3. Electrons cannot enter the electrolyte. They travel through an external circuit, producing electricity.

4. Electrons travel back to the cathode where they combine with hydrogen ions and oxygen to form water.

A fuel cell provides DC (direct current) voltage that can be used to power motors, lights or other electrical appliances. To supply electricity for homes, businesses, and buildings, however, the direct current must be changed into AC (alternating current). A device called an "inverter" makes this conversion.

Hydrogen needed by a fuel cell can be extracted from a variety of fuels. Natural gas—a chemical combination of carbon and hydrogen atoms—is perhaps the most common fuel, but other hydrocarbon fuels can also be used. For example, some fuel cells operate on gases released from wastewater digesters or from landfills. In the future, gas made from coal or biomass might be candidate fuels. Some types of fuel cells extract the hydrogen in a separate fuel processor called a "reformer"; other fuel cells incorporate reforming inside the cell stack itself.

Figure 6.5: A description of a process. In this description, the subject has been partitioned into major and minor stages.

Source: U.S. Department of Energy, http://www.fe.doe.gov/coal_power/fuelcells/fuelcells_howitworks.shtml.

The use of senses, similes, analogies, and metaphors will make your description richer and more vivid. Readers who are unfamiliar with your subject will especially benefit from these techniques, because concepts they understand are being used to describe things they don't understand.

Conclusion

The conclusion of a technical description should be short and concise. Conclusions often describe one working cycle of the object, place, or process.

> When the International Space Station is completed, it will be a center of activity. Researchers will be conducting experiments. Astronomers will study the stars. Astronauts will be working, sleeping, exercising, and relaxing. The solar power arrays will pump energy into the station, keeping it powered up and running.

> Fighting Hodgkin's disease is difficult but not impossible. After detection and diagnosis, you and your doctors will work out treatment options and staging objectives. If treatment is successful, remission can continue indefinitely.

Some technical descriptions won't include an instruction. They stop when the last item has been described.

Using Digital Photography in Descriptions

Digital cameras and scanners offer an easy way to insert visuals into your descriptions. These digitized pictures are inexpensive, alterable, and easily added to a text. Moreover, they work well in print and on-screen texts. Here are some photography basics to help you use a camera more effectively.

Resolution—Digital cameras will usually allow you to set the resolution at which you shoot pictures. If you are using your camera to put pictures on the web, you should use the 640 × 480 pixel setting. This setting will allow the picture to be downloaded quickly, because the file is smaller. If you want your picture to be a printed photograph, a minimum 1280 × 1024 pixel setting is probably needed. Online pictures are usually best saved in jpeg or png format, while print photos should be saved as tiff files.

ISO sensitivity—ISO is the amount of light the camera picks up. A high ISO number like 400 is good for shooting in dark locations or snapping something that is moving quickly. A low ISO number like 100 produces a higher-quality shot. You might leave your camera's ISO setting on "auto" unless you need to make adjustments.

(continued)

To see examples of similes, analogies, and metaphors in descriptions, go to **www.pearsonhighered.com/johnsonweb4/6.10**

Shutter speed—Some digital cameras and conventional cameras will allow you to adjust your shutter speed. The shutter speed determines how much light is allowed into the camera when you push the button. Shutter speeds are usually listed from 1/1000 of a second to 1 second. Slower speeds (like 1 second) capture more detail but risk blurring the image, especially if your subject is moving. Faster speeds (like 1/1000 second) will not capture as much detail, but they are good for moving subjects.

Cropping—Once you have downloaded your picture to your computer, you can "crop" the picture to remove things you don't want. Do you want to remove an old roommate from your college pictures? You can use the cropping tool to cut him or her out of the picture. Most word processors have a cropping tool that you can use to carve away unwanted parts of your pictures (Figure A).

Cropping a Digital Photograph

The cropping tool lets you frame the part of the picture you want.

The toolbar offers a variety of options for altering the picture.

Figure A: Using the cropping tool, you can focus the photograph on the subject. Or, you can eliminate things or people you don't want in the picture.

Retouching—One of the main advantages of digital photographs is the ease with which they can be touched up. Professional photographers make ample use of programs like Adobe Photoshop to manipulate their photographs. If the picture is too dark, you can lighten it up. If the people in the photo have "red-eye," you can remove that unwanted demonic stare.

One of the nice things about digital photography is that you can make any photograph look professional. With a digital camera, since the "bullets are free," as photographers say, you can experiment freely.

GO TO THE NET

To learn more about improving your style, go to **www.pearsonhighered.com/johnsonweb4/6.11**

Using Style and Design

The style and design of your technical description should reflect your readers' needs and characteristics, as well as the places in which they will be using the document. Your style should be simple and straightforward, and the design should clarify and support the written text.

Plain, Simple Style

Most technical descriptions are written in the plain style. Here are some suggestions for improving the style of your technical description:

Use simple words and limit the amount of jargon.

Focus on the details your readers need to know, and cut the fat.

Keep sentences short, within breathing length.

Remove any subjective qualifier words like "very," "easy," "hard," "amazing."

Use the senses to add color, texture, taste, sound, and smell.

In most cases, the best style for a technical description is an unobtrusive style. However, the style of your technical description will depend on the context in which it will be used. A technical description that will be part of your company's sales literature, for example, will usually be more persuasive than a technical specification kept in your company's files.

Page Layout That Fits

The page layout of your description or specification likely depends on your company's existing documentation or an established corporate design. If you are given a free hand to design the text, you might consider these design features:

Link

For more information on designing documents, see Chapter 18, page 482.

Use a two-column or three-column format to leave room for images.

List any minor parts in bulleted lists.

Add a sidebar that draws attention to an important part or feature.

Use headings to clarify the organization.

Put measurements in a table.

Use your imagination when you design the document. Figure 6.6, for example, shows how columns and tables can be used to pack in a solid amount of information while still presenting the information in an attractive way.

Graphics That Illustrate

Graphics are especially helpful in technical descriptions. Pictures, illustrations, and diagrams help readers visualize your subject and its parts.

Using your computer, you can collect or create a wide range of graphics. Many free-use graphics are widely available on the Internet. (Reminder: Unless the site

To see well-designed descriptions, go to **www.pearsonhighered.com/johnsonweb4/6.12**

A Specification

The title is clearly stated at the top of the text.

Photo shows product

The narrower column makes the text easier to scan.

Company name and logo are framed with white space.

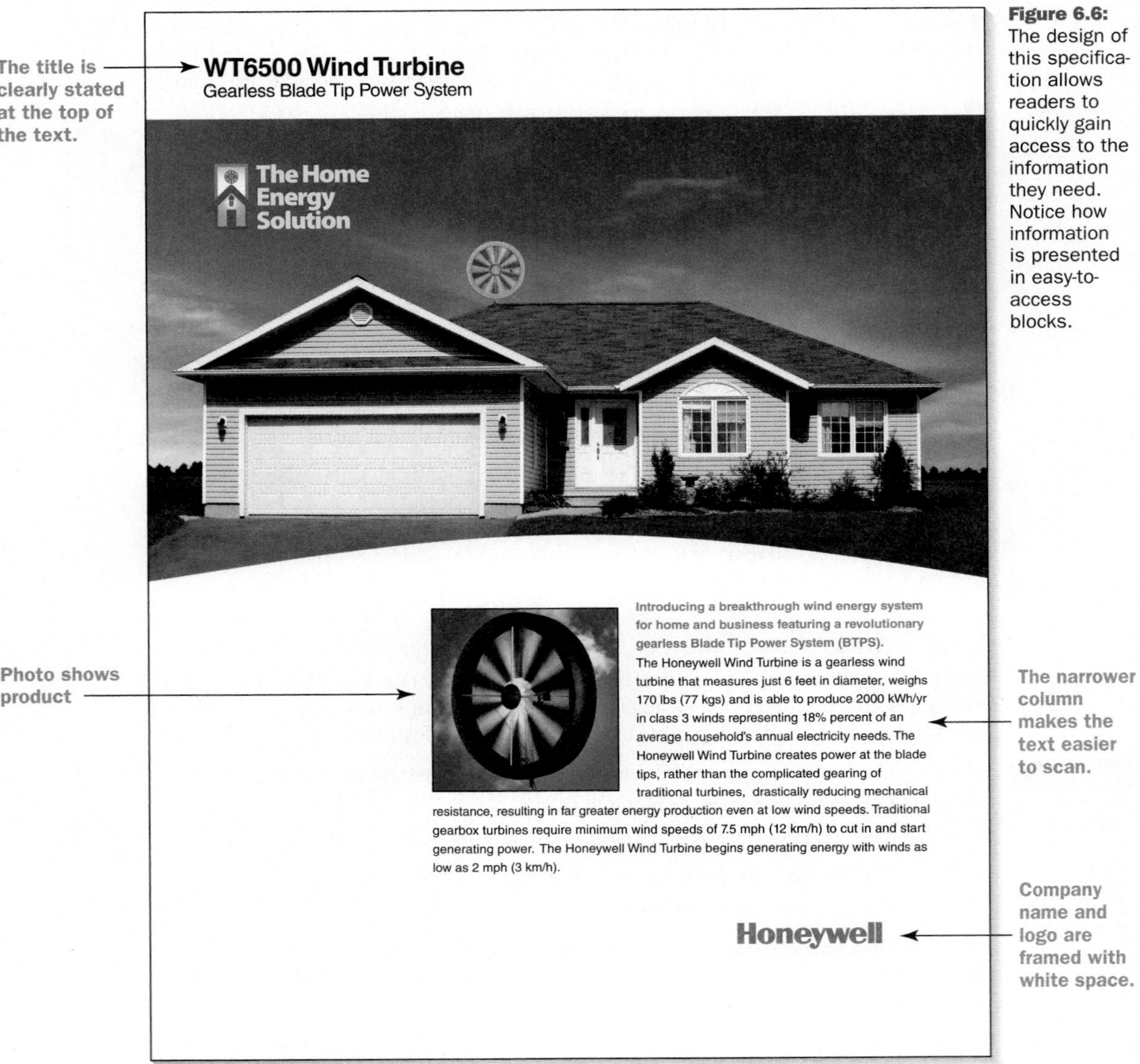

WT6500 Wind Turbine
Gearless Blade Tip Power System
The Home Energy Solution
Introducing a breakthrough wind energy system for home and business featuring a revolutionary gearless Blade Tip Power System (BTPS).
The Honeywell Wind Turbine is a gearless wind turbine that measures just 6 feet in diameter, weighs 170 lbs (77 kgs) and is able to produce 2000 kWh/yr in class 3 winds representing 18% percent of an average household's annual electricity needs. The Honeywell Wind Turbine creates power at the blade tips, rather than the complicated gearing of traditional turbines, drastically reducing mechanical resistance, resulting in far greater energy production even at low wind speeds. Traditional gearbox turbines require minimum wind speeds of 7.5 mph (12 km/h) to cut in and start generating power. The Honeywell Wind Turbine begins generating energy with winds as low as 2 mph (3 km/h).
Honeywell

Figure 6.6: The design of this specification allows readers to quickly gain access to the information they need. Notice how information is presented in easy-to-access blocks.

Source: Honeywell, Inc.

GO TO THE NET Need graphics? Go to **www.pearsonhighered.com/johnsonweb4/6.13**

Figure 6.6: *(continued)*

These annotations help readers understand the parts of the product.

Bringing wind technology to life.

We've turned wind turbines inside out.
According to the National Wind Technology Center, more than 80 percent of the U.S. residential market experience winds of less than 10 mph (16 km/h) over 90 percent of the time. Standard gear box turbines require a minimum of 7.5 mph (12 km/h) to begin producing energy. The Honeywell Wind Turbine's breakthrough technology can produce energy in as little wind as 2 mph (3 km/h) and can perform in winds up to 42 mph (67.5 km/h), enabling consumers who live in areas with low wind speeds to leverage wind as an energy source.

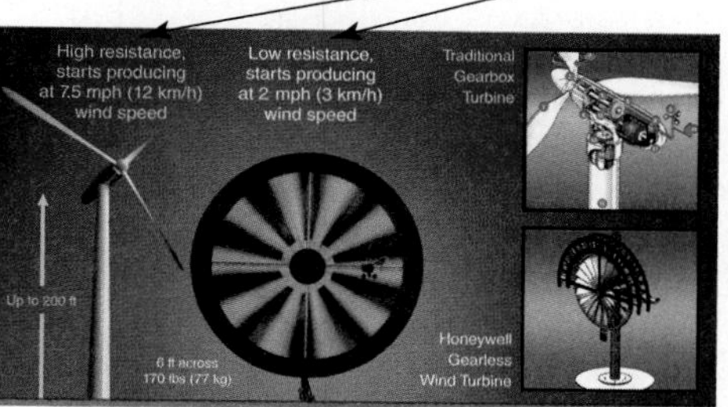

The layout balances the design features of the page.

Dealer Support
The Honeywell Wind Turbine Service Network will handle installation and services for all customers. Honeywell Wind Turbine Dealers will benefit from the following marketing and support programs.

State incentives of 30-100% for wind systems

Dealer Demo Rebate Program

One Source Certified Installation

Technical Support

Web Resource Center

Verifiable Power Output

In Store Merchandising

Training

Consumer DVD

Marketing & Public Relations

Honeywell Model WT 6500 Star Gate Specifications

Honeywell Model WT 6500 Star Gate Specifications
Honeywell Gearless Wind Turbine
Blade Tip Power System (BTPS)
BTPS Permanent Magnet Electric Generator, Patents Pending
UL Certified (Dec 2009)
Installs on Pole or Roof Mount
Lowest Cost kWh Installed Technology in Class
Enclosed Blade Tip Power System
Wide Wind Acceptance Angle
Acoustic Noise Emissions < 35dB
Tip to Tip Blade Dimension 5.7' (170 cm)
170 lbs (77.2 kg)
120 AC 60 Hz
220 AC 50 Hz (May 2010)
2.2 KW Plate Power
< 2mph (3 km/h) Cut in Speed, Shut down 42 mph (67.5 km/h)
Renewable Electric Generation 2000 kWh/yr - Class 3 Winds Renewable Electric Generation 2500 kWh/yr - Class 4 Winds (D.O.E. average US household electric 11,000 kWh/yr)
Smart Box Control System (includes) *Optimal Power Transfer Controller* *True Sine Wave Inverter* *Battery Power Management System* *Wind Direction & Speed Measurement Control System* *Standard RS485 Communication Port*
5 Year Limited Warranty
Annual CO2 Displacement 2.2 Tons
Product Design Life 20 Years

This table lists the specifications of the product.

WindTronics
380 W. Western Suite 301, Muskegon, MI 49440
toll free 866.6.EARTH.0 (866-632-7840)
local 231.332.1188
fax 231.726.5029
www.windtronics.com
info@windtronics.com

Honeywell

(continued)

Technical Descriptions Becoming More Visual

Figure 6.7: As our culture becomes more visual, even technical descriptions need to be as visual as possible. Scott McCloud drew these images for Google's Chrome browser.

For websites that discuss photography, go to **www.pearsonhighered.com/johnsonweb4/6.14**

specifies that the graphics are free to use, you must ask permission to use them.) If you cannot find graphics on the Internet, you can use a digital camera to take photographs that can be downloaded into your text. You can also use a scanner to digitize pictures, illustrations, and diagrams.

Here are some guidelines for using graphics in a technical description:

Use a title and figure number with each graphic, if possible.

Refer to the graphic by number in the written text.

Include captions that explain what the graphic shows.

Label specific features in the graphic.

Place the graphic on the page where it is referenced or soon afterward.

Link

For more information on using graphics, see Chapter 19, page 523.

It is not always possible to include titles, numbers, and captions with your graphics. In these situations, graphics should appear immediately next to or immediately after the places in the text where they are discussed.

As we move into a visual age, readers will expect technical descriptions to be more and more visual. For example, Scott McCloud, a well-known graphic artist, was brought in to illustrate the technical description of Google's Internet browser, Chrome (Figure 6.7). Technical descriptions will almost certainly become more visual and interactive in the future.

MICROGENRE

Technical Definitions

In workplace documents, you need to provide clear and precise definitions of technical terms, especially in detailed documents like technical descriptions, patents, specifications, experiments, and field observations.

Technical definitions are used a variety of ways. In most cases, they are embedded in other, larger documents. They are usually written into the main text, and they sometimes appear in the margins. In larger technical documents, definitions of technical terms will often appear before the introduction, or a *glossary of technical terms* will be added as an appendix.

A basic definition, often called a *sentence definition*, has three parts: (1) the term being defined, (2) the category in which the term belongs, and (3) the distinguishing features that differentiate it from its category.

Category

Distinguishing characteristics

An ion is an atom that has a negative or positive charge because it has more electrons or fewer electrons than usual.

An *extended definition* starts with a sentence definition like this one and then expands on it in the following ways:

Word history and etymology. Use a dictionary to figure out where a technical term originated and how its meaning has evolved (e.g., "The word *ion,* which means 'going' in Greek, was coined by physicist Michael Faraday in 1834.")

Examples. Include examples of how the term is used in a specific field (e.g., "For example, when hydrogen chloride (HCl) is dissolved in water, it forms two ions, H^+ and Cl^-.")

Negation. Define your subject by explaining what it is not (e.g., "An ion is not a subatomic particle.")

Division into parts. Divide the subject into its major parts and define each of those parts separately (e.g., "An ion has protons, electrons, and neutrons. Protons are subatomic particles with a positive charge found in the nucleus, while electrons . . .").

Similarities and differences. Compare your subject to objects or places that are similar, highlighting their common characteristics and their differences (e.g., "An ion is an atom with a nucleus and electrons, except an ion does not have an equal number of protons and electrons like a typical atom.")

Analogy. Compare your subject to something completely different but with some similar qualities (e.g., "An ion is like an unattached, single person at a dance, searching for oppositely-charged ions to dance with.")

Graphics. Use a drawing, picture, diagram, or other kind of graphic to provide an image of your subject.

Write

Write your own technical definition. List five technical terms that are important to your field of study. Write a sentence definition for each of them. Then, choose one of these sentence definitions and write a 300-word extended definition of that term.

An extended definition starts with a sentence definition. Then it uses a variety of rhetorical tools to expand and sharpen the meaning of the term.

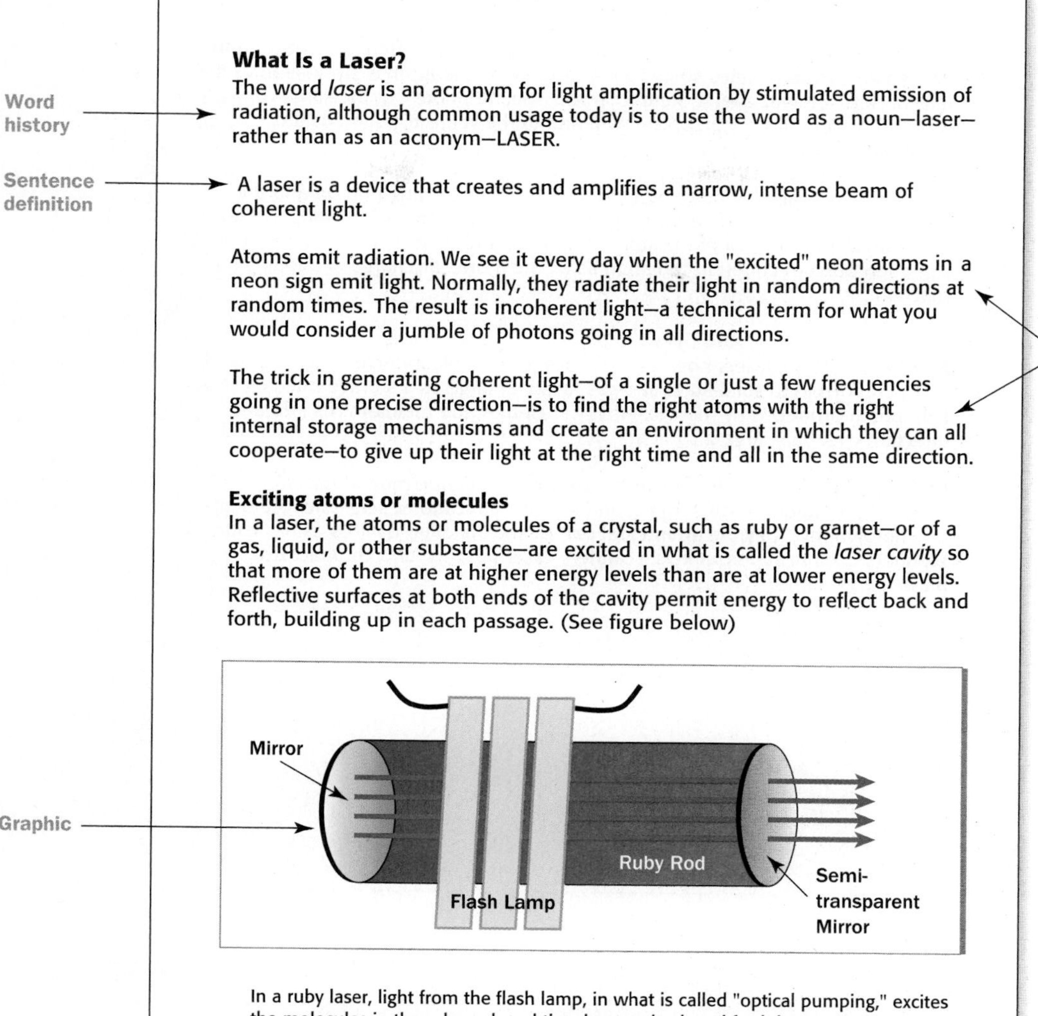

What Is a Laser?

Word history

The word *laser* is an acronym for light amplification by stimulated emission of radiation, although common usage today is to use the word as a noun—laser—rather than as an acronym—LASER.

Sentence definition

A laser is a device that creates and amplifies a narrow, intense beam of coherent light.

Atoms emit radiation. We see it every day when the "excited" neon atoms in a neon sign emit light. Normally, they radiate their light in random directions at random times. The result is incoherent light—a technical term for what you would consider a jumble of photons going in all directions.

Similarities and differences

The trick in generating coherent light—of a single or just a few frequencies going in one precise direction—is to find the right atoms with the right internal storage mechanisms and create an environment in which they can all cooperate—to give up their light at the right time and all in the same direction.

Exciting atoms or molecules

In a laser, the atoms or molecules of a crystal, such as ruby or garnet—or of a gas, liquid, or other substance—are excited in what is called the *laser cavity* so that more of them are at higher energy levels than are at lower energy levels. Reflective surfaces at both ends of the cavity permit energy to reflect back and forth, building up in each passage. (See figure below)

Graphic

In a ruby laser, light from the flash lamp, in what is called "optical pumping," excites the molecules in the ruby rod, and they bounce back and forth between two mirrors until coherent light escapes from the cavity.

Source: Reprinted with permission of Alcatel-Lucent USA Inc.

(continued)

If a photon whose frequency corresponds to the energy difference between the excited and ground states strikes an excited atom, the atom is stimulated as it falls back to a lower energy state to emit a second photon of the same (or a proportional) frequency, in phase with and in the same direction as the bombarding photon.

This process is called *stimulated emission*. The bombarding photon and the emitted photon may then each strike other excited atoms, stimulating further emission of photons, all of the same frequency and phase. This process produces a sudden burst of coherent radiation as all the atoms discharge in a rapid chain reaction.

Wide range of sizes and uses

History → First built in 1960, lasers now range in size from semiconductor lasers as small as a grain of salt to solid-state and gas lasers as large as a storage building. The light beam produced by most lasers is pencil-thin and maintains its size and direction over very large distances.

Lasers are widely used in industry for cutting and boring metals and other materials, in medicine for surgery, and in communications, scientific research, and holography. They are an integral part of such familiar devices as bar-code scanners used in supermarkets, scanners, laser printers, and compact disk players. ← Examples

CHAPTER REVIEW

- Technical descriptions and specifications are written to describe objects, places, phenomena, and processes. They are important documents in all technical workplaces.
- Basic features of a technical description or specification include a title, introduction, body, graphics, and conclusion.
- An object, place, or process can be partitioned according to its features, functions, or stages.
- Technical descriptions and specifications tend to be written in the "plain style," meaning that words and sentences are simple, direct, and concise.
- To add a visual element to the description, use the senses, similes, analogies, and metaphors to describe the subject.
- Graphics are crucial in technical descriptions, because the purpose of the description is to allow the readers to "see" the object, place, or process. You can use pictures, illustrations, and diagrams to add graphics to your text.
- The design of the description will depend on how it is being used. In sales literature, the design will probably be colorful or ornate. A specification for the company's files, on the other hand, might be rather plainly designed.

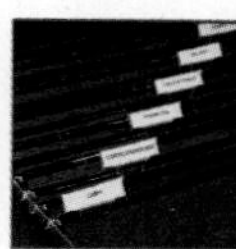

EXERCISES AND PROJECTS

Individual or Team Projects

1. Find a technical description on the Internet or in your workplace or home. First, determine the rhetorical situation (subject, purpose, readers, context) for which the description was written. Then, study its content, organization, style, and design. Write a two-page memo to your instructor in which you offer a critique of the description. What do you find effective about the description? What could be improved?

2. Your company sells a variety of products, listed below. Choose one of these items and write a one-page technical description. Your description should be aimed at a potential customer who might purchase one of these products:

    plasma-screen television
    DVD player
    MP3 player
    bicycle
    clock radio
    telescope
    washing machine

Sample descriptions are available at
www.pearsonhighered.com/johnsonweb4/6.15

baby stroller
toaster
coffeemaker
video camera

3. Find a common process that you can describe. Then, describe that process, walking your readers through its stages. In your description, you should define any jargon or technical terms that may be unfamiliar to your readers.

4. Study the description for the Canadarm or Mars Exploration Rover in this chapter. Write a two-page memo critiquing the description you chose. Do you think the description includes enough information? How might its content, organization, style, and design be improved?

Collaborative Projects

Your group has been assigned to describe a variety of renewable energy sources that might be used in your state. These energy sources might include solar, wind, geothermal, biomass generators, and fuel cells. While keeping the energy needs and limitations of your region in mind, offer a brief description of each of these renewable energy sources, showing how it works, its advantages, and its disadvantages.

In a report to your state's energy commissioner, describe these energy sources and discuss whether you think they offer possible alternatives to nonrenewable energy sources.

For additional technical writing resources, including interactive sample documents, document design tutorials and guidelines, and more, go to **www.mytechcommlab.com**.

Revision Challenge

Figure B shows a fact sheet from the Occupational Safety and Health Administration (OSHA). The description of flooding and flooding cleanup in this document is fine for an office environment. However, the size of the document makes it not particularly portable into areas that have been flooded.

Revise and redesign this document so that it fits on a 3 × 5 inch card that will be given out to first responders. This "Quickcard" will be easier to carry, and it can be stored in pockets and small storage areas. It could also be laminated, so it would hold up to severe conditions, like those found in flood zones.

Using the principles discussed in this chapter, analyze the content, organization, style, and design of this document. Then, revise this fact sheet so that it would be suitable for the kinds of emergency situations in which it would be used.

A Description That Could Be Revised

OSHA FactSheet

Flood Cleanup

Flooding can cause the disruption of water purification and sewage disposal systems, overflowing of toxic waste sites, and dislodgement of chemicals previously stored above ground. Although most floods do not cause serious outbreaks of infectious disease or chemical poisonings, they can cause sickness in workers and others who come in contact with contaminated floodwater. In addition, flooded areas may contain electrical or fire hazards connected with downed power lines.

Floodwater

Floodwater often contains infectious organisms, including intestinal bacteria such as E. coli, Salmonella, and Shigella; Hepatitis A Virus; and agents of typhoid, paratyphoid and tetanus. The signs and symptoms experienced by the victims of waterborne microorganisms are similar, even though they are caused by different pathogens. These symptoms include nausea, vomiting, diarrhea, abdominal cramps, muscle aches, and fever. Most cases of sickness associated with flood conditions are brought about by ingesting contaminated food or water. Tetanus, however, can be acquired from contaminated soil or water entering broken areas of the skin, such as cuts, abrasions, or puncture wounds. Tetanus is an infectious disease that affects the nervous system and causes severe muscle spasms, known as lockjaw. The symptoms may appear weeks after exposure and may begin as a headache, but later develop into difficulty swallowing or opening the jaw.

Floodwaters also may be contaminated by agricultural or industrial chemicals or by hazardous agents present at flooded hazardous waste sites. Flood cleanup crew members who must work near flooded industrial sites also may be exposed to chemically contaminated floodwater. Although different chemicals cause different health effects, the signs and symptoms most frequently associated with chemical poisoning are headaches, skin rashes, dizziness, nausea, excitability, weakness, and fatigue.

Pools of standing or stagnant water become breeding grounds for mosquitoes, increasing the risk of encephalitis, West Nile virus or other mosquito-borne diseases. The presence of wild animals in populated areas increases the risk of diseases caused by animal bites (e.g., rabies) as well as diseases carried by fleas and ticks.

Protect Yourself

After a major flood, it is often difficult to maintain good hygiene during cleanup operations. To avoid waterborne disease, it is important to wash your hands with soap and clean, running water, especially before work breaks, meal breaks, and at the end of the work shift. Workers should assume that any water in flooded or surrounding areas is not safe unless the local or state authorities have specifically declared it to be safe. If no safe water supply is available for washing, use bottled water, water that has been boiled for at least 10 minutes or chemically disinfected water. (To disinfect water, use 5 drops of liquid household bleach to each gallon of water and let it sit for at least 30 minutes for disinfection to be completed.) Water storage containers should be rinsed periodically with a household bleach solution.

If water is suspected of being contaminated with hazardous chemicals, cleanup workers may need to wear special chemical resistant outer clothing and protective goggles. Before entering a contaminated area that has been flooded, you should don plastic or rubber gloves, boots, and other protective clothing needed to avoid contact with floodwater.

Figure B: This fact sheet is somewhat long-winded and not easy to access in an emergency situation. Try turning it into a "Quickcard" that fits on a 3 × 5 inch card. You can use both sides of the card.

Source: Occupational Safety and Health Administration www.osha.gov/OshDoc/data_Hurricane_Facts/floodcleanup.pdf.

(continued)

Figure B: *(continued)*

Decrease the risk of mosquito and other insect bites by wearing long-sleeved shirts, long pants, and by using insect repellants. Wash your hands with soap and water that has been boiled or disinfected before preparing or eating foods, after using the bathroom, after participating in flood cleanup activities, and after handling articles contaminated by floodwater. In addition, children should not be allowed to play in floodwater or with toys that have been in contact with floodwater. Toys should be disinfected.

What to Do If Symptoms Develop

If a cleanup worker experiences any of the signs or symptoms listed above, appropriate first aid treatment and medical advice should be sought. If the skin is broken, particularly with a puncture wound or a wound that comes into contact with potentially contaminated material, a tetanus vaccination may be needed if it has been five years or more since the individual's last tetanus shot.

Tips to Remember

- Before working in flooded areas, be sure that your tetanus shot is current (given within the last 10 years). Wounds that are associated with a flood should be evaluated for risk; a physician may recommend a tetanus immunization.
- Consider all water unsafe until local authorities announce that the public water supply is safe.
- Do not use contaminated water to wash and prepare food, brush your teeth, wash dishes, or make ice.
- Keep an adequate supply of safe water available for washing and potable water for drinking.
- Be alert for chemically contaminated floodwater at industrial sites.
- Use extreme caution with potential chemical and electric hazards, which have great potential for fires and explosions. Floods have the strength to move and/or bury hazardous waste and chemical containers far from their normal storage places, creating a risk for those who come into contact with them. Any chemical hazards, such as a propane tank, should be handled by the fire department or police.
- If the safety of a food or beverage is questionable, throw it out.
- Seek immediate medical care for all animal bites.

This is one in a series of informational fact sheets highlighting OSHA programs, policies or standards. It does not impose any new compliance requirements. For a comprehensive list of compliance requirements of OSHA standards or regulations, refer to Title 29 of the Code of Federal Regulations. This information will be made available to sensory impaired individuals upon request. The voice phone is (202) 693-1999; teletypewriter (TTY) number: (877) 889-5627.

For more complete information:

OSHA Occupational Safety and Health Administration

U.S. Department of Labor
www.osha.gov
(800) 321-OSHA

DSTM 9/2005

CASE STUDY

In the Vapor

Linda Galhardy, a computer engineer at Gink, rushed to finish her technical description of FlashTime, a new handheld video game console her team had developed. She and her team had suffered through two long months of 18-hour days with no weekends off in order to get to this point. They were exhausted.

But, they succeeded in developing a prototype of the console, and they had created a couple of innovative games. The purpose of Linda's technical description was to update Gink's management on the project and persuade them to begin planning for production. If everything worked out, the game console could be released in October.

This mad rush all started when Gink's management read on gaming blogs that CrisMark, Gink's main competitor, was only six months away from developing a new handheld gaming device that would revolutionize the market. According to the bloggers, CrisMark's console had a small screen, much like a Nintendo DS, but it also allowed the gamer to project video onto any blank wall. According to reports, CrisMark had figured out a way to keep that video from shaking or turning with the controller. If true, this handheld game would revolutionize the market and take away a significant chunk of Gink's marketshare.

Fortunately, Gink had some smart engineers like Linda, and it had a reputation for more reliable and innovative consoles and games. Consumers tempted to buy CrisMark's console would probably be willing to wait a couple of extra months for a similar console from Gink. Gink's engineers and designers would have time to catch up.

Gink's management was impressed with Linda's presentation, including her technical description and the prototype. The marketing division was brought in immediately to begin working up the advertising and public relations campaign. Linda was congratulated and told to wait for management's answer.

A week later, Linda's supervisor, Thomas Hale, sent her a text message with a link to a popular consumer electronics blog. On the front page of the blog was a picture of Linda's prototype and a story taken from "leaked" sources. The story was surprisingly accurate.

She was furious, but she didn't know whom to blame. Did the marketing department leak the information? A week later, several consumer electronics blogs reported that CrisMark was abandoning its attempt to develop their revolutionary new handheld console. Anonymous sources said CrisMark abandoned the project when they heard Gink was ahead of them in developing the new product.

Soon afterward, Linda's boss, Thomas, called her into his office and told her the FlashTime project was being "slow-tracked."

(continued)

"You mean we're being killed slowly," Linda said angrily.

"Well," replied Thomas, "we need to make it look like we weren't just floating vaporware. So, you and a couple designers are going to stay on the project part-time. A year from now, we'll quietly pull the plug."

"That's garbage. We can do this, Thomas!" Linda said loudly. "We can make this console!"

Thomas said in a reassuring voice, "This thing was never going to get made anyway, Linda. Management just wanted to get something into the blogs to head off CrisMark. It worked. As soon as our 'leaked' story hit the blogs, CrisMark's stock dropped like a rock. They had to cancel the project to concentrate on their current products. Management just wanted to scare them off."

Linda couldn't believe what she was hearing. "So we were just making up some vaporware to scare off CrisMark?"

Thomas nodded. "Yeah, and it worked. Congratulations."

If you were Linda, what would you do at this point? Would you play along? Would you try to convince management to continue the project? Would you leak the truth to the blogs? Would you strike out on your own?

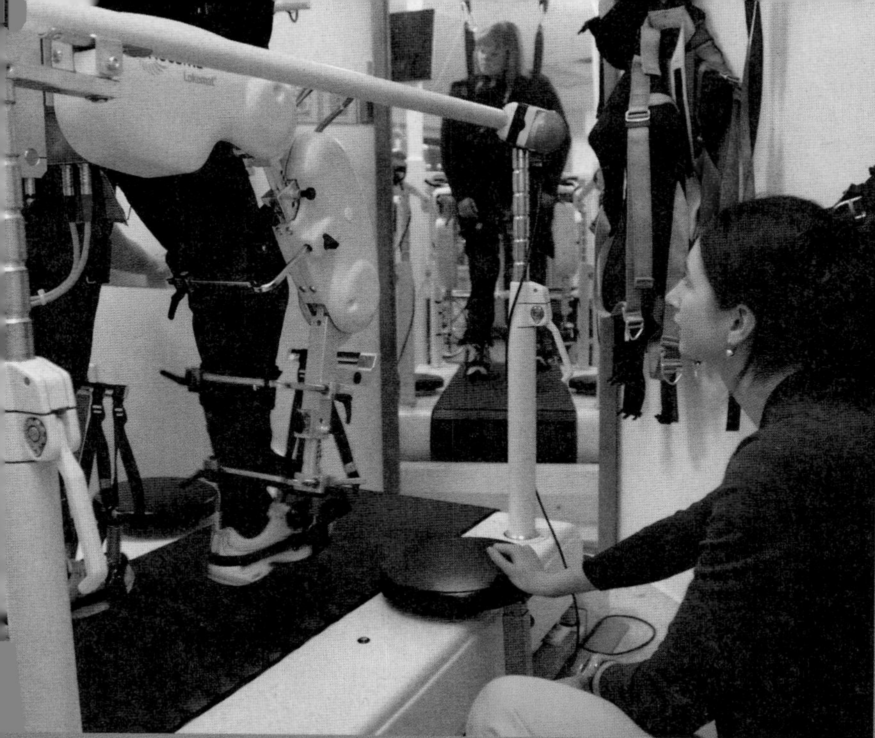

CHAPTER

7

Instructions and Documentation

Planning and Researching 166

Planning for Cross-Cultural Readers and Contexts 179

Organizing and Drafting 181

Using Style and Design 192

Help: On-Screen Documentation 194

Microgenre: Emergency Instructions 199

Chapter Review 201

Exercises and Projects 202

Case Study: The Flame 204

In this chapter, you will learn:

- The importance of instructions, specifications, and procedures in the technical workplace.
- The basic features of instructions, specifications, and procedures.
- How to plan and research instructions, specifications, and procedures.
- About the needs of cross-cultural readers.
- How to organize and draft instructions, specifications, and procedures.
- How style and design can be used to highlight and reinforce written text.

More than likely, you have read and used countless sets of instructions in your lifetime. Instructions are packaged with the products we buy, such as phones, cameras, and televisions. In the technical workplace, documentation helps people complete simple and complex tasks, such as downloading software, building an airplane engine, drawing blood from a patient, and assembling a computer motherboard.

Instructions and other kinds of documentation are among the least noticed but most important documents in the technical workplace. Documentation tends to fall into three categories:

> **Instructions**—Instructions describe how to perform a specific task. They typically describe how to assemble a product or do something step by step.
>
> **Specifications**—Engineers and technicians write specifications (often called "specs") to describe in exact detail how a product is assembled or how a routine process is completed.
>
> **Procedures/Protocols**—Procedures and protocols are written to ensure consistency and quality in a workplace. In hospitals, for example, doctors and nurses are often asked to write procedures that describe how to handle emergency situations or treat a specific injury or illness. Similarly, scientists will use protocols to ensure consistent methods in the laboratory.

To avoid confusion in this chapter, the word *documentation* will be used as a general term to mean instructions, procedures, and specifications. When the chapter discusses issues that are specific to instructions, procedures, or specifications, those terms will be used.

Link

For more information on writing descriptions, go to Chapter 6, page 133.

Documentation can take on many different forms, depending on when and where it will be used. For example, Figure 7.1 shows survival instructions from an entertaining and useful book called Worst-Case Scenarios. These instructions use simple text and visuals to explain how to jump out of a moving car—in case you ever need to.

Figure 7.2 shows an example of a protocol used at a hospital. This protocol works like most sets of instructions. A numbered list of steps explains what to do if someone is experiencing chest pain. The protocol includes helpful visuals, like the gray boxes and pictures of a doctor, to signal situations where a doctor should be called.

Figure 7.3 on pages 173–176, is a testing specification used by civil engineers. The numbers and indentation help engineers follow the procedure consistently.

Planning and Researching

When you are asked to write documentation for a product or procedure, you should first consider the situations in which it might be used. You also need to research the process you are describing so that you fully understand it and can describe it in detail.

Instructions and Documentation

These models illustrate a couple of common organizational patterns for instructions, procedures, and specifications. You should adjust these patterns to fit the process you are describing.

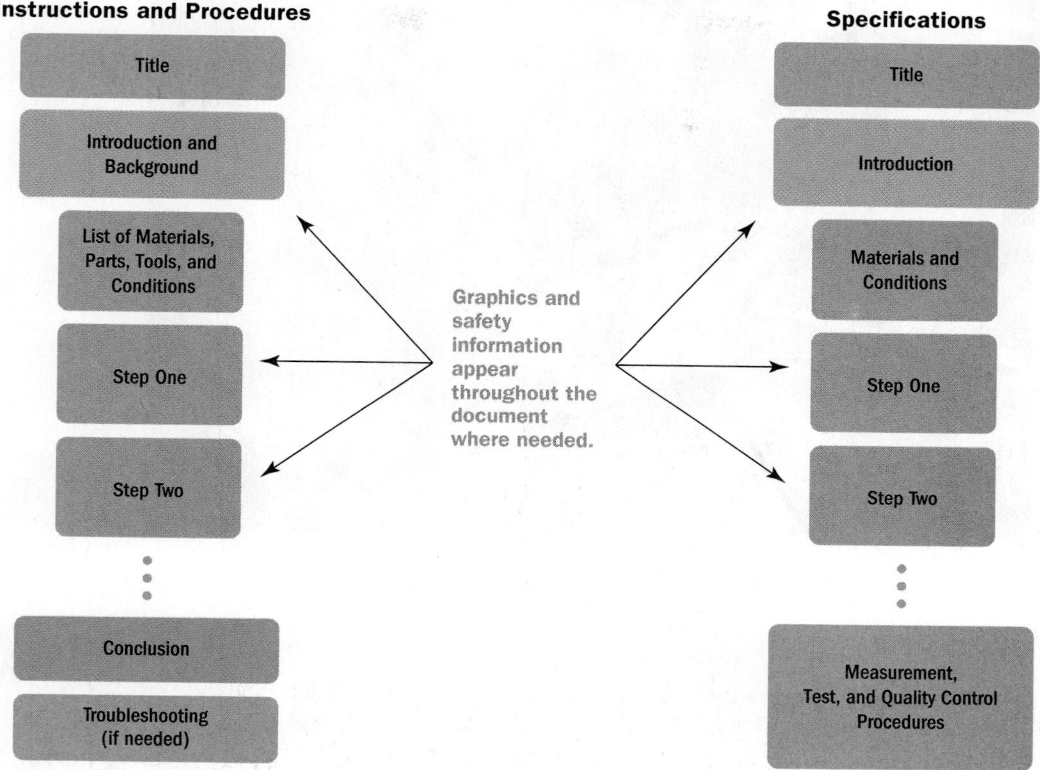

Basic Features

Documentation tends to follow a consistent step-by-step pattern, whether you are describing how to make coffee or how to assemble an automobile engine.

Here are the basic features in most forms of documentation:

- Specific and precise title
- Introduction with background information
- List of materials, parts, tools, and conditions required
- Sequentially ordered steps
- Graphics
- Safety information
- Measurement, test, and quality control procedures (for specifications)
- Conclusion that signals the completion of the task

To see samples of various kinds of instructions, go to **www.pearsonhighered.com/johnsonweb4/7.1**

Good Documentation Is Essential in the Technical Workplace

Instructions, procedures, and specifications are important, though often unnoticed, documents.

Planning

When planning, first gain a thorough grasp of your subject, your readers, and their needs. A good way to start is to answer the Five-W and How Questions:

Who might use this documentation?

Why is this documentation needed?

What should the documentation include?

Where will the documentation be used?

When will the documentation be used?

How will this documentation be used?

Once you have answered these questions, you are ready to define the rhetorical situation that will shape how you write the text.

SUBJECT Give yourself time to use the product or follow the process. What is it, and what does it do? Are there any unexpected dangers or difficulties? Where might users have trouble assembling this product or following directions?

A Set of Instructions

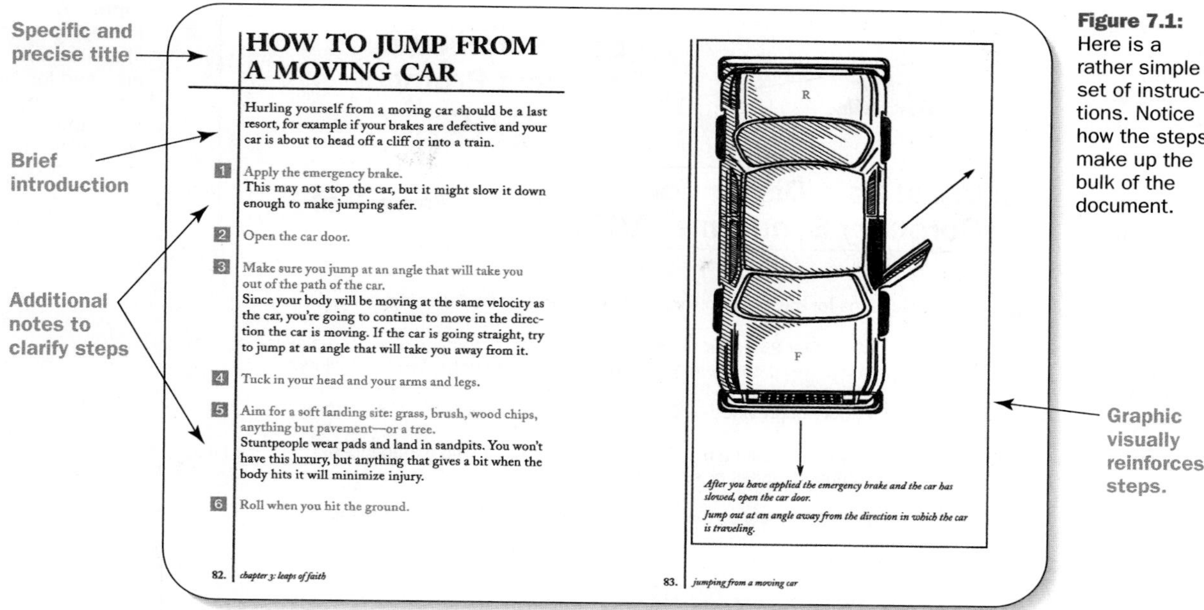

HOW TO JUMP FROM A MOVING CAR

Hurling yourself from a moving car should be a last resort, for example if your brakes are defective and your car is about to head off a cliff or into a train.

1. Apply the emergency brake.
   This may not stop the car, but it might slow it down enough to make jumping safer.
2. Open the car door.
3. Make sure you jump at an angle that will take you out of the path of the car.
   Since your body will be moving at the same velocity as the car, you're going to continue to move in the direction the car is moving. If the car is going straight, try to jump at an angle that will take you away from it.
4. Tuck in your head and your arms and legs.
5. Aim for a soft landing site: grass, brush, wood chips, anything but pavement—or a tree.
   Stuntpeople wear pads and land in sandpits. You won't have this luxury, but anything that gives a bit when the body hits it will minimize injury.
6. Roll when you hit the ground.

After you have applied the emergency brake and the car has slowed, open the car door.

Jump out at an angle away from the direction in which the car is traveling.

82. *chapter 3: leaps of faith*

83. *jumping from a moving car*

Figure 7.1: Here is a rather simple set of instructions. Notice how the steps make up the bulk of the document.

Source: From Worst-Case Scenario Survival Handbook™ *by Joshua Piven and David Borgenicht. Copyright © 1999 by Quirk Productions, Inc. Used with permission of Chronicle Books LLC, San Francisco. Visit Chronicle Books.com.*

PURPOSE Take a moment to consider and compose your purpose statement, limiting yourself to one sentence. Some key verbs for the purpose statement might include the following:

to instruct	*to guide*
to show	*to lead*
to illustrate	*to direct*
to explain	*to train*
to teach	*to tutor*

For example, here are a few purpose statements that might be used in a set of instructions:

> The purpose of these instructions is to show you how to use your new QuickTake i700 digital video camera.
>
> These procedures will demonstrate the suturing required to complete and close up a knee operation.
>
> These specifications illustrate the proper use of the Series 3000 Router to trim printed circuit boards.

Link

For more information on defining a document's purpose, see Chapter 1, page 6.

A Procedure

Header shows identification number of procedure.

Title of emergency procedure

Brief introduction defines medical condition.

Steps of the procedure

Gray areas signal situations in which a doctor needs to be involved.

WEST VIRGINIA EMS SYSTEM	EMT-Paramedic Treatment Protocol **4202**
Chest Pain/Discomfort Acute Coronary Syndrome (ACS)	**Page 1 of 3**

A. Indications for this protocol include one or more of the following:

1. Male over 25 years of age or female over 35 years of age, complaining of substernal chest pain, pressure or discomfort unrelated to an injury.

2. History of previous ACS/AMI with recurrence of "similar" symptoms.

3. Any patient with a history of cardiac problems who experiences lightheadedness or syncope.

4. Patients of any age with suspected cocaine abuse and chest pain.

B. Perform **MAMP (4201).**

C. Obtain 12 lead ECG, if available and causes no delay in treatment or transport.

D. If patient has no history of allergy to aspirin **and** has no signs of active bleeding (i.e., bleeding gums, bloody or tarry stools, etc.), then administer 4 (four) 81 mg chewable aspirin orally (324 mg total). Note: May be administered prior to establishment of IV access.

E. If blood pressure > 90 systolic and patient has **not** taken *Viagra* or *Levitra* within last 24 hours (or *Cialis* within the last 48 hours):

1. Administer nitroglycerine 0.4 mg (1/150 gr) SL. Note: May be administered prior to establishment of IV access.

2. Repeat every 3-5 minutes until pain is relieved.

3. If blood pressure falls below 90 systolic or decreases more than 30 mm Hg below patient's normal baseline blood pressure, then discontinue dosing and **contact MCP** to discuss further treatment.

West Virginia Office of Emergency Medical Services - State ALS Protocols
4602 Stroke.doc Finalized 12/1/01, Revised 9/11/07

Source: West Virginia EMS System, 2007.

Figure 7.2: Procedures like this one are used for training. They also standardize care.

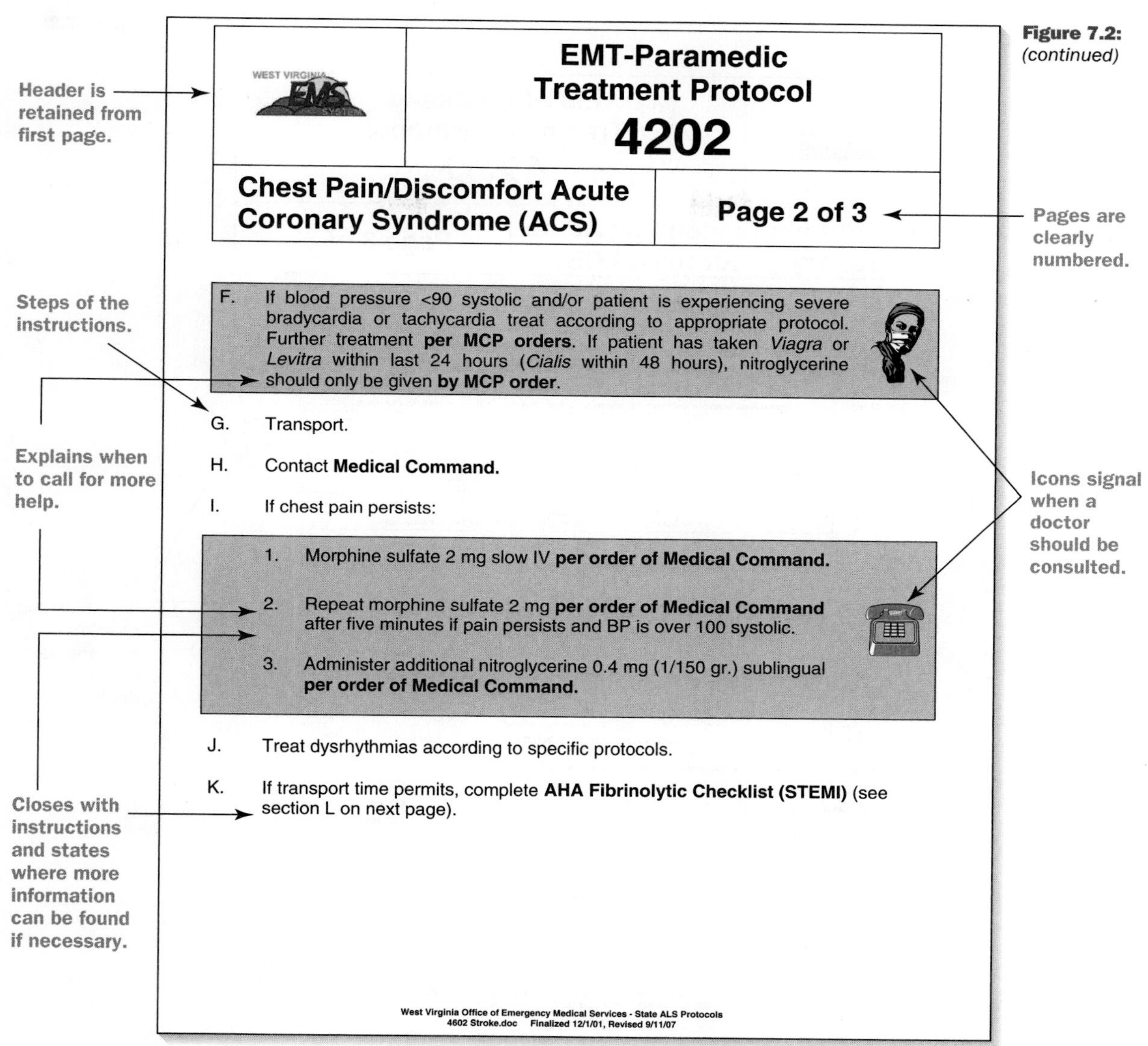

Header is retained from first page.
WEST VIRGINIA
EMS
EMT-Paramedic
Treatment Protocol
4202
Chest Pain/Discomfort Acute Coronary Syndrome (ACS)
Page 2 of 3
Pages are clearly numbered.
Steps of the instructions.
F. If blood pressure <90 systolic and/or patient is experiencing severe bradycardia or tachycardia treat according to appropriate protocol. Further treatment per MCP orders. If patient has taken Viagra or Levitra within last 24 hours (Cialis within 48 hours), nitroglycerine should only be given by MCP order.
G. Transport.
H. Contact Medical Command.
I. If chest pain persists:
Explains when to call for more help.
1. Morphine sulfate 2 mg slow IV per order of Medical Command.
2. Repeat morphine sulfate 2 mg per order of Medical Command after five minutes if pain persists and BP is over 100 systolic.
3. Administer additional nitroglycerine 0.4 mg (1/150 gr.) sublingual per order of Medical Command.
Icons signal when a doctor should be consulted.
J. Treat dysrhythmias according to specific protocols.
K. If transport time permits, complete AHA Fibrinolytic Checklist (STEMI) (see section L on next page).
Closes with instructions and states where more information can be found if necessary.
West Virginia Office of Emergency Medical Services - State ALS Protocols
4602 Stroke.doc Finalized 12/1/01, Revised 9/11/07

Figure 7.2: *(continued)*

(continued)

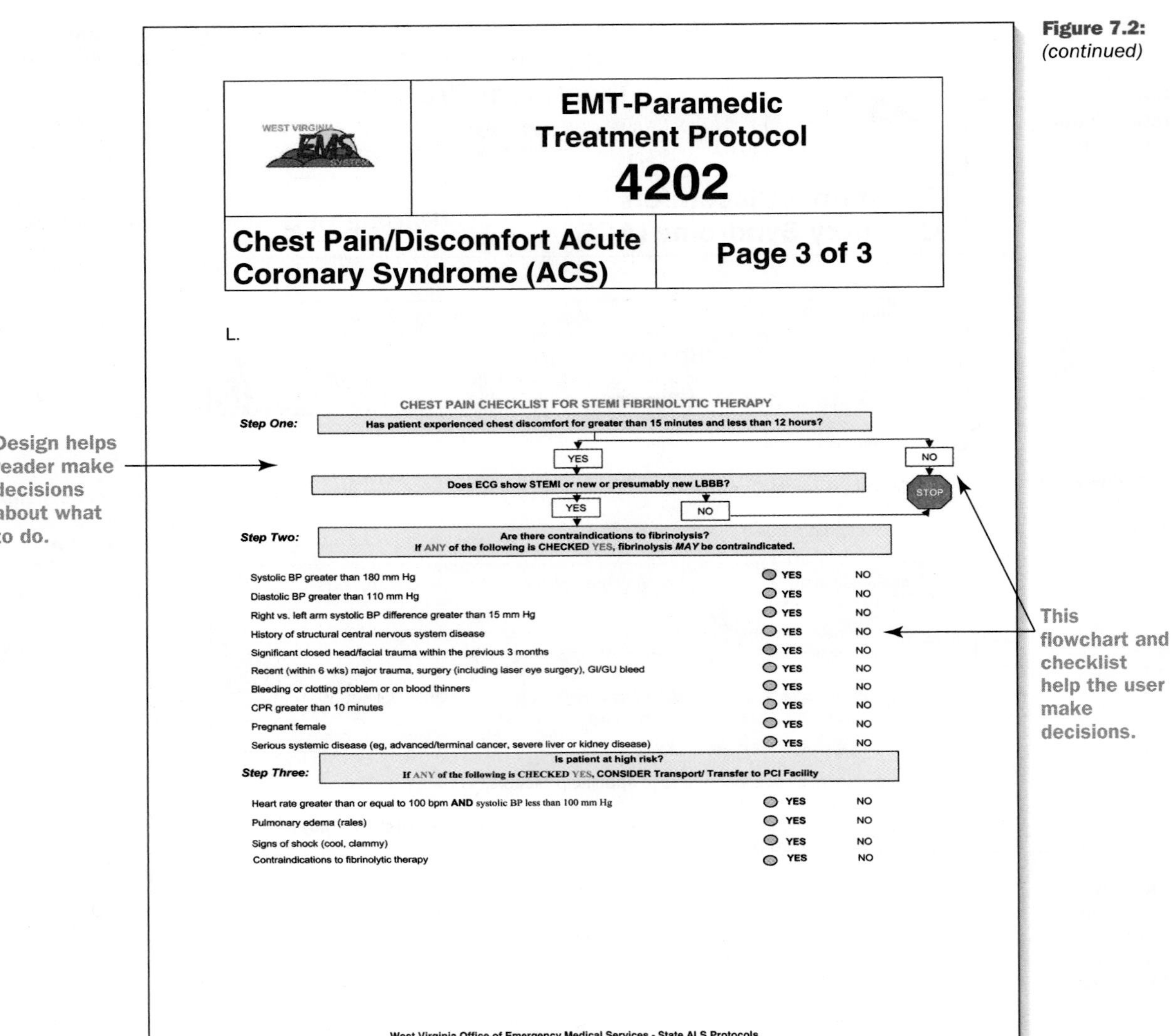
EMT-Paramedic Treatment Protocol 4202

Chest Pain/Discomfort Acute Coronary Syndrome (ACS)

Page 3 of 3

L.

CHEST PAIN CHECKLIST FOR STEMI FIBRINOLYTIC THERAPY

Step One: **Has patient experienced chest discomfort for greater than 15 minutes and less than 12 hours?**

YES → **Does ECG show STEMI or new or presumably new LBBB?**

NO → STOP

YES → Step Two; NO → STOP

Step Two: **Are there contraindications to fibrinolysis?**
If ANY of the following is CHECKED YES, fibrinolysis *MAY* be contraindicated.

Systolic BP greater than 180 mm Hg	YES	NO
Diastolic BP greater than 110 mm Hg	YES	NO
Right vs. left arm systolic BP difference greater than 15 mm Hg	YES	NO
History of structural central nervous system disease	YES	NO
Significant closed head/facial trauma within the previous 3 months	YES	NO
Recent (within 6 wks) major trauma, surgery (including laser eye surgery), GI/GU bleed	YES	NO
Bleeding or clotting problem or on blood thinners	YES	NO
CPR greater than 10 minutes	YES	NO
Pregnant female	YES	NO
Serious systemic disease (eg, advanced/terminal cancer, severe liver or kidney disease)	YES	NO

Step Three: **Is patient at high risk?**
If ANY of the following is CHECKED YES, CONSIDER Transport/ Transfer to PCI Facility

Heart rate greater than or equal to 100 bpm **AND** systolic BP less than 100 mm Hg	YES	NO
Pulmonary edema (rales)	YES	NO
Signs of shock (cool, clammy)	YES	NO
Contraindications to fibrinolytic therapy	YES	NO

West Virginia Office of Emergency Medical Services - State ALS Protocols
4602 Stroke.doc Finalized 12/1/01, Revised 9/11/07

WEST VIRGINIA
EMS
SYSTEM
Design helps reader make decisions about what to do.
This flowchart and checklist help the user make decisions.

Figure 7.2: *(continued)*

A Specification

Figure 7.3: This specification explains how a civil engineer would conduct a materials test.

Introduction explains the purpose of the specifications.

INDIANA DEPARTMENT OF TRANSPORTATION
OFFICE OF MATERIALS MANAGEMENT

DRY FLOW TESTING
OF
FLOWABLE BACKFILL MATERIALS
ITM No. 217-07T

1.0 SCOPE.

1.1 This test method covers the procedure for the determination of the flow time of dry flowable backfill materials for the purpose of verifying changes in sand sources for an approved Flowable Backfill Mix Design (FBMD).

1.2 The values stated in either acceptable English or SI metric units are to be regarded separately as standard, as appropriate for a specification with which this ITM is used. Within the text, SI metric units are shown in parentheses. The values stated in each system may not be exact equivalents; therefore, each system shall be used independently of the other, without combining values in any way.

1.3 This ITM may involve hazardous materials, operations, and equipment. This ITM may not address all of the safety problems associated with the use of the test method. The user of the ITM is responsible for establishing appropriate safety and health practices and determining the applicability of regulatory limitations prior to use.

2.0 REFERENCES.

2.1 AASHTO Standards.

M 231 Weighing Devices Used in the Testing of Materials

T 304 Uncompacted Void Content of Fine Aggregate

T 248 Reducing Samples of Aggregate to Testing Size

3.0 TERMINOLOGY. Definitions for terms and abbreviations shall be in accordance with the Department's Standard Specifications, Section 101, except as follows.

3.1 Dry flow time. The time to for a specified sample size of dry flowable materials to flow through a specified funnel

Source: Indiana Department of Transportation, http://www.in.gov/indot/div/M&T/itm/pubs/217_testing.pdf.

(continued)

Figure 7.3: *(continued)*

4.0 SIGNIFICANCE AND USE.

4.1 This ITM is used to determine the time of dry flow of loose uncompacted flowable backfill material through a flow cone. The flowable backfill material includes sand or sand and fly ash mixture. The test result is done to ensure that an alternate sand shall have the same flow characteristic as the sand in the approved FBMD.

4.2 The dry flow cone test characterizes the state of flow of dry materials on any sand of known grading that may provide information about the sand or sand and fly ash mixture angularity, spherical shape, and surface texture.

4.3 Other test procedures or test methods exist for various flow cones with different dimensions and cone tip forms and sizes that may or may not have a correlation with the AASHTO T 304 flow cone.

5.0 APPARATUS.

5.1 Cylindrical measure, in accordance with AASHTO T 304, except the nominal 100-mL cylindrical measure is replaced by a one quart glass jar

The testing apparatus is clearly described.

5.2 Metal spatula, with a blade approximately 4 in. (100 mm) long, and at least ¾ in. (20 mm) wide, with straight edges. The end shall be cut at a right angle to the edges. (The straight edge of the spatula blade is used to strike off the fine aggregate.)

This "nested" numbering system is commonly used with specifications for easy reference.

5.3 Timing device, such as a stop watch, with an accuracy to within ± 0.1 seconds

5.4 Balance, Class G2, conforming to the requirements of AASHTO M 231

5.5 Sample splitter, in accordance with AASHTO T 248 for fine aggregate

6.0 SAMPLE PREPARATION.

6.1 The sample may consist of sand or a sand and fly ash mixture proportioned according to the FBMD. The sample shall be oven dried at 230 ± 9° F (110 ± 5° C) for 24 h. Upon completion of the drying, the sample shall be split to a sample size of approximately 1,500 g using a small sample splitter for fine aggregate in accordance with AASHTO T 248.

6.2 The dry sample of sand or sand and fly ash mixture shall be thoroughly mixed with the spatula until the sample appears to be homogenous.

Figure 7.3: *(continued)*

The testing procedure is explained step by step.

7.0 PROCEDURE.

7.1 Place the dry sample into the one quart glass jar and put the lid on. Agitate the glass jar to mix the dry sample for 30 seconds.

7.2 Place a finger at the end of the funnel to block the opening of the funnel.

7.3 Pour and empty the dry sample of sand or sand and fly ash mixture from the glass jar into the Mason jar.

7.4 Level the dry sample in the Mason jar with a spatula.

7.5 Place the empty glass jar directly centered under the funnel.

7.6 Remove the finger and allow the dry sample to fall freely into the glass jar, and start timing the dry flow.

7.7 Record the time T_1 of the dry flow to an accuracy of ±0.1 second.

7.8 Repeat 7.1 through 7.6 for times T_2 and T_3.

8.0 CALCULATIONS.

8.1 Calculate the average dry flow time of the dry flowable backfill materials as follows:

$$T_{average} = \frac{T_1 + T_2 + T_3}{3}$$

where:

$T_{average}$ = Average dry flow time, s
T_1 = Dry flow time on first trial, s
T_2 = Dry flow time on second trial, s
T_3 = Dry flow time on third trial, s

9.0 REPORT.

9.1 Report the average dry flow time to within ± 0.1 seconds.

(continued)

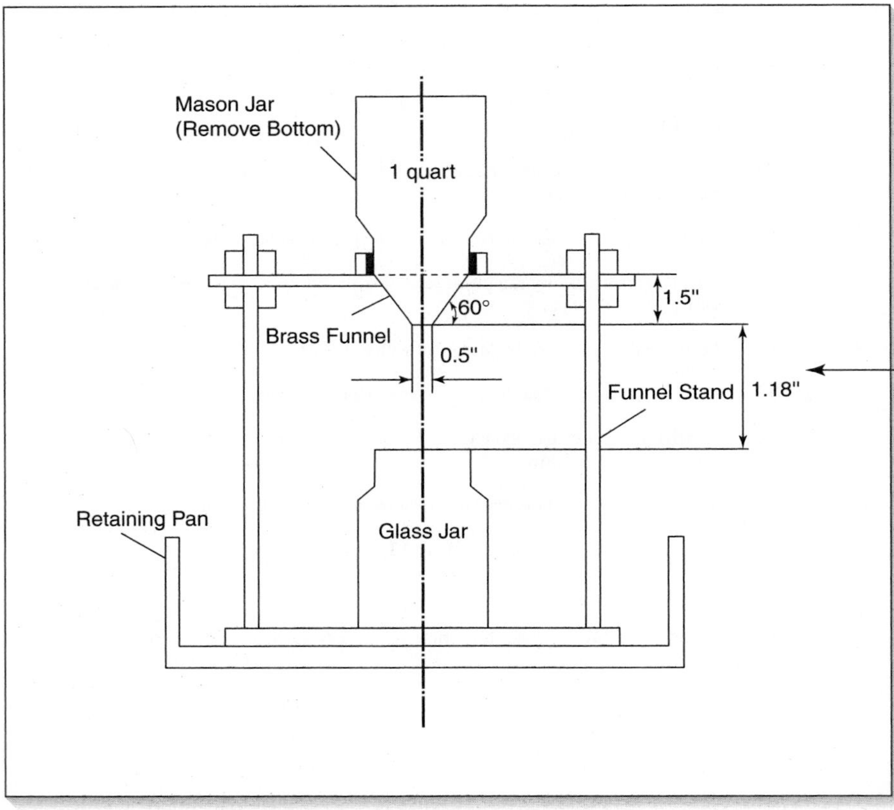

Mason Jar
(Remove Bottom)
1 quart
60°
1.5"
Brass Funnel
0.5"
Funnel Stand
1.18"
Retaining Pan
Glass Jar

Figure 7.3: *(continued)*

The illustration helps the engineer build the testing apparatus.

READERS Of course, it is difficult to anticipate all the types of people who might use your documentation. But people who decide to use a specific set of instructions, a set of specifications, or a procedure usually have common characteristics, backgrounds, and motivations that you can use to make your documentation more effective.

Primary readers (action takers) are people who will use your documentation to complete a task. What is their skill level? How well do they understand the product or process? What is their age and ability?

Secondary readers (advisors) are people who might supervise or help the primary readers complete the task. What is the skill level of these secondary readers? Are they training/teaching the primary readers? Are they helping the primary readers assemble the product or complete the process?

Tertiary readers (evaluators) often use documentation to ensure quality. Auditors and quality experts will review procedures and specifications

For reader analysis worksheets, go to **www.pearsonhighered.com/johnsonweb4/7.2**

closely when evaluating products or processes in a technical workplace. Also, sets of instructions can be used as evidence in lawsuits. So, these kinds of readers need to be considered and their needs anticipated.

Gatekeeper readers (supervisors) who may or may not be experts in your field will need to look over your documentation before it is sent out with a product or approved for use in the workplace. They will be checking the accuracy, safety, and quality of your documentation.

Take care not to overestimate your readers' skills and understanding. If your documentation is specific and uses simple language, novice users will appreciate the added help. Experienced users may find your documentation a bit too detailed, but they can just skim information they don't need. In most cases, you are better off giving your readers more information than they need.

Link

For more information on analyzing readers, see Chapter 2, page 20.

CONTEXT OF USE Put yourself in your readers' place for a moment. When and where will your readers use the documentation? In their living rooms? At a workbench? At a construction site? In an office cubicle? At night? Each of these different places and times will require you to adjust your documentation to your readers' needs.

Depending on the context of use, instructions can follow a variety of formats. They may be included in a user's manual, or they could be part of a poster explaining how to accomplish a task. Increasingly, documentation is being placed on websites for viewing or downloading. Figure 7.4, for example, shows how the documentation from *Worst-Case Scenarios* is portrayed on a website.

Context of use also involves safety and liability issues. If users of the documentation are at risk for injury or an accident, you are ethically obligated to warn them about the danger and tell them how to avoid it. You should try to anticipate all the ways users might injure themselves or experience a mishap while following the documentation you are providing. Then, use warning statements and cautions (discussed later in this chapter) to help them avoid these problems.

Researching

Once you have defined the rhetorical situation, you should spend some time doing research on the task you are describing. Research consists of gaining a thorough understanding of your subject by considering it from several angles (Figure 7.5). Here are a few research strategies that are especially useful when writing documentation.

DO BACKGROUND RESEARCH You should research the history and purpose of the product or process you are describing. If the product or process is new, find out why it was developed and study the documents that shaped its development. If the product or process is not new, determine whether it has evolved or changed. Also, collect any prior instructions, procedures, or specifications that might help you write your own documentation.

MAKE OBSERVATIONS Observe people using the product. If, for example, you are writing instructions for using a coffeemaker, observe someone making coffee with it. Pay attention to his or her experiences, especially any mistakes. Your notes from these experiments will help you anticipate some of the situations and problems your readers will experience.

For document planning worksheets, go to **www.pearsonhighered.com/johnsonweb4/7.3**

Instructions on a Website

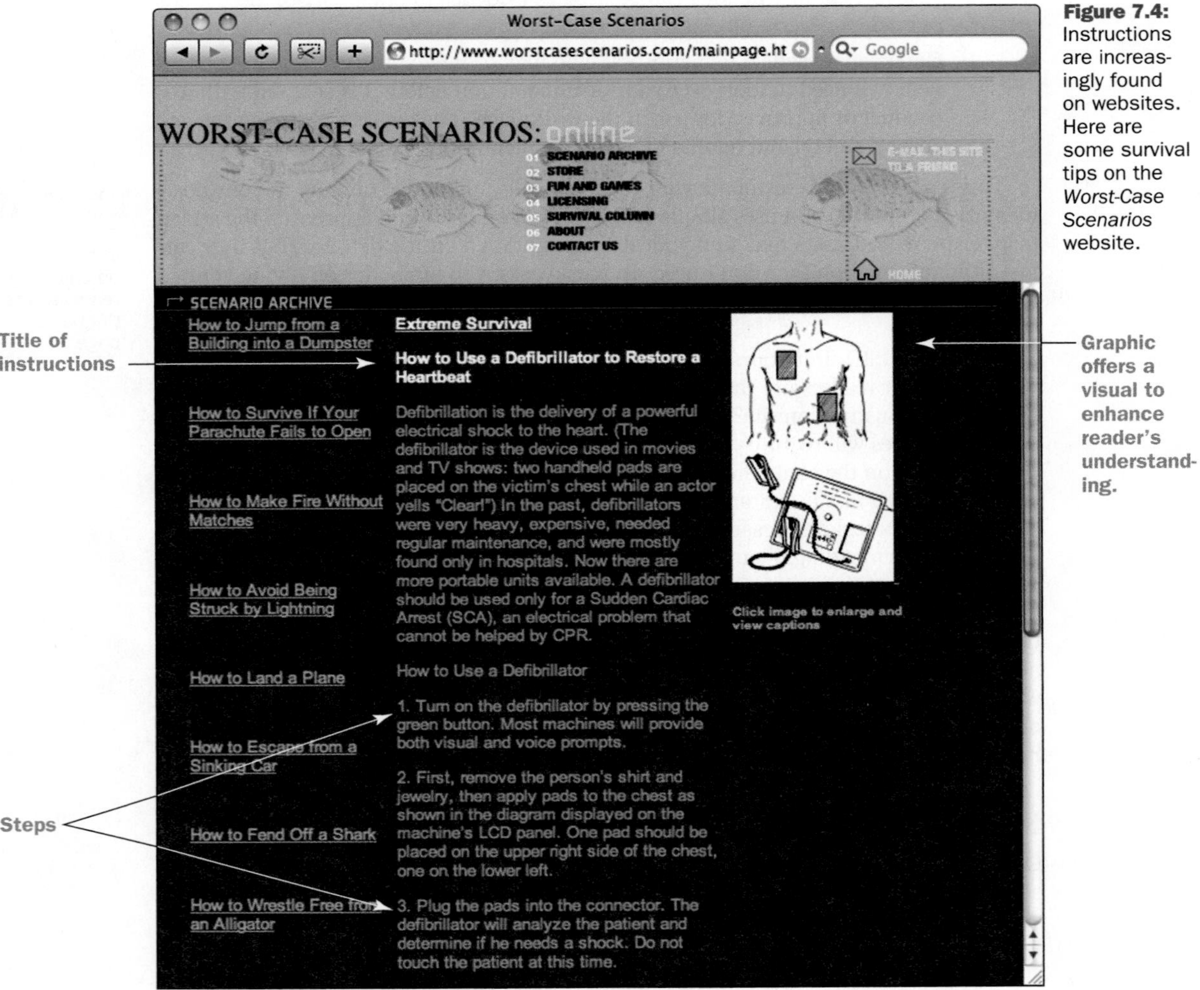

Figure 7.4: Instructions are increasingly found on websites. Here are some survival tips on the *Worst-Case Scenarios* website.

Source: © 1999–2007 by Quirk Productions, Inc. Worst Case Scenario® and The Worst-Case Scenario Survival Handbook™ are trademarks of Quirk Productions, Inc. www.worstcasescenarios .com. Used with permission. All rights reserved.

ASK SUBJECT MATTER EXPERTS (SMEs) Interview experts who are familiar with the product or have used the procedure. They may be able to give you some insight or pointers into how the product is actually used or how the procedure is completed. They might also be able to point out trouble spots where nonexperts might have problems.

GO TO THE NET — To read about blunders in which writers did not anticipate their readers' needs, go to **www.pearsonhighered.com/johnsonweb4/7.4**

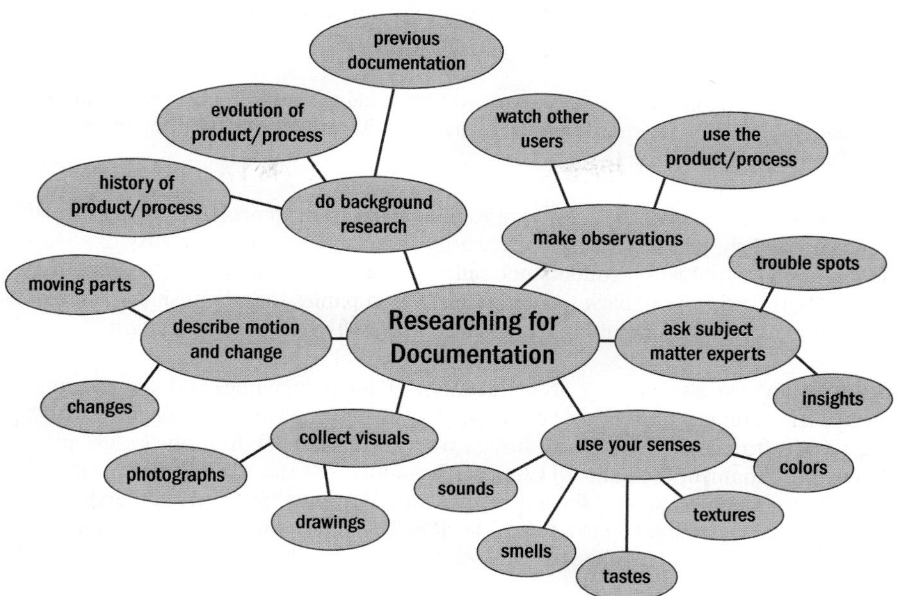

Figure 7.5: When doing research to write documentation, you should study your subject from a few different perspectives.

USE YOUR SENSES Where appropriate, take notes about appearance, sounds, smells, textures, and tastes. These details will add depth to your documentation. They will also help your readers determine if they are following the directions properly.

DESCRIBE MOTION AND CHANGE Pay special attention to the movements of your subject and the way it changes as you complete the steps. Each step will lead to some kind of motion or change. By noting these motions and changes, you will be better able to describe them.

Link

For more information on doing research, go to Chapter 14, page 386.

COLLECT VISUALS If available, collect graphics that can help you illustrate the steps you are describing in your documentation. If necessary, you can take photographs with a digital camera, or you can use drawings to illustrate your subject.

Planning for Cross-Cultural Readers and Contexts

As world trade continues to expand, documentation needs to be written with cross-cultural readers in mind. As you profile your readers, you should anticipate how your documentation should be adjusted for readers who come from a variety of cultural backgrounds.

GO TO THE NET

For more advice on cross-cultural writing, go to **www.pearsonhighered.com/johnsonweb4/7.5**

Verbal Considerations

First, figure out how the written parts of your documentation will fit the needs of cross-cultural readers.

TRANSLATE THE TEXT If the documentation is being sent to a place where another language is spoken, you should have it translated. Then, include both the translated and English versions with the product.

USE BASIC ENGLISH If the documentation might be sent to people who are not fluent in English, you should use basic words and phrases. Avoid any jargon, idioms, and metaphors that will be understood only by North Americans (e.g., "senior citizens," "bottom line," or "back to square one"). Companies that do business in cross-cultural markets often maintain lists of basic words to use in documentation.

CHECK MEANINGS OF NAMES AND SLOGANS Famously, names and slogans don't always translate well into other languages. For example, Pepsi's "Come alive with Pepsi" was translated into, "Pepsi brings your ancestors back from the grave" in Taiwan. In Spanish, the name of Chevrolet's popular car the "Nova" means "It doesn't go." Coors's slogan "Turn it loose" translated into "Suffer from diarrhea" in some Spanish-speaking countries. So, check with people who are familiar with the target culture and its language to see if your names and slogans have meanings other than those intended.

Design Considerations

Next, do some research to figure out the best way to present your documentation visually to your target readers.

USE ICONS CAREFULLY Some symbols commonly used in North America can be offensive to other cultures. A pointing finger, for example, is offensive in some Central and South American countries. An "OK" sign is offensive in Arab nations. Dogs are "unclean" animals in many parts of the world, so cartoon dogs often do not work well in documentation for multicultural readers. To avoid these problems, minimize your use of animals, human characters, or body parts whenever possible. Make sure that any icons you use will not confuse or offend your readers in some unintended way.

Link

For more information on working with cross-cultural readers, go to Chapter 2, page 28.

USE IMAGES CAREFULLY Images can convey unintended messages to readers, especially in high-context cultures like those of Asia, parts of Africa, and the Middle East. For instance, how people are dressed in photographs can signal respect or disrespect. Obvious displays of emotion in professional settings can be seen as rude. In some conservative Middle Eastern countries, meanwhile, photographs of people are used for identification only. So, before moving forward with your documentation, you might ask someone from the target culture to look over the images for any unintended meanings.

GO TO THE NET

To read about some other funny translation problems, go to **www.pearsonhighered.com/johnsonweb4/7.6**

Organizing and Drafting

Like other technical documents, documentation should include an introduction, body, and conclusion. The introduction typically offers background information on the task being described. The body describes the steps required to complete the task. The conclusion usually offers readers a process for checking their work.

Specific and Precise Title

The title of your documentation should clearly describe the specific task the reader will complete.

> **Not descriptive:** RGS-90x Telescope
>
> **Descriptive:** Setting Up Your RGS-90x Telescope
>
> **Not descriptive:** Head Wound
>
> **Descriptive:** Procedure for Treating a Head Wound

Introduction

The length of the introduction depends on the complexity of the task and your readers' familiarity with it. If the task is simple, your introduction might be only a sentence long. If the task is complex or your readers are unfamiliar with the product or process, your introduction may need to be a few paragraphs long.

Introductions should include some or all of the following moves.

STATE THE PURPOSE Simple or complex, all documentation should include some kind of statement of purpose.

> These instructions will help you set up your RGS-90x telescope.
>
> The Remington Medical Center uses these procedures to treat head wounds in the emergency room.

STATE THE IMPORTANCE OF THE TASK You may want to stress the importance of the product or perhaps the importance of doing the task correctly.

> Your RGS-90x telescope is one of the most revolutionary telescope systems ever developed. You should read these instructions thoroughly so that you can take full advantage of the telescope's numerous advanced features.
>
> Head wounds of any kind should be taken seriously. The following procedures should be followed in all head wound cases, even the cases that do not seem serious.

DESCRIBE THE NECESSARY TECHNICAL ABILITY You may want to describe the necessary technical background that readers will need to use the product or to complete the task. Issues like age, qualifications, education level, and prior training are often important considerations that should be mentioned in the introduction.

GO TO THE NET

Want to see some sample introductions for instructions? Go to **www.pearsonhighered.com/johnsonweb4/7.7**

With advanced features similar to those found in larger and more specialized telescopes, the RGS-90x can be used by casual observers and serious astronomers alike. Some familiarity with telescopes is helpful but not needed.

Because head wounds are usually serious, a trained nurse should be asked to bandage them. Head wounds should never be bandaged by trainees without close supervision.

IDENTIFY THE TIME REQUIRED FOR COMPLETION If the task is complex, you may want to estimate the time readers will need to complete all the steps.

Initially, setting up your telescope should take about 15 to 20 minutes. As you grow more familiar with it, though, setup should take only 5 to 10 minutes.

Speed is important when treating head wounds. You may have only a few minutes before the patient goes into shock.

MOTIVATE THE READER An introduction is a good place to set a positive tone. Add a sentence or two to motivate readers and make them feel positive about the task they are undertaking.

With push-button control, automatic tracking of celestial objects, and diffraction-limited imaging, an RGS-90x telescope may be the only telescope you will ever need. With this powerful telescope, you can study the rings of the planet Saturn or observe the feather structure of a bird from 50 yards away. This telescope will meet your growing interests in astronomical or terrestrial viewing.

Head wounds of any kind are serious injuries. Learn and follow these procedures so you can effectively treat these injuries without hesitation.

Procedures and specifications often also include motivational statements about the importance of doing the job right, as companies urge their employees to strive for the highest quality.

List of Parts, Tools, and Conditions Required

After the introduction, you should list the parts, tools, and conditions required for completing the task.

LIST THE PARTS REQUIRED This list should identify all the necessary items required to complete the task. Your parts list will allow readers to check whether all the parts were included in the package (Figure 7.6). Other items not included with the package, like adhesive, batteries, and paint, should be mentioned at this point so readers can collect these items before following the steps.

IDENTIFY TOOLS REQUIRED Nothing is more frustrating to readers than discovering midway through a set of instructions or a specification that they need a tool that was not previously mentioned. The required tools should be listed up front so readers can gather them before starting.

A Parts List

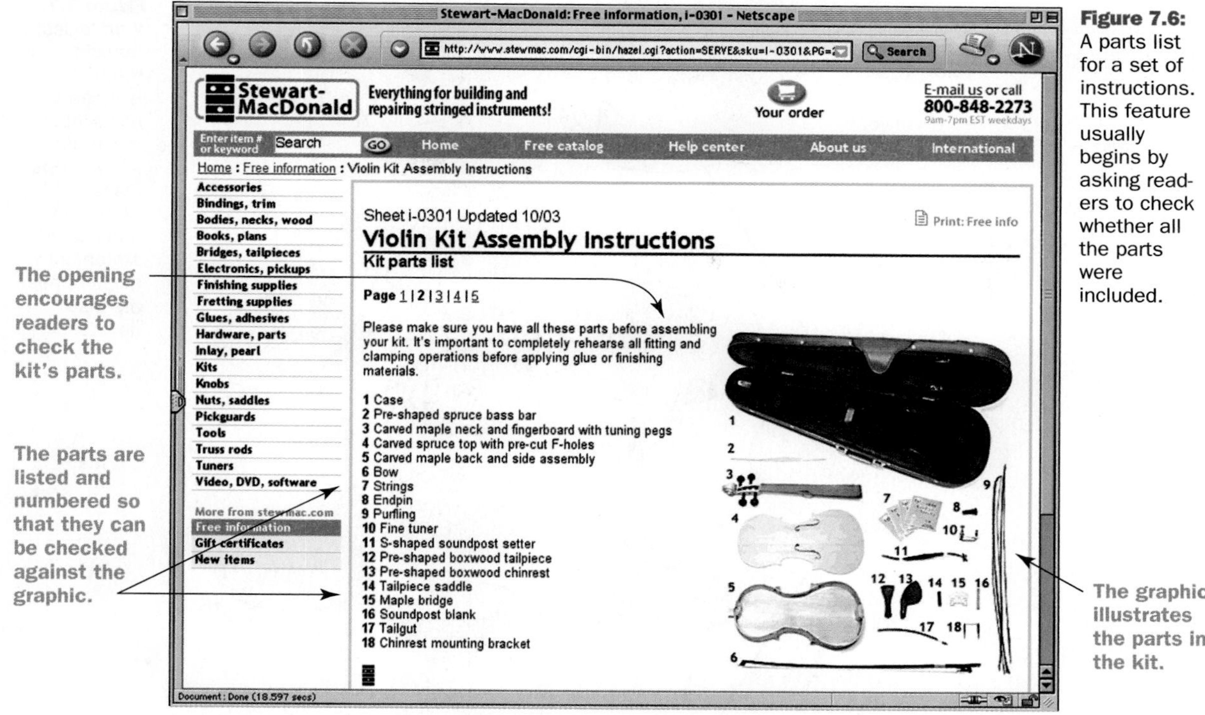

Figure 7.6: A parts list for a set of instructions. This feature usually begins by asking readers to check whether all the parts were included.

Source: Stewart-MacDonald, http://www.stewmac.com.

SPECIFY SPECIAL CONDITIONS If any special conditions involving temperature, humidity, or light are required, mention them up front.

> Paint is best applied when temperatures are between 50°F and 90°F.
>
> If the humidity is above 75 percent, do not solder the microchips onto the printed circuit board. High humidity may lead to a defective joint.

Sequentially Ordered Steps

The steps are the centerpiece of any form of documentation, and they will usually make up the bulk of the text. These steps need to be presented logically and concisely, allowing readers to easily understand them and complete the task.

As you carve the task you are describing into steps, you might use logical mapping to sort out the major and minor steps (Figure 7.7). First, put the overall task you are describing on the left-hand side of the screen or a sheet of paper. Then, break the task down into its major and minor steps. You might also, as shown in Figure 7.7, make note of any necessary hazard statements or additional comments that might be included.

Using Logical Mapping to Identify Steps in a Task

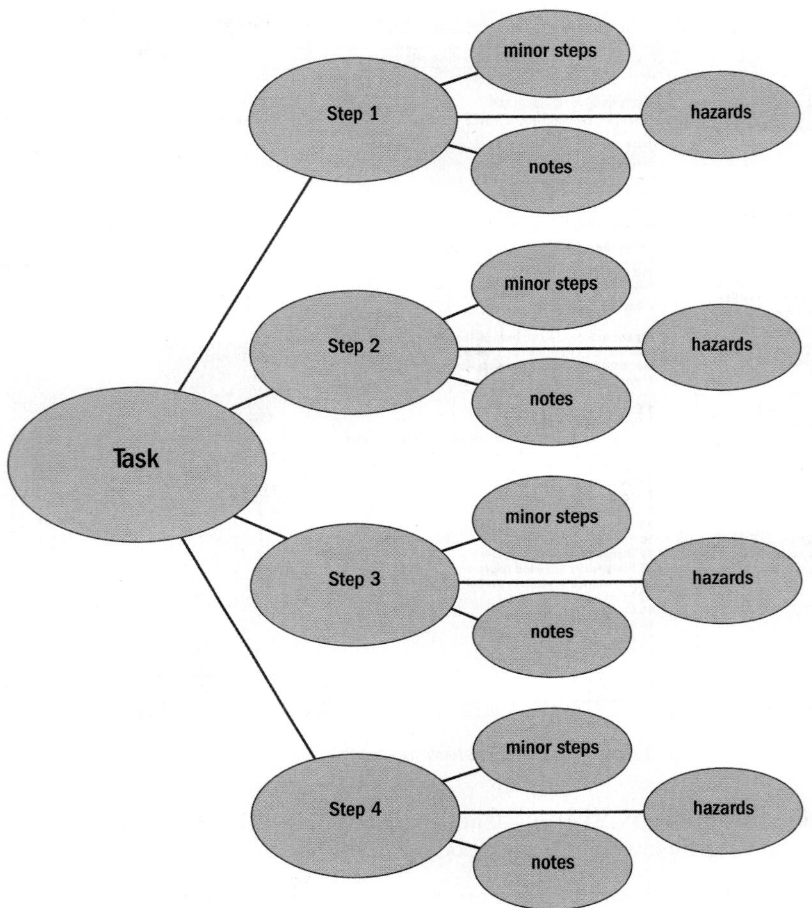

Figure 7.7: With logical mapping, the task is broken down into major and minor steps. Places where notes and hazard statements might appear are also noted.

Once you have organized the task into major and minor steps, you are ready to draft your instructions.

USE COMMAND VOICE Steps should be written in *command voice,* or imperative mood. To use command voice, start each step with an action verb.

1. Place the telescope in an upright position on a flat surface.
2. Plug the coil cord for the Electronic Controller into the HBX port (see Figure 5).

In most steps, the verb should come first in the sentence. This puts the action up front, while keeping the pattern of the steps consistent. The "you" in these sentences is not stated, but rather implied ("*You* place the telescope in an upright position").

Instructions with Sequentially Ordered Steps

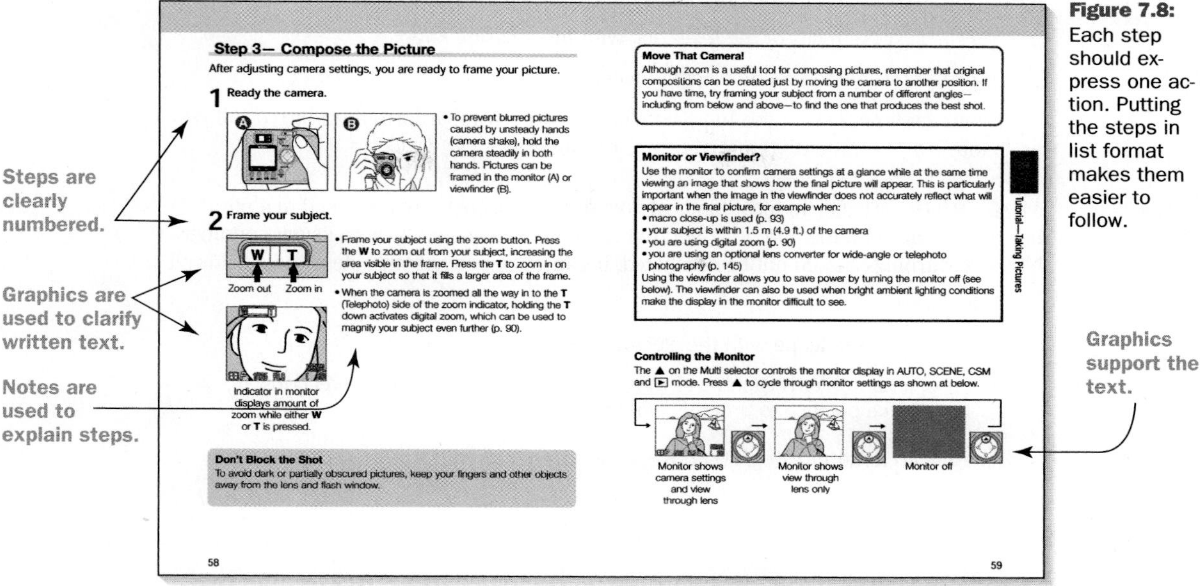
Step 3— Compose the Picture

After adjusting camera settings, you are ready to frame your picture.

1 Ready the camera.

- To prevent blurred pictures caused by unsteady hands (camera shake), hold the camera steadily in both hands. Pictures can be framed in the monitor (A) or viewfinder (B).

2 Frame your subject.

- Frame your subject using the zoom button. Press the **W** to zoom out from your subject, increasing the area visible in the frame. Press the **T** to zoom in on your subject so that it fills a larger area of the frame.
- When the camera is zoomed all the way in to the **T** (Telephoto) side of the zoom indicator, holding the **T** down activates digital zoom, which can be used to magnify your subject even further (p. 90).

Zoom out Zoom in

Indicator in monitor displays amount of zoom while either **W** or **T** is pressed.

Don't Block the Shot

To avoid dark or partially obscured pictures, keep your fingers and other objects away from the lens and flash window.

58

Move That Camera!

Although zoom is a useful tool for composing pictures, remember that original compositions can be created just by moving the camera to another position. If you have time, try framing your subject from a number of different angles—including from below and above—to find the one that produces the best shot.

Monitor or Viewfinder?

Use the monitor to confirm camera settings at a glance while at the same time viewing an image that shows how the final picture will appear. This is particularly important when the image in the viewfinder does not accurately reflect what will appear in the final picture, for example when:

- macro close-up is used (p. 93)
- your subject is within 1.5 m (4.9 ft.) of the camera
- you are using digital zoom (p. 90)
- you are using an optional lens converter for wide-angle or telephoto photography (p. 145)

Using the viewfinder allows you to save power by turning the monitor off (see below). The viewfinder can also be used when bright ambient lighting conditions make the display in the monitor difficult to see.

Controlling the Monitor

The ▲ on the Multi selector controls the monitor display in AUTO, SCENE, CSM and ▶ mode. Press ▲ to cycle through monitor settings as shown at below.

Monitor shows camera settings and view through lens

Monitor shows view through lens only

Monitor off

Tutorial—Taking Pictures

59

Figure 7.8: Each step should express one action. Putting the steps in list format makes them easier to follow.

Source: Nikon, Cool Pix 885 guide.

STATE ONE ACTION PER STEP Each step should express only one action (Figure 7.8). You might be tempted to state two smaller actions in one step, but your readers will appreciate following each step separately.

Ineffective

2. Place the telescope securely on its side as shown in Figure 4 and open the battery compartment by simultaneously depressing the two release latches.

Revised

2. Place the telescope securely on its side as shown in Figure 4.
3. Open the battery compartment by simultaneously depressing the two release latches.

However, when two actions must be completed at the same time, you should put them in the same sentence.

6. Insert a low-power eyepiece (e.g., 26mm) into the eyepiece holder and tighten the eyepiece thumbscrew.

You should state two actions in one step only when the two actions are dependent on each other. In other words, completion of one action should require the other action to be handled at the same time.

KEEP THE STEPS CONCISE Use concise phrasing to describe each step. Short sentences allow readers to remember each step while they work.

7. Adjust the focus of the telescope with the focusing knob.
8. Center the observed object in the lens.

If your sentences seem too long, consider moving some information into a follow-up "Note" or "Comment" that elaborates on the step.

NUMBER THE STEPS In most kinds of documentation, steps are presented in a numbered list. Start with the number 1 and mark each step sequentially with its own number. Notes or warnings should not be numbered, because they do not state steps to be followed.

Incorrect

9. Aim the telescope with the electronic controller.
10. Your controller is capable of moving the telescope in several directions. It will take practice to properly aim the telescope.

There is no action in step 10 above, so a number should not be used.

Correct

9. Aim the telescope with the electronic controller.

Your controller is capable of moving the telescope in several directions. It will take practice to properly aim the telescope.

An important exception to this "number only steps" guideline involves the numbering of procedures and specifications. Procedures and specifications often use an itemized numbering system in which lists of cautions or notes are "nested" within lists of steps.

10.2.1 Putting on Clean Room Gloves, Hood, and Coveralls

10.2.1.1 Put on coveralls.

10.2.1.2 Put on a pair of clean room gloves so they are fully extended over the arm and the coverall sleeves. Glove liners are optional.

10.2.1.3 Put a face/beard mask on, completely covering the mouth and nose.

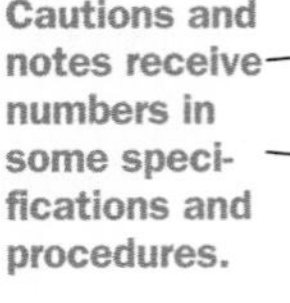

Cautions and notes receive numbers in some specifications and procedures.

10.2.1.3.1 *Caution: No exposed hair is allowed in the fab.*

10.2.1.3.2 *Caution: Keep your nose covered at all times while in the fab.*

10.2.1.3.3 *Note: Do not wear the beard cover as a facemask. A beard cover should be used with a facemask to cover facial hair.*

In specifications, comments and hazard statements are numbered because the numbers help people refer to specific lines in the document.

In some cases, you may also want to use *paragraph style* to describe the steps (Figure 7.9). In these situations, you can use headings or sequential transitions to highlight the steps. Numerical transitions ("first," "second," "third," "finally") are best in most cases. In shorter sets of steps, you might use transitions like "then," "next," "5 minutes later," and "finally" to mark the actions. In Figure 7.9, headings are used to mark transitions among the major steps.

To see good and bad examples of instructions, go to **www.pearsonhighered.com/johnsonweb4/7.8**

Paragraph Style Instructions

Introduction to instructions

Paragraphs slow the readers down.

Headings highlight steps.

List style is used when appropriate.

Numbers refer readers to the graphics.

Operating the saw.

Starting the saw is easy if you follow the instructions. But make sure you read the user's manual for your saw first, so you know how it works and are familiar with all its parts and controls.

Checking the chain
If the chain isn't new, it's probably a good idea to file it, since cutting is both easier and safer when the chain is sharp. Also make sure the chain is tensioned properly (1). Don't forget that a new chain should always be re-tensioned after operating the saw for a short period (2).

Fuel
When filling the saw with fuel and chain oil, place the saw on a stable surface. To reduce dangerous emissions, choose environmental petrol and vegetable-based chain oil. The overfill protection helps you avoid unnecessary spillage (3). And considering the risk of fires, you should always move the saw before starting it.

Safe distance
It's good to work together with someone, but make sure they are at least five metres away when you start to use the saw. Of course, when felling trees, this distance should be increased considerably.

Start
When you're ready to start, place the saw flat on the ground and clear the area around the bar.

1. Activate the chain brake by pushing the kickback protection forward, as otherwise the chain will start to rotate when the saw starts.
2. Depress the SmartStart decompression control, if the saw has this feature.
3. If the engine is cold, pull the choke out fully.
4. Put your right foot partway through the rear handle and hold the front handle firmly with your left hand. Pull the starter handle with your right hand until the engine starts (4).
5. Now push the choke in again, with the throttle on half way. Continue to pull the starter handle until the saw starts. Hit the throttle once so the engine speed drops to idle. If the engine is already warm, don't use the choke, but the other steps are the same.

If the saw is difficult to start despite being warm, pull out the choke like you do during cold starts, but push it back in right away. When you've got the saw started, don't disengage the chain brake until you're ready to saw.

Checking the chain brake
Now check that the chain brake works. Place the saw on a stable surface and squeeze the throttle. Activate the chain brake by pushing your left wrist against the kickback protection, without releasing the handle. The chain should stop straight away. (5)

Does chain lubrication work?
Also check the chain lubrication. Hold the saw above a light surface, such as a stump, and hit the throttle. A line of oil should be visible on the surface. (6)

Sawing practice
If you're not used to using a chainsaw, we recommend you first get acquainted with the saw by practising a while on a suitable log. (7)

How to operate the saw
There are some basic rules for using a chainsaw. Hold it firmly by both handles and hold your thumbs and fingers right around the handles. Make sure you hold your left thumb under the front handle, to reduce the force of a possible kickback.

Good balance
It's good to have respect for the saw, but don't be afraid of it. If you hold it close to your body it won't feel as heavy. Also, you'll be more balanced and in better control of the saw. For the best balance, stand with your feet apart. (7)

Pulling and pushing chain
You can saw with both the upper and the lower edge of the bar. When using the lower edge, you're sawing with a pulling chain, which means that the chain pulls the saw away from you. Using the upper edge of the chain, you're sawing with a pushing chain, so the chain pushes the saw towards you.

Bend your knees
Save your back by not working with a bent back. Instead, bend your knees if you're working at a low level.

Moving around
When moving around the worksite, make sure the chain is not rotating by activating the chain brake or turning off the engine. For longer distances, use the bar guard. (8)

Figure 7.9: Paragraph style often takes up less space. It is a little harder to read at a glance, but this style can often be more personal. In these instructions the pictures show helpful details to clarify the steps.

Source: Husqvarna.

(continued)

Figure 7.9: *(continued)*

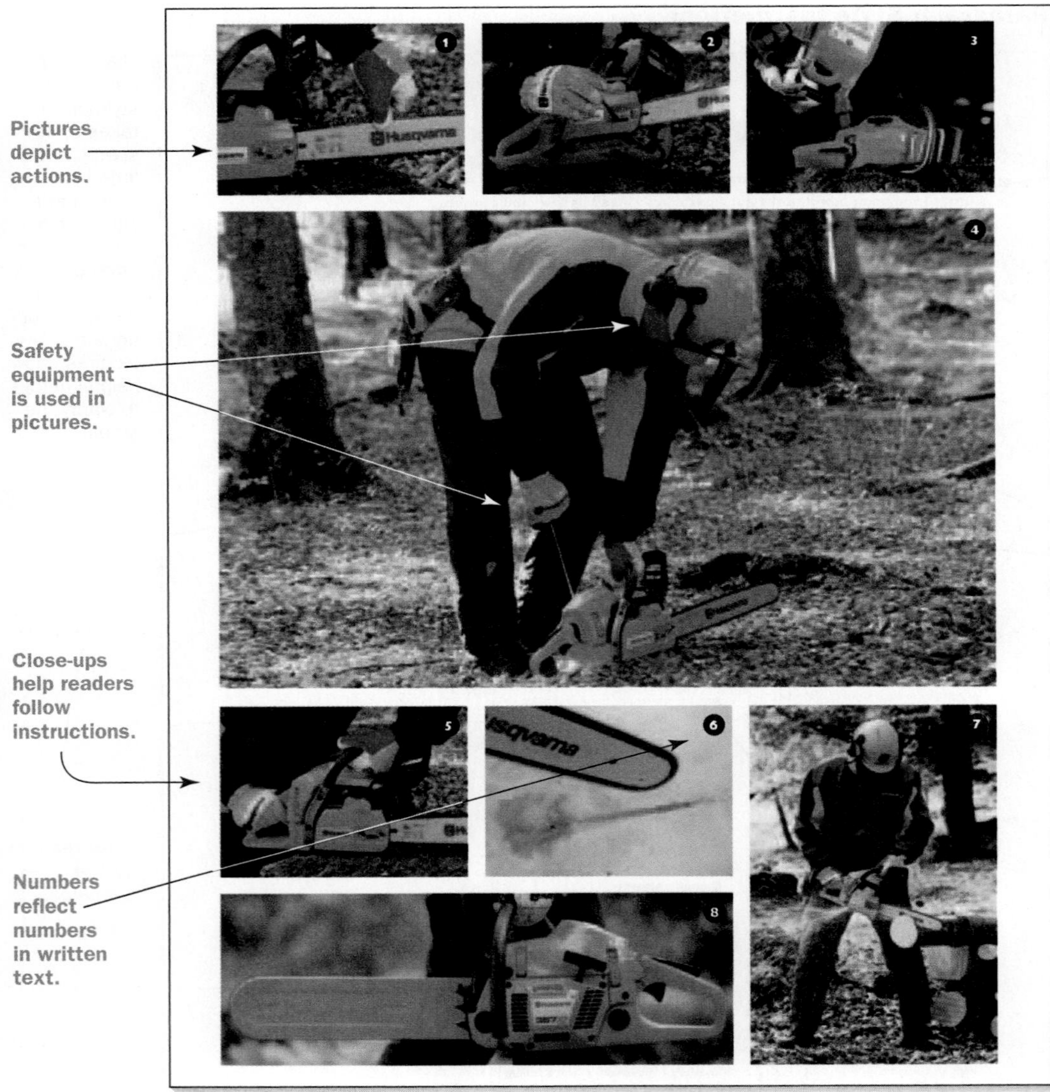

Pictures depict actions.
Safety equipment is used in pictures.
Close-ups help readers follow instructions.
Numbers reflect numbers in written text.
1
2
3
4
5
6
7
8

Readers often find paragraph style harder to follow because they cannot easily find their place again in the list of instructions. In some cases, though, paragraph style takes up less space and sounds more friendly and conversational.

AT A GLANCE

Writing Effective Steps

- Use command voice.
- State one action per step.
- Keep the steps concise.
- Number the steps.
- Add comments, notes, or examples.
- Provide feedback.
- Refer to the graphics.

ADD COMMENTS, NOTES, OR EXAMPLES After each step, you can include additional comments or examples that will help readers complete the action. Comments after steps might include additional advice or definitions for less experienced readers. Or, comments might provide troubleshooting advice in case the step did not work out.

3. Locate a place to set your telescope.

Finding a suitable place to set up your telescope can be tricky. A paved area is optimal to keep the telescope steady. If a paved area is not available, find a level place where you can firmly set your telescope's tripod in the soil.

Comments and examples are often written in the "you" style to maintain a positive tone.

PROVIDE FEEDBACK After a difficult step or group of steps, you might offer a paragraph of feedback to help readers assess their progress.

> When you finish these steps, the barrel of your telescope should be pointed straight up. The tripod should be stable so that it does not teeter when touched. The legs of the tripod should be planted firmly on the ground.

REFER TO THE GRAPHICS In the steps, refer readers to any accompanying graphics. A simple statement like, "See Figure 4" or "(Figure 4)" will notify readers that a graphic is available that illustrates the step. After reading the step, they can look at the graphic for help in completing the step properly.

If graphics are not labeled, they should appear immediately next to the step or below it so readers know which visual goes with each step.

Safety Information

Safety information should be placed early in the documentation and in places where the reader will be completing difficult or dangerous steps. A common convention in technical writing is to use a three-level rating for safety information and warnings: *Danger, Warning,* and *Caution.*

DANGER Signals that readers may be at risk for serious injury or even death. This level of warning is the highest, and it should be used only when the situation involves real danger to the readers.

> *Danger*: Do not remove grass from beneath your riding lawn mower while the engine is running (even if the blade is stopped). The blade can cause severe injury. To clear out grass, turn off the lawn mower and disconnect the spark plug before working near the blade.

Want to learn more about using safety symbols? Go to **www.pearsonhighered.com/johnsonweb4/7.9**

WARNING Signals that the reader may be injured if the step is done improperly. To help readers avoid injury, warnings are used frequently.

> *Warning*: When heated, your soldering iron will cause burns if it touches your skin. To avoid injury, always return the soldering iron to its holder between uses.

CAUTION Alerts readers that mistakes may cause damage to the product or equipment. Cautions should be used to raise readers' awareness of difficult steps.

> *Caution*: The new oil filter should be tightened by hand only. Do not use an oil filter wrench for tightening, because it will cause the filter to seal improperly. If the filter is too tight, oil will leak through the filter's rubber gasket, potentially leading to major damage to your car's engine.

AT A GLANCE

Labeling Hazards

- Danger—Risk of serious injury or death is possible.
- Warning—Injury likely if step is handled improperly.
- Caution—Damage to the product or equipment is possible.

Safety information should tell your readers the following three things: (1) the hazard, (2) the seriousness of the hazard, and (3) how to avoid injury or damage. As shown in Figure 7.10, safety information should appear in two places:

- If a hazard is present throughout the procedure, readers should be warned before they begin following the steps. In these cases, danger and warning statements should appear between the introduction and the steps.
- If a hazard relates to a specific step, a statement should appear prominently before that step. It is important for readers to see the hazard statement before the step so that they can avoid damage or injury.

You can use symbols to highlight safety information. Icons are available to reinforce and highlight special hazards such as radioactive materials, electricity, or chemicals. Figure 7.11 shows a few examples of icons commonly used in safety information.

In our litigious culture, the importance of safety information should not be underestimated. Danger, warning, and caution notices will not completely protect your company from lawsuits, but they will give your company some defense against legal action.

Safety Symbols

Source: From Geoffrey Peckham, "Safety Symbols," Compliance Engineering Magazine, *http://www.ce-mag.com/archive/02/03/peckham.html (Figure 1). Used with permission, Hazard Communication Systems, LLC. safetylabel.com.*

Figure 7.11: Here are a few examples of ISO and IEC symbols used on safety signs (hot surface, laser, radiation).

Placement of Hazard Statements

Warning statements are prominently displayed.

Symbols draw attention to warnings.

Boxes are used to capture the readers' attention.

Figure demonstrates proper use of the machine, including use of appropriate safety devices like glasses, earmuffs, and gloves.

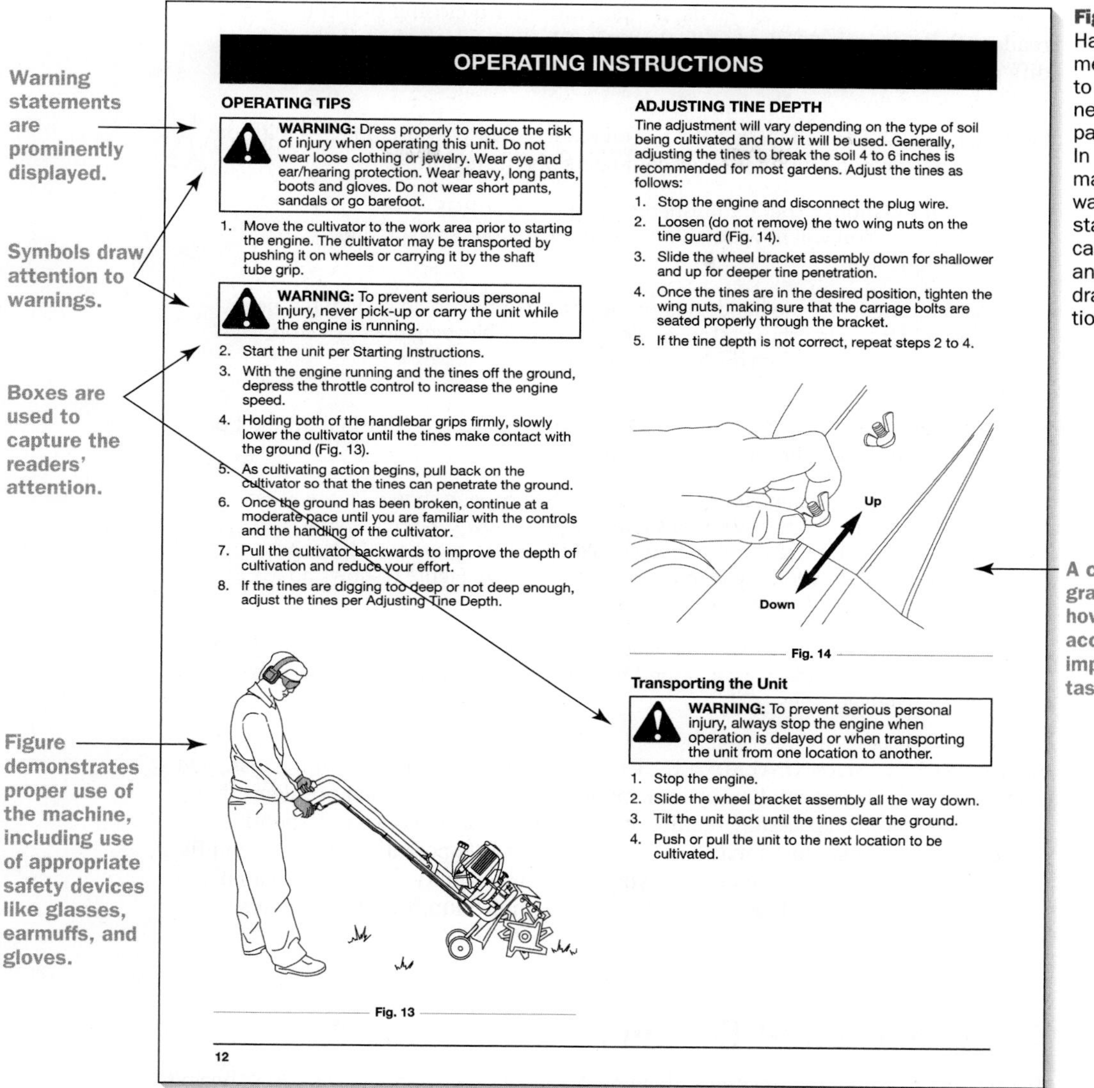
OPERATING INSTRUCTIONS

OPERATING TIPS

WARNING: Dress properly to reduce the risk of injury when operating this unit. Do not wear loose clothing or jewelry. Wear eye and ear/hearing protection. Wear heavy, long pants, boots and gloves. Do not wear short pants, sandals or go barefoot.

1. Move the cultivator to the work area prior to starting the engine. The cultivator may be transported by pushing it on wheels or carrying it by the shaft tube grip.

WARNING: To prevent serious personal injury, never pick-up or carry the unit while the engine is running.

2. Start the unit per Starting Instructions.
3. With the engine running and the tines off the ground, depress the throttle control to increase the engine speed.
4. Holding both of the handlebar grips firmly, slowly lower the cultivator until the tines make contact with the ground (Fig. 13).
5. As cultivating action begins, pull back on the cultivator so that the tines can penetrate the ground.
6. Once the ground has been broken, continue at a moderate pace until you are familiar with the controls and the handling of the cultivator.
7. Pull the cultivator backwards to improve the depth of cultivation and reduce your effort.
8. If the tines are digging too deep or not deep enough, adjust the tines per Adjusting Tine Depth.

Fig. 13

ADJUSTING TINE DEPTH

Tine adjustment will vary depending on the type of soil being cultivated and how it will be used. Generally, adjusting the tines to break the soil 4 to 6 inches is recommended for most gardens. Adjust the tines as follows:

1. Stop the engine and disconnect the plug wire.
2. Loosen (do not remove) the two wing nuts on the tine guard (Fig. 14).
3. Slide the wheel bracket assembly down for shallower and up for deeper tine penetration.
4. Once the tines are in the desired position, tighten the wing nuts, making sure that the carriage bolts are seated properly through the bracket.
5. If the tine depth is not correct, repeat steps 2 to 4.

Fig. 14

Transporting the Unit

WARNING: To prevent serious personal injury, always stop the engine when operation is delayed or when transporting the unit from one location to another.

1. Stop the engine.
2. Slide the wheel bracket assembly all the way down.
3. Tilt the unit back until the tines clear the ground.
4. Push or pull the unit to the next location to be cultivated.

12

A close-up graphic shows how to accomplish important tasks.

Figure 7.10: Hazard statements need to be prominent in the page design. In this user's manual, the warnings stand out because boxes and symbols draw attention to them.

Source: Ryobi, 2000.

Conclusion That Signals Completion of Task

When you have listed all the steps, you should offer a closing that tells readers that they are finished with the task. Closings can be handled a few different ways.

SIGNAL COMPLETION OF THE TASK Tell readers that they are finished with the steps. Perhaps you might offer a few comments about the future.

> Congratulations! You are finished setting up your RGS-90x telescope. You will now be able to spend many nights exploring the night skies.

> When completed, the bandaging of the head wound should be firm but not too tight. Bleeding should stop within a minute. If bleeding does not stop, call an emergency room doctor immediately.

DESCRIBE THE FINISHED PRODUCT You might describe the finished product or provide a graphic that shows how it should look.

> When you have completed setting up your telescope, it should be firmly set on the ground and the eyepiece should be just below the level of your eyes. With the Electronic Controller, you should be able to move the telescope horizontally and vertically with the push of a button. Figure 5 shows how a properly set up telescope should look.

> Your bandaging of the patient's head should look like Figure B. The bandaging should be neatly wound around the patient's head with a slight overlap in the bandage strips.

OFFER TROUBLESHOOTING ADVICE Depending on the complexity of the task, you might end your documentation by anticipating some of the common problems that could occur. Simple tasks may require only a sentence or two of troubleshooting advice. More complex tasks may require a table that lists potential problems and their remedies (Figure 7.12). Depending on your company's ability to provide customer service, you may also include a web address or a phone number where readers can obtain additional help.

Using Style and Design

People often assume that technical documentation should sound dry and look boring. But documentation can be—and sometimes should be—written in a more interesting style and can use an eye-catching design. There are ways, especially with consumer products, that you can use style and design to reflect readers' attitudes as they follow the steps.

To download safety symbols, go to **www.pearsonhighered.com/johnsonweb4/7.10**

Troubleshooting Guide

TROUBLESHOOTING

Your XM Reference Tuner is designed and built to provide trouble-free performance without the need for service. If it does not appear to be functioning correctly, please follow these troubleshooting tips.

1. Make sure all connectors are properly attached to the rear panel of the XM Reference Tuner and to your home audio system.
2. Make sure that the unit is plugged into a standard 120V home current. You may also plug the Power Cord of the XM Reference Tuner into the switched AC Outlets of your pre-amp/receiver, if available.
3. Make sure you have activated your XM Satellite Radio subscription. You must activate your XM Satellite Radio service in order to use this product. To activate your XM Satellite Radio service, make sure you have a major credit card and your XM Satellite Radio ID Number handy and contact XM at http://activate.xmradio.com. You can also activate your service by calling 1-800-852-9696.

If you see this on the display,	**You should:**
NO SIGNAL Cause: The XM signal is out of range.	Reposition your High-Gain Antenna.
ANTENNA	Make sure your antenna is securely attached to the rear of the unit, that the antenna cable is unkinked and undamaged. Turn the XM Reference Tuner off and then back on to reset this message.
OFF AIR Cause: The channel selected is not currently broadcasting.	Tune in to another channel.

If this happens,	**You should:**
You can only receive XM channels 0 and 1. Cause: Your XM Reference Tuner is not activated.	Contact XM as described in the ACTIVATING YOUR XM SYSTEM section.
The audio sounds distorted.	Reduce the audio output level as described in the **Menu Commands and Settings Line Level Out** section.
The audio level is too low.	Increase the audio output level as described in the **Menu Commands and Settings Line Level Out** section.
You cannot tune in to a channel. (See "Channel Skip/Add.")	1. Make sure the channel has not been "skipped." 2. Verify that you are authorized to receive that channel as part of your subscription with XM and that you have not asked XM to block that channel on your XM Reference Tuner.
No song title or artist name displays.	Nothing is wrong with your receiver. This is normal on many talk and news channels where there is no song playing. Also, this information may not yet be in the database for some music channels.

Figure 7.12: Troubleshooting guides are often provided in a table format with problems on the left and solutions on the right. Note the positive, constructive tone in this table.

Source: Polk Audio, Satellite Radio XRt 12 Tuner owner's guide.

GO TO THE NET

To see examples of troubleshooting guides, go to **www.pearsonhighered.com/johnsonweb4/7.11**

HELP

On-Screen Documentation

One of the major changes brought about by computers is the availability of on-screen documentation. Today, user's manuals and instructions for products are often available through a website, provided on a CD-ROM, or included with the Help feature in a software package.

The advantages of online documentation are numerous, and the disadvantages are few. The greatest advantage is reduced cost. After all, with some products like software, the accompanying user's manual costs more to print than the software itself. By putting the manual on a website, CD-ROM, or online Help, a company can save thousands of dollars almost immediately. Also, online documentation can be updated regularly to reflect changes in the product or revisions to the documentation.

Companies are also putting specifications and procedures online, usually on the company's intranet. That way, all employees can use their computers to easily call up the documents they need.

Several options are available for on-screen documentation:

Online Help—Increasingly, the online Help features that come with software packages are being used to present instructions (Figure A). Online Help features

Online Help

Users can use the search function to find help on specific topics.

The instructions are listed here.

A screen shot shows users how the screen will look.

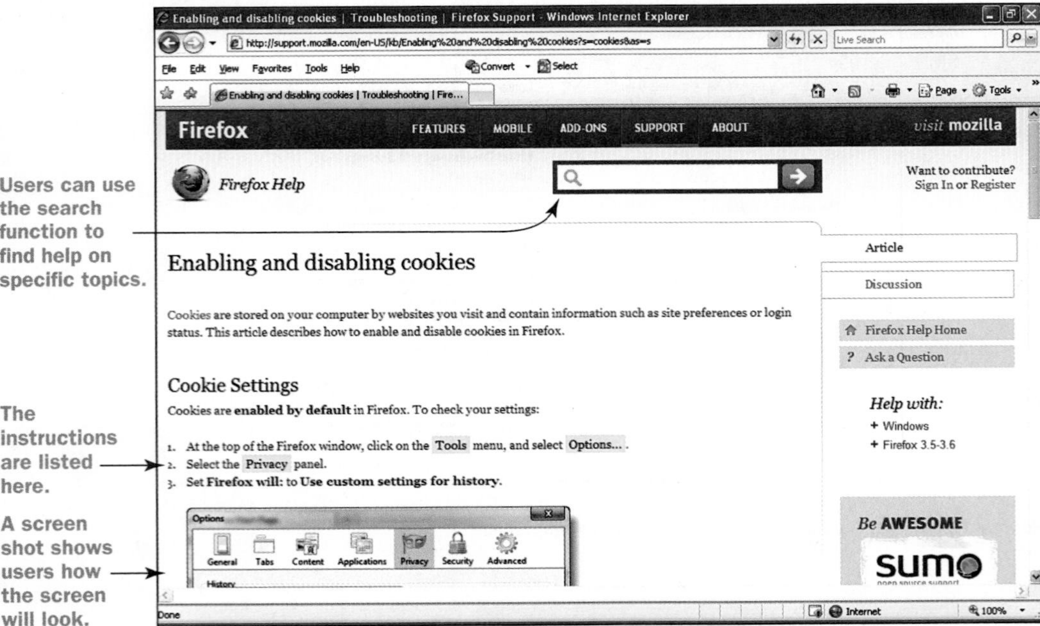

Source: Mozilla Firefox, 2011.

Figure A: Online Help features are another place where instructions are commonly found, especially for software programs. These instructions are from the online Help feature for Mozilla Firefox.

GO TO THE NET

To learn more about on-screen documentation, go to **www.pearsonhighered.com/johnsonweb4/7.12**

allow readers to access the instructions more quickly, because they do not need to hunt around for the user's manual. As more documents move online, it is likely that instructions will increasingly be offered as online Help. To write instructions as online Help, you will need Help-authoring software like RoboHelp and Doc-to-Help, which simplifies the writing of Help features.

CD-ROM—Increasingly, computers have a CD burner as a standard feature. To make a CD, you can use web development software such as Adobe Dreamweaver or Microsoft Expression Web to create the files. Then, burn the files to a CD the same way you would save to a disk. Browser programs like Firefox, Safari, or Explorer will be able to read these multimedia documents.

Website—You can put your documentation on a website for use through the Internet or a company intranet. These files can be created with web development software such as Dreamweaver or Expression Web.

Portable document format (PDF)—Software programs such as Adobe Acrobat can turn your word-processing files into PDFs, which retain the formatting and color of the original document. PDFs can be read by almost any computer, and they store information efficiently. They can also be password protected so that readers cannot tamper with the text. PDFs can be placed on a website for easy downloading.

Plain Style with a Touch of Emotion

Documentation tends to be written in the plain style. Here are some suggestions for improving the style of your technical description:

Use simple words and limit the amount of jargon.

Define any words that might not be familiar to your readers.

Keep sentences short, within breathing length.

Use the command or imperative style (verb first) for any instructional steps.

Keep it simple and don't over-explain basic steps or concepts.

How can you further improve the style of your documentation? First, look at your original analysis of your readers and the contexts in which your document will be used. Pay attention to your readers' needs, values, and attitudes. Then, try to identify the emotions and attitudes that shape how they will be reading and using the instructions. Will they be enthusiastic, frustrated, happy, apprehensive, or excited?

Identify a word that best reflects readers' feelings as they are using your documentation. Then, use logical mapping to come up with some words that are associated with that word (Figure 7.13).

Link

For more ideas about improving style, go to Chapter 17, page 455.

Want to know more about Help-authoring software? Go to **www.pearsonhighered.com/johnsonweb4/7.13**

Mapping a Tone for Instructions

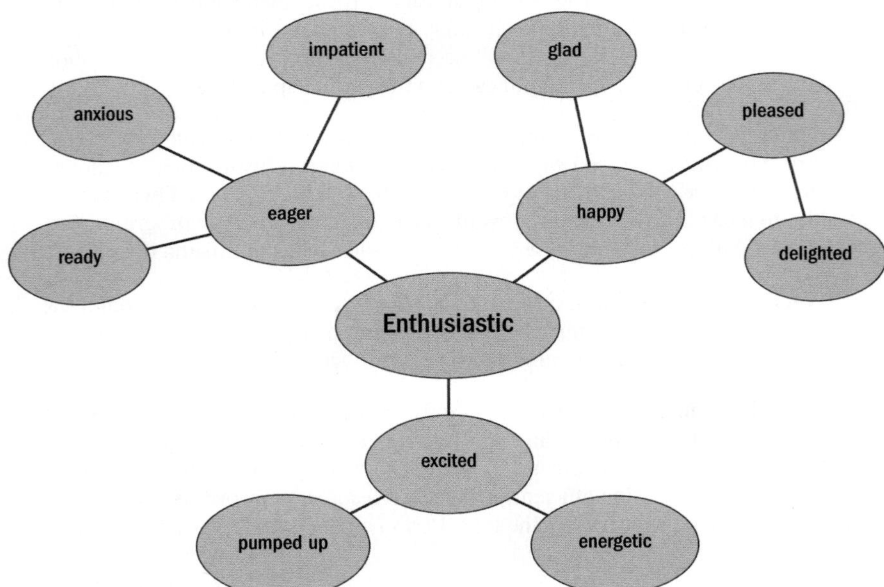

Figure 7.13: You can create an enthusiastic tone by finding synonyms associated with the word "enthusiastic."

After you have found words that are associated with the appropriate tone, use them in the introduction, notes, and conclusion. These words, when used strategically, will reflect your readers' attitudes or emotions.

If your readers have a negative attitude (perhaps they are annoyed that they need to read instructions), you can use antonyms to counteract their feelings. For example, to soothe annoyed readers, use words like *satisfy, pleasure, please, delight,* and *fulfill* to counteract their negativity. Don't overuse them, though, because angry readers may detect your attempt to soothe them and feel like you are being patronizing.

Functional, Attractive Page Layout

The page layout of your documentation should be both functional and attractive, like the example in Figure 7.14. Here are some techniques that you might try:

Incorporate graphics that illustrate and reinforce the written text.

Try using a two- or three-column format that leaves room for graphics.

Use boxes, borders, and lines to highlight important information, especially safety information.

Use headings that clearly show the levels of information in the text.

GO TO THE NET

Worksheets to help you revise your instructions are available at
www.pearsonhighered.com/johnsonweb4/7.14

If you are not sure how to design your instructions, study sample texts from your home or workplace. You can use these examples as models for designing your own documentation.

Link

For more help on document design, see Chapter 18, page 482.

Graphics That Reinforce Written Text

With computers, a variety of methods are available for you to add graphics to your documentation. You can add illustrations and diagrams (as in Figure 7.14). Or, you can use a digital camera or scanner to add graphics to your text. Here are some tips:

Number and title your graphics, so readers can locate them.

Refer to the graphics by number in the written steps.

Put each graphic next to or below the step that refers to it.

Place graphics on the same page as the step that refers to them.

Check whether the graphics you have chosen could be misunderstood by or offensive to cross-cultural readers.

Keep in mind that the graphics should reinforce, not replace, the written text.

Link

For more information on using graphics, see Chapter 19, page 523.

Design Is Important in Documentation

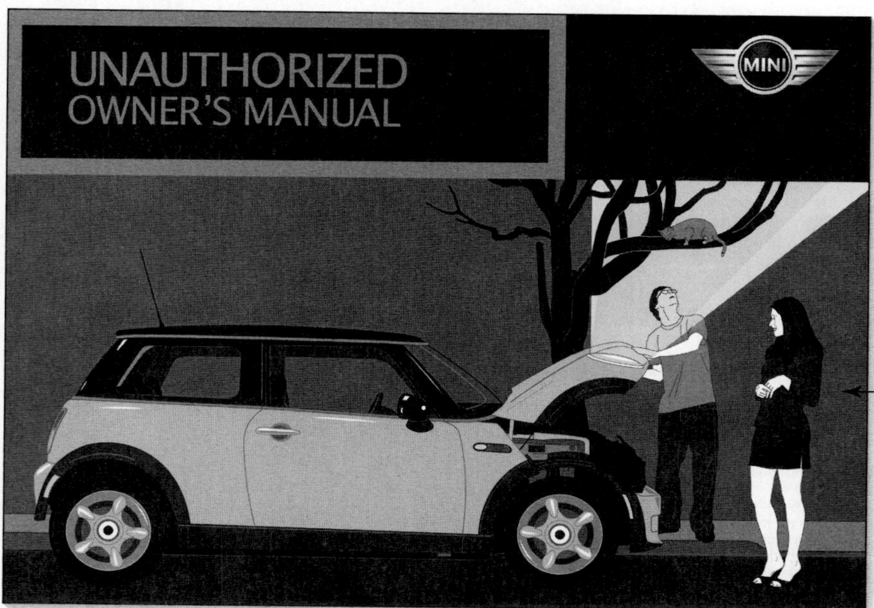

Figure 7.14: In this "unauthorized" user's guide for the Cooper Mini, the instructions use graphics to illustrate the text while making the information attractive.

Images of people help users identify with the product.

(continued)

Figure 7.14: *(continued)*

Headings make the text easy to scan.

BEST PLACES TO STASH STUFF IN YOUR MINI

First things first. This book fits snugly inside the case provided for your factory-authorized owner's manual, and is designed to be stowed away in the glove box. Or conveniently left about on the coffee table.

1. Coin Caddy
Technically engineered as the beverage holder. Reinvented by resourceful you as the handy loose change receptacle.
⚠CAUTION: When parking in low-lit urban areas, cover all exposed coinage with cup.

2. Toll Ticket Cranny
Your MINI's dashboard console features four vertical slits (two within driver's reach, two for co-pilot assistance) for securing toll receipts and parking stubs.

3. The Glove Box
Re-engineered by MINI designers. Because no one really keeps gloves in there anymore. Use it as a CD box. A toolbox. Or turn up the air conditioning, and it's a refrigerated icebox.

Fig. 5

4. Cubby Space
Two center console bins for maps, cell phone, spare change, loose French fries. One per door for sunglasses (optimists). Collapsible umbrellas (pessimists).

⚠ Cooling/Heating Feature
Turns your MINI's glove box into a mini refrigerator for chilling contents to 50° Fahrenheit. For emergency roadside candy bars and spare hero sandwiches. Your MINI goes great with mayo. Or activate the heater to warm contents from soup to nuts. (See Fig. 6.)

Hazard statements are highlighted with an icon.

Numbers and pointers highlight places discussed in text.

BEST PLACES TO STASH STUFF IN YOUR MINI

BASICS
STASH
CUSTOMIZE
ROMANCE
MUSIC
GOOD MOJO
TROUBLE

Fig. 6 Cooling/Heating Glove Box

Strategic Golf Club Placement
With both rear seats folded down, your MINI can accommodate four golf bags comfortably. Unfortunately, this leaves no room to accommodate a foursome. Recommended: with one rear seat folded down, fit one to two bags lying angled diagonally, and one to two golf buddies seated vertically.

How to Fit a Bike:

1. Adjust passenger's side seat to furthest forward non-reclining position.
2. Remove front wheel. OF BICYCLE!
3. Lay the bike on the side opposite the rear derailleur rotating the handlebars counterclockwise until they are parallel with the cargo area floor.
4. Place front wheel in cargo area and close hatch.

Instructions are easy to find in this numbered list.

Fig. 7Use floor space as parking citation crumple zone.

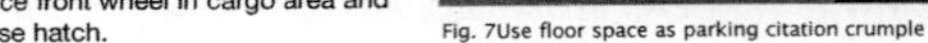

For tips on laying out pages for documentation, go to **www.pearsonhighered.com/johnsonweb4/7.15**

Figure 7.14: *(continued)*

Secret Jogging Key Nook
The undercarriage of your MINI features a handy hideaway for stashing cumbersome key sets when you take a break from motoring for a jog, bike ride, or skinny-dipping romp in the country. Place your hand just forward of either rear wheel near the rocker panel. Reach up and under the wheel well. You'll find a flat4"X4" secret spot no one knows about but you. And tens of thousands of fellow U.S.MINI owners.

⚠NOTE: Always perform a quick 360° visual scan to make sure no one is watching you. (See Fig. 8 & 9.)

Additional Storage Information
Pizza Capacity: The passenger's side floor accommodates four large pies stacked.

A little humor can be refreshing.

Fig.8 & 9 The jogging key nook.

Close-ups show users more detail.

User-Testing Your Documentation

As you work on the style and design of your documentation, you should do some user-testing to see how real readers will react to it. Chapter 20 discusses four kinds of tests that would be helpful while user-testing your documentation: (1) read-and-locate tests, (2) understandability tests, (3) performance tests, and (4) safety tests. These tests would strengthen the usability of your documentation, while helping you clarify the style and make the design more effective and appealing.

MICROGENRE

Emergency Instructions

Most of the time, emergency instructions go almost unnoticed. They are posted on a wall or put in the pocket of an airplane seat. But when an emergency happens, these instructions may be the difference between life and death. People need to take action, and they have little time to do so.

Emergency instructions need to be highly visual and brief. Here are some strategies for writing them:

Put the title in large lettering. Big lettering (at least 26-point font size) will help people locate your emergency instructions in a crisis.

For sample sets of effective and ineffective instructions, go to **www.pearsonhighered.com/johnsonweb4/7.16**

Title is in a large font size, making instructions easy to find.

This set of emergency instructions is designed to help train passengers handle a variety of potential crises.

ON-BOARD TRAIN EMERGENCY INSTRUCTIONS

ALWAYS

Use the Passenger Emergency Intercom to contact a Train Crew Member

Listen for Announcements

FIRE

Move to an unaffected car

Remain inside – tracks are electrified

Follow instructions of emergency workers

Train crews can access fire extinguishers

Do not activate emergency cord

Instructions are brief and in command voice.

MEDICAL

If a passenger is in distress, notify a crew member immediately

If you are medically qualified and able to assist, identify yourself to the crew

Familiar icons help readers find important information.

POLICE

Notify the crew of any unlawful or suspicious activity on board your train

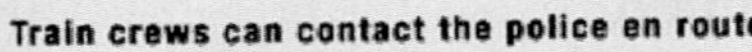

Train crews can contact the police en route

EVACUATION

Open panel above side door

Pull red handle down

Slide door open

Exit the train only when directed

Source: New York Metropolitan Transportation Authority, http://www.mta.info/lirr/safety/bilevel3.htm.

Use only one side of a page or placard. Your emergency instructions should fit on one side of a page, poster, sign, or placard, because people must be able to read them at a glance.

Use familiar icons. Highlight specialized information with icons that represent fire, water, electricity, hazardous materials, police, first aid, fire extinguishers, and so on.

Group information into visual blocks. Use lines, boxes, and white space to create frames around information that belongs together.

GO TO THE NET For first aid sites on the web, go to **www.pearsonhighered.com/johnsonweb4/7.17**

Put the safety of people first. Keeping people safe should always be your first concern. Instructions for saving property, machines, or information should be secondary.

Use command voice and keep sentences brief. Each of your commands should fit on one line and should have as few words as possible.

Minimize unnecessary explanations. In a crisis, people need to know what to do, not why to do it that way. So don't clutter up the sign with explanations.

Tell readers to call for help. Remind readers that they should call 911 or contact emergency responders as soon as possible.

Write

Write your own emergency instructions. Pick a potential emergency situation on campus or in your home. Create emergency instructions that fit on one side of a standard piece of paper. Make sure you include icons that highlight important points.

CHAPTER REVIEW

- Documentation describes step by step how to complete a task.
- Basic features of documentation include a specific and precise title; an introduction; a list of parts, tools, and conditions required; sequentially ordered steps; graphics; safety information; a conclusion; and troubleshooting information.
- Determine the rhetorical situation by asking the Five-W and How Questions and analyzing the document's subject, purpose, readers, and context of use.
- Organize and draft your documentation step by step, breaking tasks down into their major and minor actions.
- Safety information should (1) identify the hazard, (2) state the level of risk (Danger, Warning, or Caution), and (3) offer suggestions for avoiding injury or damage.
- Try to use a style that reflects or counters readers' attitudes as they follow the steps.
- Graphics offer important support for the written text. They should be properly labeled by number and inserted on the page where they are referenced.
- User-testing your documentation with sample readers is an effective way to work out any bugs and locate places for improvement. Your observations of these sample readers should help you revise the document.

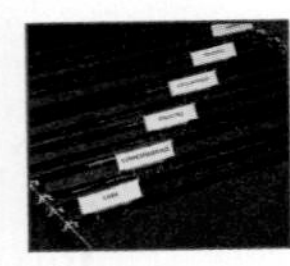

Individual or Team Projects

1. Find an example of documentation in your home or workplace. Using concepts discussed in this chapter, develop a set of criteria to evaluate its content, organization, style, and design. Then, write a two-page memo to your instructor in which you analyze the documentation. Highlight any strengths and make suggestions for improvements.

2. On the Internet or in your home, find information on first aid (handling choking, treating injuries, using CPR, handling drowning, treating shock, dealing with alcohol or drug overdoses). Then, turn this information into a text that is specifically aimed at college students living on campus. You should keep in mind that these readers will be reluctant to read this text—until it is actually needed. So, write and design it in a way that will be both appealing before injuries occur and highly usable when an injury has occurred.

3. The Case Study at the end of this chapter presents a difficult ethical decision. Pretend you are Mark Harris in this situation. As Mark, write a memo to your supervisor in which you express your concerns about the product. Tell your supervisor what you think the company should do about the problem.

4. Choose a culture that is quite different from your own. Through the Internet and your library, research how documentation is written and designed in that culture. Look for examples of documentation designed for people of that culture. Then, write a memo to your instructor in which you explain how documentation is different in that target culture.

Collaborative Projects

Have someone in your group bring to class an everyday household appliance (toaster, blender, hot air popcorn popper, clock radio, MP3 player, etc.). With your group, write and design documentation for this appliance that would be appropriate for 8-year-old children. Your documentation should keep the special needs of these readers in mind. The documentation should also be readable and interesting to these readers, so they will actually use it.

> For additional technical writing resources, including interactive sample documents, document design tutorials and guidelines, and more, go to **www.mytechcommlab.com.**

Revision Challenge

These instructions for playing Klondike are technically correct; however, they are hard to follow. Can you use visual design to revise these instructions to make them more readable?

Playing Klondike (Solitaire)

Many people know Klondike simply as solitaire, because it is such a widely played solitaire game. Klondike is not the most challenging form of solitaire, but it is very enjoyable and known worldwide.

To play Klondike, use one regular pack of cards. Dealing left to right, make seven piles from 28 cards. Place one card on each pile, dealing one fewer pile each round. When you are finished dealing, the pile on the left will have one card, the next pile on the left will have two cards, and so on. The pile farthest to the right will have seven cards. When you are finished dealing the cards, flip the top card in each pile faceup.

You are now ready to play. You may move cards among the piles by stacking cards in decreasing numerical order (king to ace). Black cards are placed on red cards and red cards are placed on black cards. For example, a red four can be placed on a black five. If you would like to move an entire stack of faceup cards, the bottom card being moved must be placed on a successive card of the opposite color. For example, a faceup stack with the jack of hearts as the bottom card can be moved only to a pile with a black queen showing on top. You can also move partial stacks from one pile to another as long as the bottom card you are moving can be placed on the top faceup card on the pile to which you are moving it. If a facedown card is ever revealed on top of a pile, it should be turned face up. You can now use this card. If the cards in a pile are ever completely removed, you can replace the pile by putting a king (or a stack with a king as the bottom faceup card) in its place.

The rest of the deck is called the "stock." Turn up cards in the stock one by one. If you can play a turned-up card on your piles, place it. If you cannot play the card, put it in the discard pile. As you turn up cards from the stock, you can also play the top card off the discard pile. For example, let us say you have an eight of hearts on top of the discard pile. You turn up a nine of spades from the stock, which you find can be played on a ten of diamonds on top of one of your piles. You can then play the eight of hearts on your discard pile on the newly placed nine of spades.

When an ace is uncovered, you may move it to a scoring pile separate from the seven piles. From then on, cards of the same suit may be placed on the ace in successive order. For example, if the two of hearts is the top faceup card in one of your piles, you can place it on the ace of hearts. As successive cards in the suit are the top cards in piles or revealed in the stock, you can place them on your scoring piles.

When playing Klondike properly, you may go through the stock only once (variations of Klondike allow you to go through the stock as many times as you like, three cards at a time). When you are finished going through the stock, count up the cards placed in your scoring piles. The total cards in these piles make up your score for the game.

GO TO THE NET

Check your solitaire instructions against ones on the net. Go to **www.pearsonhighered.com/johnsonweb4/7.18**

CASE STUDY

The Flame

Mark Harris was recently hired as a design engineer at Fun Times Toy Manufacturers, Inc. In a way, the job was a dream come true. Mark liked designing toys, and he had two small children at home.

When he began working at the company, he inherited a project from the previous engineer, who had taken a job with another company. The project was an indoor campfire made out of a gas stove. The stove produced a flame about a foot tall in a pile of fake logs. The campfire was designed to be used outside, but it could be used inside with special precautions.

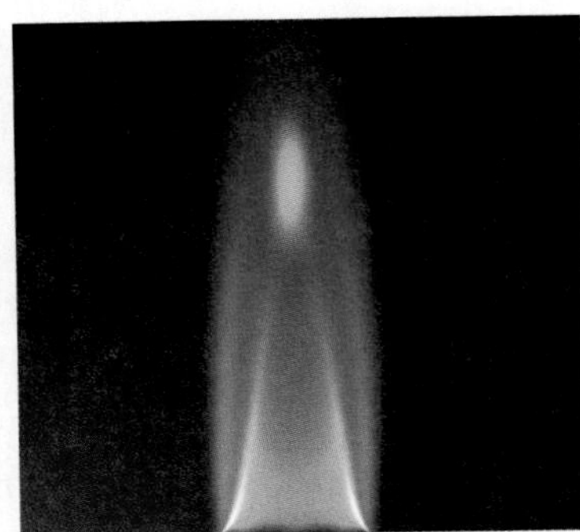

The prior engineer on the project had already built a prototype. All Mark needed to do was write the instructions for the user's manual. So, he took the campfire home to try it out with his kids.

At home, the campfire started up easily in his living room, but Mark immediately noticed that the flame was somewhat dangerous. As his kids tried to roast marshmallows over the fire, he started to become a little nervous. He wondered if the couch and rug were at risk of catching on fire. Certainly, he could imagine situations where the campfire could be used incorrectly, causing burns or fire.

The next day, he stopped by his supervisor's office to express his concerns. His supervisor, though, waved off the problems. "Yeah, the previous guy mentioned some problems, too. All you need to do is fill the user's manual with warnings and warning symbols. That should protect us from lawsuits."

As Mark walked out the door, his supervisor said, "Listen, we need that thing out the door in a couple weeks. Our manufacturers in Mexico have already retooled their plants, and we're ready to go. You need to get that manual done right now."

Mark went back to his office a bit concerned. Wouldn't his company still be negligent if people or their property were harmed? Would warnings and warning symbols really protect them from negligence? And even if the company couldn't be sued, would it be ethical to produce a potentially harmful product? What would you do if you were in Mark's place?

GO TO THE NET

To find information on product liability, go to **www.pearsonhighered.com/johnsonweb4/7.19**

CHAPTER

8 Proposals

Planning and Researching *206*

Organizing and Drafting *213*

Using Style and Design *226*

Microgenre: The Elevator Pitch *232*

Chapter Review *234*

Exercises and Projects *235*

Case Study: The Mole *245*

In this chapter, you will learn:

- The purpose of proposals and their uses in the workplace.
- The basic features and types of proposals.
- How to plan and do research for a proposal.
- How to organize and draft the major sections in a proposal.
- Strategies for using plain and persuasive style to make a proposal influential.
- How document design and graphics can enhance a proposal.

Proposals are the lifeblood of the technical workplace. Whatever your field, you will be asked to write proposals that describe new projects, present innovative ideas, offer new strategies, and promote services. The purpose of a proposal is to present your ideas and plans for your readers' consideration.

Effective proposal writing is a crucial skill in today's technical workplaces. Almost all projects begin with proposals, so you need to master this important genre to be successful.

Proposals are categorized in a couple of different ways. *Internal* proposals are used within a company to plan or propose new projects or products. *External* proposals are used to offer services or products to clients outside the company.

Proposals are also classified as *solicited* or *unsolicited,* depending on whether they were requested by the readers or not.

> **Solicited proposals** are proposals requested by the readers. For example, your company's management might request proposals for new projects. Or, your team might be "solicited" to write a proposal that answers a request for proposals (RFP) sent out by a client.
>
> **Unsolicited proposals** are proposals not requested by the readers. For example, your team might prepare an unsolicited internal proposal to pitch an innovative new idea to your company's management. Or, your team might use an unsolicited external proposal as a sales tool to offer your company's clients a product or service.

Figure 8.1 shows an internal, solicited proposal. In this example, a team within the company is pitching a plan to overhaul the company's website. This proposal is being used to persuade management to agree to the team's ideas.

Another kind of proposal is the grant proposal. Researchers and nonprofit organizations prepare grant proposals to obtain funding for their projects. For example, one of the major funding sources for grants in science and technology is the National Science Foundation (NSF). Through its website, the NSF offers funding opportunities for scientific research (Figure 8.2).

Planning and Researching

Because proposals are difficult to write, it is important to follow a reliable writing process that will help you develop your proposal's content, organization, style, and design. An important first step in this process is to start with a planning and researching phase. During this phase, you will define the rhetorical situation and start collecting content for the proposal.

Planning

A good way to start planning your proposal is to analyze the situations in which it will be used. Begin by answering the Five-W and How Questions:

> *Who will be able to say yes to my ideas, and what are their characteristics?*
>
> *Why is this proposal being written?*

To see sample proposals, go to **www.pearsonhighered.com/johnsonweb4/8.1**

Proposals

This is a basic model for organizing a proposal. This organizational pattern, though, is flexible, allowing the contents of a proposal to be arranged in a variety of ways.

QUICK START

Basic Features of Proposals

Proposals typically explain the current situation (often called the "problem") and then offer a plan for improving it (also called the "solution"). In technical workplaces, proposals tend to have the following features:

- Introduction
- Description of the current situation
- Description of the project plan
- Review of qualifications
- Discussion of costs and benefits
- Graphics
- Budget

As with other genres, though, you should not mechanically follow the pattern shown here. The proposal genre is not a formula to be followed mechanically. You should alter this pattern to suit the needs of your proposal's subject, purpose, readers, and context of use.

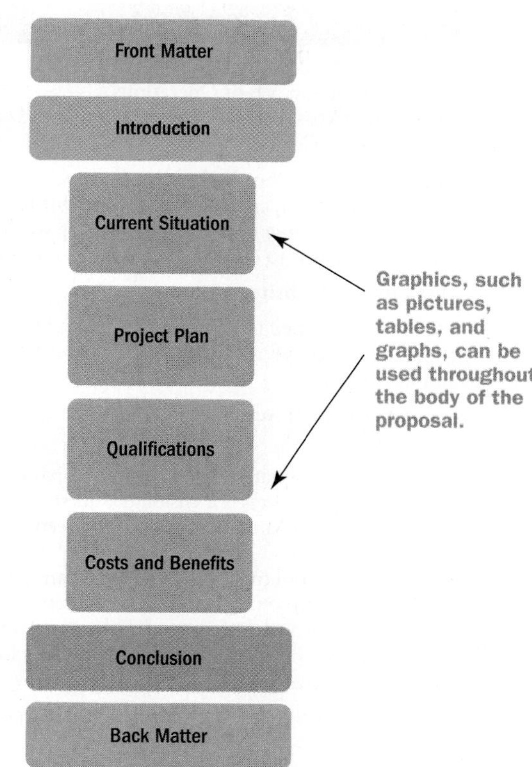

What information do the readers need to make a decision?
Where will the proposal be used?
When will the proposal be used?
How will the proposal be used?

Once you have answered these questions, you are ready to start thinking in-depth about your proposal's subject, purpose, readers, and context of use.

To see requests for proposals (RFPs), go to **www.pearsonhighered.com/johnsonweb4/8.2**

An Internal Solicited Proposal

Figure 8.1: This small proposal is an internal proposal that is pitching a new idea to a manager. After a brief introduction, it describes the current situation and offers a plan for solving a problem. It concludes by highlighting the benefits of the plan.

JumperCom

Date: April 11, 2011
To: Jim Trujillo, VP of Operations
From: Sarah Voss, Lambda Engineering Team Leader
Re: Cutting Costs

Internal proposals are often written in memo format.

At our meeting on April 4th, you asked each project team to come up with one good idea for cutting costs. Our team met on April 7th to kick around some ideas. At this meeting we decided that the best way to cut costs is to expand and enhance the company's website.

Background information signals that the proposal was solicited.

The main point of the proposal is stated up front.

Our Current Website

When we developed the current website in Spring 2005, it served our company's purposes quite well. For its time, the website was attractive and interactive.

Six years later, our website is no longer cutting edge—it's obsolete. The website

- looks antiquated, making our company seem out of touch
- does not address our customers' questions about current products
- does not address our customers' needs for product documentation
- is not a tool that our salespeople can use to provide answers and documentation to the customers
- does not answer frequently asked questions, forcing clients to call our toll-free customer service lines for answers to simple questions.

This section describes the current situation.

As a result, our outdated website is causing a few important problems. First, we are likely losing sales because our customers don't see us as cutting edge. Second, we are wasting hundreds of thousands of dollars on printed documents that the customers throw away after a glance. And, third, we are unnecessarily spending many more thousands of dollars on customer service representatives and tollfree phone lines. *A conservative estimate suggests that our outdated website could be costing us around $400,000 each year.*

GO TO THE NET
To learn more about different kinds of proposals, go to **www.pearsonhighered.com/johnsonweb4/8.3**

Figure 8.1: (continued)

This section offers a plan for the readers' consideration.

Renovating the Website

We believe a good way to cut costs and improve customer relations is to renovate the website. We envision a fully interactive site that customers can use to find answers to their questions, check on prices, and communicate with our service personnel. Meanwhile, our sales staff can use the website to discuss our products with clients. Instead of lugging around printed documents, our salespeople would use their laptop computers to show products or make presentations.

Renovating the site will require four major steps:

Step One: Study the Potential Uses of Our Website

With a consultant, we should study how our website might be better used by customers and salespeople. The consultant would survey our clients and salespeople to determine what kind of website would be most useful to them. The consultant would then develop a design for the website.

The plan is described step by step.

Step Two: Hire a Professional Web Designer to Renovate the Site

We should hire a professional web designer to implement our design, because modern websites are rather complex. A professional would provide us with an efficient, well-organized website that would include all the functions we are seeking.

Step Three: Train One of Our Employees to Be a Webmaster

We should hire or retrain one of our employees to be the webmaster of the site. We need someone who is working on the site daily and making regular updates. Being the webmaster for the site should be this employee's job description.

Step Four: User-Test the New Website with Our Customers and Salespeople

Once we have created a new version of the website, we should user-test it with our customers and salespeople. Perhaps we could pay some of our customers to try out the site and show us where it could be improved. Our salespeople will certainly give us plenty of feedback.

2

(continued)

Figure 8.1: *(continued)*

Proposal concludes by discussing costs and benefits of the plan.

At the end of this process, we would have a fully functioning website that would save us money almost immediately.

Costs and Benefits of Our Idea

Renovating the website would have many advantages:

- The new website will save us printing costs. We estimate that the printing costs at our company could be sliced in half—perhaps more—because our customers would be able to download our documents directly from the website, rather than ask us to send these documents to them. That's a potential savings of $300,000.
- The new website will provide better service to our customers. Currently, our customers go to the website first when they have questions. By providing more information in an interactive format, we can cut down dramatically on calls to our customer service center. We could save up to $120,000 in personnel costs and long-distance charges.
- Our sales staff will find the new website a useful tool when they have questions. When products change, salespeople will immediately see those changes reflected on the website. As a result, more sales might be generated because product information will be immediately available online.

A quick estimate shows that a website renovation would cost us about $40,000. We would also need to shift the current webmaster's responsibilities from part time to full time, costing us about $20,000 per year more. The savings, though, are obvious. For an initial investment of $60,000 and a yearly investment of $20,000 thereafter, we will minimally save about $400,000 a year.

Thank you for giving us this opportunity to present our ideas. If you would like to talk with us about this proposal, please call me at 555-1204, or e-mail me at sarahv@jumpercom.net.

3

The National Science Foundation Home Page

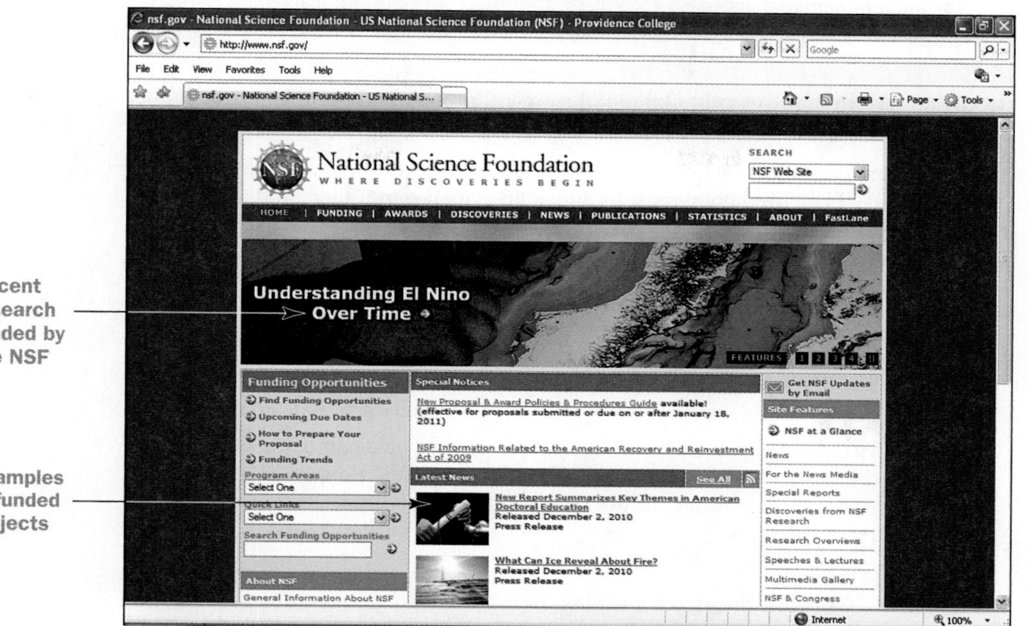

Figure 8.2: The National Science Foundation (NSF) web-site offers information on grant opportunities. The home page, shown here, discusses some of the recent research projects that have received grants.

Source: National Science Foundation, http://www.nsf.gov.

SUBJECT Define exactly what your proposal is about. Where are the boundaries of the subject? What need-to-know information must readers have if they are going to say yes to your ideas?

Link

For more help on defining need-to-know information, go to Chapter 2, page 22.

PURPOSE Clearly state the purpose of your proposal in one sentence. What should the proposal achieve?

Some key action verbs for your purpose statement might include the following:

to persuade	*to present*
to convince	*to propose*
to provide	*to offer*
to describe	*to suggest*
to argue for	*to recommend*
to advocate	*to support*

A purpose statement might look something like this:

> The purpose of this proposal is to recommend that our company change its manufacturing process to include more automation.

Want to learn more about the NSF? Go to **www.pearsonhighered.com/johnsonweb4/8.4**
Need help defining your purpose? Go to **www.pearsonhighered.com/johnsonweb4/8.5**

Link

To read more about defining your purpose, go to Chapter 1, page 6.

In this proposal, our aim is to persuade the state of North Carolina to develop a multimodal approach to protect itself from stronger hurricanes, which may be caused by climate change.

READERS More than any other kind of document, proposals require you to fully understand your readers and anticipate their needs, values, and attitudes.

Primary readers (action takers) are the people who can say yes to your ideas. They need good reasons and solid evidence to understand and agree to your ideas.

Secondary readers (advisors) are usually technical experts in your field. They won't be the people who say yes to your proposal, but their opinions will be highly valued by your proposal's primary readers.

Tertiary readers (evaluators) can be just about anyone else who might have an interest in the project and could potentially undermine the project. These readers might include lawyers, journalists, and community activists, among others.

Gatekeepers (supervisors) are the people at your own company who will need to look over your proposal before it is sent out. Your immediate supervisor is a gatekeeper, as are your company's accountants, lawyers, and technical advisors.

Link

For more strategies for analyzing your readers, see Chapter 2, page 20.

CONTEXT OF USE The document's context of use will also greatly influence how your readers will interpret the ideas in your proposal.

Physical context concerns the places your readers may read, discuss, or use your proposal.

Economic context involves the financial issues and economic trends that will shape readers' responses to your ideas.

Ethical context involves any ethical or legal issues involved in your proposed project.

Political context concerns the people inside and outside your company who will be affected by your proposal.

Link

For more help defining the context of use, turn to Chapter 2, page 26.

Proposals are legal documents that can be brought into court if a dispute occurs, so you need to make sure that everything you say in the proposal is accurate and truthful.

Researching

After defining your proposal's rhetorical situation, you should start collecting information and creating the content of your document (Figure 8.3). Chapter 14 describes how to do research, so research strategies won't be fully described here. However, here are some research strategies that are especially applicable for writing proposals:

DO BACKGROUND RESEARCH The key to writing a persuasive proposal is to fully understand the problem you are trying to solve. Use the Internet, print sources, and interviews to find as much information about your subject as you can.

Worksheets are available to help you analyze your readers and contexts of use. Go to **www.pearsonhighered.com/johnsonweb4/8.6**

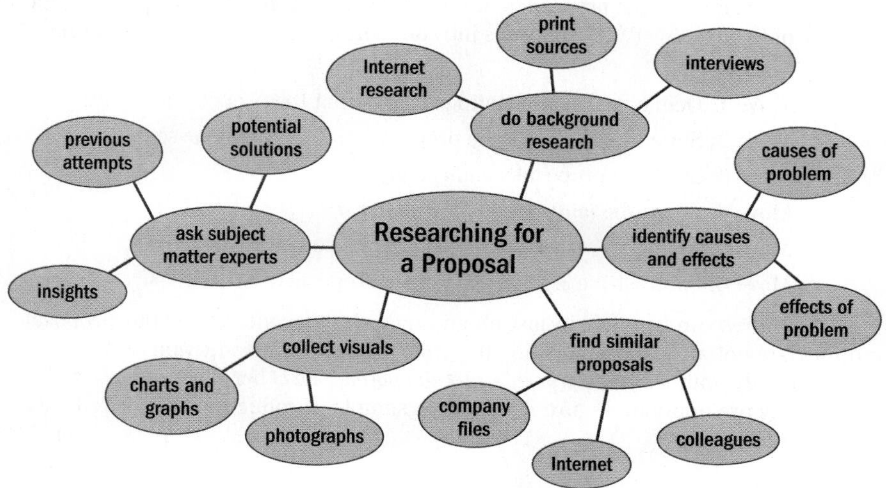

Figure 8.3: A logical map, like this one, might help you research your subject from a variety of directions. When researching the background of a proposal, collect as much information as possible.

ASK SUBJECT MATTER EXPERTS (SMEs) Spend time interviewing experts who know a great amount about your subject. They can probably give you insight into the problem you are trying to solve and suggest some potential solutions.

PAY ATTENTION TO CAUSES AND EFFECTS All problems have causes, and all causes create effects. In your observations of the problem, try to identify its causes and effects.

FIND SIMILAR PROPOSALS On the Internet or at your workplace, you can probably locate proposals that have dealt with similar problems in the past. They might give you some insight into how similar problems have been solved in the past.

Link

To learn more about doing research, turn to Chapter 14, page 386.

COLLECT VISUALS Proposals are persuasive documents, so they often include plenty of graphics, such as photographs, charts, illustrations, and graphs. Collect any materials, data, and information that will help you add a visual dimension to your proposal. If appropriate, you might use a digital camera to take pictures to include in the document.

Organizing and Drafting

Writing the first draft of a proposal is always difficult because proposals describe the future—a future that you are trying to envision for your readers and yourself.

A good way to draft your proposal is to write it one section at a time. Think of the proposal as four or five separate mini-documents that can stand alone. When you finish drafting one section, move on to the next.

For more help on researching your subject, go to **www.pearsonhighered.com/johnsonweb4/8.7**

Writing the Introduction

As with all documents, the proposal's introduction sets a context, or framework, for the body of the document. A proposal's introduction will usually include up to six moves:

Move 1: Define the *subject,* stating clearly what the proposal is about.

Move 2: State the *purpose* of the proposal, preferably in one sentence.

Move 3: State the proposal's *main point.*

Move 4: Stress the *importance of the subject.*

Move 5: Offer *background information* on the subject.

Move 6: Forecast the *organization* of the proposal.

Link

For additional help on writing introductions, see Chapter 16, page 430.

These moves can be made in just about any order, depending on your proposal, and they are not all required. Minimally, your proposal's introduction should clearly identify your *subject, purpose,* and *main point.* The other three moves are helpful, but they are optional. Figure 8.4 shows a sample introduction that uses all six moves.

Describing the Current Situation

The aim of the *current situation* section—sometimes called the *background* section—is to define the problem your plan will solve. You should accomplish three things in this section of the proposal:

- Define and describe the problem.
- Discuss the causes of the problem.
- Discuss the effects of the problem if nothing is done.

For example, let us say you are writing a proposal to improve safety at your college or workplace. Your current situation section would first define the problem by proving there is a lack of safety and showing its seriousness. Then, it would discuss the causes and effects of that problem.

AT A GLANCE

The Current Situation

- Define and describe the problem.
- Discuss the causes of the problem.
- Discuss the effects if nothing is done about the problem.

MAPPING OUT THE SITUATION Logical mapping is a helpful technique for developing your argument in the current situation section. Here are some steps you can follow to map out the content:

1. Write the problem in the middle of your screen or piece of paper. Put a circle around the problem.
2. Write down the two to five major causes of that problem. Circle them, and connect them to the problem.
3. Write down some minor causes around each major cause, treating each major cause as a separate problem of its own. Circle the minor causes and connect them to the major causes.

GO TO THE NET

To see sample proposal introductions, go to **www.pearsonhighered.com/johnsonweb4/8.8**

Introduction to a Proposal

Growth and Flexibility with Telecommuting

A Proposal to Northside Design

Offers background information.

Founded in 1982, Northside Design is one of the classic entrepreneurial success stories in architecture. Today, this company is one of the leading architectural firms in the Chicago market with over 50 million dollars in annual revenue. With growth, however, comes growing pains, and Northside now faces an important decision about how it will manage its growth in the near future. The right decision could lead to more market share, increased sales, and even more prominence in the architectural field. However, Northside also needs to safeguard itself against over-extension in case the Chicago construction market unexpectedly begins to recede.

Defines the subject.

Stresses the importance of the subject.

"Northside needs to safeguard itself against over-extension in case the Chicago construction market unexpectedly begins to recede."

To help you make the right decision, this proposal offers an innovative strategy that will support your firm's growth while maintaining its flexibility. Specifically, we propose Northside implement a telecommuting network that allows selected employees to work a few days each week at home. Telecommuting will provide your company with the office space it needs to continue growing. Meanwhile, this approach will avoid a large investment in new facilities and disruption to the company's current operations.

States the purpose of the proposal.

States the main point.

In this proposal, we will first discuss the results of our research into Northside's office space needs. Second, we will offer a plan for using a telecommuting network to free up more space at Northside's current office. Third, we will review Insight Systems' qualifications to assist Northside with its move into the world of telecommuting. And finally, we will go over some of the costs and advantages of our plan.

Forecasts the body of the proposal.

1

Figure 8.4: This introduction makes all six "moves." As a result, it is somewhat lengthy. Nevertheless, this introduction prepares readers to understand the information in the body of the proposal.

Mapping Out the Current Situation

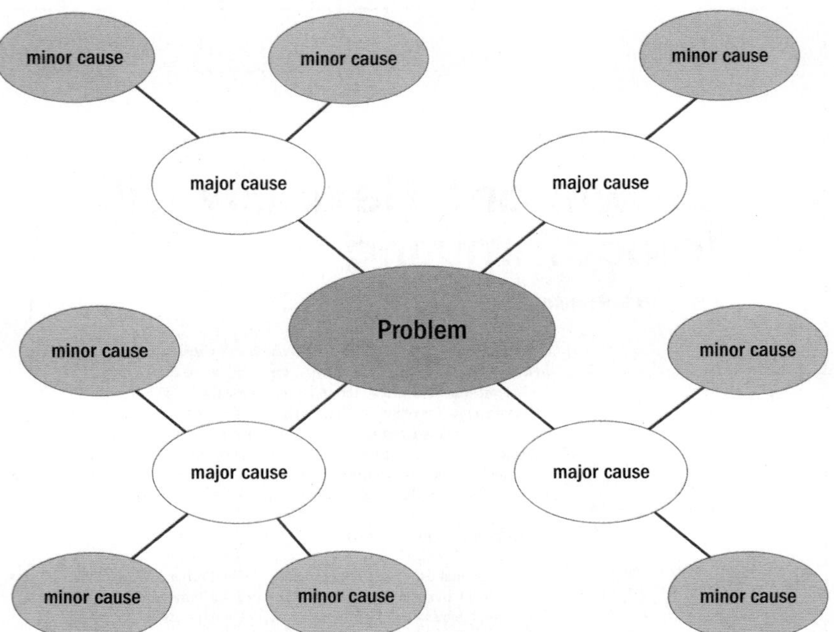

Figure 8.5: Logical mapping helps you figure out what caused the problem that you are trying to solve.

Figure 8.5 illustrates how your logical map for the current situation section might look.

DRAFTING THE CURRENT SITUATION SECTION Your proposal's current situation section should include an opening, a body, and a closing:

> **Opening**—Identify and define the problem you will describe.
>
> **Body**—Discuss the *causes* of the problem, showing how these causes brought about the problem.
>
> **Closing**—Discuss the *effects* of not doing anything about the problem.

The length of the current situation section depends on your readers' familiarity with the problem. If readers are new to the subject, then several paragraphs or even pages might be required. However, if they fully understand the problem already, maybe only a few paragraphs are needed.

Figure 8.6 demonstrates one possible pattern for organizing this section of the proposal, though this section can be organized a variety of ways.

Example Current Situation Section

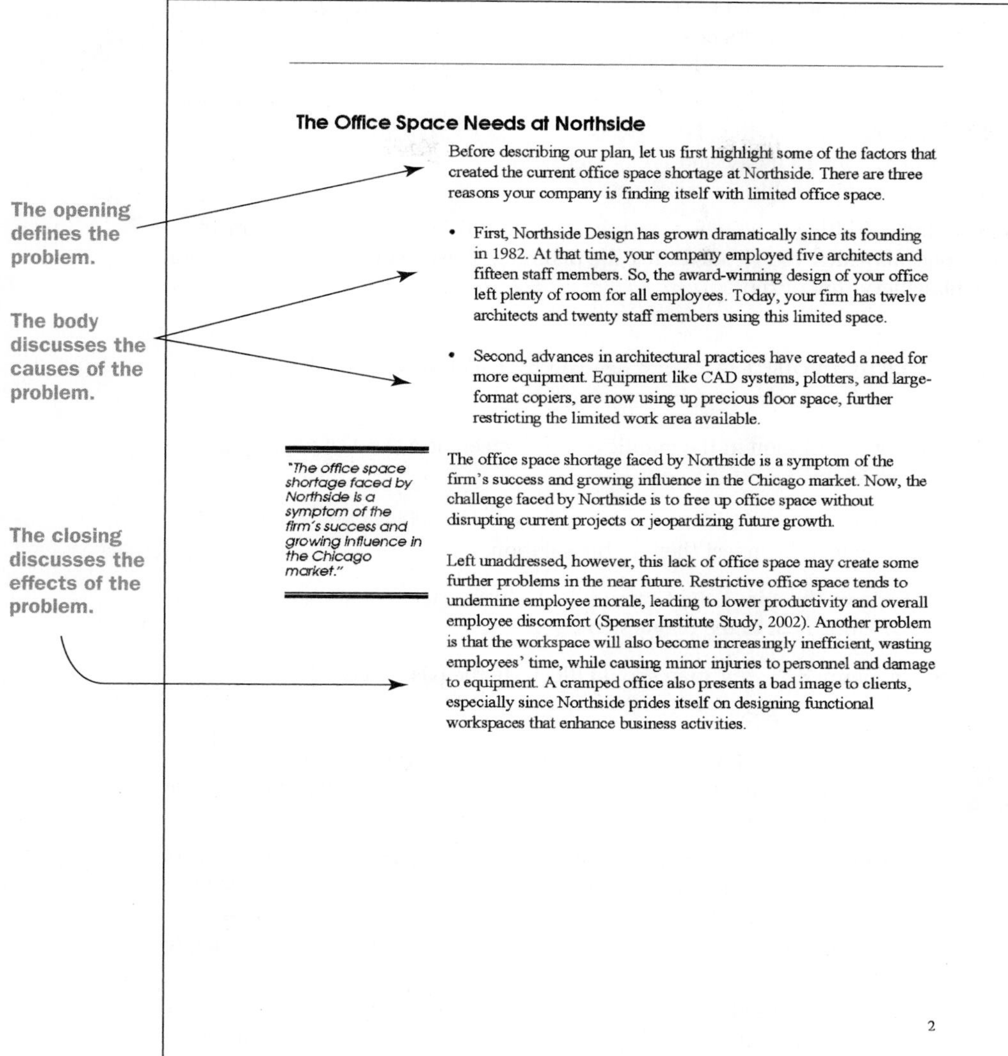

The opening defines the problem.
The body discusses the causes of the problem.
The closing discusses the effects of the problem.

The Office Space Needs at Northside

Before describing our plan, let us first highlight some of the factors that created the current office space shortage at Northside. There are three reasons your company is finding itself with limited office space.

- First, Northside Design has grown dramatically since its founding in 1982. At that time, your company employed five architects and fifteen staff members. So, the award-winning design of your office left plenty of room for all employees. Today, your firm has twelve architects and twenty staff members using this limited space.
- Second, advances in architectural practices have created a need for more equipment. Equipment like CAD systems, plotters, and large-format copiers, are now using up precious floor space, further restricting the limited work area available.

"The office space shortage faced by Northside is a symptom of the firm's success and growing influence in the Chicago market."

The office space shortage faced by Northside is a symptom of the firm's success and growing influence in the Chicago market. Now, the challenge faced by Northside is to free up office space without disrupting current projects or jeopardizing future growth.

Left unaddressed, however, this lack of office space may create some further problems in the near future. Restrictive office space tends to undermine employee morale, leading to lower productivity and overall employee discomfort (Spenser Institute Study, 2002). Another problem is that the workspace will also become increasingly inefficient, wasting employees' time, while causing minor injuries to personnel and damage to equipment. A cramped office also presents a bad image to clients, especially since Northside prides itself on designing functional workspaces that enhance business activities.

2

Figure 8.6: The current situation section includes an opening, a body, and a closing. The causes of the problem are discussed mainly in the body paragraphs, while the effects are usually discussed at the end of the section.

To see documents that analyze problems and causes, go to **www.pearsonhighered.com/johnsonweb4/8.9**

Describing the Project Plan

A proposal's *project plan* section offers a step-by-step method for solving the problem. Your goal is to tell your readers *how* you would like to handle the problem and *why* you would handle it that way. In this section, you should do the following:

- Identify the solution.
- State the objectives of the plan.
- Describe the plan's major and minor steps.
- Identify the deliverables or outcomes.

As you begin drafting this section, look back at your original purpose statement for the proposal, which you wrote during the planning phase. Now, imagine a solution that might achieve that purpose.

MAPPING OUT THE PROJECT PLAN When you have identified a possible solution, you can again use logical mapping to turn your idea into a plan:

1. Write your solution in the middle of your screen or a sheet of paper. Circle this solution.
2. Write down the two to five major steps needed to achieve that solution. Circle them and connect them to the solution.
3. Write down the minor steps required to achieve each major step. Circle them and connect them to the major steps.

As shown in Figure 8.7, your map should illustrate the basic steps of your plan.

AT A GLANCE

The Project Plan

- Identify the solution.
- State the objectives of the plan.
- Describe the plan's major and minor steps.
- Identify the deliverables or outcomes.

DRAFTING THE PROJECT PLAN SECTION Your project plan section should have an opening, a body, and a closing. This section will describe step by step how you will achieve your project's purpose.

Opening—Identify your overall solution to the problem. You can even give your plan a name to make it sound more real (e.g., the "Restore Central Campus Project"). Your opening might also include a list of project objectives so readers can see what goals your plan is striving to achieve.

Body—Walk the readers through your plan step by step. Address each major step separately, discussing the minor steps needed to achieve that major step. It is also helpful to tell readers *why* each major and minor step is needed.

Closing—Summarize the final *deliverables,* or outcomes, of your plan. The deliverables are the goods and services that you will provide when the project is finished.

As shown in Figure 8.8, the project plan section explains *why* each step is needed and identifies *what* will be delivered when the project is finished.

GO TO THE NET

For more advice about writing project plans in proposals, go to **www.pearsonhighered.com/johnsonweb4/8.10**

Mapping Out a Project Plan

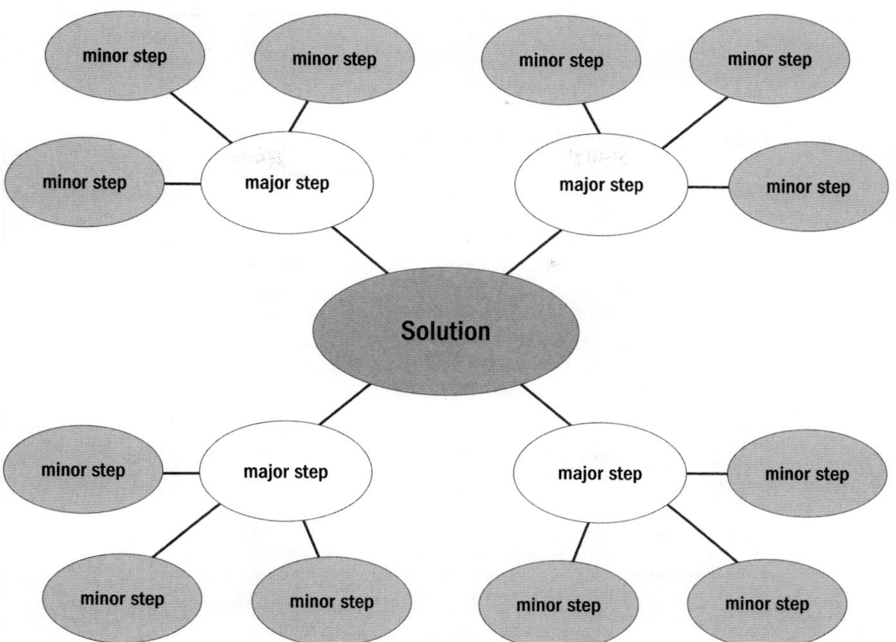

Figure 8.7: Logical mapping will help you figure out how to solve the problem. In a map, like the one shown here, you can visualize your entire plan by writing out the major and minor steps.

Describing Qualifications

The qualifications section presents the credentials of your team or company, showing why your team is qualified to carry out the project plan. Minimally, the aim of this section is to show that your team or company is able to do the work. Ideally, you also want to prove that your team or company is *best qualified* to handle the project.

As you begin drafting this section, keep the following saying in mind: *What makes us different makes us attractive.* In other words, pay attention to the qualities that make your team or company different from your competitors. What are your company's strengths? What makes you better than the others?

A typical qualifications section offers information on three aspects of your team or company:

Description of personnel—Short biographies of managers who will be involved in the project; demographic information on the company's workforce; description of support staff.

Description of organization—Corporate mission, philosophy, and history of the company; corporate facilities and equipment; organizational structure of the company.

Previous experience—Past and current clients; a list of similar projects that have been completed; case studies that describe past projects.

Project Plan Section

Figure 8.8: An effective project plan section includes an opening, a body, and a closing. The opening states the solution and offers some objectives. The body walks the readers through the plan's steps. The closing identifies the major deliverables of the plan.

Our Plan: Flexibility and Telecommuting

Managing Northside's limited office space requires a solution that allows the company to grow but does not sacrifice financial flexibility. Therefore, we believe a successful solution must meet the following objectives:

Objectives

- minimize disruption to Northside's current operations
- minimize costs, preserving Northside's financial flexibility
- retain Northside's current office on Michigan Avenue
- foster a dynamic workplace that will be appealing to Northside's architects and staff.

Our Objectives:
- *minimize disruption*
- *minimize costs*
- *retain Northside's current office*
- *foster a dynamic workplace*

To meet these objectives, Insight Systems proposes to collaborate with Northside to develop a telecommunication network that allows selected employees to work at home.

Solution to problem

The primary advantage of telecommuting is that it frees up office space for the remaining employees who need to work in the main office. Telecommuting will also avoid overextending Northside's financial resources, so the firm can quickly react to the crests and valleys of the market.

Our plan will be implemented in four major phases. First, we will study Northside's telecommuting options. Second, we will design a local area network (LAN) that will allow selected employees to telecommute from a home office. Third, we will train Northside's employees in telecommuting basics. And finally, we will assess the success of the telecommuting program after it has been implemented.

Phase One: Analyzing Northside's Telecommuting Needs

We will start out by analyzing the specific workplace requirements of Northside's employees and management. The results of this analysis will allow us to work closely with Northside's management to develop a telecommuting program that fits the unique demands of a dynamic architecture firm.

3

Figure 8.8: (continued)

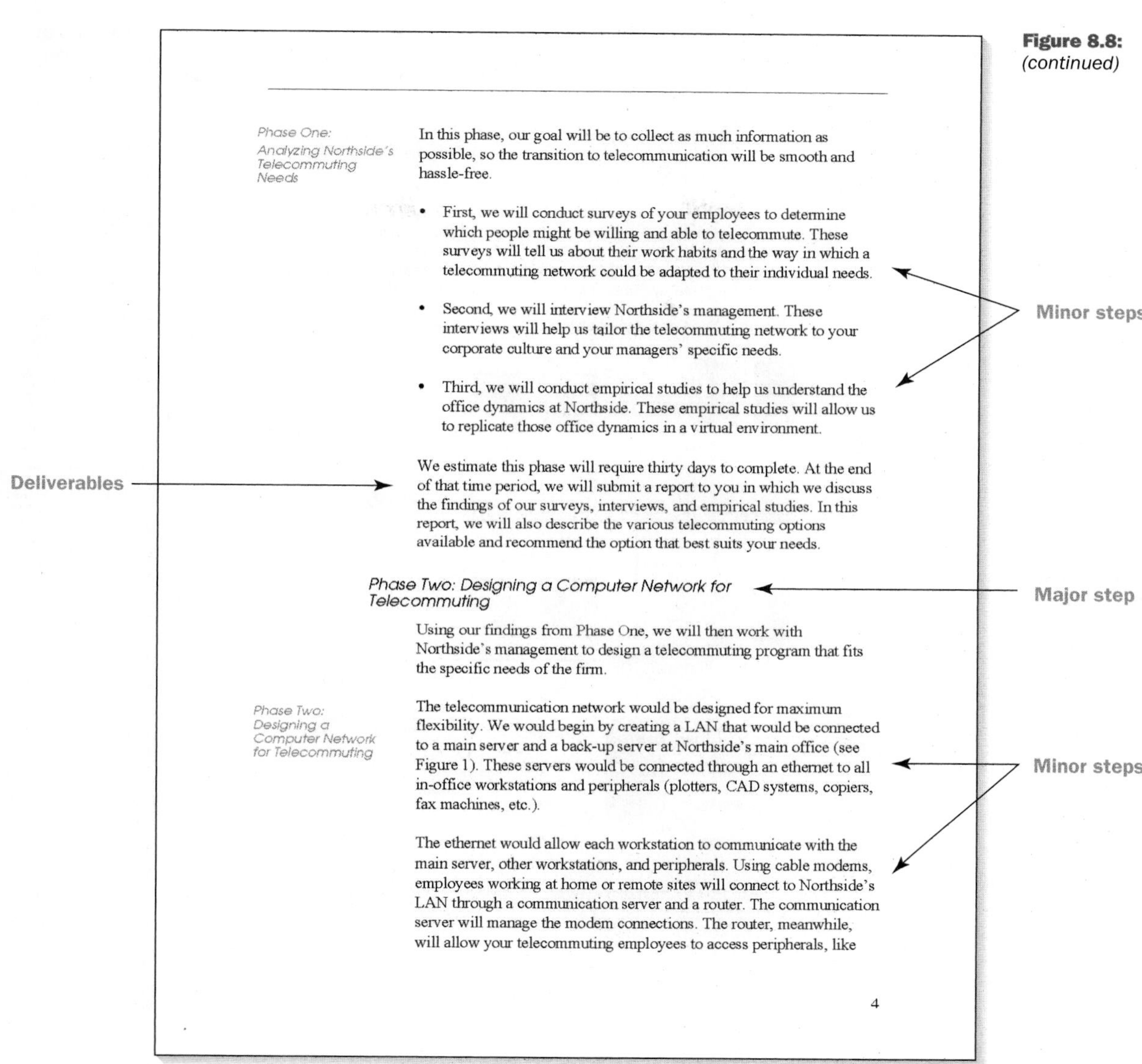

Phase One: Analyzing Northside's Telecommuting Needs
In this phase, our goal will be to collect as much information as possible, so the transition to telecommunication will be smooth and hassle-free.
• First, we will conduct surveys of your employees to determine which people might be willing and able to telecommute. These surveys will tell us about their work habits and the way in which a telecommuting network could be adapted to their individual needs.
• Second, we will interview Northside's management. These interviews will help us tailor the telecommuting network to your corporate culture and your managers' specific needs.
• Third, we will conduct empirical studies to help us understand the office dynamics at Northside. These empirical studies will allow us to replicate those office dynamics in a virtual environment.
We estimate this phase will require thirty days to complete. At the end of that time period, we will submit a report to you in which we discuss the findings of our surveys, interviews, and empirical studies. In this report, we will also describe the various telecommuting options available and recommend the option that best suits your needs.
Phase Two: Designing a Computer Network for Telecommuting
Using our findings from Phase One, we will then work with Northside's management to design a telecommuting program that fits the specific needs of the firm.
Phase Two: Designing a Computer Network for Telecommuting
The telecommunication network would be designed for maximum flexibility. We would begin by creating a LAN that would be connected to a main server and a back-up server at Northside's main office (see Figure 1). These servers would be connected through an ethernet to all in-office workstations and peripherals (plotters, CAD systems, copiers, fax machines, etc.).
The ethernet would allow each workstation to communicate with the main server, other workstations, and peripherals. Using cable modems, employees working at home or remote sites will connect to Northside's LAN through a communication server and a router. The communication server will manage the modem connections. The router, meanwhile, will allow your telecommuting employees to access peripherals, like
4
Minor steps
Deliverables
Major step
Minor steps

(continued)

Figure 8.8: *(continued)*

the plotters and copiers, through the ethernet. The router will also allow Northside's main office to connect easily with future branch offices and remote clients.

To ensure the security of the LAN, we will equip the network with the most advanced security hardware and software available. The router (hardware) will be programmed to serve as a "firewall" against intruders. We will also install the most advanced encryption and virus software available to protect your employees' transmissions.

Figure 1: The Local Area Network

The graphic illustrates a complex concept.

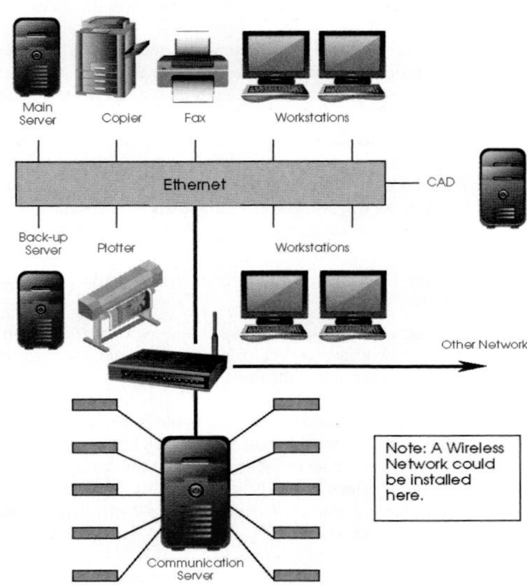

Figure 1: An ethernet allows you to interconnect the office internally and externally.

5

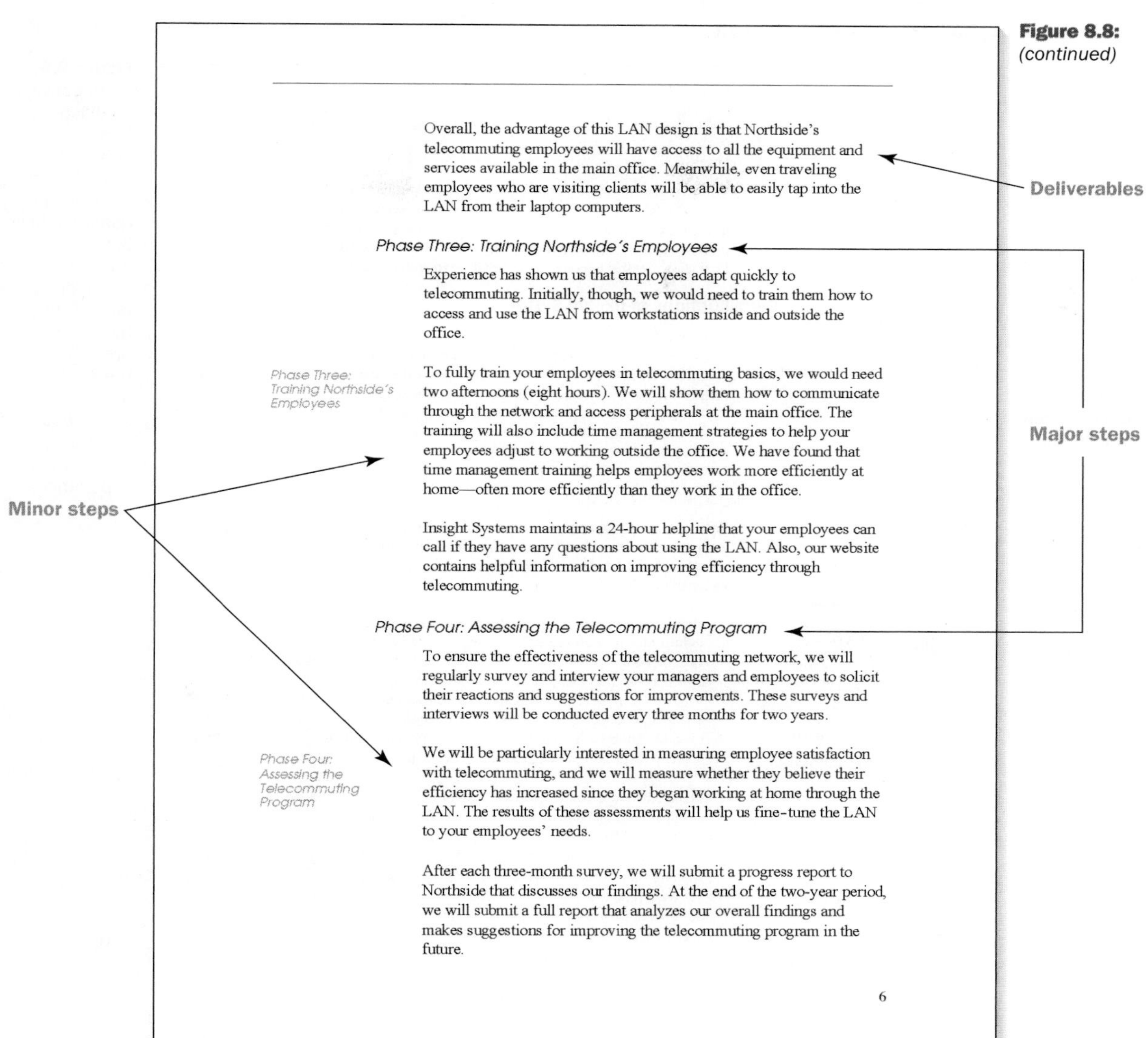

Overall, the advantage of this LAN design is that Northside's telecommuting employees will have access to all the equipment and services available in the main office. Meanwhile, even traveling employees who are visiting clients will be able to easily tap into the LAN from their laptop computers.

Phase Three: Training Northside's Employees

Experience has shown us that employees adapt quickly to telecommuting. Initially, though, we would need to train them how to access and use the LAN from workstations inside and outside the office.

Phase Three: Training Northside's Employees

To fully train your employees in telecommuting basics, we would need two afternoons (eight hours). We will show them how to communicate through the network and access peripherals at the main office. The training will also include time management strategies to help your employees adjust to working outside the office. We have found that time management training helps employees work more efficiently at home—often more efficiently than they work in the office.

Insight Systems maintains a 24-hour helpline that your employees can call if they have any questions about using the LAN. Also, our website contains helpful information on improving efficiency through telecommuting.

Phase Four: Assessing the Telecommuting Program

To ensure the effectiveness of the telecommuting network, we will regularly survey and interview your managers and employees to solicit their reactions and suggestions for improvements. These surveys and interviews will be conducted every three months for two years.

Phase Four: Assessing the Telecommuting Program

We will be particularly interested in measuring employee satisfaction with telecommuting, and we will measure whether they believe their efficiency has increased since they began working at home through the LAN. The results of these assessments will help us fine-tune the LAN to your employees' needs.

After each three-month survey, we will submit a progress report to Northside that discusses our findings. At the end of the two-year period, we will submit a full report that analyzes our overall findings and makes suggestions for improving the telecommuting program in the future.

6

Deliverables
Major steps
Minor steps

Figure 8.8: *(continued)*

Qualifications Section

When we complete this plan, Northside Design will have a fully functional telecommuting network that will allow selected employees to work from home. You should see an immediate improvement in productivity and morale. Meanwhile, you will be able to stay financially flexible to compete in the Chicago architectural market.

Qualifications at Insight Systems

At Insight Systems, we know this moment is a pivotal one for Northside Design. To preserve and expand its market share, Northside needs to grow, but it cannot risk overextending itself financially. For these reasons, Insight Systems is uniquely qualified to handle this project, because we provide flexible, low-cost telecommuting networks that help growing companies stay responsive to shifts in their industries.

Management and Labor

With over seventy combined years in the industry, our management team offers the insight and responsiveness required to handle your complex growth needs. (The résumés of our management team are included in Appendix B).

"With over seventy combined years in the industry, our management team offers the insight and responsiveness required to handle your complex growth needs."

Hanna Gibbons, our CEO, has been working in the telecommuting industry for over 20 years. After she graduated from MIT with a Ph.D in computer science, she worked at Krayson International as a systems designer. Ten years later, she had worked her way up to Vice President in charge of Krayson's Telecommuting Division. In 2002, Dr. Gibbons took over as CEO of Insight Systems. Since then, Dr. Gibbons has built this company into a major industry leader with gross sales of $15 million per year.

Frank Roberts, Chief Engineer at Insight Systems, has 30 years of experience in the networked computer field. He began his career at Brindle Labs, where he worked on artificial intelligence systems using analog computer networks. In 1995, he joined the Insight Systems team, bringing his unique understanding of networking to our team. Frank is very detail oriented, often working long hours to ensure that each computer network meets each client's exact specifications and needs.

7

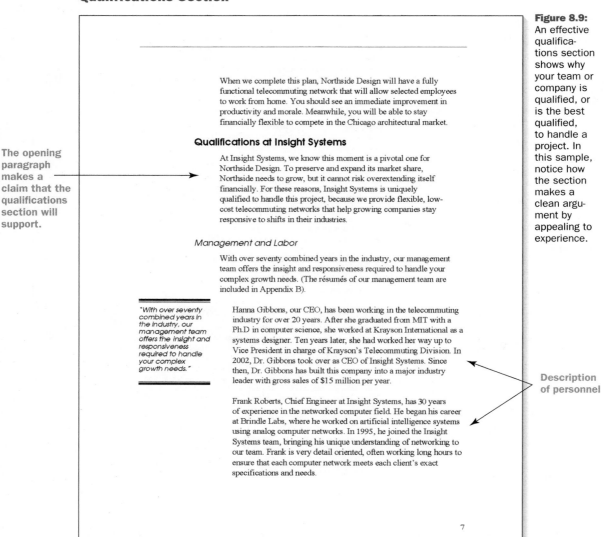

The opening paragraph makes a claim that the qualifications section will support.

Description of personnel

Figure 8.9: An effective qualifications section shows why your team or company is qualified, or is the best qualified, to handle a project. In this sample, notice how the section makes a clean argument by appealing to experience.

Figure 8.9 shows a sample qualifications section that includes these three kinds of information. Pay attention to how this section does more than describe the company—it makes an argument that the bidders are uniquely qualified to handle the project.

For more help writing a qualifications section, go to **www.pearsonhighered.com/johnsonweb4/8.11**

Figure 8.9: *(continued)*

Lisa Miller, Insight System's Senior Computer Engineer, has successfully led the implementation of thirty-three telecommuting systems in companies throughout the United States. Earning her computer science degree at Iowa State, Lisa has won numerous awards for her innovative approach to computer networking. She believes that clear communication is the best way to meet her clients' needs.

Our management is supported by one of most advanced teams of high technology employees. Insight Systems employs twenty of the brightest engineers and technicians in the telecommunications industry. We have aggressively recruited our employees from the most advanced universities in the United States, including Stanford, MIT, Illinois, Iowa State, New Mexico, and Syracuse. Several of our engineers have been with Insight Systems since it was founded.

Corporate History and Facilities

Insight Systems has been a leader in the telecommuting industry from the beginning. In 1982, the company was founded by John Temple, a pioneer in the networking field. Since then, Insight Systems has followed Dr. Temple's simple belief that computer-age workplaces should give people the freedom to be creative.

Description of organization

"Insight Systems earned the coveted '100 Companies to Watch' designation from Business Outlook Magazine."

Recently, Insight Systems earned the coveted "100 Companies to Watch" designation from *Business Outlook Magazine* (May 2010). The company has worked with large and small companies, from Vedder Aerospace to the Cedar Rapids Museum of Fine Arts, to create telecommuting options for companies that want to keep costs down and productivity high.

Insight Systems' Naperville office has been called "a prototype workspace for the information age" (*Gibson's Computer Weekly*, May 2006). With advanced LAN systems in place, only ten of Insight System's fifty employees actually work in the office. Most of Insight Systems' employees telecommute from home or on the road.

Experience You Can Trust

Our background and experience give us the ability to help Northside manage its needs for a more efficient, dynamic office space. Our key to success is innovation, flexibility, and efficiency.

8

You should never underestimate the importance of the qualifications section in a proposal. In the end, your readers will not accept the proposal if they do not believe that your team or company has the personnel, facilities, or experience to do the work.

Concluding with Costs and Benefits

Most proposals end by summarizing the benefits of the project and identifying the costs of the project. The conclusion of a proposal usually makes most of these five moves:

Move 1: Make an obvious *transition* that signals the conclusion.

Move 2: State the *costs* of the project.

Move 3: Summarize the *benefits* of the project.

Move 4: Briefly *describe the future* if the readers say yes.

Move 5: *Thank* the readers and offer contact information.

AT A GLANCE

Concluding a Proposal

- Restate the proposal's main point (the solution).
- Say thank you.
- Describe the next step.
- Provide contact information.

Start out this section by making an obvious transition that will wake up your readers. Say something like, "Let us conclude by summarizing the costs and benefits of our plan." Then, tell them the costs in a straightforward way without an apology or sales pitch.

As shown in our budget, this renovation will cost $287,000.

We anticipate that the price for retooling your manufacturing plant will be $5,683,000.

After the costs, summarize or list the two to five major benefits of saying yes to your project. Usually, these benefits are the deliverables you mentioned in the project plan section earlier in your proposal. By putting these benefits right after the costs, you can show readers exactly what they will receive for their investment.

Then, add a paragraph in which you describe the better future that will happen if the readers agree to your ideas. Show them how their investment of funds will make an improvement.

Finally, thank the readers for their consideration of the proposal and offer contact information (e.g., phone number and e-mail address) where they can reach you if they have more questions or need more information.

In most cases, your discussion of the benefits should not add new ideas to the proposal. In Figure 8.10, for example, notice that this proposal's costs and benefits section really doesn't add anything new to the proposal, except the cost of the project.

Link

For more information on writing conclusions, go to Chapter 16, page 443.

Using Style and Design

Style and design are crucial to a proposal's success. A clear, persuasive proposal that is attractive and easy to navigate will be much more competitive than one that sounds boring and looks plain. Plus, good style and design helps inspire trust in your readers.

GO TO THE NET

To see discussions of costs and benefits, go to **www.pearsonhighered.com/johnsonweb4/8.12**

Costs and Benefits Section

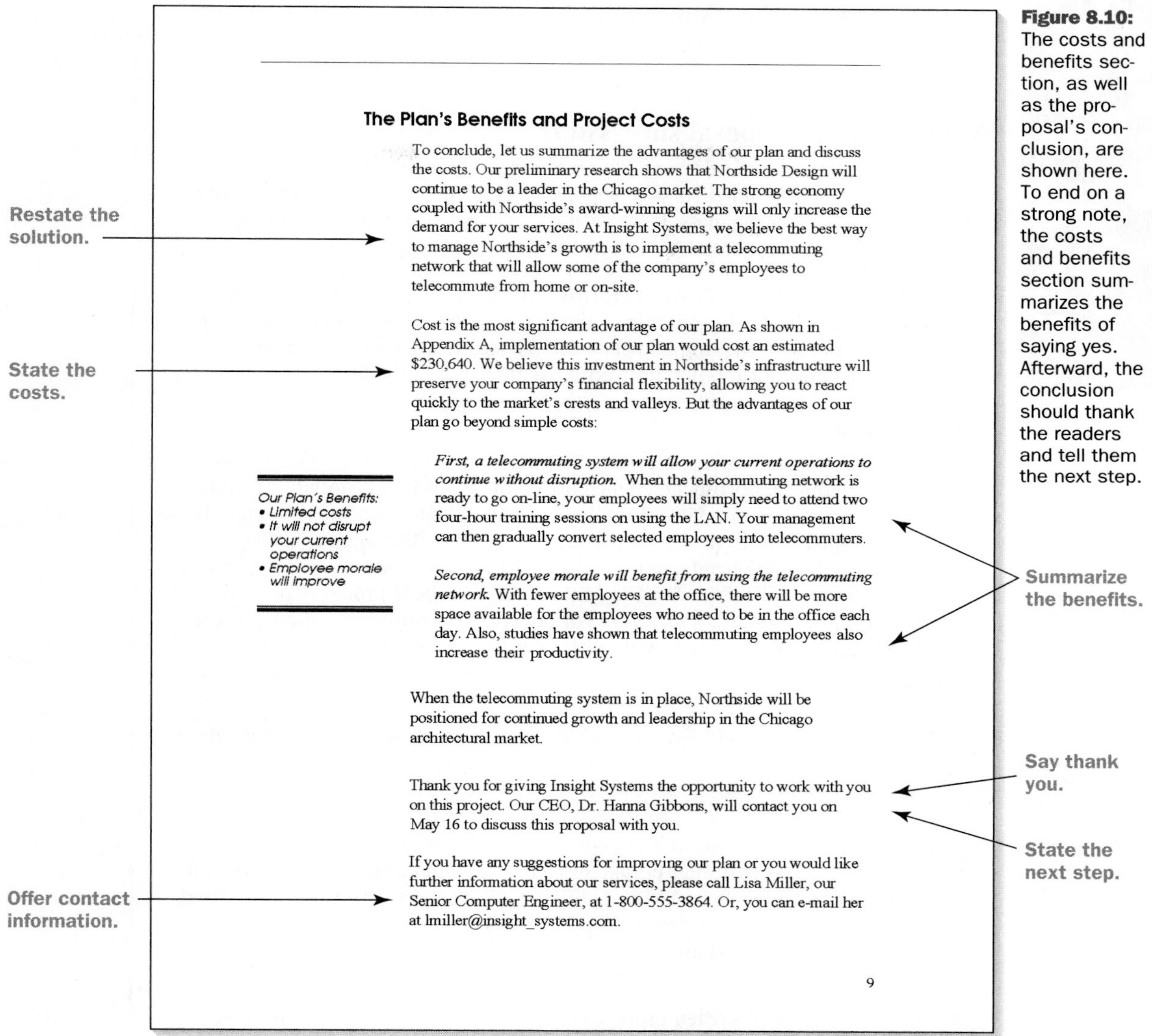

The Plan's Benefits and Project Costs
To conclude, let us summarize the advantages of our plan and discuss the costs. Our preliminary research shows that Northside Design will continue to be a leader in the Chicago market. The strong economy coupled with Northside's award-winning designs will only increase the demand for your services. At Insight Systems, we believe the best way to manage Northside's growth is to implement a telecommuting network that will allow some of the company's employees to telecommute from home or on-site.
Cost is the most significant advantage of our plan. As shown in Appendix A, implementation of our plan would cost an estimated $230,640. We believe this investment in Northside's infrastructure will preserve your company's financial flexibility, allowing you to react quickly to the market's crests and valleys. But the advantages of our plan go beyond simple costs:
Our Plan's Benefits:
• Limited costs
• It will not disrupt your current operations
• Employee morale will improve
First, a telecommuting system will allow your current operations to continue without disruption. When the telecommuting network is ready to go on-line, your employees will simply need to attend two four-hour training sessions on using the LAN. Your management can then gradually convert selected employees into telecommuters.
Second, employee morale will benefit from using the telecommuting network. With fewer employees at the office, there will be more space available for the employees who need to be in the office each day. Also, studies have shown that telecommuting employees also increase their productivity.
When the telecommuting system is in place, Northside will be positioned for continued growth and leadership in the Chicago architectural market.
Thank you for giving Insight Systems the opportunity to work with you on this project. Our CEO, Dr. Hanna Gibbons, will contact you on May 16 to discuss this proposal with you.
If you have any suggestions for improving our plan or you would like further information about our services, please call Lisa Miller, our Senior Computer Engineer, at 1-800-555-3864. Or, you can e-mail her at lmiller@insight_systems.com.
9
Restate the solution.
State the costs.
Summarize the benefits.
Say thank you.
State the next step.
Offer contact information.

Figure 8.10: The costs and benefits section, as well as the proposal's conclusion, are shown here. To end on a strong note, the costs and benefits section summarizes the benefits of saying yes. Afterward, the conclusion should thank the readers and tell them the next step.

Jane Perkins

GLOBAL WEB STRATEGIST AND PRODUCER, A.T. KEARNEY, CHICAGO, ILLINOIS

A.T. Kearney is a global management consulting firm.

What makes a proposal successful?

Writing proposals is critical work at A.T. Kearney, a global management consulting firm made up primarily of MBAs with engineering backgrounds. It is how we obtain a significant part of our business. Our success in writing proposals depends on key aspects of our proposal process:

- *Web-based processes for data and expertise.* We have formal processes in place to proactively gather important data that is necessary for providing potential clients with decision-making information. We also rely on an informal and collaborative network for tapping into expertise throughout the firm. These processes help us write proposals that include the necessary level of detail to persuade stakeholders.
- *A cohesive proposal and selling process.* For many opportunities, we develop a series of proposal documents, which are components of a negotiation/selling process, often extending over months. For example, an opportunity might begin with a formal RFP to down-select competitors. The RFP often specifies providing both online and print versions with strict requirements on content, length, and software application. For the next round, we would probably prepare a PowerPoint presentation for the client buyers. This round would open the door for refinements on the scope and initiate negotiations. The proposal process might then conclude with a proposal letter to summarize the aspects we had discussed.
- *Innovation and internal education.* Our proposal writing spurs much innovation and forward thinking, as we respond to evolving client needs and opportunities, and we build on recent project work and research. Proposal teams are not only educating clients through proposals but also important, internal stakeholders, such as those responsible for ensuring financials and quality delivery of the project. The use of SharePoint sites that are maintained by practices and collaborative applications facilitate this learning process. Additionally, expert reviews at specified stages of the proposal process are part of our culture—leadership insights and approvals help the proposal team, while the latest thinking is spread across the firm.

As you can see, a well-written proposal is only part of a much larger process.

Want more help developing your proposal's style? Go to **www.pearsonhighered.com/johnsonweb4/8.13**

A Balance of Plain and Persuasive Styles

Proposals need to educate and persuade, so they tend to use a mixture of plain and persuasive styles.

Link
For more strategies for using plain and persuasive style, see Chapter 17, page 445.

> **Plain style**—Use plain style in places where description is most important, such as the current situation section, the project plan section, and the qualifications section.
>
> **Persuasive style**—Use persuasive style in places where readers are expected to make decisions, such as the proposal's introduction and the costs and benefits section.

One persuasive style technique, called "setting a tone," is particularly effective when writing a proposal.

1. Determine how you want your proposal to sound. Do you want it to sound exciting, innovative, or progressive? Choose a word that best reflects the tone you want your readers to hear as they are looking over your proposal.
2. Put that word in the middle of your screen or sheet of paper. Circle it.
3. Write words associated with that word around it. Circle them also.
4. Keep mapping farther out until you have filled the page or screen.

Figure 8.11 shows a logical map of the word *progressive.* As shown in this map, you can quickly develop a set of words that are related to this word.

Once you have finished mapping out the tone, you can weave these words into your proposal to create a theme. As discussed in Chapter 17, if you use these words carefully and strategically in your proposal, your readers will hear this tone as they are reading your document.

An Attractive, Functional Design

The design of your proposal needs to be visually appealing, while helping readers find the information they need. Figure 8.12, for example, shows the first page of a well-designed proposal. Notice how the design of the proposal sets a professional, progressive tone for the whole document—even before you start reading.

When designing your proposal, you should consider three components: graphics, page design, and medium:

Link
For more information on using graphics, see Chapter 19, page 523.

> **Graphics**—In proposals, it is common to include charts, graphs, maps, illustrations, photographs, and other kinds of graphics to reinforce important points.

Mapping to Set a Tone

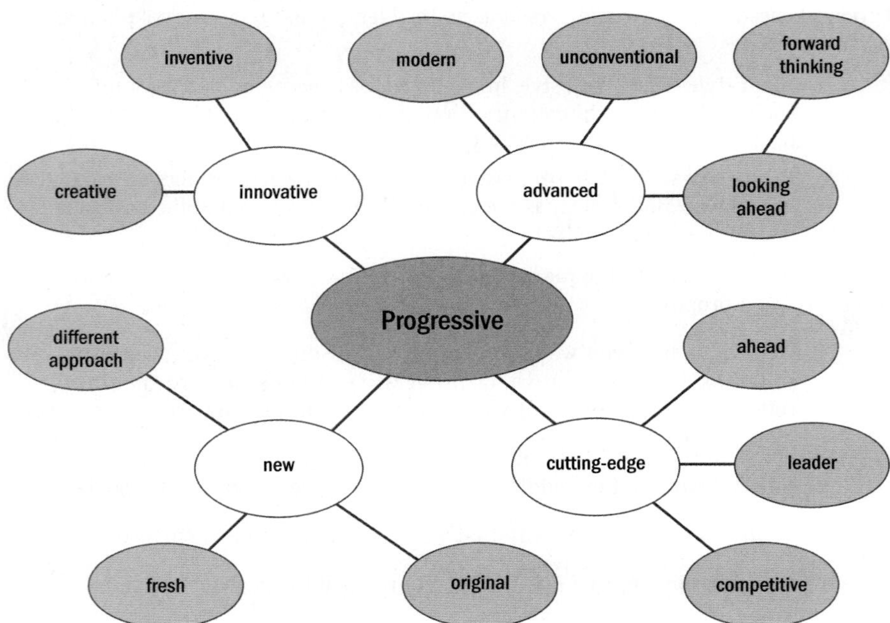

Figure 8.11: Logical mapping can help you develop a tone for your proposal. Weaving these words into a text would make the document sound "progressive."

Page design—Page layouts for proposals vary from simple to elaborate. At a minimum, you should use headings, lists, and graphics. More elaborate page layouts might include multiple columns, margin comments, pull quotes, and sidebars. Choose a page design that suits your readers and the context in which they will use the proposal.

Medium—The appropriate medium is also an important choice. Paper is still the norm for most proposals, but increasingly, companies are using multimedia, websites, and presentation software to deliver their ideas.

Link

For more information on page and screen layout, see Chapter 18, page 486.

Link

For more information on developing oral presentations with presentation software, see Chapter 21, page 579.

In the rush to get the proposal out the door, you might be tempted to skip the design phase of the writing process. You might even convince yourself that visual design doesn't matter to readers. Don't fool yourself. Good design is very important in proposals. Design makes the proposal more attractive while helping readers locate important information in the document.

To see samples of well-designed and badly designed proposals, go to **www.pearsonhighered.com/johnsonweb4/8.15**

Sample Proposal Design

Header adds color and anchors the top of the page.

Title of proposal is bold and easy to read.

Headings provide access points into text.

Two-column format makes text more scannable.

Figure 8.12: The page design of this proposal sets a professional tone while helping readers scan for important information. The tone is set with positive words and phrasings, while the graphics and headings make the information easy to access.

Graphic supports text while adding color.

A Proposal for Upgrading the National-Scale Soil Geochemical Database for the United States

The most requested data from the U.S. Geological Survey's (USGS) National Geochemical Database is a set of 1,323 soil samples. Why? Consider the following examples:

Example 1—Imagine for a moment that you are employed by an environmental regulatory agency of either the Federal or a State Government. Your assignment is to establish a "remediation value" for arsenic in soil at a contaminated site where a wood preservative facility once operated. Current arsenic values in soils at the facility range from 15 to 95 ppm and you must decide the concentration of arsenic that is acceptable after remediation efforts are completed. Scientists refer to the natural or native concentration of an element in soils as the "background concentration." Given the fact that arsenic occurs naturally in all soils, how would you determine the background concentration of arsenic in soils for this particular area?

Example 2—Your environmental consulting firm has been assigned to work with a team of specialists conducting a risk-based assessment of land contaminated with lead, zinc, and cadmium from a metal foundry. The assessment would determine the likelihood of adverse health or ecological effects caused by the contaminants. Again, an important part of this determination is, "What is the background concentration of these elements in the soil?"

What data are available for persons responsible for making the determinations of background concentrations for soils contaminated with potentially toxic metals? The most-often-quoted data set for background concentrations of metals and other trace elements in soil of the conterminous United States consists of only 1,323 samples collected during the 1960s and 1970s by the U.S. Geological Survey (Boerngen and Shacklette, 1981; Shacklette and Boerngen, 1984). (There is a similar data set for Alaska (Gough and others, 1984, 1988)). Samples for the "Shacklette data" were collected from a depth of about 1 ft, primarily from noncultivated fields having native vegetation, and samples were analyzed for more than 40 elements. Data in this study represent about one sample per 2,300 mi^2, indicating that very few samples were collected in each State. For example, the State of Arizona is covered by only 47 samples, and Pennsylvania has only 16. Despite the low number of samples, this data set is still being used on a regular basis to determine background concentrations of metals in soil to aid in remediation or risk-based assessments of contaminated land.

The only other national-scale soil geochemical data set for the United States was generated by the Natural Resources Conservation Service (NRCS), formerly the Soil Conservation Service (Holmgren and others, 1993). This data set consists of 3,045 samples of agricultural soil collected from major crop-producing areas of the conterminous United States. The primary purpose of this study was to assess background levels of lead and cadmium in major food crops and in soils on which these crops grow. Thus, the samples were only analyzed for five metals—lead, cadmium, copper, zinc, and nickel.

ARSENIC

Figure 1. Map of arsenic distribution in soils and other surficial materials of the conterminous United States based on 1,323 sample localities as represented by the black dots.

The Shacklette data set allows us to produce geochemical maps for specific elements, such as that shown on figure 1 for arsenic (Gustavsson and others, 2001). A map produced from such sparse data points obviously carries a large degree of uncertainty with it and does not have the resolution needed to answer many of the questions raised by land-management and regulatory agencies, earth scientists, and soil scientists. An example of the poor data set resolution is illustrated for Pennsylvania (fig. 2). The State is divided into major soil taxonomic units referred to as Suborders (Soil Survey Staff, 1999). Suborders group similar soil types in any region. The dots represent the sample points from the Shacklette data set. The few sample points shown on figure 2 illustrate that this data set would be inadequate for someone who must define the arsenic content of a given soil. At this time, no data set exists that will allow us to make these kinds of determinations.

The USGS and NRCS are currently studying the feasibility of a national-scale soil geochemical survey that will increase the sample density of the Shacklette data set by at least a factor of 10. This project, called Geochemical Landscapes, began in October 2002. The first 3 years will be devoted to determining how such a survey should be conducted. Therefore, we are actively soliciting input from potential customers of the new data. Interested members of the private sector, government, or academic communities

U.S. Department of the Interior
U.S. Geological Survey

Printed on recycled paper

USGS Fact Sheet FS-015-03
March 2003

Source: U.S. Geological Survey, 2003.

MICROGENRE

The Elevator Pitch

An elevator pitch is a one- or two-minute proposal that pitches a new idea, project, or service to potential investors or clients. If you type "Elevator Pitch Competition" into an Internet search engine, you will find videos of competitions in which entrepreneurs or college students compete to give the best elevator pitches. The best elevator pitches sell a good idea quickly. The elevator pitch shown here is for graFighters, an online game created by Dave Chenell and Eric Cleckner, two students at Syracuse.

Here is how to create your own elevator pitch:

Introduce yourself and establish your credibility. Remember that people invest in other *people,* not in projects. So tell them *who* you are and *what* you do.

Grab them with a good story. You need to grab your listeners' attention right away, so ask them, "What if _____?" or explain, "Recently, _____ happened and we knew there must be a better way."

Present your big idea in one sentence. Don't make them wait. Hit them with your best idea up front in one sentence.

Give them your best two or three reasons for doing it. The secret is to sell your idea, not explain it. List your best two or three reasons with minimal explanation.

Mention something that distinguishes you and your idea from the others. What is unique about your idea? How does your idea uniquely fit your listener's prior investments?

Offer a brief cost-benefits analysis. Show them very briefly that your idea is worth their investment of time, energy, or money.

Make sure they remember you. End your pitch by telling them something memorable about you or your organization. Make sure you put your contact information in their hand (e.g., a business card or résumé). If they allow it, leave them a written version of your pitch.

Write

Make your own elevator pitch. Think of a new product or service that might be useful to students at your university. Then, with a team of other students, write a persuasive one-minute elevator pitch. If time allows, present your pitch to your class.

Elevator Pitch: graFighters

graFighters is the first online fighting game for your hand drawn characters. Upload your drawings to the site, challenge your friends, and watch as they battle it out from any computer or mobile device.

The idea for the game was spawned by the countless number of hours we spent drawing characters in our notebooks during class. At one point, we were sitting next to each other arguing about whose drawing would win in a fight. After wasting a healthy amount of class time on the discussion, we decided that it would be awesome if there were a way to let the characters decide for themselves. That moment is what inspired us to develop the game.

What makes graFighters really different than any other fighting game is that you don't control your character during the fight. When you upload your drawing, the computer analyzes how that character has been drawn and then determines its strengths, weaknesses, and fighting style. The algorithm that makes all of these decisions is something we've code named "Cornelius."

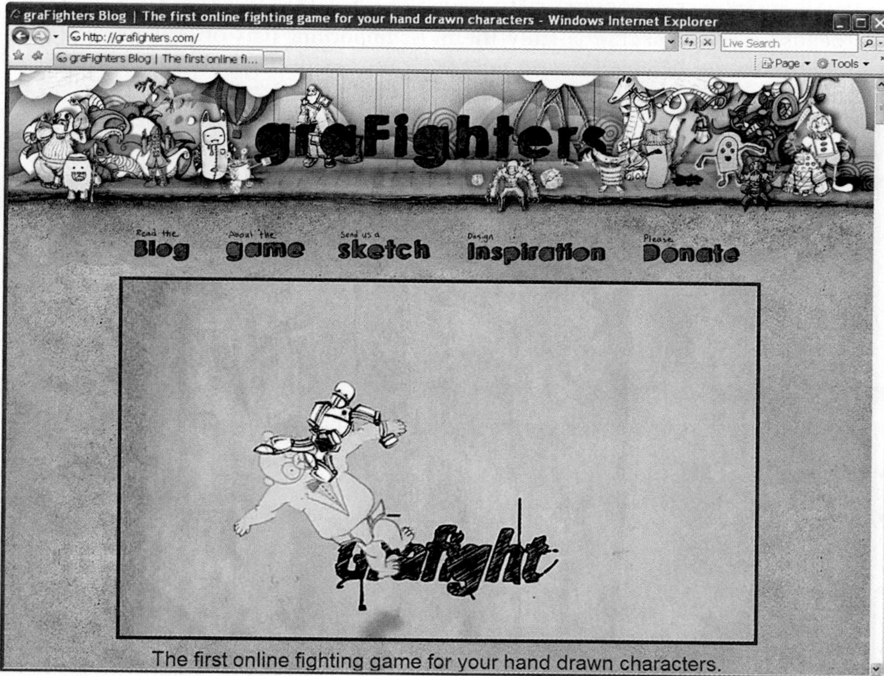

graFighters is a business idea pitched by Syracuse students Dave Chenell and Eric Cleckner.

Source: Dave Chenell and Eric Cleckner, graFighters.com.

(continued)

So what we're trying to get across is that the most important part of the game happens with your pencil and paper. That's where your fights are really won or lost. Once you have uploaded your character to the site, your drawing has essentially become a living, breathing, fighting machine. At this point, the player's role has changed from creator to coach, setting up matches and determining what changes they could make to their fighter to develop them into a better warrior. In essence, we're letting players "Lose Control" (graFighters tag-line) in exchange for a more creative type of gameplay.

We are excited about creating a game with art and design right at the core of it. For us, it's all about the creativity. The game is played with your pencil and paper—the rest, everything that happens on the site, is the aftermath of your creation.

To date, with the help of our awesome team of programmers, we have been able to develop the game engine, which is the part that actually animates the characters and allows them to fight. We are also in the midst of refining the character uploading process. But the reason that we've reached out to the kickstarter community is because we need money to build "Cornelius," which truly is the most important part of the game.

We are asking for $20,000, all of which will go toward the development of "Cornelius." This algorithm is truly the most important part of graFighters. With it in place, we will be able to accurately portray your drawings in the game just as you have imagined them. This means that if your characters have the ability to fly and breathe fire it will actually do that in-game. Or if you gave your character a double-barrel shotgun, you can expect to see paper bullets wreaking havoc on your opponent.

Please help us raise the money to build this algorithm. Drawings all over the world will be eternally grateful for your donation.

CHAPTER REVIEW

- Proposals are documents that present ideas or plans for consideration.
- Proposals can be internal or external and solicited or unsolicited.
- Proposals usually include five sections: (1) introduction, (2) current situation, (3) project plan, (4) qualifications, and (5) costs and benefits.
- When planning the proposal, start out by defining the Five-W and How Questions. Then define the rhetorical situation (subject, purpose, readers, context of use).
- When organizing and drafting the proposal, each major section should be written separately with an opening, a body, and a closing.
- Proposals should use a combination of plain and persuasive styles to motivate readers.
- Design is absolutely essential in professional proposals. You should look for places where page layout and graphics will improve the readability of your proposal.

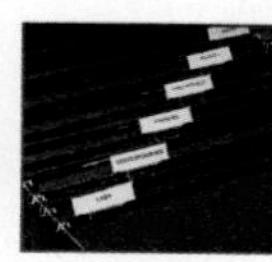

EXERCISES AND PROJECTS

Individual or Team Projects

1. Analyze the proposal "'Meaningful' Outdoor Experiences for Students," at the end of this chapter. Study the content, organization, style, and design of the proposal. Does it cover the four areas (current situation, project plan, qualifications, costs and benefits) that were discussed in this chapter? Where does it include more or less information than you would expect in a proposal?

   Write a two-page memo to your instructor in which you discuss whether you believe the proposal was effectively written and designed. Discuss the content, organization, style, and design in specific detail. Highlight the proposal's strengths and suggest ways in which it could be improved.

2. Find a Request for Proposals (RFP) at www.fedbizopps.gov or in the classifieds of your local newspaper. Analyze the RFP according to the Who, What, Where, When, Why, and How Questions. Then, prepare a presentation for your class in which you (a) summarize the contents of the RFP, (b) discuss why you believe the RFP was sent out, and (c) explain what kinds of projects would be suitable for this RFP.

3. Find a proposal that demonstrates weak style and/or design. Using the style and design techniques discussed in this chapter and in Chapters 17–19, revise the document so that it is more persuasive and more visual. Locate places where blocks of text could be turned into lists, and identify places where graphics would reinforce the text.

Collaborative Project: Improving Campus

As students, you are aware—perhaps more so than the administrators—of some of the problems on campus. Write a proposal that analyzes the causes of a particular problem on campus and then offers a solution that the administration might consider implementing. The proposal should be written to a *named* authority on campus who has the power to say yes to your proposal and put it into action.

Some campus-based proposals you might consider include the following:

- improving lighting on campus to improve safety at night.
- improving living conditions in your dorm or fraternity/sorority.
- creating a day-care center on campus.
- creating an adult commuter room in the student union.
- improving campus facilities for the handicapped.
- improving security in buildings on campus.
- creating a university public relations office run by students.
- increasing access to computers on campus.
- improving the parking situation.
- reducing graffiti and/or litter on campus.

To find sample proposals for analysis, go to **www.pearsonhighered.com/johnsonweb4/8.16**

- improving food service in the student union.
- helping new students make the transition to college.
- changing the grading system at the university.
- encouraging more recycling on campus.
- reducing the dependence on cars among faculty and students.

Use your imagination to come up with a problem to which you can offer a solution. You don't need to be the person who implements the program. Just offer some guidance for the administrators. In the qualifications section of your proposal, you will likely need to recommend that someone else should do the work.

For additional technical writing resources, including interactive sample documents, document design tutorials and guidelines, and more, go to **www.mytechcommlab.com.**

Revision Challenge

This proposal in Figure A is written well, but it could be strengthened in organization, style, and design. Read through the proposal and identify places where you could make the proposal stronger.

Sample Proposal

Figure A: This proposal is well written, but not perfect. How might you revise this document to make its content, organization, style, and design even stronger?

Project Summary

Organization:	**Monterey Bay High School**
Applicant Name:	Ted Watson
P.I.:	Ariel Jane Lichwart
Address:	590 Foam Street, Monterey, CA 93940
Telephone Number:	Ted: (831) 244-7843, Ariel: (831) 244-7842
Email Address:	Ted: Twatson@mbhs.edu, Ariel: Ajl@mbhs.edu
Partner:	Carmel River Watershed Council

Area of Interest: **"Meaningful" Outdoor Experiences for Students**

Project Title: A Meaningful Outdoor Experience for Students: To Monitor the Health of the Carmel River and Carmel Watershed

Project Period: 09/01/2003 – 10/31/2004

Project Objectives:

- To provide a meaningful outdoor experience for all of the 650 students at Monterey Bay High School;
- To assess the environmental quality and health of the Carmel River and the impact these problems have on the entire watershed and the larger ecological system;
- To teach the students how to monitor fish traps, conduct water quality tests, measure stream flow, operate plant surveys, perform bird density measurements (point counts), sample invertebrate and plankton tows, and perform herpetology surveys along the River;
- To assist local agencies that are concerned with the health of the Carmel River, to maintain existing populations and to assist in restoring those that are in decline through habitat restoration work, community-based education programs and increased local involvement in environmental projects;
- To raise the students' social awareness, stimulate observation, motivate critical thinking, and develop problem-solving skills.

Summary of Work: To provide high school students with Meaningful Outdoor Experiences by assessing the environmental quality and health of the Carmel River and Watershed.

Total Federal Funds:	**$50,000**
Cost Sharing:	**$25,643**
Total Project Costs:	**$75,643**

Source: National Marine Sanctuaries, http://sanctuaries.noaa.gov/news/bwet.

(continued)

Figure A: *(continued)*

Grant Proposal

Purpose of Project

Monterey Bay High School is requesting $50,000 from NOAA for Environmental Education Projects in the Monterey Bay Watershed to provide high school students with **Meaningful Outdoor Experiences** by assessing the environmental quality and health of the Carmel River and Watershed. The Carmel River Watershed Council will be our collaborative working partner during the course of the grant.

How

The project will take place over a one-year period of time, beginning September 1, 2003. It will consist of three phases: the preparation phase, the action phase, and the reflection phase. The preparation and reflection stages will each be 3 months long; the action phase will take place over a 6 month period.

Project Objectives

- To provide a meaningful outdoor experience for all of the 650 students at Monterey Bay High School;
- To assess the environmental quality and health of the Carmel River and the impact these problems have on the entire watershed and the larger ecological system;
- To teach the students how to monitor fish traps, conduct water quality tests, measure stream flow, operate plant surveys, perform bird density measurements (point counts), sample invertebrate and plankton tows, and perform herpetology surveys along the River;
- To assist local agencies that are concerned with the health of the Carmel River, to maintain existing populations and to assist in restoring those that are in decline through habitat restoration work, community-based education programs and increased local involvement in environmental projects;
- To raise the students' social awareness, stimulate observation, motivate critical thinking, and develop problem-solving skills.

Preparation Phase

All of the 650 students at Monterey Bay High School (grades 9-12) will participate in the preparation phase. At the outset of the preparation phase, we will propose the question: What is the present health of the Carmel River and how does the river's health affect the entire watershed?

During this phase the students and the teachers will do the initial background research on the Carmel River. The students will hear guest lectures delivered by our partner the Carmel River Watershed Council, and other state and local organizations that will be collaborating with us on this project (Carmel River Steelhead Association, Monterey Peninsula Audubon Society, Ventana Wilderness Society and the Monterey County Water Management District, California State Parks). These agencies will also supply us

Figure A:
(continued)

with data previously collected that charts the Carmel River's vegetation, birds, wildlife, and fisheries, invertebrates, and water quality. Certain problems will quickly become apparent to the students such as erosion, vegetation loss, coliform and other water pollutants, loss of wildlife densities and use of habitat, steelhead declines, and invertebrate declines. After discussion and analysis students will make their predictions about other potential problems in the Carmel River Watershed. We will then set up our goals to investigate in greater depth some of the problems associated with the River and the Watershed.

Action Phase
All of the students will participate in this field research phase. After a site visit to four selected Carmel River locations (upper Watershed, middle River site, site adjacent to the High School and Carmel River mouth), we will set up a phenological schedule (timeline) for sampling the sites with the students and the teachers. Each site will be sampled at least once per month over a six month period. Some of the activities and projects conducted at each site will include the monitoring of fish traps, conducting water quality tests, measuring stream flow, operating plant surveys, performing bird density measurements (point counts) and a bird banding program, sampling invertebrate and plankton tows, and performing herpetology surveys. All students will keep extensive journals, and record their observations and experiments at the River sites.

Reflection Phase
Following the six-month research project at the Carmel River Watershed sites, the students and teachers will discuss their findings, refocus on the initial question, analyze the conclusions reached, and evaluate the results. We will also conduct appropriate assessment activities to evaluate the project and the learning achieved. The students will be divided into teams, and each team will write up a different aspect of each finding. The students will present this data to all interested stakeholders, including the Carmel River Steelhead Association, the Monterey County Water Management District, Monterey Regional Parks, California State Parks, the Ventana Wilderness Society and the Monterey Peninsula Audubon Society. Their conclusions will be used by these agencies to help evaluate further watershed studies, to maintain existing populations, and to assist in restoring those that are in decline through habitat restoration work, community-based education programs, and increased local involvement in environmental projects. Local media will report on our results and the projects that will take place as a result of the students' work.

Why

1 - Demonstrate to students that local actions can impact the greater water environment (i.e., Monterey Bay). An intentional connection needs to be made to water quality, the watershed, and the larger ecological system.

Through this project, the students will learn that the Carmel River is besieged with problems such as erosion, vegetation loss, coliform and other water pollutants, loss of wildlife densities and use of habitat, steelhead declines, and invertebrate declines. Part of

(continued)

Figure A:
(continued)

the Reflection phase will be used to study the impact these problems in the River have on the entire Watershed and the larger ecological system. As with all of the lessons we teach at our habitat, the students will learn about the connectedness of all living creatures and all actions, both large and small.

2 - Experiences should include activities where questions, problems, and issues are investigated through data collection, observation, and hands-on activities.

Much of the work done throughout this project involves data collection, observation, and hands-on activities. In the preparation phase, the students and teachers will be analyzing data previously collected by our partner organization, the Carmel River Watershed Council, on the vegetation, birds, wildlife and fisheries, invertebrates, and water quality. In the action phase, the students and teachers will perform hands-on activities at four sites along the river, at least once a month at each site, over a six-month period. These activities include the monitoring of fish traps, conducting water quality tests, measuring stream flow, operating plant surveys, performing bird density measurements (point counts) and banding birds, sampling invertebrate and plankton tows, and performing herpetology surveys. In the reflection phase, the students will examine the questions asked in light of the data they collected. They will then share their data with supporting local and state organizations so that some of the problems discovered can begin to be eradicated.

3 - Experiences should stimulate observation, motivate critical thinking, develop problem solving skills, and instill confidence in students.

There is no question but that this project will stimulate observation, motivate critical thinking, and develop problem solving skills. As a result, the students will acquire confidence, not only in their investigative skills, but also to be active participants in the public debate on many environmental issues. It is our experience that projects such as this one not only raise social awareness, but academic skills as well. It is a well known fact that students learn better by doing; by aligning the work done for this project with California state Standards (see number 7) the academic performance of the students will clearly be enhanced.

4 - Activities should encourage students to assist, share, communicate, and connect directly with the outdoors. Experiences can include: (1) Investigative or experimental design activities where students or groups of students use equipment, take measurements and make observations for the purpose of making interpretations and reaching conclusions; (2) Project-oriented experiences, such as restoration, monitoring, and protection projects, that are problem solving in nature and involve many investigative skills. These experiences should involve fieldwork, data collection, and analysis.

As stated above, all of the activities for this project involve investigative or experimental design activities where the students use equipment, take measurements, and make observations for the purpose of making interpretations and reaching conclusions. In

Figure A: *(continued)*

addition, many of the activities in the action phase involve monitoring projects that are problem solving in nature and involve many investigative skills. All of the experiences involve fieldwork, data collection, and analysis.

5- The "Meaningful" outdoor experiences need to be part of a sustained activity; the total duration leading up to and following the activity should involve a significant investment of instructional time.

Since many of the activities that will take place during all three phases will be aligned with the California State Standards, the "Meaningful" outdoor experiences will be embedded into the science curriculum (see number 7), and will encompass at least 10% of instructional time over the year period of the grant. This sustained activity will involve a significant outlay of instructional time, not just in the classroom, but in the field as well. As the budget narrative sets forth, all teachers, the principal investigator, and the field coordinators will invest many hours preparing and teaching the lessons as well as helping the students to perform the tasks at the site.

6- An experience should consist of three general parts: a preparation phase—which focuses on a question, problem, or issue and involves students in discussions about it; an action phase—which includes one or more outdoor experiences sufficient to conduct the project, make the observations, or collect the data required; and the reflection phase—which refocuses on the question, problem, or issue, analyzes the conclusion reached, evaluates the results, and assesses the activity and the learning.

This project will consist of a preparation phase, an action phase, and a reflection phase. The details of these are set forth above.

7- "Meaningful" outdoor experiences must be an integral part of the instructional program and clearly part of what is occurring concurrently in the classroom; be aligned with the California academic learning standards; and make appropriate connections among subject areas and reflect an integrated approach to learning. Experiences should occur where and when they fit into the instructional sequence.

All of the sampling and research will be aligned with the California State Standards. For example, water sampling fits into our 9th grade chemistry curriculum; erosion aligns with the earth science requirement for 10th and 11th grade; wildlife monitoring is part of the scientific investigation component that is required for all grade levels. In addition, this Project is a perfect complement to the River of Words program, where the students submit poetry and art to the international contest that is based on their experiences and observations of their Watershed (see description, above).

8- Project should demonstrate partnerships that form a collaborative working relationship, with all partners taking an active role in the project.

The Carmel River Watershed Council (CRWC) will be a collaborative working partner during the course of this grant. The CRWC is a nonprofit, community-based organization

(continued)

Figure A: *(continued)*

founded in 1999 to work with local, state, and federal agencies for improved management of the Carmel River Watershed. The primary mission of the CRWC is the protection of the natural resources that form the Carmel River Watershed.

The CRWC will take an active role in this Project. During the course of this Project, they will provide the students and teachers at Monterey Bay High School with data that has been collected by them and other agencies concerning the vegetation, birds, wildlife and fisheries, invertebrates, and water quality so that the students can assess and analyze the potential problems in the Carmel River. The principal investigator, who is a consultant for the CRWC (and is also a teacher at the High School), will organize and coordinate the project, and provide, on an ongoing basis, expertise, teaching skills, and leadership to guide the staff in working with the students in the sampling of the river, monitoring of fish traps, conducting water quality tests, measuring stream flow, operating plant surveys, performing bird density measurements and bird banding, sampling invertebrate and plankton tows, and performing herpetology surveys. In all of these activities, he and other consultants at CRWC will work with the staff at the Biological Sciences Project and the participating teachers at the Monterey Bay High School to provide the students with a meaningful outdoor experience that will encourage them to have greater pride and ownership of their environment and become more effective and thoughtful community leaders and participants. The students will be doing important research that will benefit not only the Carmel River Watershed Council, but other local and state agencies, such as the Carmel River Steelhead Association, the Monterey County Water Management District, Monterey Regional Parks, California State Parks, the Ventana Wilderness Society, and the Monterey Peninsula Audubon Society. All of the data that is gathered will be shared with these groups through student presentations and be used to evaluate future watershed activities that will help to maintain existing populations and assist in restoring ones that are in decline through habitat restoration work, community-based education programs, and increased local involvement in community projects.

Who

The principal organization is the Monterey Bay High School. The Carmel River Watershed Council will be collaborating as our partner. Both organizations have been described in great detail, above.

The target audience is high school students, grades 9–12. All of the 650 students in the school will participate in all three phases of the project. The students and teachers at Monterey Bay High School, as well as the consultants that work with the Carmel River Watershed Council, have a great deal of experience using outdoor, hands-on activities as an integral part of the educational process. This project is an ideal and natural complement to the activities that take place at our habitat. We are committed to using the environment as a context for learning and improving the understanding of environmental stewardship of students and teachers. Many of the teachers at the school have extensive experience in teaching bird banding, data analysis, and data measurement, and they understand how to teach these types of scientific techniques to this age group of students. For those that do not have expertise in the subject matter that will be emphasized in this Project, the

Figure A:
(continued)

Principal Investigator and other consultants from local agencies with expertise in these areas will provide **Professional Development** for the teachers.

The Project Manager for the Habitat will be coordinating this project and the Project Intern will assist in the gathering and computation of the data. Four teachers, all from the science department, will participate in all three phases of the Project, and will serve as field coordinators.

A consultant from the Carmel River Watershed Council, who is also a science teacher at the High School, will be directing the study and serve as the Principal Investigator. He will coordinate the three phases, oversee the teachers, and synchronize the visits and data collection during the action phase (see number 8, above).

Where

The Project will take place at Monterey Bay High School and on the Carmel River. The Carmel River is 36 miles long. It drains about 255 square miles while flowing NW out of the valley between the Santa Lucia Mountains on the South and the Sierra del Salinas to the North and East. The river empties into the Pacific Ocean near Carmel, California. The lower river flows from the mouth to the narrows, about 9 miles up stream; the middle river flows from the narrows to Camp Stephani; the upper river flows through rugged canyons.

We will set up four study sites: at the Carmel River mouth; at a site close to the High School; in the mid-Valley area; and at Cachagua Community Park below Los Padres Dam.

Need

This important project could not take place without government financial assistance. The financial health of the state and county budgets preclude funding for projects such as this one; in addition private grant funding is becoming increasingly difficult to secure. Without the financial aid of the federal government, these types of meaningful outdoor experiences for students could not take place.

Benefits or Results Expected

There are a number of benefits and results that will be derived from the proposed activities. First, the students will gather important data that will help to clarify what can be done to help bring the Carmel River and the Carmel Watershed to a state of good health. The students will be doing significant research that will benefit not only the Carmel River Watershed Council, but other local and state agencies, such as the Carmel River Steelhead Association, the Monterey County Water Management District, Monterey Regional Parks, California State Parks, the Ventana Wilderness Society, and the Monterey Peninsula Audubon Society. All of the data that is gathered will be shared with these groups through student presentations and be used to evaluate future watershed

(continued)

Figure A: *(continued)*

activities that will help to maintain existing populations and assist in restoring ones that are in decline through habitat restoration work, community-based education programs, and increased local involvement in community projects.

In addition, the process of progressing through the preparation phase, the action phase, and the reflection phase will provide the students with a meaningful outdoor experience that will encourage them to have greater pride and ownership of their environment and become more effective and thoughtful community leaders and participants. This project will stimulate observation, motivate critical thinking, and develop problem solving skills. As a result, the students will acquire confidence, not only in their investigative skills, but also to be active participants in the public debate on many environmental issues.

Project Evaluation

We will employ various methodologies to ensure that we are meeting the goals and objectives of our project. First, we will give pre and post tests to the students before the project begins and at its conclusion to determine if they have learned the science and skills that we want to teach through this project. In addition, students will be tested periodically throughout the course of the project as part of their coursework requirements.

Second, all students will keep extensive journals in which they will evaluate the data they have collected. At the end of the project, they will be asked to summarize their findings and submit them in a report. In addition, each student will be required to perform an independent science experiment that relates to the work being done at the River (i.e., measure the amount of nitrates in the water).

Third, the results of the data that are gathered will be presented by the students to the local and state agencies that are stakeholders in the Carmel River Watershed. These reports will be evaluated and critiqued by these agencies, along with suggestions for future studies.

Fourth, the students will use their experiences at the river as the basis for the art and poetry they submit to the River of Words competition that takes place each spring.

Finally, interested classes of incoming 9th graders will be invited to take a trip to one of the sites where the High School students can mentor or teach them some of the skills that they have learned.

CASE STUDY

The Mole

Henry Espinoza knew this proposal was important to his company, Valen Industries. If the proposal was successful, it could be worth millions of dollars in short-term and long-term projects for his employer. For Henry, winning the project meant a likely promotion to vice president and a huge raise in salary. The company's CEO was calling him daily to check up and see if it was going well. That meant she was getting stressed out about the project.

Through the grapevine, Henry knew his company really had only one major competitor for the project. So the odds were good that Valen Industries would win the contract. Of course, Henry and his team needed to put together an innovative and flawless proposal to win, because their major competitor was going all out to get this project.

Then, one day, something interesting happened. One of Henry's team members, Vera Houser, came to him with an e-mail from one of her friends, who worked for their main competitor. In the e-mail, this person offered to give Valen Industries a draft of its competitor's proposal. Moreover, this friend would pass along any future drafts as the proposal evolved. In the e-mail, Vera's friend said he was very frustrated working at his "slimy" company, and he was looking for a way out. He also revealed that his company was getting "inside help" on the proposal, but Henry wasn't sure what that meant.

Essentially, this person was hinting that he would give Henry's team a copy of the competitor's proposal and work as a mole if they considered hiring him at Valen Industries. Henry knew he would gain a considerable advantage over his only competitor if he had a draft of its proposal to look over. The ability to look at future drafts would almost ensure that he could beat his competitor. Certainly, he thought, with this kind of money on the line, a few rules could be bent. And it seemed that his competitor was already cheating by receiving inside help.

Henry went home to think it over. What would be the ethical choice in this situation? How do you think Henry should handle this interesting opportunity?

GO TO THE NET

Discussions of proprietary information are available at
www.pearsonhighered.com/johnsonweb4/8.17

CHAPTER

9

Activity Reports

Types of Activity Reports *247*

Planning and Researching *254*

Organizing and Drafting *260*

Using Style and Design *263*

Microgenre: The Status Report *264*

Chapter Review *266*

Exercises and Projects *266*

Case Study: Bad Chemistry *268*

In this chapter, you will learn:

- The basic features of activity reports.
- About the different kinds of activity reports and how they are used in the workplace.
- How to determine the rhetorical situation for an activity report.
- How to organize and draft an activity report.
- Strategies for using an appropriate style.
- How to design and format activity reports.

Today, companies are using computer networks to create management structures that are less hierarchical. As a result, companies require fewer levels of managers than before, because computer networks help top executives better communicate with employees throughout the company.

These "flatter" management structures require more communication, quicker feedback, and better accountability among employees in the company. As a result, activity reports are more common than ever in the technical workplace.

Activity reports are used to objectively present ideas or information within a company. This genre has many variations, making it adaptable to many situations that you will encounter in the technical workplace.

Types of Activity Reports

Even though the various types of activity reports are similar in most ways, they are called a variety of names that reflect their different purposes. Activity reports share one goal—to objectively inform readers about (1) what happened, (2) what is happening, and (3) what will happen in the near future.

Progress Reports

A progress report, also called a status report, is written to inform management about the progress or status of a project. These reports are usually written at regular intervals—weekly, biweekly, or monthly—to update management on what has happened since the last progress report was submitted. Your company's management may also periodically request a progress report to stay informed about your or your team's activities.

A typical progress report will provide the following information:

- a summary of completed activities
- a discussion of ongoing activities
- a forecast of future activities

Figure 9.1, for example, shows a progress report that is designed to update management on a project.

Briefings and White Papers

Briefings and white papers are used to inform management or clients about an important issue. Typically, briefings are presented verbally, while white papers are provided in print. Occasionally, briefings will also appear as "briefs" in written form.

Briefings and white papers typically present gathered facts in a straightforward and impartial way. They include the following kinds of information:

- a summary of the facts
- a discussion of the importance of these facts
- a forecast about the importance of these facts in the future

QUICK START

Activity Reports

This is a basic model for organizing an activity report. There are many different types of activity reports, so this pattern can and should be altered to fit the content and purpose of your document.

Basic Features of Activity Reports

An activity report usually includes the following features, which can be modified to suit the needs of the situations in which the report will be used:

- Introduction
- Summary of activities
- Results of activities or research
- Future activities or research
- Incurred or future expenses
- Graphics

Keep in mind that activity reports are used for a variety of purposes. You should adjust this pattern to suit your needs.

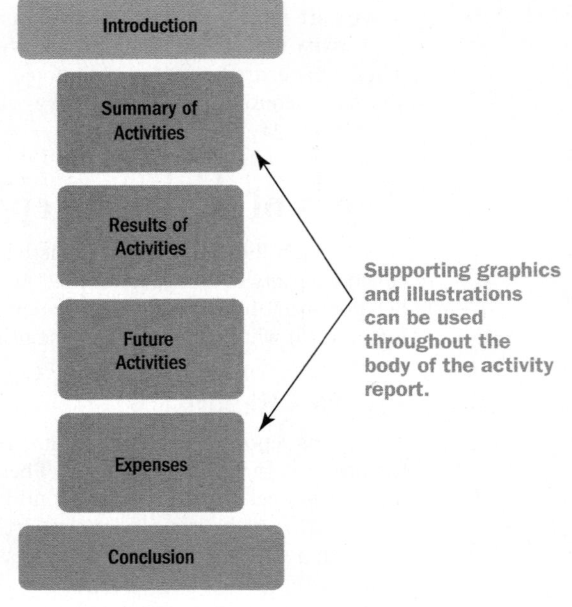

An effective briefing presents the facts as concisely as possible, leaving time for questions and answers. When you brief an audience on your subject, try to do so as objectively as possible. Then, interpret the importance of those facts based on evidence, not on speculation.

Briefings and white papers typically do not advocate for any specific side or course of action. Instead, they present the facts in a straightforward way and offer an objective assessment of what those facts mean. It is up to the readers to decide what actions are appropriate.

Figure 9.2 on pages 250–252 shows the executive summary of a white paper written by senior officers at PayPal. The white paper discusses PayPal's attempts to manage "phishing," a fraudulent form of e-mail spam.

GO TO THE NET

To see other sample progress reports, go to **www.pearsonhighered.com/johnsonweb4/9.1**

A Progress Report

Figure 9.1: A progress report describes the team's activities and discusses future activities.

Hanson Engineering

March 14, 2011
To: Charlie Peterson, Director
From: Sue Griego, Iota Team Manager
Subject: Progress Report, March 14

Subject, purpose, and main point are identified up front.

This month, we made good progress toward developing a new desalinization method that requires less energy than traditional methods.

Ongoing activities are described objectively.

Our activities have centered around testing the solar desalinization method that we discussed with you earlier this year. With solar panels, we are trying to replicate the sun's natural desalinization of water (Figure A). In our system, electricity from the photovoltaic solar panels evaporates the water to create "clouds" in a chamber, similar to the way the sun makes clouds from ocean water. The salt deposits are then removed with reverse osmosis, and freshwater is removed as steam.

The graphic supports the text.

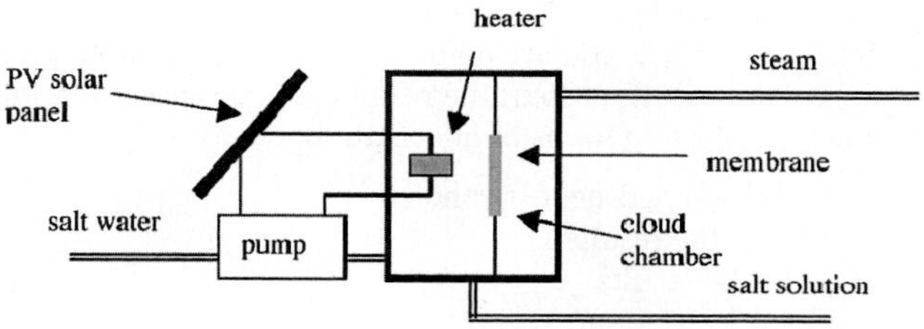

Figure A: The Desalinator

Results are presented.

We are succeeding on a small scale. Right now, our solar desalinator can produce an average of 2.3 gallons of freshwater an hour. Currently, we are working with the system to improve its efficiency, and we soon hope to be producing 5 gallons of freshwater an hour.

Report ends with a look to the future and a brief conclusion.

We are beginning to sketch out plans for a large-scale solar desalinization plant that would be able to produce thousands of gallons of freshwater per hour. We will discuss our ideas with you at the April 18 meeting.

Our supplies and equipment expenses for this month were $8,921. Looks like things are going well. E-mail me if you have questions. (Suegriego@hansoneng.net)

Incident Reports

Incident reports describe an event, usually an accident or irregular occurrence, and they identify what corrective actions have been taken. As with other kinds of activity reports, incident reports present the facts as objectively as possible. They provide the following information:

- a summary of what happened (the facts)
- a discussion of why it happened
- a description of how the situation was handled
- a discussion of how the problem will be avoided in the future

It is tempting, especially when an accident was your fault, to make excuses or offer apologies, but an incident report is not the place to do so. As with other activity reports, you should concentrate on the facts. Describe what happened as honestly and clearly as possible. You can make excuses or apologize later.

Figure 9.3 on pages 253–254 shows a typical incident report in which management is notified of an accident in a laboratory.

Laboratory Reports

Laboratory reports are written to describe experiments, tests, or inspections. If you have taken a laboratory class, you are no doubt familiar with lab reports. These reports describe the experiment, present the results, and discuss the results. Lab reports typically include the following kinds of information:

- a summary of the experiment (methods)
- a presentation of the results
- a discussion of the results

A White Paper

A Practical Approach to Managing Phishing

Michael Barrett, Chief Information Security Officer

Dan Levy, Senior Director of Risk Management – Europe

April 2008

Figure 9.2: A white paper presents technical information objectively, allowing readers to make decisions based on the facts.

Source: From A Practical Approach to Managing Phishing *by Michael Barrett, Chief Information Security Officer, and Dan Levy, Director of Risk Management-Europe. PayPal, April 2008. These materials have been reproduced with the permission of PayPal, Inc.*

To see white papers, go to **www.pearsonhighered.com/johnsonweb4/9.2**

Figure 9.2: *(continued)*

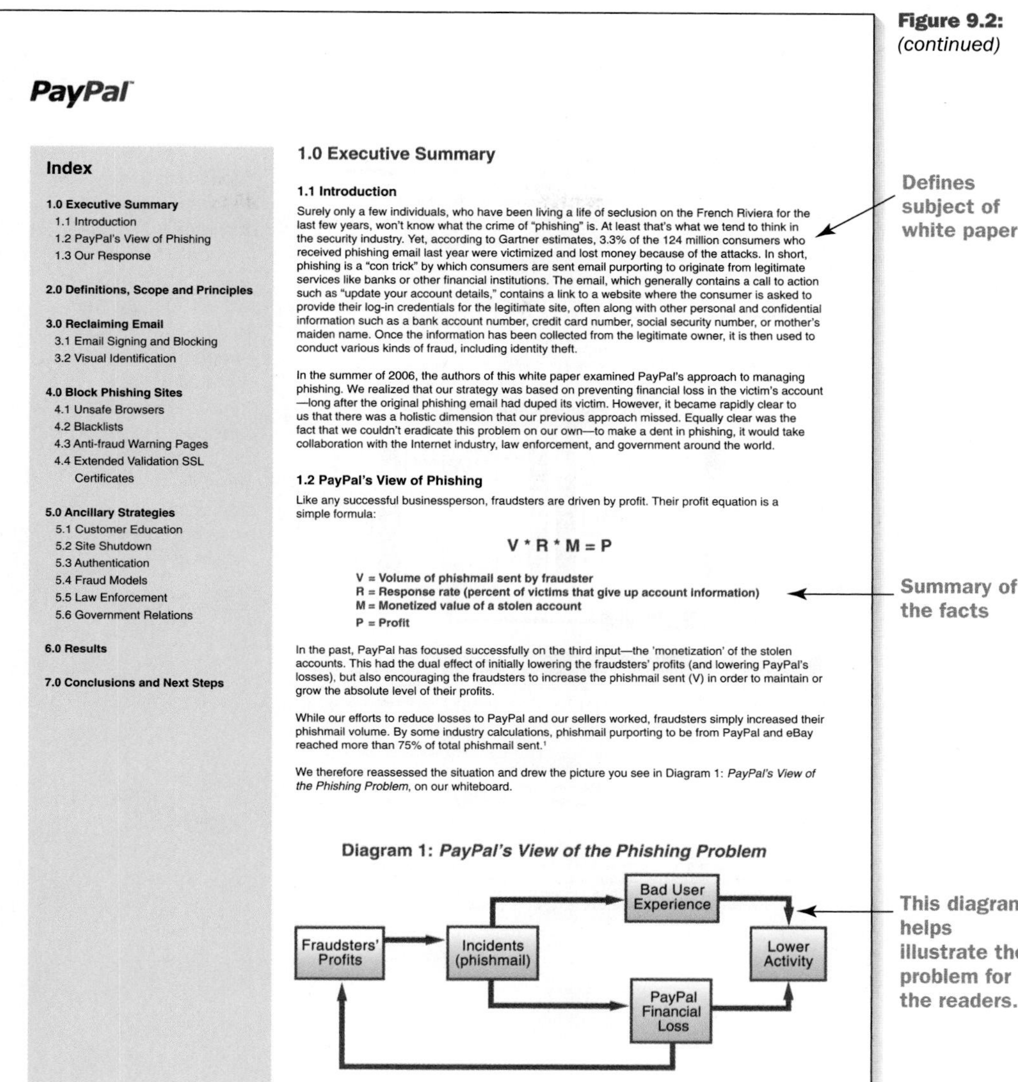

PayPal

Index

1.0 Executive Summary
- 1.1 Introduction
- 1.2 PayPal's View of Phishing
- 1.3 Our Response

2.0 Definitions, Scope and Principles

3.0 Reclaiming Email
- 3.1 Email Signing and Blocking
- 3.2 Visual Identification

4.0 Block Phishing Sites
- 4.1 Unsafe Browsers
- 4.2 Blacklists
- 4.3 Anti-fraud Warning Pages
- 4.4 Extended Validation SSL Certificates

5.0 Ancillary Strategies
- 5.1 Customer Education
- 5.2 Site Shutdown
- 5.3 Authentication
- 5.4 Fraud Models
- 5.5 Law Enforcement
- 5.6 Government Relations

6.0 Results

7.0 Conclusions and Next Steps

1.0 Executive Summary

1.1 Introduction

Surely only a few individuals, who have been living a life of seclusion on the French Riviera for the last few years, won't know what the crime of "phishing" is. At least that's what we tend to think in the security industry. Yet, according to Gartner estimates, 3.3% of the 124 million consumers who received phishing email last year were victimized and lost money because of the attacks. In short, phishing is a "con trick" by which consumers are sent email purporting to originate from legitimate services like banks or other financial institutions. The email, which generally contains a call to action such as "update your account details," contains a link to a website where the consumer is asked to provide their log-in credentials for the legitimate site, often along with other personal and confidential information such as a bank account number, credit card number, social security number, or mother's maiden name. Once the information has been collected from the legitimate owner, it is then used to conduct various kinds of fraud, including identity theft.

In the summer of 2006, the authors of this white paper examined PayPal's approach to managing phishing. We realized that our strategy was based on preventing financial loss in the victim's account —long after the original phishing email had duped its victim. However, it became rapidly clear to us that there was a holistic dimension that our previous approach missed. Equally clear was the fact that we couldn't eradicate this problem on our own—to make a dent in phishing, it would take collaboration with the Internet industry, law enforcement, and government around the world.

1.2 PayPal's View of Phishing

Like any successful businessperson, fraudsters are driven by profit. Their profit equation is a simple formula:

$$V * R * M = P$$

V = Volume of phishmail sent by fraudster
R = Response rate (percent of victims that give up account information)
M = Monetized value of a stolen account
P = Profit

In the past, PayPal has focused successfully on the third input—the 'monetization' of the stolen accounts. This had the dual effect of initially lowering the fraudsters' profits (and lowering PayPal's losses), but also encouraging the fraudsters to increase the phishmail sent (V) in order to maintain or grow the absolute level of their profits.

While our efforts to reduce losses to PayPal and our sellers worked, fraudsters simply increased their phishmail volume. By some industry calculations, phishmail purporting to be from PayPal and eBay reached more than 75% of total phishmail sent.[1]

We therefore reassessed the situation and drew the picture you see in Diagram 1: *PayPal's View of the Phishing Problem*, on our whiteboard.

Diagram 1: *PayPal's View of the Phishing Problem*

2

Defines subject of white paper

Summary of the facts

This diagram helps illustrate the problem for the readers.

(continued)

Figure 9.2: *(continued)*

Explains the team's solution for addressing the problem

We knew that fraudsters' profits drove them to send phishmail. In turn, phishmail drove two responses from our customers: a bad user experience and/or financial loss. Even though the vast majority of users would immediately delete the mail or not click on the phishing links, the negative experience caused residual concern about whether or not PayPal had given up their email addresses or otherwise caused them to be a target.

A much smaller group of users would give up their private financial data (including their passwords), leading to financial loss for PayPal and our customers. And of course, PayPal's financial loss was the fraudsters' profit—thus encouraging even more phishmail and perpetuating the cycle. No matter which path the customer followed, PayPal observed consistently lower activity from these users.

"...phishing is only the first step in a series of crimes that also frequently includes unauthorized account access, identity theft, financial fraud, money laundering, and others. While all of these crimes need to be considered together, this paper focuses primarily on the first step of this crime – phishing, specifically via email."

1.3 Our Response

We determined five blocking points (illustrated in Diagram 2: Five Blocking Points) where PayPal could break the business model of fraudsters, and we applied a strategy against each point. There would not be one silver bullet, but perhaps five would do the trick.

Diagram 2: *Five Blocking Points*

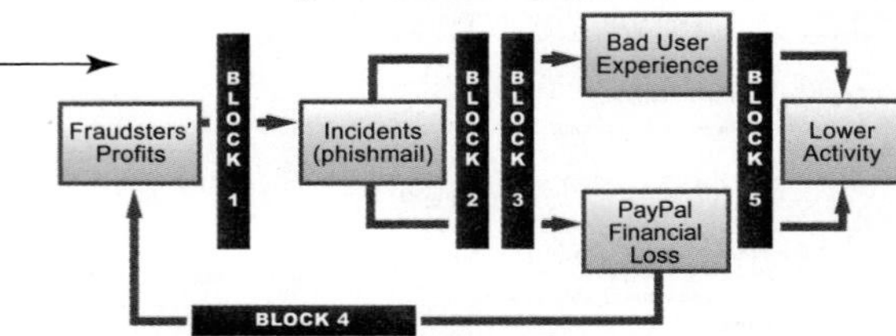

Block / Strategy	Short Explanation
1. Reclaim email	Prevent phishmail from ever entering customers' inboxes. To be judged by the adoption of email authentication by ISPs who agree not to deliver unsigned email.
2. Block phishing sites	Prevent phishing sites from being displayed to customers.
3. Authenticate users	Prevent stolen login/password combinations from being used on PayPal.com.
4. Prosecute	Create a disincentive by pursuing legal prosecution (in partnership with government and law enforcement).
5. Brand & Customer Recovery	Ensure that targeted customers would still use PayPal.

This paper concentrates on our top two strategies of reclaiming email and blocking phishing sites. For each, we will address both our "passive" and our "active" solutions, corresponding to the different segments of our user base. There are also a few other strategies that we believe are important, and while they may be ancillary to those described above, we'll give some detail about them nonetheless.

3

An Incident Report

Figure 9.3: An incident report is not the place to make apologies or place blame. You should state the facts as objectively as possible.

Red Hills Health Sciences Center (RH)

Testing and Research Division
201 Hospital Drive, Suite A92
Red Hills, CA 92698

March 10, 2010

To: Brian Jenkins, Safety Assurance Officer

From: Hal Chavez, Testing Laboratory Supervisor

Subject: Incident Report: Fire in Laboratory

Subject and purpose are stated up front.

I am reporting a fire in Testing Laboratory 5, which occurred yesterday, March 9, 2010, at 3:34 p.m.

What happened is described objectively.

The fire began when a sample was being warmed with a bunsen burner. A laboratory notebook was left too close to the burner, and it caught fire. One of our laboratory assistants, Vera Cather, grabbed the notebook and threw it into a medical waste container. The contents of the waste container then lit on fire, filling the room with black smoke. At that point, another laboratory assistant, Robert Jackson, grabbed the fire extinguisher and emptied its contents into the waste container, putting out the fire. The overhead sprinklers went off, dousing the entire room.

What was done about it is noted.

Even though everyone seemed fine, we decided to send all lab personnel down to the emergency room for an examination. While we were in the waiting room, Vera Cather developed a cough and her eyes became red. She was held for observation and released that evening when her condition was stable. The rest of us were looked over by the emergency room doctors, and they suggested that we stay out of the laboratory until it was thoroughly cleaned.

I asked the hospital's HazMat team to clean up the mess that resulted from the fire. We had been working with samples of *Borrelia burgdorferi* bacteria, which causes Lyme disease. I was not sure if the

(continued)

Want to see other incident reports? Go to **www.pearsonhighered.com/johnsonweb4/9.3**

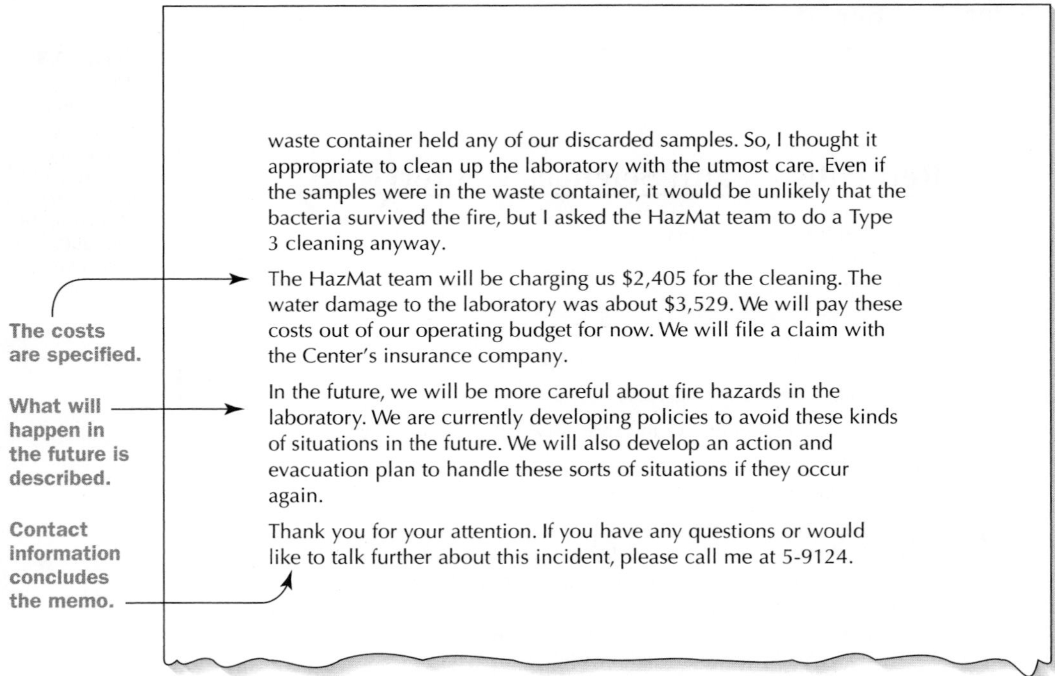
waste container held any of our discarded samples. So, I thought it appropriate to clean up the laboratory with the utmost care. Even if the samples were in the waste container, it would be unlikely that the bacteria survived the fire, but I asked the HazMat team to do a Type 3 cleaning anyway.

The HazMat team will be charging us $2,405 for the cleaning. The water damage to the laboratory was about $3,529. We will pay these costs out of our operating budget for now. We will file a claim with the Center's insurance company.

In the future, we will be more careful about fire hazards in the laboratory. We are currently developing policies to avoid these kinds of situations in the future. We will also develop an action and evacuation plan to handle these sorts of situations if they occur again.

Thank you for your attention. If you have any questions or would like to talk further about this incident, please call me at 5-9124.

The costs are specified.
What will happen in the future is described.
Contact information concludes the memo.

Figure 9.3: *(continued)*

Lab reports, like other activity reports, emphasize the facts and data. Here is not the place to speculate or to develop a new theory. Instead, your lab report should present the results as objectively as possible and use those results to support the reasoned discussion that follows.

Figure 9.4 shows an example of a laboratory report. In this report, the writer describes the results of the testing as objectively as possible.

Planning and Researching

One of the nice things about writing activity reports is that you have already developed most of the content. You are probably familiar with your readers. So, minimal planning is required, and the research has been mostly completed. These internal reports, after all, are supposed to describe your activities.

A good workplace practice you might adopt is keeping an *activity journal* or *work log* on your computer or in a notebook. In your journal, start out each day by jotting down the things you need to accomplish. As you complete each of these activities, note the dates and times they were completed and the results.

GO TO THE NET Other lab reports can be viewed at **www.pearsonhighered.com/johnsonweb4/9.4**

A Lab Report

Figure 9.4: A lab report walks readers through the methods, results, and discussion. Then, it offers any conclusions, based on the facts.

FEND-LAB, INC.
2314 Universal St., Suite 192
San Francisco, CA 94106
(325) 555-1327
www.fendlabcal.com

Test Address
NewGen Information Technology, LLC
3910 S. Randolph
Slater, CA 93492

Client
Brian Wilson
Phone: 650-555-1182
Fax: 650-555-2319
e-mail: brian_wilson@cssf.edu

Mold Analysis Report
Report Number: 818237-28
Date of Sampling: 091310
Arrival Date: 091510
Analysis Date: 092010
Technician: Alice Valles

Lab Report: Mold Test

The introduction states the subject, purpose, and main point.

In this report, we present the results of our testing for mold at the offices of NewGen Information Technology, at 3910 S. Randolph in Slater, California. Our results show above-normal amounts of allergenic mold, which may lead to allergic reactions among the residents.

Testing Methods

On 13 September 2010, we took samples from the test site with two common methods: Lift Tape Sampling and Bulk Physical Sampling.

Methods are described, explaining how the study was done.

Lift Tape Sampling. We located 10 areas around the building where we suspected mold or spores might exist (e.g., water stains, dusty areas, damp areas). Using 8-cm-wide strips of transparent tape, we lifted samples and pressed them into the nutrient agar in petri dishes. Each sample was sealed and sent to our laboratory, where it was allowed to grow for one week.

(continued)

Figure 9.4: *(continued)*

Bulk Physical Sampling. We located 5 additional areas where we observed significant mold growth in ducts or on walls. Using a sterilized scraper, we removed samples from these areas and preserved them in plastic bags. In one place, we cut a 1-inch-square sample from carpet padding because it was damp and contained mold. This sample was saved in a plastic bag. All the samples were sent to our laboratory.

At the laboratory, the samples were examined through a microscope. We also collected spores in a vacuum chamber. Mold species and spores were identified.

Results of Microscopic Examination

The following chart lists the results of the microscope examination:

Mold Found	Location	Amount
Trichoderma	Break room counter	Normal growth
Geotrichum	Corner, second floor	Normal growth
Cladosporium	Air ducts	Heavy growth
Penicillium spores	Corkboard in bathroom	Normal growth

Descriptions of molds found:

Results are presented objectively, without interpretation.

Trichoderma: Trichoderma is typically found in moistened paper and unglazed ceramics. This mold is mildly allergenic in some humans, and it can create antibiotics that are harmful to plants.

Geotrichum: Geotrichum is a natural part of our environment, but it can be mildly allergenic. It is usually found in soil in potted plants and on wet textiles.

Cladosporium: Cladosporium can cause serious asthma and it can lead to edema and bronchiospasms. In chronic cases, this mold can lead to pulmonary emphysema.

Penicillium: Penicillium is not toxic to most humans in normal amounts. It is regularly found in buildings and likely poses no threat.

2

Figure 9.4: *(continued)*

Results are interpreted and discussed.

Discussion of Results

It does not surprise us that the client and her employees are experiencing mild asthma attacks in their office, as well as allergic reactions. The amount of Cladosporium, a common culprit behind mold-caused asthma, is well above average. More than likely, this mold has spread throughout the duct system of the building, meaning there are probably no places where employees can avoid coming into contact with this mold and its spores.

The other molds found in the building could be causing some of the employees' allergic reactions, but it is less likely. Even at normal amounts, Geotrichum can cause irritation to people prone to mold allergies. Likewise, Trichoderma could cause problems, but it would not cause the kinds of allergic reactions the client reports. Penicillium in the amounts found would not be a problem.

The results of our analysis lead us to believe that the Cladosporium is the main problem in the building.

The conclusion restates the main point and recommends action.

Conclusions

The mold problem in this building will not go away over time. Cladosporium has obviously found a comfortable place in the air ducts of the building. It will continue to live there and send out spores until it is removed.

We suggest further testing to confirm our findings and measure the extent of the mold problem in the building. If our findings are confirmed, the building will not be safely habitable until a professional mold remover is hired to eradicate the mold.

Ignoring the problem would not be wise. At this point, the residents are experiencing mild asthma attacks and occasional allergic reactions. These symptoms will only grow worse over time, leading to potentially life-threatening situations.

Contact us at (325) 555-1327 if you would like us to further explain our methods and/or results.

3

Page from an Activity Journal

The "To Do" list keeps track of what needs to be handled each day.

Color-coding allows users to prioritize items on the "To Do" list.

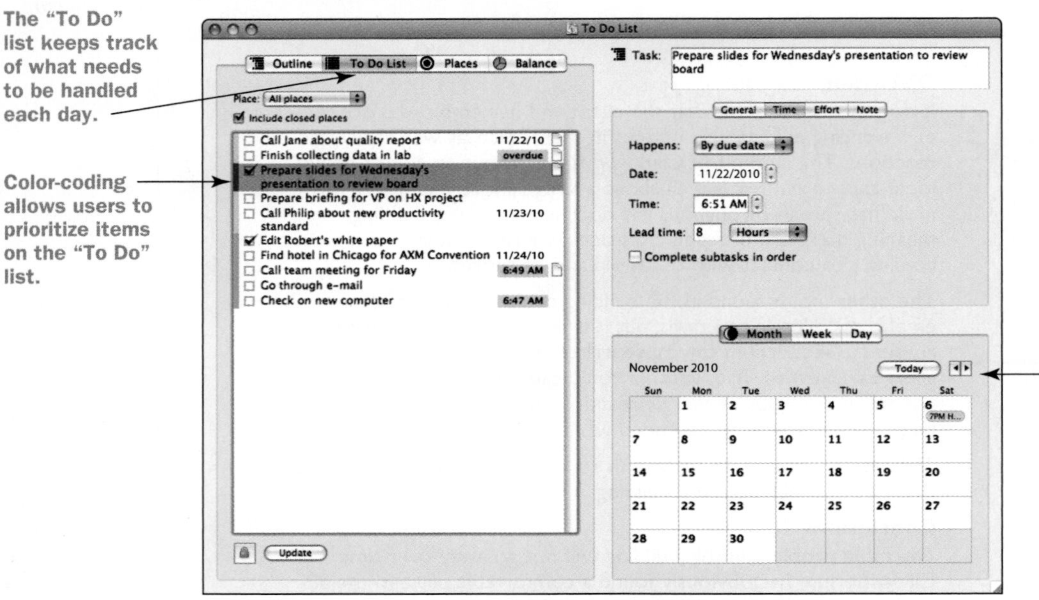

Figure 9.5: A variety of software and groupware programs have the ability to keep a "To Do" list and calendar for you. Here is a program by Llamagraphics called Life Balance.

The calendar can be used to keep track of day-to-day activities.

Source: Llamagraphics, Life Balance.

Computers can also help you keep track of your activities. Software like Llamagraphic's Life Balance and groupware programs like Lotus Notes and Microsoft Outlook can maintain "to do" lists while keeping a calendar of your activities (Figure 9.5). Similarly, mobile phones can also maintain a "to do" list and your calendar. When you complete a task, you can cross it off your list. The mobile phone will keep track of the time and date.

At first, keeping an activity journal will seem like extra work. But you will soon realize that your journal keeps you on task and saves you time in the long run. Moreover, when you need to report on your activities for the week or month, you will have a record of all the things you accomplished.

Analyzing the Rhetorical Situation

With your notes in front of you, you are ready to plan your activity report. You should begin by briefly answering the Five-W and How Questions:

Who might read or use this activity report?

Why do they want the report?

For links to software packages that help you keep activity journals, go to **www.pearsonhighered.com/johnsonweb4/9.5**

What information do they need to know?

Where will the report be used?

When will the report be used?

How might the report be used?

After considering these questions, you can begin thinking about the rhetorical situation that will shape how you write the activity report or present your briefing.

SUBJECT The subject of your report includes your recent activities. Include only information your readers need to know.

PURPOSE The purpose of your report is to describe what happened and what will happen in the future. In your introduction, state your purpose directly:

> In this memo, I will summarize our progress on the Hollings project during the month of August 2010.
>
> The purpose of this briefing is to update you on our research into railroad safety in northwestern Ohio.

You might use some of the following action verbs to describe your purpose:

to explain	*to show*	*to demonstrate*
to illustrate	*to present*	*to exhibit*
to justify	*to account for*	*to display*
to outline	*to summarize*	*to inform*

Link

To learn more about defining a purpose, go to Chapter 1, page 6.

READERS Think about the people who will need to use your report. The readers of activity reports tend to be your supervisors. Occasionally, though, these kinds of reports are read by clients (lab reports or briefings) or used to support testimony (white papers). An incident report, especially when it concerns an accident, may have a range of readers who plan to use the document in a variety of ways.

Link

For more ideas about reader analysis, turn to Chapter 2, page 20.

CONTEXT OF USE The context of use for your activity report will vary. In most cases, your readers will simply scan and file your report. Similarly, oral briefings are not all that exciting. Your listeners will perk up for the information that interests them, but they will mostly be checking to see if you are making progress.

Nevertheless, take a moment to decide whether your activity report discusses any topics that involve troublesome ethical or political issues. When mistakes happen, auditors and lawyers will go through your activity reports, looking for careless statements or admissions of fault. So, your statements need to reflect your actual actions and the results of your work.

GO TO THE NET

Worksheets to help you define your purpose, readers, and context of use can be found at **www.pearsonhighered.com/johnsonweb4/9.6**

Link

For more help defining the context, go to Chapter 2, page 26.

Moreover, if you are reporting expenses in your activity report, they need to be accurate. Auditors and accountants will look at these numbers closely. If your numbers don't add up, you may have some explaining to do.

Organizing and Drafting

Remember, organizing and drafting activity reports should not take too much time. If you find yourself taking more than an hour to write an activity report, you are probably spending too much time on this routine task.

To streamline your efforts, remember that all technical documents have an introduction, a body, and a conclusion. Each of these parts of the document makes predictable moves that you can use to guide your drafting of the report.

Writing the Introduction

Readers of your activity report are mostly interested in the facts. So, your introduction should give them only a brief framework for understanding those facts. To provide this framework, you should concisely

- define your subject.
- state your purpose.
- state your main point.

Figures 9.1, 9.3 and 9.4 show examples of concise introductions that include these three common introductory moves.

Link

For more advice on writing introductions, turn to Chapter 16, page 430.

If your readers are not familiar with your project (e.g., you are giving a demonstration to clients), you might want to expand the introduction by also offering background information, stressing the importance of the subject, and forecasting the body of the report. For example, Figure 9.6 shows the introduction of a document that would accompany a demonstration for people who would not be familiar with micromachines.

Writing the Body

In the body of the activity report, you should include some or all of the following:

Summary of activities—In chronological order, summarize the project's two to five major events since your previous activity report. Highlight any advances or setbacks in the project.

Results of activities or research—In order of importance, list the two to five most significant results or outcomes of your project. To help a reader scan, you might even use bullets to highlight these results.

Future activities or research—Tell readers what you will be doing during the next work period.

To see more examples of briefs, go to **www.pearsonhighered.com/johnsonweb4/9.7**

Full Introduction for an Activity Report

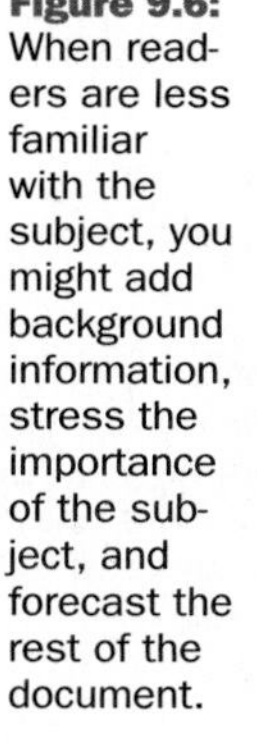

Figure 9.6: When readers are less familiar with the subject, you might add background information, stress the importance of the subject, and forecast the rest of the document.

Wilson National Laboratory

Always Moving Forward

Nanotech Micromachines Demonstration for Senators Laura Geertz and Brian Hanson
Presented by Gina Gould, Head Engineer

Background information is offered for readers unfamiliar with the topic. → Nanotechnology is the creation and utilization of functional materials, devices, and systems with novel properties and functions that are achieved through the control of matter, atom by atom, molecule by molecule, or at the macromolecular level. A revolution has begun in science, engineering, and technology, based on the ability to organize, characterize, and manipulate matter systematically at the nanoscale.

Purpose and main point are mentioned here. → In this demonstration, we will show you how the 5492 Group at Wilson National Laboratory is applying breakthroughs in nanotechnology science toward the development of revolutionary new micromachines. Our work since 2002 has yielded some amazing results that might dramatically expand the capacity of these tiny devices.

Forecasting shows the structure of the briefing. → Today, we will first show you a few of the prototype micromachines we have developed with nanotechnology principles. Then, we will present data gathered from testing these prototypes. And finally, we will discuss future uses of nanotechnology in micromachine engineering.

Expenses—If asked, you should state the costs incurred over the previous week or month. Highlight any places where costs are deviating from the project's budget.

The body of the activity report shown in Figure 9.7 includes these four items.

Writing the Conclusion

The conclusion should be as brief as possible. You should

- restate your main point.
- restate your purpose.

To see more examples of activity report memos, go to **www.pearsonhighered.com/johnsonweb4/9.8**

Progress Report

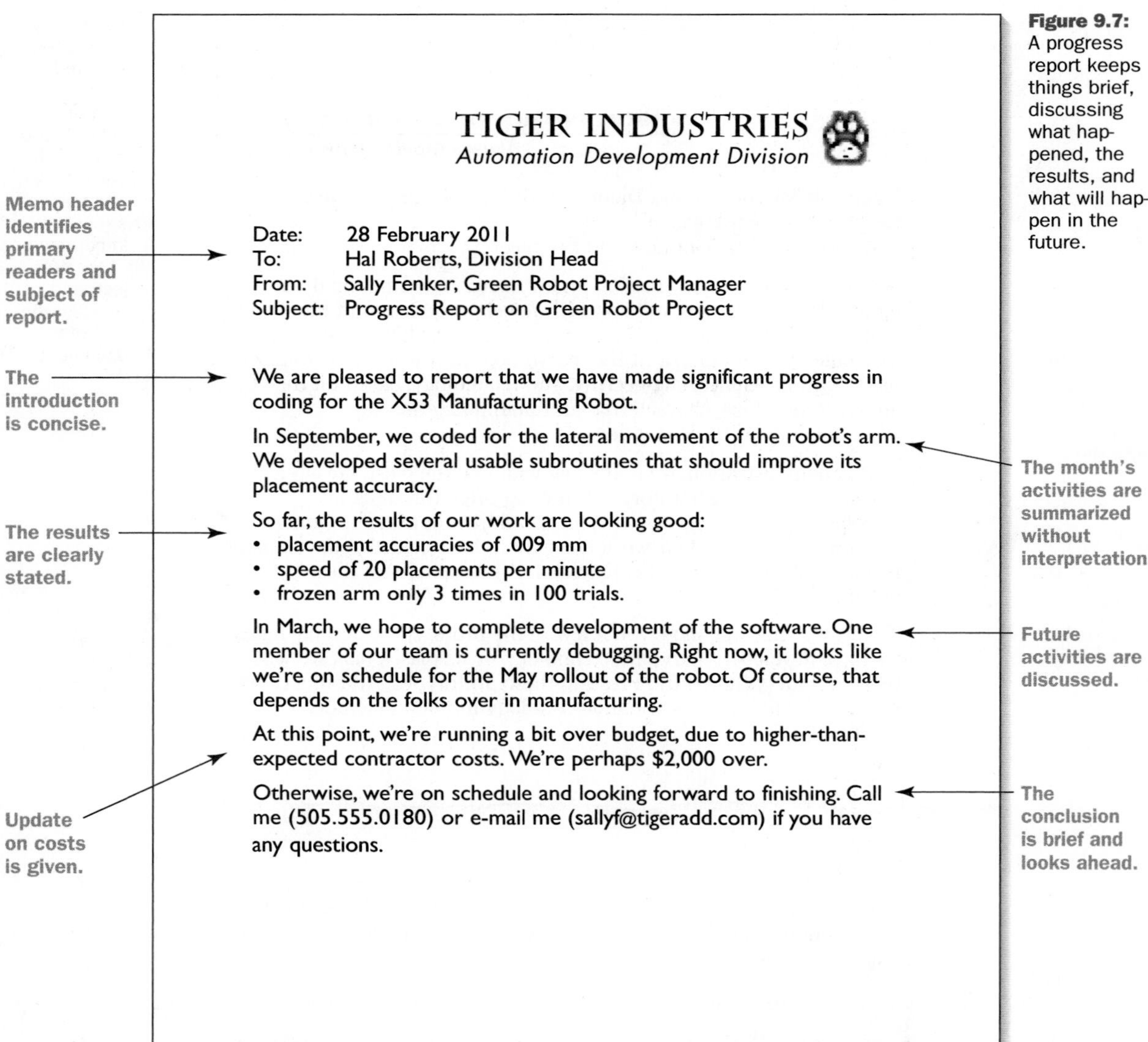

TIGER INDUSTRIES

Automation Development Division

Date: 28 February 2011
To: Hal Roberts, Division Head
From: Sally Fenker, Green Robot Project Manager
Subject: Progress Report on Green Robot Project

We are pleased to report that we have made significant progress in coding for the X53 Manufacturing Robot.

In September, we coded for the lateral movement of the robot's arm. We developed several usable subroutines that should improve its placement accuracy.

So far, the results of our work are looking good:

- placement accuracies of .009 mm
- speed of 20 placements per minute
- frozen arm only 3 times in 100 trials.

In March, we hope to complete development of the software. One member of our team is currently debugging. Right now, it looks like we're on schedule for the May rollout of the robot. Of course, that depends on the folks over in manufacturing.

At this point, we're running a bit over budget, due to higher-than-expected contractor costs. We're perhaps $2,000 over.

Otherwise, we're on schedule and looking forward to finishing. Call me (505.555.0180) or e-mail me (sallyf@tigeradd.com) if you have any questions.

Memo header identifies primary readers and subject of report.
The introduction is concise.
The results are clearly stated.
Update on costs is given.
The month's activities are summarized without interpretation.
Future activities are discussed.
The conclusion is brief and looks ahead.

Figure 9.7: A progress report keeps things brief, discussing what happened, the results, and what will happen in the future.

- make any recommendations, if appropriate.
- look to the future.

These concluding moves should be made in a maximum of two to four sentences.

> To conclude, in this demonstration, our goal was to update you on our progress toward developing nanotechnology micromachines. Overall, it looks like we are making solid progress toward our objectives, and we seem to be on schedule. Over the next couple of months, we will be facing some tough technical challenges. At that point, we will know if micromachines are feasible with current technology.

Link
For more ideas about writing conclusions, go to Chapter 16, page 443.

The conclusion shown in Figure 9.7 is probably more typical than the example above. Most analytical reports have a one or two sentence conclusion.

Using Style and Design

Generally, activity reports follow a plain style and use simple design. These documents are mostly informative, not overly persuasive, so you should try to keep them rather straightforward.

Using a Plain Style

As you revise your document with style in mind, pay attention to the following elements:

Sentences—Using plain style techniques, make sure that (1) the subject is the "doer" of most sentences and (2) the verb expresses the action in most sentences. Where appropriate, eliminate any prepositional chains.

Paragraphs—Each paragraph should begin with a topic sentence that makes a direct statement or claim that the rest of the paragraph will support. This topic sentence will usually appear as the first or second sentence of each paragraph.

Tone—Since activity reports are often written quickly, you should make sure you are projecting an appropriate tone. It might be tempting to be sarcastic or humorous, but this is not the place. After all, you never know how the activity report might be used in the future. While giving briefings, you want to project a professional tone. If you have negative information to convey, state it candidly with no apologies.

Link
For more information on using plain style, see Chapter 17, page 455.

Link
For more information on using an appropriate tone, go to Chapter 17, page 472.

Using Design and Graphics

The design of your activity report should also be straightforward. Usually, the design of these documents is governed by a standard format, like a memo format or perhaps a standardized form for lab reports. Your company will specify the format for activity reports. Otherwise, you might use the templates available with your word-processing program.

Link
For help using templates, go to Chapter 18, page 487.

For more advice on plain style, go to
www.pearsonhighered.com/johnsonweb4/9.9

If you want to add any visuals, you should center them in the text and place them after the point where you refer to them. Even though activity reports tend to be short, you should still label the graphic and refer to it by number in the text.

If you are presenting a briefing orally, you should look for ways to include graphics to support your presentation. Photographs can help the audience visualize what you are talking about, so use your digital camera to snap some pictures. Graphs are always helpful for showing trends in the data.

Overall, your activity report should be clear and straightforward without any stylistic or visual embellishment. You want to state your points as quickly and simply as possible.

MICROGENRE

The Status Report

A status report is a short, barebones e-mail that highlights your or your team's recent activities. In some ways, they are similar to the "status updates" on Facebook and other social networking sites, because they are intended to keep people up to date on what you are doing. Status reports in the workplace, though, are longer and they are submitted weekly, sometimes daily. Here is how to write one:

Identify the purpose and date in the subject line. You can type something like "Status Report for T. Jennings: 9/22/11" in the subject line.

Put your name, project name, and date at the top of the e-mail message. These items will help readers identify the report if the text of the e-mail is separated from the subject line.

Describe the project's status. Use brief phrases to list your activities. Sometimes, you will be asked to estimate the amount of time devoted to each task.

Record any tasks you have completed. If you have finished something, tell the readers it's done.

Identify tasks for the next reporting period. List the things you plan to accomplish before the next status report is due.

Highlight any problems or concerns. List any short-term or long-term issues that might sidetrack the project.

Identify any costs. If you spent any money on the project, identify those costs. Normal operating costs usually don't need to be reported.

Increasingly, microblogs like Twitter are being used as status update tools, so all this information may need to be crammed into 140-word posts.

For worksheets to help you revise, go to **www.pearsonhighered.com/johnsonweb4/9.10**

Write

Write your own status report. While working on your current project, e-mail two status reports each week to your instructor and your other team members.

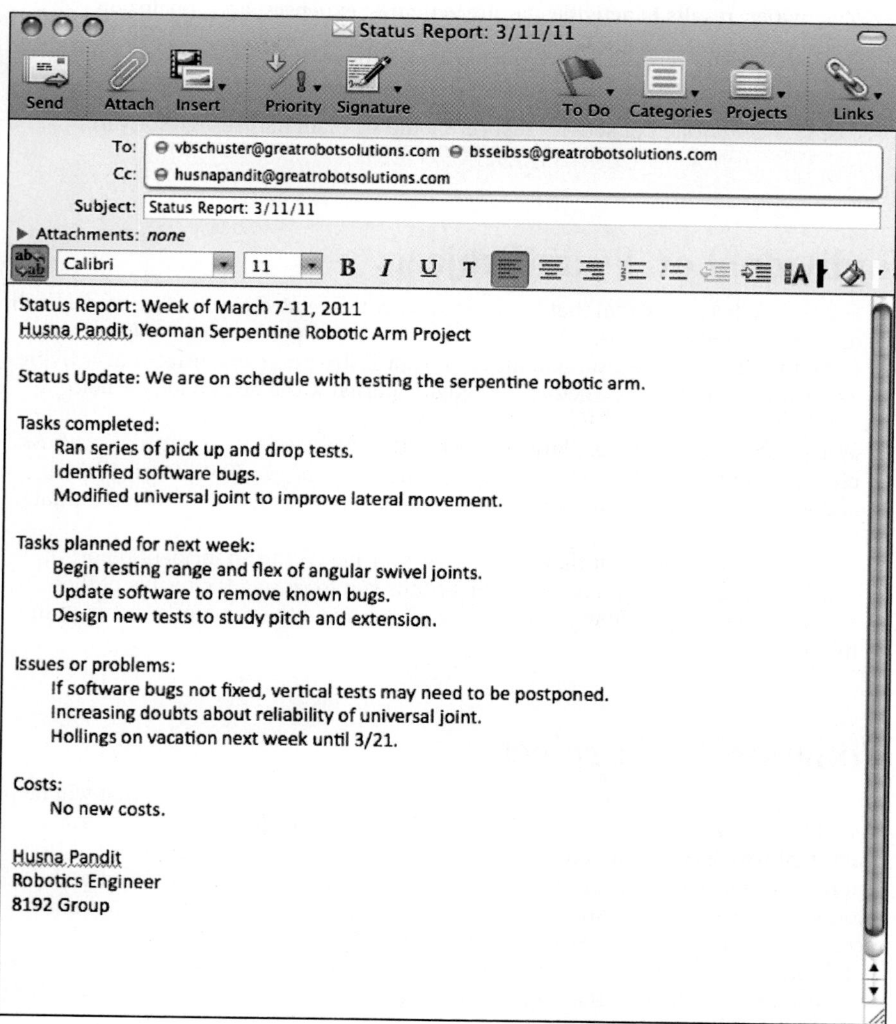
Status Report: 3/11/11

Send Attach Insert Priority Signature To Do Categories Projects Links

To: vbschuster@greatrobotsolutions.com bsseibss@greatrobotsolutions.com
Cc: husnapandit@greatrobotsolutions.com
Subject: Status Report: 3/11/11
Attachments: *none*

Status Report: Week of March 7-11, 2011
Husna Pandit, Yeoman Serpentine Robotic Arm Project

Status Update: We are on schedule with testing the serpentine robotic arm.

Tasks completed:
Ran series of pick up and drop tests.
Identified software bugs.
Modified universal joint to improve lateral movement.

Tasks planned for next week:
Begin testing range and flex of angular swivel joints.
Update software to remove known bugs.
Design new tests to study pitch and extension.

Issues or problems:
If software bugs not fixed, vertical tests may need to be postponed.
Increasing doubts about reliability of universal joint.
Hollings on vacation next week until 3/21.

Costs:
No new costs.

Husna Pandit
Robotics Engineer
8192 Group

CHAPTER REVIEW

- Activity reports include progress reports, briefings, white papers, reports that accompany demonstrations, incident reports, and lab reports.
- An activity report typically includes the following sections: introduction, summary of activities, results of activities, future activities, expenses, and conclusion.
- While preparing to write an activity report, analyze the rhetorical situation by anticipating the readers and the context in which the report will be used.
- The style and design of activity reports should be plain and straightforward.

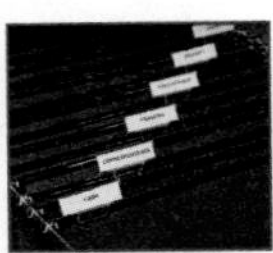

EXERCISES AND PROJECTS

Individual or Team Projects

1. For a week, keep a journal that tracks your activities related to school or work. Each day, make up a "to do" list. Then, as you complete each task, cross it off and write down the results of the task. In a memo to your instructor, summarize your activities for the week and discuss whether the activity journal was a helpful tool or not.

2. While you are completing a large project in this class or another, write a progress report to your instructor in which you summarize what you have accomplished and what you still need to complete. Submit the progress report in memo format.

3. Think back to an accident that occurred in your life. Write an incident report in which you explain what happened, the actions you took, and the results of those actions. Then, discuss how you made changes to avoid that kind of accident in the future.

Collaborative Project

Your group has been asked to develop a standardized information sheet that will help students report accidents on your campus. Think of all the different kinds of accidents that might happen on your campus. Your information sheet should explain how to report an accident to the proper authorities on campus. Encourage the users of the information sheet to summarize the incident in detail, discuss the results, and make recommendations for avoiding similar accidents in the future.

Of course, numerous potential accidents could occur on campus. Your group may need to categorize them so that readers contact the right authorities.

For additional technical writing resources, including interactive sample documents, document design tutorials and guidelines, and more, go to **www.mytechcommlab.com.**

To view information worksheets, go to **www.pearsonhighered.com/johnsonweb4/9.11**

Revision Challenge

The activity report in Figure A is intended to notify students of the recent changes in computer use policies at their small college. How would you revise this report to help it achieve its purpose?

Smith College

Office of the Provost

Date: August 4, 2010
To: Smith College students
From: Provost George Richards
Subject: File-Sharing

Smith College has worked hard to develop a top-notch computer network on campus to provide access to the Internet for a variety of purposes. As a Smith student, you are welcome to use these computers for legal purposes, You will find the college's computer usage policies explained in the *Computer Policy for Smith College,* version 03.29.10, which is in effect until eclipsed by a revision.

Illegal downloading and sharing of copyrighted materials is a problem, especially music files, over our network. Violating our computer usage policies puts the college at risk for copyright infringement lawsuits. Our information technology experts also tell me that these activities slow down our network because these files require large amounts of bandwidth.

Recently, we have installed network tools and filters that allow us to detect and block illegal file sharing. We are already warning students about illegal use. After October 1, 2010, we will begin disciplining people who share files illegally by suspending their computer privileges. Repeated violations will be referred to the college's Academic Integrity Review Board.

You should know that if a complaint is filed by copyright owners, Smith College must provide your name and address to prosecutors. Our computer usage policy explicitly states that using college computers for illegal activities is forbidden and those who do will be prosecuted. Thank you for giving us this opportunity to stress the seriousness of this situation. We expect your compliance.

Figure A: This activity report needs help. How might you revise it?

GO TO THE NET

To download an electronic copy of this revision challenge file, go to **www.pearsonhighered.com/johnsonweb4/9.12**

CASE STUDY

Bad Chemistry

Amanda Jones works as a chemical engineer at BrimChem, one of the top plastics companies in the country. Recently, her division had hired a bright new chemical engineer named Paul Gibson. Paul was tall and good-looking, and he was always polite. At lunch during Paul's first week, Amanda and a co-worker teased him about being a "Chippendales guy." Paul laughed a little, but it was apparent that the comment offended him. So, Amanda was careful from then on about her comments regarding his appearance.

A few months after starting at BrimChem, Paul went to a convention and came back somewhat agitated. Amanda asked him what was wrong. After a pause, Paul told her that one of the managers, Linda Juno, had made a pass at him one evening at the convention, suggesting he come up to her room "for a drink." When he declined, she became angry and said, "Paul, you need to decide whether you want to make it in this company." She didn't speak to him for the rest of the convention.

Paul told Amanda he was a bit worried about keeping his job with the company, since he was still on "probationary" status for his first year. Being on probation meant Linda or anyone else could have him fired for the slightest reason.

Later that day, though, Linda came down to Paul's office and seemed to be patching things up. After Linda left his office, Paul flashed Amanda a thumbs-up signal to show that things were all right.

The next week, Amanda was working late and passed by Paul's office. Linda was in his office giving him a backrub. He was obviously not enjoying it. He seemed to be making the best of it, though, and he said, "OK, thank you. I better finish up this report."

Linda was clearly annoyed and said, "Paul, I let you off once. You better not disappoint me again." A minute later, Linda stormed out of Paul's office.

The next day, Paul stopped Amanda in the parking lot. "Amanda, I know you saw what happened last night. I'm going to file a harassment complaint against Linda. If I'm fired, I'll sue the company. I'm tired of being harassed by her and other women in this company."

Amanda nodded. Then Paul asked, "Would you write an incident report about what you saw last night? I want to put some materials on file." Amanda said she would.

A week later, Paul was fired for a minor mistake. Amanda hadn't finished writing up the incident report.

If you were Amanda, what would you do at this point? If you would finish writing the incident report, what would you say and how would you say it?

To find links on sexual harassment in the workplace, go to **www.pearsonhighered.com/johnsonweb4/9.13**

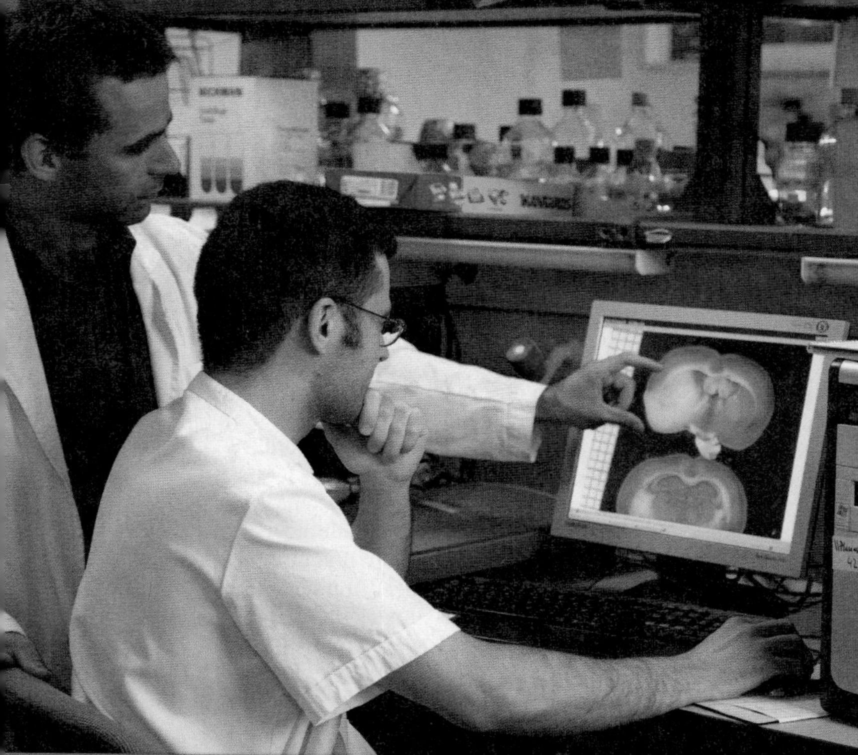

CHAPTER 10 Analytical Reports

Types of Analytical Reports 270
Planning and Researching 272
Organizing and Drafting 282
Help: Using Google Docs to Collaborate with International Teams 298
Drafting Front Matter and Back Matter 299
Using Style and Design 303
Microgenre: The Poster Presentation 309
Chapter Review 310
Exercises and Projects 311
Case Study: The X-File 313

In this chapter, you will learn:

- About the various kinds of analytical reports used in technical workplaces.
- How to use the IMRaD pattern for organizing reports.
- How to determine the rhetorical situation for your reports.
- How to develop a methodology for collecting and analyzing information.
- Strategies for organizing and drafting an analytical report.
- How to use style and design to highlight important information and make it understandable.

Link

For more strategies on organizing information in large documents, see Chapter 16, page 428.

Analytical reports are some of the most common large documents produced in the technical workplace. An analytical report is usually a formal response to a research question. It typically describes a research methodology, presents results, discusses those results, and makes recommendations.

Research is the foundation of a good analytical report. To assist you with the research process, you can use computers to access incredible amounts of information. Even a modest search on the Internet quickly unearths a small library of information on any given subject. Meanwhile, the number-crunching capabilities of computers can help you highlight subtle trends in data collected from experiments and observations.

The diagram in the Quick Start shows a general pattern for organizing an analytical report. To help you remember this generic report pattern, you might do what many researchers do—memorize the acronym IMRaD (Introduction, Methods, Results, and Discussion). This acronym outlines the main areas that most reports address.

The IMRaD pattern for reports is flexible and should be adapted to the specific situation in which you are writing. For example, in some reports, the methodology section can be moved into an appendix at the end of the report, especially if your readers do not need a step-by-step description of your research approach. In other reports, you may find it helpful to combine the results and discussion sections into one larger section.

Link

For more information on the scientific method, turn to Chapter 14, page 389.

Like patterns for other documents, the IMRaD pattern is not a formula to be followed strictly. Rather, it is a guide to help you organize the information you have collected. The IMRaD pattern actually reflects the steps in the scientific method: (1) identify a research question or hypothesis, (2) create a methodology to study the question, (3) generate results, and (4) discuss those results, showing how they answer the research question or support the hypothesis.

Types of Analytical Reports

Analytical reports are formal documents that present findings and make recommendations. Here are a few types:

Research reports—The purpose of a research report is to present the findings of a study. Research reports often stress the causes and effects of problems or trends, showing how events in the past have developed into the current situation.

Empirical research reports—Empirical research reports are written when a scientific project is finished. They first define a research question and hypothesis. Then they describe the methods of study and the results of the research project. And finally, they discuss these results and draw conclusions about what the project discovered. These reports often start out as laboratory reports, but they can also evolve into published scientific articles.

Completion reports—Most projects in technical workplaces conclude with a completion report. These documents are used to report back to management or to the client, assessing the outcomes of a project or initiative.

GO TO THE NET

To read more about the IMRaD pattern, go to **www.pearsonhighered.com/johnsonweb4/10.1**

Analytical Report

Analytical reports tend to follow the IMRaD pattern as shown here. However, reports can be organized a variety of ways. You should adjust this pattern to suit your subject and purpose.

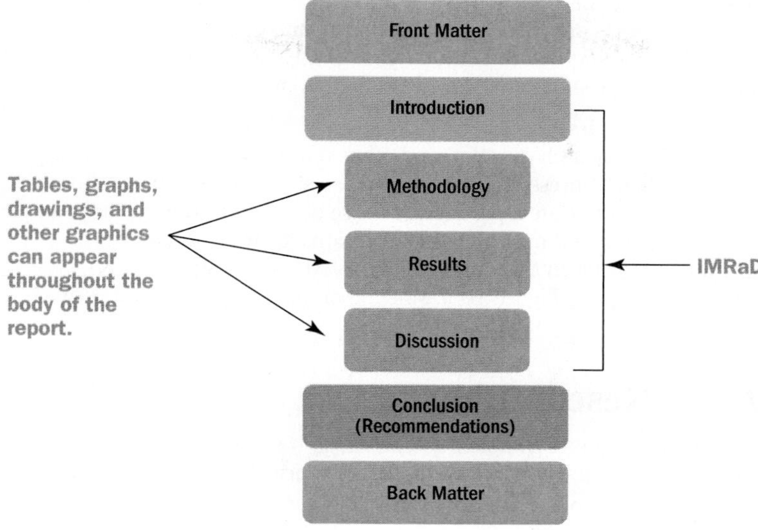

Basic Features of Analytical Reports

Because analytical reports are used in so many different ways, it is hard to pin down a basic pattern that they tend to follow. Nevertheless, you will find that analytical reports typically include the following basic features:

- Introduction
- Methodology or research plan
- Results
- Discussion of the results
- Conclusions or recommendations

The pattern shown above is a common one, but the sections of analytical reports can be arranged and combined in a variety of ways. You should organize your report to fit the needs of your project.

For examples of analytical reports, go to **www.pearsonhighered.com/johnsonweb4/10.2**

Recommendation reports—Recommendation reports are often used to make suggestions about the best course of action. These reports are used to study a problem, present possible solutions, and then recommend what actions should be taken.

Feasibility reports—Feasibility reports are written to determine whether developing a product or following a course of action is possible or sensible. Usually, these reports are produced when management or the clients are not sure whether something can be done. The feasibility report helps determine whether the company should move forward with a project.

Figure 10.1 shows a research report on alcohol use among young adults. Like any research report, it defines a research question and offers a methodology for studying that research question. Then, it presents the results of the study and discusses those results. The report does not advocate a particular course of action. If this report were a recommendation report, it would make recommendations for ways to address alcohol use in young adults. (Later in this chapter, Figure 10.6, starting on p. 286, is an example of a feasibility report. Figure 10.9, starting on p. 306, is an example of a recommendation report.)

Planning and Researching

An analytical report can be a large, complex document, so it is important that you plan properly with a full understanding of the report's rhetorical situation.

Planning

You should start planning the document by first identifying the elements of the rhetorical situation. Begin by answering the Five-W and How Questions:

Who might read this report?

Why was this report requested?

What kinds of information or content do readers need?

Where will this report be read?

When will this report be used?

How will this report be used?

With the answers to these questions fresh in your mind, you can begin defining the rhetorical situation in which your report will be used.

SUBJECT What exactly will the report cover, and what are the boundaries of its subject? What information and facts do readers need to know to make a decision?

PURPOSE What should the report accomplish, and what do the readers expect it to accomplish? What is its main goal or objective?

Figure 10.1: A research report does not typically recommend a specific course of action. It concentrates on presenting and discussing the facts.

National Survey on Drug Use and Health

The NSDUH Report

Issue 31 — 2006

A clear title for the report is placed up front.

Underage Alcohol Use among Full-Time College Students

Main points are placed up front in an easy-to-access box.

In Brief

- The rates of past month, binge, and heavy alcohol use among full-time college students aged 18 to 20 remained steady from 2002 to 2005
- Based on 2002 to 2005 combined data, 57.8 percent of full-time college students aged 18 to 20 used alcohol in the past month, 40.1 percent engaged in binge alcohol use, and 16.6 percent engaged in heavy alcohol use
- Based on 2002 to 2005 combined data, male full-time students in this age group were more likely to have used alcohol in the past month, engaged in binge alcohol use, and engaged in heavy alcohol use than their female counterparts

During the past decade, increased attention has been directed toward underage alcohol use and binge drinking among college students and the negative consequences related to these behaviors.[1-5] Binge drinking refers to the "consumption of a sufficiently large amount of alcohol to place the drinker at increased risk of experiencing alcohol-related problems and to place others at increased risk of experiencing secondhand effects" (p. 287).[2]

Background information stresses the importance of the subject.

The National Survey on Drug Use and Health (NSDUH) asks respondents aged 12 or older to report their frequency and quantity of alcohol use during the month before the survey. NSDUH defines binge alcohol use as drinking five or more drinks on the same occasion (i.e., at the same time or within a couple of hours of each other) on at least 1 day in the past 30 days. NSDUH defines heavy alcohol use as drinking five or more drinks on the same occasion on each of 5 or more days in the past 30 days. All heavy alcohol users are also binge alcohol users.

The subject is defined.

The NSDUH Report (formerly **The NHSDA Report**) is published periodically by the Office of Applied Studies, Substance Abuse and Mental Health Services Administration (SAMHSA). All material appearing in this report is in the public domain and may be reproduced or copied without permission from SAMHSA. Additional copies of this report or other reports from the Office of Applied Studies are available online: http://www.oas.samhsa.gov. Citation of the source is appreciated. For questions about this report, please e-mail: shortreports@samhsa.hhs.gov.

Source: Office of Applied Studies Substance Abuse and Mental Health Services Administration [SAMHSA], 2006.

(continued)

Figure 10.1:
(continued)

NSDUH REPORT: UNDERAGE ALCOHOL USE AMONG FULL-TIME COLLEGE STUDENTS   Issue 31, 2006

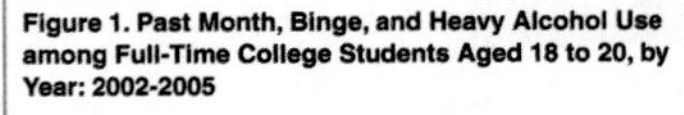
Figure 1. Past Month, Binge, and Heavy Alcohol Use among Full-Time College Students Aged 18 to 20, by Year: 2002-2005

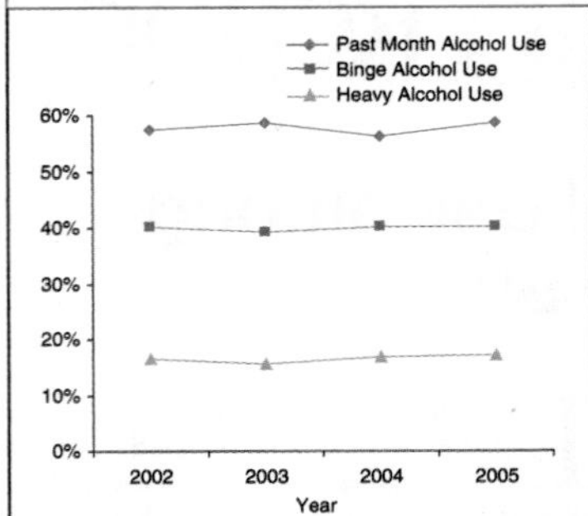

Source: SAMHSA, 2002-2005 NSDUHs.

Figure 2. Past Month, Binge, and Heavy Alcohol Use among Full-Time College Students Aged 18 to 20, by Gender: 2002-2005

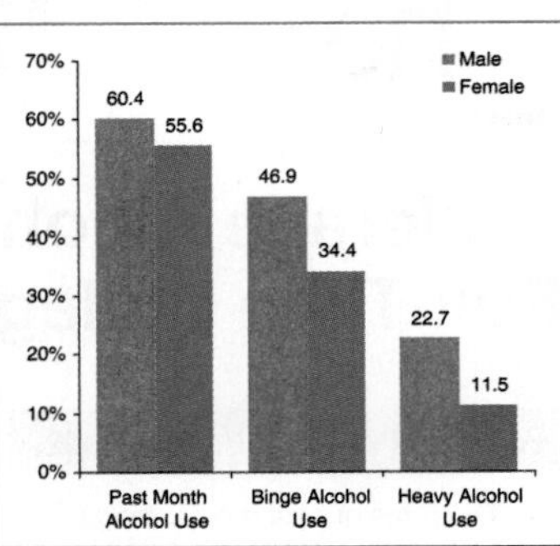

Source: SAMHSA, 2002-2005 NSDUHs.

Graphs show trends in data while supporting written text.

The methodology is described in detail.

NSDUH also asks young adults aged 18 to 22 about college attendance. For this analysis, respondents were classified as college students if they reported that they were in their first through fourth year (or higher) at a college or university and that they were a full-time student. Respondents who were on break from college were considered enrolled if they intended to return to college or university when the break ended.[6]

Results are presented and discussed.

Data from the 2005 survey indicate that young adults aged 18 to 22 enrolled full time in college were more likely than their peers not enrolled full time (i.e., part-time college students and persons not currently enrolled in college) to use alcohol in the past month, binge drink, and drink heavily.[7]

The purpose is stated in a direct way.

This report examines trends and patterns in the rates of alcohol use among full-time college students who have not yet reached the legal drinking age (i.e., college students aged 18 to 20) based on data from the 2002, 2003, 2004, and 2005 NSDUHs.

Headings make the text easy to scan.

Demographic Characteristics of Full-Time College Students Aged 18 to 20

From 2002 to 2005, an average of 5.2 million young adults aged 18 to 20 were enrolled full time in college each year. This represents 41.3 percent of young adults in this age range. Full-time college students included an average of 2.8 million women aged 18 to 20 (46.0 percent of women in this age group) and 2.4 million men aged 18 to 20 years (36.9 percent of men in this age group) each year. Over half of full-time college students aged 18 to 20 (58.2 percent) lived in the same household with a parent, grandparent, or parent-in-law, while 41.8 percent lived independently of a parental relative.[8]

Past Month Alcohol Use

From 2002 to 2005, the rates of past month alcohol use among full-time college students aged 18 to 20 remained steady (Figure 1), with an annual average of 57.8 percent (3.0 million

Figure 10.1: *(continued)*

Issue 31, 2006 NSDUH REPORT: UNDERAGE ALCOHOL USE AMONG FULL-TIME COLLEGE STUDENTS

Figure 3. Past Month, Binge, and Heavy Alcohol Use among Full-Time College Students Aged 18 to 20, by Age: 2002-2005

Source: SAMHSA, 2002-2005 NSDUHs.

Charts are used to present results visually in ways that are easy to scan.

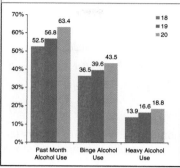

More results and discussion complete the report.

students) using alcohol in the past month. Male full-time students in this age group were more likely to have used alcohol in the past month than their female counterparts (60.4 vs. 55.6 percent) (Figure 2). Rates of past month alcohol use among this group increased with increasing age (Figure 3). Among full-time college students aged 18 to 20, those living with a parent, grandparent, or parent-in-law were less likely to have used alcohol in the past month than those who were not living with a parental relative (51.2 vs. 67.0 percent).

Binge Alcohol Use

Rates of past month binge alcohol use among full-time college students aged 18 to 20 also remained steady from 2002 to 2005 (Figure 1), with an annual average of 40.1 percent (2.1 million students) engaging in binge alcohol use. In this group of young adult students, males were more likely to have engaged in binge alcohol use than females (46.9 vs. 34.4 percent) (Figure 2).

Rates of binge alcohol use among this group also increased with increasing age (Figure 3). Full-time college students aged 18 to 20 living with a parent, grandparent, or parent-in-law were less likely to have engaged in binge alcohol use than full-time college students aged 18 to 20 who were not living with a parental relative (34.0 vs. 48.5 percent).

Heavy Alcohol Use

From 2002 to 2005, rates of heavy alcohol use among full-time college students aged 18 to 20 also remained steady (Figure 1), with an annual average of 16.6 percent (866,000 students) engaging in heavy drinking. As is true for past month and binge alcohol use, rates of heavy alcohol use were higher among males than females and increased with increasing age (Figures 2 and 3). Among full-time college students aged 18 to 20, those living with a parent, grandparent, or parent-in-law were less likely to have engaged in heavy alcohol use than those who were not living with a parental relative (12.3 vs. 22.5 percent).

The authors choose not to include a conclusion because the report is brief and does not make recommendations.

End Notes

1 Reifman, A., & Watson, W. K. (2003). Binge drinking during the first semester of college: Continuation and desistance from high school patterns. *Journal of American College Health, 52,* 73-81.

2 Wechsler, H., & Nelson, T. F. (2001). Binge drinking and the American college student: What's five drinks? *Psychology of Addictive Behaviors, 15,* 287-291.

3 Turrisi, R., Wiersma, K. A., & Hughes, K. K. (2000). Binge-drinking-related consequences in college students: Role of drinking beliefs and mother-teen communications. *Psychology of Addictive Behaviors, 14,* 342-355.

4 Weingardt, K. R., Baer, J. S., Kivlahan, D. R., Roberts, L. J., Miller, E. T., & Marlatt, G. A. (1998). Episodic heavy drinking among college students: Methodological issues and longitudinal perspectives. *Psychology of Addictive Behaviors, 12,* 155-167.

5 Wechsler, H., Davenport, A., Dowdall, G., Moeykens, B., & Castillo, S. (1994). Health and behavioral consequences of binge drinking in college: A national survey of students at 140 campuses. *Journal of the American Medical Association, 272,* 1672-1677.

6 Respondents whose current college enrollment status was unknown were excluded from the analysis.

7 Office of Applied Studies. (2006). *Results from the 2005 National Survey on Drug Use and Health: National findings* (DHHS Publication No. SMA 06-4194, NSDUH Series H-30). Rockville, MD: Substance Abuse and Mental Health Services Administration.

8 Living with a parental relative is defined as currently living in the same household with a parent, grandparent, or parent-in-law. Respondents who did not live with a parental relative and had unknown information on one or more household relationships were excluded from this analysis.

Sources are properly listed.

You should be able to express the purpose of your report in one sentence. A good way to begin forming your purpose statement is to complete the phrase "The purpose of my report is to . . . ." You can then use some of the following action verbs to express what the report will do:

to analyze	*to develop*	*to determine*
to examine	*to formulate*	*to recommend*
to investigate	*to devise*	*to decide*
to study	*to create*	*to conclude*
to inspect	*to generate*	*to offer*
to assess	*to originate*	*to resolve*
to explore	*to produce*	*to select*

READERS Who are the primary readers (action takers), secondary readers (advisors), and tertiary readers (evaluators)? Who are the gatekeeper readers (supervisors) for this report?

Primary readers (action takers) are the people who need the report's information to make some kind of decision. Anticipate the decision they need to make and provide the information they require. If they need recommendations, present your suggestions in a direct, obvious way.

Secondary readers (advisors) are usually experts or other specialists who will advise the primary readers. More than likely, they will be most concerned about the accuracy of your facts and the validity of your reasoning.

Link

For more help defining your readers and their characteristics, go to Chapter 2, page 20.

Tertiary readers (evaluators) might be people you didn't expect to read the report, like reporters, lawyers, auditors, and perhaps even historians. Anticipate these kinds of audiences and avoid making unfounded statements that might be used to harm you or your company.

Gatekeeper readers (supervisors) will probably include your immediate supervisor. Other gatekeepers, like your company's legal or technical experts, might need to review your report before it is sent to the primary readers.

CONTEXT OF USE Where, when, and how will the report be used? What are the economic, political, and ethical factors that will influence the writing of the report and how readers will interpret it?

Physical context—Consider the various places where your report might be used, such as in a meeting or at a conference. What adjustments will be needed to make the report more readable/usable in these situations?

Economic context—Anticipate the financial issues that may influence how your readers will interpret the results and recommendations in your report, especially if you are recommending changes.

Political context—Think about the politics involved with your report. On a micropolitical level (i.e., office politics), you should determine who stands to

To find worksheets to help you analyze readers and contexts of use, go to **www.pearsonhighered.com/johnsonweb4/10.3**

gain or lose from the information in your report. On a macropolitical level, you should consider the larger political trends that will shape the reception of your report.

Ethical context—Consider any legal or ethical issues that might affect your report and the methods you will use to collect information. For example, if you are doing a study that involves people or animals, you may need to secure permission release forms.

Link

For more information on defining the context of use, go to Chapter 2, page 28.

If you are writing a report with a team, collaborate on your answers to these questions about the rhetorical situation. If your team begins the project with a clear understanding of the subject, purpose, readers, and context of use, you will likely avoid unnecessary conflict and wasted time.

Researching

With the rhetorical situation fresh in your mind, you can start collecting information for your report. It is important that you define a research question and develop a plan (a methodology) for conducting research on your subject.

Research in technical fields typically follows a predictable process:

1. Define a research question.
2. State a hypothesis.
3. Develop a research methodology.
4. Collect information by following the research methodology.
5. Analyze gathered information and compare it to the hypothesis.

The most effective methodologies are the ones that collect information from a variety of sources (Figure 10.2).

Researching a Subject

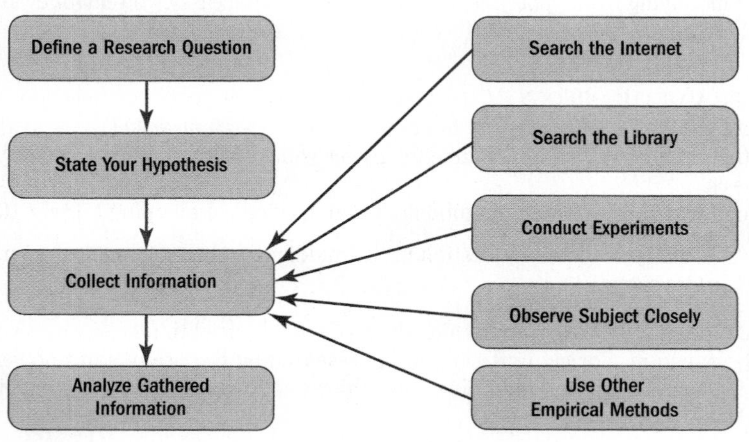

Figure 10.2: A research methodology is a plan for gathering information, preferably from a variety of sources.

For more information about legal issues in research, go to **www.pearsonhighered.com/johnsonweb4/10.4**

DEFINE A RESEARCH QUESTION AND HYPOTHESIS Reports are usually written to answer a specific *research question* or test a *hypothesis.* So, you should begin by defining the question you are trying to answer. Write down the question in one sentence.

> Could we convert one of our campus buildings to a renewable heating source, like solar?
>
> Why are the liver cancer rates in Horn, Nevada, higher than the national average?
>
> How much would it cost to automate our factory in Racine, Wisconsin?
>
> Is it feasible to reintroduce wolves into the Gila Wilderness Area?

Link

To learn more about forming a hypothesis, turn to Chapter 14, page 389.

At this point, you should also write down a one-sentence hypothesis. A hypothesis is essentially an educated guess or tentative explanation that answers your research question. You don't *need* a hypothesis at this point; however, some people like to begin with their best try at answering the research question. With your hypothesis stated, you will have a better idea about what direction your research will take and what you are trying to prove or disprove.

> We believe we could convert a building like Engineering Hall to a solar heating source.
>
> Our hypothesis is that liver cancer rates in Horn, Nevada, are high because of excessive levels of arsenic in the town's drinking water.
>
> Automating our Racine plant could cost $2 million, but the savings will offset that figure in the long run.
>
> Reintroducing wolves to the Gila Wilderness Area is feasible, but there are numerous political obstacles and community fears to be overcome.

Remember, though, that a hypothesis is just a possible answer (i.e., your best guess) to your research question. As you move forward with your research, you will likely find yourself modifying this hypothesis to fit the facts. In some cases, you might even need to abandon your original hypothesis completely and come up with a new one.

DEVELOP A METHODOLOGY Once you have defined your research question and stated a hypothesis, you are ready to develop your research *methodology.* A methodology is the series of steps you will take to answer your research question or test your hypothesis.

A good way to invent your methodology is to use logical mapping (Figure 10.3).

1. Write your research question in the middle of a sheet of paper or your computer screen.
2. Identify the two to five major steps you would need to take to answer that question. For example, if you are researching the use of solar power to generate heat for a building, one major step in your methodology might be

Want help defining a hypothesis? Go to **www.pearsonhighered.com/johnsonweb4/10.5**

Using Logical Mapping to Develop a Methodology

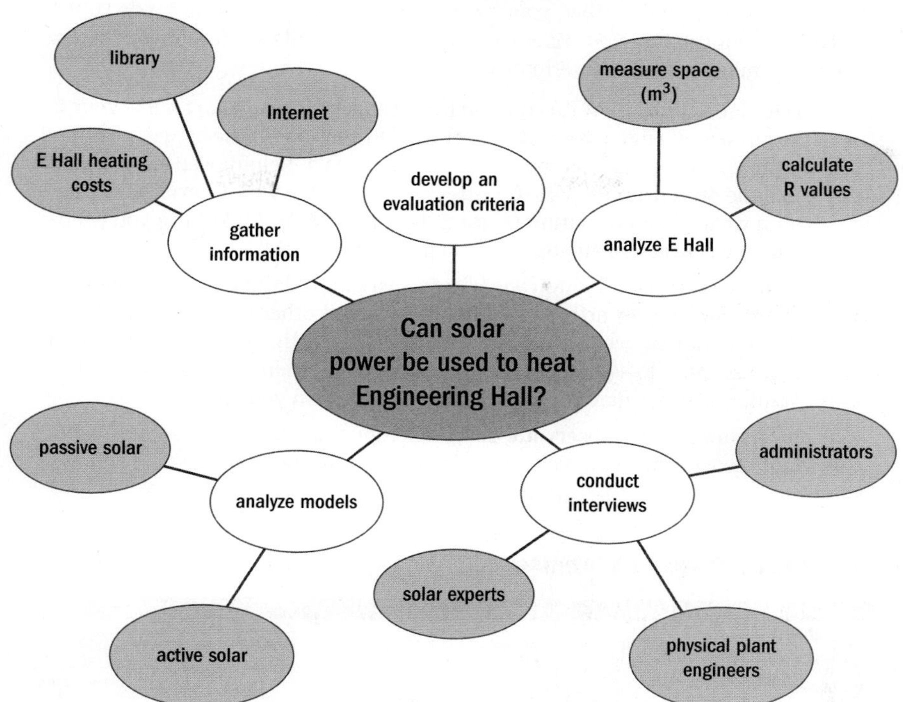

Figure 10.3: When mapping a methodology, ask yourself how you might answer the research question. Then decide on the major and minor steps in your methodology.

to use the Internet to collect information on solar power and its potential applications.

3. Identify the two to five minor steps required to achieve each major step. In other words, determine which smaller steps are needed to achieve each major step.

4. Keep filling out and revising your map until you have fully described the major and minor steps in your methodology.

Your logical map will help you answer the How question about your methodology. By writing down your major and minor steps, you are describing how you would go about studying your subject.

Allow yourself to be creative at this point. As you keep mapping out farther, your methodology will likely evolve in front of you. You can cross out some steps and combine others. In the end, a reasonable methodology is one that is "replicable," meaning that readers can obtain the same results if they redo your research. As you invent and devise your methodology, ask yourself whether someone in your research area could duplicate your work.

COLLECT INFORMATION Your methodology is your road map to collecting information. Once you have described your methodology, you can use it to guide your research. The information you collect will become the results section of your report.

There are many places to find information:

Link

For more strategies for using search engines, see Chapter 14, page 394.

Internet searches—With some well-chosen keywords, you can use your favorite search engine to start collecting information on the subject. Cut and paste the materials you may need and bookmark helpful websites. Websites like the ones from the U.S. Census Bureau and the Pew Charitable Trust offer a wealth of information (Figures 10.4 and 10.5). Make sure you properly cite any sources when you enter them into your notes.

Library research—Using your library's catalogs, databases, and indexes, start searching for articles, reports, books, and other print documents that discuss your subject. You should make copies of the materials you find or use a scanner to save them on your computer. Again, keep track of where you found these materials so you can cite them in your report.

Experiments and observations—Your report might require experiments and measurable observations to test your hypothesis. Each field follows

U.S. Census Bureau Website

Figure 10.4: The U.S. Census Bureau website is an excellent resource for statistics. The website offers statistics on income, race, trade, and lifestyle issues.

Source: www.census.gov.

An Archive on the Internet

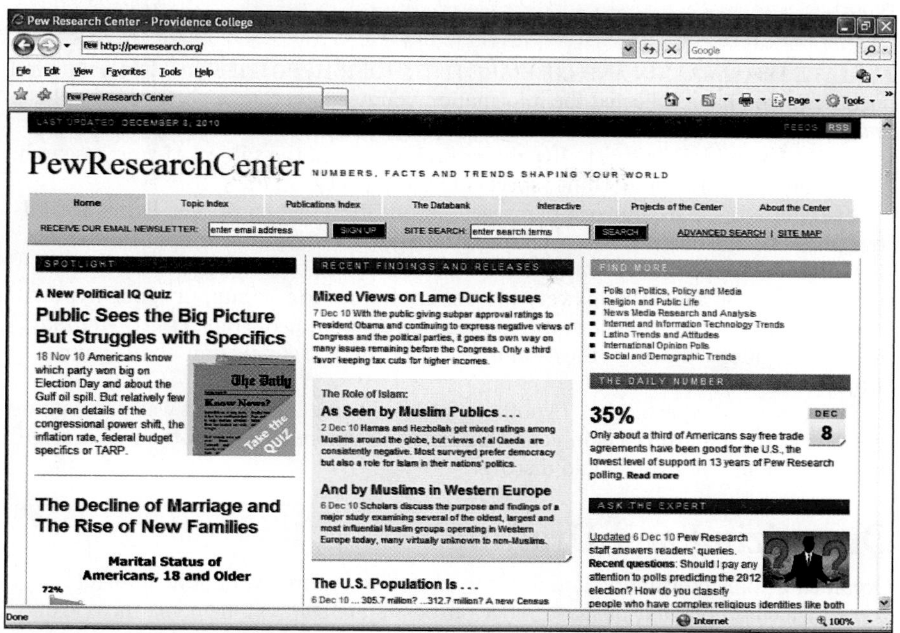

Source: *http://www.pewinternet.org/reports.asp.*

Figure 10.5: The Pew Research Center's Pew Internet & American Life Project's archive offers reports on the Internet and American life. Pew collects data on religion, consumption, lifestyle issues, and many other topics.

different methods for conducting experiments and observations. Learn about research methodologies used in your field and use them to generate empirical data.

Other empirical methods—You might conduct interviews, pass out surveys, or do case studies to generate empirical information. A report should include some empirical material, preferably generated by you or your company. This kind of data will strengthen the credibility of your report.

To avoid any problems with plagiarism or copyright, you need to carefully cite your sources in your notes. It's fine to take material from the Internet or quote passages from print sources, as long as you quote, paraphrase, and cite them properly when you write your report. Chapter 15 offers note-taking strategies that will help you avoid any issues of plagiarism or copyright. Chapter 4 discusses ethical issues involving copyright.

Link

For more information on plagiarism and copyright, go to Chapter 15, page 420.

While collecting information, you will likely find that your methodology will evolve. These kinds of changes in the research plan are not unusual. Keep track of

Link

Chapter 14 offers a full discussion of research. For more research strategies, see page 386.

any changes, because you will need to note them in the methodology section of your report.

ANALYZE INFORMATION AND COMPARE IT TO YOUR HYPOTHESIS Here's the hard part. Once you have collected the information you need for your report, you should analyze your materials closely to identify any major issues and themes. From your materials, try to identify two to five major findings about the subject of your research. These findings will be important aspects of your report.

Link

For more information on developing and modifying research questions and hypotheses, turn to Chapter 14, page 389.

Also, look back at your original research question or hypothesis. Do your findings answer your original research question and/or support your hypothesis? If they do, you are probably ready to start drafting your report. However, if your findings don't sufficiently answer your research question or support your hypothesis, you still need to collect more information and data. In some cases, you may need to abandon your hypothesis and reach a different conclusion about your subject.

Don't worry if your original hypothesis ends up being incorrect. It was just a guess anyway. In the end, the facts in your final report are more important than your original guess about what you would discover.

Organizing and Drafting

Organizing your information and drafting analytical reports will not be difficult if you stay focused on your purpose. Your purpose statement will help you include only need-to-know information in the report. Although you might be tempted to include everything you collected, don't do this. Anything beyond need-to-know information will only muddle your document, making the most important ideas harder for your readers to find.

AT A GLANCE

Moves in an Introduction

- Define the subject.
- State the purpose.
- State the main point.
- Stress the importance of the subject.
- Offer background information.
- Forecast the remainder of the report.

Writing the Introduction

Let's be honest. Reports are not the most interesting documents to read. Given the slightest excuse, your readers will start thumbing through your report's pages, looking for the main points. That's why your introduction needs to grab their attention and give them good reasons to read your report closely.

Your report should begin with an introduction that sets a framework, or context, for the rest of the document (Figure 10.6 on pp. 286–297). Typically, the introduction will include some or all of the following six moves, though not necessarily in this order:

Move 1: Define the *subject* of the report.

Move 2: State the *purpose* of the report, preferably in one sentence.

Move 3: State the report's *main point,* which is likely your main conclusion or recommendation.

Dan Small

SOFTWARE ENGINEER, INTELLIGENT SYSTEMS AND ROBOTICS CENTER, SANDIA NATIONAL LABORATORIES

The Sandia National Laboratories in Albuquerque, New Mexico, develop advanced technology for national defense and progress.

AT WORK

What is the most efficient way to write a report?

As a software engineering professional working at a major R&D lab like Sandia, a large part of my duties includes documenting the research performed by my team. I regularly write analytical reports to our management and our external customers. It's really important that these documents concisely communicate both the technical details and the "big picture" implications of the results.

When writing an analytical report, the first thing I always do is make an outline, which usually follows this general form (much like IMRaD):

> **Introduction**—Make sure that you keep the intended audience in mind here. This needs to be the section where you give the overview and scope of the work you have done. (Believe me, Dilbert has it right with respect to writing for managers versus writing for a technical audience.)
>
> **Data**—Present the data that you have collected, preferably in a graphical form. Make sure that there is an appropriate legend and markings that support your analysis.
>
> **Analysis**—Speak directly to the data that you present. Point out those aspects that support your conclusions and those that don't. (An analytical report is not designed to persuade, only to present.)
>
> **Conclusions**—Draw your conclusions from the analysis presented above. Make sure that the major points that need to be communicated are handled here and are supported by your data and analysis. Make sure that the larger implications (if any) of the results are communicated here as well.

I cannot stress enough the importance of effective writing in the technical workplace. The main difference I see between technical professionals who advance and those who stagnate is their ability to communicate. Given that you are reading this book, I can tell you that you are on the right track.

Move 4: Stress the *importance of the subject,* especially to the readers.

Move 5: Offer *background information* on the subject.

Move 6: Forecast the *organization* of the report.

It is fine to be straightforward in the introduction of your report, because these documents are not read for pleasure. You can make statements like, "The purpose of

Link

For more information on writing introductions, see Chapter 16, page 430.

Have writer's block? Go to **www.pearsonhighered.com/johnsonweb4/10.9**

this report is to . . ." and "In this report, we demonstrate that . . . ." Your readers will appreciate your clarity and forthright approach.

Link

For more information on writing a purpose statement, see Chapter 1, page 6.

Describing Your Methodology

Following the introduction, reports typically include a methodology section that describes step by step how the study was conducted. The aim of the methodology section is to describe the steps you went through to collect information on the subject. This section should include an opening, body, and closing.

OPENING In the opening paragraph, start out by describing your overall approach to collecting information (Figure 10.6). If you are following an established methodology, you might mention where it has been used before and who used it.

Link

For help organizing sections in larger documents, see Chapter 16, page 428.

BODY In the body of the methodology section, walk your readers step by step through the major parts of your study. As you highlight each major step, you should also discuss the minor steps that were part of it.

CLOSING To close the methodology section, you might discuss some of the limitations of the study. For example, your study may have been conducted with a limited sample (e.g., college students at a small Midwestern university). Perhaps time limitations restricted your ability to collect comprehensive data. All methodologies have their limitations. By identifying your study's limitations, you will show your readers that you are aware that other approaches may yield different results.

Summarizing the Results of the Study

In the results section, you should summarize the major *findings* of your study. A helpful guideline is to discuss only the two to five major results. That way, you can avoid overwhelming your readers with a long list of findings, especially ones that are not significant.

This section should include an opening, a body, and perhaps a closing.

- In your opening paragraph for the results section, briefly summarize your major results (Figure 10.6).
- In the body of this section, devote at least one paragraph to each of these major results, using data to support each finding.
- In the closing (if needed), you can again summarize your major results.

Link

For help making tables, charts, and graphs, see Chapter 19, starting on page 523.

Your aim in this section is to present your findings as objectively and clearly as possible. To achieve this aim, state your results with minimal interpretation. You should wait until the discussion section to offer your interpretation of the results.

If your study generated numerical data, you should use tables, graphs, and charts to present your data in this section. As discussed in Chapter 19, these graphics should support the written text, not replace it.

To see sample reports with good introductions, go to **www.pearsonhighered.com/johnsonweb4/10.10**

Discussing Your Results

The discussion section is where you will analyze the results of your research. As you review your findings, identify the two to five major conclusions you might draw from your information or data.

The discussion section should start out with an opening paragraph that briefly states your overall conclusions about the results of your study (Figure 10.6). Then, in the body of this section, you should devote a paragraph or more to each of your conclusions. Discuss the results of your study, detailing what you think your results show.

Stating Your Overall Conclusions and Recommendations

The conclusion of a report should be concise. You should make some or all of the following six moves, which are typical in a larger document:

MAKE AN OBVIOUS TRANSITION A heading like "Our Recommendations" or "Summary" will cue readers that you are concluding. Or, you can use phrases like "In conclusion" or "To sum up" to signal that the report is coming to an end.

RESTATE THE MAIN POINTS Boiled down to its essence, what did your study show or demonstrate? Tell your readers what you proved, disproved, or did not prove.

STATE YOUR RECOMMENDATIONS If you have been asked to make recommendations, your conclusion should identify two to five actions that your readers should consider. A good way to handle recommendations is to present them in a bulleted list (Figure 10.6).

AT A GLANCE

Moves in the Conclusion

- Make an obvious transition.
- Restate the main points.
- State your recommendations.
- Re-emphasize the importance of the study.
- Look to the future.
- Say thank you and offer contact information.

RE-EMPHASIZE THE IMPORTANCE OF THE STUDY Tell your readers why you believe your study is important. Tell them why you think the results and recommendations should be taken seriously.

LOOK TO THE FUTURE You might discuss future research paths that could be pursued. Or, you could describe the future you envision if readers follow your recommendations.

SAY THANK YOU AND OFFER CONTACT INFORMATION After thanking your readers for their interest, you might also provide contact information, such as a phone number and e-mail address. Readers who have questions or want to comment on the report will then be able to contact you.

Like the introduction, your conclusion doesn't need to make these moves in this order, nor are all the moves necessary. Minimally, you should restate your main point and state any recommendations. You can use the other moves to help end your report on a positive note.

Link

For more ideas on writing conclusions, see Chapter 16, page 443.

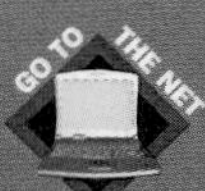

To see samples of well-written methodologies, go to **www.pearsonhighered.com/johnsonweb4/10.11**

A Feasibility Report

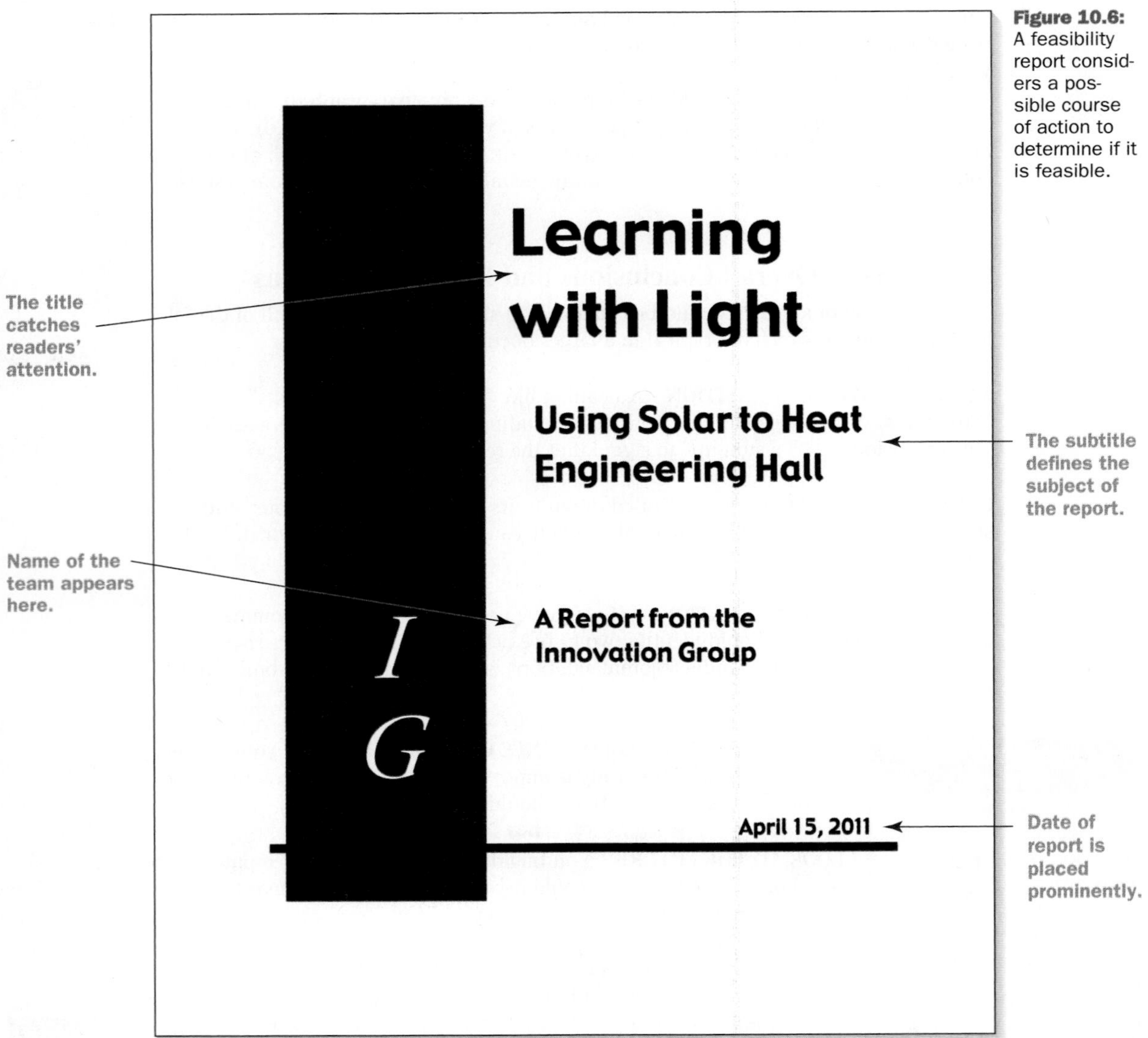

Learning with Light
Using Solar to Heat Engineering Hall
A Report from the Innovation Group
I
G
April 15, 2011
The title catches readers' attention.
The subtitle defines the subject of the report.
Name of the team appears here.
Date of report is placed prominently.

Figure 10.6: A feasibility report considers a possible course of action to determine if it is feasible.

To see reports with well-written conclusions, go to **www.pearsonhighered.com/johnsonweb4/10.12**

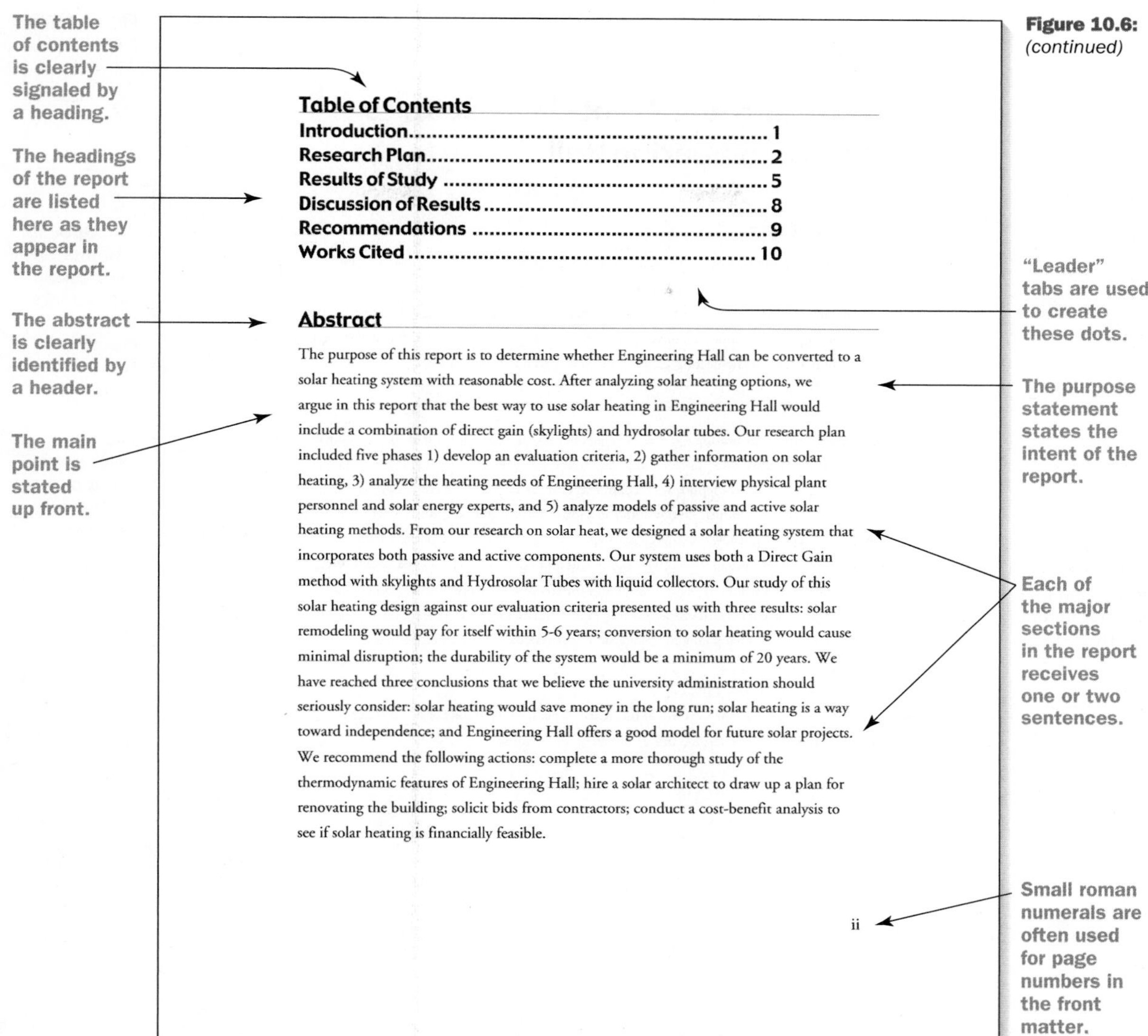

The table of contents is clearly signaled by a heading.
The headings of the report are listed here as they appear in the report.
The abstract is clearly identified by a header.
The main point is stated up front.
"Leader" tabs are used to create these dots.
The purpose statement states the intent of the report.
Each of the major sections in the report receives one or two sentences.
Small roman numerals are often used for page numbers in the front matter.

Table of Contents

Introduction.. 1
Research Plan... 2
Results of Study 5
Discussion of Results 8
Recommendations 9
Works Cited 10

Abstract

The purpose of this report is to determine whether Engineering Hall can be converted to a solar heating system with reasonable cost. After analyzing solar heating options, we argue in this report that the best way to use solar heating in Engineering Hall would include a combination of direct gain (skylights) and hydrosolar tubes. Our research plan included five phases 1) develop an evaluation criteria, 2) gather information on solar heating, 3) analyze the heating needs of Engineering Hall, 4) interview physical plant personnel and solar energy experts, and 5) analyze models of passive and active solar heating methods. From our research on solar heat, we designed a solar heating system that incorporates both passive and active components. Our system uses both a Direct Gain method with skylights and Hydrosolar Tubes with liquid collectors. Our study of this solar heating design against our evaluation criteria presented us with three results: solar remodeling would pay for itself within 5-6 years; conversion to solar heating would cause minimal disruption; the durability of the system would be a minimum of 20 years. We have reached three conclusions that we believe the university administration should seriously consider: solar heating would save money in the long run; solar heating is a way toward independence; and Engineering Hall offers a good model for future solar projects. We recommend the following actions: complete a more thorough study of the thermodynamic features of Engineering Hall; hire a solar architect to draw up a plan for renovating the building; solicit bids from contractors; conduct a cost-benefit analysis to see if solar heating is financially feasible.

ii

Figure 10.6: *(continued)*

(continued)

Figure 10.6: *(continued)*

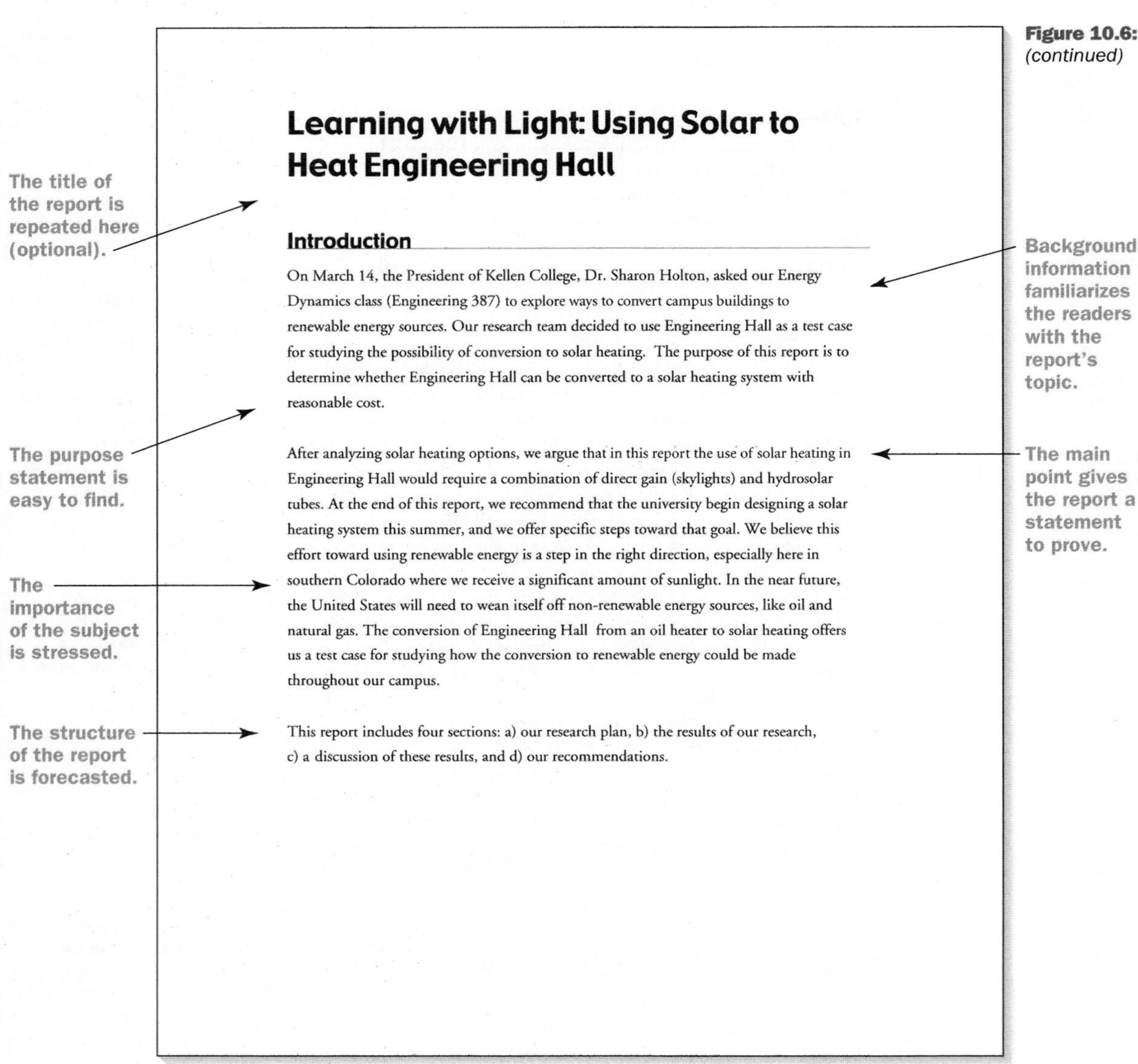

The title of the report is repeated here (optional).
The purpose statement is easy to find.
The importance of the subject is stressed.
The structure of the report is forecasted.
Background information familiarizes the readers with the report's topic.
The main point gives the report a statement to prove.

Learning with Light: Using Solar to Heat Engineering Hall

Introduction

On March 14, the President of Kellen College, Dr. Sharon Holton, asked our Energy Dynamics class (Engineering 387) to explore ways to convert campus buildings to renewable energy sources. Our research team decided to use Engineering Hall as a test case for studying the possibility of conversion to solar heating. The purpose of this report is to determine whether Engineering Hall can be converted to a solar heating system with reasonable cost.

After analyzing solar heating options, we argue that in this report the use of solar heating in Engineering Hall would require a combination of direct gain (skylights) and hydrosolar tubes. At the end of this report, we recommend that the university begin designing a solar heating system this summer, and we offer specific steps toward that goal. We believe this effort toward using renewable energy is a step in the right direction, especially here in southern Colorado where we receive a significant amount of sunlight. In the near future, the United States will need to wean itself off non-renewable energy sources, like oil and natural gas. The conversion of Engineering Hall from an oil heater to solar heating offers us a test case for studying how the conversion to renewable energy could be made throughout our campus.

This report includes four sections: a) our research plan, b) the results of our research, c) a discussion of these results, and d) our recommendations.

Figure 10.6: *(continued)*

The opening of the research section includes a summary of the methodology.

Research Plan

To study whether heating Engineering Hall with solar is feasible, we followed a five-part research plan:

Phase 1: Develop evaluation criteria
Phase 2: Gather information
Phase 3: Study the heating needs of Engineering Hall
Phase 4: Interview physical plant personnel and solar energy experts
Phase 5: Analyze models of passive and active solar heating methods

Phase 1: Develop Evaluation Criteria

The heading states a major step.

In consultation with President Holton and Carlos Riley, Director of Campus Planning, we determined that a successful conversion to solar heating would need to meet three criteria:

The evaluation criteria are listed and defined.

Cost-Effectiveness—the solar heating system would need to be cost-neutral in the long run. In other words, any costs of converting to solar heating would need to be offset by the savings of the system.

Minimal Disruption—conversion needs to cause only minimal disruption to the use of the building. Any construction would need to occur mostly over the summer or in a way that allowed the building to be used.

Durability—the heating system would need to be reliable over time. It would need to last at least 20 years with routine maintenance.

These criteria were the basis of our evaluation of solar heating systems in the Results and Discussion sections later in this report.

2

(continued)

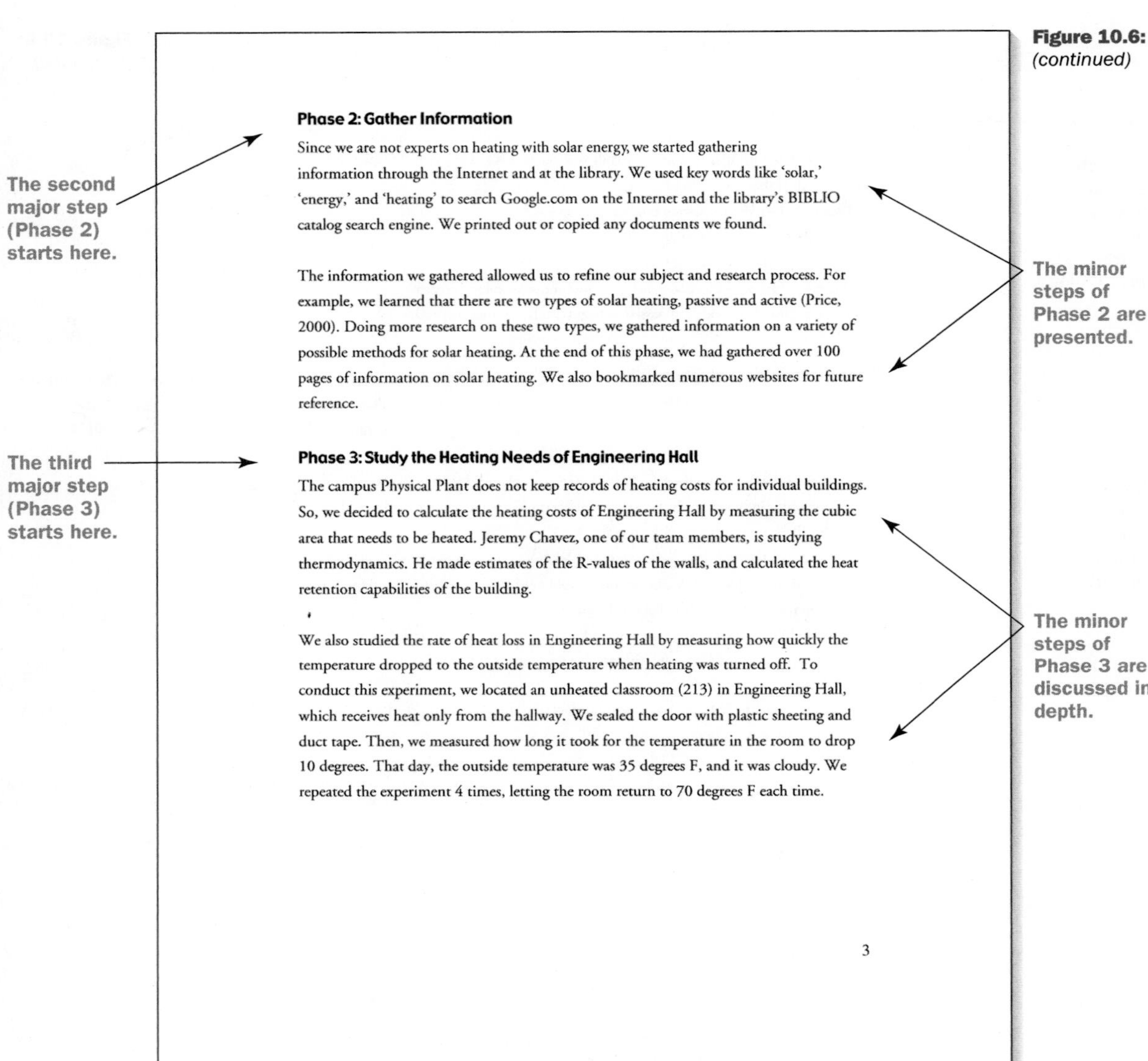
The second
major step
(Phase 2)
starts here.
The minor
steps of
Phase 2 are
presented.
The third
major step
(Phase 3)
starts here.
The minor
steps of
Phase 3 are
discussed in
depth.

Phase 2: Gather Information

Since we are not experts on heating with solar energy, we started gathering information through the Internet and at the library. We used key words like 'solar,' 'energy,' and 'heating' to search Google.com on the Internet and the library's BIBLIO catalog search engine. We printed out or copied any documents we found.

The information we gathered allowed us to refine our subject and research process. For example, we learned that there are two types of solar heating, passive and active (Price, 2000). Doing more research on these two types, we gathered information on a variety of possible methods for solar heating. At the end of this phase, we had gathered over 100 pages of information on solar heating. We also bookmarked numerous websites for future reference.

Phase 3: Study the Heating Needs of Engineering Hall

The campus Physical Plant does not keep records of heating costs for individual buildings. So, we decided to calculate the heating costs of Engineering Hall by measuring the cubic area that needs to be heated. Jeremy Chavez, one of our team members, is studying thermodynamics. He made estimates of the R-values of the walls, and calculated the heat retention capabilities of the building.

We also studied the rate of heat loss in Engineering Hall by measuring how quickly the temperature dropped to the outside temperature when heating was turned off. To conduct this experiment, we located an unheated classroom (213) in Engineering Hall, which receives heat only from the hallway. We sealed the door with plastic sheeting and duct tape. Then, we measured how long it took for the temperature in the room to drop 10 degrees. That day, the outside temperature was 35 degrees F, and it was cloudy. We repeated the experiment 4 times, letting the room return to 70 degrees F each time.

3

Figure 10.6: *(continued)*

Figure 10.6: *(continued)*

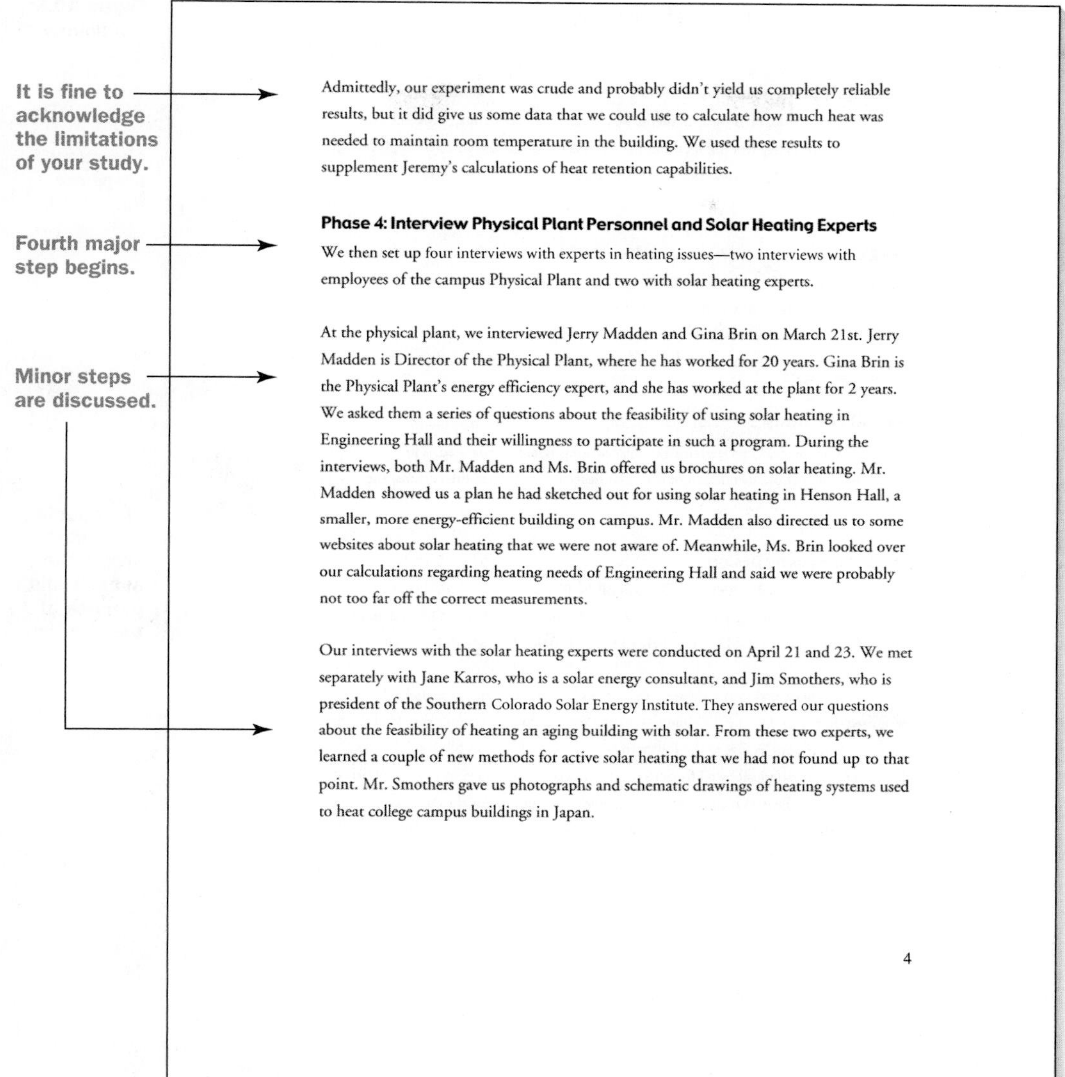

It is fine to acknowledge the limitations of your study.
Fourth major step begins.
Minor steps are discussed.

Admittedly, our experiment was crude and probably didn't yield us completely reliable results, but it did give us some data that we could use to calculate how much heat was needed to maintain room temperature in the building. We used these results to supplement Jeremy's calculations of heat retention capabilities.

Phase 4: Interview Physical Plant Personnel and Solar Heating Experts

We then set up four interviews with experts in heating issues—two interviews with employees of the campus Physical Plant and two with solar heating experts.

At the physical plant, we interviewed Jerry Madden and Gina Brin on March 21st. Jerry Madden is Director of the Physical Plant, where he has worked for 20 years. Gina Brin is the Physical Plant's energy efficiency expert, and she has worked at the plant for 2 years. We asked them a series of questions about the feasibility of using solar heating in Engineering Hall and their willingness to participate in such a program. During the interviews, both Mr. Madden and Ms. Brin offered us brochures on solar heating. Mr. Madden showed us a plan he had sketched out for using solar heating in Henson Hall, a smaller, more energy-efficient building on campus. Mr. Madden also directed us to some websites about solar heating that we were not aware of. Meanwhile, Ms. Brin looked over our calculations regarding heating needs of Engineering Hall and said we were probably not too far off the correct measurements.

Our interviews with the solar heating experts were conducted on April 21 and 23. We met separately with Jane Karros, who is a solar energy consultant, and Jim Smothers, who is president of the Southern Colorado Solar Energy Institute. They answered our questions about the feasibility of heating an aging building with solar. From these two experts, we learned a couple of new methods for active solar heating that we had not found up to that point. Mr. Smothers gave us photographs and schematic drawings of heating systems used to heat college campus buildings in Japan.

4

(continued)

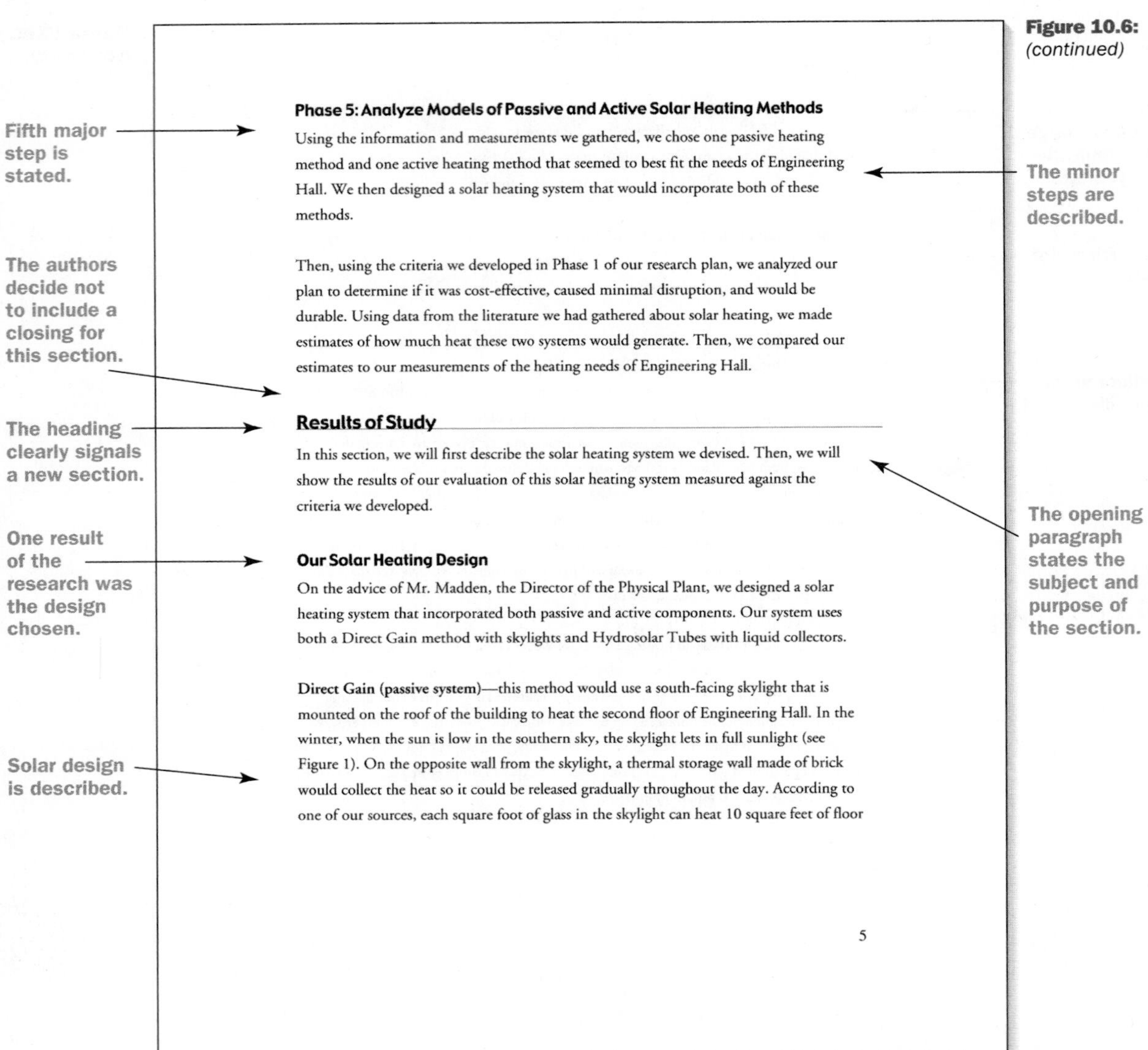

Figure 10.6: *(continued)*
Fifth major step is stated.
The minor steps are described.
The authors decide not to include a closing for this section.
The heading clearly signals a new section.
One result of the research was the design chosen.
The opening paragraph states the subject and purpose of the section.
Solar design is described.

Phase 5: Analyze Models of Passive and Active Solar Heating Methods

Using the information and measurements we gathered, we chose one passive heating method and one active heating method that seemed to best fit the needs of Engineering Hall. We then designed a solar heating system that would incorporate both of these methods.

Then, using the criteria we developed in Phase 1 of our research plan, we analyzed our plan to determine if it was cost-effective, caused minimal disruption, and would be durable. Using data from the literature we had gathered about solar heating, we made estimates of how much heat these two systems would generate. Then, we compared our estimates to our measurements of the heating needs of Engineering Hall.

Results of Study

In this section, we will first describe the solar heating system we devised. Then, we will show the results of our evaluation of this solar heating system measured against the criteria we developed.

Our Solar Heating Design

On the advice of Mr. Madden, the Director of the Physical Plant, we designed a solar heating system that incorporated both passive and active components. Our system uses both a Direct Gain method with skylights and Hydrosolar Tubes with liquid collectors.

Direct Gain (passive system)—this method would use a south-facing skylight that is mounted on the roof of the building to heat the second floor of Engineering Hall. In the winter, when the sun is low in the southern sky, the skylight lets in full sunlight (see Figure 1). On the opposite wall from the skylight, a thermal storage wall made of brick would collect the heat so it could be released gradually throughout the day. According to one of our sources, each square foot of glass in the skylight can heat 10 square feet of floor

5

Figure 10.6: *(continued)*

space (Solar Thermal Energy Group 2003). If so, we would need 200 square feet of glass to heat the 2000 sq. ft. on the second floor of Engineering Hall. In other words, we would need the equivalent of a 5 ft. by 40 ft. skylight across the roof of the building.

Graphics help illustrate complex concepts.

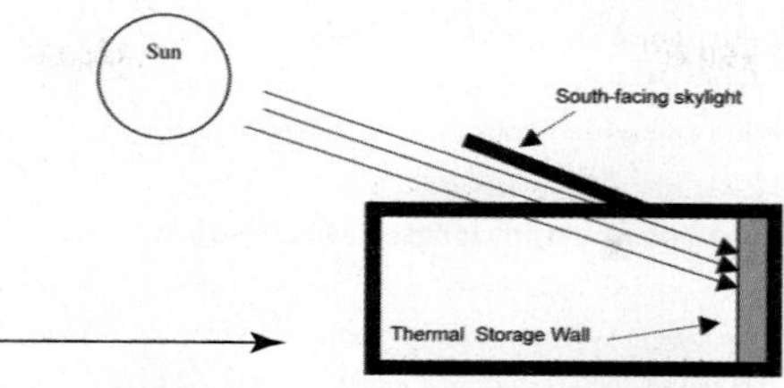

Figure 1: Direct Gain Method with Skylights and a Thermal Storage Wall

Hydrosolar Tubes (active system)—this method would place liquid solar collectors by the southern-facing base of the building to gather heat from the sun (Langa, 1981; Clive, 2007). The heated water would then be piped into hot water registers inside the first-floor rooms, with a pump that runs on solar electricity (Figure 2). The circulating water would heat each room through registers placed along the walls (Meeker & Boyd, 1983; Eklund, et al., 1979). We decided we would need six of these systems—one for each south-facing room in Engineering Hall. Each system can heat a room of 600 square feet on a sunny day, allowing us to heat the 3000 sq. ft. of space on the first floor of the building.

A graphic helps the readers visualize the project.

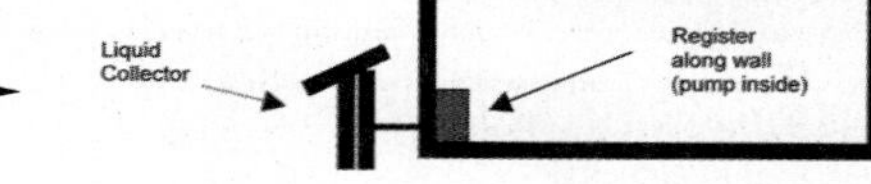

Figure 2: Hydrosolar Tubes with Liquid Collector and a Register

We also determined that the solar heating system would not be able to stand alone. A small backup electric heating system would need to be installed for the occasional cloudy days (which are rare here in south-central Colorado).

6

(continued)

Figure 10.6: *(continued)*

Here begins a section that describes the specific results generated by the researchers.

Evaluation of Solar Heating Design

Our evaluation of this solar heating design against our criteria presented us with three results:

- Solar remodeling would pay for itself within 5-6 years.
- Conversion to solar heating would cause minimal disruption.
- The durability of the system would be a minimum of 20 years.

The first result is described objectively.

Result: Solar remodeling would pay for itself within 5–6 years

Our measurements show that a combination of passive and active systems like the ones we designed would pay for itself in 5-6 years. With help from Mr. Madden, we estimate that installing the skylights and adding a thermal wall would cost about $24,000. Installing the liquid collectors with tanks, pumps, and registers would cost $23,000. The backup electric heating system would cost about $5,000 and cost about $500 a year to run. The total cost of the system would be $57,000 to heat the building for ten years. This compares favorably with our estimate that the current heating system costs about $4,500 per year for a total of $45,000 over ten years. Renovating the oil heater, Ms. Brin from the Physical Plant estimates, would cost around $20,000. Thus, converting to solar saves us $8000 over ten years.

Note how numbers are used to support the presentation of results.

The second result is presented with support.

Result: Conversion to solar heating would cause minimal disruption

Jane Karros, one of the solar consultants we interviewed, estimated that installation of a solar heating system would take approximately 3 months. If the work was begun late in the spring semester and completed by the end of the summer, the disruption to the operations of the building would be minimal. Certainly, the disruption would be no more than a typical renovation of a building's heating system (Ramlow & Nusz, 2006).

7

Figure 10.6:
(continued)

The third result is presented.

Result: The durability of the system would be a minimum of 10 years

The literature we gathered estimates that the durability of this solar heating system would be a minimum of 20 years (Price 2000). The direct gain system, using the skylight and thermal storage wall, could be used indefinitely with minor upkeep. The hydrosolar tube system, using liquid solar collectors and a solar pump, would likely need to be renovated in 20 years. These systems have been known to go 25 years without major renovation.

Discussion of Results

The discussion section starts out with an opening paragraph to redirect the discussion.

Based on our research and calculations, we have reached three conclusions, which suggest that a solar heating system would be a viable option for Engineering Hall.

The results of the research are discussed.

Solar heating would save money in the long run

Engineering Hall is one of the oldest, least efficient buildings on campus. Yet, our calculations show that putting in skylights and liquid collectors would be sufficient to heat the building on most days. According to our estimates, the solar remodeling would more than pay for itself in 10 years. Moreover, we would eliminate the need for the current oil heating system in the building, which will likely need to be replaced in that time period.

Solar heating is a way toward independence

Our dependence on imported oil and natural gas puts our society at risk a few different ways (US DOE 2004). Engineering Hall's use of oil for heating pollutes our air, and it contributes to our nation's dependence on other countries for fuel. By switching over to solar heating now, we start the process of making ourselves energy independent. Engineering Hall's heating system will need to be replaced soon anyway. Right now would be a good time to think seriously about remodeling to use solar.

8

(continued)

Figure 10.6: *(continued)*

Note how the authors are offering their opinions in this section.

Engineering Hall offers a good model for future projects

Engineering Hall is one of the older buildings on campus. Consequently, it presents some additional renovation challenges that we would not face with the newer buildings. We believe that Engineering Hall provides an excellent model for conversion to solar energy because it is probably one of the more difficult buildings to convert. If we can make the conversion with this building, we can almost certainly make the conversion with other buildings.

It might also be noteworthy that our discussions with Mr. Madden and Ms. Brin at the Physical Plant showed us that there is great enthusiasm for making this kind of renovation to buildings like Engineering Hall.

The recommendations section is clearly signaled here.

Recommendations

In conclusion, we think the benefits of remodeling Engineering Hall to use solar heating clearly outweigh the costs. We recommend the following actions:

- Complete a more thorough study of the heating needs of Engineering Hall.
- Hire a solar architect to draw up a plan for renovating the building.
- Solicit bids from contractors.
- Conduct a formal cost-benefit analysis to see if solar heating is financially feasible for Engineering Hall.

The recommendations are put in bullet form to make them easy to read.

Here is the main point and a look to the future.

If all goes well, remodeling of Engineering Hall could be completed in the summer of 2011. When the remodeling is complete, the college should begin saving money within ten years. Moreover, our campus would house a model building that could be studied to determine whether solar heating is a possibility for other buildings on campus.

9

Figure 10.6: *(continued)*

Offering contact information is a good way to end the report.

The materials cited in the report are listed here in APA format.

Thank you for your time and consideration. After you have looked over this report, we would like to meet with you to discuss our findings and recommendations. Please call Dan Garnish at 555-9294.

Works Cited

Clive, K. (2007). *Build your own solar heating system.* Minneapolis, MN: Lucerno.

Eklund, K. (1979). *The solar water heater workshop manual* (2nd ed.). Seattle, WA: Ecotape Group.

Langa, F. (1981). *Integral passive solar water heating book.* Davis, CA: Passive Solar Institute.

Meeker, J. & Boyd, L. (1983). Domestic hot water installations: The great, the good, and the unacceptable. *Solar Age 6,* 28-36.

Price, G. (2000). *Solar remodeling in southern New Mexico.* Las Cruces, NM: NMSU Energy Institute.

Ramlow, B. & Nusz, B. (2006). *Solar water heating.* Gabriola Island, BC: New Society Publishers.

Solar Thermal Energy Group.(2003). *Solar home and solar collector plans.* Retrieved from http://www.jc-solarhomes.com

US. Department of Energy. (2004). Residential solar heating retrofits. Retrieved from http://www.eere.energy.gov/consumerinfo/factsheets/ac6.html

10

Using Google Docs to Collaborate with International Teams

In the universe of Google products, there is a helpful collaboration tool called Google Docs (http://docs.google.com). Google Docs is useful in two ways. First, it includes a free, web-based suite of online software that can be used for writing documents, creating spreadsheets, and making presentations. It's similar to the Microsoft Office software package.

Second, and more importantly, Google Docs allows you to collaborate with others by storing and "sharing" documents (Figure A). When a file is placed in Google Docs, the creator can share it with other members of the team. They can then work on the document, too.

Figure A shows the Google Docs interface. In the center of the screen, you can see a list of files that are being shared among different collaborative groups. On the left-hand side of the screen, the user names of regular team members are listed, allowing the creator of a document to designate quickly who can read or edit each file.

The Google Docs Interface

Buttons to upload and share files

Types of documents

User names of team members

Files shared among team members

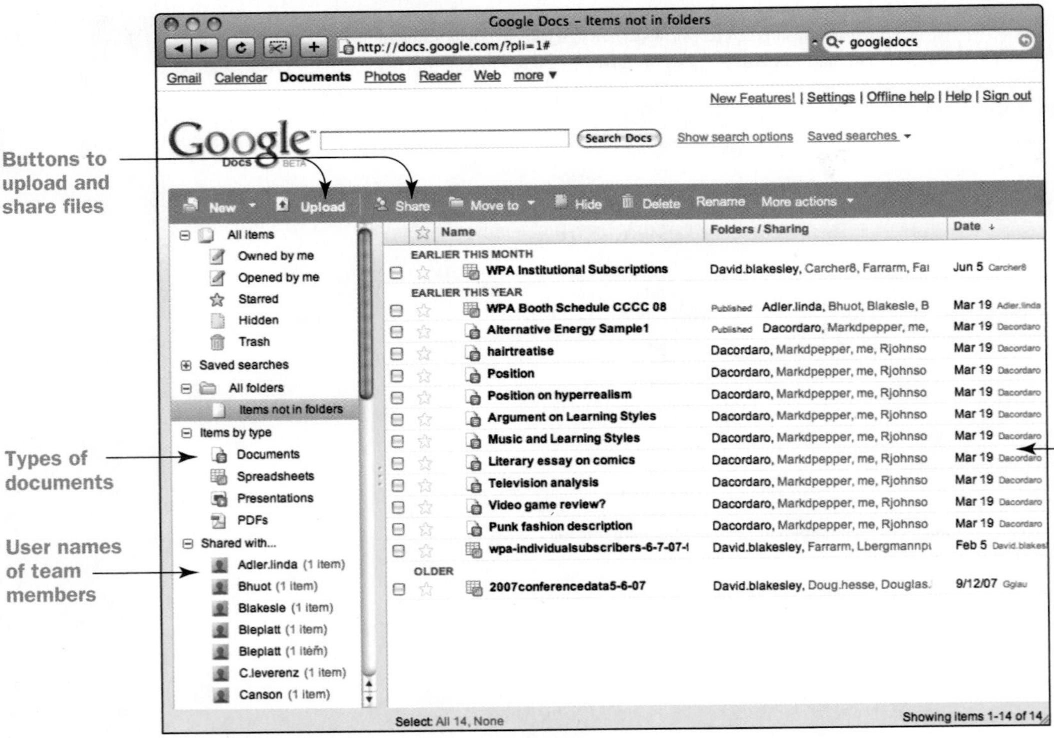

Figure A: Google Docs provides a common space for sharing documents with international teams.

Source: GOOGLE is a trademark of Google Inc.

Want to know more about virtual teaming? Go to **www.pearsonhighered.com/johnsonweb4/10.13**

People working on collaborative projects, especially with international teams, find Google Docs to be an amazingly helpful workspace. With Google Docs, everyone can access and edit the most recent versions of any document. This allows teams to avoid the usual questions about whether everyone has the latest version of the file.

Another advantage is that this free software is available to everyone. One of the problems with working internationally is that team members in different parts of the world are often using a variety of software for word processing, spreadsheets, and presentations on different operating systems. So document sharing among members of an international team can get bogged down with tricky conversions between software packages and operating systems. But if the whole team is using Google Docs, everyone has access to the same software.

Even if your team agrees to use a common software package (e.g., MS Word), members may be using different versions or running old versions of Windows, Linux, or Mac OS. Google Docs solves that problem, too, because everyone will be using the up-to-date versions of its software. Since the software is free, you won't hear the common complaints from your team members about how their management won't pay for software or operating system upgrades.

The main drawback to Google Docs is that the software suite is not as advanced as Microsoft or Adobe products (despite Google's claims to the contrary). Also, when documents from other software applications, like MS Word, are placed on Google Docs, they can lose some of their formatting and design. Plus, you should always remember that with Google Docs, your files are being stored on a server outside your company. So, any high-security files should not be stored or shared on Google Docs.

Google Docs isn't perfect, but for international teams, its benefits far outweigh its shortcomings. The ability to share documents with Google Docs and use common software makes collaborating internationally much easier.

Link

For more advice about working in teams, see Chapter 3, page 59.

Drafting Front Matter and Back Matter

Most reports also include front matter and back matter. Front matter includes the letter of transmittal, title page, table of contents, and other items that are placed before the first page of the main report. Back matter includes appendixes, glossaries, and indexes that are placed after the main report.

Developing Front Matter

Front matter may include some or all of the following items:

LETTER OR MEMO OF TRANSMITTAL Typically, reports are accompanied by a letter or memo of transmittal. A well-written letter or memo gives you an opportunity to make positive personal contact with your readers before they read your report.

TITLE PAGE Title pages are an increasingly common feature in analytical reports. A well-designed title page sets a professional tone while introducing readers to the

Link

For more information on writing letters and memos of transmittal, see Chapter 5, page 94.

To learn more about groupware, go to **www.pearsonhighered.com/johnsonweb4/10.14**

Designing the Title Page

Learning with Light:
Using Solar to Heat
Engineering Hall

A Report from the
Innovation Group

April 15, 2011

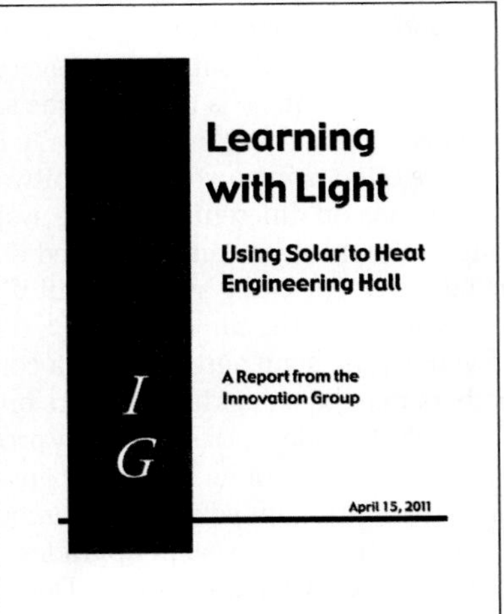

Figure 10.7: Designing an effective title page takes only a few moments. Those few moments of work, though, can make a solid first impression on the readers. Most readers would be attracted to the report on the right.

subject of the report. The title page should include all or some of the following features:

- a specific title for the report
- the names of the primary readers, their titles, and the name of their company or organization
- the names of the writers, their titles, and the name of their company or organization
- the date on which the report was submitted
- company logos, graphics, or rules (lines) to enhance the design

Link

For graphic design strategies, see Chapter 18, page 482.

The title of your report should give readers a clear idea of what the report is about (Figure 10.7). A title like "Solar Heating" probably is not specific enough. A more descriptive title like "Learning with Light: Using Solar to Heat Engineering Hall" gives readers a solid idea about what the report will discuss.

ABSTRACT OR EXECUTIVE SUMMARY If your report is longer than ten pages, you should consider including an abstract or executive summary.

An abstract is a summary of the report that uses the phrasing in the report and follows its organizational structure. When writing an abstract, you should draw key sentences directly from the report itself. Start out with the purpose statement of the report, and then state the main point. From there, draw one or two key sentences from each major section. In an abstract for a report, for example, you would probably include these items in the following order:

- Purpose statement (one sentence)
- Main point (one sentence)

Want to see some sample texts that use good and bad design? Go to **www.pearsonhighered.com/johnsonweb4/10.15**

- Methodology (one or two sentences)
- Results (one or two sentences)
- Discussion (one or two sentences)
- Recommendations (one or two sentences)

You should modify the sentences in places to make the abstract readable, but try to retain the phrasing of the original report as much as possible.

Abstract

Purpose of report leads the abstract.

The main point comes second.

The remainder of the abstract mirrors the structure of the report (methodology, results, discussion, recommendations).

The purpose of this report is to determine whether the Engineering Hall's heating system can be converted to solar with reasonable cost. After analyzing solar heating options, we argue in this report that the best way to add solar heating to Engineering Hall would include a combination of direct gain (skylights) and hydrosolar tubes. Our research plan included five phases: (1) develop evaluation criteria, (2) gather information on solar heating, (3) analyze the heating needs of Engineering Hall, (4) interview physical plant personnel and solar energy experts, and (5) analyze models of passive and active solar heating methods. From our research, we designed a solar heating system that incorporates both passive and active components. Our system uses both a direct gain method with skylights and hydrosolar tubes with liquid collectors. Our evaluation of this solar heating design using our evaluation criteria yielded three results: Solar remodeling would pay for itself within 5–6 years; conversion to solar heating would cause minimal disruption; the durability of the system would be a minimum of 10 years. We have reached three conclusions that we believe the university administration should seriously consider: Solar heating would save money in the long run; solar heating is a way toward independence; and Engineering Hall offers a good model for future solar projects. We recommend the following actions: Complete a more thorough study of the thermodynamic features of Engineering Hall; hire a solar architect to draw up a plan for renovating the building; solicit bids from contractors; and conduct a cost-benefit analysis to see if solar heating is financially feasible.

An executive summary is a concise, *paraphrased* version of your report (usually one page) that highlights the key points in the text. The two main differences between an abstract and an executive summary are that (1) the summary does not follow the organization of the report, and (2) the summary does not use the exact phrasing of the report. In other words, a summary paraphrases the report and organizes the information to highlight the key points.

Report Summary

The purpose of the report is placed early in the summary.

The main point is also placed up front.

This report was written in response to a challenge to our Energy Dynamics class (Engineering 387) from Dr. Sharon Holton, President of Kellen College. She asked us to develop options for converting campus buildings to renewable energy sources. In this report, we discuss the possibility of converting Engineering Hall's heating system to solar. We conclude that heating Engineering Hall with solar sources would require a combination of direct gain (skylights) and hydrosolar tubes. The combination of these two solar technologies would ensure adequate heating for almost all the building's heating needs. A backup heater could be retained for sustained cold spells.

To see sample abstracts and summaries, go to **www.pearsonhighered.com/johnsonweb4/10.16**

> To develop the information for this report, we followed a five-step research plan: (1) develop an evaluation criteria, (2) gather information on solar heating, (3) analyze the heating needs of Engineering Hall, (4) interview physical plant personnel and solar energy experts, and (5) analyze models of passive and active solar heating that would be appropriate for this building.
>
> The results of our research are mostly anecdotal, but they show that solar heating is possible, even for an older building on campus. We believe that our results show that Engineering Hall can be a model for developing solar heating systems around campus, because it is truly one of the more difficult buildings at Kellen to convert to solar heating. Newer buildings on campus would almost certainly be easier to convert. We conclude by pointing out that solar heating would save money in the long run. In the case of Engineering Hall, solar remodeling would pay for itself in 5–6 years.
>
> We appreciate your taking time to read this report. If you have any questions or would like to meet with us, please call Dan Garnish at 555–9294.

The remainder of the summary organizes information in order of importance.

The executive summary will often duplicate the contents of the introduction, but it should not replace the introduction. Instead, it should be written so that it can stand alone, apart from the rest of the report.

TABLE OF CONTENTS If your report runs over ten pages, you should consider adding a table of contents. A table of contents is helpful to readers in two ways. First, it helps them quickly access the information they need in the report. Second, it offers an overall outline of the contents of the report. Since reports tend to be larger documents, your readers will appreciate a quick summary of the report's contents.

In the table of contents, the headings should be the same as the ones used in your report. Then, use tabs or leader tabs to line up the page numbers on the right side. Leader tabs are used to insert a line of dots or dashes from the heading to the page number.

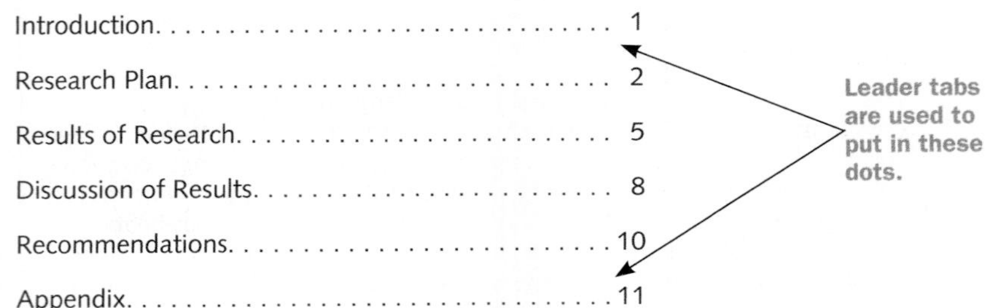
Table of Contents

Introduction. 1

Research Plan. 2

Results of Research. 5

Discussion of Results. 8

Recommendations. 10

Appendix. 11

Leader tabs are used to put in these dots.

Developing Back Matter

Back matter in a report might include appendixes, a glossary of terms, and calculations. Keep in mind, though, that most readers will never look at the back matter. So, if you have something important to say, do not say it here.

AT A GLANCE

Front Matter and Back Matter

Front matter—items that appear before the main report:
- Letter or memo of transmittal
- Title page
- Abstract or executive summary
- Table of contents

Back matter—items that appear after the main report:
- Appendixes
- Glossary of terms
- Calculations

APPENDIXES Appendixes are storage areas for information that may or may not be useful to certain readers. For example, additional data tables or charts not needed in the body of the report might be displayed in an appendix. You might include clippings from newspapers or magazines.

GLOSSARY OF TERMS Depending on the complexity of your report and your readers' familiarity with the subject, you may want to include a short glossary of terms. When creating a glossary, look back over your report and highlight words that may not be familiar to your nonexpert readers. Then, in a glossary at the back of the report, list these terms and write sentence definitions for each.

Link

For more information on writing sentence definitions, see Chapter 6, page 155.

CALCULATIONS In highly technical reports, you may want to include your calculations in the back matter. Here is where you can demonstrate how you arrived at the figures in the report.

Using Style and Design

Your reports don't need to be difficult to read, and they don't need to look boring. The most readable reports are written in plain style, with limited use of persuasion. Meanwhile, the page design is usually straightforward, but graphics, page layout, and color can make a report more attractive and readable.

Link

For more help with using plain style, see Chapter 17, page 455.

Using Plain Style in a Persuasive Way

Reports are persuasive in an unstated way, putting the emphasis on the soundness of the methodology, the integrity of the results, and the reasonableness of the discussion. Plain style will make your reports sound straightforward and clear to readers.

While revising your report, you might pay close attention to the following plain style techniques:

> **Make "doers" the subjects of sentences**—Reports tend to overuse the passive voice, which makes them harder to read than necessary. If the passive voice is required in your field, use it. But if you want your writing to be more effective, make your sentences active by making the "doers" the subjects of the sentences.
>
> **Passive:** Saplings had been eaten during the winter by the deer we monitored, because they were desperate for food.

Want to learn about passive voice? Go to **www.pearsonhighered.com/johnsonweb4/10.17**

> **Active:** The deer we monitored ate saplings to survive the winter because they were desperate for food.

Use breathing-length sentences—Reports are notorious for using sentences that are far too long to be understood. As you write and revise your report, look for places where your sentences are too long to be stated in one breath. These sentences should be shortened to breathing length or divided into two breathing-length sentences.

Link

For more information on eliminating nominalizations, see Chapter 17, page 458.

Eliminate nominalizations—Reports often include nominalizations that cloud the meaning of sentences. You should revise these sentences for clarity:

> **Nominalization:** This report offers *a presentation* of our findings.
>
> **Revised:** This report *presents* our findings.
>
> **Nominalization:** We made *a decision* to initiate *a replacement* of the Collings CAD software.
>
> **Revised:** We *decided to replace* the Collings CAD software.
>
> **Revised further:** We *replaced* the Collings CAD software.

AT A GLANCE

Improving Style in Analytical Reports

- Make "doers" the subjects of sentences.
- Use breathing-length sentences.
- Eliminate nominalizations.
- Define jargon and specialized terms.

Nominalizations may make your report sound more formal, but eliminating them will increase clarity, which readers will appreciate.

Define jargon and specialized terms—Jargon should not be completely eliminated from reports. Instead, these words should be defined for nonexperts. When you need to use a specialized term, use a sentence definition or parenthetical definition to clarify what the word means.

Link

For more information on writing sentence and parenthetical definitions, see Chapter 6, page 133.

> **Sentence definition:** A gyrocompass is a directional finding device that uses a gyroscope to compensate for the earth's rotation and thus points to true north.
>
> **Parenthetical definition:** Spotting a blue grouse, *a plump, medium-sized bird with feathered legs and bluish gray plumage,* is especially difficult because they are well camouflaged and live in higher mountain areas.

A Straightforward Design

Link

For more help with document design, see Chapter 18, page 482.

People rarely read reports word for word. Instead, they scan these documents. Therefore, you should use document design and graphics to highlight your main points and offer readers access points to start reading.

Choose a functional document design—Your report's design should reflect the subject of the report and the preferences of your readers. So, you

Need a definition for a jargon term? Go to **www.pearsonhighered.com/johnsonweb4/10.18**

Page Layouts for Reports

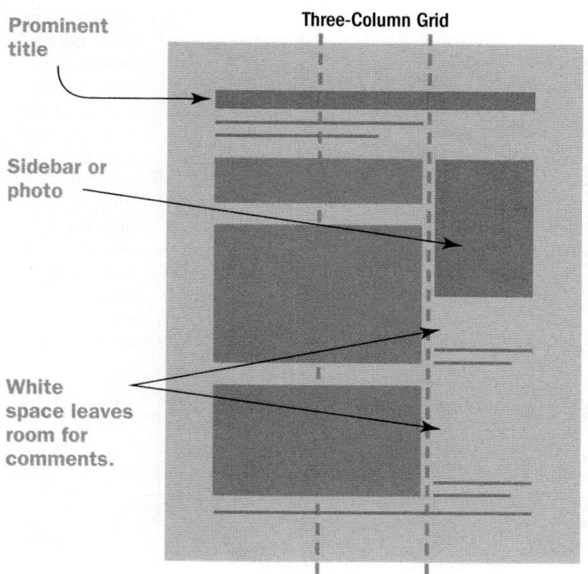

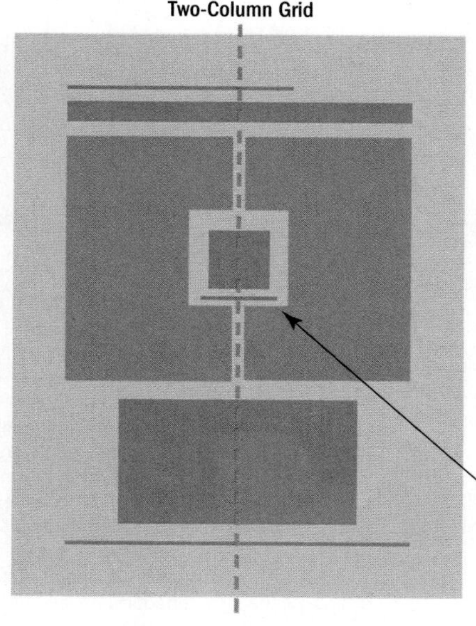

Figure 10.8: Reports don't have to look boring. A little attention to design will make your report inviting and easier to read. In these page layouts, grids have been used to balance the text, leaving room for margin text and other access points.

might experiment with page layout, the design of headings, and uses of graphics. You might look for ways to use multicolumn formats, which allow you to add pull quotes, sidebars, and other page layout enhancements (Figure 10.8).

Using graphics to clarify and enhance meaning—Tables and graphs are especially helpful tools for displaying data. As you write the report, you should actively look for places where tables might be used to better display your data or information. Also, look for places where a graph could show trends in the data.

Link

For more information on making and using graphs, see Chapter 19, page 523.

Your design and graphics should fit your subject, readers, and the document's context of use. Often, as a project comes to a finish, writers do not put enough time into the design of their report. As a result, the report makes a bad first impression on readers. With a little effort, you can create a design that ensures a positive first impression, while making need-to-know information easier to find.

Figure 10.9, for example, shows a well-designed report. The writers have designed this document to make it highly scannable. The two-column format and large headings make the information more accessible. Meanwhile, the larger font and ample white space help readers to look quickly through the document.

Figure 10.9: This recommendation report uses visual design features, such as a double-column format, headings, and lists, to make the information more accessible.

NIJ SEXUAL ASSAULT ON CAMPUS

Heather M. Karjane, Bonnie S. Fisher, and Francis T. Cullen

The title is large and easy to locate.

Sexual Assault on Campus:
What Colleges and Universities Are Doing About It

Campus crime in general and sexual assault in particular have been receiving more attention than in the past, and concern has been expressed at the highest levels of government. On the Federal level, Congress responded by enacting several laws requiring institutions of higher education to notify students about crime on campus, publicize their prevention and response policies, maintain open crime logs, and ensure sexual assault victims their basic rights.[3] The Clery Act, the most notable of these laws, mandates an annual security report from each Federally funded school (see "Recent Federal Laws on Campus Crime").

In 1999, Congress asked the National Institute of Justice to find out what policies and procedures schools use to prevent and respond to reports of sexual assault.[4] The resulting study revealed that schools are making strides

About the Authors

Heather M. Karjane, Ph.D., is coordinator for gender issues at the Commonwealth of Massachusetts Administrative Office of the Trial Court. Bonnie S. Fisher, Ph.D., and Francis T. Cullen, Ph.D., are faculty in the Division of Criminal Justice at the University of Cincinnati. The Police Executive Research Forum conducted some of the field research.

Recent Federal Laws on Campus Crime

Starting in 1990, Congress acted to ensure that institutions of higher education have strategies to prevent and respond to sexual assault on campus and to provide students and their parents accurate information about campus crime. The major Federal laws pertaining to this study are:

Student Right-to-Know and Campus Security Act of 1990 (the "Clery Act"*) (20 U.S.C. § 1092). This law, Title II of Public Law 101–542, requires that schools annually disclose information about crime, including specific sexual crime categories, in and around campus.

Campus Sexual Assault Victims' Bill of Rights of 1992. This amendment to the 1990 act requires that schools develop prevention policies and provide certain assurances to victims. The law was amended again in 1998 to expand requirements, including the crime categories that must be reported.

*The act was renamed in 1998 the "Jeanne Clery Disclosure of Campus Security Policy and Campus Crime Statistics Act" in honor of a student who was sexually assaulted and murdered on her campus in 1986.

This "sidebar" offers background information.

1

Source: U.S. Department of Justice, National Institute of Justice, 2005.

Figure 10.9:
(continued)

The two-column format makes the report easy to scan.

in some areas but must continue efforts to increase student safety and accountability. After summarizing what is known about the nature and extent of sexual assault on campus, the researchers highlighted findings regarding response policies and procedures; reporting options; barriers and facilitators; reporter training and prevention programming; victim resources; and investigation, adjudication, and campus sanctions. The study's baseline information can be used to measure progress in how institutions of higher education respond to sexual assault.

These large, bold headings help readers find the information they need.

The scope of the problem

Administrators want their campuses to be safe havens for students as they pursue their education and mature intellectually and socially. But institutions of higher education are by no means crime-free; women students face a high risk for sexual assault.

Just under 3 percent of all college women become victims of rape (either completed or attempted) in a given 9-month academic year. On first glance, the risk seems low, but the percentage translates into the disturbing figure of 35 such crimes for every 1,000 women students. For a campus with 10,000 women students, the number could reach 350. If the percentage is projected to a full calendar year, the proportion rises to nearly 5 percent of college women. When projected over a now-typical 5-year college career, one in five young women experiences rape during college.[5]

Counter to widespread stranger-rape myths, in the vast majority of these crimes—between 80 and 90 percent—victim and assailant know each other.[6] In fact, the more intimate the relationship, the more likely it is for a rape to be completed rather than attempted.[7] Half of all student victims do not label the incident "rape."[8] This is particularly true when no weapon was used, no sign of physical injury is evident, and alcohol was involved—factors commonly associated with campus acquaintance rape.[9] Given the extent of non-stranger rape on campus, it is no surprise that the majority of victimized women do not define their experience as a rape.

These reasons help explain why campus sexual assault is

(continued)

Figure 10.9: (continued)

SEXUAL ASSAULT ON CAMPUS NIJ

The header at the top of the page makes the text feel consistent.

not well reported. Less than 5 percent of completed and attempted rapes of college students are brought to the attention of campus authorities and/or law enforcement.[10] Failure to recognize and report the crime not only may result in underestimating the extent of the problem, but also may affect whether victims seek medical care and other professional help. Thus, a special concern of the study was what schools are doing to encourage victims to come forward.

Federal law and the schools' response

Institutions of higher education vary widely in how well they comply with Clery Act mandates and respond to sexual victimization. Overall, a large proportion of the schools studied—close to 80 percent—submit the annual security report required by the Act to the U.S. Department of Education; more than two-thirds include their crime statistics in the report. Yet, according to a General Accounting Office study, schools find it difficult to consistently interpret and apply the Federal reporting requirements, such as deciding which incidents to cite in the annual report, classifying crimes, and the like.[11]

For definitions and explanation of terminology such as "acquaintance rape," see Karjane et al., *Campus Sexual Assault: How America's Institutions of Higher Education Respond,* Oct. 2002, NCJ 196676: 2–3; for an analysis of how colleges and universities define sexual assault, see chapter 3.

This screened box offers a source for more information if the reader wants it.

Definitions, even of such terms as "campus" and "student," are often a challenge and contribute to inconsistency in calculating the number of reported sexual assaults. Only 37 percent of the schools studied report their statistics in the required manner; for example, most schools failed to distinguish forcible and nonforcible sex offenses in their reports as required by the Clery Act.

Ample white space gives the text a less intimidating feel.

The issues and the findings

Congress specified the issues to be investigated (see "Study Design"). Key areas of concern were whether schools have a written sexual assault response policy; whether and how they define sexual misconduct; who on campus is trained to respond to reports of sexual assault; how students can report sexual victimization; what

3

MICROGENRE

The Poster Presentation

Poster presentations have become one of the principal ways to display technical and scientific research. Poster presentations are typically 1 × 1.5 meters (3.5 × 5 feet), and they are often exhibited at conventions and conferences. Readers can study the researcher's methodology and the results of an experiment or research study. Poster presentations are designed to stand alone, but the researcher is often nearby to answer questions. Here is how to make one:

Choose the appropriate software. PowerPoint or Presentation will work fine if you are creating a poster with slides. InDesign or QuarkXPress are better for large, single-page posters, because they allow more flexibility with text and graphics.

Use a prominent title that attracts the reader. Your title should be large and prominent, 70 points or higher, so it catches the reader's attention. The best titles ask a question or hint at an interesting discovery.

Divide your presentation into major sections. Poster presentations usually follow the organization of an analytical report (e.g., IMRaD). Each major section of your report should be represented by a space on the poster. Label each section with a prominent heading.

Keep each section brief and to the point. Each section of the poster will usually have 500 words or less. So, keep the text simple and straightforward. Focus on the facts and need-to-know information.

Use text that is readable from 4 to 5 feet away. Most readers will be standing this distance from the poster, so the font size of the body text should be about 16 points. Headings should be 36 points or higher.

Design an interesting layout. Generally, the flow of information should be left to right and top to bottom. Poster presentations usually use a three-column or four-column layout.

Include images, graphs, and tables. Readers may not have time to read all your written text, but they will look at the visuals. So, where possible, put your data into visual forms that can be scanned quickly.

Write

Make your own poster presentation. You can use your poster presentation as a way to organize your ideas and data before you write your analytical report. Or, you can turn an existing report into a poster presentation.

GO TO THE NET

For worksheets to help you revise and edit your report, go to
www.pearsonhighered.com/johnsonweb4/10.20

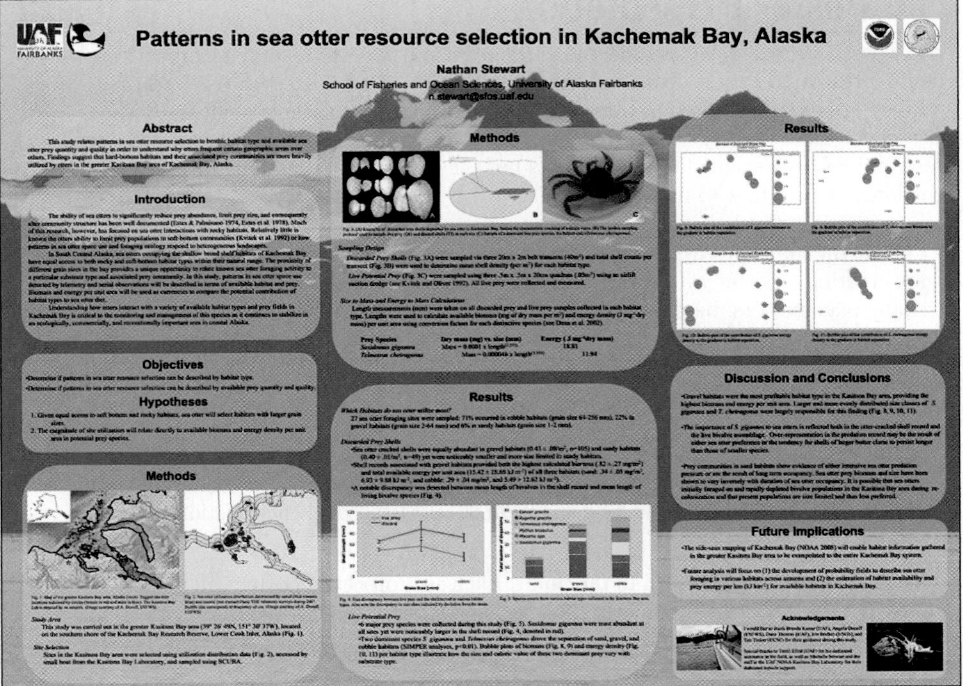

This poster presentation has all the major parts of an analytical report. It also uses graphics well to reinforce the written text.

Source: Used with permission of Nathan Stewart.

CHAPTER REVIEW

- Various types of analytical reports are used in the workplace, including research reports, completion reports, recommendation reports, feasibility reports, and empirical research reports.
- Use the IMRaD acronym to remember the generic report pattern: Introduction, Methods, Results, and Discussion.
- Determine the rhetorical situation for your report by considering the subject, purpose, readers, and context of use of your report.
- Define your research question, and formulate a hypothesis to be tested. Then, develop a methodology and identify the two to five major steps you will need to take to establish or refute your hypothesis.
- Collect information using library, Internet, and empirical sources.

GO TO THE NET

Sample reports for analysis are available at **www.pearsonhighered.com/johnsonweb4/10.21**

- Organize and draft your report following the separate "moves" for an introduction, body, and conclusion.
- Both the style and the design of analytical reports should serve to clarify the contents of the report.

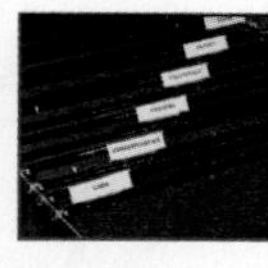

EXERCISES AND PROJECTS

Individual or Team Projects

1. Write a small report (two or three pages) in which you conduct a preliminary study of a scientific or technical topic that interests you. In your report, create a short methodology that will allow you to collect some overall information about the subject. Then, present the results of your study. In the conclusion, talk about how you might conduct a larger study on the subject and the limitations you might face as you enlarge the project.

2. Devise a research question of interest to you. Then, using logical mapping, sketch out a methodology that would allow you to study that research question in depth. When you have finished outlining the methodology, write a one-page memo to your instructor in which you describe the methodology and discuss the kinds of results your research would produce.

3. Write an executive summary of an article from a scientific or technical journal in your field. You should be able to find journals in your campus library. Your summary should paraphrase the article, highlighting its main points, methodology, results, discussion, and conclusions.

Collaborative Project: Problems in the Community

With a team of classmates, choose a local problem in your community about which you would like to write an analytical report. The readers of your report should be people who can take action on your findings, like the mayor or the city council.

After defining the rhetorical situation in which your report will be used, develop a step-by-step methodology for studying the problem. Then, collect information on the topic. Here are some ideas for topics that might interest you:

- Driving under the influence of alcohol or drugs
- Violence
- Underage drinking
- Illegal drug use
- Homelessness
- Teen pregnancy
- Water usage
- Pollution
- Graffiti
- Environmental health

In your report, present the results of your study and discuss those results. In the conclusion of your report, make recommendations about what local authorities can do to correct the problem or improve the situation.

For additional technical writing resources, including interactive sample documents, document design tutorials and guidelines, and more, go to **www.mytechcommlab.com.**

Revision Challenge

The introduction to a report shown here is somewhat slow and clumsy. How could you revise this introduction to make it sharper and more interesting?

Introduction

Let us introduce ourselves. We are the Putnam Consulting Firm, LLC, based out of Kansas City, KS. We specialize in the detection and mitigation of lead-related problems. We have been in business since 1995, and we have done many studies much like the one we did for you.

Our CEO is Lisa Vasquez. She has been working with lead-related issues for nearly two decades. In her opinion, lead poisoning is one of the greatest epidemics facing the United States today. The effects of lead on children, especially poor children, is acute. Dr. Vasquez has devoted her life to addressing the lead problem so today and tomorrow's children won't be damaged by continued neglect of the issue.

One of today's silent villains is lead poisoning. It is one of the most prevalent sources of childhood health problems. Each year, several hundred children in Yount County are treated for serious cases of lead poisoning. There are countless others who are suffering silently because their symptoms are either too minor or they go unnoticed by parents, teachers, and other authorities. For example, consider the case of Janice Brown in northwestern Yount County. She showed the effects of lead poisoning in her cognitive development, leading to a lower IQ rating. When her house was tested for lead poisoning, it was found that her parents had sandblasted the sides of the home to remove the old paint. Lead had saturated the ground around Janice's home and there was residual lead dust everywhere in the home. The levels of lead in Janice's blood were nearly 10 times the amount that causes problems.

You will find other troubling stories like this one in this report. We were hired by the Yount County Board of Directors to complete this study on the amounts of lead in the county. Unfortunately, we have found that Yount County is in particular trouble. Much of the county's infrastructure and housing was built when lead use was at its peak. Housing in the county relies heavily on lead pipes. Most of the houses were painted inside and out with lead-based paint.

Ironically, lead poisoning is also one of the most preventable problems that face children today. The continual persistence of the problem is a stain on our nation, and it is a particular problem that Yount County must face. In this report, we offer some recommendations.

To see a revision of this introduction, go to **www.pearsonhighered.com/johnsonweb4/10.22**

CASE STUDY

The X-File

George Franklin works as an environmental engineer for Outdoor Compliance Associates. He and his team are responsible for writing analytical reports called environmental impact statements (EISs) on sites that might be developed for housing or businesses. An EIS is required by the government before any work can begin.

George's responsibility on the team is to track down any historical uses of the site. For example, if a gas station had once been on the site, his job is to make sure the old underground holding tanks were removed. If the site once held a factory, his job is to make sure that chemicals had not been dumped in the area.
George loves being an environmental detective.

A year ago, George's team had written an EIS for a site where a new apartment complex was planned. The site was about a mile from a major research university. While researching the site, George discovered that it had housed part of the city's waste treatment center until the early 1950s. George's discovery wasn't a problem, though, because any contaminants would have disappeared long ago. So, George's team wrote a favorable EIS, clearing the site for development. The building plan was approved by the city, and construction started soon after.

Then, yesterday, an old, yellowed file mysteriously appeared in George's office mailbox. The file was marked "Confidential" and had no return address. George looked inside.

In the file was a report that nuclear weapons research had once been done at the university during the 1940s. Not recognizing the potential harm of the nuclear waste, the scientists had sent tons of radioactive waste down the drain. The nuclear waste ended up at the city's old waste treatment center. The waste, including the nuclear waste, was then spread around the grounds of the waste treatment plant.

George grabbed his Geiger counter and went out to the building site. The apartment development was now half built. When he pulled out his Geiger counter, it immediately began detecting significant levels of radiation.

The radiation wasn't high enough to violate government standards, but it was close. George knew that the building permit still would have been granted even if knowledge of this site's nuclear past had been known. However, he also knew that as soon as people heard about the radiation, no one would want to live in the apartments. That would be a disaster for the developer building them.

If he reported the radiation, there was a good chance George's company would be sued by the builder for missing this important problem with the property. Moreover, other builders would likely never again hire his company to write an EIS. George would almost certainly lose his job, and the company he worked for might be forced out of business. But then, he could keep it quiet. The radiation, after all, was not above government standards.

If you were George, how would you handle this situation?

More information on nuclear waste disposal is available at **www.pearsonhighered.com/johnsonweb4/10.23**

CHAPTER

11

Starting Your Career

Setting Goals, Making a Plan 315

Preparing a Résumé 320

Writing Effective Application Letters 331

Help: Designing a Scannable/Searchable Résumé 333

Creating a Professional Portfolio 339

Interviewing Strategies 343

Microgenre: The Bio 347

Chapter Review 348

Exercises and Projects 349

Case Study: The Lie 351

In this chapter, you will learn:

- How to set goals and make a plan for finding a career position.
- Methods for using both web-based and traditional networking.
- How to design and prepare a résumé in both electronic (scannable) and traditional formats.
- Techniques for writing a persuasive application letter.
- How to create a targeted, professional portfolio that highlights your background and experience.
- Strategies for effective interviewing.

Finding a good job requires energy, dedication, and optimism—and readiness to hear the word "no." You will probably look for a job at least a few times during your lifetime. U.S. Department of Labor statistics show that people tend to change careers four to six times in their lives, including twelve to fifteen job changes. In other words, being able to find a job is now becoming a necessary skill in a successful career.

Setting Goals, Making a Plan

From the beginning, you should adopt a professional attitude about looking for a job. To approach the task professionally, you should start your search by first setting some specific goals and developing a plan for reaching those goals.

Setting Goals

Before you begin sending your résumé to every company you can think of, spend some time setting goals for your search. Begin by answering the Five-W and How Questions:

What are my needs and wants in a job/career?

Who would I like to work for?

Where would I like to live?

When do I need to be employed?

Why did I choose this career path in the first place?

How much salary, vacation, and benefits do I need?

Type your answers into a file, or write them on a piece of paper. Be specific about things like location, salary, vacation, and the amount of time you have to find a position. If you don't know the answers to specific questions, like salary, then use Internet search engines to help you find these answers. If you are still in school, you can ask your instructors to give you a good estimate of salaries in your field.

Using a Variety of Job-Seeking Paths

With your goals in mind, you can begin developing a plan for finding a good job. Fortunately, your computer offers a doorway to an amazing number of job-seeking pathways.

JOB SEARCH ENGINES The numerous job search engines available on the Internet are good places to start looking for jobs. By entering a few keywords, you should be able to locate a variety of job opportunities. Most search engines will allow you to limit your search by region, job title, or industry. Some search engines will even allow you to post your résumé so employers can find you. For example, Monster.com, shown in Figure 11.1, has a link on its home page that takes you to its résumé-posting area.

Career Materials

These models show typical organizational patterns for a résumé and application letter. These patterns should be altered to fit your background and the kinds of jobs you are applying for.

Résumé

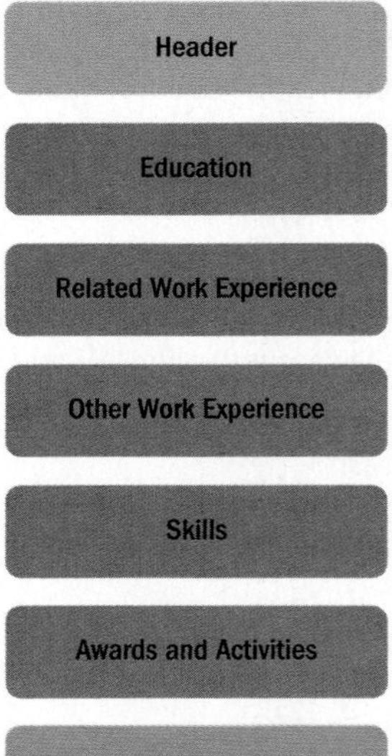

Application Letter

Your Address

Date

Employer's Address

Dear X,

Introduction

Description of Educational Background

Description of Work Experience

Description of Skills or Talents

Conclusion

Sincerely,

Your Name

Basic Features

Minimally, your career materials will include a résumé and an application letter. These documents will usually have the following features:

Résumé

- Header with your name and contact information
- Career objective/career summary
- Educational background
- Related work experience
- Other work experience
- Skills
- Awards and activities
- References

Application Letter

- Header with addresses
- Introduction that identifies position being applied for
- Description of educational background
- Description of work experience
- Description of specialized skills or talents
- Conclusion that thanks readers

To see employment trends, go to **www.pearsonhighered.com/johnsonweb4/11.1**

An Internet Job Search Engine

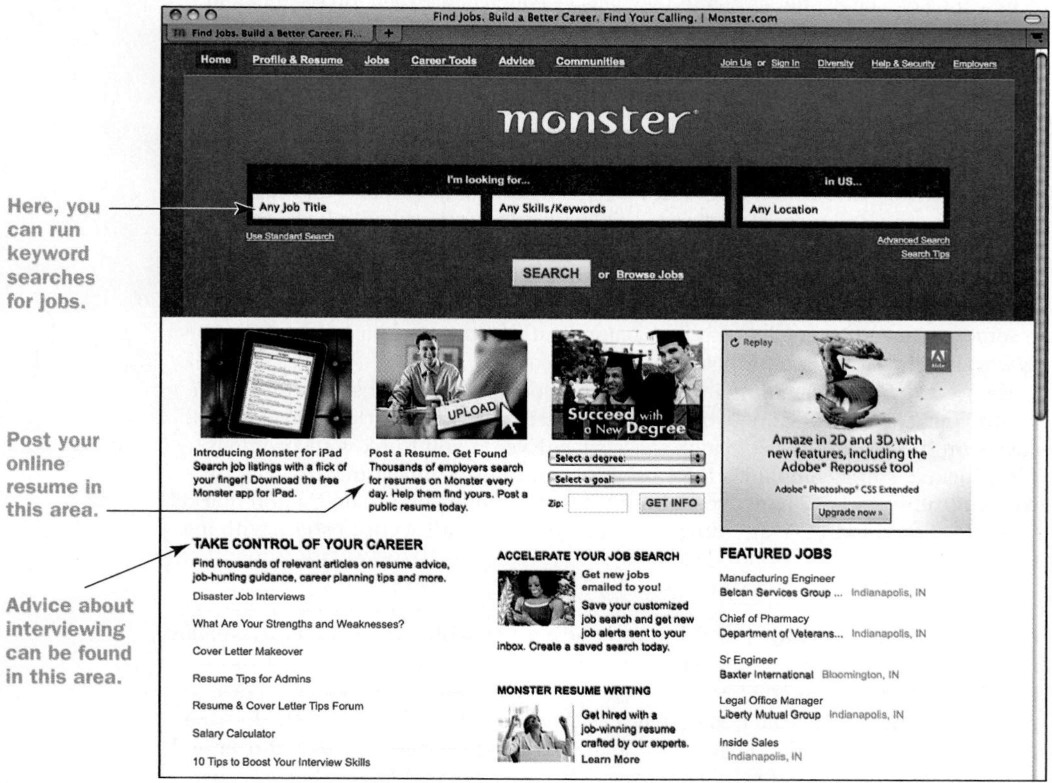

Figure 11.. Monster.com is one of the more popular job search engines. It offers a variety of tools to aid your search.

Source: Monster.com. Reprinted by permission.

Here are some of the more popular job search engines:

4Jobs.com	*Usajobs.opm.gov*
CollegeRecruiter.com	*Hotjobs.yahoo.com*
Indeed.com	*SimplyHired.com*
Careerbuilder.com	*Dice.com*
Jobfox.com	*Monster.com*
Jobster.com	*Worktree.com*
Collegegrad.com	

For websites that will help you answer these goal-setting questions, go to **www.pearsonhighered.com/johnsonweb4/11.2**

Job search engines are fast becoming the best way to find a job, especially in technical fields. They offer easy and inexpensive contacts between job seekers and employers. Employers are beginning to prefer search engines because these tools can be used to run a nationwide search for the best people. You should check these sites regularly and post your résumé on at least a few of them.

SOCIAL NETWORKING WEBSITES More than ever, people are finding jobs through their connections on social networking sites like Facebook, MySpace, and Second Life. There are also business-centered social networking sites like LinkedIn and Ryze that you should join because they are designed to help people do career networking. You should turn your social networking site(s) into a job finding tool by including your résumé and any other information you would like potential employers to see. In some cases, employers may find you before you find them because they use social networking sites to search for potential new employees.

Before you go on the job market, however, you should spend some time cleaning up your current social networking sites. Interviewers regularly do background checks on their candidates through these sites. Even if you think your profile is safely locked, some companies have been known to use connections like alumni from your university to gain access to profiles. So, while you are on the job market, your site should not have anything on it that you wouldn't want to share with a potential employer.

PERSONAL NETWORKING Someone you know is probably aware of a job available in your field. Or, they know someone who knows about a job. Make a list of your friends, relatives, and instructors who might be able to help you find a job. Then, send each of these people an e-mail that tells them you are "on the market," looking for a job. You might even attach a résumé to your e-mail so they can look it over and perhaps forward it to a potential employer.

More than likely, you will be pleasantly surprised by the response to these e-mails. Your friends and relatives know more people than you realize. Meanwhile, your instructors are often aware of opportunities available in your area.

PROFESSIONAL NETWORKING Most career tracks have professional groups associated with them. Engineers, for example, have the Institute of Electrical and Electronics Engineers (IEEE), while medical practitioners have the American Medical Association (AMA). Technical writers have the Society for Technical Communication (STC). These professional groups are especially helpful for networking with people who are already employed in your field.

Most large professional groups have a local chapter you can join. Chapter meetings offer great opportunities to contact people who have jobs similar to the one you want. Also, local chapters usually have websites that post job openings in your area.

You should become involved with these groups as soon as possible, even if you have not graduated from college yet. It takes a while to become a regular at meetings, but once people get to know you, they can be very helpful with finding job opportunities.

For links to job search sites, go to
www.pearsonhighered.com/johnsonweb4/11.3

COLLEGE PLACEMENT OFFICE Most colleges have a placement office that is available to students. The placement office may have jobs posted on its website, or you can visit the office itself. There, you can sign up for interviews and speak with a counselor about improving your job-searching skills.

TARGETING Make up a list of ten to twenty "target" companies for which you might want to work. Then, look at their websites, paying special attention to each company's human resources office. From each website, write down notes about the company's mission, products, history, and market.

If one of your targeted companies does not have a job available, send its human resources department a copy of your résumé and an application letter. In your letter, tell the target company's human resources officer that you are sending materials for their files in case a position becomes available.

CLASSIFIED ADVERTISEMENTS In the classifieds section of a newspaper, especially the Sunday edition, you will find job advertisements. Keep in mind, though, that newspapers carry advertisements for only a few jobs in any given area. *Most jobs are not advertised in the paper.* But these ads are worth checking once a week.

Most major newspapers now have companion websites that list online classifieds. If you are looking for a position in a specific city or state, these online classifieds may be helpful. They include all the jobs that are printed in the classified advertisements in the newspaper.

The secret to effective job searching is to set clear goals and have a strategy for reaching those goals. With the variety of electronic tools available, you have many different paths to follow to find a position (Figure 11.2). The most successful job seekers use them all.

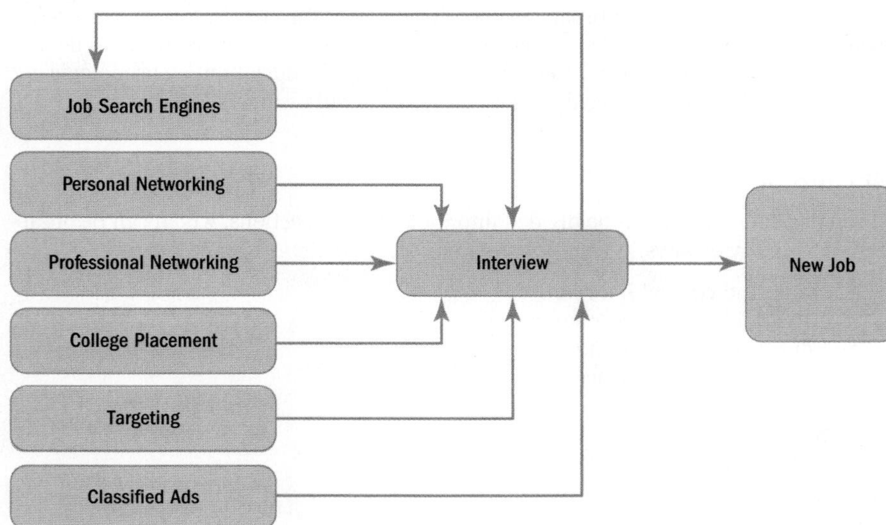

Figure 11.2: There are many tools available for finding a job and building a career. You should take advantage of all of them.

Preparing a Résumé

A résumé is a summary of your background, experience, and qualifications. Usually, résumés for entry-level jobs fit on one page. Résumés for advanced positions might extend to two pages. You need to spend some serious time and effort on your résumé, because it may be one of the most important documents you will ever write.

Your résumé has to make a good first impression because it is usually the first item that employers will look at. If your résumé reflects the qualifications and experience they are looking for, they may read your other materials and give you a call. If your résumé is poorly written or poorly designed, chances are slim that you will ever get your foot in the door.

Types of Résumés

Résumés tend to follow one of two organizational approaches: the *chronological* approach and the *functional* approach.

> **Chronological approach**—Organizes the résumé according to education and work experience, highlighting a job seeker's qualifications in a few areas. A chronological résumé might be organized into sections such as Education, Work Experience, Skills, and Awards/Activities.
>
> **Functional approach**—Organizes the résumé according to your talents, abilities, skills, and accomplishments. A functional résumé would be organized into sections such as Leadership, Design Experience, Communication Skills, and Training Abilities.

By far, the chronological approach is the most common in technical fields, especially for entry-level jobs. It allows you to highlight the details of your experience. Generally, the functional approach is not advantageous for new graduates because their experiences are limited and have not given them enough opportunities to demonstrate their abilities. For most entry-level jobs, you should submit a chronological résumé unless you are specifically asked for a functional résumé.

Chronological Résumé

Chronological résumés can be divided into the following sections, which can be organized in a variety of ways:

- Name and contact information
- Career objective/career summary
- Educational background
- Related work experience
- Other work experience
- Skills
- Awards and activities
- References

For links to several major newspapers' classified ads, go to **www.pearsonhighered.com/johnsonweb4/11.5**

The headings in this list are the ones commonly used in chronological résumés, but other headings are available. Use headings that best highlight your qualifications and set the specific tone you desire (Figures 11.3, 11.4, and 11.5).

NAME AND CONTACT INFORMATION At the top of the page, your résumé should include a heading that states your name, address, phone number, e-mail address, and perhaps a website address. To catch the reader's eye, your name should appear in a larger font size. The heading can be left justified, centered, or right justified, depending on the design of your résumé. Some people like to add a rule (a line) to set off the heading from the rest of the information in the résumé.

Link

For more information on using fonts effectively, go to Chapter 18, page 500.

CAREER OBJECTIVE OR CAREER SUMMARY (OPTIONAL) A *career objective* is a phrase or sentence that describes the career you are seeking. Your career objective should specify the type of position you are applying for and the industry in which you want to work.

> Seeking a position as a Physician's Assistant in a research hospital.
>
> A computer programming career working with mainframes in the defense industry.
>
> Looking for a career as a school psychologist working with children who have behavioral problems.
>
> Mechanical Engineer developing nanotechnological solutions for biotech applications.

Avoid writing a career objective that is too narrow or too broad. A career objective that is too narrow might eliminate you from some potential jobs. For example, if you specify that you are looking for a job at a "large engineering firm," the human resources officer at a small engineering firm might assume you are not interested in her company's job. On the other hand, an objective that is too broad, like "a job as an electrical engineer," might give the impression that you are not sure about what kind of career would suit your talents.

A *career summary* is a brief sentence or paragraph that describes your career to this point. Career summaries are typically used by people with years of experience.

> I have been employed in the hospitality industry for 10 years, working my way up from desk clerk to assistant manager at a major downtown hotel. My specialty is coordinating catering services for large conferences.
>
> My experience as a webmaster includes designing and managing a variety of interactive sites that have been used by large companies to promote their products and solicit new business.

Your career objective or career summary should convey a sense of what makes you unique or interesting as an applicant. A career objective or career summary is not required in a résumé. In fact, some résumé experts suggest that these statements unnecessarily take up space on a résumé, because they rarely do more than describe the position being applied for. If you are looking for room to put more information on your résumé, you might leave out the career objective or career summary.

GO TO THE NET

To find other headings for your résumé, go to **www.pearsonhighered.com/johnsonweb4/11.6**

Chronological Résumé

Name and contact information are prominently displayed.

Career objective describes position sought.

Education is highlighted by placement early in the résumé.

Work experience lists duties with bullets.

Skills are listed separately for emphasis.

Awards and memberships are placed later in the résumé.

Anne Franklin

834 County Line Rd.
Hollings Point, Illinois 62905

Home: 618-555-2993
Mobile: 618-555-9167
e-mail: afranklin@unsb5.net

Career Objective

A position as a naturalist, specializing in agronomy, working for a distribution company that specializes in organic foods.

Educational Background

Bachelor of Science, Southern Illinois University, expected May 2011.
Major: Plant and Soil Science
Minor: Entomology
GPA: 3.2/4.0

Work Experience

Intern Agronomist, December 2009–August 2010
Brighter Days Organic Cooperative, Simmerton, Illinois

- Consulted with growers on organic pest control methods. Primary duty was sale of organic crop protection products, crop nutrients, seed, and consulting services.
- Prepared organic agronomic farm plans for growers.
- Provided crop-scouting services to identify weed and insect problems.

Field Technician, August 2007–December 2009
Entomology Department, Southern Illinois University

- Collected and identified insects.
- Developed insect management plans.
- Tested organic and nonorganic pesticides for effectiveness and residuals.

Skills

Computer Experience: Access, Excel, Outlook, PowerPoint, and Word. Global Positioning Systems (GPS). Database Management.
Machinery: Field Tractors, Combines, Straight Trucks, and Bobcats.
Communication Skills: Proposal Writing and Review, Public Presentations, Negotiating, Training, Writing Agronomic and Financial Farm Plans.

Awards and Memberships

Awarded "Best Young Innovator" by the Organic Food Society of America, 2009.
Member of Entomological Society of America.

References Available Upon Request

Figure 11.3: Anne Franklin's résumé uses a list style. She achieves a classic look by using a serif font (Garamond) and left justification.

Chronological Résumé

Name and contact information are prominently displayed.

A career summary is used to summarize experience.

Work experience is featured because it is placed early in the résumé.

Other work experience is listed concisely.

A paragraph lists classes taken.

Résumé ends by mentioning where a dossier can be obtained.

James L. Mondragon

576 First Avenue, Rolla, Missouri 65408
Phone: 573-555-4391, e-mail: bigmondy12@umr.edu

Career Summary

My interests in chemical engineering started with a childhood fascination with plastic products. I enrolled at the University of Missouri–Rolla because of their strong chemical engineering program, especially in the area of applied rheology and polymeric materials. I have also completed a co-op with Vertigo Plastics in St. Louis. My background includes strong computer modeling skills, especially involving polymeric materials.

Work Experience

Vertigo Plastics, Inc., St. Louis, Missouri, 5/09–1/10, 5/10–1/11
Co-op Chemical Engineer
Performed inspections of chemical equipment and plant equipment affected by chemical systems (such as boilers and condensers). Monitored the performance of chemical systems at various sites throughout the Vertigo Plastics system. Performed calculations and wrote reports summarizing the performance of those systems. Helped troubleshoot problems.

Other Work Experience

To Go Pizza, *Server,* Springfield, Missouri, 2/03–7/05

Educational Background

University of Missouri–Rolla, BSE, Expected May 2011
Major: Chemical Engineering
Advanced Coursework included Chemical Engineering Fluid Flow, Chemical Engineering Heat Transfer, Chemical Engineering Thermodynamics I & II, Process Dynamics and Control, Chemical Engineering Reactor Design, Chemical Engineering Economics, Chemical Process Safety, Chemical Process Design, Chemical Process Materials.

Activities and Awards

Awarded Stevenson Scholarship in Engineering, 2009
Treasurer, UMR Student Chapter, American Institute of Chemical Engineers (AIChE), 2009–Present
Member, Tau Beta Pi, 2009

Dossier with References Available at UMR Placement Services (573-555-2941)

Figure 11.4: James Mondragon's résumé uses a paragraph style. It uses a sans serif font for headings (Myriad) and a serif font (Times) for the body text. The headings are centered.

Chronological Résumé of Student Who Returned to College

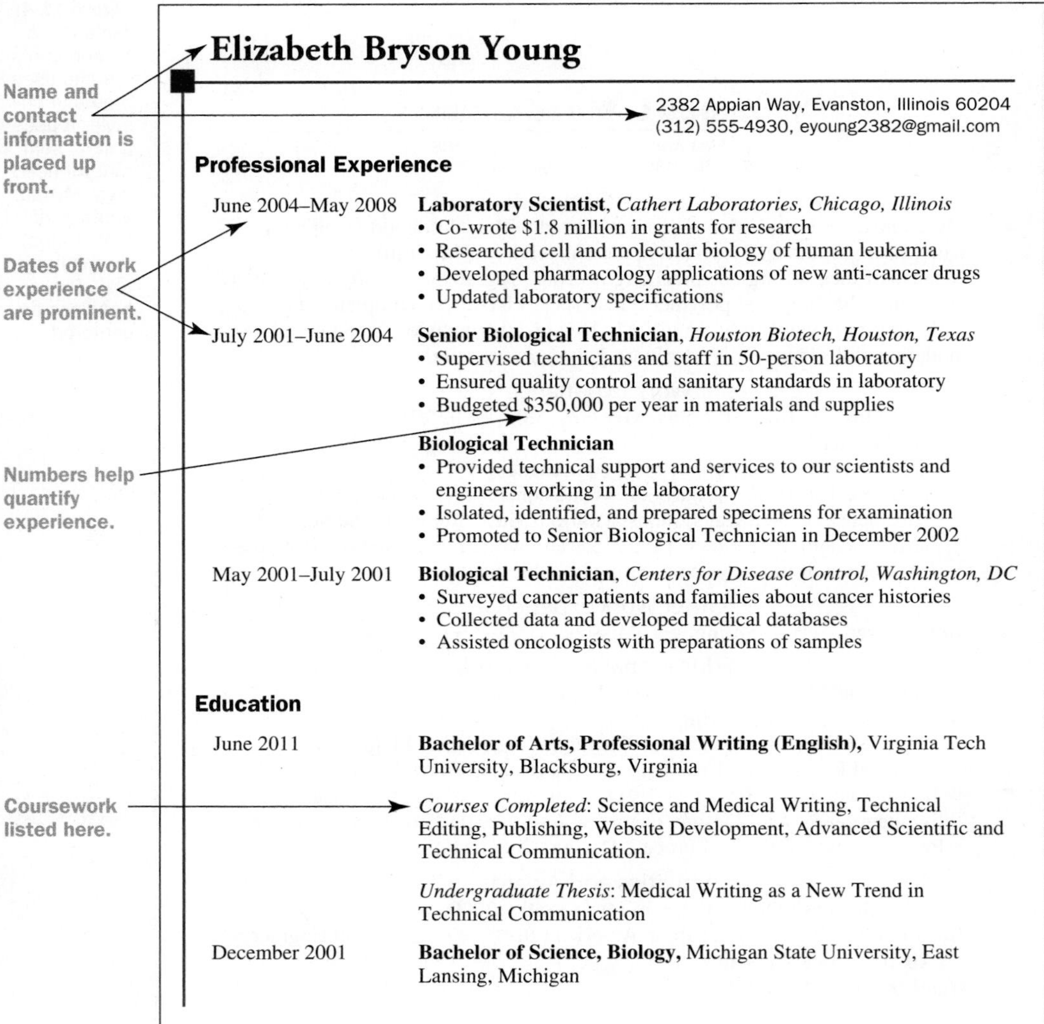

Elizabeth Bryson Young

2382 Appian Way, Evanston, Illinois 60204
(312) 555-4930, eyoung2382@gmail.com

Professional Experience

June 2004–May 2008 **Laboratory Scientist**, *Cathert Laboratories, Chicago, Illinois*
- Co-wrote $1.8 million in grants for research
- Researched cell and molecular biology of human leukemia
- Developed pharmacology applications of new anti-cancer drugs
- Updated laboratory specifications

July 2001–June 2004 **Senior Biological Technician**, *Houston Biotech, Houston, Texas*
- Supervised technicians and staff in 50-person laboratory
- Ensured quality control and sanitary standards in laboratory
- Budgeted $350,000 per year in materials and supplies

Biological Technician
- Provided technical support and services to our scientists and engineers working in the laboratory
- Isolated, identified, and prepared specimens for examination
- Promoted to Senior Biological Technician in December 2002

May 2001–July 2001 **Biological Technician**, *Centers for Disease Control, Washington, DC*
- Surveyed cancer patients and families about cancer histories
- Collected data and developed medical databases
- Assisted oncologists with preparations of samples

Education

June 2011 **Bachelor of Arts, Professional Writing (English),** Virginia Tech University, Blacksburg, Virginia

Courses Completed: Science and Medical Writing, Technical Editing, Publishing, Website Development, Advanced Scientific and Technical Communication.

Undergraduate Thesis: Medical Writing as a New Trend in Technical Communication

December 2001 **Bachelor of Science, Biology,** Michigan State University, East Lansing, Michigan

Name and contact information is placed up front.
Dates of work experience are prominent.
Numbers help quantify experience.
Coursework listed here.

Figure 11.5: This résumé from a student who returned to college features work experience ahead of educational experience.

Figure 11.5: *(continued)*

Computer Skills

Word, InDesign, Photoshop, Illustrator, Excel, Dreamweaver, PowerPoint

Awards

AMWA Best Biotech Article in 2006 for "Turning a Corner on Leukemia," *Biotech News*, March 2006

Researcher of the Year, Cathert Laboratories, 2005

Professional Organizations

American Medical Writers Association (AMWA)

Society of Technical Communication (STC)

American Association for the Advancement of Science (AAAS)

Volunteer work adds depth to experience and shows maturity.

Volunteer Work

Health Volunteers Overseas, 2003–2007

Habitat for Humanity, 1998–present

Northern Chicago Urban Ministries, 2010–present

References listed on second page.

References

Thomas VanArsdale, Chief Scientist
Cathert Laboratories
1208 Discovery Lane
Chicago, IL 60626
(312) 654-3864
tvanarsdale@cathertlabs.com

Franklin Charleston, Assistant Professor
Department of English
North Carolina State University
Tompkins Hall, Box 8105
Raleigh, NC 27695-8105
(919) 515-3866
franklcharleston@ncsu.edu

Christina Smith, Associate Professor
Department of English
Virginia Tech University
323 Shanks Hall (0112)
Blacksburg, VA 24061
(540) 231-6501
chrsmith34@vt.edu

Want more advice on writing résumés? Go to **www.pearsonhighered.com/johnsonweb4/11.7**

Kim Isaacs

EXECUTIVE DIRECTOR, ADVANCED CAREER SYSTEMS, INC.

Advanced Careers Systems is a résumé-writing and career development company.

How has the Internet changed the job search process?

The Internet has changed the job search dramatically, and while the web makes the search easier in many ways, it also requires the seekers to be savvier than ever. Because job seekers can search the Internet for positions, they can now apply for jobs that they might not have even heard about in the past. They can also apply for international jobs as easily as national ones. Seekers can go to major online career sites and set up "job agents" that enable them to enter information about their goals and receive notifications when matching jobs become available.

Because the candidate pool is so large, many employers report that they're constantly inundated with new résumés, and employers have the luxury of seeking a "perfect match." That's why it's so important, for example, to make sure that electronic résumés use keywords that will be found on a résumé-tracking system. Even if your résumé is excellent, it can be overlooked without the right wording and phrasing.

Despite the increase in using the Internet to search for positions, online recruiting still accounts for a small fraction of actual hires (the figure keeps changing, but it's around 5 percent). Of course job seekers should use the Internet to research companies, scour for opportunities, and take advantage of the career tools available, but they still need to focus most of their effort on old-fashioned job search methods such as networking.

EDUCATIONAL BACKGROUND Your educational background should list your most recent college degree and other degrees in reverse chronological order—most recent to least recent. You can list the degree you are working on now as "expected" with the month and year when you should graduate.

Name each college and list your major, minor, and any distinctions you earned (e.g., scholarships or summa cum laude or distinguished-scholar honors) (Figure 11.3). You might also choose to mention any coursework you have completed that is related to your career (Figures 11.4 and 11.5).

Any specialized training, such as welding certification, experience with machinery, or military training, might also be listed here with dates of completion.

If you are new to the technical workplace, you should place your educational background early in your résumé. Your degree is probably your most prominent achievement, so you want to highlight it. If you have years of professional work experience, you may choose to put your educational background later in your résumé, allowing you to highlight your years of experience.

For sample résumés, go to **www.pearsonhighered.com/johnsonweb4/11.8**

RELATED WORK EXPERIENCE Any career-related jobs, internships, or co-ops that you have held should be listed, starting with the most recent and working backward chronologically. For each job, include the title of the position, the company, and dates of employment (month and year).

Below each position, list your workplace responsibilities. As demonstrated in Figures 11.3, 11.4, and 11.5, you can describe these responsibilities in a bulleted list or in a brief paragraph. Use action verbs and brief phrases to add a sense of energy to your work experience. Also, where possible, add any numbers or details that reflect the importance of your responsibilities.

> Coordinated a team of 15 student archaeologists in the field.
>
> Participated in the development of UNIX software for a Cray supercomputer.
>
> Worked with over 100 clients each year on defining, updating, and restoring their water rights in the Wilkins Valley.

Verb-first phrases are preferable to full sentences, because they are more scannable and require less space. Some action verbs you might consider using on your résumé include the following:

adapted	*devised*	*organized*
analyzed	*directed*	*oversaw*
assisted	*equipped*	*planned*
collaborated	*examined*	*performed*
collected	*exhibited*	*presented*
compiled	*implemented*	*proposed*
completed	*increased*	*recorded*
conducted	*improved*	*researched*
constructed	*instructed*	*studied*
coordinated	*introduced*	*supervised*
corresponded	*investigated*	*taught*
designed	*managed*	*trained*
developed	*observed*	*wrote*

You might be tempted to exaggerate your responsibilities at a job. For example, a cashier at a fast food restaurant might say he "conducted financial transactions." In reality, nobody is fooled by these kinds of puffed-up statements. They simply draw attention to a lack of experience and, frankly, a mild lack of honesty. There is nothing wrong with simply and honestly describing your experiences.

OTHER WORK EXPERIENCE Almost everyone has worked at jobs that were not related to his or her desired career. If you have worked at a pizza place, waited tables, or

Helpful services for managing your letters of reference can be found at **www.pearsonhighered.com/johnsonweb4/11.9**

painted houses in the summer, those jobs can be listed in your résumé. But they should not be placed more prominently than your related work experience, nor should they receive a large amount of space. Instead, simply list these jobs in reverse chronological order, with names, places, and dates.

> Pizza Chef. Giovanni's. Lincoln, Nebraska. September 2010–August 2011.
>
> Painter. Campus Painters. Omaha, Nebraska. May 2010–September 2010.
>
> Server. Crane River Brewpub and Cafe. Omaha, Nebraska. March 2008–May 2010.

Do not offer any additional description. After all, most people are well aware of the responsibilities of a pizza chef, painter, and server. Any more description will only take up valuable space on your résumé.

Don't underestimate the importance of non-career-related jobs on your résumé. If you do not have much professional work experience, any kind of job shows your ability to work and hold a job. As your career progresses, these items should be removed from your résumé.

SKILLS Résumés will often include a section that lists career-related skills. In this area, you may choose to mention your abilities with computers, including any software or programming languages you know how to use. If you have been trained on any specialized machines or know how to do bookkeeping, these skills might be worth mentioning. If you have proven leadership abilities or communication skills (technical writing or public speaking) you might also list those.

> *Computer Skills:* Word processing (Word, WordPerfect), desktop publishing (InDesign, Quark), web design (Dreamweaver, FrontPage), and data processing (Excel, Access).
>
> *Leadership Abilities:* President of Wilkins Honor Society, 2010–2011. Treasurer of Lambda Kappa Kappa Sorority, 2008–2011. Volunteer Coordinator at the Storehouse Food Shelter, 2009 to present.

The skills section is a good place to list any training you have completed that does not fit under the "Educational Background" part of your résumé.

AWARDS AND ACTIVITIES List any awards you have won and any organized activities in which you have participated. For example, if you won a scholarship for your academic performance, list it here. If you are an active member of a club or fraternity, show those activities in this part of your résumé. Meanwhile, volunteer work is certainly worth mentioning, because it shows your commitment to the community and your willingness to take the initiative.

REFERENCES Your references are the three to five people who employers can call to gather more information about you. Your references could include current or former supervisors, instructors, colleagues, and professionals who know you and your work.

You can download résumé templates at **www.pearsonhighered.com/johnsonweb4/11.10**

AT A GLANCE

Sections in a Chronological Résumé

- Name and contact information
- Career objective/ career summary
- Educational background
- Related work experience
- Other work experience
- Skills
- Awards and activities
- References

They should be people you trust to offer a positive account of your abilities. Each reference listing should include a name, title, address, phone number, and e-mail address. Before putting someone in your list of references, though, make sure you ask permission. Otherwise, he or she may be surprised when a recruiter or human resources officer calls to ask questions about you.

References can take up a large amount of space on a résumé, so they are typically not listed on the résumé itself. Instead, a line at the bottom of the résumé states, "References available upon request." Then, the references—listed on a separate sheet of paper under the heading "References"—can be sent to any employer who requests them.

If the employer asks for references to appear on the résumé, you can add them at the bottom of your résumé, usually on a second page (Figure 11.5). To avoid causing your résumé to go over two pages, you may need to list your references in two or three columns on the second page.

Functional Résumé

The functional résumé is less common than the chronological résumé, especially for new college graduates. This type of résumé is designed to highlight the job applicant's abilities and skills by placing them up front in the résumé. The advantage of this type of résumé is its ability to boil years of experience down to a few strengths that the job applicant would like to highlight.

Figure 11.6 shows an example of a functional résumé. In this example, note how the résumé places the applicant's strengths, including awards, early in the document. Then, the remainder of the résumé concisely lists details about the applicant's employment background, educational background, and professional memberships.

Designing the Résumé

The design of your résumé should reflect your personality and the industry in which you want to work. A résumé for an engineering firm, for example, will probably be somewhat plain and straightforward. A résumé for a graphic artist position at a magazine, on the other hand, should demonstrate some of your skills as a designer. There are, of course, exceptions. Some progressive engineering firms, for example, might prefer a layout that reflects your innovative qualities.

Most word-processing programs include résumé templates that you can use to lay out your information. If you use one of these templates, alter the design in some way, because unfortunately, many thousands of people have access to the same templates. So, employers often receive several résumés that look identical. You want yours to stand out.

If you decide to design your own résumé, Chapter 18 offers design principles that are helpful toward creating a design for your résumé. These principles are balance, alignment, grouping, consistency, and contrast. All of these principles should be used to design your résumé.

A Functional Résumé

Name and contact information are placed up front.

The objective describes the position sought.

Qualities are summarized here, including awards.

Employment and education histories are very concise, listing only the details.

Walter David Trimbal

818 Franklin Drive
Atlanta, Georgia 30361
404-555-2915

Objective: Senior architect position in a firm that specializes in urban revitalization projects.

Leadership Experience

- Managed design team for additions/alterations for small-scale commercial buildings in downtown Atlanta.
- Led planning charette for Bell Hill Neighborhood renovation.
- Awarded a 2008 "Archi" for design of Delarma Commerce Center.

Technical Expertise

- Experienced with the latest developments in computer-aided drafting hardware and software.
- Able to resolve conflicting building and zoning codes, while preparing site plans.

Community Involvement

- Founding Member of the Better Atlanta Commission in 2001 and served as board member until 2009.
- Served on the Architecture Public Involvement Committee for Atlanta Metropolitan Council of Governments from 2008–2009.

Employment Background

Vance & Lipton—Architects, Senior Architect, Atlanta, GA, 2005 to present.
Fulton County Planning Department, Planning Architect, Atlanta, GA, 2001–2005.
Ronald Alterman—Architect, Intern, Boston, MA, 2000–2001.

Education Background

B.A. in Architecture, Boston College, Boston, MA, 2001.
A.A. in Computer-Aided Drafting, Augusta Technical College, Augusta, GA, 1998.

Professional Memberships

National Council of Architectural Registration Boards
Greater Atlanta Architects Guild

References Available Upon Request

Figure 11.6: A functional résumé puts the applicant's abilities and skills up front where an employer will see them. Other features, such as employment history and education, are minimized.

Balance—Pay attention to the vertical and horizontal balance of the page. Your résumé should not be weighted too heavily toward the left or right, top or bottom.

Alignment—Different levels of information should be consistently indented to make the résumé easy to scan. Don't just align everything at the left margin; instead, use vertical alignment to create two or three levels in the text.

Grouping—Use white space to frame groups of information. For example, a job listed on your résumé with its responsibilities should be identifiable as a chunk of text. Sometimes using rules, especially horizontal lines, is a good way to carve a résumé into quickly identifiable sections (Figure 11.3).

Consistency—The design of your résumé should be internally consistent. Use boldface, italics, and font sizes consistently. Bullets or other symbols should also be used consistently throughout the résumé.

Contrast—Titles and headings should be noticeably different from the body text. To contrast with the body text, you might choose a different serif or sans serif font for your titles and headings. You can increase the font sizes and/or use boldface to make the résumé more scannable.

Link

For more information on page design, turn to Chapter 18, page 482.

A helpful strategy for designing your résumé is to collect résumés from other people. There are also numerous books and websites available that offer ideas about designing résumés. You can use these sample résumés as models for designing your own.

Writing Effective Application Letters

Your résumé will provide the employer with facts and details about your education, work experience, and skills. An effective application letter strives to prove that you are uniquely qualified for the available position. Two common mistakes are made in application letters. First, applicants simply restate the information available on the résumé, failing to demonstrate why they are the right person for the job. Second, they discuss why the position would be good *for them* (e.g., "A job at Gurson Industries would help me reach my goal to become an electrical engineer working with sensors"). To put it bluntly, employers don't really care whether their job is good for you.

Instead, your letter should prove to potential employers that your education, experience, and skills will allow you to make a valuable contribution to *their* company. Put the emphasis on *their* needs, not yours. Your letter should fit on one page for an entry-level position. Two pages would be a maximum for any job.

Content and Organization

Like any letter, an application letter will have an introduction, body, and conclusion (Figure 11.7). It will also include common features of a letter such as the header (your address and the employer's address), a greeting ("Dear"), and a closing salutation ("Sincerely") with your signature.

Link

For more information on writing letters, go to Chapter 5, page 94.

The Basic Pattern of an Application Letter

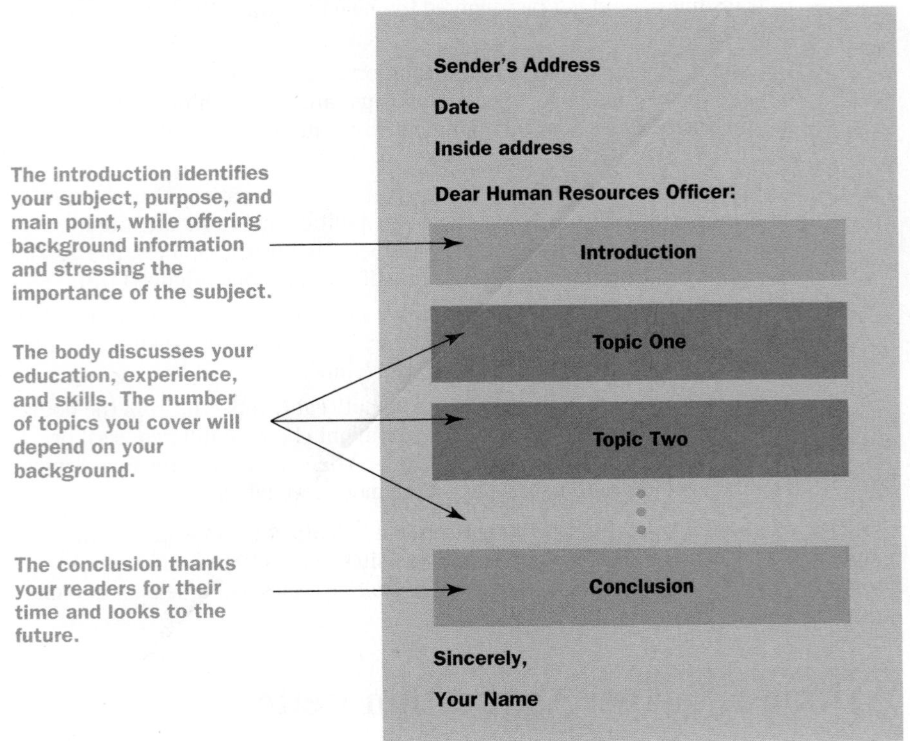

Sender's Address
Date
Inside address
Dear Human Resources Officer:
Introduction
Topic One
Topic Two
Conclusion
Sincerely,
Your Name
The introduction identifies your subject, purpose, and main point, while offering background information and stressing the importance of the subject.
The body discusses your education, experience, and skills. The number of topics you cover will depend on your background.
The conclusion thanks your readers for their time and looks to the future.

Figure 11.7: An application letter includes the common features of a letter, like the sender's address, date, inside address, introduction, body, and closing salutation.

INTRODUCTION You should begin your letter by making up to five moves commonly found in the introduction of any document: Identify your subject, state your purpose, state your main point, stress the importance, and offer background information.

Subject and purpose

Background information

Main point and importance

> Dear Ms. Sims:
>
> I would like to apply for the Organic Agronomist position you advertised through HotJobs.com on March 17. My experience with organic innovations in plant and soil science as well as my minor in entomology would allow me to make an immediate contribution to your company.

This introduction makes all five introductory moves, but you don't need to use all five moves for a successful introduction. Minimally, an introduction should identify the subject (the position being applied for), state your purpose ("I would like to apply for your job"), and state your main point ("I am uniquely qualified for your position").

GO TO THE NET Need more help making a scannable résumé? Go to **www.pearsonhighered.com/johnsonweb4/11.11**

Designing a Scannable/Searchable Résumé

HELP

Companies will sometimes ask for a *scannable* résumé (Figure A). These kinds of résumés are scannable by computers, which sort and rank candidates. Also, résumés posted on job search engines need to be searchable through keywords.

How are scannable/searchable résumés used by employers? Usually, after all the résumés are scanned by the computer, a human resources officer or recruiter will enter ten keywords that describe the position. Then, the computer returns a ranked list of the applicants who matched the most keywords. To survive the cut, you need to find a way to anticipate the keywords that will be entered. Here's a hint: The job advertisement probably contains many of the keywords the employer will be looking for.

Here are some ideas for developing a scannable/searchable résumé:

- Use well-known keywords to describe your skills and experience.
- Use terms in predictable ways. For example, you should write "Managed a team of technicians" rather than "Responsible for guiding a contingent of technical specialists."
- Include acronyms specific to your field (e.g., CAD, TQM, APA, IEEE).
- Use common headings found in résumés: Career Objective, Work Experience, Skills, Qualifications, Education, Honors, Publications, Certifications.
- At the end of your résumé, make a list of any additional traits or skills you possess: time management, dependability, efficiency, leadership, responsibility. These may be used as keywords.

If you are making a paper-based scannable résumé:

- Use white 8½-by-11-inch paper, printed on one side only.
- Do not fold or staple the paper.
- Place your name on its own line at the top of the page.
- Use a standard address format below your name.
- List each phone number on its own line.
- Use standard typefaces like Arial, Helvetica, Times, New York, or Garamond. The computer may have trouble reading other fonts.
- Don't use a font size smaller than 10 points for any of the text.
- Don't use italics, underlining, shadows, or reverse type (white type on a black background) because scanners have trouble reading them.
- Don't use vertical and horizontal lines, graphics, boxes, or shading.
- Don't use two- or three-column formats. One column is easier for the computer to scan.

Fortunately, scanning machines and computers do not care about the length or design of your résumé. Your scannable résumé should be plain in design, and it can be longer than your regular résumé, if needed.

If you suspect your résumé will be scanned, your best strategy is to make two résumés. One should be your regular résumé and the other should be a scannable résumé. Your scannable résumé should be clearly identified for the employer with a cover note. That way, the employer won't be confused by the submission of two résumés.

A Scannable Résumé

Name and contact information is plainly presented.

Headings are predictable and easy to locate.

Italics have been removed for easier scanning.

Bullets have been removed to simplify text.

Anne Franklin
834 County Line Rd.
Hollings Point, IL 62905
Home: 618-555-2993
Mobile: 618-555-9167
e-mail: afranklin@unsb5.net

Career Objective: A position as a naturalist, specializing in agronomy, working for a distribution company that specializes in organic foods.

Educational Background
Bachelor of Science, Southern Illinois University, expected May 2011.
Major: Plant and Soil Science
Minor: Entomology
GPA: 3.2/4.0

Work Experience
Intern Agronomist, December 2009–August 2010
Brighter Days Organic Cooperative, Simmerton, IL
Consulted with growers on organic pest control methods. Primary duty was sale of organic crop protection products, crop nutrients, seed, and consulting services.
Prepared organic agronomic farm plans for growers.
Provided crop-scouting services to identify weed and insect problems.

Field Technician, August 2007–December 2009
Entomology Department, Southern Illinois University
Collected and identified insects.
Developed insect management plans.
Tested organic and nonorganic pesticides for effectiveness and residuals.

Skills
Computer Experience: Access, Excel, Outlook, PowerPoint, and Word. Global Positioning Systems (GPS). Database Management.
Machinery: Field Tractors, Combines, Straight Trucks, and Bobcats.
Communication Skills: Proposal Writing and Review, Public Presentations, Negotiating, Training, Writing Agronomic and Financial Farm Plans.

Awards and Memberships
Awarded "Best Young Innovator" by the Organic Food Society of America (OFSA), 2009.
Member of Entomological Society of America (ESA).

Figure A: A scannable résumé removes much of the formatting and design, allowing the computer to more easily locate keywords. Compare this résumé with the regular résumé in Figure 11.3.

GO TO THE NET

To see sample application letters, go to **www.pearsonhighered.com/johnsonweb4/11.12**

BODY In the body of the letter, you should include two to three paragraphs that show how your educational background, work experience, and skills fit the employer's needs. You should organize the body of your letter to highlight your strengths. If your educational background is your best asset, put that paragraph right after the letter's introduction (Figure 11.8). If your work experience is stronger than your education, then put that information up front (Figure 11.9).

Remember that you are making an argument, so each paragraph should start out with a claim, and the rest of the paragraph should support that claim with examples, facts, and reasoning. Here is a sample paragraph that supports a claim about an agronomist's educational background.

A clear claim

Support for that claim

> My education and research as an organic agronomist would benefit your company significantly. As a Plant and Soil Science major at Southern Illinois University, I have been studying and researching environmentally safe alternatives to pesticides. Specifically, my mentor, Professor George Roberts, and I have been working on using benevolent insects, like ladybird beetles (*Coleomegilla maculata*), to control common pests on various vegetable plants. We have also developed several varieties of organic insecticidal soaps that handle the occasional insect infestation.

In the body, you need to back up your claims with facts, examples, details, and reasoning—you need proof. You also need to do more than simply restate information that can be found on your résumé. Instead, you should breathe life into your letter by telling stories about yourself.

CONCLUSION Your conclusion should make three moves: thank the reader, offer contact information, and look to the future. Your goal is to leave a positive impression.

Thank you statement

Look to the future

Contact information

> Thank you for this opportunity to apply for your opening. I look forward to hearing from you about this exciting position. I can be contacted at home (618–555–2993) or through e-mail (afranklin@unsb5.net).

Keep the conclusion concise, and avoid any pleading for the position. Employers will not look favorably on someone who is begging for the job.

Style

Another way an application letter differs from a résumé is in its style. You want to adopt a style that conveys a sense of your own personality and your interest in the position available.

"YOU" ATTITUDE Put the emphasis of the letter on your readers by using the "*you*" *attitude*. By strategically using the words "you" and "your" in the letter, you can discuss your qualifications from your readers' point of view.

> Your company would benefit from my mechanical engineering training and hands-on experience as a certified welder.

Letter of Application Emphasizing Education

Figure 11.8: A letter of application should make an argument, not simply restate items that can be found on the résumé.

834 County Line Rd.
Hollings Point, Illinois 62905

April 1, 2011

Valerie Sims, Human Resources Manager
Sunny View Organic Products
1523 Cesar Chavez Lane
Sunny View, California 95982

Dear Ms. Sims:

Opening paragraph states the subject, purpose, and main point. →

I would like to apply for the Organic Agronomist position you advertised through HotJobs.com on March 17. My experience with organic innovations in plant and soil science as well as my minor in entomology would allow me to make an immediate contribution to your company.

Education is discussed up front with examples. →

My education and research as an organic agronomist would benefit your company significantly. As a Plant and Soil Science major at Southern Illinois University, I have been studying and researching environmentally safe alternatives to pesticides. Specifically, my mentor, Professor George Roberts, and I have been working on using benevolent insects, like ladybird beetles (*Coleomegilla maculata*), to control common pests on various vegetable plants. We have also developed several varieties of organic insecticidal soaps that handle the occasional insect infestation.

Work experience is used to show potential contributions to employer. →

I also worked as an intern for Brighter Days Organic Cooperative, a group of organic farmers, who have an operation similar to Sunny View. From your website, I see that you are currently working toward certification as an organic farm. At Brighter Days, I wrote eleven agronomic plans for farmers who wanted to change to organic methods. My work experience in the organic certification process would be helpful toward earning certification for Sunny View in the shortest amount of time.

Other skills are highlighted to show unique abilities. →

Finally, I would bring two other important skills to your company: a background in farming and experience with public speaking. I grew up on a farm near Hollings Point, Illinois. When my father died, my mother and I kept the farm going by learning how to operate machinery, plant the crops, and harvest. We decided to go organic in 1999, because we always suspected that my father's death was due to chemical exposure. Based on our experiences with going organic, I have given numerous public speeches and workshops to the Farm Bureau and Future Farmers of America on organic farming. My farming background and speaking skills would be an asset to your operation.

Conclusion ends on a positive note and offers contact information. →

Thank you for this opportunity to apply for your opening. I look forward to hearing from you about this exciting position. I can be contacted at home (618-555-2993) or through e-mail (afranklin@unsb5.net).

Sincerely,

Anne Franklin

Anne Franklin

Letter of Application Emphasizing Work Experience

Figure 11.9: The organization of the application letter should highlight strengths. Here, work experience is being highlighted by appearing immediately after the introduction.

576 First Avenue
Rolla, Missouri 65408

March 10, 2011

Mr. Harold Brown, Human Resources Director
Farnot Plastic Solutions
4819 Renaissance Lane
Rochester, New York 14608

Dear Mr. Brown:

Opening paragraph uses background information to make personal connection. →

Last week, you and I met at the University of Missouri–Rolla engineering job fair. You mentioned that Farnot Plastics might be interviewing entry-level chemical engineers this spring to work on applications of polymeric materials. If a position becomes available, I would like to apply for it. With my experience in applied rheology and polymeric materials, I would be a valuable addition to your company.

Work experience with co-op is highlighted. →

My work experience includes two summers as a co-op at Vertigo Plastics, a company similar in size, products, and services to Farnot. My responsibilities included inspecting and troubleshooting the plant's machinery. I analyzed the production process and reported on the performance of the plant's operations. While at Vertigo, I learned to work with other chemical engineers in a team-focused environment.

Paragraph on education makes connections to employer's needs. →

My education in chemical engineering at University of Missouri–Rolla would allow me to contribute a thorough understanding of plastics engineering to Farnot. In one of the best programs in the country, I have excelled at courses in thermodynamics, chemical process design, and chemical process materials. In addition, my work in the university's state-of-the-art chemical laboratories has prepared me to do the prototype building and vacuum forming that is a specialty of your company. I also have experience working with the CAD/CAM systems that your company uses.

Conclusion indirectly requests the interview. →

The enclosed résumé highlights my other qualifications. I enjoyed speaking with you at the job fair, and I would appreciate an opportunity to talk to you again about opportunities at Farnot. If you would like more information or you would like to schedule an interview, please call me at 573-555-4391 or e-mail me at bigmondy12@umr.edu.

Sincerely,

J.L. Mondragon

James L. Mondragon

Enclosure: Résumé

Your group is one of the fastest growing medical practices in Denver. As a member of your medical staff, I would be reliable and hardworking.

Link

For more information on using the "you" style in letters, go to Chapter 5, page 116.

To put emphasis on the readers rather than on yourself, you might also change "I" sentences to "my" sentences. The first sentence above, for example, uses the phrase "my mechanical engineering training" rather than "I have been trained in mechanical engineering." This subtle change lessens the overuse of "I" in the letter, allowing you to put more focus on your readers.

ACTIVE VOICE In stressful situations, like writing application letters, you might be tempted to switch to passive voice. Passive voice in application letters will make you sound detached and even apathetic.

Passive: The senior project on uses of nanotechnology in hospitals was completed in November 2010.

Active: I completed my senior project on the uses of nanotechnology in hospitals in November 2010.

Passive: The proposal describing the need for a new bridge over the Raccoon River was completed and presented to the Franklin City Council.

Active: My team completed the proposal for a new bridge over the Raccoon River and presented it to the Franklin City Council.

Link

For more help on using active voice, see Chapter 17, page 455.

Why active voice? The active voice shows that you took action. You were in charge, not just a passive observer of the events around you.

AT A GLANCE

Style in an Application Letter

- "You" attitude—Put the emphasis on the employer.
- Active voice—Put yourself, not events, in charge.
- Nonbureaucratic tone—No one wants to hire bureaucrats, so don't sound like one.
- Themes—Point out what makes you different or attractive.

NONBUREAUCRATIC TONE Avoid using business clichés to adopt an artificially formal tone. Phrases such as "per your advertisement" or "in accordance with your needs" only make you sound stuffy and pretentious.

Bureaucratic: Pursuant to your advertisement in the *Chicago Sun-Times*, I am tendering my application toward your available position in pharmaceutical research.

Nonbureaucratic: I am applying for the pharmaceutical researcher position you advertised in the *Chicago Sun-Times*.

Employers are not interested in hiring people who write in such a pretentious way—unless they are looking for a butler.

THEMES Think of one quality that sets you apart from others. Do you work well in teams? Are you a self-starter? Are you able to handle pressure? Are you quality minded? These are some themes that you might mention throughout your letter.

While working on the Hampton County otter restoration project, I was able to communicate clearly and effectively, helping us win over a skeptical public that was concerned about the effects of introducing these animals to local streams.

GO TO THE NET

For more advice about letter-writing style, go to **www.pearsonhighered.com/johnsonweb4/11.13**

Beyond simply telling readers that you have a special quality, use examples to show them how you have used this quality to solve problems in the past. Some qualities you might mention include the ability to

- make public presentations.
- be a leader.
- manage time effectively.
- motivate yourself and others.
- follow instructions.
- meet deadlines.
- work well in a team.
- write clearly and persuasively.
- handle stressful situations.

Weave one of these qualities into your letter or address it directly. One theme should be enough, because using more than one theme will sound like you are boasting.

Revising and Proofreading the Résumé and Letter

When you are finished writing your résumé and application letter, you should spend a significant amount of time revising and proofreading your materials. These documents need to be revised just like any other documents.

Content—Did you provide enough information? Did you provide too much? Do the résumé and letter answer the interviewers' basic questions about your qualities and abilities? Do your materials match up with the job advertisement?

Organization—Are your materials organized in ways that highlight your strengths? Can readers easily locate information about your education, work experience, and skills?

Style—Do your materials set a consistent theme? Is the tone projected in the letter of application appropriate?

Design—Is the résumé highly scannable? Does it follow the design principles of balance, alignment, grouping, consistency, and contrast?

When you are finished revising, proofread everything carefully. Have as many people as possible look over your materials, especially your résumé. For most jobs, even the smallest typos can become good excuses to pitch an application into the recycle bin. Your materials need to be nearly flawless in grammar and spelling.

Creating a Professional Portfolio

Increasingly, it is becoming common for employers to ask applicants to bring a *professional portfolio* to their interview. Sometimes, interviewers will ask that a portfolio be sent ahead of the interview so they can familiarize themselves with the applicant and his or her abilities.

A portfolio is a collection of materials that you can use to demonstrate your qualifications and abilities. Portfolios include some or all of the following items:

- Résumé
- Samples of written work
- Examples of presentations
- Descriptions and evidence of projects
- Diplomas and certificates
- Awards
- Letters of reference

These materials can be placed in a three-ring binder with dividers and/or put on a website or CD-ROM.

Your personal portfolio should not be left with the interviewer, because it may be lost or not returned. Instead, you should create a "giveaway" portfolio with copies of your work that you can leave with the employer.

Even if interviewers do not ask you to bring a portfolio, putting one together is still well worth your time. You will have a collection of materials to show your interviewers, allowing you to make the strongest case possible that you have the qualifications and experience they need.

Collecting Materials

If you have not made a portfolio before, you are probably wondering if you have enough materials to put one together.

Much of your written work in college is suitable for your portfolio. Your class-related materials do not need to be directly applicable to the job you are applying for. Instead, potential employers are interested in seeing evidence of your success and your everyday abilities. Documents written for class will show evidence of success.

A Print Portfolio

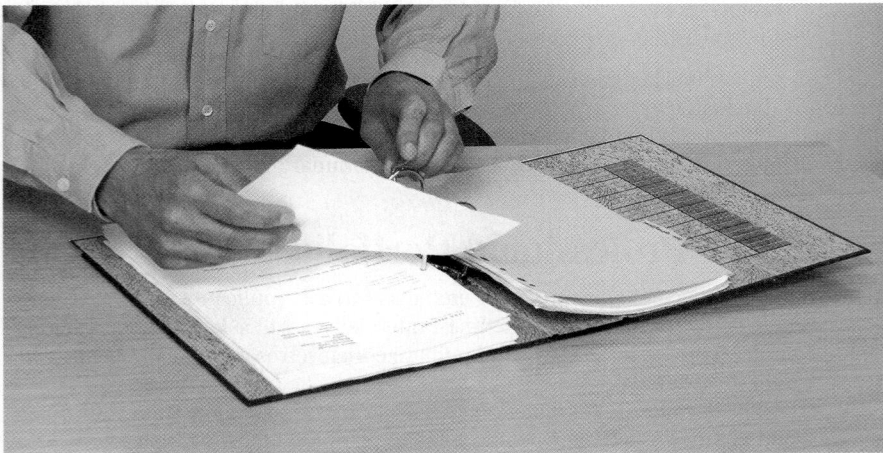

A portfolio is a useful tool to bring to an interview. It holds all your work in an accessible package so that you can support your claims during your interview.

Where can materials for your portfolio be found? Go to **www.pearsonhighered.com/johnsonweb4/11.14**

Even nontechnical materials—like your critical analysis of Beethoven's Ninth Symphony that you wrote for a course in classical music—show your ability to do research, adopt a critical perspective, and write at length. These items are appropriate if you are looking for materials to fill a portfolio.

Organizing Your Portfolio

A particularly good way to organize a portfolio is to follow the categories in a typical résumé:

Cover sheet—A sheet that includes your name, address, phone number, and e-mail address.

Educational background—Diplomas you have received, workshops you have attended, a list of relevant courses, and college transcripts.

Related work experience—Printed materials from your previous jobs, internships, co-ops, and volunteer work. You might include performance reviews, news articles that mention you, or brochures about specific projects. You might even include photographs of places you have worked, projects you have worked on, and people you have worked with.

Specialized skills—Certificates of completion for training, including any coursework you have completed outside of your normal college curriculum.

Writing samples and publications—Examples of your written work, presentations, or websites.

Awards—Any award certificates or letters of congratulation. You might include letters that mention any scholarships you have received.

Other interests—Materials that reflect your other activities, such as volunteer work, sports, or hobbies. Preferably, these materials would be relevant to the kinds of jobs you are seeking.

References—Letters of reference from instructors, colleagues, co-workers, or employers. Also, it is helpful to keep a list of references with phone numbers and addresses where they can be contacted.

AT A GLANCE

Organizing Your Portfolio

- Cover sheet
- Educational background
- Related work experience
- Specialized skills
- Writing samples and publications
- Awards
- Other interests
- References

Assembling the Portfolio in a Binder

Your portfolio needs to be easy to use at an interview. If your portfolio is nothing more than a hodgepodge of paper stuffed in a three-ring binder, you will find it difficult to locate important information.

So, go to an office supply store and purchase the following items:

Three-ring binder—The binder should be at least 2 inches wide to hold your materials, but it should not be too large to be comfortably carried to an interview. Find one that is suitable for a formal situation like an interview.

Dividers with tabs—Tabbed dividers are helpful for separating your materials into larger categories. The tabs are helpful for finding materials quickly, especially in a stressful situation like an interview.

Pocketed folders or clear plastic sleeves—Put your materials in pockets or clear plastic sleeves. That way, you can easily insert and remove your materials when you want.

As you assemble the portfolio, keep professionalism and ease of use in mind:

- Copies of your résumé can usually be placed in a pocket on the inside cover of the three-ring binder. You may need them if interviewers do not have a copy of your résumé in front of them.
- Labels can be used to provide background information on each item in the portfolio. These labels are helpful in jogging your memory during a stressful interview.

Creating an Electronic Portfolio

If you know how to create a basic website, you can create an electronic portfolio for yourself (Figure 11.10). An electronic portfolio has several advantages. It can

- be accessed from anywhere there is a networked computer, including an interviewer's office.

Electronic Portfolio

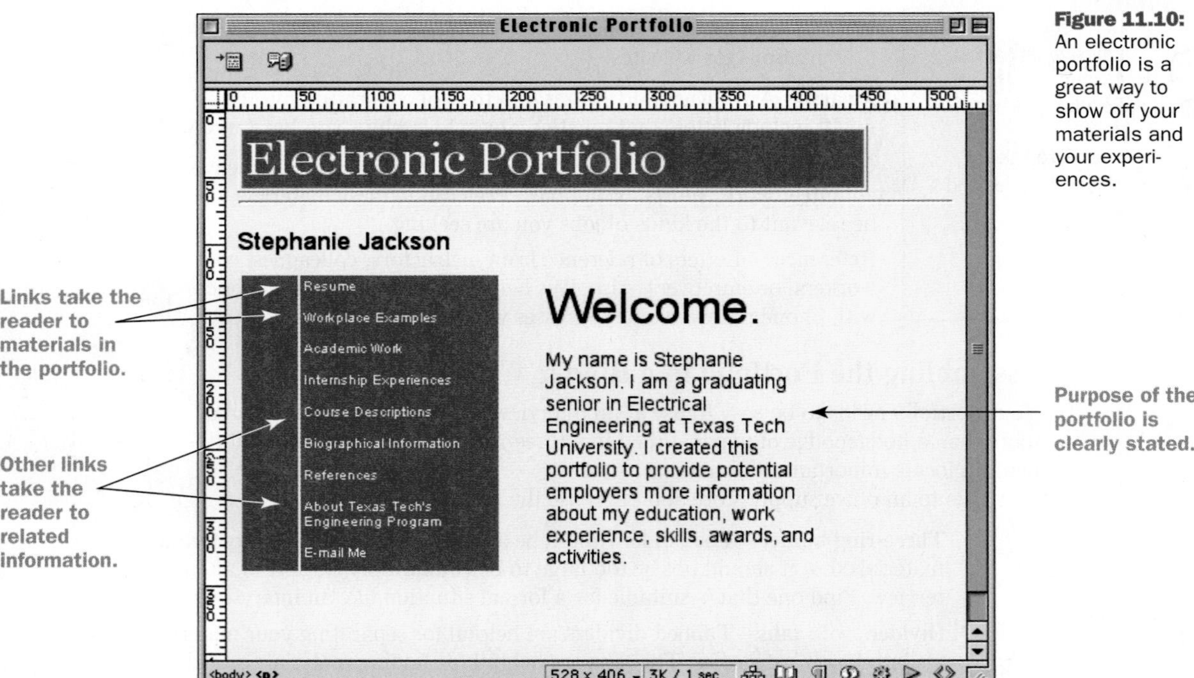

Figure 11.10: An electronic portfolio is a great way to show off your materials and your experiences.

To see samples of online portfolios, go to **www.pearsonhighered.com/johnsonweb4/11.15**

- include multimedia texts such as movies, presentations, and links to websites you have created.
- include materials and links to information that would not typically be found in a nonelectronic portfolio. For example, you might put links to your university and academic department to help interviewers learn about your educational background.

You should keep your electronic portfolio separate from any personal websites you create. The materials you include in your electronic portfolio should be appropriate for discussion in a professional interview. (Your vacation photos from that Aspen skiing trip with your friends probably aren't appropriate.)

Interviewing Strategies

When you are called for an interview, an employer is already telling you that you are qualified for the position. Now you just need to compete with the other qualified candidates who are also interviewing for the position.

Preparing for the Interview

In many ways, interviewing is a game. It's not an interrogation. The interviewers are going to ask you some questions or put you in situations that will test your problem-solving abilities. To answer the interviewers' questions appropriately, you need to do some preparation.

The Interview

Interviewing is more like a game than an interrogation. Once you know how the game is played, you can win it.

To visit sites that discuss interviewing strategies, go to **www.pearsonhighered.com/johnsonweb4/11.16**

RESEARCH THE COMPANY Before the interview, find out as much as possible about the company and the people who will be interviewing you. The Internet, especially the company's website, is a great place to start. But you should also look for magazine and newspaper articles on the company in your university or local library. While researching the company, locate facts about the size of the company, its products, and its competitors. You should also be aware of major trends in the company's market.

DRESS APPROPRIATELY The interview game begins with your appearance. There is an old saying: "Dress for the job you want, not for the job you have." When you are interviewing, you should be dressed in a suitably formal manner, avoiding flashy jewelry or too much cologne or perfume. Appropriate clothing usually depends on the kind of job you are seeking. During your interview, you want to be one of the best-dressed people in the room.

At the Interview

When you are at the interview, try to relax and present yourself as someone the interviewer would like to hire. Remember that each question from an interviewer is a move in the game, and there are always appropriate countermoves available.

GREET PEOPLE WITH CONFIDENCE When you meet people at the interview, you should greet them with confidence. Most North Americans will expect you to shake their hand firmly and make eye contact. Let the interviewer indicate where you are going to sit. Then, set your briefcase and/or portfolio at your side on the floor. Don't put things on the interviewer's desk, unless he or she asks you to.

When time permits, write down the name and title of each interviewer you meet. You will need these names and titles when you are writing thank you e-mails and letters after the interview.

Link
For more information on making a professional presentation, see Chapter 21, page 574.

Link
For more strategies on answering questions, see Chapter 21, page 593.

ANSWER QUESTIONS Interviewers will usually work from a script of questions. However, most interviews go off the script as interesting topics come up. You should be prepared with answers to some of the following questions:

"Tell me about yourself"—Spend about two minutes talking about your work experience, education, and skills, relating them to the position. Don't start with, "I was born in New York in 1986 . . . ."

"What about this position attracted you?"—Talk about the strengths of the company and what qualities of the position you find interesting.

"Why should we hire you?"—Stress qualifications and skills appropriate to the job that set you apart from the other candidates.

"Where do you want to be in three to five years?"—Without being too ambitious, talk about doing your job well and moving up in the company.

"What are your salary requirements?"—This question is uncommon, but you should have a salary figure in your head in case it is asked. You don't want to fumble this question or ask for too little.

"What is your greatest strength?"—Discuss a strong qualification, skill, or knowledge area relevant to the job.

"What is your greatest weakness?"—Discuss something you would like to learn that would enhance your ability to do your job (e.g., a new language, more advanced computer skills, greater communication skills). The "weakness" question is not the time to admit your shortcomings. If you are a procrastinator or you have a bad temper, the interview is not the time to confess your sins. Also, answers like, "I just work too hard" or "I am too committed to doing an excellent job" don't really fool anyone. Instead, mention something you want to learn to improve yourself.

USE YOUR PORTFOLIO No doubt you have been told numerous times, "Show, don't just tell." In your interview, use your professional portfolio to back up your answers to interview questions. The materials in your portfolio should be used to reinforce your statements about your background, qualifications, and skills.

ASK QUESTIONS As the interview comes to an end, the interviewer will usually ask if you have any questions. You should be ready to ask two or three insightful questions. Here are a few examples of questions that will demonstrate your interest in the company and the job:

Where do you see the company going in the next five years?

What can you tell me about your customers/clients?

What kinds of additional learning opportunities are available?

What happens in a normal day at this position?

Avoid asking questions at this point about salary, vacation, and benefits. Usually these items are discussed after a job offer has been made.

After the interviewer answers each question, you might offer a follow-up statement that reinforces one of your strengths.

AT A GLANCE

Interviewing Strategies

- Research the company.
- Dress appropriately.
- Greet people with confidence.
- Answer questions.
- Use your portfolio.
- Ask questions.
- Leave with confidence.

LEAVE WITH CONFIDENCE When the interview is finished, thank the interviewers for their time and say that you are looking forward to hearing from them. Also, ask if they would like you to send them any other information. Then, shake each interviewer's hand firmly and go.

As soon as possible after the interview, find a place where you can write down everything you can remember about what you and the interviewers talked about. These notes may be helpful later, especially if a week or two lapses before you hear about the job. Your notes should mention any important discussion points that developed during the interview.

Writing Thank You Letters and/or E-Mails

After an interview, it is polite to write a thank you letter to the people who interviewed you. A basic thank you letter shows your appreciation for the interviewers' time while expressing continued interest in the job. A more sophisticated letter could reinforce one or more of your strengths, in addition to saying thank you and expressing continued interest in the job (Figure 11.11).

Want some strategies for answering tough questions? Go to **www.pearsonhighered.com/johnsonweb4/11.17**

A Thank You Letter

Figure 11.11: A thank you letter can be used to reinforce an important point that came out during the interview.

834 County Line Rd.
Hollings Point, Illinois 62905

May 13, 2011

Inside address

Valerie Sims, Human Resources Manager
Sunny View Organic Products
1523 Cesar Chavez Lane
Sunny View, California 95982

Dear Ms. Sims:

A thank you statement leads off the letter.

Thank you for interviewing me for the Organic Agronomist position at Sunny View Organic Products. I enjoyed meeting you and the others at the company. Now, after speaking with you, I am more interested than ever in the position.

The main point is clearly stated in the introduction.

An important point is reinforced with details from the interview.

I noticed during the interview that my experience with benevolent insects was a recurring topic of discussion. This area of pest control is indeed very exciting, and my work with Professor George Roberts is certainly cutting edge. By paying attention to release times and hatching patterns, we have been able to maximize the effectiveness of predator insects. I would bring the latest research in this area to Sunny View's crops.

Again, thank you for the opportunity to interview with you for this position. I can be contacted at home (618-555-2993) or through e-mail (afranklin@unsb5.net). Please call or e-mail me if you need more information or would like to speak with me further about this position.

The conclusion thanks the reader and offers contact information.

Sincerely,

Anne Franklin

The signature includes the full name.

If you send a thank you through e-mail, follow it with a letter through the mail. An e-mail is nice for giving the interviewers immediate feedback, but the letter shows more professionalism.

If all goes well, you will be offered the position. At that point, you can decide if the responsibilities, salary, and benefits fit your needs.

MICROGENRE

The Bio

In the workplace, you will need to write your professional biography or "bio." A bio is a brief description of your career that highlights your background and accomplishments. Bios are typically included in job-finding materials, social networking sites, "About Us" pages on websites, qualifications sections, and marketing materials. Also, career consultants often ask people to write their own "retirement bio" that describes the careers they hope to have. Writing your own retirement bio might seem a bit odd when you're still in college, but it's a great way to lay out your long-term goals and think about what you want to do with your life.

Here are some strategies for writing your bio:

Be concise. You may be asked to keep your bio to a specific length: one sentence, 100 words, 250 words, or 500 words. Try to say as much as you can in the space available.

Hook the readers. Tell the readers what you do in the first sentence, and highlight what makes you different. For example, "Lisa Geertz is a biomedical engineer who specializes in developing mobility devices for children."

Write in the third-person. You should refer to yourself by name, even though you are writing about yourself. A formal bio will use your last name throughout, while an informal bio will use your first name.

Identify your three to five major achievements. Describe each of your major accomplishments in one sentence or a maximum of two sentences.

Be specific about your accomplishments. Where possible, use numbers, facts, dates, and other figures to quantify the *impact* of your achievements.

Offer personal information. If space allows, you can talk about your family, where you live, or your favorite activities. Personal information adds a human touch.

Contact information. If appropriate, include contact information, such as your e-mail address and phone number.

Write

Write your own retirement bio. Imagine yourself at the end of your career, getting ready to retire. Write a 250-word bio that reviews your career and accomplishments. Go ahead and dream. Set your goals high.

welch architecture

Portfolio Bio Client Access Contact Blog

CLIFF WELCH ARCHITECT

BIOGRAPHY

Cliff is a Dallas Architect whose work has been honored at the local and national levels. His background includes working with the late Dallas modernist Bud Oglesby, was a principal with Design International, and now has his own practice. His firm's focus is modern architecture, concentrating on residential, interiors, and small scale commercial work. He has been a leading resource and proponent for the restoration and preservation of post-war modernism in Dallas.

In addition to his practice, he is past President of the Dallas Architectural Foundation, and has taught graduate level design at the University of Texas, Arlington. Cliff is a past Executive Board member for the Dallas Chapter AIA, served two years as Commissioner of Design, and has Chaired several chapter events such as the Ken Roberts Memorial Delineation Competition, Retrospect, and Home Tours. He has also served as a design awards juror for other chapters around the state.

Cliff was featured in Texas Architect as one of five young design professionals leading the way into the coming century and has been honored as Dallas American Institute of Architects' Young Architect of the Year.

Source: Welch Architecture, http://www.welcharchitecture.com.

CHAPTER REVIEW

- From the beginning, you should adopt a professional approach to job seeking: Set goals and make a plan.
- Your job-searching plan should include job search engines, personal networking, professional networking, targeting, and classified advertisements.
- Two types of résumés are commonly used, the chronological résumé and the functional résumé. Either should summarize your background, experience, and qualifications.
- A well-designed résumé should use the basic principles of design; a scannable résumé incorporates keywords so that it can be sorted electronically.

- An effective application letter can be more individual and specifically targeted than the résumé. It shares the common features of a letter, including the appropriate "moves" for the introduction, body, and conclusion.
- An appropriate style for an application letter uses the "you" attitude, active voice, and nonbureaucratic language; you can also work in a theme that will make your résumé stand out.
- A professional portfolio is a helpful tool for an interview because it can demonstrate your qualifications and abilities; material for your "master" portfolio can be assembled from classroom work, volunteer projects, related work experience, awards, certificates, or letters of reference.
- When you go to your interview, be prepared with information about the company and appropriate questions to ask; dress appropriately for the position; greet people with confidence; answer questions thoughtfully; and obtain the name and title of each interviewer (and the correct spellings) for a follow-up note to be sent that day.

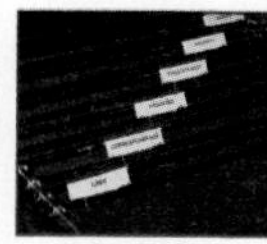

EXERCISES AND PROJECTS

Individual or Team Projects

1. Imagine that you are looking for an internship in your field. Write a one-page résumé that summarizes your education, work experience, skills, awards, and activities. Then, pay attention to issues of design, making sure the text uses principles of balance, alignment, grouping, consistency, and contrast.

2. Using an Internet job search engine, use keywords to find a job for which you might apply after college. Underline the qualifications required and the responsibilities of the position. Then, write a résumé and application letter suitable for the job. In a cover memo addressed to your instructor, discuss some of the reasons you would be a strong candidate for the position. Then, discuss some areas where you might need to take more courses or gain more experience before applying for the position.

3. Contact a human resources manager at a local company. Request an "informational interview" with this manager (preferably in person, but an e-mail interview is sufficient). If the interview is granted, ask him or her the best way to approach the company about a job. Ask what kinds of qualifications the company is usually looking for in a college graduate. Ask what you can do now to enhance your chances of obtaining a position at the company. After your interview, present your findings to your class.

4. Make an electronic portfolio of your materials. Besides the materials available in your regular portfolio, what are some other links and documents you might include in your electronic portfolio?

For examples of good and bad résumés, go to **www.pearsonhighered.com/johnsonweb4/11.18**

Collaborative Project

With a group of people pursuing similar careers, develop some career goals and a job search plan for reaching those goals. Then, send each member of the group out to collect information. One group member should try out job search engines, while another should explore personal and professional networking opportunities. One member should create a list of potential employers in your area. Another should explore newspapers' online classified advertisements.

After your group has collected the information, write a small report for your class and instructor in which you discuss the results of your research. What did your group discover about the job market in your field? Where are the hottest places to find jobs? How can professional groups and personal networking help you make contacts with potential employers?

Individuals in your group should then choose one job that seems interesting. Each member of the group should write an application letter for that job and create a résumé that highlights qualifications and strengths.

Then the group should come up with five questions that might be asked at an interview, four questions that are usually asked at an interview, and one question that is meant to trip up an interviewee.

Finally, take turns interviewing each other. Ask your questions and jot down good answers and bad answers to the questions. Discuss how each member of the group might improve his or her interviewing skills based on your experiences with this project.

For additional technical writing resources, including interactive sample documents, document design tutorials and guidelines, and more, go to **www.mytechcommlab.com**.

CASE STUDY

The Lie

Henry Romero had wanted to be an architect his whole life. Even as a child, he was fascinated by the buildings in nearby Chicago. So, he spent years preparing to be an architect, winning top honors at his university in design.

He was thrilled when one of the top architectural firms, Goming and Cooper, announced it would be visiting his campus looking to interview promising new talent. This firm was certainly one of the top firms in the country. It had designed some of the most innovative buildings in recent years. Henry was especially interested in the firm's international projects, which would allow him to work and live abroad. Recently, in *Architecture Times,* he had read about the firm's successful relationships with companies in France.

So, Henry gave his résumé and a letter of application to the university's placement office personnel and told them to send it to Goming and Cooper. A month later, his placement counselor called him and said he had an interview with the firm. Henry was very pleased.

There was only one problem. On his résumé, Henry had put down that he "read and spoke French fluently." But, in all honesty, he had taken French for only a few years in high school and one year in college. He hadn't really mastered the language. If necessary, he could muddle his way through a conversation in French, but he was hardly fluent.

To make things worse, last week Henry's roommate, Paul, had had a nightmarish interview. He too had reported on his résumé that he spoke a language fluently—in this case, German. When he arrived at the interview, the interviewer decided to break the ice by talking in German. Paul was stunned. He stammered in German for a few minutes. Then the interviewer, clearly angry, showed him the door.

Henry was worried. What if the same thing happened to him? He knew the ability to speak French was probably one of the reasons he had received the interview in the first place. If the interviewer found out he was not fluent in French, there was a good chance the interview would go badly.

If you were Henry, how would you handle this touchy situation?

CHAPTER

12

Strategic Planning, Being Creative

Using Strategic Planning 353

Help: Planning with Online Calendars 356

Generating New Ideas 357

Chapter Review 363

Exercises and Projects 363

Case Study: Getting Back to Crazy 365

In this chapter, you will learn:

- To plan a project in an organized way.
- Strategies for identifying your objectives and "top-rank objective."
- To develop a list of tasks and a timeline for your project.
- How to use the technical writing process as a planning tool.
- How to generate new ideas in the technical workplace.
- Strategies for being creative and trusting your instincts.
- Five techniques for generating new ideas.

The writing process for any technical document should begin with a "planning phase" in which you think about how you are going to best inform and persuade your readers. Good planning will save you time while helping you write more efficiently and effectively.

Using Strategic Planning

Effective strategic planning will save you time, while helping you produce higher-quality documents and presentations that are informative and persuasive to your readers. A time-tested method for strategic planning includes three steps: (1) setting objectives, (2) creating a list of tasks or "task list," and (3) setting a timeline (Figure 12.1).

Step 1: Set Your Objectives

To begin the planning process, you first need to figure out what you want your project to achieve.

LIST PROJECT OBJECTIVES On your computer or a sheet of paper, make a brainstorming list of the objectives of your project. For a smaller project, you may list only a few objectives. For a larger project, your list of objectives will probably include many items that vary in importance. At this point, as you brainstorm, you should list any objectives that come to mind. You can prioritize and condense the list later.

IDENTIFY THE TOP-RANK OBJECTIVE When your list is complete, rank your objectives from the most important to the least important. Identify your "top-rank

Steps of Strategic Planning

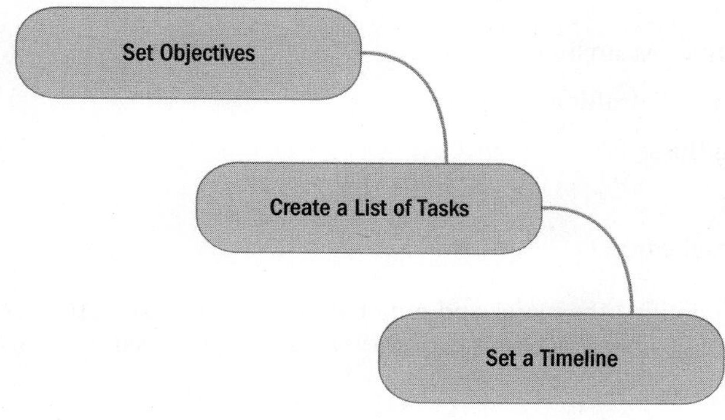

Figure 12.1: Good project planning involves identifying your objective and then breaking the project down into tasks that are set on a timeline.

For more project planning tips, go to **www.pearsonhighered.com/johnsonweb4/12.1**

objective" (or what a marketing guru would call your "TRO"). That's the main goal that your project will strive to reach. More than likely, your top-rank objective is going to be almost identical to the *purpose* of your project.

Then, express your project's top-rank objective in one sentence:

> Our main objective is to persuade the university's vice president of information technology to upgrade the wireless network on campus.

> The primary goal of this project is to develop a solar car that will be competitive in the American Solar Challenge race.

Link
For more help with identifying your purpose go to Chapter 1, page 6.

If you are having trouble expressing your top-rank objective or purpose in one sentence, you probably need to narrow the scope of your project. A top-rank objective that requires more than one sentence is probably too complicated to guide your strategic planning.

Step 2: Create a List of Tasks (or Task List)

Once you have identified your top-rank objective, you should then convert the remainder of your objectives into tasks that you will need to perform. Logical mapping and developing a "task list" are helpful ways to make this conversion from objectives to tasks.

Link
For more tips on using logical maps, go to Chapter 14, page 388.

MAP OUT THE PROJECT TASKS Put your top-rank objective (purpose) in the middle of your screen or a piece of paper, and ask yourself, "What are the two to five major steps necessary to achieve this goal?" Once you have identified your major steps, then identify the two to five minor steps that will help you achieve each major step.

You shouldn't reinvent the wheel with every new project. For example, if your project involves writing a document, you can use the "technical writing process" described in Chapter 1 to help you figure out the major steps of your project (Figure 12.2). Here are the stages of the writing process again:

Link
For more information about the technical writing process, go to Chapter 1, page 2.

Planning and researching

Organizing and drafting

Improving the style

Designing

Revising and editing

Once you have identified the major and minor steps of your project, put each of these steps on your calendar along with the other tasks that you need to accomplish to finish the project.

Not sure how to set your objective? Go to **www.pearsonhighered.com/johnsonweb4/12.2**

Mapping Out a Plan

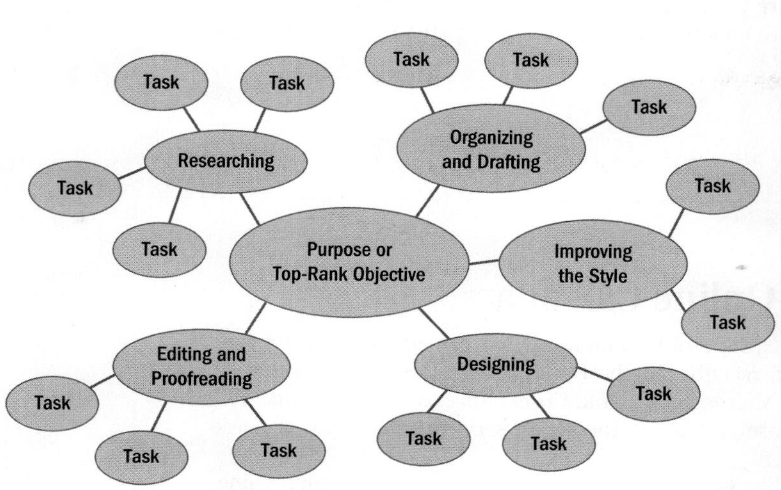

Figure 12.2: To create a project plan, map out the two to five major steps. Then, add two to five minor tasks for each major step. Your plan can then be converted into a task list.

Project Planning Steps

AT A GLANCE

- Step 1: Set your objectives.
- Step 2: Create a list of objectives.
- Step 3: Set a timeline.

CREATE A TASK LIST When your logical map is finished, you can transform it into a list of tasks, or a *task list* (Figure 12.2). The major steps in your map will become the larger stages of the project. Meanwhile, the minor steps will become individual tasks that you need to complete at each stage.

Step 3: Set a Timeline

In technical workplaces, setting timelines is essential. A timeline allows you to keep track of your progress toward completing the project. If you are working alone, the timeline will help you avoid procrastination (and a mad rush to the finish). If you are working with a team, the timeline will help everyone work together to reach the same milestones and deadlines.

ASSIGN A DATE TO EACH TASK Working backward from your project's deadline, identify how much time each part of the project will require. Then, on your task list, write down specific dates when each part of the project should be completed. Online calendars and project planning software are available to help you fill out your timeline (see the Help box). These calendaring programs are widely used in technical workplaces, because they allow team members to check each other's calendars and the project calendar.

Link

For more information on creating a project calendar, go to Chapter 3, page 48.

For software that will help you create a "to do" or task list, go to **www.pearsonhighered.com/johnsonweb4/12.3**

SCHEDULE MEETINGS OR CHECKPOINTS At regular intervals on your timeline (each week, every two weeks, or each month), schedule meetings with your team. Meetings can be boring, but people tend to use them as deadlines to get their tasks completed. If you are working alone, you can use the major steps in your project as "checkpoints" to ensure that you are making steady progress toward finishing the project.

Planning with Online Calendars

An online calendar is a helpful tool for planning your project and coordinating with team members. Until recently, the best online calendars have been tied to large software suites like Microsoft Outlook, Corel Office, and IBM Lotus. These calendar programs are closely linked to the e-mail services included with these software suites.

When search engines like Google and Yahoo! jumped into the e-mail game, online calendars took an interesting leap forward. Now, your calendar (like your e-mail) can be accessed through any computer, mobile phone, or PDA that gives you access to Google or Yahoo! portals (Figure A). So, your calendar is no longer tied to your personal computer. You can access it anywhere. That's a significant advantage.

Online calendars are helpful because you can easily schedule events, including any deadlines and meetings (and social activities). Then, you can set up your preferences, and the calendar will send you reminder messages through your e-mail. That way, you won't miss an important meeting, and you will be fully aware when you blow past the deadline for a project—even if there's nothing you can do about it.

You can set up your calendar to let others check your schedule to find times when they can meet with you. If you scheduled a meeting or an after-work gathering, your calendar program can remind people with e-mail messages ("Hey, don't forget we're meeting at Cy's Roost on Thursday at 5:30").

Online calendars are especially useful for project planning. Once you have created your list of project tasks, you should enter the items into your calendar. Work backward from the project deadline, as shown in Figure A. Put something like "Proposal Due" on the day of the deadline. From last to first, start entering the other tasks into the calendar. Once all the tasks are entered, you can move them around to create a project timeline.

Finally, set up your preferences so your online calendar sends you reminders about when parts of the project need to be completed. The reminders should keep you on schedule.

Online calendaring is easy, and it's a great way to stay organized. Give it a try.

Want to learn more about creating a schedule? Go to **www.pearsonhighered.com/johnsonweb4/12.4**

Scheduling with an Online Calendar

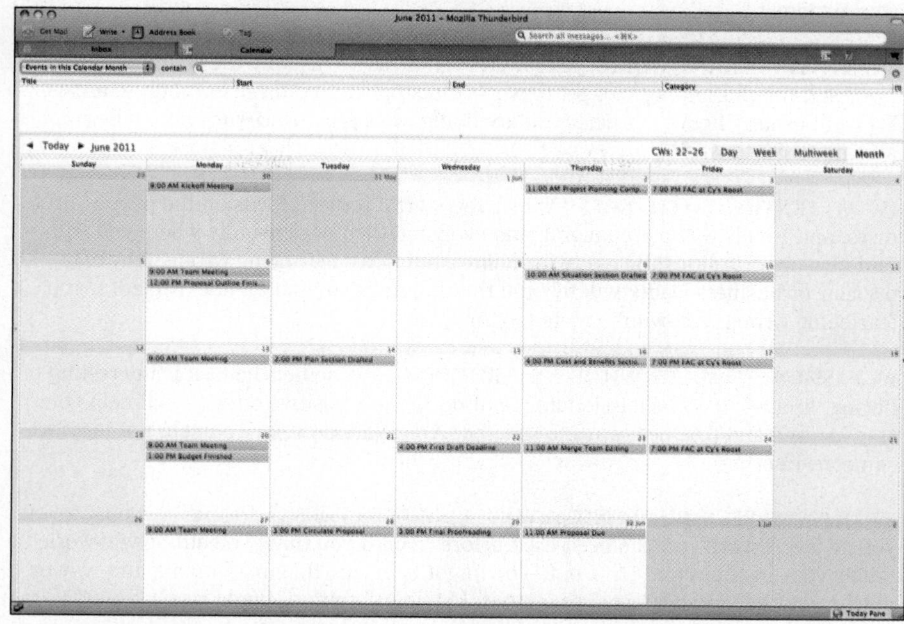

Figure A: Online calendars are widely used in technical workplaces. They are usually linked to e-mail programs. The screen shown here is Mozilla Thunderbird with the Lightning add-on for calendaring.

Source: Mozilla.

Generating New Ideas

In today's technical workplace, it is difficult to overestimate the importance of innovation and creative thinking. The ability to "think outside the box" has become a tiresome cliché, but this overused phrase highlights the importance of being creative in the high-tech community. Your company's new products and services will only have a short lead time before competitors have answered with their own versions. So, the ability to generate new ideas and solutions is highly valued.

When you finish strategic planning, it's time to start being creative and "inventing" new ideas that will become the content of your technical documents.

Tips for Being More Creative

Being creative is the ability to come up with new ideas or alternatives for solving problems. Everyone values creativity, but creativity can also be threatening. After all, when you develop something completely new or you do something differently, you're shifting the power balance (MacLoed, 2009). More often than not, the people around you will react with skepticism to new ideas or new ways of doing things. People rarely embrace new ideas right away.

GO TO THE NET

Use a free online calendar program. Go to **www.pearsonhighered.com/johnsonweb4/12.5**

Here are a few guidelines to keep in mind as you begin inventing the content of your document.

CREATIVITY IS HARD WORK. Breakthrough ideas are usually the result of hard work and tough thinking. So, get to work and stop waiting for inspiration to arrive. You will usually figure out what you are doing *while you are working,* not while you are waiting to get to work.

PAY ATTENTION TO CHANGE. When you start a project, focus on the people, processes, and trends that are changing and evolving. Change is usually where you will find new opportunities. Be ready for failure, but don't fear being wrong. Any entrepreneur or business leader will tell you that failure is common and expected. If you fear being wrong, you won't create anything new.

BE PASSIONATE ABOUT WHAT YOU ARE DOING. Whether the task is interesting or boring, find a way to be passionate about doing it. A positive outlook will help you stay focused and find new ways to succeed. Above all, do it for yourself, not just for someone else.

HEAR NAYSAYERS, BUT DON'T ALWAYS LISTEN TO THEM. People are going to tell you "it won't work" or "it's been tried before." But if you think something will work, follow your instincts and try it out. You might be seeing the problem in a new way or maybe your good idea's time has arrived. Let your intuitions guide you.

Inventing Ideas

All right, let's put that creativity to work. Sometimes the hardest part about starting a new project is just putting ideas and words on the screen. Fortunately, several *invention techniques* can help you get your ideas out there. Five of the best techniques for technical communication are logical mapping, brainstorming, freewriting, outlining/boxing, and using the journalist's questions. Try them all to see which one works best for you.

LOGICAL MAPPING Logical mapping is a visual way to invent your ideas, helping you to discover their logical relationships.

To map the content of your document, start by putting your subject in the middle of the screen or a piece of blank paper. Put a circle or box around it. Then, start typing or writing your other ideas around the subject, and put circles or boxes around them (Figure 12.3).

Now, fill the screen or page with words and phrases related to the subject. Start connecting related ideas by drawing lines among them. As you draw lines, you will begin to identify the major topics, concepts, or themes that will be important parts of the document you are writing. These major issues can be found in the clusters of your map.

Software programs such as Inspiration (shown in Figure 12.3), Visio, MindManager, and IHMC Concept Mapping Software can help you do logical mapping on screen. Otherwise, you can use the Draw function of your word processor to create "text boxes" and draw lines among them. With a little practice, you will find that you can create logical maps on the screen with little effort.

To find links to logical mapping software, go to **www.pearsonhighered.com/johnsonweb4/12.6**

Logical Mapping Software

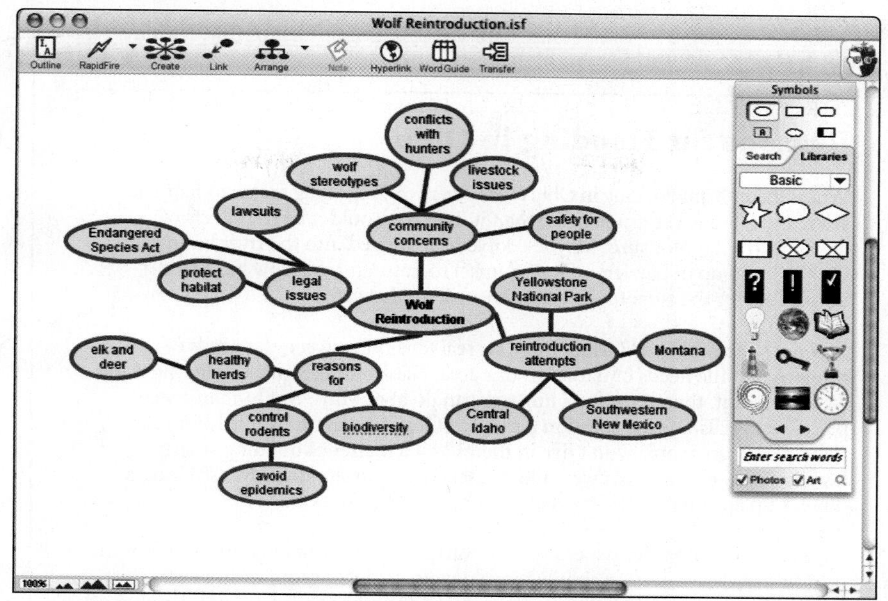

Figure 12.3: A variety of software packages, some of them free, are available for doing logical mapping. The program shown here, Inspiration, will help you map out your ideas. Then, it will turn them into an outline.

Source: Diagram created in Inspiration® by Inspiration Software®, Inc.

BRAINSTORMING OR LISTING Some people like to make lists of their ideas rather than drawing concept maps. Make a quick list of everything you know or believe about your topic. One page or one screen is probably enough. Just write down any words or phrases that come to mind. You're brainstorming.

Then, pick out the best two or three ideas from your list. Make a second list in which you concentrate on these key ideas. Again, write down all the words and phrases that you can think up. Making two lists will force you to think more deeply about your subject while narrowing the scope of your project.

You can continue this brainstorming process indefinitely with a third or fourth list, but eventually you will find it difficult to come up with new ideas. At that point, you should be able to sort your lists into clusters of ideas. These clusters can then be mined for the major topics that will become the content of your text.

FREEWRITING Freewriting is easy. Simply put your fingers on the keyboard and start typing into a document file in your word processor. Type for 5 to 10 minutes before you stop to look over your work. Don't worry about the usual constraints of writing such as sentences, paragraphs, grammatical correctness, or citations. Just keep typing. Eventually, you will find that you have filled one or more screens with words, sentences, and fragments of sentences (Figure 12.4).

GO TO THE NET Having trouble with writer's block? For more help, go to **www.pearsonhighered.com/johnsonweb4/12.7**

Freewriting

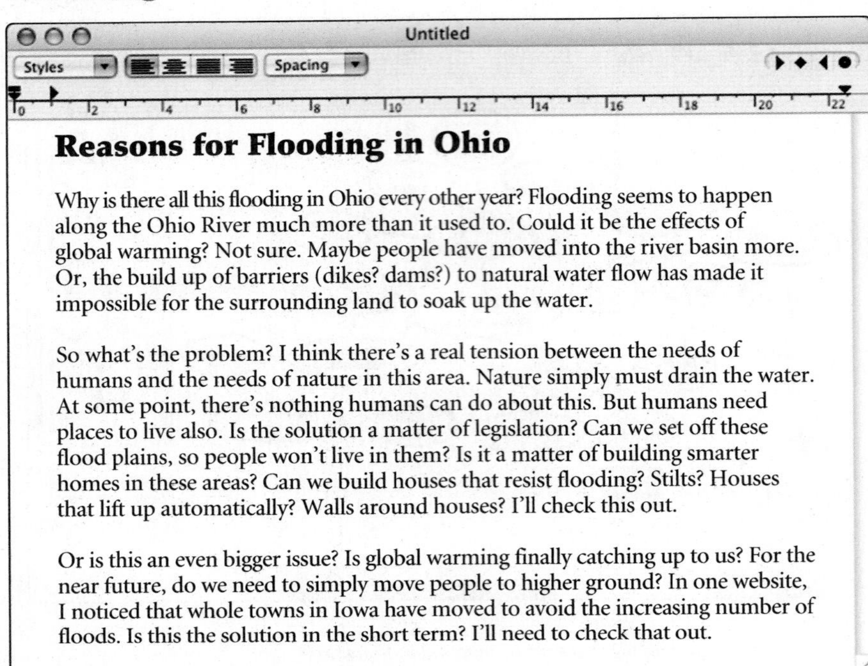

Reasons for Flooding in Ohio

Why is there all this flooding in Ohio every other year? Flooding seems to happen along the Ohio River much more than it used to. Could it be the effects of global warming? Not sure. Maybe people have moved into the river basin more. Or, the build up of barriers (dikes? dams?) to natural water flow has made it impossible for the surrounding land to soak up the water.

So what's the problem? I think there's a real tension between the needs of humans and the needs of nature in this area. Nature simply must drain the water. At some point, there's nothing humans can do about this. But humans need places to live also. Is the solution a matter of legislation? Can we set off these flood plains, so people won't live in them? Is it a matter of building smarter homes in these areas? Can we build houses that resist flooding? Stilts? Houses that lift up automatically? Walls around houses? I'll check this out.

Or is this an even bigger issue? Is global warming finally catching up to us? For the near future, do we need to simply move people to higher ground? In one website, I noticed that whole towns in Iowa have moved to avoid the increasing number of floods. Is this the solution in the short term? I'll need to check that out.

Figure 12.4: While freewriting, just get your ideas on the screen. Simply writing ideas down will help you locate important ideas and directions for research.

You may or may not end up using many of the words and sentences in your freewriting draft, but the purpose of freewriting is to put your ideas on the screen. It helps you fight through writer's block.

When you're done freewriting, identify the two to five major items in your text that seem most important. Then, spend 5 to 10 minutes freewriting about each of these items separately. Like magic, within half an hour to an hour, your freewriting will probably give you the material you need to write your text.

OUTLINE OR BOXING Outlines can be used throughout the drafting process. Most word-processing programs will allow you to draft in *Outline* mode or *Document Map* mode (Figure 12.5). Sometimes it helps to sketch an outline before you start drafting. That way, you can see how the document will be structured.

Boxing is less formal. As you plan your document, draw boxes on the screen or a piece of paper that show the major ideas or topics in your document (Figure 12.6). Then, type or write your ideas into the boxes. If you want to make multiple levels in your text, simply create boxes within boxes. You can use the Table function in your word-processing software to make boxes. When using the

GO TO THE NET

These invention tools don't work for you? To learn about other invention techniques, go to **www.pearsonhighered.com/johnsonweb4/12.8**

Outlining or Document Mapping

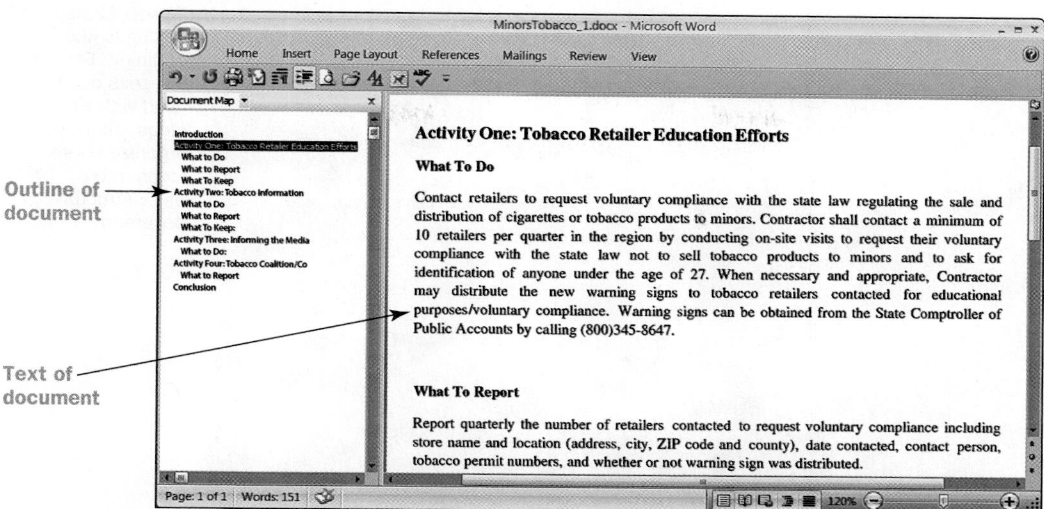

Figure 12.5: In Document Map mode, the computer automatically outlines your document on the left. This feature allows you to keep the whole structure of the document in mind as you work on individual parts.

Table function, start out with a few boxes. Then add cells to the table as you need more boxes.

THE JOURNALIST'S QUESTIONS The journalist's questions focus on the who, what, where, when, why, and how of an issue. They are also called the "Five-W and How questions." Separately, for each question, write down any words, phrases, and sentences that come to mind about your topic. These six questions will help you view your subject from a variety of viewpoints and perspectives.

Who was involved?

What happened?

Where did it happen?

When did it happen?

Why did it happen?

How did it happen?

When using the journalist's questions, pay special attention to anything about your subject that is changing or has changed. If you ask what has changed recently about your subject, you will likely focus in on what is most important about it.

Boxing

Introduction: Report on Flooding in Ohio

Purpose Statement: This report will provide strategies for managing flooding in the Ohio River Valley.

Main Point: Solving flooding means restoring wetlands and slowing development.

Importance of Subject: If we don't do something now, it will only become worse as the effects of global warming are felt.

Section One: The Problem

Development in Ohio River Valley

Increased water due to global warming

Additional dams and retaining walls

Section Two: The Plan

Restore wetlands

Limit development along rivers

Create holding reservoirs for water

Remove some retaining walls

Conclusions: We're Running Out of Time

We need to restore wetlands and lessen development on river

Advantages of these recommendations

The future

Figure 12.6: Boxing is like outlining. Each of the cells can be filled with your ideas. Then, you organize these ideas into a more structured document.

In the workplace, people use a variety of strategies to be creative. The five described here are especially useful for writing technical documents, but you may have your own ways of generating new ideas. Whatever you do, don't be discouraged if your good ideas aren't always accepted. Creativity usually means change, and change can be intimidating to others. Every once in a while, though, an idea catches fire. Those moments make the effort of being creative worth it.

CHAPTER REVIEW

- Developing a project plan is a process of identifying tasks and setting a timeline for completing them.
- Start out your project planning by listing your objectives and identifying your "top-rank objective."
- Create a task list and then put those tasks on a timeline.
- Strategic planning will lead to the need to generate new ideas and be creative.
- Being creative is hard work, but you should pay attention to change and trust your instincts.
- Logical mapping, brainstorming, freewriting, outlining/boxing, and the journalist's questions are good ways to generate new ideas.
- Creativity means change, so don't be discouraged if your good ideas are not accepted immediately.

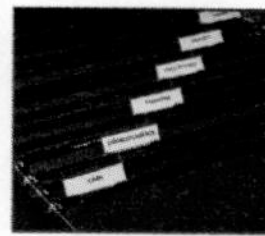

EXERCISES AND PROJECTS

Individual or Team Projects

1. For a project you are working on right now, go through the strategic planning process described in this chapter. First, list your objectives and identify a top-rank objective. Then, express that top-rank objective as a one-sentence purpose statement for the project. Second, create a task list of items that will need to be completed for the project. Third, put those tasks on a timeline or calendar and schedule any deadlines or meetings that will be part of the project. If you are working with a group, make sure everyone is following the same objectives, task list, and calendar.

2. In this chapter, you learned five different invention techniques to be creative and generate new ideas. Pick a technical subject. Then, use two different invention techniques to generate ideas about that subject. Compare and contrast the results. Which of the techniques worked better for this subject? Why? Which technique felt more comfortable to you? Is there something about the way you think that might make some kinds of invention techniques more effective than others?

3. On the Internet, use a search engine to find advice about being creative. Find five ways to be creative that aren't mentioned in this chapter. Do you think these websites are offering good advice? What criticisms or skepticisms might you have about the advice they offer for being more creative? What are your criticisms and skepticisms about the advice offered in this chapter?

Collaborative Project

Imagine that you and your group have been hired to write a travel book about Quebec, Canada, and you need to go there for four days this spring to do your

To learn more about working in teams, go to **www.pearsonhighered.com/johnsonweb4/12.9**

research. More than likely, most of you have never been to Quebec. It's a historic city in eastern Canada that reminds many people of an older European city. Quebec has great food, a fun nightlife, music, and arts and culture. It is also not far from great hiking, whale watching, and other outdoor experiences.

With your group, use listing or brainstorming to come up with a list of "objectives" that you would like to achieve during your trip. Check out Quebec on the Internet to decide on some of the things you would like to explore while you are there. Then, narrow your list to one top-rank objective and a few other major objectives that you would like to reach while you are on the trip.

Now, create a task list of major activities that you will do when you arrive in Quebec. Remember, you only have four days, so you will need to prioritize what you can experience while you are there. Also, the members of your group don't need to do everything together. You can go your separate ways, giving you a chance to explore more of Quebec.

Finally, use an online calendaring program to schedule your visit. On a four-day timeline, assign a time to each activity (task). Identify who will be going where and doing what each hour of the day. Also, schedule daily meetings or checkpoints in which your group will meet to catch up and make adjustments to the plan.

When you are finished, give your plan to your boss (your instructor) for approval.

Extra Challenge: Try to do this collaborative exercise virtually, without meeting face to face.

For additional technical writing resources, including interactive sample documents, document design tutorials and guidelines, and more, go to **www.mytechcommlab.com.**

CASE STUDY

Getting Back to Crazy

Lisa Stewart had been working for Fluke!, an Internet search company based in Silicon Valley, California. Her boss, Jack Hansen, would tell her stories about the exciting days when Fluke! began. He said new ideas were always welcome, no matter how crazy or far-fetched. He said there were only a dozen employees working long hours, so they could create just about anything they could imagine. Those were the good days.

One of those crazy ideas became an Internet search engine that made Fluke! a successful company. The search engine used social networking to rank information, products, and services. Advertisers loved it. Fluke! grew into a multi-million dollar company in one year.

Over the years, though, Fluke! had lost its edge. While other search engine companies, like Google, were always coming up with new products and services, the software engineers at Fluke! kept trying to improve its existing search engine. Fluke! was still well regarded, but revenues had dropped off sharply over the last few years.

So, the company's Board of Directors hired a new CEO, Amanda Jackson, who had a reputation for bringing established Silicon Valley companies back to life. The new CEO immediately put everyone in the company through a three-day workshop with "Creativity Engineers," who showed them how to be innovative and inventive. Most employees, including Lisa, were skeptical at first, but they learned a great deal. They were excited about creating some new products and getting back some of that Silicon Valley magic that Lisa's boss, Jack, was always talking about.

On a darker note, though, the new CEO also began circulating a rumor that the company would be going through a dramatic shake up if creative new products weren't put in the pipeline. Whole teams of software engineers would be fired if they kept doing business as usual.

The Monday after the creativity workshop, Lisa came to work with some new ideas. She was ready to throw out some crazy thoughts to see if her team could come up with something new. Some of her other team members came in with new ideas, too.

Unfortunately, the creativity workshop didn't seem to have any effect on her boss, Jack. Their Monday morning team meeting was the same old boring discussion about how to adjust the Fluke! search engine to improve its speed. There was no talk of new products or crazy ideas.

After the meeting, Lisa was frustrated, and so were some of her fellow team members. Not only did they not have an opportunity to share new ideas, but Lisa was worried about the lack of innovation. If the new CEO was looking for people to fire for lack of creativity, Lisa's team was putting itself on the chopping block. Her boss, Jack, obviously didn't see the threat.

If you were Lisa, what would you do?

CHAPTER

13

Persuading Others

Persuading with Reasoning 367
Persuading with Values 372
Help: Persuading Readers Online 373
Persuasion in High-Context Cultures 378
Chapter Review 382
Exercises and Projects 382
Case Study: Leapfrogging with Wireless 384

In this chapter, you will learn:

- How people are persuaded by reasoning and values.
- How to reason with logic, examples, and evidence to support your views.
- How to use values to appeal to common goals and ideals while using language familiar to your readers.
- About persuasion in high-context cultures.

When you have finished planning your project and generating ideas, you should spend some time developing your "persuasion strategy." Persuasion is not only about changing other people's minds. It is also about giving people good reasons to do things they might already want to do. In some cases, persuasion is about building someone else's confidence in you, your company, or your company's products and services.

Effective technical documents typically include a blend of reasoning-based and values-based persuasion strategies (Figure 13.1).

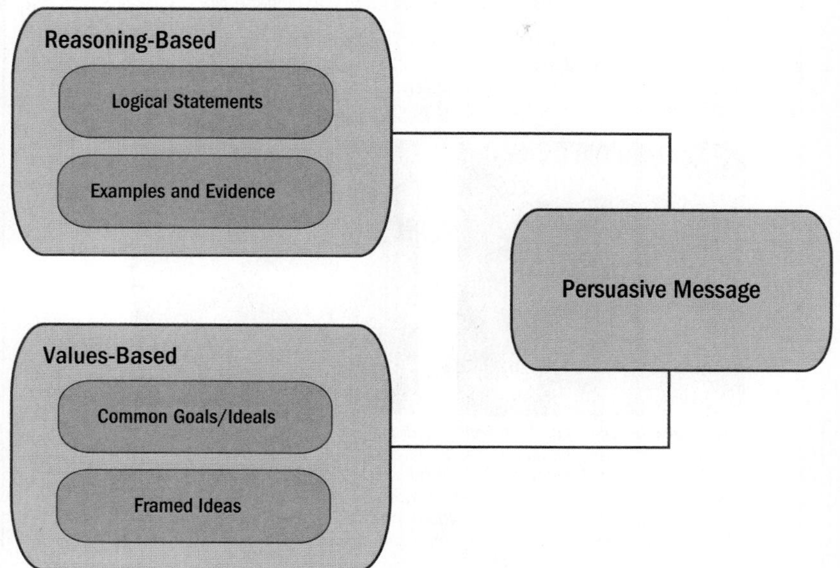

Figure 13.1: Technical documents and presentations typically include a blend of reasoning-based and values-based statements.

Persuading with Reasoning

Let's look at reasoning first. Reasoning is the use of logic and examples to show others the strengths and merits of your ideas. Reasoning has two basic forms:

Reasoning with logic—Using logically constructed statements such as *if . . . then, either . . . or, cause and effect, effect . . . because, costs and benefits,* and *better and worse*

Reasoning with examples and evidence—Using real or realistic statements, such as examples, prior experiences, facts, data, observations, and quotes from experts

Figure 13.2 shows a briefing from NASA that uses an assortment of logical statements, examples, and evidence to support its points. In this example, notice how the writer from NASA weaves logic, examples, and evidence together to present a persuasive document.

To learn more about rhetoric and persuasion, go to **www.pearsonhighered.com/johnsonweb4/13.1**

A Document That Uses Reasoning

Figure 13.2: In this brief, scientists at NASA Goddard Institute try to explain their findings about global warming. Due to political resistance, scientists often need to be persuasive in explaining the results of their research.

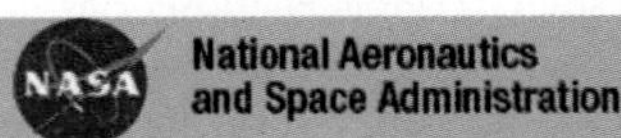

Scientists Confirm Earth's Energy Is Out of Balance

04.28.05

Scientists have concluded more energy is being absorbed from the sun than is emitted back to space, throwing the Earth's energy "out of balance" and warming the globe.

Scientists from NASA, Columbia University, New York, and Lawrence Berkeley National Laboratory, Berkeley, Calif., used satellites, data from buoys and computer models to study the Earth's oceans. They confirmed the energy imbalance by using precise measurements of increasing ocean heat content over the past 10 years.

These images are visual observations that show the changes in the Earth's climate.

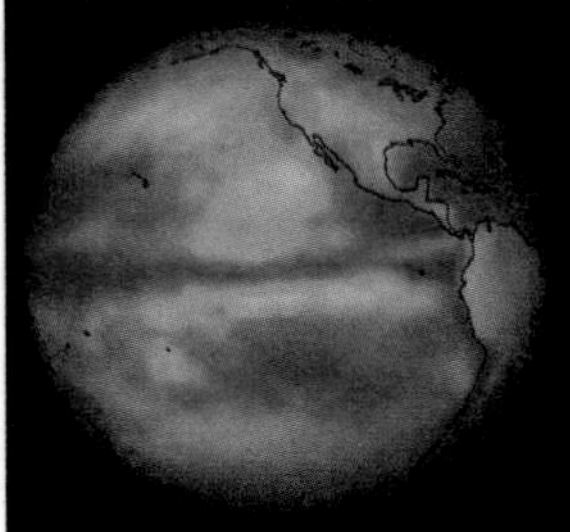

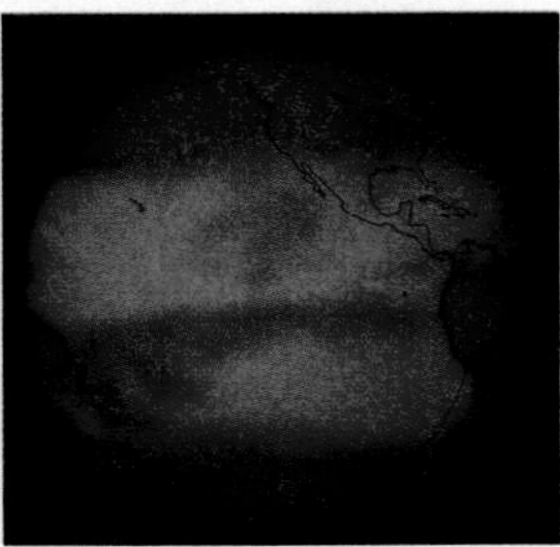

Image above: *Clouds and the Earth's Radiant Energy System (CERES) measurements show the reflected solar radiation (left) and emitted heat radiation (right) for January 1, 2002. In both images, the lightest areas represent thick clouds, which both reflect radiation from the Sun and block heat rising from the Earth's surface. Notice the clouds above the western Pacific Ocean, where there is strong uprising of air, and the relative lack of clouds north and south of the equator.*

Data is used to support factual claims.

The study reveals Earth's energy imbalance is large by standards of the planet's history. The imbalance is 0.85 watts per meter squared. That will cause an additional warming of 0.6 degrees Celsius (1 degree Fahrenheit) by the end of this century.

The main point is expressed as a cause and effect statement.

To understand the difference, think of a one-watt light bulb shining over an area of one square meter (10.76 square feet). Although it doesn't seem like much, adding up the number of feet around the world creates a big effect. To put this number into perspective, an imbalance of one-watt per square meter, maintained for the past 10,000 years is enough to melt ice equivalent to one kilometer (0.6 mile) of sea level, if there were that much ice.

Source: Rani Chohan, Goddard Space Flight Center, NASA, http://www.nasa.gov/vision/earth/environment/earth_energy.html.

For websites that discuss using logic, go to **www.pearsonhighered.com/johnsonweb4/13.2**

Figure 13.2: *(continued)*

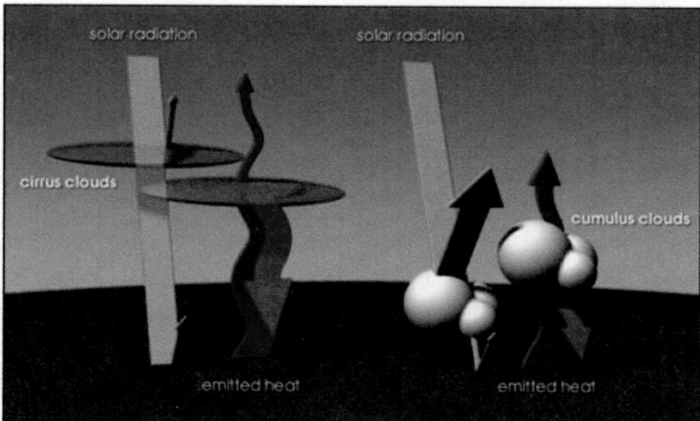

Image above: *The Earth's energy balance refers to the amount of energy received from the sun (yellow arrows) minus the energy reflected and emitted from the Earth (red arrows). Clouds play an important role in regulating this balance. Thin cirrus clouds permit sunlight to pass through them, while blocking a significant amount of the heat radiating from the surface. Thick cumulus clouds reflect most sunlight, and block the majority of heat radiating from the surface. Credit: NASA*

"The energy imbalance is an expected consequence of increasing atmospheric pollution, especially carbon dioxide, methane, ozone, and black carbon particles. These pollutants block the Earth's heat radiation from escaping to space, and they increase absorption of sunlight," said Jim Hansen of NASA's Goddard Institute for Space Studies, New York. He is the lead author of the new study, which is in this week's Science Magazine Science Express.

This quote from an expert supports and clarifies the message.

As the Earth warms it emits more heat. Eventually the Earth will be back in balance, if the greenhouse gas emissions are kept at the same level of today. Scientists know it takes the ocean longer to warm than the land. The lag in the ocean's response has practical consequences. It means there is an additional global warming of about one degree Fahrenheit that is already in the pipeline. Even if there were no further increase of human-made gases in the air, climate would continue to warm that much over the next century.

These are cause and effect statements.

Here are "if . . . then" statements.

Warmer world-wide water temperatures also affect other things. "Warmer waters increase the likelihood of accelerated ice sheet disintegration and sea level rise during this century," Hansen said. Since 1993, sea levels have been measured by satellite altimeters. Data has shown they have risen by approximately 3.1 centimeters or 1.26 inches per decade.

Although 3.1 centimeters is a small change, the rate of increase is twice as large as in the preceding century. There are positive feedbacks that come into play, as the area of ice melt increases. The researchers agree monitoring ice sheets and sea level is necessary to best ensure the system is in balance.

The data used here helps explain the magnitude of the problem.

Reasoning with Logic

Logic offers you a variety of ways to persuade your readers. When using logic, you are trying to appeal to your readers' common sense or beliefs. Logic allows you to build more complex ideas from simpler facts. When using logical reasoning, you are essentially saying things like, "If you believe X, then you should also believe Y," or "Either you believe X, or you believe Y," or "X is happening because Y happened."

IF . . . THEN Perhaps the most common logical pattern is the *if . . . then* statement. When using *if . . . then* statements, you persuade your readers by demonstrating that something they already believe leads logically to something else they should also accept. You are basically saying, "If you believe X, then you should also believe Y."

> If we are going to be ready for next summer's hurricane season, then we cannot wait until next spring to begin planning.

> Internet thieves will be able to steal your identity if you don't take steps to protect yourself.

Link

For more on using *if . . . then* and *either . . . or* statements, go to Chapter 16, page 440.

EITHER . . . OR When using *either . . . or* statements, you are offering your readers a choice between two paths. You are telling them, "Either you believe X or you believe Y."

> Either we take steps to control crime in our area or we risk handing over our streets to criminals.

> We need to either start redesigning the car to use a hybrid engine or take the risky path of hoping oil prices drop dramatically over the next few years.

Either . . . or statements can be risky because readers may choose the path you didn't want them to take, or they may reject both choices. So, you need to make the "correct" path obvious to them if you want them to go in a particular direction.

Link

For more help with using cause and effect and costs and benefits statements, go to Chapter 16, page 437.

CAUSE AND EFFECT When using *cause and effect* statements, you are demonstrating to your readers how specific causes lead to specific effects. You are showing them that "X is caused by Y."

> Gradually, desertification causes a dryland, such as the Sonoran Desert, to lose its ability to support plants and animals.

> The effects of this problem can be sobering. Last year, intoxicated drivers caused 83 accidents in Holt County, killing four people and costing taxpayers $1.1 million.

Similarly, the word *because* can also signal a cause and effect relationship:

> The Stonyridge Windfarm project should be approved because it will generate electricity and revenue without further polluting our area.

> The Internet went down repeatedly over the summer because the server kept crashing.

COSTS AND BENEFITS When trying to persuade people, you might find it helpful to show them the *costs and benefits* of your ideas. In most cases, you will want to show them that the benefits outweigh the costs by saying something like "The benefits of doing X will be worth the cost, Y." In some cases, though, you might point out that the costs would be too high for the few benefits gained.

> Building a wireless network in the fraternity house will require an up-front investment, but we would save money because each member would no longer have to pay the phone company for a separate DSL line.
>
> Since St. Elizabeth Hospital's main building has become obsolete in almost every way, the benefits of remodeling it would not justify the costs.

BETTER AND WORSE Another persuasive strategy is to show that your ideas are better than the alternatives. You are arguing that "X is better/worse than Y."

Link

For more help with using empirical observations, go to Chapter 14, page 398.

> In 2008, we decided to go with AMD's Athlon 64 FX-57 microprocessor for our gaming-dedicated computers. The other chips on the market just couldn't match its balance of speed and reliability.
>
> In the long run, we would be better off implementing our automation plan right now, while we are retooling the manufacturing plant. If we wait, automating the lines will be almost impossible as we return to full capacity.

Reasoning with Examples and Evidence

Examples and evidence allow you to reason with your readers by showing them real or realistic support for your claims.

FOR EXAMPLE Using an example is a good way to clarify and support a complex idea while making it seem more realistic to readers. You should say something like, "For example, X happened, which is similar to what we are experiencing now."

> For example, some parasitoid wasps inject polydnaviruses into the egg or larva of the moth host. The wasp eggs survive in the host's body because the virus suppresses the immunity of the host.
>
> If, for instance, a high-speed railway were to be built between Albuquerque and Santa Fe, commuters would cut their commuting times in half while avoiding the dangerous 50-mile drive.

Phrases like *for example*, *for instance, in a specific case, to illustrate* and *such as* signal that you are using an example.

EXPERIENCES AND OBSERVATIONS Personal experiences and observations can often be persuasive as long as readers trust the credibility of the source. You are telling your readers, "I have seen/experienced X before, so I know Y is likely true."

AT A GLANCE

Reasoning-Based Persuasion

- *If . . . then*
- *Either . . . or*
- Cause and effect
- Costs and benefits
- Better and worse
- *For example,*
- Experiences and observations
- Facts and data
- Quotes from experts

While we were in the Arctic Circle, we observed a large male polar bear kill and devour a cub, turning to cannibalism to survive.

When our team began closely monitoring the medications of schizophrenic prisoners at the Oakwood Correctional Facility, we observed a dramatic decline in the number of hallucinations and delusional episodes among the prisoners.

FACTS AND DATA Empirically proven facts and data generated from experiments and measurements can offer some of the strongest forms of evidence. People generally trust observed facts and numbers.

The facts about influenza epidemics on college campuses are amazing. A 2004 study on college campuses showed that 5 out of 10 students will become infected when the virus finds its way into a dormitory. (Venn, p. 15)

Recently published data shows that for every child who may experience a prolonged benefit from the hemophilus vaccine, two to three children may develop vaccine-induced diabetes. (Akers & Wilson, p. 126)

Always remember to cite any sources when referring to facts and data.

Link

For more ideas about using interviews and observations to collect information, go to Chapter 14, page 398.

QUOTES FROM EXPERTS A recognized authority on a subject can also be used to support your points. You can use quotes as evidence to back up your claims.

Dr. Jennifer Xu, a scientist at Los Alamos National Laboratory, recently stated, "Our breakthroughs in research on edge localized modes (ELMs) demonstrate that we are overcoming the hurdles to fusion nuclear power" (Xu, 2006).

The lead biologist for the study, Jim Filks, told us that mercury levels in fish from the Wildcat Creek had dropped 18 percent over the last ten years, but levels were still too high for the fish to be safely consumed.

Persuading with Values

Values-based persuasion can be more subtle than reasoning-based persuasion. Values-based persuasion is effective because people prefer to say yes to someone who holds the same values and beliefs as they do (Lakoff, 2004). Moreover, confidence and trust go a long way toward convincing people what to believe and what to do.

Values-based persuasion uses two forms:

Goals and ideals—The use of goals, needs, values, and attitudes that you share with your readers

Frames—The use of words, phrases, and themes that reflect your readers' point of view and values

For more about values-based persuasion, go to **www.pearsonhighered.com/johnsonweb4/13.4**

HELP

Persuading Readers Online

The Internet offers some interesting persuasion challenges because influencing people is usually best done in person. So, if you need to persuade someone through e-mail, websites, or text messages, you might need to be a little more creative. Noah Goldstein and Steve Martin are researchers who have compiled persuasive strategies into a book called *Yes! 50 Scientifically Proven Ways to Be Persuasive.* Here are ten of their strategies that work well in online situations.

CREATE AN IMPRESSION OF SCARCITY. Suggest that the product or service will be available in limited amounts or for a short time.

DECREASE THE NUMBER OF OPTIONS. Give clients only a few options, because having too many choices could actually cause them to hesitate and seek a simpler solution.

LABEL THE READER INTO A SPECIFIC GROUP. If you accurately label the person you are trying to persuade (e.g., "engineering major," "doctor," or "instructor") they are more likely to respond favorably.

TELL PEOPLE THEIR PRIOR BELIEFS WERE CORRECT. People don't like to be inconsistent, so they will often cling to prior beliefs for the sake of consistency. So, assure them that their prior beliefs were correct under the old conditions, but new conditions call for them to think and act differently than before.

ASK THEM TO HELP. Readers respond more positively if they think you are asking them to help in some way. Instead of explaining what you need or what you want them to do, ask them for their help in figuring out how to solve the problem.

ADMIT YOU WERE WRONG (IF YOU WERE). If a product or service didn't work as expected, taking responsibility will actually build trust in readers. In other words, an admission that the team or company came up short is persuasive because it shows you're working toward improvement.

USE "YOU" TO REFER TO READERS. People from Western cultures respond favorably to the word "you" in persuasive situations (e.g., "You will receive the following services."). Interestingly, though, people from Eastern cultures tend to respond more favorably to statements that signal a service or product will be best for all.

USE THE WORD "BECAUSE" TO MAKE YOUR ARGUMENTS SOUND RATIONAL. Simply adding the word "because" to your explanations will signal that you have thought things through and are being reasonable.

RHYME PHRASES TO MAKE THEM MORE CONVINCING. You probably remember rhymes like "An apple a day keeps the doctor away," or "If it doesn't fit, you must acquit." Rhymed phrases are easier to agree with, even if they sound hokey.

FACE TIME BEATS E-MAIL TIME. Studies show that people reach agreement and resolve problems better face to face than through e-mail. So, ask yourself whether a meeting or a phone call would be more effective.

When using values-based persuasion, you are trying to convince readers to *identify* with you, your company, or your company's products and services. Advertisers spend billions each year appealing to consumers' sense of values to develop product identification (Figure 13.3). When a company succeeds in associating itself with a particular set of positive values, its written materials will be much more effective.

Link

For more information about profiling readers, turn to Chapter 2, page 20.

Appealing to Common Goals and Ideals

To appeal to common goals and ideals, you should begin by looking closely at the profile you developed of your readers, which was discussed in Chapter 2. This profile will contain clues about your readers' goals, needs, values, and attitudes.

GOALS Almost all people have personal and professional goals they are striving to reach, and so do most companies. If possible, discuss your readers' goals with them. Then, in your document, you can show them how your product or services will help them reach those goals.

> Solar power will maintain the appearance of your home while adding value. From the street, photovoltaic shingles look just like normal roofing materials, but they will give you energy independence, save you money, and increase the value of your home.
>
> By renovating Centennial Park, Mason City will go a long way toward reaching its goal of becoming a town that promotes a lifestyle of learning and leisure.

NEEDS A fundamental difference exists between needs and goals. *Needs* are the basic requirements for survival.

> Your just-in-time manufacturing process requires suppliers to have parts ready to ship within hours. We can guarantee that parts will be shipped within one hour of your order.
>
> Equipment in the operating room must be sterilized with alcohol to ensure our patients' safety.

If you aren't sure what your readers need, ask them. They will usually be very specific about what they require.

To learn more about what people think they need, go to **www.pearsonhighered.com/johnsonweb4/13.5**

Appealing to Values

Readers identify with these elderly people who are still apparently active.

Notice how words like "approved," "science," and "common sense" reflect readers' values.

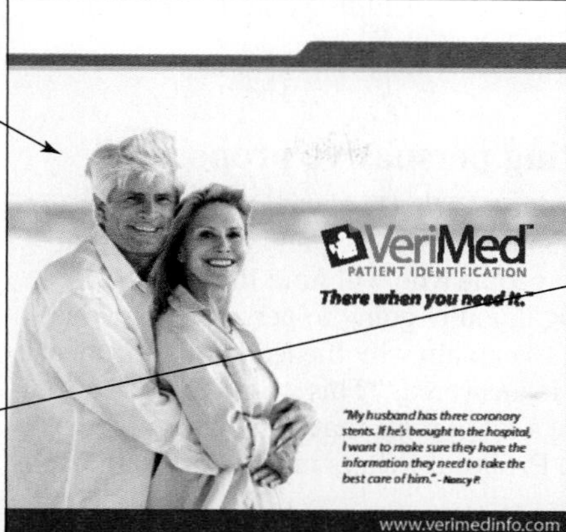

unfortunately,
you may not be able to prevent a medical emergency

fortunately, you can prepare

VeriMed™ is the **first and only FDA-approved patient identification system that uses an implantable microchip.** While that may sound like science fiction, it's really down-to-earth, common sense when it comes to your life.

About the size of a grain of rice and inserted just under your skin, each VeriMed microchip contains a unique identification number that emergency personnel, equipped with an appropriate radio frequency identification (RFID) scanner, may scan to **immediately identify you and access your personal health information** — facilitating appropriate treatment with less delay. VeriMed is there when you need it.

the VeriMed Microchip (actual size)

Figure 13.3: The product described in this booklet is an RFID (Radio Frequency Identification) device that is injected under the skin—a potentially troubling product. The authors appeal to values to persuade readers of the merits of the product.

The shared goal of quick service is regularly mentioned.

The fear of being unable to communicate (an attitude) is brought out here and other places in the text.

Use of "you" appeals directly to readers.

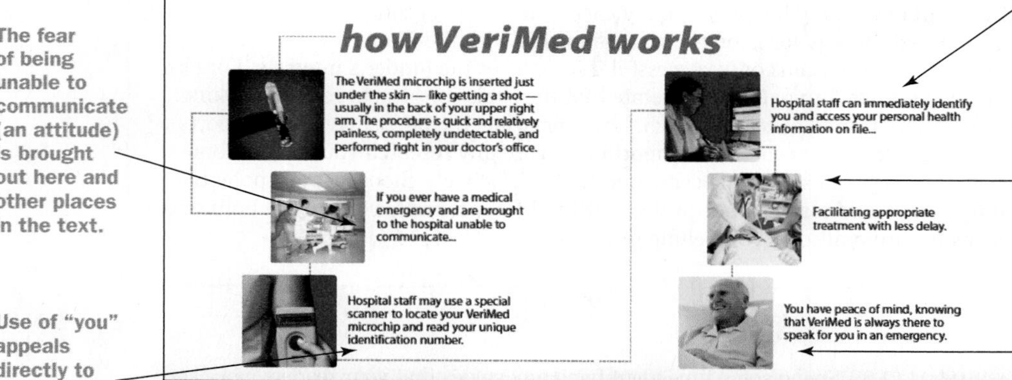

Images are used to put readers in the place of a patient.

The image shows the achievement of a common goal.

Source: VeriMed, http://www.verimedinfo.com.

AT WORK

Mary Nauman

FREELANCE WRITER-EDITOR OF GRANTS AND PROPOSALS

Mary Nauman specializes in writing grant proposals, including ones for scientific, medical, and technical research.

Can you give us some tips for writing persuasive proposals?

Two key methods for crafting a persuasive proposal are: (1) always provide your reader with a compelling yet accurate story about your project; and (2) ensure that your goals are aligned with the mission of the people who will fund the project.

First, your proposal needs to be compelling if you're going to persuade someone to fund the project. Recently, I asked a client to explain why the federal government should fund his research on mammograms. He answered, "This would provide a tool that doesn't currently exist." Sorry. There are a lot of tools that don't currently exist, but that doesn't mean these tools are needed. Proposal writers must take a step back and look at the full impact of the project.

To describe the full impact, my client's proposal immediately stated that his research is a first step toward reducing the number of false-positive mammograms. False-positive results are a painful and expensive problem experienced by 49 percent of all women at least once within their first ten mammograms. By providing this context without overselling his research, my client conveyed his comprehensive vision. He also paved the way for a long-term relationship with the funder.

Secondly, a proposal is only successful if it matches the funder's interests. For example, two private foundations may state that their missions are to reduce smoking. One may support lung cancer research, however, while the other may only support anti-smoking advertisements. It is important to carefully research your target donor and, if necessary, call and ask specific questions. Selectively choosing your potential funding agency and placing your project within the appropriate context are both critical steps toward crafting a compelling proposal.

Link

For more strategies on identifying shared values through the Internet, go to Chapter 2, page 25.

SHARED VALUES Spend some time identifying any values that your readers, you, and your company have in common. For example, if your readers stress high-quality service, more than likely, they will want to work with suppliers who also value high quality.

> If you have any questions or problems with your custom CAD software, we have engineers available 24/7 through our website and over the phone. When you call, a person, not a machine, will answer the phone.
>
> Like your organization, our company has a "People First" policy, which we believe is essential in keeping our employees satisfied and productive.

You can usually find shared values by looking on your readers' websites.

For more information about using identification to sell products, go to **www.pearsonhighered.com/johnsonweb4/13.6**

ATTITUDES A reader's attitude toward your subject, you, or your company can greatly determine whether your message is persuasive. So, you might use your reader's positive attitude to your advantage or show understanding when the reader has a negative attitude.

> Purchasing your first new car can be exciting and just a little bit stressful. This guide was created to help you survive your first purchase.

> Like you, we're always a bit nervous when our company upgrades operating systems on all its computers. The process, though, is mostly painless, and the improved speed and new features are worth the effort.

Words alone will rarely change someone's attitude, but you can show that you empathize with your readers' point of view. Perhaps your readers will give you a chance if they think you understand how they feel about the subject.

AT A GLANCE

Values-Based Persuasion

- Goals
- Needs
- Shared values
- Attitudes
- Framing
- Reframing

Framing Issues from the Readers' Perspective

Being persuasive often means seeing and describing an issue from your readers' perspective. Linguists and psychologists call this *framing* an issue (Lakoff, 2004, p. 24). By properly framing or reframing an issue, you can appeal to your readers' sense of values.

FRAMING To frame an issue, you should look closely at the profile you developed of the readers. Locate the one or two words or phrases that best characterize your readers' perspective on the issue. For example, let's say your readers are interested in *progress.* Your profile shows that they see the world in terms of *growth* and *advancement.* You can use logical mapping to develop a frame around that concept.

In Figure 13.4, for example, the word "progress" was put in the middle of the screen to create a logical map of ideas. Then, words and phrases that cluster around "progress" were added. As you fill in the logical map, it should show you the frame from which your readers understand the issue you are discussing. Knowing their frame will help you choose content and phrasing that support or reinforce their point of view.

Link

For more help on using style persuasively, go to Chapter 17, page 472.

REFRAMING In some cases, your readers might not see an issue in a way that is compatible with your or your company's views. In these situations, you may need to "reframe" the issue for them. To reframe an issue, look a little deeper into your reader profile to find a value that you and your readers share. Once you locate a common value, you can use logical mapping to reframe the issue in a way that appeals to your readers. For example, the poster in Figure 13.5 shows how NASA is trying to reframe the debate over humans traveling to the moon. They recognize that some people may be skeptical about the need to go to the moon. So, they reframe the argument around common values that most readers will share.

GO TO THE NET

Want to learn more about framing? Go to **www.pearsonhighered.com/johnsonweb4/13.7**

Using Logical Mapping to Create a Frame

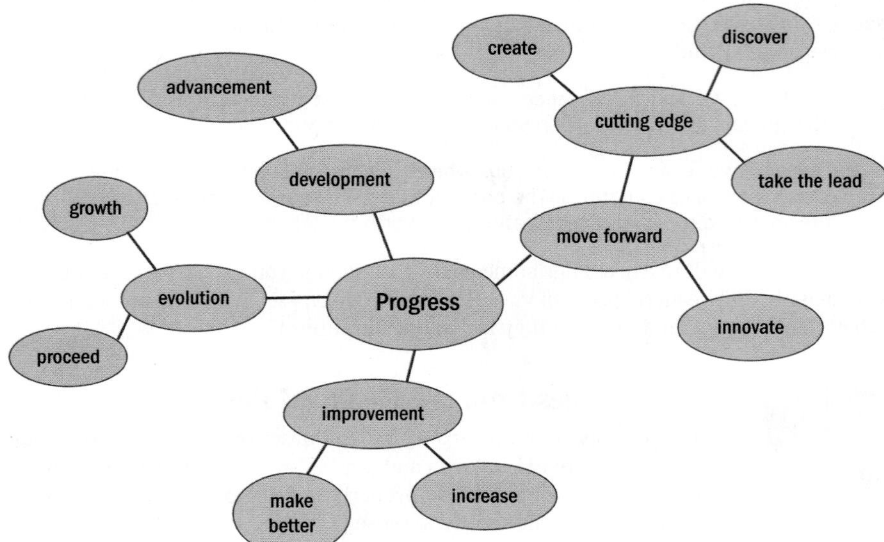

Figure 13.4: Framing is a useful way to describe an idea through the readers' point of view. Here, a logical map shows how the word "progress" would be framed.

Persuasion in High-Context Cultures

Cross-cultural communication can be difficult, but persuasion is especially challenging when the audience or readers are from a "high-context" culture. High-context cultures, which include many in Asia, the Middle East, and sub-Saharan Africa, tend to put more emphasis on community than on individuals. These cultures also tend to put a high value on consensus, interpersonal harmony, hierarchy, and rituals. They tend to stress long-term goals over short-term gains.

What is not said in a conversation is often very significant, because people from high-context cultures rely on *contextual cues* to interpret what a speaker means. As a result, negotiators from low-context cultures, such as Europe, the Americas, and Australia, tend to mistakenly believe that high-context negotiators are inefficient, indirect, vague, and ambiguous. Meanwhile, negotiators from high-context cultures, such as Japan, China, Korea, Indonesia, and parts of Africa, often find their low-context counterparts, especially Americans, to be abrupt, aggressive, and far too emotional.

Persuasion and negotiation are still important in high-context cultures, but the interactions are more subtle and indirect. Here are some guidelines to help you navigate such interactions:

Guideline 1: Develop long-term relationships—High-context cultures put a high value on existing relationships and reputations. Spend time familiarizing your readers with your company and explaining its reputation in the field. This will take time, because companies and governments from high-context cultures

GO TO THE NET: Want to learn more about etiquette in Asian cultures? Go to **www.pearsonhighered.com/johnsonweb4/13.8**

Reframing by Using Common Values

Argues that "real benefits" are to be gained from the mission.

Identifies six benefits (values) that are held by most people.

Uses an image with a child to encourage readers to think about the future.

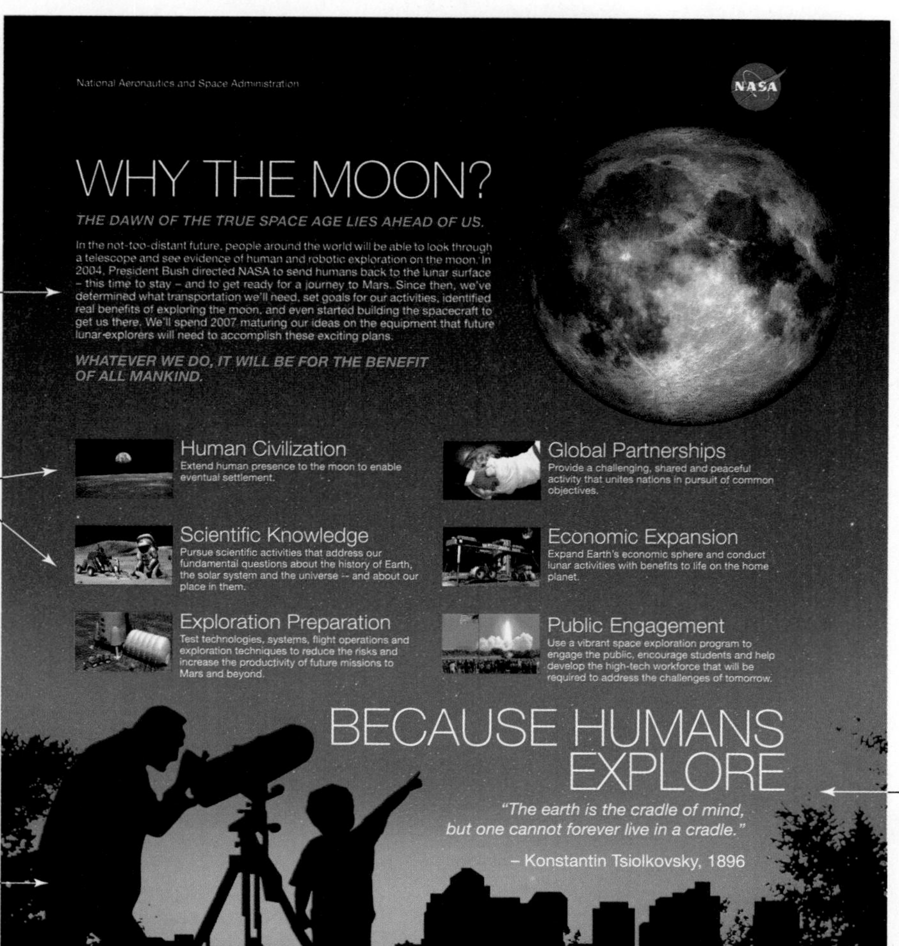

Figure 13.5: In this poster, NASA uses several common values to argue that travel to the moon is important. NASA's goal is to overcome public skepticism about the cost and necessity of the program.

The answer to the question "Why the Moon?" appeals to an assumed trait of humanity—a desire to explore.

Working in High-Context Cultures

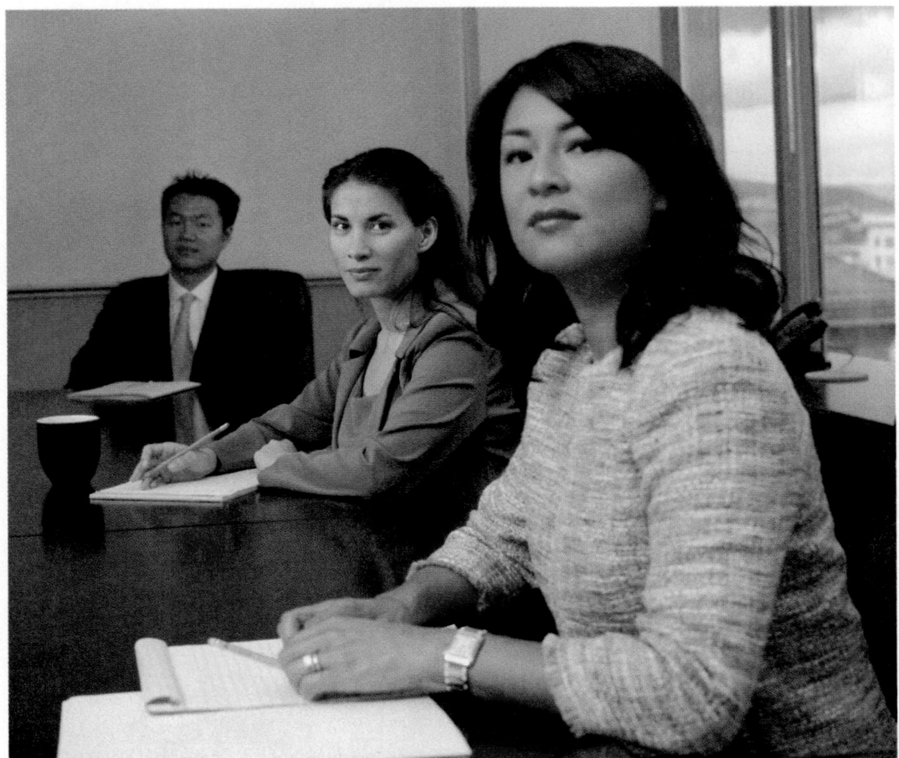

Readers from high-context cultures often use different forms of persuasion.

tend to think long term. Your company may need to invest years of effort to build a strong relationship with a company from a high-context culture.

Guideline 2: Use intermediaries to build relationships—In high-context cultures, strangers tend to be treated skeptically. So, you might look for a trusted intermediary who knows both parties to help you make connections.

Guideline 3: Rely on facts and reasoning—In high-context cultures, attempts to persuade directly are often viewed negatively. So, a fact-based presentation with solid reasoning tends to be much more effective.

Guideline 4: Avoid arguing strongly for or against—Outward argumentativeness can be viewed as threatening and disrespectful and is often counterproductive. Moreover, arguing directly against someone else's position might be perceived as an attack on that person. Instead, if you disagree with someone, restate the facts of your ideas and use reasoning to explain them.

To learn more about high-context cultures, go to **www.pearsonhighered.com/johnsonweb4/13.9**

Guideline 5: Strive to reach consensus—Social harmony is greatly valued in high-context cultures. So, you should strive for consensus with your high-context counterparts. They will tend to react skeptically to plans in which one side seems to benefit more than another. Instead, your ideas will be most persuasive if you present them as a win–win for both sides.

Guideline 6: Speak collectively, not individually—To maintain social harmony, you should tend toward speaking collectively ("we" or "us") rather than individually ("I" or "me"). Avoid saying something like, "Here's my opinion." Instead, strive to characterize what is best for all.

Guideline 7: Be patient and wait for the "point"—People from high-context cultures often approach complex issues holistically, discussing all issues at the same time. To a low-context person, it can seem as though nothing is being decided. People from low-context cultures tend to be eager to "get to the point," while people from high-context cultures usually assume the point is obvious and need not be stated. The point of a discussion or document will tend to arrive near the end, so you should be patient and wait for it.

Guideline 8: Remember that "no" is rarely used—Direct refusal, rejection, and the answer "no" tend to be avoided in high-context cultures, especially in professional settings. A direct refusal can be considered an insult. So, refusals are handled with some care. If you are being refused, you might hear a deferral like, "We will consider your ideas." In some cases, as in Indonesia, you may receive an initial "yes" with a later "but" that signals the refusal. When a "no" needs to be conveyed, often an intermediary will be asked to deliver the bad news.

Guideline 9: Don't be informal—In high-context cultures, hierarchy and rituals are important, and respect for social status is expected. In professional settings, people from high-context cultures tend to address each other formally. The American tendency to quickly become informal and familiar (e.g., "You can just call me Jim") can be viewed as disrespectful or aggressive. Even when you know someone well, you should address that person by title and demonstrate respect for his or her position.

Guideline 10: Defer to hierarchy—High-context cultures tend to put great emphasis on hierarchy and social standing. A person of lower standing is expected to defer to someone of higher standing. Meanwhile, causing someone of higher standing to become embarrassed or agitated, even accidentally, will usually undermine or scuttle any negotiations.

Link

For more strategies on communicating with cross-cultural readers, go to Chapter 2, page 36.

Guideline 11: Minimize emotions—Smiles are welcome in high-context cultures, but obvious signs of emotion, like anger, hilarity, annoyance, or bemusement, will usually be taken far more seriously than they would in low-context cultures. Losing your composure is almost a sure way to end a professional relationship.

Research is always helpful if you need to communicate with people from another culture. A surprising amount of helpful information can be found on the web.

As technical fields globalize, North Americans are finding it helpful to learn how other cultures operate. To be persuasive, you should learn how people from the target culture tend to think and negotiate. Listen carefully and don't become too frustrated when you make mistakes—because you will.

CHAPTER REVIEW

- All technical documents are persuasive in some way, even those that are intended to be strictly "informational."
- Persuasion is about giving people good reasons to do things, while building their confidence in you and your company.
- Technical workplaces tend to rely on reasoning-based persuasion and values-based persuasion.
- Reasoning-based persuasion relies on logic, examples, and evidence to support claims.
- Values-based persuasion uses shared ideals, mutual values, common goals, and credibility to build and strengthen relationships.
- Cross-cultural communication can be difficult, but persuasion is especially challenging when the audience or readers are from a "high-context" culture.

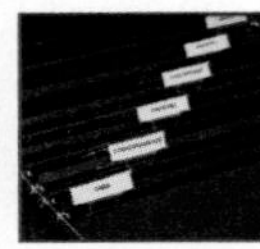

Individual or Team Projects

1. Compare and contrast two technical documents. In a PowerPoint presentation, show why one document is more persuasive than the other. How does the more persuasive document use reasoning and values to persuade readers? What would make the other document more persuasive?

2. Find websites that use images and graphics to set a particular tone. How persuasive are the images? Why do you think they persuade the readers, or don't persuade the readers? If you were the intended reader of the website, what would make the website more persuasive to you? What would make you do what the authors of the website want you to do?

3. Write a white paper to your class that studies the persuasion strategies of a culture other than your own. For example, you might explore persuasion strategies in China or France. What would typical people from these cultures find persuasive? And what would be the most effective way to persuade them, without offending them? You can learn more about white papers in Chapter 9.

Collaborative Project

Ask one of your team members to bring in a common household product (e.g., toaster, popcorn popper, video game, mobile phone, etc.). Using reasoning-based and values-based persuasion strategies, develop a strategy to promote this item to college students—in other words, people much like your classmates. What would be some of your best arguments in favor of the product? How would you downplay any weaknesses? How might you use the buyers' values to urge them to identify with the product?

Now, let's change directions. Imagine that you need to promote this same product to people who are 60 years old or older. How might your persuasion strategies change? What kinds of reasoning would be more appropriate for this audience? How might you encourage them to value the product by identifying with it?

In a 5-minute presentation to your class, compare and contrast your team's strategies for marketing the product to college students and older people. How would your strategies be similar? How would they differ? How might these differences affect the organization, style, and design of the documentation that goes with the product?

For additional technical writing resources, including interactive sample documents, document design tutorials and guidelines, and more, go to **www.mytechcommlab.com.**

CASE STUDY

Leapfrogging with Wireless

Tim Jenks knew this was going to be an interesting project. Tim was an electrical engineer who worked for Seal Beach Communications, a company that built wireless networks in southern California. Most of the company's customers lived and worked in the suburbs of Los Angeles.

Recently, his company won a contract to install wireless networks on Native American reservations in New Mexico. The tribes had received a $6 million grant to build wireless networks that would bring phone and Internet services to their members. The tribes were leapfrogging the technology, in some cases going from almost no services in some areas straight to high-speed wireless services in all areas. The grant was designed to help tribal members access educational opportunities and perhaps even work from home.

The project sounded fascinating, and Tim was eager to work with Native Americans, with whom he had almost no experience. Plus, this project was completely different from his normal work in southern California. It sounded like something he could be proud of.

His boss, Tina Guerrero, visited him at a worksite. "Tim, I have some good news. You're going to be our reconnaissance guy on the New Mexico project. Next week, we're sending you on a fact-finding mission to the future job site."

Tina explained that Tim would be meeting with tribal leaders to talk about where towers would be placed, where wireless clouds should be created, and what public facilities would be covered (e.g., libraries, cultural centers, schools).

"There is one complication—and it's a big one," cautioned Tina. "One of my friends who has worked with reservations in New Mexico says the tribes are all different. The Navajo sometimes handle issues very differently than the Pueblo tribes along the Rio Grande. We know very little about the Apache tribes in the north and south of the state."

A bit concerned, Tim conceded he didn't know anything about these cultural differences.

Tina continued, "The last thing we need is one of our engineers offending people and making things harder on us. So, I want you to find out whatever you can about the tribes in New Mexico and how they operate in business settings."

Imagine that you are Tim. Using the Internet and any other sources, discover what he should know about the cultural customs and business practices of Native Americans in New Mexico. How should he handle the cross-cultural issues involved in this project, especially when meeting tribal leaders for the first time?

GO TO THE NET

For some websites that discuss Native American culture and practices, go to **www.pearsonhighered.com/johnsonweb4/13.10**

CHAPTER

14 Researching and Research Methods

Beginning Your Research 387

Defining Your Research Subject 388

Formulating a Research Question or Hypothesis 389

Developing a Research Methodology 391

Triangulating Materials 393

Chapter Review 404

Exercises and Projects 404

Case Study: The Life of a Dilemma 406

In this chapter, you will learn:

- How to define your research subject using logical mapping to define its boundaries.
- How to formulate a research question or hypothesis that will guide your research.
- To develop a research methodology and revise it as needed.
- Methods for triangulating information to ensure reliability.
- To use the many available electronic research tools.
- To find electronic, print, and empirical sources for your research.

Computers and computer networks have made research both easier and more challenging. Not long ago, finding *enough* information was the hard part when doing research. Today, with access to the Internet, you will find seemingly endless amounts of information available on any given topic.

The problem caused by this overwhelming amount of facts, data, and opinions is an *information glut.* An information glut exists when there is more information than time available to collect, interpret, and synthesize that information.

What should you do about this overwhelming access to information? You should view "research" as a form of *information management.* Research is now a process of shaping the flow of information, so you can locate and use the information you need. As an information manager, you need to learn how to evaluate, prioritize, interpret, and store that information so you can use it effectively.

Empirical research is a critical part of working in technical disciplines.

Of course, in addition to collecting existing information, *primary research* (empirical research) is important in the technical workplace. Primary research involves observing and/or directly experiencing the subject of your study. By conducting experiments, doing field studies, using surveys, and following other empirical methods, you can make your own observations and collect your own data. The most effective research usually blends these kinds of empirical observations with the existing information available through computer networks and libraries.

For websites that discuss the information glut, go to **www.pearsonhighered.com/johnsonweb4/14.1**

Beginning Your Research

In technical fields, researchers typically use a combination of primary and secondary sources to gain a full understanding of a particular subject.

Primary sources—Information collected from observations, experiments, surveys, interviews, ethnographies, and testing

Secondary sources—Information drawn from academic journals, magazine articles, books, websites, research databases, DVDs, CD-ROMs, and reference materials

Most researchers begin their research by first locating the secondary sources available on their subject. Once they have a thorough understanding of their subject, they use primary research to expand on these existing materials.

Your research with primary and secondary sources should follow a process similar to this:

1. Define the research subject.
2. Formulate a research question and hypothesis.
3. Develop a research methodology.
4. Triangulate electronic, print, and empirical sources of information.
5. Appraise collected information to determine reliability.

A good research process begins by clearly defining the research subject. Then, it follows a research methodology in which a variety of sources are located and appraised for reliability (Figure 14.1).

A Research Process

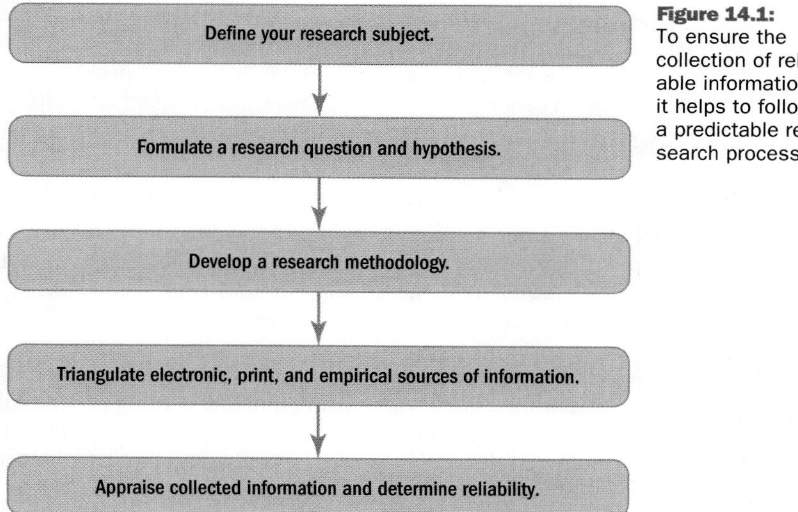

Define your research subject.
Formulate a research question and hypothesis.
Develop a research methodology.
Triangulate electronic, print, and empirical sources of information.
Appraise collected information and determine reliability.

Figure 14.1: To ensure the collection of reliable information, it helps to follow a predictable research process.

Defining Your Research Subject

Your first task is to define your research subject as clearly as possible. You should begin by identifying what you already know about the subject and highlighting areas where you need to do more research.

A reliable way to start is to first develop a *logical map* of your research subject (Figure 14.2). To create a logical map, write your subject in the middle of your screen or a piece of paper. Then, around that subject, begin noting everything you already know or believe about it. As you find relationships among these ideas, you can draw lines connecting them into clusters. In places where you are not sure of yourself, simply jot down your thoughts and put question marks (?) after them.

Using Mapping to Find the Boundaries of a Subject

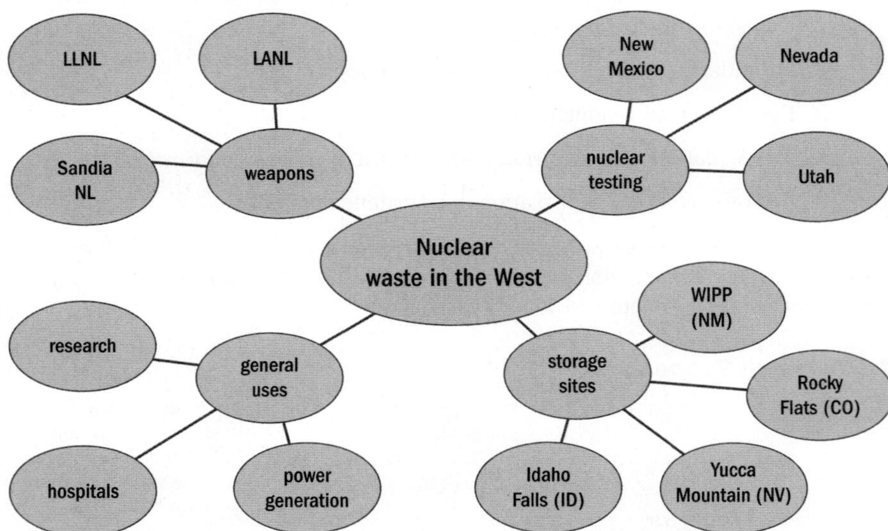

Figure 14.2: A logical map can help you generate ideas about your subject. It can also show you where you need to do research.

As you make your logical map, you will notice that some ideas will lead to other, unexpected ideas—some seemingly unrelated to your subject. When this happens, just keep writing them down. Don't stop. These unexpected ideas are evidence that you are thinking creatively by tapping into your visual-spatial abilities. You may end up crossing out many of these ideas, but some may offer you new insights into the subject.

Mapping is widely used in technical disciplines, and it is gaining popularity in highly scientific and technical research where visual thinking is being used to enhance creativity. You might find it strange to begin your research by drawing circles and lines, but mapping will reveal relationships that you would not otherwise discover.

GO TO THE NET

Want to learn more about being creative? Go to **www.pearsonhighered.com/johnsonweb4/14.2**

Narrowing Your Research Subject

After defining your subject, you also need to look for ways to narrow and focus your research. Often, when people start the research process, they begin with a very broad subject (e.g., nuclear waste, raptors, lung cancer). Your logical map and a brief search on the Internet will soon show you that these kinds of subjects are too large for you to handle in the time available.

General Subject (too broad)	Angled Research Area (narrowed)
Nuclear waste	Transportation of nuclear waste in western states
Eagles	Bald eagles on the Mississippi
Lung cancer	Effects of secondhand smoke
Water usage	Water usage on the TTU Campus
Violence	Domestic abuse in rural areas

To help narrow your subject, you need to choose an *angle* on the subject. An angle is a specific direction that your research will follow. For example, "nuclear waste" may be too large a subject, but "the hazards of transporting nuclear waste in the western United States" might be a good angle for your research. Likewise, research on raptors is probably too large a subject, but "the restoration of bald eagles along the Mississippi River" might be a manageable project.

By choosing an angle, you will be able to narrow your research subject to a manageable size.

Formulating a Research Question or Hypothesis

Once you have narrowed your subject, you should then formulate a *research question* and *hypothesis*.

The purpose of a research question is to guide your empirical or analytical research. Your research question does not need to be very specific when you begin your research. It simply needs to give your research a direction to follow.

Try to devise a research question that is as specific as possible:

> Why do crows like to gather on our campus during the winter?
>
> What are the effects of violent television on boys between the ages of 10 and 16?
>
> Is solar power a viable energy source for South Dakota?

Your hypothesis is your best guess about an answer to your research question:

> **Hypothesis:** The campus is the best source of available food in the wintertime, because students leave food around. Crows naturally congregate because of the food.

GO TO THE NET

Want to see other workplaces that use mapping? Go to **www.pearsonhighered.com/johnsonweb4/14.3**

> **Hypothesis:** Boys between the ages of 10 and 16 model what they see on violent television, causing them to be more violent than boys who do not watch violent television.

> **Hypothesis:** Solar power is a viable energy source in the summer, but cloudiness in the winter makes it less economical than other forms of renewable energy.

As you move forward with your research, you will probably need to refine or sharpen your original research question and hypothesis. For now, though, ask the question that you would most like to answer. Then, to form your hypothesis, answer this question to the best of your knowledge. Your hypothesis should be your best guess for the moment.

AT WORK

Kristy Raine

REFERENCE LIBRARIAN AND ARCHIVIST, MOUNT MERCY UNIVERSITY

Mount Mercy University is located in Cedar Rapids, Iowa.

How can a reference librarian help me do research in my field?

Reference librarians make the perfect guides for helping you with your research. Reference librarians assist students in formulating questions, considering sources for their inquiries, and finding the relevant materials they need. Some librarians are generalists, knowledgeable in many fields, and often work at colleges and small universities. Large campuses offer subject specialists for students and faculty in particular disciplines. Good reference librarians are familiar with their institution's print and electronic collections and understand how to access the most appropriate sources for a successful result.

Helping students narrow the search to a single question is a first step; librarians refer to this process as the "reference interview." The librarian typically asks about the task (its requirements and expectations) and most importantly, the steps already taken in the search. With those answers in hand, the librarian may know of several viable research paths and might recommend two or more choices, all the while refining options as the reference conversation continues. During this stage, the librarian is internally formulating a research plan, one that meets the student's needs. The reference interview also enables the librarian to assess the student's abilities in conveying an idea, in expressing his or her understanding of the assignment, and in demonstrating his or her knowledge of library use. The conversation between the librarian and student is, perhaps, the most crucial part of library service.

As the interview continues, the librarian discusses search options with the student, and with mutual understanding of the goals, embarks on finding sources. This journey

Having trouble refining your hypothesis? Go to **www.pearsonhighered.com/johnsonweb4/14.4**

may involve demonstrating how to use a periodical database, how to interpret its search results, and how to organize and cite sources for a references list. The student may explore the library to find books, journals, and media of interest. Ultimately, the librarian encourages the student to use the library's resources so that he or she leaves not only with relevant information but also with new skills for approaching future research tasks.

Developing a Research Methodology

With your research question and hypothesis formed, you are ready to start developing your *research methodology.* A research methodology is a step-by-step procedure that you will use to study your subject. As you and your team consider how to do research on your subject, begin thinking about all the different ways you can collect information.

Mapping Out a Methodology

Logical mapping can help. Put the purpose of your research in the middle of your screen or a piece of paper. Ask, "*How* are we going to achieve this purpose?" Then, answer this question by formulating the two to five major steps you will need to take in your research. Each of these major steps can then be broken down into minor steps (Figure 14.3).

Using the map in Figure 14.3, for example, a team of researchers might devise the following methodology for studying their research question:

Methodology for Researching Nuclear Waste Transportation:

- Collect information off the Internet from sources for and against nuclear waste storage and transportation.
- Track down news stories in the print media and collect any journal articles available on nuclear waste transportation.
- Interview experts and survey members of the general public.
- Study the Waste Isolation Pilot Plant (WIPP) in New Mexico to see if transportation to the site has been a problem.

Note that these researchers are planning to collect information from a range of electronic, print, and empirical sources.

Describing Your Methodology

After mapping out your research methodology, begin describing it in outline form (Figure 14.4).

Sometimes, as shown in Figure 14.4, it is also helpful to identify the kinds of information you expect to find in each step. By clearly stating your *expected findings* before you start collecting information, you will know if your research methodology is working the way you expected.

Need help developing a methodology? Go to **www.pearsonhighered.com/johnsonweb4/14.5**

Mapping Out a Methodology

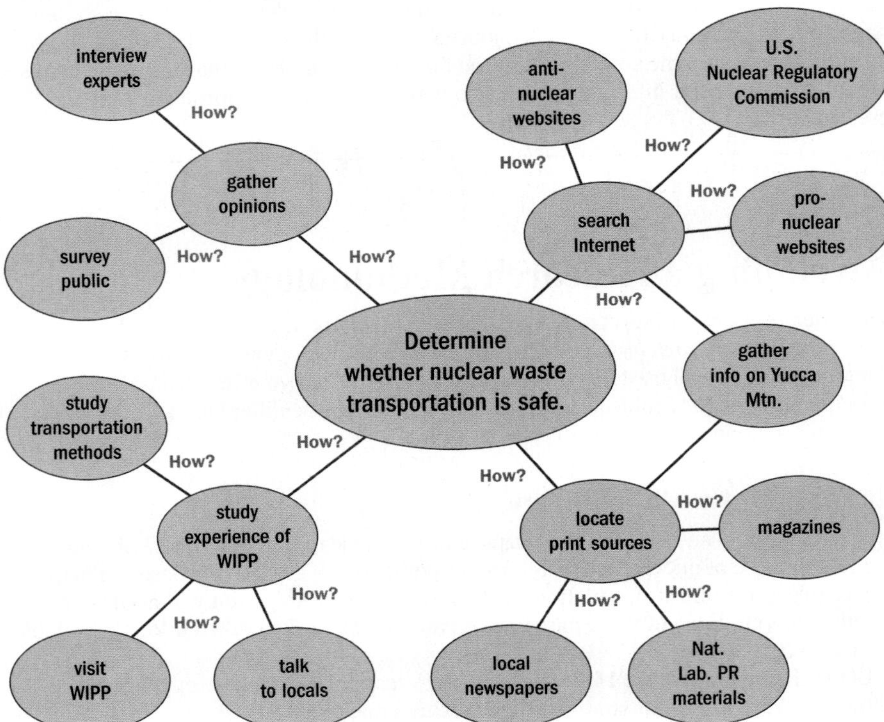

Figure 14.3: Logical mapping can help you sketch out a methodology. Keep asking the *How?* question as you consider the steps needed to complete your project.

At the end of your methodology, add a step called "Analysis of Findings." If you collected data, you will need to do some statistical analysis. If you conducted interviews or tracked down information on the Internet, you will need to spend some time checking and verifying your sources.

Using and Revising Your Methodology

A good methodology is like a treasure map. You and your research team can use it as a guide to uncover answers to questions that intrigue you.

Almost certainly, you will deviate from your methodology while doing your research. Sometimes you will find information that takes you down an unexpected path. Sometimes information you expected to find is not available. In other cases, experiments and surveys return unexpected findings.

When you deviate from your methodology, note these changes in direction. A change in the methodology is not a failure. It is simply a recognition that research is not formulaic and can be unpredictable. Research is a process of discovery. Sometimes the most important discoveries are made when we deviate from the plan.

Outlining a Research Methodology

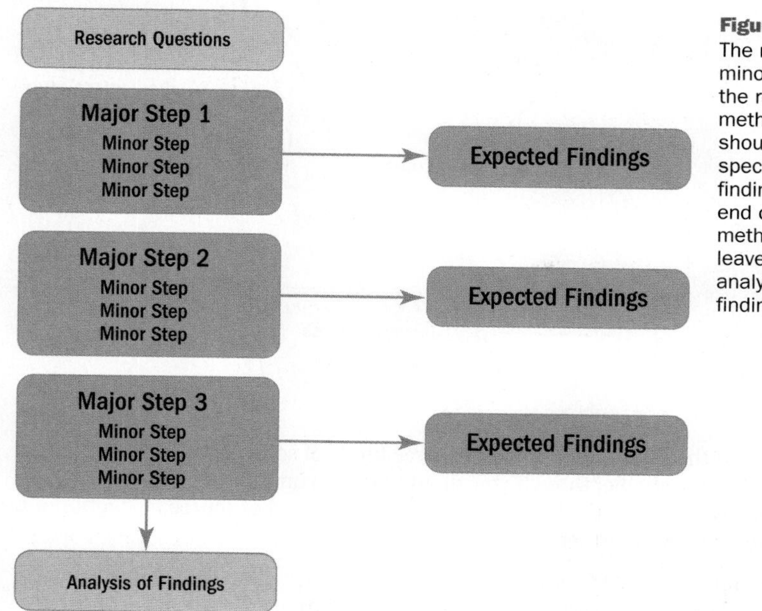

Figure 14.4: The major and minor steps in the research methodology should result in specific kinds of findings. At the end of the methodology, leave time for analyzing your findings.

Triangulating Materials

To ensure that your methodology is reliable, you should draw information from a variety of sources. Specifically, you should always try to *triangulate* your materials by collecting information from electronic, print, and empirical sources (Figure 14.5). Triangulation allows you to compare and contrast sources, thereby helping you determine which information is reliable and which is not. Plus, triangulation gives your readers confidence in your research, because you will have collected information from a variety of sources.

Solid research draws from three kinds of information:

- **Electronic sources:** Websites, DVDs, CD-ROMs, listservs, research databases, television and radio, videos, podcasts, blogs
- **Print sources:** Books, journals, magazines and newspapers, government publications, reference materials, microform/microfiche
- **Empirical sources:** Experiments, surveys, interviews, field observations, ethnographies, case studies

By drawing information from all three kinds of sources, you will be able to verify the facts you find:

- If you find similar facts in all three kinds of sources, you can be reasonably confident that the information is reliable.

Tools for helping you do research electronically are available at **www.pearsonhighered.com/johnsonweb4/14.6**

The Research Triangle

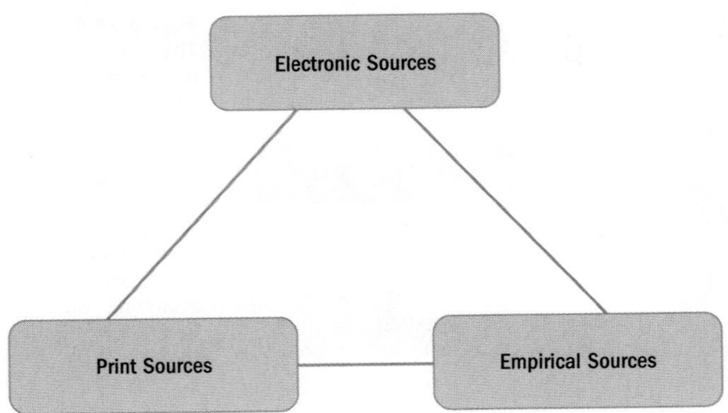

Figure 14.5: In any research project, try to draw information from electronic, print, and empirical sources.

- If you find the information in two of three kinds of sources, the information is probably still reliable, though you should be less confident.
- If, however, you find the information in only one kind of source, it might not be reliable and needs further confirmation.

Keep in mind that "truth" and "facts" are more slippery than we want to admit. A source may claim it is providing the truth, but until you can confirm that source's facts through triangulation, you should always treat any claims skeptically.

Also remember that there are always two sides to every issue. So, do not restrict your research to only one side. If you look only for sources that confirm what you already believe, you will probably not gain a deeper understanding of the subject. After all, even when you absolutely disagree with someone else, his or her argument may give you additional insight into the issue you are researching. Keep an open mind.

Using Electronic Sources

Because electronic sources are so convenient, a good place to start collecting information is through your computer.

Websites—Websites are accessible through browsers like Firefox, Explorer, or Safari. When using search engines like MetaCrawler.com, Google.com, Yahoo.com, and Ask.com, among many others, you can run keyword searches to find information on your subject (Figures 14.6 and 14.7).

CD-ROMs—A compact disc (CD-ROM) can hold a library's worth of text and images. Often available at your local or campus library, CD-ROMs are usually searchable through keywords or subjects. Encyclopedias and databases are also available on CD-ROM.

Listservs—Listservs are ongoing e-mail discussions, usually among specialists in a field. Once you find a listserv on your subject, you can usually subscribe to the discussion. A politely phrased question may return some helpful answers from other subscribers to the listserv.

For links to Internet search engines, go to **www.pearsonhighered.com/johnsonweb4/14.7**

Internet Search Engine

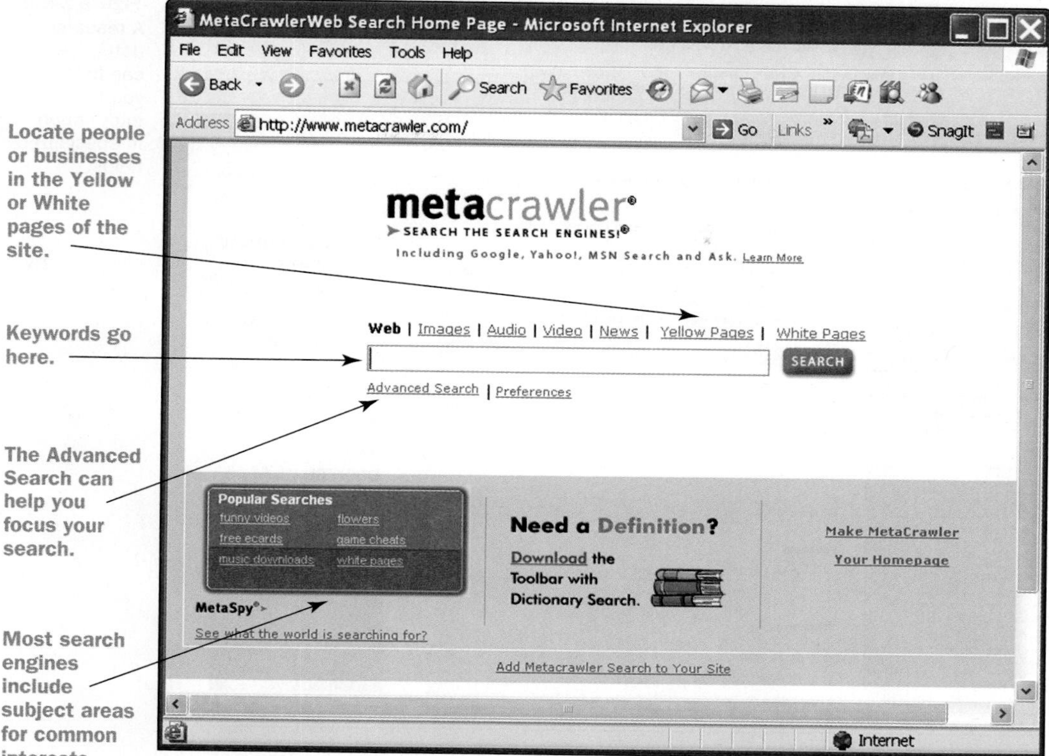

Figure 14.6: MetaCrawler is one of the more useful Internet search engines. By typing in some keywords, you can locate limitless amounts of information.

Source: metacrawler, http://www.metacrawler.com.

Television and radio—You can locate television and radio documentaries or news programs that address your subject. In some cases, copies of these materials will be available at your library on a DVD.

Research databases—If you are looking for scientific and technical articles on your subject, you might first locate a research database that collects materials about your subject (Figure 14.7). Your campus library likely subscribes to a variety of databases that can be searched electronically.

Podcasts—Increasingly, information is being put on websites as podcasts. Podcasts can be played on your computer or any MP3 player (not only on an iPod). They often sound like radio broadcasts. Video podcasts are also becoming more popular.

Videos—Increasingly, documentaries and training videos are available on websites like You Tube, digital videodiscs (DVDs), or videotapes (VHS). Your library or even video rental stores may have these kinds of materials available.

GO TO THE NET

For links to solid electronic sources of information, go to **www.pearsonhighered.com/johnsonweb4/14.8**

Research Databases

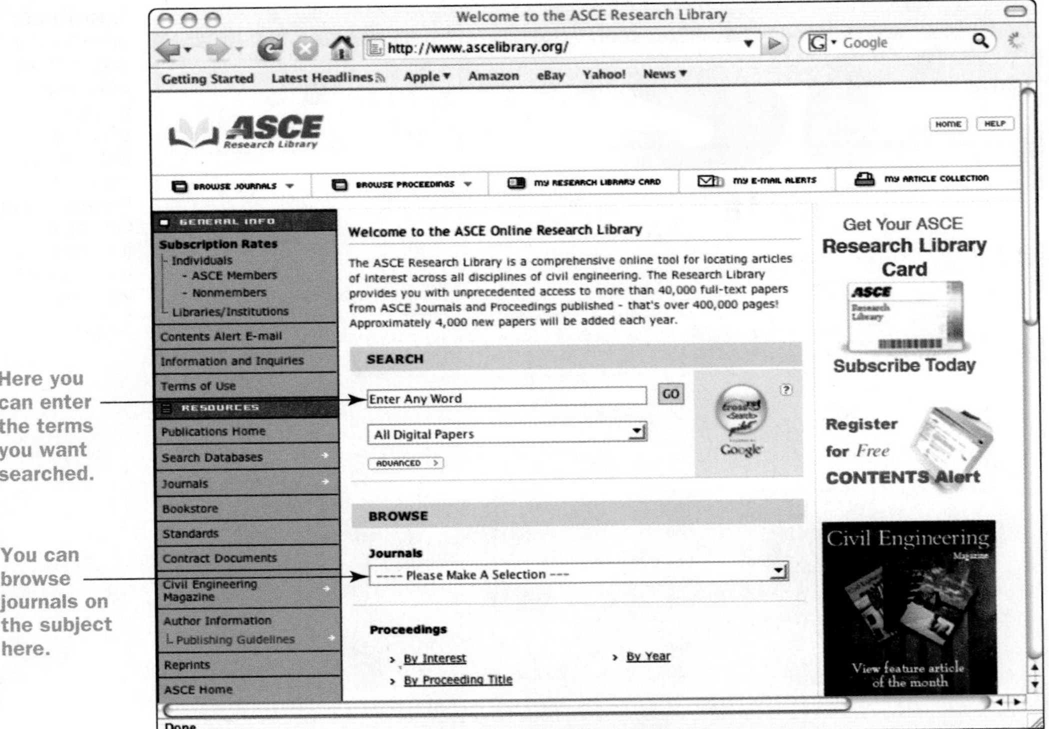

Figure 14.7: A research database can help you target information in specialized fields.

Source: American Society of Civil Engineers, http://www.ascelibrary.org.

Blogs—Blogs are Internet sites where a commentator or group of commentators often "publish" raw information, opinions, and hearsay. Blogs can be a good source for keeping up with the cutting edge of a research area.

Using Print Sources

With easy access to information through the Internet, you may be tempted to forego using the print sources available at your library. Researchers who neglect print sources are making a serious mistake.

Printed documents are still the most abundant and reliable sources of information. In the rush to use electronic sources, many people have forgotten that their nearby library is loaded with books and periodicals on almost any subject. These print sources can usually be located by using your computer to access the library's website (Figure 14.8).

GO TO THE NET Want to know more about blogs? Go to **www.pearsonhighered.com/johnsonweb4/14.9**

A Library's Search Engine

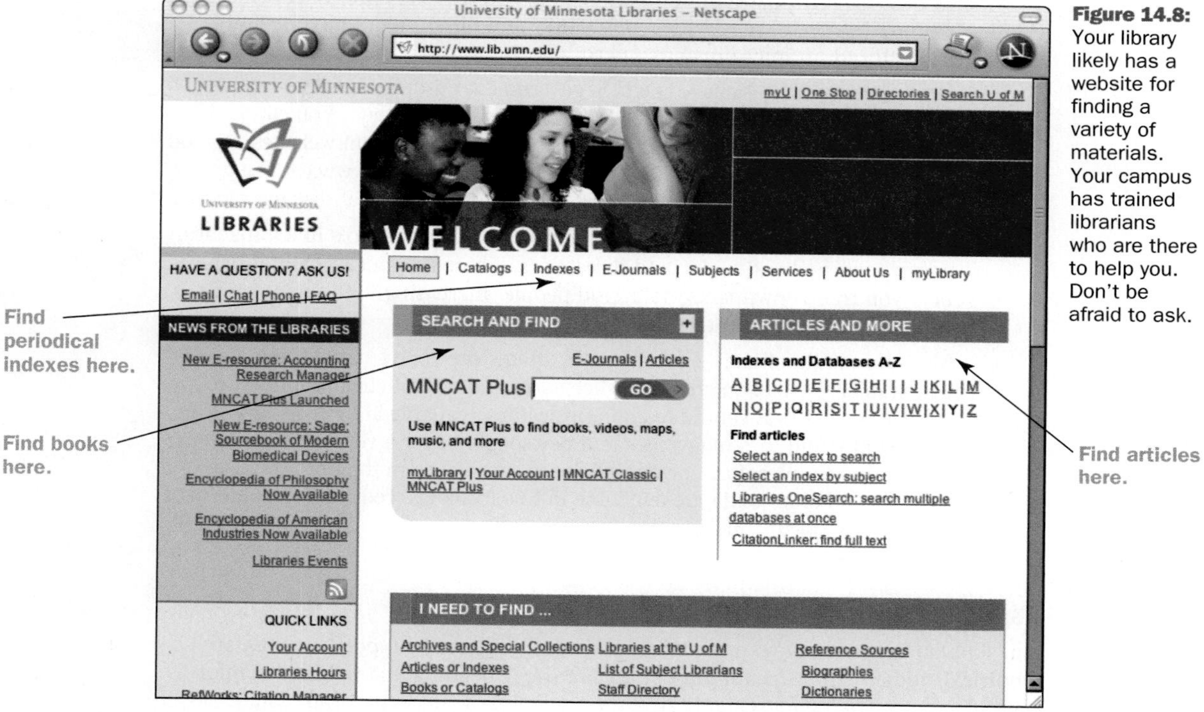

Figure 14.8: Your library likely has a website for finding a variety of materials. Your campus has trained librarians who are there to help you. Don't be afraid to ask.

Source: © 2008 Regents of the University of Minnesota. All rights reserved.

Here are a few of the many kinds of print materials that you can use:

Books—Almost all libraries have electronic cataloging systems that allow you to use author name, subject, title, and keywords to search for books on your subject. Once you have located a book on your subject, look at the books shelved around it to find other useful materials.

Journals—Using a *periodical index* at your library, you can search for journal articles on your subject. Journal articles are usually written by professors and scientists in a research field, so the articles can be rather detailed and hard to understand. Nevertheless, these articles offer some of the most exact research on any subject. Periodical indexes for journals are usually available online at your library's website, or they will be available as printed books in your library's reference area.

Magazines and newspapers—You can also search for magazine and newspaper articles on your subject by using the *Readers' Guide to Periodical Literature* or a newspaper index. The *Readers' Guide* and

GO TO THE NET For access to online news and legal and business information, go to **www.pearsonhighered.com/johnsonweb4/14.10**

newspaper indexes are likely available online at your library's website or in print form. Recent editions of magazines or newspapers might be stored at your library. Older magazines and newspapers have usually been stored on microform or microfiche.

Government publications—The U.S. government produces a surprising number of useful books, reports, maps, and other documents. You can find these documents through your library or through government websites. A good place to start is *The Catalog of U.S. Government Publications* (www.catalog.gpo .gov), which offers a searchable listing of government publications and reports.

Reference materials—Libraries contain many reference tools like almanacs, encyclopedias, handbooks, and directories. These reference materials can help you track down facts, data, and people. Increasingly, these materials can also be found online in searchable formats.

Microform/microfiche—Libraries will often store copies of print materials on microform or microfiche. Microform and microfiche are miniature transparencies that can be read on projectors available at your library. You will usually find that magazines and newspapers over a year old have been transferred to microform or microfiche to save space in the library. Also, delicate and older texts are available in this format to reduce the handling of the original documents.

Using Empirical Sources

You should also generate your own data and observations to support your research. Empirical studies can be *quantitative* or *qualitative*, depending on the kinds of information you are looking for. Quantitative research allows you to generate data that you can analyze statistically to find trends. Qualitative research allows you to observe patterns of behavior that cannot be readily boiled down into numbers.

Experiments—Each research field has its own experimental procedures. A controlled experiment allows you to test a hypothesis by generating data. From that data, you can confirm or dispute the hypothesis. Experiments should be *repeatable*, meaning the results can be replicated by another experimenter.

Field observations—Researchers often carry field notebooks to record their observations of their research subjects. For example, an ornithologist might regularly note the birds she observes in her hikes around a lake. Her notebook would include her descriptions of birds and their activities.

Interviews—You can ask experts to answer questions about your subject. On almost any given college campus, experts are available on just about any subject. Your well-crafted questions can draw out very useful information and quotes. Figure 14.9, for example, shows a classic interview with Bill Gates and Steve Jobs.

Surveys and questionnaires—You can ask a group of people to answer questions about your subject. Their answers can then be scored and ana-

For The Catalog of U.S. Government Publications, go to **www.pearsonhighered.com/johnsonweb4/14.11**

Doing Empirical Research

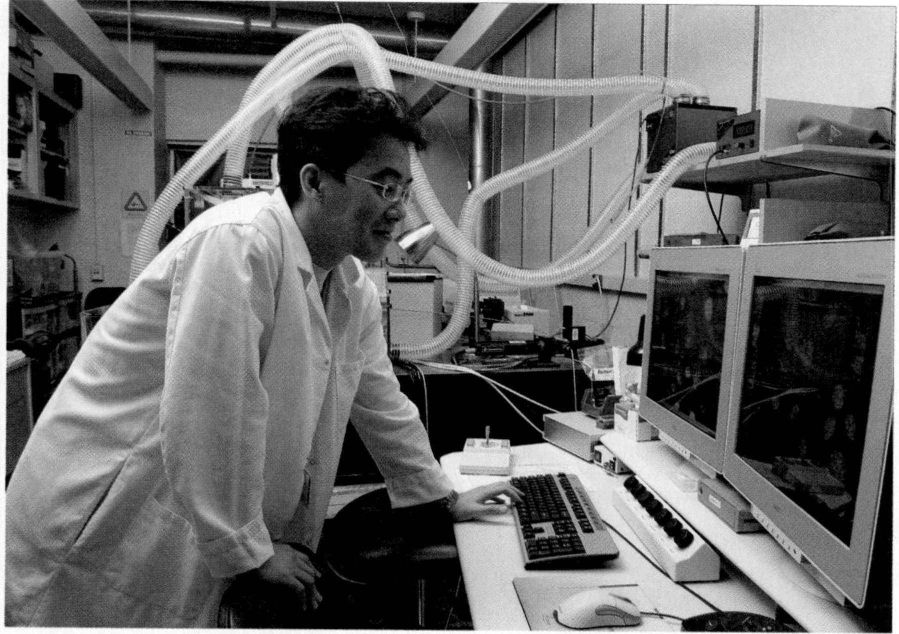

Empirical research requires you to observe your subject directly.

Triangulating Research

AT A GLANCE

Solid research draws from three kinds of information:

- Electronic sources—Internet, DVDs, CD-ROMs, listservs, television and radio, videos, blogs
- Print sources—books, journals, magazines and newspapers, government publications, reference materials, microform/microfiche
- Empirical sources—experiments, surveys, interviews, field observations, ethnographies, case studies

lyzed for trends. Survey questions can be *closed-ended* or *open-ended.* Closed-ended questions ask respondents to choose among preselected answers. Open-ended questions allow respondents to write down their views in their own words. Figure 14.10 shows pages from a survey with both closed-ended and open-ended questions.

Ethnographies—An ethnography is a systematic recording of your observations of a defined group or culture. Anthropologists use ethnographies to identify social or cultural trends and norms.

Case studies—Case studies typically offer in-depth observations of specific people or situations. For example, a case study might describe how a patient reacted to a new treatment regimen that manages diabetes.

When conducting empirical research, you should follow the *scientific method* (Figure 14.11). The concept of a scientific method was first conceived by Francis Bacon, a seventeenth-century English philosopher. Later in the seventeenth century, the London Royal Society, a club of scientists, gave the scientific method the form we recognize now.

For more information on conducting qualitative empirical studies, go to **www.pearsonhighered.com/johnsonweb4/14.12**

Following is a transcript of the interview Kara Swisher and Walt Mossberg conducted with Microsoft Chairman Bill Gates and Apple CEO Steve Jobs at the Wall Street Journal's *D: All Things Digital conference on May 30, 2007.*

Scripted question →

Kara: So let's get started. I wanted to ask, there's been a lot of mano-a-mano/catfight kind of thing in a lot of the blogs and the press and stuff like that, and we wanted to–the first question I was interested in asking is what you think each has contributed to the computer and technology industry, starting with you, Steve, for Bill, and vice versa.

Steve: Well, you know, Bill built the first software company in the industry and I think he built the first software company before anybody really in our industry knew what a software company was, except for these guys. And that was huge. That was really huge. And the business model that they ended up pursuing turned out to be the one that worked really well, you know, for the industry. I think the biggest thing was, Bill was really focused on software before almost anybody else had a clue that it was really the software.

Uses follow-up prompt for more detail. →

Kara: Was important?

Steve: That's what I see. I mean, a lot of other things you could say, but that's the high order bit. And I think building a company's really hard, and it requires your greatest persuasive abilities to hire the best people you can and keep them at your company and keep them working, doing the best work of their lives, hopefully. And Bill's been able to stay with it for all these years.

Scripted question →

Walt: Bill, how about the contribution of Steve and Apple?

Bill: Well, first, I want to clarify: I'm not Fake Steve Jobs. What Steve's done is quite phenomenal, and if you look back to 1977, that Apple II computer, the idea that it would be a mass-market machine, you know, the bet that was made there by Apple uniquely–there were other people with products, but the idea that this could be an incredible empowering phenomenon, Apple pursued that dream. Then one of the most fun things we did was the Macintosh and that was so risky. People may not remember that Apple really bet the company. Lisa hadn't done that well, and some people were saying that general approach wasn't good, but the team that Steve built even within the company to pursue that, even some days it felt a little ahead of its time–I don't know if you remember that Twiggy disk drive and...

Steve: One hundred twenty-eight K.

Kara: Oh, the Twiggy disk drive, yes.

Bill: Steve gave a speech once, which is one of my favorites, where he talked about, in a certain sense, we build the products that we want to use ourselves. And so he's really pursued that with incredible taste and elegance that has had a huge impact on the industry. And his ability to always come around and figure out where that next bet should be has been phenomenal. Apple literally was failing when Steve went back and re-infused the innovation and risk-taking that have been phenomenal. So the industry's benefited immensely from his work. We've both been lucky to be part of it, but I'd say he's contributed as much as anyone.

Steve: We've also both been incredibly lucky to have had great partners that we started the companies with and we've attracted great people. I mean, so everything that's been done at Microsoft and at Apple has been done by just remarkable people, none of which are sitting up here today.

Source: D: All Things Digital, http://d5.allthingsd.com/200705311/d5-gates-jobs-transcript.

Figure 14.9: In this interview transcript, notice how the interviewers are asking scripted questions and then following up with unscripted questions.

For websites that discuss empirical research methods, go to **www.pearsonhighered.com/johnsonweb4/14.13**

Pages from a Questionnaire on Campus Safety

Introduction explains how to complete the survey.

These closed-ended questions yield numerical data.

Open-ended questions allow participants to elaborate on their answers.

Campus Survey 75

Campus Perception Survey

The following questions are about how safe you feel or don't feel on campus. For each situation please tell us if you feel: very safe, reasonably safe, neither safe nor unsafe, somewhat unsafe, very unsafe, or if this situation does not apply to you. (Please circle the number that best represents your answer or **NA** if the situation does not apply to you.)

How safe do you feel...

	Very Unsafe	Somewhat Unsafe	Neither Safe Nor Unsafe	Reasonably Safe	Very Safe	
walking alone on campus during daylight hours?	1	2	3	4	5	NA
waiting alone on campus for public transportation during daylight hours?	1	2	3	4	5	NA
walking alone in parking lots or garages on campus during daylight hours?	1	2	3	4	5	NA
walking alone on campus after dark?	1	2	3	4	5	NA
waiting alone on campus for public transportation after dark?	1	2	3	4	5	NA
walking alone in parking lots or garages on campus after dark?	1	2	3	4	5	NA
working in the library stacks late at night?	1	2	3	4	5	NA
while alone in classrooms?	1	2	3	4	5	NA
Student Activity Center during the day?	1	2	3	4	5	NA
Student Activity Center at night?	1	2	3	4	5	NA

Are there any specific areas on campus where you do not feel safe? Please specify which areas, which campus, and when; for example, evenings only or any time. ____________________

____

Do you have any special needs related to safety on campus? ____________________

Have you ever used services related to safety issues, sexual harassment, or sexual assault that are provided on campus by the following?

How satisfied were you with help from this source?

			Very Dissatisfied	Somewhat Dissatisfied	Neither Satisfied Nor Dissatisfied	Somewhat Satisfied	Very Satisfied
Campus Police	YES	NO	1	2	3	4	5
Women's Center	YES	NO	1	2	3	4	5
Campus ministry	YES	NO	1	2	3	4	5
Campus counseling (Belknap)	YES	NO	1	2	3	4	5
Student Health Services (Belknap)	YES	NO	1	2	3	4	5
Campus counseling (Health Science)	YES	NO	1	2	3	4	5
Student Health Services (Health Science)	YES	NO	1	2	3	4	5
Psychological Services Ctr (Psychology Clinic)	YES	NO	1	2	3	4	5
Affirmative Action Office	YES	NO	1	2	3	4	5
Security escort services after dark	YES	NO	1	2	3	4	5
Residence Hall staff	YES	NO	1	2	3	4	5
Office of Student Life	YES	NO	1	2	3	4	5
Disability Resource Center	YES	NO	1	2	3	4	5
Access Center	YES	NO	1	2	3	4	5
Faculty member	YES	NO	1	2	3	4	5
Other ________ (Please specify.)	YES	NO	1	2	3	4	5

Source: *Bledsoe & Sar, 2001.*

Figure 14.10: A survey is a good way to generate data for your research. In this example, both closed-ended and open-ended questions are being used to solicit information.

(continued)

Figure 14.10: *(continued)*

The survey uses statements to measure participants' reactions to specific situations or opinions.

Campus Survey 78

The following are some beliefs that may be held about the role of women and men in today's society. There are no right or wrong answers. (Please circle the response that best describes your opinion.)

	Strongly Disagree		Somewhat Agree		Strongly Agree
A man's got to show the woman who's boss right from the start.	1	2	3	4	5
Women are usually sweet until they've caught a man, but then they let their true self show.	1	2	3	4	5
In a dating relationship a woman is largely out to take advantage of a man.	1	2	3	4	5
Men are out for only one thing.	1	2	3	4	5
A lot of women seem to get pleasure from putting a man down.	1	2	3	4	5
A woman who goes to the home or apartment of a man on their first date implies that she is willing to have sex.	1	2	3	4	5
Any female can get raped.	1	2	3	4	5
Any healthy woman can successfully resist a rapist if she really wants to.	1	2	3	4	5
Many women have an unconscious wish to be raped, and may then unconsciously set up a situation in which they are likely to be attacked.	1	2	3	4	5
If a woman gets drunk at a party and has intercourse with a man she's just met there, she should be considered "fair game" to other males at the party who also want to have sex with her whether she wants to or not.	1	2	3	4	5

What **percentage of women** who report a rape would you say are lying because they are angry and want to get back at the man they accuse? ______%

What **percentage of reported rapes** would you guess were merely invented by women who discovered they were pregnant and wanted to protect their own reputation? ______%

Did you attend any type of student orientation conducted by University of Louisville during Summer 2000 or at the beginning of this term? (Please circle your answer). **YES NO**

If your answer was "No, I did not attend orientation", please continue on next page.

	None		Some		A Lot
At the orientation you attended, how much information about violence against women issues did you receive? (Please circle your answer).	1	2	3	4	5

Using the Scientific Method

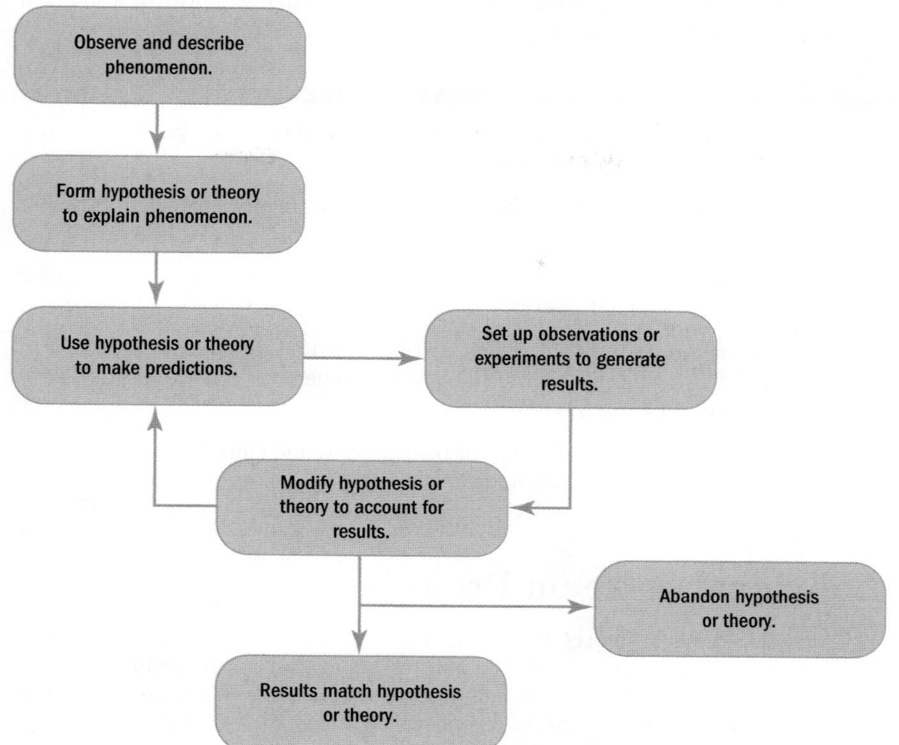

Figure 14.11: The "scientific method" is a systematic way to study a phenomenon by forming a hypothesis and developing empirical methods to test it.

The Scientific Method:

1. Observe and describe a phenomenon.
2. Formulate a hypothesis or theory that explains the phenomenon.
3. Use the hypothesis or theory to make predictions.
4. Use observations and experiments to generate results that confirm or deny your predictions.
5. Modify the hypothesis or theory to account for your results.
6. Repeat steps 3 through 5 until results match your hypothesis or theory OR you abandon the hypothesis or theory.

The scientific method can be used with quantitative or qualitative forms of empirical research. Whether you are doing an experiment in a laboratory or making field

To learn more about the scientific method, go to **www.pearsonhighered.com/johnsonweb4/14.14**

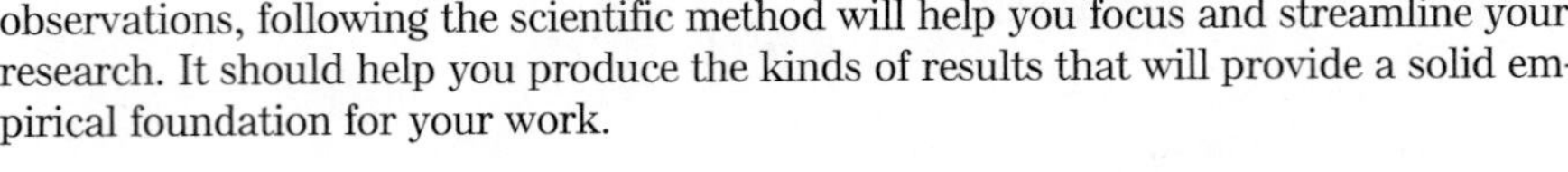

observations, following the scientific method will help you focus and streamline your research. It should help you produce the kinds of results that will provide a solid empirical foundation for your work.

CHAPTER REVIEW

- Research today involves collecting information from diverse sources that are available in many media, including the Internet.
- Effectively managing existing information is often as important as creating new information.
- Logical mapping can be used to define a subject and highlight places where information needs to be found.
- A research methodology is a planned, step-by-step procedure that you will use to study the subject. Your research methodology can be revised as needed as your research moves forward.
- Triangulation is a process of using electronic, print, and empirical sources to obtain and evaluate your findings and conclusions.

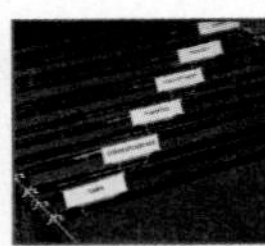

EXERCISES AND PROJECTS

Individual or Team Projects

1. Think of a technical subject that interests you. Then, collect information from electronic and print sources. Write a progress report to your instructor in which you highlight themes in the materials you've found. Discuss any gaps in the information that you might be able to fill with more searching or empirical study. Some possible topics might include the following:

   Wildlife on campus
   Surveillance in America
   Hybrid motor cars
   The problems with running red lights on or near campus
   Safety on campus at night
   The effects of acid rain in Canada
   Migration of humpback whales

2. On the Internet, find information on a subject that you think is junk science or is influenced by junk science. Pay close attention to the reputations of the researchers and their results. Can you find any information to back up their claims? Pay special attention to where they receive their funding for the research. When you are finished searching the web, make a report to your class on your findings. Show your audience how junk science influences the debate on your subject.

   Here are a few possible topics:

   Evolution versus creation science
   Genetically engineered foods
   Managing forests to prevent fires

Mobile phones and cancer
Experimentation on animals
Herbicides and insecticides
Diets and dietary supplements
Global warming
Smoking and secondhand smoke
Air and water pollution
Transporting nuclear waste
Welfare abuse

3. Survey your class on a campus issue that interests you. Write five questions and let your classmates select among answers like "strongly agree," "agree," "disagree," and "strongly disagree." Then, tabulate the results of the survey. Write a memo to your instructor in which you discuss the trends revealed by your findings. In your memo, also point out places where your methodology might be challenged by someone who doubts your findings. Discuss how you might strengthen your survey if you wanted to do a larger study on this subject.

Collaborative Project

With a group, develop a methodology for studying substance abuse (alcohol abuse or abuse of prescription drugs or illegal drugs) on campus. First, use logical mapping to identify what you already know or believe about substance abuse on your campus. Second, formulate a research question that your research will answer. Third, use logical mapping to sketch out a methodology that would help you generate results to answer your research question.

Your methodology should use triangulation to gather information from a broad range of sources. In other words, you should plan to gather information from electronic, print, and empirical sources.

Finally, write up your methodology, showing the step-by-step procedures you will use to study substance abuse on campus. Your methodology should be written in such a way that others can duplicate it. It should also clearly identify the kinds of results you expect your research to generate.

Give your methodology to your instructor. At this point, your instructor may ask you to continue your research, following your methodology. As you do your research, note places where you changed your methodology or found information you did not expect.

For additional technical writing resources, including interactive sample documents, document design tutorials and guidelines, and more, go to **www.mytechcommlab.com**.

GO TO THE NET

For websites that discuss substance abuse on college campuses, go to **www.pearsonhighered.com/johnsonweb4/14.15**

CASE STUDY

The Life of a Dilemma

Jen Krannert was a third-year student in the biomedical engineering program at North Carolina State University. She was looking forward to doing a co-op semester in the spring, so she could gain some valuable experience (and earn a little money).

She applied to several co-op programs, including programs at Baxter, Biogen, and Boston Scientific. The one that caught her interest most, though, was at GenBenefits, a small biotech laboratory in California. At the interview, GenBenefits' recruiters talked with her about their work on embryonic stem cell research and how they were on the cusp of some major breakthroughs. She was very excited about the possibility of being part of that kind of research.

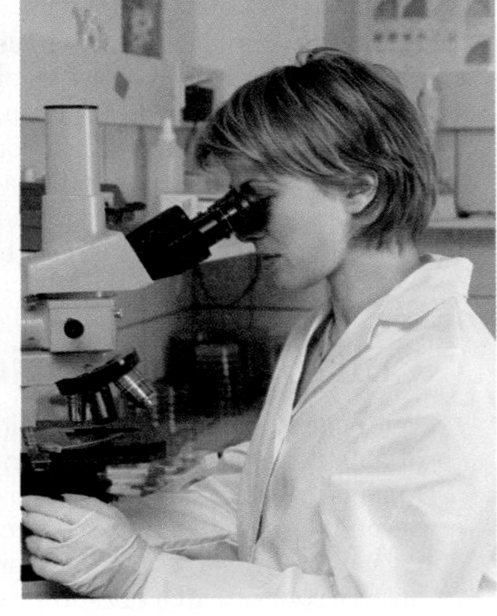

Two weeks after the interview, Allen Marshall, GenBenefits' Vice President of Research, called Jen personally to offer her the co-op position. She accepted right away and spent the next hour e-mailing her other co-op opportunities to tell them she had accepted a position. She also called her parents, who were thrilled for her. They were very excited about her being part of this cutting-edge research.

The next day, she called her best friend Alice Cravitz, who was a student at Duke University. At first, Alice was enthusiastic, but when Jen told her that she would be working with embryonic stem cells, Alice grew quiet.

"What's wrong?" Jen asked.

Alice said, "I just think doing research on embryonic stem cells is unethical. Those are human lives you will be messing around with and ultimately destroying."

Jen became a little defensive. "First, they are embryos, not people. Second, these embryos are the leftover products of fertilization clinics. They will never be implanted and will probably be destroyed anyway. Third, we could save many, many lives with this research."

Sensing Jen's defensiveness, Alice changed the subject. After Jen hung up the phone, she wasn't sure how she felt about her co-op now. She realized she wasn't sure how she felt about embryonic stem cell research, so she didn't know if she believed it was ethical or not. The co-op was a great opportunity. If she turned it down, she would likely not find another for the spring. Plus, she really believed that this research could lead to some incredible medical breakthroughs.

If you were Jen, how might you use the research methods described in this chapter to sort out this ethical dilemma?

Want to learn more about embryonic stem cell research? Go to **www.pearsonhighered.com/johnsonweb4/14.16**

CHAPTER

15

Using Sources and Managing Information

Taking Useful Notes *408*

Documenting Sources *413*

Help: Using a Citation Manager *416*

Appraising Your Information *417*

Avoiding Plagiarism *420*

Chapter Review *421*

Exercises and Projects *422*

Case Study: The Patchwriter *423*

In this chapter, you will learn:

- How to take helpful notes from your sources.
- How to quote, paraphrase, and summarize information.
- To document sources using APA and other bibliographic styles.
- Methods for determining if a source is reliable or too biased.
- How to avoid plagiarism and using the ideas of others without giving them proper credit.

The previous chapter showed you how to do effective research in the technical workplace. Today, networked computers make gathering information easier than ever. You can quickly access seemingly unlimited amounts of information on almost any subject through the Internet, searchable databases, and your library's website. Meanwhile, contacting people for interviews and surveys is also becoming simpler.

The problem with having so much access to information is keeping track of it all. When a basic search turns up hundreds of useful hits on your subject, your challenge isn't finding enough information. Your challenge is managing that information in ways that allow you to maximize your use of it. Good research, after all, isn't only about uncovering the truth about a subject. It is also about showing your readers how you explored your subject and gathered your information. They will be much more confident in your research if you can show them exactly how you arrived at your conclusions.

In this chapter, you will learn some strategies for managing the large amounts of information that is generated by your research.

Taking Useful Notes

On almost any subject, you are going to find a wealth of information. At this point, you need to start thinking like an information manager. After all, only some of the information that you collected will be important to your readers. Most of the information you find will be unnecessary for your readers to take action or make a decision (Figure 15.1).

Need-to-Know Versus Want-to-Tell Information

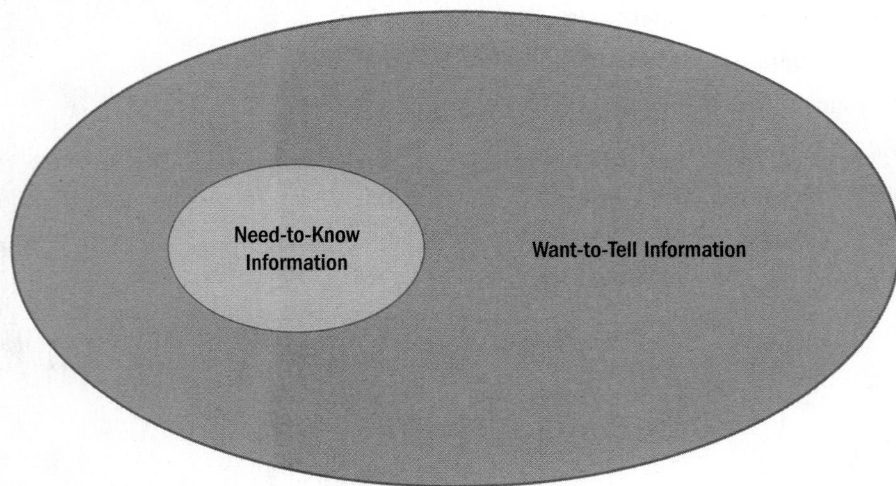

Figure 15.1: While researching, you will find much more information than your readers need. Give them only the information they need to make a decision.

GO TO THE NET: For some great ideas on note taking, go to **www.pearsonhighered.com/johnsonweb4/15.1**

Managing Information

As you decide what to include in the document you are writing, you need to distinguish between *need-to-know* information and *want-to-tell* information.

- *Need-to-know information* includes material that your readers require to take action or make a decision.
- *Want-to-tell information* includes material that you would like to tell your readers but that is not necessary for them to take action or make a decision.

After you have gone through all the effort to collect information, you will want to tell the readers about everything you found. But your readers don't need (or want) all that information. They want only the information they need to take action or make an informed decision. Any extra want-to-tell information will just cloud their ability to understand your document.

Careful Note Taking

Reliable note taking is essential when you do research. If you are organized when you take notes, you will find the information you collected easy to use in the document you are writing.

Note-organizing software and database programs can help you keep track of the information you find. Many researchers write their notes exclusively on laptops or their mobile phones. A pen and pad of paper is also still a good way to keep track of information.

What is most important, though, is to have a workable system for taking notes. Here are some note-taking strategies you might consider using:

RECORD EACH SOURCE SEPARATELY Make sure you clearly identify the author, title of the work, and the place where you found the information (Figure 15.2). For information off the Internet, write down the webpage address (URL) and the date and time you found the information. For a print document, write down where the information was published and who published it. Also, record the library call number of the document.

For large research projects, you might consider making a separate word-processing file for each of your authors or sources, like the one shown in Figure 15.2. That way, you can more easily keep your notes organized.

TAKE DOWN QUOTATIONS When taking down a quotation from a source, be sure to copy the exact wording of the author. If you are taking a quote from a website, you might avoid errors by using the Copy and Paste functions of your computer to copy the statement directly from your source into your notes.

In your notes, you should put quotation marks around any material you copied word for word from a source.

> According to Louis Pakiser and Kaye Shedlock, scientists for the Earthquake Hazards Program at the U.S. Geological Survey, "the assumption of random occurrence with time may not be true." (1997, para. 3)

Keeping Notes on Your Computer

Bibliographic information on the source

Summary of source

Direct quote from source

Commentary from the researcher (you)

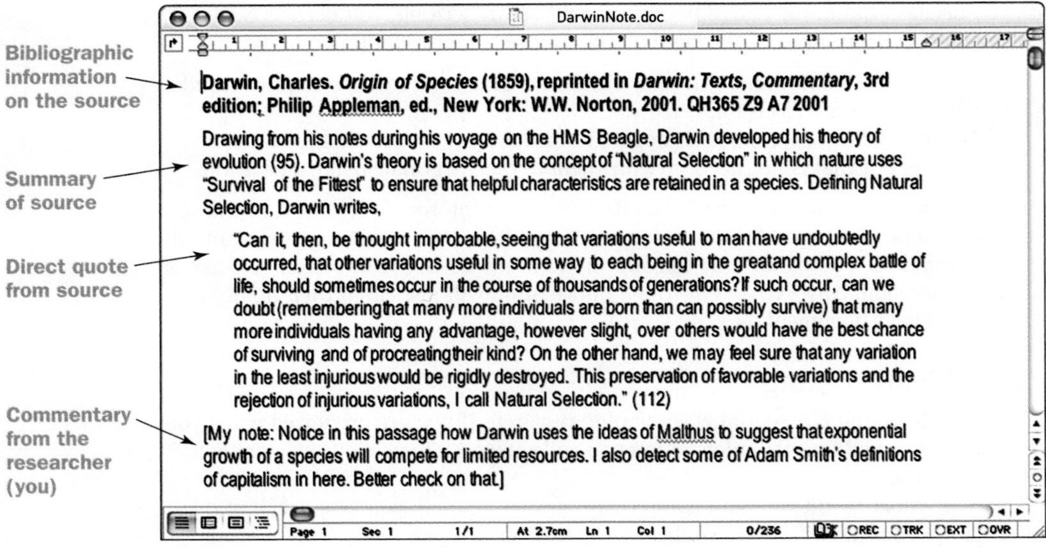

DarwinNote.doc
Darwin, Charles. Origin of Species (1859), reprinted in Darwin: Texts, Commentary, 3rd edition; Philip Appleman, ed., New York: W.W. Norton, 2001. QH365 Z9 A7 2001
Drawing from his notes during his voyage on the HMS Beagle, Darwin developed his theory of evolution (95). Darwin's theory is based on the concept of "Natural Selection" in which nature uses "Survival of the Fittest" to ensure that helpful characteristics are retained in a species. Defining Natural Selection, Darwin writes,
"Can it, then, be thought improbable, seeing that variations useful to man have undoubtedly occurred, that other variations useful in some way to each being in the great and complex battle of life, should sometimes occur in the course of thousands of generations? If such occur, can we doubt (remembering that many more individuals are born than can possibly survive) that many more individuals having any advantage, however slight, over others would have the best chance of surviving and of procreating their kind? On the other hand, we may feel sure that any variation in the least injurious would be rigidly destroyed. This preservation of favorable variations and the rejection of injurious variations, I call Natural Selection." (112)
[My note: Notice in this passage how Darwin uses the ideas of Malthus to suggest that exponential growth of a species will compete for limited resources. I also detect some of Adam Smith's definitions of capitalism in here. Better check on that.]
Page 1
Sec 1
1/1
At 2.7cm
Ln 1
Col 1
0/236
REC
TRK
EXT
OVR

Figure 15.2: Most notes include a combination of summaries, paraphrases, direct quotes, and personal comments.

If the quoted material runs more than three lines in your text, you should set off the material by indenting it in the text.

> Louis Pakiser and Kaye Shedlock, scientists for the Earthquake Hazards Program at the U.S. Geological Survey, make the following point:
>
> > When plate movements build the strain in rocks to a critical level, like pulling a rubber band too tight, the rocks will suddenly break and slip to a new position. Scientists measure how much strain accumulates along a fault segment each year, how much time has passed since the last earthquake along the segment, and how much strain was released in the last earthquake. (1997, para. 4)
>
> If we apply this rubber band analogy to the earthquake risk here in California . . .

Link

For more information on citing sources, go to Appendix C, which begins on page A-23.

When you are quoting a source, you also need to include an in-text citation at the end of the quote. In the two examples above, the in-text citation is the information in the parentheses.

Overall, you should use direct quotes sparingly in your technical writing. You might be tempted to use several quotes from a source, because the authors "said it right." If you use too many quotes, though, your writing will sound fragmented and patchy, because the quotes disrupt the flow of your text.

For links to note-taking software, go to **www.pearsonhighered.com/johnsonweb4/15.2**

PARAPHRASE IDEAS A better way to incorporate someone else's ideas into your writing is to paraphrase them. When paraphrasing, you are presenting another person's ideas in your own words. You still need to give the original author credit for the ideas, but you do not need to use quotation marks around the text. To paraphrase something, you should:

- reorganize the information to highlight important points.
- use plain language, replacing jargon and technical terms with simpler words.
- include an in-text citation.

In the following example, a quote from an original document is paraphrased:

Original Quote

"But in many places, the assumption of random occurrence with time may not be true, because when strain is released along one part of the fault system, it may actually increase on another part. Four magnitude 6.8 or larger earthquakes and many magnitude 6–6.5 shocks occurred in the San Francisco Bay region during the 75 years between 1836 and 1911. For the next 68 years (until 1979), no earthquakes of magnitude 6 or larger occurred in the region. Beginning with a magnitude 6.0 shock in 1979, the earthquake activity in the region increased dramatically; between 1979 and 1989, there were four magnitude 6 or greater earthquakes, including the magnitude 15.1 Loma Prieta earthquake. This clustering of earthquakes leads scientists to estimate that the probability of a magnitude 6.8 or larger earthquake occurring during the next 30 years in the San Francisco Bay region is about 67 percent (twice as likely as not)."

Effective Paraphrase

In-text citation

Pakiser and Shedlock (1997) report that large earthquakes are mostly predictable, because an earthquake in one place usually increases the likelihood of an earthquake somewhere nearby. They point out that the San Francisco area—known for earthquakes—has experienced long periods of minor earthquake activity (most notably from 1911 to 1978, when no earthquakes over magnitude 6 occurred). At other times in San Francisco, major earthquakes have happened with more frequency, because large earthquakes tend to trigger other large earthquakes in the area.

Simple language is used.

Some of the more technical details have been removed to enhance understanding.

Improper Paraphrase

Pakiser and Shedlock (1997) report the assumption of random occurrence of earthquakes may not be accurate. Earthquakes along one part of a fault system may increase the frequency of earthquakes in another part. For example, the San Francisco Bay region experienced many large earthquakes between 1836 and 1911. For the next six decades until 1979, only smaller earthquakes (below magnitude 6) occurred in the area. Then, there was a large rise in earthquakes between 1979 and 1989. Scientists estimate that the probability of an earthquake of magnitude 6.8 or larger is 67 percent in the next 30 years in the Bay area.

Much of the original wording is retained.

Language is still overly technical for the readers.

Link

For more information on plagiarism, see page 420.

The "effective" paraphrase shown here uses the ideas of the original quote, while reordering information to highlight important points and simplifying the language. The "improper" paraphrase on the previous page duplicates too much of the wording from the original source and does not effectively reorder information to highlight important points. In fact, this improper paraphrase is so close to the original, it could be considered plagiarism.

In many ways, paraphrasing is superior to using direct quotes. A paraphrase allows you to simplify the language of a technical document, making the information easier for readers to understand. Also, you can better blend the paraphrased information into your writing because you are using your writing style, not the style of the source.

Warning: Make sure you are paraphrasing sources properly. Do not use the author's original words and phrases. Otherwise, when you draft your document, you may forget that you copied some of the wording from the original text. These duplications may leave you vulnerable to charges of plagiarism or copyright violation.

SUMMARIZE SOURCES When summarizing, your goal is to condense the ideas from your source into a brief passage. Summaries usually strip out many of the examples, details, data, and reasoning from the original text, leaving only the essential information that readers need to know. Like a paraphrase, summaries should be written in your own words. When you are summarizing a source for your notes:

- Read the source carefully to gain an overall understanding.
- Highlight or underline the main point and other key points.
- Condense key points into lists, where appropriate.
- Organize information from most important to least important.
- Use plain language to replace any technical terms or jargon in the original.
- Use in-text citations to identify important ideas from the source.

To see an example of summarizing, consider the passage about predicting earthquakes shown in Figure 15.3. When summarizing this text, you would first need to identify the main point and key points in the text. The main point is that scientists are increasingly able to estimate the probability of an earthquake in a specific area in the near future.

Now, locate the other key points in the text, of which there are three: (1) the frequency of earthquakes in the past helps scientists predict them in the future; (2) earthquakes are not random events, and they tend to occur in clusters; and (3) measurements of the strain on the earth can help scientists measure the probability of a future earthquake.

In the summary shown in Figure 15.4, the details in the original text have been stripped away, leaving only a condensed version that highlights the main point and a few other key issues. As shown here, the summary uses the writer's own words, not the words from the original source.

WRITE COMMENTARY In your notes, you might offer your own commentary to help interpret your sources. Your commentary might help you remember why you collected the information and how you thought it could be used. To avoid plagiarism, it is important to visually distinguish your commentary from summaries,

For more tips on paraphrasing and summarizing, go to **www.pearsonhighered.com/johnsonweb4/15.3**

Original Text to Be Summarized

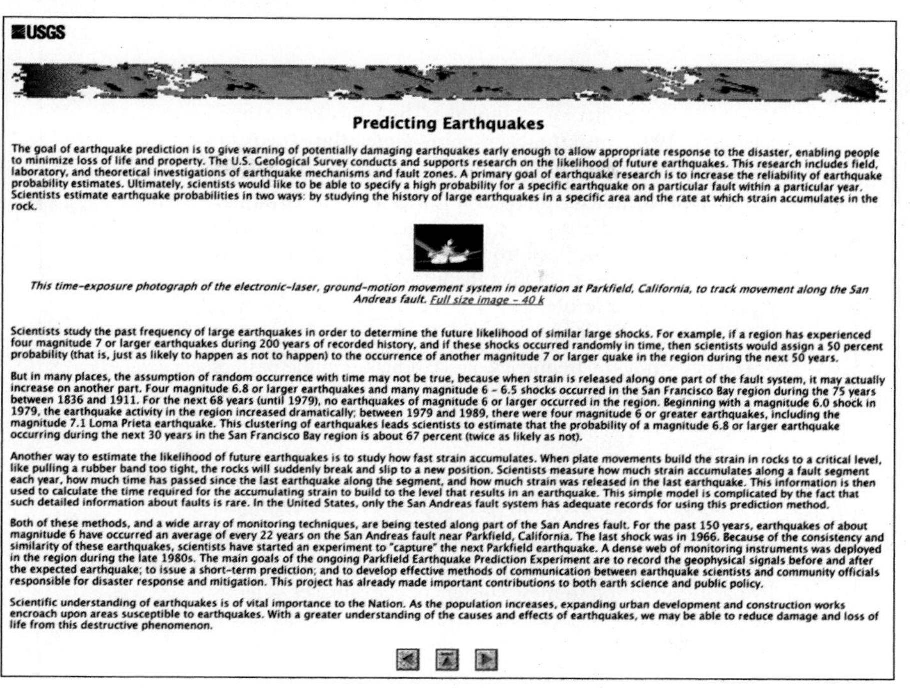

USGS

Predicting Earthquakes

The goal of earthquake prediction is to give warning of potentially damaging earthquakes early enough to allow appropriate response to the disaster, enabling people to minimize loss of life and property. The U.S. Geological Survey conducts and supports research on the likelihood of future earthquakes. This research includes field, laboratory, and theoretical investigations of earthquake mechanisms and fault zones. A primary goal of earthquake research is to increase the reliability of earthquake probability estimates. Ultimately, scientists would like to be able to specify a high probability for a specific earthquake on a particular fault within a particular year. Scientists estimate earthquake probabilities in two ways: by studying the history of large earthquakes in a specific area and the rate at which strain accumulates in the rock.

This time-exposure photograph of the electronic-laser, ground-motion movement system in operation at Parkfield, California, to track movement along the San Andreas fault. Full size image - 40 k

Scientists study the past frequency of large earthquakes in order to determine the future likelihood of similar large shocks. For example, if a region has experienced four magnitude 7 or larger earthquakes during 200 years of recorded history, and if these shocks occurred randomly in time, then scientists would assign a 50 percent probability (that is, just as likely to happen as not to happen) to the occurrence of another magnitude 7 or larger quake in the region during the next 50 years.

But in many places, the assumption of random occurrence with time may not be true, because when strain is released along one part of the fault system, it may actually increase on another part. Four magnitude 6.8 or larger earthquakes and many magnitude 6 - 6.5 shocks occurred in the San Francisco Bay region during the 75 years between 1836 and 1911. For the next 68 years (until 1979), no earthquakes of magnitude 6 or larger occurred in the region. Beginning with a magnitude 6.0 shock in 1979, the earthquake activity in the region increased dramatically; between 1979 and 1989, there were four magnitude 6 or greater earthquakes, including the magnitude 7.1 Loma Prieta earthquake. This clustering of earthquakes leads scientists to estimate that the probability of a magnitude 6.8 or larger earthquake occurring during the next 30 years in the San Francisco Bay region is about 67 percent (twice as likely as not).

Another way to estimate the likelihood of future earthquakes is to study how fast strain accumulates. When plate movements build the strain in rocks to a critical level, like pulling a rubber band too tight, the rocks will suddenly break and slip to a new position. Scientists measure how much strain accumulates along a fault segment each year, how much time has passed since the last earthquake along the segment, and how much strain was released in the last earthquake. This information is then used to calculate the time required for the accumulating strain to build to the level that results in an earthquake. This simple model is complicated by the fact that such detailed information about faults is rare. In the United States, only the San Andreas fault system has adequate records for using this prediction method.

Both of these methods, and a wide array of monitoring techniques, are being tested along part of the San Andres fault. For the past 150 years, earthquakes of about magnitude 6 have occurred an average of every 22 years on the San Andreas fault near Parkfield, California. The last shock was in 1966. Because of the consistency and similarity of these earthquakes, scientists have started an experiment to "capture" the next Parkfield earthquake. A dense web of monitoring instruments was deployed in the region during the late 1980s. The main goals of the ongoing Parkfield Earthquake Prediction Experiment are to record the geophysical signals before and after the expected earthquake; to issue a short-term prediction; and to develop effective methods of communication between earthquake scientists and community officials responsible for disaster response and mitigation. This project has already made important contributions to both earth science and public policy.

Scientific understanding of earthquakes is of vital importance to the Nation. As the population increases, expanding urban development and construction works encroach upon areas susceptible to earthquakes. With a greater understanding of the causes and effects of earthquakes, we may be able to reduce damage and loss of life from this destructive phenomenon.

Figure 15.3: The original text contains many details that can be condensed into a summary.

Source: U.S. Geological Survey, http://pubs.usgs.gov/gip/earthg1/predict.html.

paraphrases, and quotations drawn from other sources. You might put brackets around your comments or use color, italics, or bold type to set them off from your other notes.

Documenting Sources

As you draft your text, you will need to *document* your sources. Documentation involves (1) naming each source with an *in-text citation* and (2) recording your sources in the *References* list at the end of the document. Documenting your sources offers the advantages of:

- supporting your claims by referring to the research of others.
- helping build your credibility with readers by showing them the support for your ideas.
- reinforcing the thoroughness of your research methodology.
- allowing your readers to explore your sources for more information.

Summary of Original Text

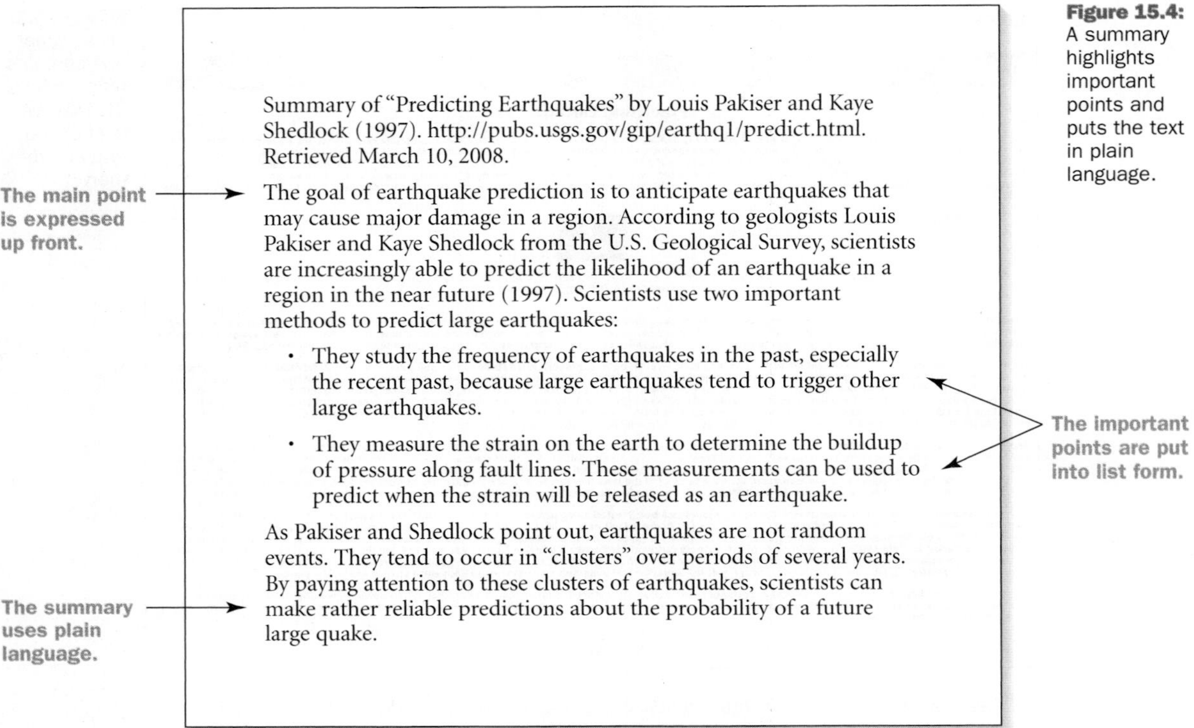

Summary of "Predicting Earthquakes" by Louis Pakiser and Kaye Shedlock (1997). http://pubs.usgs.gov/gip/earthq1/predict.html. Retrieved March 10, 2008.

The goal of earthquake prediction is to anticipate earthquakes that may cause major damage in a region. According to geologists Louis Pakiser and Kaye Shedlock from the U.S. Geological Survey, scientists are increasingly able to predict the likelihood of an earthquake in a region in the near future (1997). Scientists use two important methods to predict large earthquakes:

- They study the frequency of earthquakes in the past, especially the recent past, because large earthquakes tend to trigger other large earthquakes.
- They measure the strain on the earth to determine the buildup of pressure along fault lines. These measurements can be used to predict when the strain will be released as an earthquake.

As Pakiser and Shedlock point out, earthquakes are not random events. They tend to occur in "clusters" over periods of several years. By paying attention to these clusters of earthquakes, scientists can make rather reliable predictions about the probability of a future large quake.

The main point is expressed up front.
The important points are put into list form.
The summary uses plain language.

Figure 15.4: A summary highlights important points and puts the text in plain language.

When should you document your sources? Any ideas, text, or images that you draw from a source need to be properly acknowledged. If you are in doubt about whether you need to cite someone else's work, you should go ahead and cite it. Citing sources will help you avoid any questions about the integrity and soundness of your work.

In Appendix C at the end of this book, you will find a full discussion of three documentation systems (APA, CSE, and MLA) that are used in technical fields. Each of these systems works differently.

The most common documentation style for technical fields is offered by the American Psychological Association (APA). The APA style, published in the *Publication Manual of the American Psychological Association,* is preferred in technical fields because it puts emphasis on the year of publication. As an example, let's briefly look at the APA style for in-text citations and full references.

GO TO THE NET

Need help citing sources? Go to **www.pearsonhighered.com/johnsonweb4/15.4**

Elements of an APA Full Reference

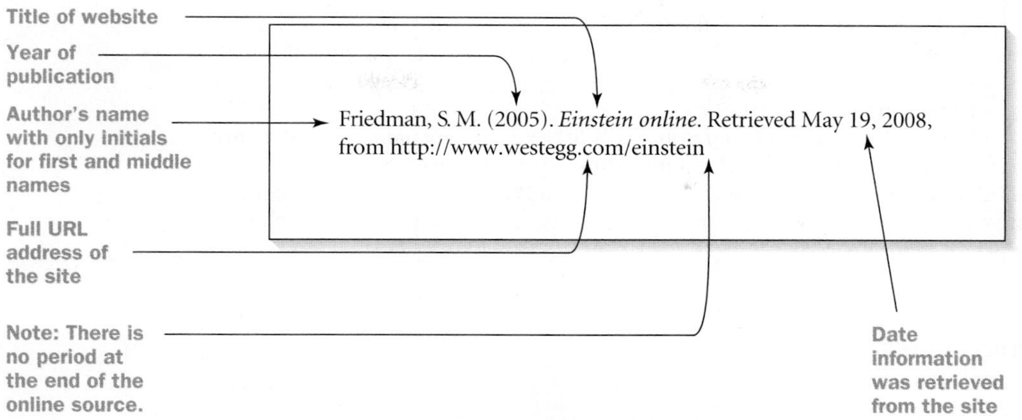

Figure 15.5: The full reference for an APA citation contains some standard elements. Here is an example of a reference for a website.

APA IN-TEXT CITATIONS In APA style, in-text citations can include the author's name, the publication year, and the page number where the information was found.

> One important study showed that physicians were regularly misusing antibiotics to treat viruses (Reynolds, 2003, p. 743).
>
> According to Reynolds (2003), physicians are regularly misusing antibiotics to treat viruses.
>
> According to Reynolds, "Doctors are creating larger problems by mistakenly treating viruses with antibiotics" (2003, p. 743).

These in-text citations are intended to refer the readers back to the list of full references at the end of the document.

Link

For a full discussion of documentation, including models for documenting references, turn to Appendix C, page A-23.

APA FULL REFERENCES The full references at the end of the document provide readers with the complete citation for each source (Figure 15.5).

> Friedman, S. M. (2005). *Einstein online.* Retrieved May 19, 2008, from http://www.westegg.com/einstein
>
> Pauling, L., & Wilson, E. B. (1935). *Introduction to quantum mechanics.* New York, NY: Dover Publications.

As you take notes, you should keep track of the information needed to properly cite your sources. That way, when you draft the document and create a references list, you will have this important information available. It can be very difficult to locate the sources of your information again after you finish drafting the document.

Using a Citation Manager

In today's technical workplace, finding information on a subject is rarely a problem. The real problem is managing all the information that is available through the Internet and print sources. You can often find yourself struggling to keep track of the sources you have located.

A good way to organize your sources is to use a *citation manager,* a software program that will help you format and store your references. Some of the more popular citation managers include EndNote, RefWorks, Procite, and NoodleBib. Firefox also offers a free web-based citation manager called Zotero, which is shown in Figure A.

Using a Citation Manager

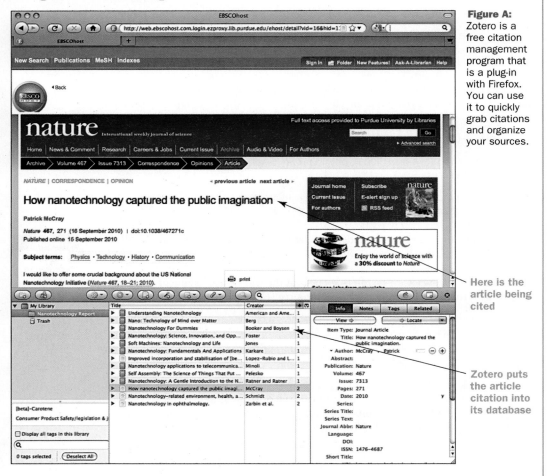

Figure A: Zotero is a free citation management program that is a plug-in with Firefox. You can use it to quickly grab citations and organize your sources.

Here is the article being cited

Zotero puts the article citation into its database

Using a citation manager isn't difficult, but it takes some practice. Here's how it works. When you find a source that looks useful, you should enter its details into the citation manager as soon as you can. The software will cue you for information about the author, title, journal title, publisher, pages, date, and so on. Then, the citation manager will automatically format all your sources into APA, MLA, CSE, and a variety of other citation styles. You just need to tell it which citation style you are using.

Some citation managers will allow you to click on an icon in the address bar of your Internet browser, which will automatically put the citation in your database of sources (Figure A).

Then, when you need to cite a specific source in your document, select "Add Citation" from your word processor's menu. The citation manager will create an in-text citation for that source. It will also add the source to your document's References list in the appropriate bibliographic format.

Another helpful feature in many citation managers is the ability to include keywords or comments about each source. You can then electronically search your citations to locate the ones that are best for your research project.

When you are working on a team project, citation managers can be especially helpful. You and your team can quickly collect all your sources into a common database. Then, the database can be searched to find the most useful sources for the project. This ability to organize a collection of sources will help your team find the information it needs while identifying any trends or gaps in the research.

Something to keep in mind, though, is that citation managers can make formatting mistakes. So, you should proofread the items in your References list before sending your document to your instructor or a client.

Appraising Your Information

All information is not created equal. In fact, some information is downright wrong or misleading. Keep in mind that even the most respected authorities usually have agendas that they are pursuing with their research. Even the most objective experiment will include some tinge of bias.

To avoid misleading information and researcher biases, you need to appraise the information you have collected and develop an overall sense of what the truth might be (Figure 15.6).

Is the Source Reliable?

Usually, the most reliable sources of information are those that have limited personal, political, or financial stakes in the subject. For example, claims about the safety of pesticides from a company that sells pesticides need to be carefully verified. Meanwhile, a study on pesticides by a university professor should be less biased, because the professor is not selling the product.

To ensure that your sources are reliable, you should always do some checking on their authors. Use an Internet search engine like Ask.com or Google.com to check out

Questions for Appraising Your Information

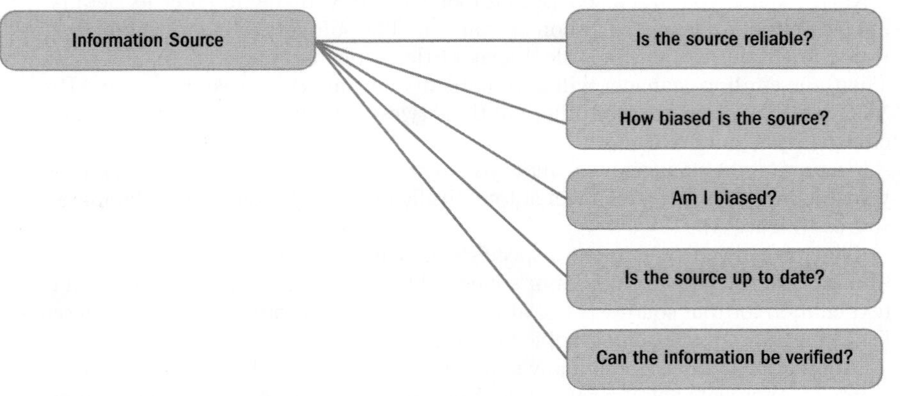

Figure 15.6: Challenge your sources by asking questions about their biases and validity.

the authors, company, or organization that produced the materials. If the researchers have a good reputation, the information is probably reliable. If you can find little or no information about the researchers, company, or organization, you should be skeptical about their research.

How Biased Is the Source?

It is safe to say that all sources of information have some bias. There is no such thing as a completely objective source. So, you need to assess the amount of bias in your source. For example, facts on creation science websites that are used to dispute Darwinian evolution are usually biased toward theories that reinforce the biblical creation story. Their information is still usable in some situations—you might even accept it as true—but you need to recognize the inherent bias in such material.

Even the most reliable sources have some bias. Researchers, after all, very much want their hypotheses to be true, so irregularities in their results might be overlooked. Bias is a natural part of research. So, when you are assessing bias, consider how much the researchers want their results to be true. If the researchers indicate that at the beginning of their research they were open to a range of answers, then the bias of the material is probably minimal. If only one answer was acceptable to the researchers (e.g., smoking does not cause lung cancer), then the material should be considered heavily biased.

Am I Biased?

As a researcher, you need to carefully examine your own biases. We all go into a research project with our own beliefs and expectations of what we will find. Our own biases can cause us to overlook evidence that contradicts our beliefs or expectations. For example, our beliefs about gender, race, sexuality, poverty, or religion, among other social issues, can strongly influence the way we conduct research and interpret our findings. These influences cannot be completely avoided, but they can be identified and taken into consideration.

David B. Resnik, JD, PhD

NATIONAL INSTITUTE OF ENVIRONMENTAL HEALTH SCIENCES,
NATIONAL INSTITUTES OF HEALTH

Why is ethics important in research?

Ethics is different from law, which also guides human conduct. First, unlike law, ethics is not enforced by government sanctions. If you cheat on your income taxes, you may be fined or jailed, but not if you cheat at cards. Second, we often appeal to our sense of ethics in deciding whether laws should be enacted or changed. Leaders of the Civil Rights Movement in the 1960s argued that laws supporting racial discrimination were unjust and must be changed. Many expressed their opposition to discriminatory laws through civil disobedience, that is, intentionally breaking the law to support their ethical principles.

Ethics is important in research for several reasons.

- First, ethical rules promote the aims of research, such as knowledge, truth, and avoidance of error. Prohibitions against fabricating, falsifying, or misrepresenting research data promote the truth and help avoid bias and error.
- Second, ethic rules promote values, such as trust, fairness, mutual respect, and accountability, which are essential to collaborative enterprises such as research. Guidelines for authorship, prohibitions against plagiarism, data sharing policies, and confidentiality rules in journal peer review encourage collaboration by protecting intellectual property interests. Most researchers want to receive credit for their contributions and do not want to have their ideas stolen or disclosed prematurely.
- Third, ethical rules hold researchers accountable to the public and help to build support for research. People and institutions that fund research projects want to be assured of the quality and integrity of the work, and they expect researchers to behave ethically.
- Fourth, ethical rules help to protect human rights, social justice, animal welfare, public health and safety, and the environment. Thus, ethical guidance is an important part of research involving animal or human subjects, biomedical products, agriculture, and engineering or public works projects.

To keep your own biases in check, consider your research subject from an alternative or opposing perspective. At a minimum, considering alternative views will only strengthen your confidence in your research. But, in some cases, you may actually gain a new perspective that can help you further your research.

Is the Source Up to Date?

Depending on the field of study, results from prior research can become obsolete rather quickly. For instance, three-year-old research on skin cancer might already be considered outdated. On the other hand, climate measurements that are over 100 years old are still usable today.

Try to find the most recent sources on your subject. Reliable sources will usually offer a *literature review* that traces research on the subject back at least a few years. These literature reviews will show you how quickly the field is changing, while allowing you to judge whether the information you have located is current.

Can the Information Be Verified?

You should be able to locate more than one independent source that verifies the information you find. If you locate the same information from a few different independent sources, chances are good that the information is reliable. If you find the information in only one or two places, it is probably less reliable.

Triangulation is the key to verifying information. If you can find the information in diverse electronic and print sources, it is probably information you can trust. You might also use empirical methods to confirm or challenge the results of others.

Avoiding Plagiarism

One thing to watch out for is plagiarism in your own work, whether it is intentional or unintentional.

Plagiarism is the use of others' words, images, or ideas without acknowledgment or permission. In most cases, plagiarism is unintentional. While researching, a person might cut and paste information off websites or duplicate passages from a book. Later, he or she might use the exact text, forgetting that the information was copied directly from a source.

In rare cases, plagiarism is intentional and therefore a form of academic dishonesty. In these cases, teachers and colleges will often punish plagiarizers by having them fail the course, putting them on academic probation, or even expelling them. Intentional plagiarism is a serious form of dishonesty.

To avoid plagiarizing, keep careful track of your sources and acknowledge where you found your information.

Keep track of sources—Whenever you are gathering information from a source, carefully note where that information came from. If you are cutting and pasting information from an online source, make sure you put quotation marks around that material and clearly identify where you found it.

Acknowledge your sources—Any words, sentences, images, data, or unique ideas that you take from another source should be properly cited. If you are taking a direct quote from a source, use quotation marks to set it off from your writing. If you are paraphrasing the work of others, make sure you cite them with an in-text citation and put a full-text citation in a references list.

Ask permission—If you want to include others' images or large blocks of text in your work, write them an e-mail to ask permission. Downloading pictures

To learn more about plagiarism, go to **www.pearsonhighered.com/johnsonweb4/15.5**

AT A GLANCE

Assessing Your Information

- Is the source reliable?
- How biased is the source?
- Am I biased?
- Is the source up to date?
- Can the information be verified?

and graphics off the Internet is really easy. But those images are usually someone's property. If you are using them for educational purposes, you can probably include them without asking permission. But, if you are using them for any other reason, you likely need to obtain permission from their owners.

Link

For more information on obtaining permission, go to Chapter 4, page 89.

You do not need to cite sources that offer information that is "common knowledge." If you find the same information in a few different sources, you probably do not need to document that information. But, if you have any doubts, you might want to cite the sources anyway to avoid any plagiarism problems.

Unfortunately, cases of plagiarism are on the rise. One of the downsides of online texts, such as websites, is the ease of plagiarism. Some students have learned techniques of "patchwriting," in which they cut and paste text from the Internet and then revise it into a document. This kind of writing is highly vulnerable to charges of plagiarism, so it should be avoided.

In the end, plagiarism harms mostly the person doing it. Plagiarism is kind of like running stoplights. People get away with it for only so long. Then, when they are caught, the penalties can be severe. Moreover, whether intentional or unintentional, plagiarizing reinforces some lazy habits. Before long, people who plagiarize find it difficult to do their own work, because they did not learn proper research skills. Your best approach is to avoid plagiarism in the first place.

CHAPTER REVIEW

- Taking good notes will help you shorten your research time, because your notes will be searchable and more useful.
- Quoting is a good way to use the words of another researcher to add credibility to your own text.
- Paraphrasing and summarizing are useful ways to explain the research of others in ways that fit the tone of your text and the technical level of your readers.
- Plagiarism can be accidental. When you are doing research, make sure you carefully keep track of where you found what information. Give proper credit where needed for specific ideas.
- After collecting information, you should carefully assess whether it is biased or outdated. Also, be aware of your own biases.

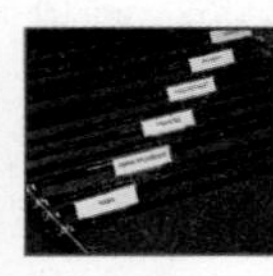

EXERCISES AND PROJECTS

Individual or Team Projects

1. For a research project you are working on, collect three electronic sources and three print sources. Create electronic notecards for each source in which you use APA format to create a full reference. Then, add three to five keywords for each source and summarize the key points in the source. Add at least one quote from each source. When you are finished, show your electronic notecards to your instructor and your research group.

2. Using an Internet search engine, find two sources on the same subject that you would consider "heavily biased." Your sources should be about a technical subject, and they should be on opposing sides of the issue. How do you know these sources are biased? What are some of the obvious markers of a biased source? On what points do the two sources disagree? Do they agree on any points? As you assess these two sources, do you feel biased yourself in some way? If so, how are you biased about this subject?

3. In the age of the Internet, plagiarism seems to have become a more significant problem. Obviously, text is easier than ever to cut and paste from electronic sources, especially the web. On the other hand, it's also easier than ever to detect plagiarism, even minor lapses. Write a two-page white paper to your instructor in which you explain whether plagiarism is a problem and whether it has grown since the invention of the Internet. In a white paper, you should not express your own opinion. Instead, stay close to the facts.

Collaborative Project

With a group, choose a technical topic that interests you. Using a citation manager, like EndNote or Zotero, collect ten sources apiece on that topic. Then, merge your databases into one larger database.

Looking closely at your sources, identify three major themes that run through these sources. What are some common issues they discuss? What issues do they seem to be arguing about? Then, identify one major gap that seems to lack coverage in the sources your group collected. Why does this gap exist? Can you find sources that would fill this gap? If not, do you think this gap is a potential place for research? If so, what kind of research is needed to fill this gap? If not, why do you think researchers have chosen not to pursue research in this area?

When you have identified three themes and one gap, write an e-mail to your instructor in which you explain your subject, identify the themes you identified, and explain why a gap in the research exists.

> For additional technical writing resources, including interactive sample documents, document design tutorials and guidelines, and more, go to **www.mytechcommlab.com.**

CASE STUDY

The Patchwriter

John Krenshaw was a new supervisor for Group 435 at Howell Laboratories. He had worked as a civil engineer for ten years in another group, and he was looking forward to being a manager in charge of other engineers. His new group was working on some very interesting road building projects that would start in India next year.

The projects sounded interesting, but he quickly realized that he would need to learn a great amount to catch up with his group. He knew the civil engineering part as well as anyone, but he didn't know what had been accomplished to this point and what would be happening in the near future. Plus, he needed to be brought up to speed on Indian engineering and construction practices.

So, he asked a member of the group, Holly Gibbs, to write a white paper that would give him the basics about the group's activities and projects. He gave her a two-week deadline, which seemed a bit tight, but he needed that information as soon as possible.

Within a week, she sent him the white paper. He was a bit surprised. When John started to read it, though, he thought the phrasing was strange and uneven. It had sentences like, "India offers you a prime opportunity to get in on the ground floor of one of the growing world economies. Don't miss your chance to invest!" Another sentence read, "India is a fractured nation still addressing some of the tensions involved with its colonial past."

"Where the heck did she get this stuff from?" John muttered as he read. He turned to his computer and started typing in keywords from Holly's white paper. Sure enough, she had taken much of the text from sources available on the web. She had changed many of the words and altered most of the sentences and paragraphs. Overall, though, the bulk of the report was taken from sources that were available on the web. Some of the information was even from Howell Laboratories' competitors.

Holly had gathered some information from engineers in Group 435, but that information was buried in the text she had lifted from other sources. John wasn't sure he could trust that information, especially considering much of the rest of the white paper had been plagiarized.

John was scheduled to meet with Holly after lunch to talk about her report. How do you think he should handle this situation?

CHAPTER

16

Organizing and Drafting

Basic Organization for Any Document 425

Using Genres to Organize Information 426

Outlining the Document 428

Organizing and Drafting the Introduction 430

Organizing and Drafting the Body 433

Organizing and Drafting the Conclusion 446

Organizing Cross-Cultural Documents 448

Chapter Review 451

Exercises and Projects 451

Case Study: The Bad News 453

In this chapter, you will learn:

- A basic organizational pattern that any document can follow.
- How genres offer patterns for organizing documents.
- How to outline the organization of a document.
- How to use presentation software to organize your information.
- How to organize and draft a document's introduction.
- Patterns of arrangement for organizing and drafting sections of a document's body.
- Basic moves used in a document's conclusion.
- Strategies for organizing documents for international readers.

Computers are great tools for helping us manage large amounts of information. Today, effective "information management" is one of the most significant challenges to communicating in the technical workplace. After all, computer networks put amazing amounts of information at our fingertips. But, we still need to organize that information in ways that make it accessible and understandable to our readers.

When writing a technical document or presentation, you will need to organize the information you've collected into patterns that are familiar to your readers. Your readers, after all, are interested in the information you have gathered. But they need you to present that information in a predictable and usable way. Otherwise, they won't be able to take full advantage of your thoughts and research on the subject.

Basic Organization for Any Document

Despite their differences, technical documents usually have something important in common. They typically include an *introduction, body,* and *conclusion.*

> **Introduction**—The introduction of your document needs to tell readers *what* you are writing about and *why* you are writing about it.
>
> **Body**—The body of your document presents the content that your readers need to know to take action or make a decision.
>
> **Conclusion**—The conclusion of your document wraps up your argument by restating your main point(s).

Sometimes it helps to remember the familiar speechwriters' saying, "Tell them what you are going to tell them. Tell them. Then, tell them what you told them."

This beginning-middle-end pattern might seem rather obvious, but people regularly forget to include distinct introductions, bodies, and conclusions in their documents. All too often, they toss readers into the details without first telling them the subject and purpose of their documents. In some cases, their documents end abruptly without summing up the major points.

Introductions and conclusions are especially important in technical documents, because they provide a *context,* or framework, for understanding the *content* in the body of the text (Figure 16.1). Without "contextual" information at the beginning and end of the document, readers find it very difficult to figure out what the author is telling them. Have you ever read a document that didn't seem to have an obvious point? More than likely, it was lacking an effective introduction and/or conclusion.

To see an example of a good introduction, body, and conclusion, consider the classic memo in Figure 16.2. This memo is referred to as the "smoking gun" memo that demonstrated that NASA and the management at Morton Thiokol were ignoring erosion problems with the shuttle's O-rings.

The problem described in this memo is clear, and its author, Roger Boisjoly, does his best to stress the importance of the problem in the introduction and conclusion. Unfortunately, his and other engineers' warnings were not heeded by higher-ups.

Why do documents need introductions and conclusions? For answers, see **www.pearsonhighered.com/johnsonweb4 /16.1**

Standard Organization of a Document

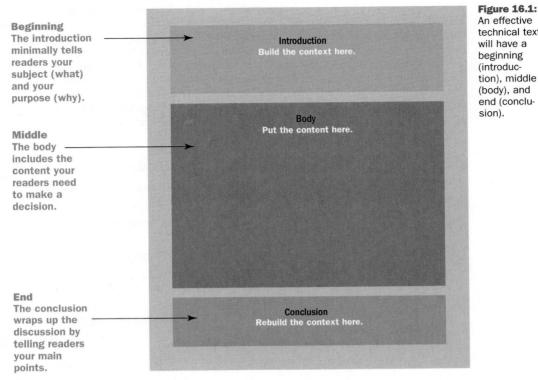

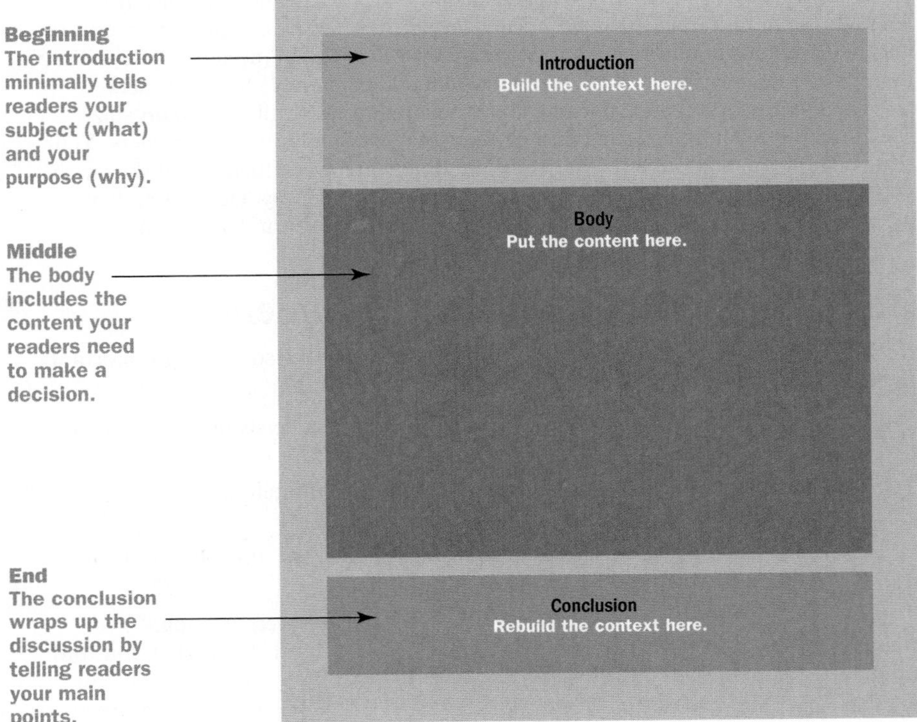
Beginning
The introduction minimally tells readers your subject (what) and your purpose (why).
Introduction
Build the context here.
Middle
The body includes the content your readers need to make a decision.
Body
Put the content here.
End
The conclusion wraps up the discussion by telling readers your main points.
Conclusion
Rebuild the context here.

Figure 16.1: An effective technical text will have a beginning (introduction), middle (body), and end (conclusion).

Using Genres to Organize Information

Once you move beyond the introduction-body-conclusion pattern, technical documents follow a variety of organization patterns. An analytical report, for example, is very different from a set of instructions, and a proposal shows little resemblance to a technical specification. These documents follow different patterns of organization, which are called *genres.*

A genre is a predictable pattern for organizing information to achieve specific purposes. For instance, here are two genres, for the report and a procedure, side by side:

Analytical Report
- Introduction
- Methods
- Results
- Discussion
- Conclusion

Procedure
- Introduction
- Materials
- List of parts/tools
- Sequentially ordered steps
- Conclusion

Both genres have an introduction and conclusion, but otherwise they are very different in organization.

The "Smoking Gun" *Challenger* Memo

Figure 16.2: Roger Boisjoly's classic memo warning of imminent disaster with the *Challenger* shuttle.

MORTON THIOKOL, INC

Wasatch Division Interoffice Memo

July 31, 1985

2870:FY86:073

TO:	R. K. Lund Vice President, Engineering
CC:	B. C. Brinton, A. J. McDonald, L. H. Sayer, J. R. Kapp
FROM:	R. M. Boisjoly Applied Mechanics - Ext. 3525
SUBJECT:	SRM O-Ring Erosion/Potential Failure Criticality

The introduction contains the purpose and main point.

This letter is written to insure that management is fully aware of the seriousness of the current O-ring erosion problem in the SRM joints from an engineering standpoint.

The body supports the argument.

The mistakenly accepted position on the joint problem was to fly without fear of failure and to run a series of design evaluations which would ultimately lead to a solution or at least a significant reduction of the erosion problem. This position is now drastically changed as a result of the SRM 16A nozzle joint erosion which eroded a secondary O-ring with the primary O-ring never sealing.

If the same scenario should occur in a field joint (and it could), then it is a jump ball as to the success or failure of the joint because the secondary O-ring cannot respond to the clevis opening rate and may not be capable of pressurization. The result would be a catastrophe of the highest order—loss of human life.

An unofficial team (a memo defining the team and its purpose was never published) with leader was formed on July 19, 1985 and was tasked with solving the problem for both the short and long term. This unofficial team is essentially nonexistent at this time. In my opinion, the team must be officially given the responsibility and the authority to execute the work that needs to be done on a non-interference basis (full time assignment until completed.)

The conclusion restates the main point.

It is my honest and very real fear that if we do not take immediate action to dedicate a team to solve the problem with the field joint having the number one priority, then we stand in jeopardy of losing a flight along with all the launch pad facilities.

R. M. Boisjoly

Concurred by: J. R. Kapp, Manager

Applied Mechanics

Source: Report of the Presidential Commission on the Space Shuttle Challenger Accident, 1986.

Indeed, each genre achieves a different purpose and therefore requires a different organizational pattern. Once you know your purpose, you should be able to easily figure out what kind of document you are being asked to write and how it should be organized (Figure 16.3).

Keep in mind that genres are patterns, not formulas. There are countless ways to organize analytical reports and procedures, depending on the needs of your readers and the context in which the document will be used. You can use these genres as starting places from which to begin organizing and drafting your documents.

Chapters 5 through 11 discuss a range of workplace genres in depth. When you begin writing one of these documents, you should turn to the chapter that describes it.

Choosing a Genre

Which genre fits my writing task?

Purpose	Genre	Type of Document
I need to tell others about a decision or event.	Correspondence (Chapter 5)	E-mail, Letter, or Memo
I need to describe an item, product, or service.	Description (Chapter 6)	Technical Description, White Paper
I need to explain how to do something.	Instructions (Chapter 7)	Instructions, Specifications, Procedures
I need to make a suggestion or propose a new project.	Proposal (Chapter 8)	Research Proposal, Planning Proposal, Implementation Proposal
I need to present information or make a recommendation.	Report (Chapters 9 and 10)	Research Report, Feasibility Report, Recommendation Report, Lab Report, Progress Report
I need to provide online information to others.	Hypertext (Chapter 22)	Website, Multimedia Document, CD-ROM
I need a job.	Résumé (Chapter 11)	Portfolio, Résumé, Scannable Résumé

Figure 16.3: Once you know the purpose of your document, you can decide which genre fits your need. Each genre provides a pattern for you to follow as you organize and draft your document.

Outlining the Document

Once you have identified the genre of your document, you can begin developing a rough outline of its organization.

Outlining may seem a bit old-fashioned, but it is very helpful when you are trying to sort out your ideas, especially as you prepare to write a large technical document. In the workplace, most people sketch out a rough outline to help them organize their ideas. In their outline, they type the document's main headings on the screen (Figure 16.4). Then, they list the contents of each section separately. The outline will usually change as new ideas, evidence, or issues emerge.

If you are unsure how to outline your document, let the genre you are following guide your thought process. The genre should give you a good sense of the larger

GO TO THE NET

Want to learn more about genres? Go to **www.pearsonhighered.com/johnsonweb4/16.2**

A Rough Outline

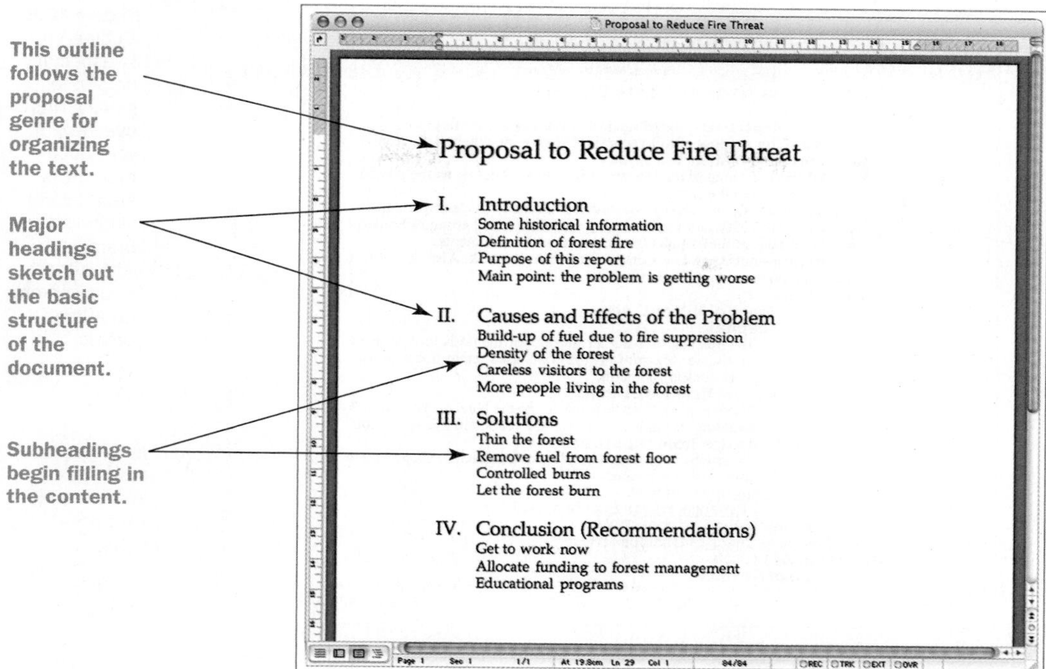

Figure 16.4: An outline doesn't need to be formal, and it should always be open to change. Here is a rough outline with some guesses about what kinds of topics will be discussed in the proposal.

sections of the document. From there, you can begin filling in the smaller topics in each section.

When outlining, you might also consider using the Outline View in your word processor. In Outline View, you can begin arranging your information by listing the headings for the document's major sections and subsections (Figure 16.5). Then, you can decide if:

- smaller sections should be merged or incorporated into larger sections.
- larger sections should be divided into separate sections.
- any need-to-know information is missing from sections.
- there is too much want-to-tell information.

While you are revising your document, Outline View also helps you quickly move information around. You can "collapse" a whole part of a document under a heading. Then, you can cut and paste to move that part somewhere else in the document.

Outline View is a great tool when you are working with a team. When you and your co-workers are brainstorming about a project or document, just toss your ideas into Outline View. Then, when you're finished brainstorming, you can start organizing

Link

For more information on working with a team, see Chapter 3, page 45.

For websites that offer outlining strategies, go to **www.pearsonhighered.com/johnsonweb4/16.3**

Outline View

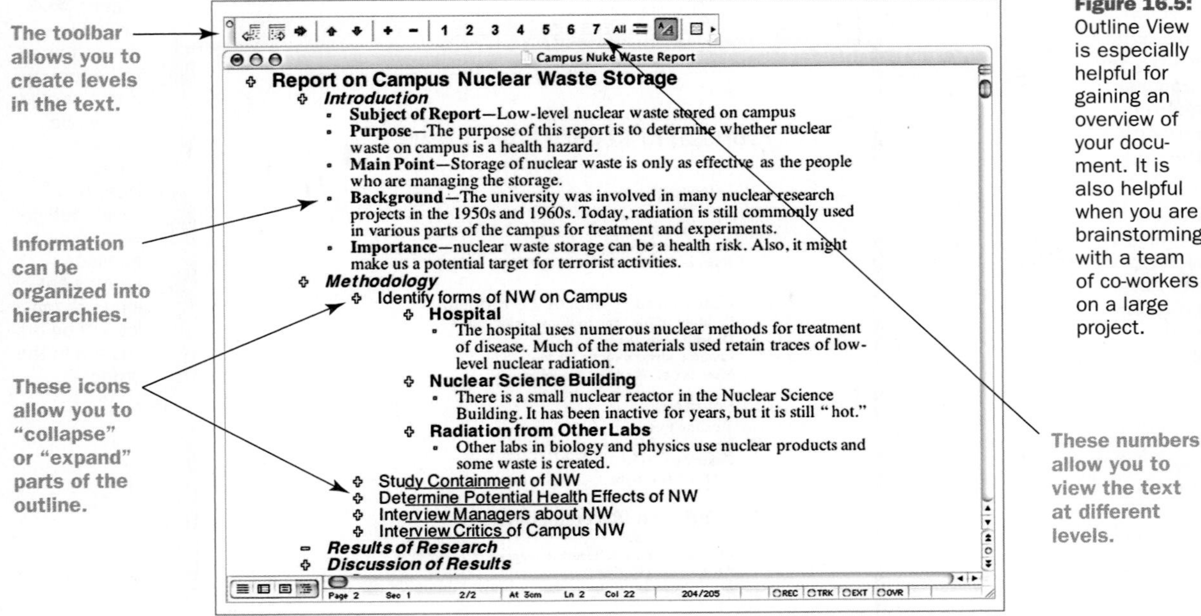

Figure 16.5: Outline View is especially helpful for gaining an overview of your document. It is also helpful when you are brainstorming with a team of co-workers on a large project.

all that information into a more structured outline. At the end of the meeting, you can print out the outline or e-mail it to the whole group to guide the drafting of the document.

Overall, an outline should be as flexible as the document itself. A computer-generated outline can be a helpful tool for planning, drafting, and revising your work.

Organizing and Drafting the Introduction

When organizing and drafting your document, you should put yourself in your readers' place. If you were the reader, what information would you want to know up front?

What is this document about?

Why did someone write this for me?

What is the main point?

Is this information important?

How is this document organized?

When an introduction answers these questions up front, readers are better able to understand the rest of the document.

For more help using Outline View, go to **www.pearsonhighered.com/johnsonweb4/16.4**

Six Opening Moves in an Introduction

These kinds of questions translate into six opening "moves" made in an introduction:

MOVE 1: DEFINE YOUR SUBJECT Tell readers what your document is about by defining the subject.

> Flooding has become a recurring problem in Darbey, our small town nestled in the Curlew Valley south of St. Louis.

In some cases, to help define the boundaries of your subject, you might also tell readers what your document is *not* going to cover.

MOVE 2: STATE YOUR PURPOSE Tell readers what you are trying to achieve. Your purpose statement should be clear and easy to find in the introduction. It should plainly tell your readers what the document will do.

Link

For more information on crafting a purpose statement, go to Chapter 1, page 6.

> This proposal offers some strategies for managing flooding in the Darbey area.

You should be able to articulate your purpose in one sentence. Otherwise, your purpose may not be clear to your readers—and perhaps not even to you.

MOVE 3: STATE YOUR MAIN POINT Tell your readers the key idea or main point that you would like them to take away from the document. Usually your main point is your overall decision, conclusion, or solution that you would like the readers to accept.

> The only long-term way to control flooding around Darbey is to purchase and restore the wetlands around the Curlew River, while enhancing some of the existing flood control mechanisms like levees and diversion ditches.

Are you giving away the ending by telling the readers your main point up front? Yes. But technical documents are not mystery stories. Just tell readers your main point in the introduction. That way, as they read the document, they can see how you came to your decision.

AT A GLANCE

Six Moves in an Introduction

- Move 1: Define your subject.
- Move 2: State your purpose.
- Move 3: State your main point.
- Move 4: Stress the importance of the subject.
- Move 5: Provide background information.
- Move 6: Forecast the content.

MOVE 4: STRESS THE IMPORTANCE OF THE SUBJECT Make sure you give your readers a reason to care about your subject. You need to answer their "So what?" questions if you want them to pay attention and continue reading.

> If development continues to expand between the town and the river, the flooding around Darbey will only continue to worsen, potentially causing millions of dollars in damage.

MOVE 5: PROVIDE BACKGROUND INFORMATION Typically, background information includes material that readers already know or won't find controversial. This material could be historical, or it could stress a connection with the readers.

GO TO THE NET

To learn more about how people read introductions, go to **www.pearsonhighered.com/johnsonweb4/16.5**

> As we mentioned in our presentation to the city council last month, Darbey has been dealing with flooding since it was founded. Previously, the downtown was flooded three times (1901, 1922, and 1954). In recent years, Darbey has experienced flooding with much more frequency. The downtown was flooded in 1995, 1998, 2000, and 2003.

MOVE 6: FORECAST THE CONTENT Forecasting describes the structure of the document for your readers by identifying the major topics it will cover.

> In this proposal, we will first identify the causes of Darbey's flooding problems. Then we will offer some solutions for managing future flooding. And finally, we will discuss the costs and benefits of implementing our solutions.

Forecasting helps readers visualize the organization of the rest of the document. It gives them a map to anticipate the topics the document will cover.

Drafting with the Six Moves

In an introduction, these moves can be made in just about any order. Figure 16.6, for example, shows how the introductory moves can be used in different arrangements.

Two Versions of an Introduction

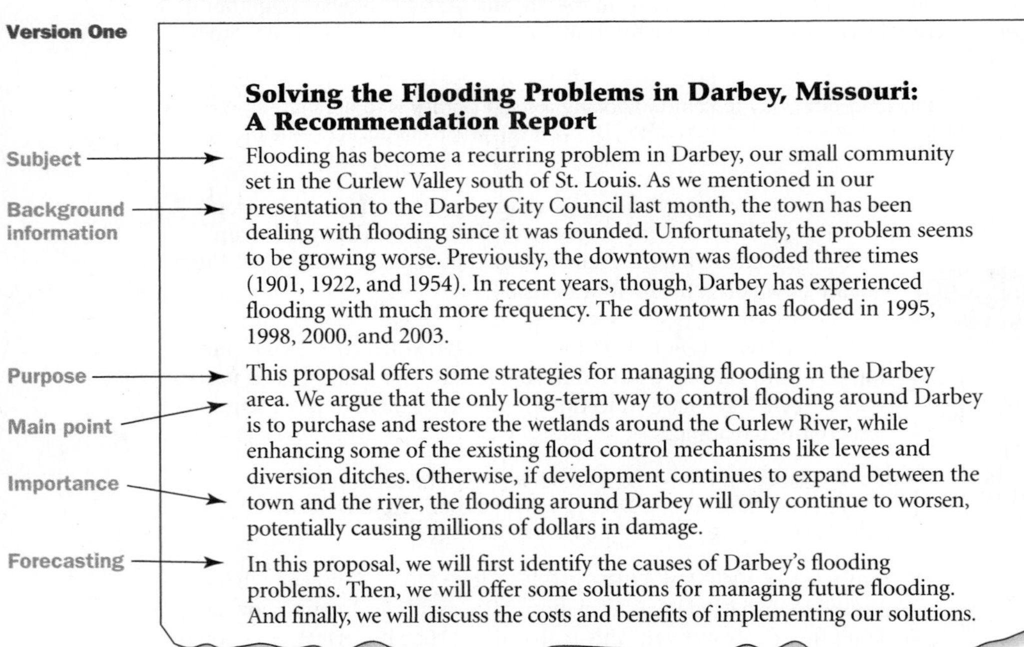

Solving the Flooding Problems in Darbey, Missouri: A Recommendation Report

Flooding has become a recurring problem in Darbey, our small community set in the Curlew Valley south of St. Louis. As we mentioned in our presentation to the Darbey City Council last month, the town has been dealing with flooding since it was founded. Unfortunately, the problem seems to be growing worse. Previously, the downtown was flooded three times (1901, 1922, and 1954). In recent years, though, Darbey has experienced flooding with much more frequency. The downtown has flooded in 1995, 1998, 2000, and 2003.

This proposal offers some strategies for managing flooding in the Darbey area. We argue that the only long-term way to control flooding around Darbey is to purchase and restore the wetlands around the Curlew River, while enhancing some of the existing flood control mechanisms like levees and diversion ditches. Otherwise, if development continues to expand between the town and the river, the flooding around Darbey will only continue to worsen, potentially causing millions of dollars in damage.

In this proposal, we will first identify the causes of Darbey's flooding problems. Then, we will offer some solutions for managing future flooding. And finally, we will discuss the costs and benefits of implementing our solutions.

Figure 16.6: As shown in these two sample introductions, the six introductory moves can be arranged in just about any order. Some arrangements, however, are more effective than others, depending on the subject and purpose of the document.

GO TO THE NET For sample introductions, see **www.pearsonhighered.com/johnsonweb4/16.6**

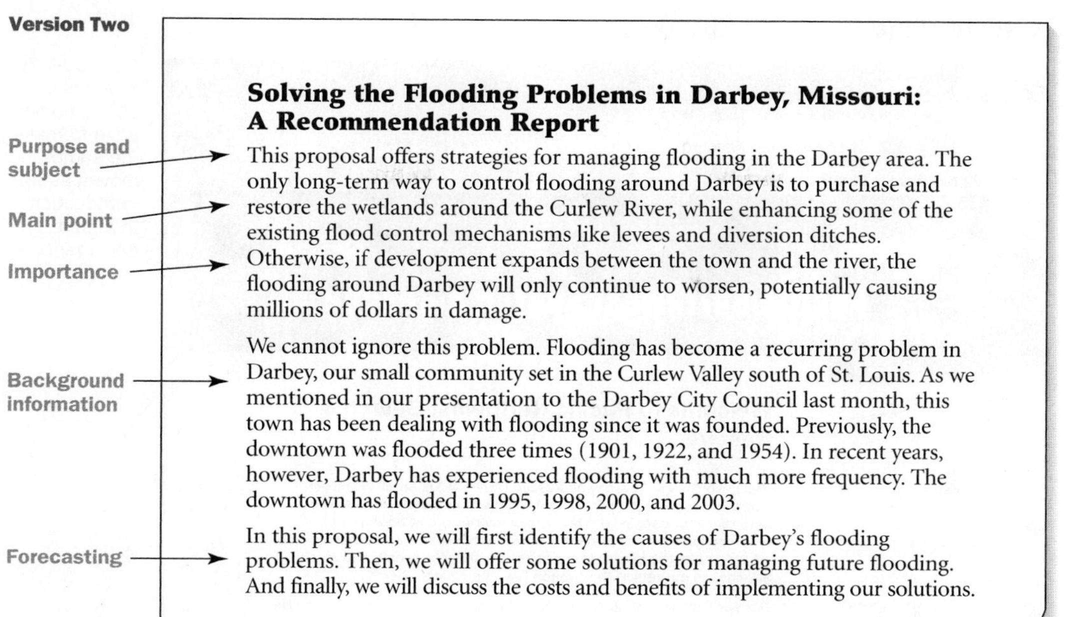
Version Two
Solving the Flooding Problems in Darbey, Missouri:
A Recommendation Report
Purpose and subject
This proposal offers strategies for managing flooding in the Darbey area. The only long-term way to control flooding around Darbey is to purchase and restore the wetlands around the Curlew River, while enhancing some of the existing flood control mechanisms like levees and diversion ditches. Otherwise, if development expands between the town and the river, the flooding around Darbey will only continue to worsen, potentially causing millions of dollars in damage.
Main point
Importance
We cannot ignore this problem. Flooding has become a recurring problem in Darbey, our small community set in the Curlew Valley south of St. Louis. As we mentioned in our presentation to the Darbey City Council last month, this town has been dealing with flooding since it was founded. Previously, the downtown was flooded three times (1901, 1922, and 1954). In recent years, however, Darbey has experienced flooding with much more frequency. The downtown has flooded in 1995, 1998, 2000, and 2003.
Background information
In this proposal, we will first identify the causes of Darbey's flooding problems. Then, we will offer some solutions for managing future flooding. And finally, we will discuss the costs and benefits of implementing our solutions.
Forecasting

Figure 16.6: *(continued)*

Any information that goes beyond these six moves should be removed from the introduction. After all, this extra information will only make it more difficult for readers to locate the subject, purpose, and main point of your document.

As discussed more thoroughly in Chapter 22, home pages in websites are also introductions. For example, the home page shown in Figure 16.7 also makes the typical moves found in an introduction.

Organizing and Drafting the Body

The body of the document is where you are going to provide the *content* that readers need to know. Here is where you will give them the information they need (facts, details, examples, and reasoning) to understand your subject and/or take action.

Carving the Body into Sections

With the exception of small documents (e.g., small memos or letters), the bodies of technical documents are typically carved into *sections*. In many ways, sections are like miniature documents, needing their own beginning, middle, and end. They typically include an *opening, body,* and *closing* (Figure 16.8).

GO TO THE NET

For more help with writing introductions, go to **www.pearsonhighered.com/johnsonweb4/16.7**

A Home Page

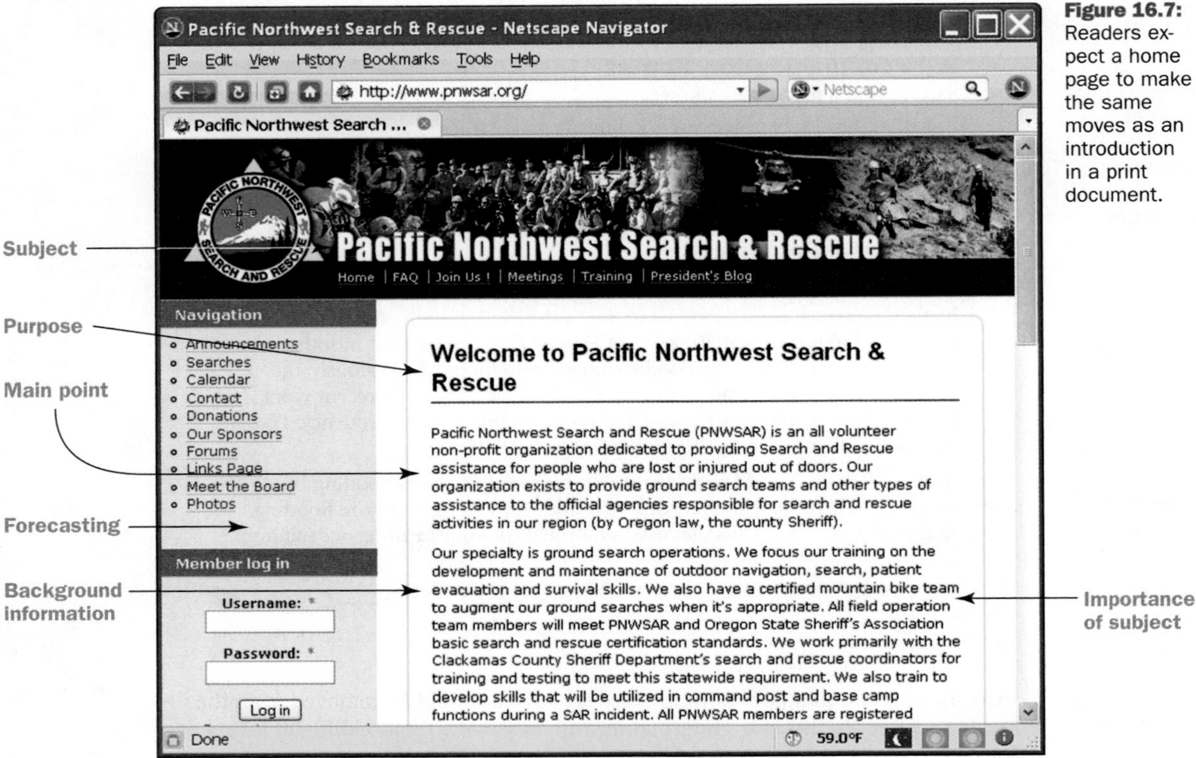

Figure 16.7: Readers expect a home page to make the same moves as an introduction in a print document.

Source: Pacific Northwest Search and Rescue, http://www.pnwsar.org.

OPENING An opening is usually a sentence or small paragraph that identifies the subject and purpose of the section. The opening usually includes a claim or set of claims that the rest of the section will support.

> **Results of Our Study**
>
> The results of our study allow us to draw two conclusions about the causes of flooding in Darbey. First, Darbey's flooding is mostly due to the recent construction of new levees by towns farther upriver. Second, development around the river is taking away some of the wetlands that have protected Darbey from flooding in the past. In this section, we will discuss each of these causes in depth.

BODY The body of a section is where you will offer support for the claim you made in the opening. The body of a section can run anywhere from one paragraph to many

A Section with an Opening, Body, and Closing

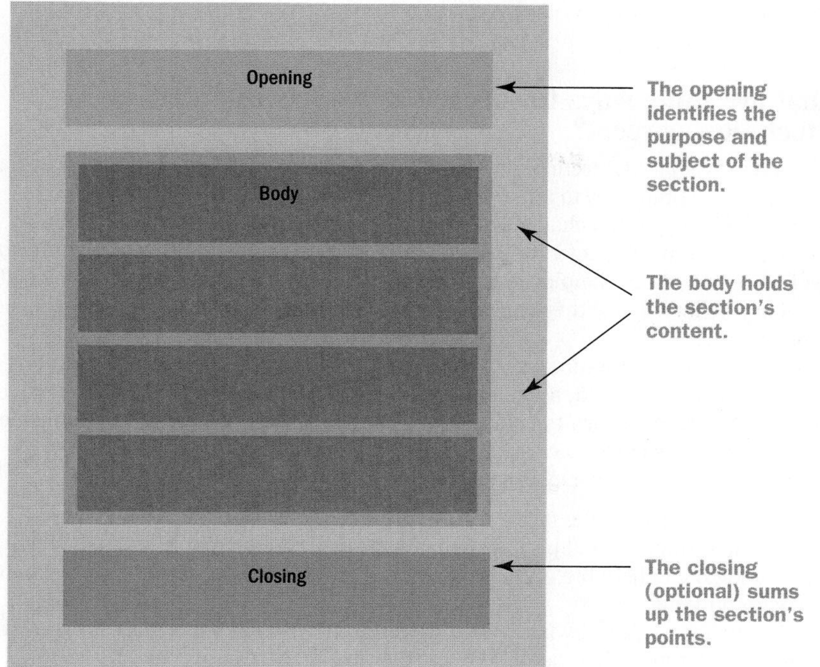

Figure 16.8: A section includes an opening and body. The closing is optional, but it can be helpful to sum up the section's point or points.

paragraphs, depending on the purpose of the section. For example, if you are discussing the results of a research study, your Results section may require three or more paragraphs in the body—one paragraph per major result.

CLOSING (OPTIONAL) A section that is large and/or complex might need a closing paragraph or sentence to wrap up the discussion. A closing usually restates the claim you made in the opening of the section. It might also look forward to the next section.

> In sum, these two causes will likely become only more significant over time. As upriver towns grow in population, there will be more pressure than ever to build more levees to protect them. Meanwhile, if development continues in the available wetlands around the Curlew River, Darbey will find that some of its last defenses against flooding have disappeared.

Overall, a section should be able to work as a stand-alone unit in the document. By having its own beginning, middle, and end, a well-written section feels like a miniature document that makes a specific point.

GO TO THE NET For sample sections that you can model, go to **www.pearsonhighered.com/johnsonweb4/16.8**

AT WORK

Brian Burridge

INTERNET SOFTWARE ARCHITECT, DESIGNER, AND DEVELOPER

What are some ways to succeed as a telecommuter in technical careers?

Although we've had the technology to telecommute for many years, it has taken many more years for companies to adapt their methodologies and workplace paradigms. Human psychology often changes much more slowly than technology. Even today, we continue to have managers who think and manage as though it were the industrial age. They expect to see employees in their seats, arriving and leaving at a specified time. They believe that if they can't see what the employee is doing, he or she must not be working.

But this is the information age and, beyond that, the age of creativity. Employee expectations have changed, as well as increasing fuel costs, extended commutes, concerns about carbon footprints, and the ever-increasing cost of real estate for business. So, more and more companies are turning to telecommuting as a solution.

Here are the keys to being as successful telecommuter:

- Schedule a trip to the facility at the beginning of the project. A few shared lunches, dinners, and perhaps other activities (catch a movie with fellow co-workers) can help break the ice and open the door of opportunity for relationship building.
- Get to know people by finding out what they like, whether they have families, what they do for fun, what they read, what they like to eat, career goals, and so forth.
- Use social websites, like LinkedIn, MySpace, and FaceBook to blog and share your personal and professional experiences. This can help build relationships on a deeper level.
- Send daily reports that provide task status and offer updates any time a problem affects the schedule. An end-of-day report should be sufficient in most cases.
- Be available via multiple forms of communication so that people feel they can easily get in touch with you, especially during normal working times.
- Look for ways to be helpful to other teammates on their tasks. Take on new responsibilities when possible.
- Finish tasks in advance of scheduled completion times to help build trust.

The successful telecommuter needs to put a special effort toward building trust with managers and team members. They cannot see you working, so you need to show them evidence by communicating with them regularly.

Finally, quality communication depends on the telecommuter's ability to use the written word (in e-mail and instant messages) to convey the point in a succinct, yet effective manner. This is the number one problem with telecommuting; simply that so many workers are really very bad at communicating their thoughts, defining problems, and proposing solutions with the written word. This is what frustrates team members and then makes them complain to their boss about the off-site employee who can't communicate. To communicate clearly, get to the point; leave out too many side notes and details that don't directly support the purpose of your communication.

Patterns of Arrangement

When writing each section, you can usually follow a *pattern of arrangement* to organize your ideas. These patterns are based on logical principles, and they can help you organize your information so that your views will be presented in a reasoned way. Major patterns of arrangement are:

- Cause and effect
- Comparison and contrast
- Better and worse
- Costs and benefits
- If . . . then
- Either . . . or
- Chronological order
- Problem/needs/solution
- Example

You will find that each section in your document likely follows one of these patterns of arrangement. One section, for example, may discuss the causes and effects of a problem. A later section might use a discussion of the costs and benefits of doing something about the problem. So, as you are organizing and drafting each section, decide which pattern of arrangement best fits your needs.

CAUSE AND EFFECT In a sense, all events are subject to *causes* and *effects.* For example, if a bridge suddenly collapses, investigators would immediately try to determine the causes for the collapse.

> In 2002, the I-40 bridge over the Arkansas River collapsed for a few different reasons. The actual collapse occurred when a runaway barge on the river rammed into one of the bridge's supports. But other causes were evident. The bridge was already weakened by erosion around the pilings, deteriorated concrete, and attrition due to recent seismic activity.

A cause can also have various effects.

> Leaving a wound untreated can be dangerous. The wound may become contaminated with dirt and germs, thus requiring more healing time. In some cases, the wound may grow infected or even gangrenous, requiring much more treatment at a hospital. Infections can be life-threatening.

As you discuss causes and effects, show how effects are the results of specific causes. For example, the white paper shown in Figure 16.9 explains how flooding occurs when specific events (causes) occur.

COMPARISON AND CONTRAST You can *compare and contrast* just about anything. When comparing and contrasting two things, first identify all the features that make them similar. Then, contrast them by noting the features that make them different (Figure 16.10). By comparing and contrasting two similar things, you can give your readers a deeper understanding of both.

Want to see examples of cause and effect sections? Go to **www.pearsonhighered.com/johnsonweb4/16.9**

A Section Using Cause and Effect

Buckinghamshire Flooding Facts

Flooding—Cause and Effect

In Buckinghamshire, as with other inland counties, there are three main causes of flooding, which are river (also known as riparian), flash, and groundwater. In all three cases the ability of the ground to absorb rainfall, like a sponge, plays a major part. This varies according to weather conditions.

River flooding occurs when rivers cannot cope with the amount of water draining into them off the land. In the winter months this is usually because the ground has become saturated and can no longer absorb water properly. When rainfall is heavy and/or prolonged, runoff reaches the rivers faster and eventually floods over their banks.

Flash flooding can happen anywhere at any time, although it is more likely in the summer months when the ground is hard and dry. Sudden downpours, such as those associated with thunderstorms, cannot soak in fast enough and the water runs off quickly into drains, ditches, and culverts that cannot cope with the volume.

Groundwater flooding is rare, occurring only when the underground water table rises to an unusually high level. Groundwater rose significantly following the very wet winter of 2002–03. Fortunately the drier spring weather meant that levels did not reach a critical point.

Whatever the cause, the effects are much the same. If floodwater enters your property it will ruin carpets, furniture, household goods, decor, and electrics. Often floodwater will be mixed with raw sewage, as drains overflow. When the flood water leaves, your home will take a long time to dry out and may smell for weeks. Plaster may have to be stripped off walls. If you are fully insured there should be no problem in getting the necessary repairs paid for, but the whole process will probably take months, and the experience is severely depressing. People who have been flooded will often say that it is far worse than being burgled.

Figure 16.9: A section that uses the cause and effect pattern of arrangement.

Opening paragraph lists the causes of the problem.

Body paragraphs discuss each cause separately.

Closing paragraph discusses the effects of the problem.

Source: Buckinghamshire County Council, http://www.southbucks.gov.uk/documents/Flooding%20FACTS3.doc.

BETTER AND WORSE In technical workplaces, you are often faced with moments in which you need to choose among different paths. In these cases, you may need to play the advantages off the disadvantages.

The "better" is discussed in terms of advantages.

Automating our assembly line with robotic workstations has clear advantages. With proper maintenance, robots can work around the clock, every day of the week. They don't take vacations, and they don't require benefits. Moreover, after an initial up-front investment, they are less expensive per unit than human labor.

A Section Using Comparison and Contrast

Figure 16.10: Comparison and contrast puts two things side by side to show their similarities and differences.

Comparison Between Hemyock and Bodiam Castles

Opening states the claim for the section to follow.

The plan of Hemyock Castle has similarities with Bodiam Castle in Kent, built some five years later.

Comparisons are made by noting similarities.

Both Hemyock and Bodiam are typical of small late medieval castles: a rectangular site with high, round corner towers and central interval towers, connected by a high curtain wall; all topped with crenellations; surrounded by a water-filled moat. Both had massive fortified gatehouses. The two castles were roughly the same size. Even the detail of Hemyock's NE Tower appears similar to Bodiam's Well Tower.

There are obvious differences:

- Hemyock Castle was built around the family's old manor house, whereas Bodiam Castle was built on a new site.
- Judging by the remains, Bodiam Castle appears to have been more lavish.
- Bodiam's interval towers were rectangular rather than round.
- Bodiam has a huge moat.

Contrasts are made by noting differences.

Presumably, Hemyock Castle was more functional. Much of the accommodations would have been provided by the old manor house, so the defensive outer walls could be simpler. This would also have allowed the use of stronger, more functional round towers. The rectangular towers at Bodiam were required to provide comfortable accommodations. Further accommodation was built into Bodiam's outer walls.

Bodiam had a huge, if easily drained, moat and complex entrance causeways.

Hemyock's moat was more functional but not easily drained. Hemyock's western entrance (now lost) may have included a short defensive causeway.

The massive gatehouse at Hemyock's eastern entrance was protected by a drawbridge across the moat and by outer bastions (now the "Guard Houses" holiday cottages). In recent centuries there have been great changes around Hemyock's eastern entrance, including diversion of the river (St. Margaret's Brook) and extension of St. Mary's Church. It is just possible that there was a more complex series of defenses and water obstacles.

Source: Hemyock Castle, http://www.hemyockcastle.co.uk/bodiam.htm.

The "worse" is shown in a less favorable light.

Our alternative to automation is to stay with human labor. Increasingly, we will become less profitable, because our competitors are moving their operations to offshore facilities, where labor is much cheaper and environmental laws are routinely ignored. Meanwhile, if we do not automate, the increasing costs of health care and an aging workforce will eventually force us to close some of our manufacturing plants in North America.

COSTS AND BENEFITS By directly weighing the *costs and benefits,* you can show readers that the price is ultimately worth the benefits of moving forward with a project.

Why College Education Is Still Worth It

How good an investment finishing college is depends on both earnings and costs—the earnings of college graduates relative to high school graduates and the costs of attending college (both tuition and foregone earnings). Tuition and fees for a four-year college for the 2003–2004 academic year averaged $7,091; the average net price—tuition and fees net of grants—was $5,558 (both amounts in 2003 dollars). If we assume that tuition and fees continue to rise as they did between the 1999–2000 and 2003–2004 school years, and conservatively look at sticker rather than net prices, the average full-time student entering a program in the fall of 2003 who completes a bachelor's degree in four years will pay $30,325 in tuition and fees. If we assume an opportunity cost equal to the average annual earnings of a high school graduate (from the March 2004 Current Population Survey) and a 5 percent discount rate for time preference, the total cost of attending college rises to $107,277. In other words, college is worthwhile for an average student if getting a bachelor's degree boosts the present value of her lifetime earnings by at least $107,277.

What is the boost to the present value of wages? At a 5 percent annual discount rate, it is $402,959. The net present value of a four-year degree to an average student entering college in the fall of 2003 is roughly $295,682—the difference between $402,959 in earnings and $107,277 in total costs. A student entering college today can expect to recoup her investment within 10 years of graduation.

It still pays to go to college—very much so, at least as much as ever before.

Source: L. Barrow, C. E. Rouse, (2005). "Does college still pay?" The Economists' Voice *2 (4), 1–8.*

Stressing the benefits to your readers is always a good way to reason with them. By putting the costs in contrast to these benefits, you can show how the advantages ultimately outweigh the costs.

IF . . . THEN Perhaps the most common way to reason is using *if . . . then* arguments. Essentially, you are saying, "If you believe in X, then you should do Y" or perhaps "If X happens, then Y is likely to happen also" (Figure 16.11). When using *if . . . then* arguments, you are leveraging something readers already believe or consider possible to convince them that they should believe or do something further.

EITHER . . . OR When using *either . . . or* statements, you are offering readers a choice or showing them two sides of the issue. You are saying, "Either you believe X, or you believe Y" or perhaps "Either X will happen, or Y will happen" (Figure 16.12). Statements using *either . . . or* patterns suggest that there is no middle ground, so you should

To see other documents that argue costs and benefits, go to **www.pearsonhighered.com/johnsonweb4/16.10**

A Section Using *If . . . Then*

Figure 16.11: *If . . . then* patterns are used to build arguments on uncertainties.

The opening sets a premise with two *if . . . then* statements.

The body of the section explores the implications of those *if . . . then* statements.

The closing poses *if . . . then* statements that are more speculative.

Effects of Climate Change on Forests

The projected 2°C (3.6°F) warming could shift the ideal range for many North American forest species by about 300 km (200 mi.) to the north. If the climate changes slowly enough, warmer temperatures may enable the trees to colonize north into areas that are currently too cold, at about the same rate as southern areas became too hot and dry for the species to survive. If the earth warms 2°C (3.6°F) in 100 years, however, the species would have to migrate about 2 miles every year.

Trees whose seeds are spread by birds may be able to spread at that rate. But neither trees whose seeds are carried by the wind, nor nut-bearing trees such as oaks, are likely to spread by more than a few hundred feet per year. Poor soils may also limit the rate at which tree species can spread north. Thus, the range over which a particular species is found may tend to be squeezed as southern areas become inhospitably hot. The net result is that some forests may tend to have a less diverse mix of tree species.

Several other impacts associated with changing climate further complicate the picture. On the positive side, CO^2 has a beneficial fertilization effect on plants, and also enables plants to use water more efficiently. These effects might enable some species to resist the adverse effects of warmer temperatures or drier soils. On the negative side, forest fires are likely to become more frequent and severe if soils become drier. Changes in pest populations could further increase the stress on forests. Managed forests may tend to be less vulnerable than unmanaged forests, because the managers will be able to shift to tree species appropriate for the warmer climate.

Perhaps the most important complicating factor is uncertainty (see U.S. Climate in the Future Climate section) whether particular regions will become wetter or drier. If climate becomes wetter, then forests are likely to expand toward rangelands and other areas that are dry today; if climate becomes drier, then forests will retreat away from those areas. Because of these fundamental uncertainties, existing studies of the impact of climate change have ambiguous results.

Source: U.S. Environmental Protection Agency, http://yosemite.epa.gov/oar/globalwarming.nsf/content/ImpactsForests.html.

use this strategy only when appropriate. When used appropriately, an *either . . . or* argument can prompt your readers to make a decision. When used inappropriately, these arguments can invite readers to make a choice you didn't expect—or perhaps to make no choice at all.

A Section Using *Either . . . Or*

Figure 16.12: The *either . . . or* pattern is helpful when only one belief or action is the right one.

Shapley-Curtis Debate

During the 1920s, Harlow Shapley and Heber Curtis debated whether observed fuzzy "spiral nebulae," which are today known as galaxies, exist either in our Milky Way Galaxy or beyond it. Shapley took the stance that these "spiral nebulae" observed in visible light do indeed exist in our galaxy, while the more conservative Curtis believed in the converse. While both sides of this debate seemed to either make too many suppositions or supported their claims with flawed data, the main points of both stances made the debate fairly close at the time.

The opening sets up two sides with *either . . . or* statements.

Shapley's Argument

Shapley argued that these "spiral nebulae" are actually within our galaxy's halo. He presented faulty data which seemed to confirm that M101, a large angular-diameter galaxy, changed angular size noticeably. We know now that for M101 to change angular size appreciably (0.02 arcsecond/year), it would have to recede from us at warp speeds based upon our current knowledge of its diameter and distance. To account for all the galaxies' observed recession from earth, not knowing of the universe's expansion, Shapley invented a special repulsion force to explain this strange phenomenon. He was not cogently able to explain why galaxies observed from earth are less densely distributed along the galactic equator, though.

The remainder of the section discusses the two sides.

Curtis's Argument

Curtis remained more conservative in his argument by not introducing outlandish suppositions like the special repulsion forces of Shapley. Curtis, although not able to explain galaxies' redshifts well either, made a convincing claim for the distribution of galaxies in the sky. Because observed galaxies seen edge-on often contain opaque gas lanes, and if our galaxy is similar to those observed galaxies, then we should not see extragalactic objects near our equator because of our own opaque dust lane. Hubble's observation of Cepheid variables in the Andromeda Galaxy confirmed the distance to a galaxy: millions of light-years, not thousands! Curtis was obviously correct.

Here, the section ends by saying which side won the debate.

Source: A. Aversa, 2003.

CHRONOLOGICAL ORDER Time offers its own logic because events happen in chronological order. You can arrange information logically according to the sequence of events.

> Three things happen inside the bronchial tubes and airways in the lungs of people with asthma. The first change is inflammation: The tubes become red, irritated, and swollen. This inflamed tissue "weeps," producing thick mucus. If the inflammation persists, it can lead to permanent thickening in the airways.
>
> Next comes constriction: The muscles around the bronchial tubes tighten, causing the airways to narrow. This is called *bronchospasm* or *bronchoconstriction.*
>
> Finally, there's hyperreactivity. The chronically inflamed and constricted airways become highly reactive to so-called triggers: things like allergens (animal dander, dust mites, molds, pollens), irritants (tobacco smoke, strong odors, car and factory emissions), and infections (flu, the common cold). These triggers result in progressively more inflammation and constriction.
>
> *Source:* G. Shapiro, 1998.

PROBLEM/NEEDS/SOLUTION The *problem/needs/solution* pattern is a common organizational scheme in technical documents, because technical work often involves solving problems. When using this pattern, you should start by identifying the problem that needs to be solved. Then, state what is needed to solve the problem. Finally, end the section by stating the solution (Figure 16.13). The three-part structure leads readers logically from the problem to the solution.

EXAMPLE Using an *example* is a good way to support your claims. Your readers might struggle with facts and data, but an example can help make all those technical details more realistic and familiar.

> To fool predators, most butterflies have evolved colors and patterns that allow them to survive. For example, the delicious Red-spotted Purple is often not eaten by birds, because it looks like a Pipevine Swallowtail, a far less appetizing insect. Other butterflies, like angel wings, look like tree bark or leaves when seen from below. If a predator comes near, though, angel wings can surprise it with a sudden burst of color from the topside of their wings. (Pyle, 1981, p. 23)

GO TO THE NET

Want to see other models that use chronological order? Go to **www.pearsonhighered.com/johnsonweb4/16.11**

A Section Describing Problems, Needs, and Solutions

Figure 16.13: The problem/needs/solution pattern is a good way to lead people logically toward taking action.

North Creek Water Cleanup Plan (TMDL)

Opening paragraph identifies the problem.

North Creek does not meet state standards for swimming and wading because there is too much bacteria in the water. Also, the federal government has determined that Chinook salmon are threatened, and other salmon species face continuing pressure from urban development.

In the 1960s, much of the watershed was home to small ranches and hobby farms. Over the past 40 years, much of the land has been redeveloped with a trend towards more urban, commercial, and suburban residential development. The basin's hydrology, how water is stored and managed throughout the basin, has also changed.

Why Are These Waters Polluted?

The section's body elaborates on the problem.

Pollution in the North Creek watershed comes from thousands of sources that may not have clearly identifiable emission points; this category of pollution is called "nonpoint" pollution. These nonpoint sources can contribute a variety of pollutants that may come from failing septic systems, livestock and pet wastes, at-home car washing, lawn and garden care, leaky machinery, and other daily activities. Some of these nonpoint sources create fecal coliform bacterial pollution that indicate the presence of fecal wastes from warm-blooded animals. Ecology has confirmed that high levels of fecal coliform bacteria exist in North Creek. For this reason, Total Maximum Daily Loads for fecal coliform bacteria were subsequently established at multiple locations through each watershed.

Although wildlife can also contribute bacteria, such sources are not defined as pollution; however, when such natural sources combine with nonpoint pollution, the result can cause the kind of problems found in North Creek.

North Creek became polluted because of the way we do things, not the activities themselves. For example, having dogs, cats, horses, and other animals as part of our life is not a problem; rather, it is the way that we care for these animals. Similarly,roads and parking lots are a necessity of our modern society, but the way we build centers is causing our local streams and creeks to be polluted. There are solutions that can be undertaken by local governments, businesses, organizations, and citizens to solve the problem.

What Can You Do to Improve North Creek?

If cleaning up local waters is important to you, think about what you can do on your own first. Do you always pick up after your pet? Can you use organic

Source: Washington State Department of Ecology, http://www.ecy.wa.gov/programs/wq/tmdl/watershed/north_creek/solution.html.

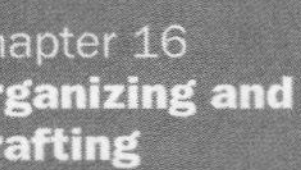

Want to see people using examples in their documents? See **www.pearsonhighered.com/johnsonweb4/16.12**

These questions are designed to highlight what is needed.

Here are the recommended solutions to the problem.

Figure 16.13: *(continued)*

fertilizer? Do you wash your car on your lawn or take it to a car wash? Can you reduce the amount of stormwater runoff from your property? Can you develop a farm plan to ensure your horse's manure is not reaching local streams? Do you practice good on-site septic system maintenance? The draft Action Plan includes information about current and future activities to clean up local waters.

There are many things residents can do now to reduce pollution reaching water bodies and to improve water quality. Here are some ways that you can help:

- Be responsible for proper septic tank maintenance or repair. If you have questions about your on-site septic system (exactly where it is located, how to maintain it), you can call the Snohomish Health District for technical assistance.
- Can you reduce stormwater leaving your property? To have a free survey of your property for ways to reduce potential water quality problems and improve stormwater management, contact Craig Young, your North Creek Basin steward.
- Keep pet and other animal wastes out of your local streams. Pick up after your pet and work with your community, association, or local government to get a pet waste collection station installed where it is needed most.
- Use landscaping methods that eliminate or reduce fertilizers and pesticides. If fertilizers are needed, organic products break down more slowly and help prevent big flushes of pollution when we have heavy rains; they also improve soil structure.
- Join local volunteers in planting trees and performing other activities that help local streams. Snohomish County, the Stilly/Snohomish Task Force and North Creek Streamkeepers, and the City of Bothell water quality volunteer program improve water quality by helping with stream restoration activities. They provide help or other opportunities to plant trees on your property or at other needed locations to help water quality (you can also volunteer to plant trees in other areas that need help too).
- Get involved in your local government's programs. Folks interested in sampling their local waters can contact Snohomish County, North Creek Streamkeepers, or your local city to explore the availability of volunteer monitoring opportunities. If you live outside of a city, be a Salmon Watcher or Watershed Keeper, or get involved in other individual or group activities to improve local waters; these are coordinated by Snohomish County staff.

(continued)

Figure 16.13: *(continued)*

The section closes by restating the main point and looking to the future.

How else can you be involved? The solution to polluting our local waters is to do some things a little differently. In this way, we can still live a normal 21st-century lifestyle, have animals as a close part of our lives, and have clean water. Citizen involvement in deciding what needs to be done is essential to making our water bodies safe places for people and fish, and you may be part of helping to design future watershed activities that haven't even been thought of yet! Check out the Related Links page for more ideas.

Organizing and Drafting the Conclusion

An effective conclusion rounds out the discussion by bringing readers back to the subject, purpose, and main point of your document. Conclusions should be concise and to the point.

Again, put yourself in your readers' place. As a document concludes, what do you want to know? More than likely, you want answers to these questions:

What is the main point?

Why is this information important to me?

Where do we go from here?

Five Closing Moves in a Conclusion

These questions translate into five moves typically found in a conclusion, which will help round off your document and reestablish the context that you created in the introduction.

MOVE 1: MAKE AN OBVIOUS TRANSITION After reading the body of your document, your readers need to wake up a bit. By using a heading such as "Final Points" or a transitional phrase such as "To sum up," you will signal to the readers that you are going to tell them your main points. Your readers will sit up and begin reading closely again. Here are some transitions that will wake up your readers:

In conclusion,	*Put briefly,*	*Overall,*
To sum up,	*In brief,*	*As a whole,*
Let us sum up,	*Finally,*	*In the end,*
In summary,	*To finish up,*	*On the whole,*
In closing,	*Ultimately,*	

For sample conclusions that you can model, see **www.pearsonhighered.com/johnsonweb4/16.13**

The readers' heightened attention will last only a little while, perhaps for a paragraph or a page. If your conclusion runs more than a page or two, your readers will lose interest and even become annoyed by its length.

Link

For more information on using transitions, see Chapter 17, page 464.

MOVE 2: RESTATE YOUR MAIN POINT In the conclusion, you need to restate your main point one more time to drive it home. After all, your readers now have all the facts, so they should be ready to make a final decision.

> If Darbey is to survive and thrive, we need to take action now to address its increasing flooding problem. By restoring wetlands, developing greenways, and building levees, we can begin preparing for the flooding problems that are almost certainly a risk in the future.

Your main point should be absolutely obvious to readers as they finish the document. You may need to say something direct, such as, "Our main point is . . ." to ensure that your readers don't miss it.

MOVE 3: RE-EMPHASIZE THE IMPORTANCE OF THE SUBJECT Sometimes readers need to be reminded of why the subject of your document is important to them.

> If we can reduce or eliminate flooding in Darbey, we will save our citizens millions of dollars in lost revenues and reconstruction. Moreover, Darbey will be viewed as a place with a future, because flooding will not continually undo all our hard work.

Don't try to scare the readers at this point. Stay positive. Stress the benefits of accepting your ideas or agreeing to your recommendations. Scaring your readers will take the shine off your document at this crucial moment.

MOVE 4: LOOK TO THE FUTURE Looking to the future is a good way to end any document (Figure 16.14). A sentence or paragraph that looks to the future will leave your readers with a positive image.

> When we have effectively managed the Curlew River, Darbey will likely see steady growth in population and industry. Once its reputation for flooding has been removed, people and businesses will likely move to this area for its riverside charm and outdoor activities. The town will experience a true revival.

AT A GLANCE

Five Moves in a Conclusion

- Move 1: Make an obvious transition.
- Move 2: Restate your main point.
- Move 3: Re-emphasize the importance of the subject.
- Move 4: Look to the future.
- Move 5: Say thank you and offer contact information.

MOVE 5: SAY THANK YOU AND OFFER CONTACT INFORMATION You might end your document by saying thank you and offering contact information.

> We appreciate your time and consideration. If you have any questions or would like to meet with us about this report, please contact the task force leader, Mary Subbock, at 555–0912 or e-mail her at msubbock@cdarbey.gov.

This kind of thank you statement leaves readers with a positive feeling and invites them to contact you if they need more information.

For more help with writing conclusions, go to **www.pearsonhighered.com/johnsonweb4/16.14**

Sample Conclusion

An obvious transition

The main point

Stresses the importance of the subject

A look to the future

A thank you with contact information

Conclusion

In conclusion, if Darbey is to survive and thrive, we need to take action now to address its increasing flooding problem. By restoring wetlands, developing greenways, and building levees, we can begin preparing for the flooding problems that are almost certainly a risk in the future.

The benefits clearly outweigh the costs. If we can reduce or eliminate flooding in Darbey, we will save our citizens millions of dollars in lost revenues and reconstruction. Moreover, Darbey will be viewed as a place with a future, because flooding will not continually undo all our hard work.When we have effectively managed the Curlew River, Darbey will likely see steady growth in population and industry. Once its reputation for flooding has been removed, people and businesses will likely move to this area for its riverside charm and outdoor activities. The town will experience a true revival.

We appreciate your time and consideration. If you have any questions or would like to meet with us about this report, please contact the task force leader, Mary Subbock, at 555-0912 or e-mail her at msubbock@cdarbey.gov.

Figure 16.14: This conclusion demonstrates all five closing moves that might be found in a document.

Organizing Cross-Cultural Documents

When communicating with international readers, you may need to alter the organization of a document to suit their different expectations. Most international readers know that North American documents tend to be direct and to the point. Even with this awareness, though, the bluntness might still distract or even irritate them. So, your message may be more effective if you consider your readers' cultural preferences.

When organizing a cross-cultural document or presentation, perhaps the greatest concern is whether you are communicating with people from a high-context culture like many in Asia, the Middle East, and Africa or a low-context culture like those in North America or Europe. In a high-context culture, people expect a writer or speaker to provide a significant amount of contextual information before expressing the point or stating the purpose of the document. To most North Americans and Europeans, this up-front contextual information may make the document seem unfocused and its message sound distracted or vague.

Link

For more information on writing cross-cultural texts, see Chapter 2, page 28.

To someone from a high-context culture, however, putting most or all of the contextual information up front signals that the writer is being careful, polite, and deliberate (Figure 16.15). This "indirect approach" signals sensitivity to the importance of the message and the refinement of the writer. In some cultures, this indirectness is intended to allow the reader to save face when problems are mentioned or criticisms are made later in the document.

An Indirect Approach Letter

Figure 16.15: This example letter demonstrates the indirect approach common in many Asian cultures. This kind of letter would be written to a reader who had met the writer, but the relationship is still new.

10 February, 2011

Mr. Wu Choa
Northern Construction Company, Ltd.
2550 Hongqiao Road
Shanghai, 200335, China

Dear Mr. Wu Choa,

Weather, health, and safety are common opening topics.

We hope you are in good health and that you returned home safely from our meeting in Shanghai. Our representatives in Shanghai tell us the weather has been mild and rainy. In our state of Missouri, the weather continues to be cold with more snow than usual, so we are expecting a late spring. We hope spring will bring less moisture.

Reinforces relationships with historical information

Hammond Industries greatly appreciates your visit with our company's representatives in Shanghai. Our company was started 25 years ago by Charles Hammond in Missouri. Hammond Industries has been supplying solar panels to Asia for 15 years. We have relationships with 8 construction companies in China. Our products are recognized for their reliability and affordability. Our sales and service team in Shanghai is dependable and happy to respond to customers' needs.

Becomes more specific by explaining corporate goals

China is a growing market for our products. Our goal is to strengthen and expand our connections in the Shanghai province. We hope our products will continue to meet the needs of large construction companies in your province.

Main point: Products are available for purchase.

We have included a catalog and brochures that describe our products. Our representatives in Shanghai are also available to provide more information. Calls can be made to Mr. Sun, our sales representative. Any orders can be made through our website.

Sincerely,

Sally Gualandi

Sally Gualandi
Special Representative for Asia
Hammond Industries

Indirect Approach Introductions

The *indirect approach* primarily affects the introductions and conclusions of documents. In Asian cultures, including China, Japan, Korea, and India, writers and speakers often work from the most general information (placed in the introduction) toward the main point (found in the conclusion).

Usually, the first aim in an indirect approach document is to establish a relationship or call attention to the existing relationship. A Japanese introduction may discuss the weather or the changing of the seasons. A Chinese or Korean introduction may refer to any prior relationships between the writer's and reader's companies.

Link

For more information on writing introductions in letters, see Chapter 5, page 102.

The example letter in Figure 16.15 plays it safe by mentioning the weather, health, and safety. Notice how the letter starts with general information and works toward more specific information. A letter like this one would be appropriate when the reader has met the writer but the relationship is still new. If the relationship were more established, the second paragraph describing the company would be removed.

For many Asian readers, the preservation of harmony is important, so introductions will likely be subtle, striving to draw attention to common values and experiences. The individual is de-emphasized, while the community and situation are emphasized.

In Arab nations and many Islamic cultures, introductions may include calls for Allah's favor on the reader, family, or business. It is common to find highly ornate statements of appreciation, almost to the point of exaggeration. Arab writers may mention their own accomplishments and the high standing of their acquaintances, while complimenting the reader's achievements and high status. This exaltation of self and acquaintances may seem unnecessary to the North American reader, but the intention is to build a relationship between writer and reader.

Letters in Mexico and some South American countries might start with polite inquiries about family members, especially spouses. They then usually move to the main point, much like North American or European documents.

Indirect Approach Conclusions

Link

For more information on writing conclusions in letters, see Chapter 5, page 106.

An indirect approach conclusion usually states the main point of the document but not in an overly direct way. The most effective conclusion is one in which writer and reader reach a common understanding, often without the main point being stated directly.

For example, in the letter in Figure 16.15, the writer is trying to persuade the reader to buy her company's products. But, she does not write, "We would like to sell you our products." Instead, she only makes the catalog and brochures available and explains where orders can be made. The reader in China will recognize that she is trying to sell something.

CHAPTER REVIEW

- The beginning of a document (introduction) builds a context. The middle (body) provides the content. And the end (conclusion) rebuilds the context.
- A genre is a predictable pattern for organizing information to achieve specific purposes.
- Outlining may seem old-fashioned, but it is a very effective way to sketch out the organization of a document.
- Presentation software can help you organize complicated information.
- Introductions usually include up to six opening moves: (1) define your subject, (2) state your purpose, (3) state your main point, (4) stress the importance of the subject, (5) provide background information, and (6) forecast the content.
- The body of larger documents is usually carved into sections. Each section has an opening, a body, and perhaps a closing.
- Sections usually follow patterns of arrangement to organize information.
- Conclusions usually include up to five closing moves: (1) make an obvious transition, (2) restate your main point, (3) re-emphasize the importance of the subject, (4) look to the future, and (5) say thank you and offer contact information.
- Some international readers, especially those from Asia, Africa, and the Middle East, will feel more comfortable with an "indirect approach" document that moves from the general to the specific.

EXERCISES AND PROJECTS

Individual and Team Projects

1. Find a document that interests you at www.pearsonhighered.com/johnsonweb4/16.16 (or you can find one on your own). Write a memo to your instructor in which you critique the organization of the document. Does it have a clear beginning, middle, and end? Does the introduction make some or all of the six opening moves? Is the body divided into sections? Does the conclusion make some or all of the five closing moves? Explain to your instructor how the document's organization might be improved.

2. Find an introduction that you think is ineffective. Then, rewrite the introduction so that it includes a clear subject, purpose, and main point. Also, stress the importance of the document's subject, provide some background information, and forecast the structure of the document's body. Sample introductions are available at www.pearsonhighered.com/johnsonweb4/16.16.

3. On the Internet or in a print document, find a section that uses one of the following organizational patterns:

    - Cause and effect
    - Comparison and contrast
    - Better and worse
    - Costs and benefits
    - *If . . . then*
    - *Either . . . or*
    - Problem/needs/solution
    - Example

    Prepare a brief presentation to your class in which you show how the text you found follows the pattern.

    You can find sample documents at www.pearsonhighered.com/johnsonweb4/16.16 that will use some or all of these organizational patterns.

Collaborative Project

Around campus or at www.pearsonhighered.com/johnsonweb3/16.16, find a large document, perhaps a report or proposal. With presentation software, outline the document by creating a slide for each section or subsection. Use the headings of the sections as titles on your slides. Then, on each slide, use bulleted lists to highlight the important points in each section.

When you have finished outlining, look over the presentation. Are there any places where too much information is offered? Are there places where more information is needed? Where could you rearrange information to make it more effective?

Present your findings to your class. Using the presentation software, show the class how the document you studied is organized. Then, discuss some improvements you might make to the organization to highlight important information.

For additional technical writing resources, including interactive sample documents, document design tutorials and guidelines, and more, go to **www.mytechcommlab.com.**

Sample documents for these exercises can be found at **www.pearsonhighered.com/johnsonweb4/16.15**

CASE STUDY

The Bad News

In most ways, the project had been a failure. Lisa Franklin was on a chemical engineering team developing a polymer that would protect lightweight tents against desert elements like extreme sun, sandstorms, and chewing insects. Hikers and campers were interested in tents with this kind of protection. But the company Lisa worked for, Outdoor Solutions, wanted to begin selling tents to the military. Bulk sales to the U.S. Army and Marines would improve the company's bottom line, as well as open new opportunities to sell other products.

Despite a promising start, Lisa and the other researchers had not yet developed a stable polymer that would provide the desired protection. The best polymers they had developed were highly flammable. Tent materials covered with these polymers went up like an inferno when exposed to flame. The nonflammable polymers, meanwhile, seemed to break down within three months of use in desert conditions. Lisa's boss, Jim Franklin, was convinced they were close to a breakthrough, but they just couldn't find the right combination to create a nonflammable polymer that would last.

Unfortunately, Lisa's boss had been giving the company president the impression that the polymer was already a success. So, the president had secured a million-dollar loan to retool a factory to produce new lightweight tents coated with the polymer. Renovation was due to begin in a month.

Before starting the factory retooling, the president of the company asked for a final update on the polymer. So Jim wrote a progress report. Then he gave the report to Lisa for final revisions, because he was going out of town on a business trip. He said, "Do whatever you want to revise the report. I'm so frustrated with this project, I don't want to look at this report anymore. When you're done, send the report to the company president. I don't need to see it again."

In the report, Lisa noticed, Jim's facts were all true, but the organization of the report hid the fact that they had not developed a workable polymer. For example, when Jim mentioned that their best polymers were highly flammable, he did so at the end of a long paragraph in the middle of the report. It was highly unlikely that the president would be reading closely enough to see that important fact.

Lisa knew Jim was trying to hide the research team's failure to develop a workable polymer. She didn't want to admit that they had failed either. But she also thought it was important that the company president understand that they had not developed a successful polymer. After all, there was a lot of money on the line.

If you were in Lisa's place, how would you handle this situation?

CHAPTER

17

Using Plain and Persuasive Style

What Is Style? 455

Writing Plain Sentences 455

Help: Intercultural Style and Translation Software 462

Writing Plain Paragraphs 464

When Is It Appropriate to Use Passive Voice? 470

Persuasive Style 472

Balancing Plain and Persuasive Style 477

Chapter Review 478

Exercises and Projects 478

Case Study: Going Over the Top 480

In this chapter, you will learn:

- The importance of style in technical communication.
- The differences between plain and persuasive style.
- Strategies for writing plain sentences and paragraphs.
- How to tell when it's appropriate to use persuasive style.
- How to use techniques of persuasive style, including tone, similes, analogies, metaphors, and pace.

How you say something is often what you say. Your document's style expresses your attitude toward the subject. It reflects your character by embodying the values, beliefs, and relationships you want to share with the readers. In a word, style is about *quality.* Style is about your and your company's commitment to excellence.

One of the great advantages of writing with computers is the ability to move text around. By learning some rather simple techniques, you can make your writing clearer and more persuasive.

What Is Style?

In technical communication, style is not embellishment or ornamentation. Style is not artificial flavoring added to a document to make the content more "interesting." A few added adjectives or exclamation marks won't do much to improve your style.

Good style goes beyond these kinds of superficial cosmetic changes. It involves

- choosing the right words and phrases.
- structuring sentences and paragraphs for clarity.
- using an appropriate tone.
- adding a visual sense to the text.

Historically, rhetoricians have classified style into three levels: plain style, persuasive style, and grand style.

AT A GLANCE

Two Common Styles in Technical Communication

- Plain style—stresses clear wording and simple prose.
- Persuasive style—is used to influence people to accept your ideas and take action.

Plain style—Plain style stresses clear wording and simple prose. It is most often used to instruct, teach, or present information. Plain style works best in documents like technical descriptions, instructions, and activity reports.

Persuasive style—There are times when you will need to influence people to accept your ideas and take action. In these situations, persuasive style allows you to add energy and vision to your writing and speaking. This style works best with proposals, letters, articles, public presentations, and some reports.

Grand style—Grand style stresses eloquence. For example, Martin Luther King, Jr., and President John F. Kennedy often used the grand style to move their listeners to do what was right, even if people were reluctant to do it.

In this chapter, we will concentrate solely on plain and persuasive style, because they are most common in technical documents. Grand style is rarely used in technical communication because it often sounds too ornate or formal for the workplace.

Writing Plain Sentences

In the past, you have probably been told to "write clearly" or "write in concrete language" as though making up your mind to do so was all it took. In reality, writing plainly is a skill that requires practice and concentration. Fortunately, once you have mastered a few basic guidelines, plain style will become a natural strength in your writing.

To see further discussions of style by classical rhetoricians, go to **www.pearsonhighered.com/johnsonweb4/17.1**

Basic Parts of a Sentence

To start, let's consider the parts of a basic sentence. A sentence in English typically has three main parts: a subject, a verb, and a comment.

Subject—What the sentence is about

Verb—What the subject is doing

Comment—What is being said about the subject

English is a flexible language, allowing sentences to be organized in a variety of ways. For example, consider these three variations of the same sentence:

Subject	Verb	Comment
The Institute	provided	the government with accurate crime statistics.
The government	was provided	with accurate crime statistics by the Institute.
Crime statistics	were provided	to the government by the Institute.

Notice that the *content* in these sentences has not changed; only the order of the words has changed. However, the focus of each sentence changes when the subject is changed. The first sentence is *about* the "Institute." The second sentence is *about* the "government." The last sentence is *about* "crime statistics." By changing the subject of the sentence, you essentially shift its focus, drawing your readers' attention to different issues.

Eight Guidelines for Plain Sentences

This understanding of the different parts of a sentence is the basis for eight guidelines that can be used to write plainer sentences in technical documents.

GUIDELINE 1: THE SUBJECT OF THE SENTENCE SHOULD BE WHAT THE SENTENCE IS ABOUT. Confusion often creeps into texts when readers cannot easily identify the subjects of the sentences. For example, what is the subject of the following sentence?

1. Ten months after the Hartford Project began in which a team of our experts conducted close observations of management actions, our final conclusion is that the scarcity of monetary funds is at the basis of the inability of Hartford Industries to appropriate resources to essential projects that have the greatest necessity.

This sentence is difficult to read for a variety of reasons, but the most significant problem is the lack of a clear subject. What is this sentence about? The word "conclusion" is currently in the subject position, but the sentence might also be about "our experts," "the Hartford Project," or "the scarcity of monetary funds."

For more information on how sentences are constructed, see **www.pearsonhighered.com/johnsonweb4/17.2**

When you run into a sentence like this one, first decide what the sentence is about. Then, cut and paste to move that subject into the subject slot of the sentence. For example, when this sentence is restructured around "our experts," readers will find it easier to understand:

The subject (underlined) is what this sentence is "about."

1a. Ten months after the Hartford Project began, our experts have concluded through close observations of management actions that the scarcity of monetary funds is at the basis of the inability of Hartford Industries to appropriate resources to essential projects that have the greatest necessity.

This sentence is still rather difficult to read. Nevertheless, it is easier to understand because the noun in the subject slot ("our experts") is what the sentence is about. We will return to this sentence later in this chapter.

GUIDELINE 2: THE SUBJECT SHOULD BE THE "DOER" IN THE SENTENCE. Readers tend to focus on who or what is doing something in a sentence. For example, which of the following sentences is easier to read?

2a. On Saturday morning, the paperwork was completed in a timely fashion by Jim.

2b. On Saturday morning, Jim completed the paperwork in a timely fashion.

Most readers would say that sentence 2b is easier to read because Jim, the subject of the sentence, is actually doing something. In the first sentence, the paperwork is merely sitting there. People make especially good subjects of sentences, because they are usually active.

GUIDELINE 3: THE VERB SHOULD STATE THE ACTION, OR WHAT THE DOER IS DOING. Once you have determined who or what is doing something, ask yourself what that person or thing is actually doing. Find the *action* and turn it into the verb of the sentence. For example, consider these sentences:

In these sentences, it becomes harder and harder to figure out the action.

3a. The detective investigated the loss of the payroll.

3b. The detective conducted an investigation into the loss of the payroll.

3c. The detective is the person who is conducting an investigation of the loss of the payroll.

Sentence 3a is easy to understand because the action of the sentence is expressed in the verb. Sentences 3b and 3c are increasingly more difficult to understand, because the action, "investigate," is further removed from the verb of the sentence.

GUIDELINE 4: THE SUBJECT OF THE SENTENCE SHOULD COME EARLY IN THE SENTENCE. Subconsciously, your readers start every sentence looking for the subject. The subject anchors the sentence, because it tells readers what the sentence is about.

For some interesting discussions of plain style, go to **www.pearsonhighered.com/johnsonweb4/17.3**

Consider these two sentences:

It's easier to locate the subject of this sentence because it is up front.

4a. If deciduous and evergreen trees experience yet another year of drought like the one observed in 1997, the entire Sandia Mountain ecosystem will be heavily damaged.

4b. The entire Sandia Mountain ecosystem will be heavily damaged if deciduous and evergreen trees experience yet another year of drought like the one observed in 1997.

The problem with sentence 4a is that it forces readers to hold all those details about trees, drought, and 1997 in short-term memory before it identifies the subject of the sentence. Readers almost feel a sense of relief when they come to the subject, because until that point they cannot figure out what the sentence is about.

Quite differently, sentence 4b tells readers what the sentence is about up front. With the subject early in the sentence, readers immediately know how to connect the comment with the subject.

The first sentence is harder to read because the action is in a nominalization.

GUIDELINE 5: ELIMINATE NOMINALIZATIONS. Nominalizations are perfectly good verbs and adjectives that have been turned into awkward nouns. For example, look at these sentences:

5a. Management has an expectation that the project will meet the deadline.

5b. Management expects the project to meet the deadline.

Here are a whole bunch of nominalizations.

In sentence 5a, "expectation" is a nominalization. Here, a perfectly good verb is being used as a noun. Sentence 5b is not only shorter than sentence 5a, but it also has more energy because the verb "expects" is now an action verb.

Consider these two sentences:

See how reducing them improves the readability?

6a. Our discussion about the matter allowed us to make a decision on the acquisition of the new x-ray machine.

6b. We discussed the matter and decided to acquire the new x-ray machine.

Sentence 6a includes three nominalizations—"discussion," "decision," and "acquisition"—making the sentence hard to understand. Sentence 6b is clearer, because the nominalizations "discussion" and "decision" have been turned into action verbs.

Why do writers use nominalizations in the first place? They use nominalizations for two reasons:

- First, humans generally think in terms of people, places, and things (nouns), so our first drafts are often filled with nominalizations, which are nouns. While revising, an effective writer will turn those first-draft nominalizations into action verbs.
- Second, some people mistakenly believe that using nominalizations makes their writing sound more formal or important. In reality, though, nominalizations only make sentences harder to read. The best way to sound important is to write sentences that readers can understand.

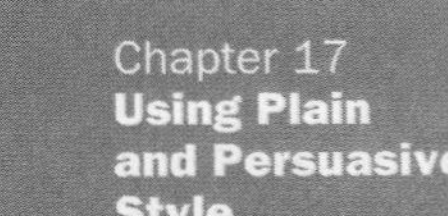

Want to learn other methods for improving style at the sentence level? Go to **www.pearsonhighered.com/johnsonweb4/17.4**

GUIDELINE 6: AVOID EXCESSIVE PREPOSITIONAL PHRASES. Prepositional phrases are necessary in writing, but they are often overused in ways that make text too long and too tedious. Prepositional phrases follow prepositions like *in, of, by, about, over,* and *under.* These phrases are used to modify nouns.

Prepositional phrases become problematic when used in excess. For example, sentence 7a is difficult to read because it links too many prepositional phrases together (the prepositional phrases are underlined and the prepositions are italicized). Sentence 7b shows the same sentence with fewer prepositional phrases:

The prepositional phrases have been reduced, clarifying the sentence's meaning.

7a. The decline *in* the number *of* businesses owned *by* locals *in* the town *of* Artesia is a demonstration *of* the increasing hardship faced *in* rural communities *in* the southwest.

7b. Artesia's declining number of locally owned businesses demonstrates the increased hardship faced by southwestern rural communities.

You should never feel obligated to eliminate all the prepositional phrases in a sentence. Rather, look for long chains of prepositional phrases. Then, try to condense the sentence by turning some of the prepositional phrases into adjectives.

For example, the phrase "in the town of Artesia" from sentence 7a was reduced in sentence 7b to the adjective "Artesia's." The phrase "in rural communities in the southwest" was reduced to "southwestern rural communities." As a result, sentence 7b is much shorter and easier to read.

Figure 17.1 shows a website in which the authors write plainly. Notice how they have used minimal prepositional phrases.

AT A GLANCE

Eight Guidelines for Plain Sentences

- Guideline 1: The subject of the sentence should be what the sentence is about.
- Guideline 2: The subject should be the "doer" in the sentence.
- Guideline 3: The verb should state the action, or what the doer is doing.
- Guideline 4: The subject of the sentence should come early in the sentence.
- Guideline 5: Eliminate nominalizations.
- Guideline 6: Avoid excessive prepositional phrases.
- Guideline 7: Eliminate redundancy in sentences.
- Guideline 8: Write sentences that are "breathing length."

GUIDELINE 7: ELIMINATE REDUNDANCY IN SENTENCES. In your effort to get your point across, you may use redundant phrasing. For example, you might write "unruly mob" as though some kinds of mobs might be orderly. Or, you might talk about "active participants" as though someone can participate without doing anything.

Sometimes buzzwords and jargon lead to redundancies like, "We should collaborate together as a team" or "Empirical observations will provide a new understanding of the subject." In some cases, you might use a synonym to modify a synonym by saying something like, "We are demanding important, significant changes."

You should try to eliminate redundancies because they use two or more words to do the work of one word. As a result, readers need to work twice as hard to understand one basic idea.

GUIDELINE 8: WRITE SENTENCES THAT ARE "BREATHING LENGTH." You should be able to read a sentence out loud in one breath. At the end of the sentence, the period (.) signals, "Take a breath." Of course, when reading silently, readers do not actually breathe when they see a period, but they do take a mental pause at the end of each sentence. If a sentence runs on and on, it forces

For more practice eliminating prepositional phrases, go to **www.pearsonhighered.com/johnsonweb4/17.5**

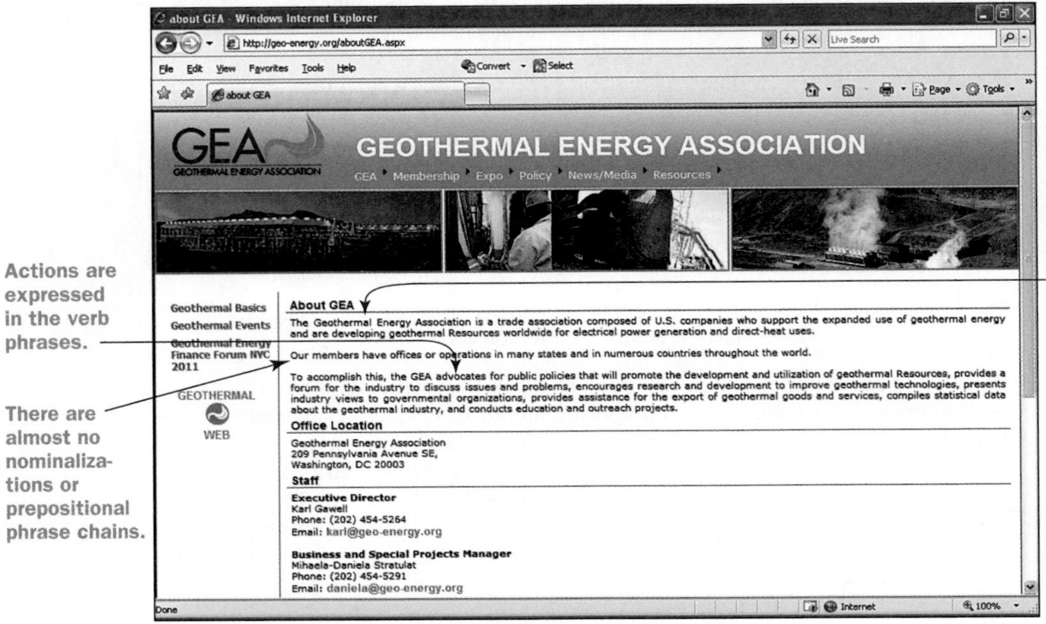

Figure 17.1: In this website, the Geothermal Energy Association uses plain language to convey its facts about geothermal energy.

Subjects are placed up front in sentences

Source: Geothermal Energy Association, http://www.geo-energy.org.

readers to mentally hold their breath. By the end of an especially long sentence, they are more concerned about when the sentence is going to end than what the sentence is saying.

Link

For examples of different breathing-length sentences, see the discussion of pace on page 475.

The best way to think about sentence length is to imagine how long it takes to comfortably say a sentence out loud.

- If the written sentence is too long to say comfortably, it probably needs to be shortened or cut into two sentences. Avoid asphyxiating the readers with sentences that go on forever.
- If the sentence is very short, perhaps it needs to be combined with one of its neighbors to give it a more comfortable breathing length. You want to avoid making readers hyperventilate over a string of short sentences.

Creating Plain Sentences with a Computer

Computers give us an amazing ability to manipulate sentences. So, take full advantage of your machine's capabilities. First, write out your draft as you normally would,

GO TO THE NET

For websites that discuss bureaucratic language, go to **www.pearsonhighered.com/johnsonweb4/17.6**

not paying too much attention to the style. Then, as you revise, identify difficult sentences and follow these seven steps:

1. Identify who or what is doing something in the sentence.
2. Turn that who or what into the subject of the sentence.
3. Move the subject to an early place in the sentence.
4. Identify what the subject is doing, and move that action into the verb slot.
5. Eliminate prepositional phrases, where appropriate, by turning them into adjectives.
6. Eliminate unnecessary nominalizations and redundancies.
7. Shorten, lengthen, combine, or divide sentences to make them breathing length.

With these seven steps in mind, let's revisit sentence 1, the example of weak style earlier in this chapter:

Original

1. Ten months after the Hartford Project began in which a team of our experts conducted close observations of management actions, our final conclusion is that the scarcity of monetary funds is at the basis of the inability of Hartford Industries to appropriate resources to essential projects that have the greatest necessity.

Now, let's apply the seven-step method for revising sentences into the plain style. First, identify who or what is doing something in the sentence, make it the subject, and move it to an early place in the sentence.

The experts are doing something.

The action is now in the verb.

1a. After a ten-month study, our experts have concluded that the scarcity of monetary funds is at the basis of the inability of Hartford Industries to appropriate resources to essential projects that have the greatest necessity.

Now eliminate the prepositional phrases, nominalizations, and redundancies.

1b. After a ten-month study, our experts concluded that Hartford Industries' budget shortfalls limit its support for priority projects.

In the revision, the "doers" ("our experts") were moved into the subject position and then moved to an early place in the sentence. Then, the action of the sentence ("concluded") was moved into the verb slot. Prepositional phrases like "after the Hartford Project" and "to appropriate resources to essential projects" were turned into adjectives. Nominalizations ("conclusion," "necessity") were turned into verbs or adjectives. And finally, the sentence was shortened to breathing length.

The resulting sentence still sends the same message—just more plainly.

Need more examples of revised sentences? Go to **www.pearsonhighered.com/johnsonweb4/17.7**

HELP

Intercultural Style and Translation Software

In the *Hitchhiker's Guide to the Galaxy* by Douglas Adams, aliens are able to communicate with each other by using a Babel fish, which is described as a "yellow and leech-like" creature that can be slipped into the ear.

Unfortunately, these Babel fish are fiction, so we need to translate our own materials or have them translated for us. Professional translators can be expensive, especially if you are just trying to translate an e-mail from one of your overseas clients or figure out a website written in another language.

Several translation software programs are available for purchase, such as Power Translator, Systran, and Translution. The accuracy of these programs has improved greatly over the last decade. A free online translation tool is the Babel Fish website (babelfish.yahoo.com), available through Yahoo! (Figure A). Another free online translator, @promt is available through PROMT (www.online-translator.com).

Translation software is far from perfect. Certainly, you should not rely exclusively on these tools to translate an important proposal or report into another language. To handle these kinds of projects, you will likely need to hire a translation service. However, for smaller texts, some helpful tips can minimize problems when using translation software:

Use basic sentences—Translation software typically tries to translate whole sentences rather than translate word by word. So, the software will struggle with complex sentences. Cut longer, more complex sentences into smaller, simpler sentences.

Use standard punctuation—Periods, commas, and question marks are fine. Less common punctuation marks like ellipses, dashes, colons, and semicolons often create difficulties for translation software. You can usually replace these uncommon marks with basic marks.

Use consistent words—In texts that will be translated, use the same word to mean the same thing. For example, if you called something an "animal" earlier in the document, you should avoid calling it a "mammal," "living being," "critter," or "quadruped" later in the document.

Avoid metaphors, sayings, and clichés—These kinds of phrases require knowledge of the culture of the source language. So, if you write something like "Our product slaughters the competition" or "She couldn't see the forest for the trees," the translation software will come back with some very strange results.

Remove any cultural, historical, or sports-related references—The translation software will not know what you mean if you say something like "That project was our battle of Gettysburg" or "He hit a home run with his presentation."

Back translate all text—After you translate something, have the software translate it back into English. Each time through, the translator will magnify the

To find out more about translation software, go to **www.pearsonhighered.com/johnsonweb4/17.8**

Free Online Translation Tool

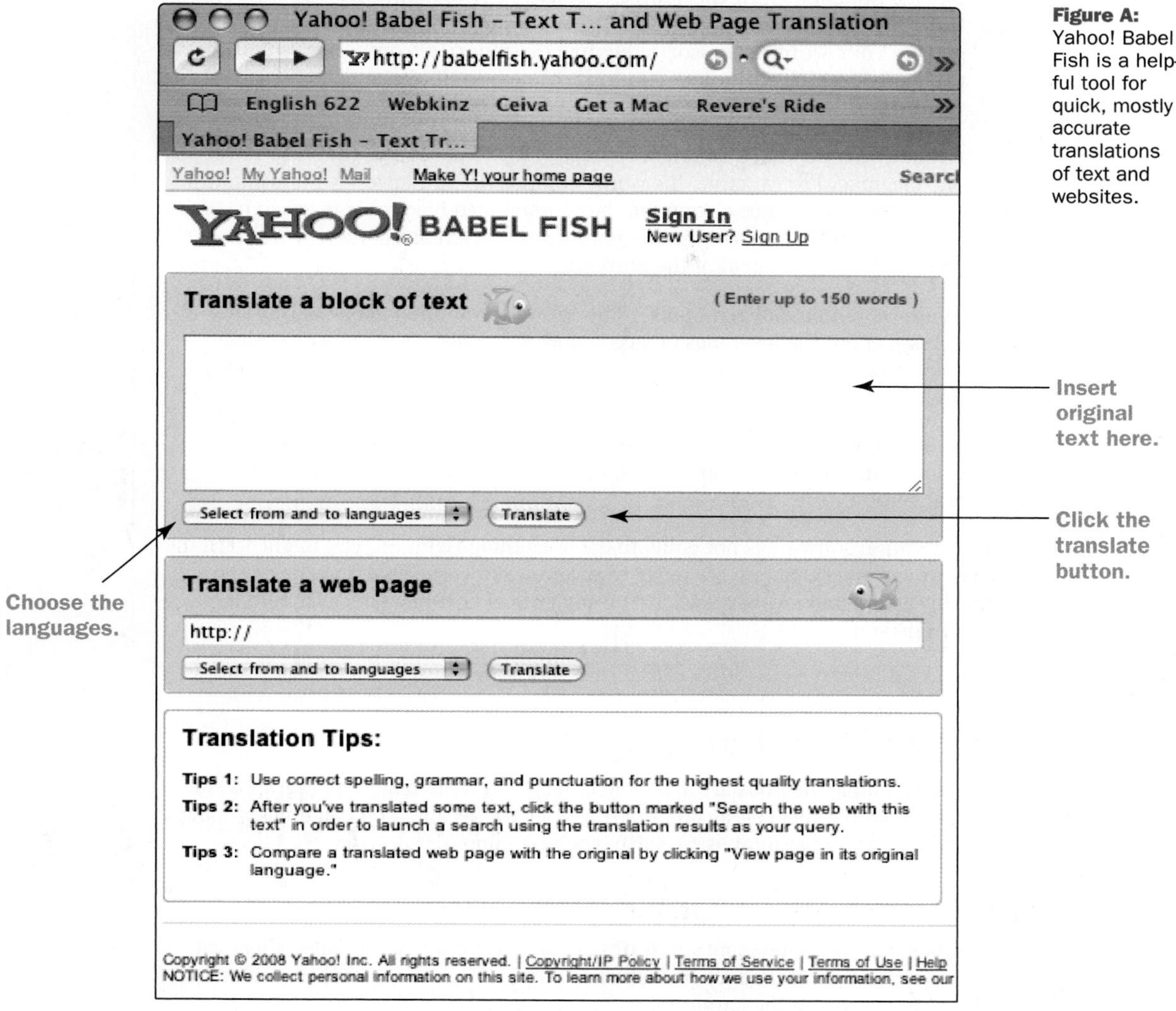

Figure A: Yahoo! Babel Fish is a helpful tool for quick, mostly accurate translations of text and websites.

Source: Reproduced with permission of Yahoo! Inc.® 2008 by Yahoo! Inc. YAHOO! and the YAHOO! logo are trademarks of Yahoo! Inc.

differences in meaning, so the results are often funny. However, the back translation can signal places where the translation software completely missed the mark.

Use the spellchecker on the original *and* translated texts—Since translation software usually tries to translate the whole sentence, any spelling errors

(continued)

or typos will cause it to make unnecessary guesses about what a sentence means. So, spellcheck the source text carefully. Then, after the text is translated, run it through the spellchecker again, using the target language. Your word processor likely has dictionaries from other languages preloaded, which will allow you to locate misspellings or other errors in the translation.

Avoid words that have double meanings—Try to use words that only have one meaning. For example, the word "compound" can have several meanings: "Working in the office compound, he compounded his problems when he used the compound to connect the compound joint." Here is another: "I was not content with the content of the proposal."

Minimize jargon and acronyms—The software will often make inaccurate guesses about the meanings of jargon and acronyms, unless you are using translation software that is specific to your field. So, simplify jargon words, and check how any acronyms were translated.

Avoid puns or other plays on words—The translation software won't understand the double meanings, and they will seem very odd in the translation.

Your translation software is not going to get everything right. So, you might warn the receiver of your text that you are using translation software. That way, he or she will forgive the errors and inconsistencies. For any critical documents, you should hire a human translator.

Writing Plain Paragraphs

Some rather simple methods are available to help you write plainer paragraphs. As with writing plain sentences, a computer really helps, because you can quickly move sentences around in paragraphs to clarify your meaning.

The Elements of a Paragraph

Paragraphs tend to include four kinds of sentences: a *transition* sentence, a *topic* sentence, *support* sentences, and a *point* sentence (Figure 17.2). Each of these sentences plays a different role in the paragraph.

TRANSITION SENTENCE The purpose of a transition sentence is to make a smooth bridge from the previous paragraph to the present paragraph. For example, a transition sentence might state the following:

> Keeping these facts about West Nile virus in mind, let us consider the actions we should take.

To learn more about paragraphing, go to **www.pearsonhighered.com/johnsonweb4/17.9**

Types of Sentences in a Typical Paragraph

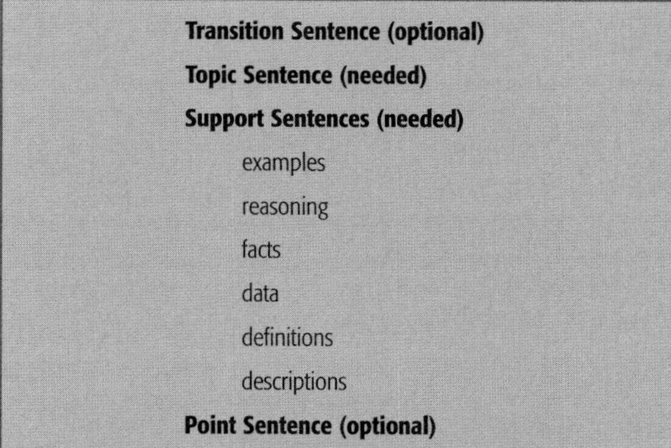
Transition Sentence (optional)
Topic Sentence (needed)
Support Sentences (needed)
examples
reasoning
facts
data
definitions
descriptions
Point Sentence (optional)

Figure 17.2: Paragraphs usually contain up to four types of sentences: transition, topic, support, and point sentences.

> What should we do to help these learning disabled children in our classrooms?

Transition sentences are typically used when the new paragraph handles a significantly different topic than the previous paragraph. They help close the gap between the two paragraphs or redirect the discussion.

TOPIC SENTENCE The topic sentence is the claim or statement that the rest of the paragraph is going to prove or support:

> To combat the spread of the West Nile virus in Pennsylvania, we recommend three immediate steps: vaccinate all horses, create a public relations campaign to raise public awareness, and spray strategically.

> Children with learning disabilities struggle to cope in the normal classroom, so teachers need to be trained to recognize their special needs.

In technical documents, topic sentences typically appear in the first or second sentence of each paragraph. They are placed up front in each paragraph for two reasons.

- The topic sentence sets a goal for the paragraph to reach. It tells readers the claim the writer is trying to prove. Then, the remainder of the paragraph proves that claim.
- The topic sentence is the most important sentence in any given paragraph. Since readers, especially scanning readers, tend to pay the most attention to the beginning of a paragraph, placing the topic sentence up front ensures that they will read it.

GO TO THE NET

More examples of these kinds of sentences are available at **www.pearsonhighered.com/johnsonweb4/17.10**

SUPPORT SENTENCES The bulk of any paragraph is typically made up of support sentences. These sentences contain examples, reasoning, facts, data, anecdotes, definitions, and descriptions.

> First, we recommend that all horses be vaccinated against West Nile virus, because they are often the most significant victims of the virus. Last year, nearly 2000 horses died from the virus. Second, we believe a public relations campaign . . .

> Learning disabled children often struggle to understand their teacher, especially if the teacher lectures for periods longer than ten minutes.

Support sentences are used to prove the claim made in the paragraph's topic sentence.

POINT SENTENCES Point sentences restate the paragraph's main point toward the end of the paragraph. They are used to reinforce the topic sentence by restating the paragraph's original claim in new words. Point sentences are especially useful in longer paragraphs where readers may not fully remember the claim stated at the beginning of the paragraph. They often start with transitional devices like "Therefore," "Consequently," or "In sum," to signal that the point of the paragraph is being restated.

AT A GLANCE

Four Kinds of Sentences in a Paragraph

- Transition sentence (optional)
- Topic sentence
- Support sentence
- Point sentence (optional)

> These three recommendations represent only the first steps we need to take. Combating West Nile virus in the long term will require a more comprehensive plan.

> Again, learning disabled children often struggle undetected in the classroom, so teachers need to be trained to recognize the symptoms.

Point sentences are optional in paragraphs, and they should be used only occasionally when a particular claim needs to be reinforced. Too many point sentences will cause your text to sound repetitious and even condescending.

Using the Four Types of Sentences in a Paragraph

Of these four kinds of sentences, only the topic and support sentences are needed to construct a good paragraph. Transition sentences and point sentences are useful in situations where bridges need to be made between paragraphs or specific points need to be reinforced.

Here are the four kinds of sentences used in a paragraph:

> 8a. How can we accomplish these five goals (transition sentence)? Universities need to study their core mission to determine whether distance education is a viable alternative to the traditional classroom (topic sentence). If universities can maintain their current standards when moving their courses online, then distance education may provide a new medium through which nontraditional students can take classes and perhaps earn a degree (support sentence). Utah State, for example,

Want some more advice about organizing paragraphs? Go to **www.pearsonhighered.com/johnsonweb4/17.11**

is reporting that students enrolled in its online courses have met or exceeded the expectations of their professors (support sentence). If, however, standards cannot be maintained, we may find ourselves returning to the traditional on-campus model of education (support sentence). In sum, the ability to meet a university's core mission is the litmus test to measure whether distance education will work (point sentence).

Here is the same paragraph with the transition sentence and point sentence removed:

8b. Universities need to study their core mission to determine whether distance education is a viable alternative to the traditional classroom (topic sentence). If universities can maintain their current standards when moving their courses online, then distance education may provide a new medium through which nontraditional students can take classes and perhaps earn a degree (support sentence). Utah State, for example, is reporting that students enrolled in its online courses have met or exceeded the expectations of their professors (support sentence). If, however, standards cannot be maintained, we may find ourselves returning to the traditional on-campus model of education (support sentence).

As you can see in paragraph 8b, some paragraphs are fine without transition and point sentences. Nevertheless, transition and point sentences can make texts easier to read while amplifying important points.

Aligning Sentence Subjects in a Paragraph

Now let's discuss how you can make paragraphs flow by weaving sentences together effectively. Have you ever read a paragraph in which each sentence seemed to go off in a new direction? Have you ever run into a paragraph that actually felt "bumpy" as you read it? More than likely, the problem was a lack of alignment of the paragraph's sentence subjects. For example, consider this paragraph:

Notice how the subjects of these sentences (underlined) are not in alignment, making the paragraph seem rough to readers.

9. The lack of technical knowledge about the electronic components in automobiles often leads car owners to be suspicious about the honesty of car mechanics. Although they might be fairly knowledgeable about the mechanical workings of their automobiles, car owners rarely understand the nature and scope of the electronic repairs needed in modern automobiles. For instance, the function and importance of a transmission in a car are generally well known to all car owners, but the wire harnesses and printed circuit boards that regulate the fuel consumption and performance of their car are rarely familiar. Repairs for these electronic components can often run over $400—a large amount to a customer who cannot even visualize what a wire harness or printed circuit board looks like. In contrast, a $400 charge for the transmission on the family car, though distressing, is more readily understood and accepted.

There is nothing really wrong with this paragraph—it's just hard to read. Why? The paragraph is difficult to read because the subject of the sentences changes with each new sentence. Look at the underlined subjects of the sentences in this paragraph.

They are all different, causing each sentence to strike off in a new direction. As a result, each sentence forces readers to shift focus to concentrate on a new subject.

With your word processor, you can easily revise paragraphs to avoid this bumpy, unfocused feeling. The secret is lining up the subjects in the paragraph.

To line up subjects, first ask yourself what the paragraph is about. Then, cut and paste words to restructure the sentences to line up on that subject. Here is a revision of paragraph 9 that focuses on the "car owners" as subjects:

Here, the paragraph has been revised to focus on people.

> 9a. Due to their lack of knowledge about electronics, some car owners are skeptical about the honesty of car mechanics when repairs involve electronic components. Most of our customers are fairly knowledgeable about the mechanical features of their automobiles, but they rarely understand the nature and scope of the electronic repairs needed in modern automobiles. For example, most people recognize the function and importance of a transmission in an automobile, but the average person knows very little about the wire harnesses and printed circuit boards that regulate the fuel consumption and performance of his or her car. So, for most of our customers, a $400 repair for these electronic components seems like a large amount, especially when these folks cannot even visualize what a wire harness or printed circuit board looks like. In contrast, most car owners think a $400 charge to fix the transmission on the family car, though distressing, is more acceptable.

In this revised paragraph, you should notice two things:

- First, the words "car owners" are not always the exact two words used in the subject slot. Synonyms and pronouns are used to add variety to the sentences.
- Second, not all the subjects need to be "car owners." In the middle of the paragraph, for example, "$400 repair" is the subject of a sentence. This deviation from "car owners" is fine as long as the majority of the subjects in the paragraph are similar to each other. In other words, the paragraph will still sound focused, even though an occasional subject is not in alignment with the others.

Of course, the subjects of the paragraph could be aligned differently to stress something else in the paragraph. Here is another revision of paragraph 9 in which the focus of the paragraph is "repairs."

This paragraph focuses on repairs.

> 9b. Repairs to electronic components often lead car owners, who lack knowledge about electronics, to doubt the honesty of car mechanics. The nature and scope of these repairs are usually beyond the understanding of most nonmechanics, unlike the typical mechanical repairs with which customers are more familiar. For instance, the importance of fixing the transmission in a car is readily apparent to most car owners, but adjustments to electronic components like wire harnesses and printed circuit boards are foreign to most customers—even though these electronic parts are crucial in regulating their car's fuel consumption and performance. So, a repair to these electronic components that costs $400 seems excessive, especially when the repair can't even be visualized by the customer. In contrast, a $400 replacement of the family car's transmission, though distressing, is more readily accepted.

GO TO THE NET

Want to practice revising a few paragraphs? Go to **www.pearsonhighered.com/johnsonweb4/17.12**

You should notice that paragraph 9a is easier for most people to read than paragraph 9b. Paragraph 9a is more readable because sentences that have "doers" in the subject slots are easier for people to read. In paragraph 9a, the car owners are active subjects, while in paragraph 9b, the car repairs are inactive subjects.

The Given/New Method

Another way to write plain paragraphs is to use the "given/new" method to weave sentences together. Described by Susan Haviland and Herbert Clark in 1974, the given/new method is based on the assumption that readers try to process new information by comparing it to information they already know. Therefore, every sentence in a paragraph should contain something the readers already know (the given) and something that the readers do not know (the new).

Consider these two paragraphs:

10a. Santa Fe has many beautiful places. Some artists choose to strike off into the mountains to work, while others enjoy working in local studios. The landscapes are wonderful in the area.

10b. Santa Fe offers many beautiful places for artists to work. Some artists choose to strike off into the mountains to work, while others enjoy working in local studios. Both the mountains and the studios offer places to savor the wonderful landscapes in the area.

Both of these paragraphs are readable, but paragraph 10b is easier to read because words are being repeated in a given/new pattern. Paragraph 10a is a little harder to read, because there is nothing "given" that carries over from each sentence to its following sentence.

Using the given/new method is not difficult. In most cases, the given information should appear early in the sentence and the new information should appear later in the sentence. The given information will provide a familiar anchor or context, while the new information will build on that familiar ground. Consider this larger paragraph:

Notice how the chaining of words in this paragraph makes it smoother to read.

11. Recently, an art gallery exhibited the mysterious paintings of Irwin Fleminger, a modernist artist whose vast Mars-like landscapes contain cryptic human artifacts. One of Fleminger's paintings attracted the attention of some young schoolchildren who happened to be walking by. At first, the children laughed, pointing out some of the strange artifacts in the painting. Soon, though, the strange artifacts in the painting drew the students into a critical awareness of the painting, and they began to ask their bewildered teacher what the artifacts meant. Mysterious and beautiful, Fleminger's paintings have this effect on many people, not just schoolchildren.

In this paragraph, the beginning of each sentence provides something given—typically an idea, word, or phrase drawn from the previous sentence. Then, the remainder of each sentence adds something new to that given information. By chaining together given and new information, the paragraph builds readers' understanding gradually, adding a little more information with each sentence.

GO TO THE NET For more information on the given/new method, go to **www.pearsonhighered.com/johnsonweb4/17.13**

In some cases, however, the previous sentence in a paragraph does not offer a suitable subject for the sentence that follows it. In these cases, transitional phrases can be used to provide readers given information in the beginning of the sentence. To illustrate,

12. This public relations effort will strengthen Gentec's relationship with leaders of the community. With this new relationship in place, the details of the project can be negotiated in terms that are fair to both parties.

In this sentence, the given information in the second sentence appears in the transitional phrase, not in the subject slot. A transitional phrase is a good place to include given information when the subject cannot be drawn from the previous sentence.

When Is It Appropriate to Use Passive Voice?

Before discussing the elements of persuasive style, we should expose a writing boogeyman as a fraud. Since childhood, you have probably been warned against using *passive voice.* You have been told to write in *active voice.*

One problem with this prohibition on passive voice is that passive voice is very common in technical communication. In some scientific fields, passive voice is the standard way of writing. So, when is it appropriate to use passive voice?

Passive voice occurs when the subject of the sentence is *being acted upon,* so the verb is in passive voice. Active voice occurs when the subject of the sentence is *doing the acting,* so the verb is an action verb. Here is an example of a sentence written in passive voice and a sentence written in active voice:

13a. The alloy was heated to a temperature of 300°C. (passive)

13b. Andy James heated the alloy to a temperature of 300°C. (active)

The passive voice sentence (sentence 13a) lacks a doer because the subject of the sentence, the alloy, is being acted upon. As a result, sentence 13b might be a bit easier to understand because a doer is the subject of the sentence. But, as you can see, sentence 13a is easy to read and understand, too.

Passive voice does have a place in technical documents. A passive sentence is fine if:

- the readers do not need to know who or what is doing something in the sentence.
- the subject of the sentence is what the sentence is about.

For example, in sentence 13a, the person who heated the alloy might be unknown or irrelevant to the readers. Is it really important that we know that Andy James heated the alloy? Or do we simply need to know that the alloy was heated? If the alloy is what the sentence is about and who heated the alloy is unimportant, then the passive voice is fine.

Do you want more information on using passive voice? See **www.pearsonhighered.com/johnsonweb4/17.14**

If you are wondering whether to use the passive voice, consider your readers. Do they expect you to use the passive voice? Do they need to know who did something in a sentence? Go ahead and use the passive voice if your readers expect you to use passive voice or they don't need to know who did what.

Consider these other examples of passive and active sentences:

14a. The shuttle bus will be driven to local care facilities to provide seniors with shopping opportunities. (passive)

14b. Jane Chavez will drive the shuttle bus to local care facilities to provide seniors with shopping opportunities. (active)

15a. The telescope was moved to the Orion system to observe the newly discovered nebula. (passive)

15b. Our graduate assistant, Mary Stewart, moved the telescope to the Orion system to observe the newly discovered nebula. (active)

In both sets of sentences, the passive voice may be more appropriate, unless there is a reason Jane Chavez or Mary Stewart needs to be singled out for special consideration.

When developing a focused paragraph, passive sentences can often help you align the subjects and use given/new strategies (i.e., make the paragraph more cohesive). For example, compare the following paragraphs, one written in passive voice and the other in active voice:

Paragraph written in passive voice

16a. We were shocked by the tornado damage. The downtown of Wheaton had been completely flattened. Brick houses were torn apart. A school bus was tossed into a farm field, as though some fairy-tale giant had crumpled it like a soda can. The town's fallen water tower had been dragged a hundred yards to the east. Amazingly enough, no one was killed by the storm. Clearly, lives were saved by the early warning that the weather bureau had sent out.

Same paragraph written in active voice

16b. The tornado damage shocked us. The tornado completely flattened the downtown of Wheaton. It had torn apart brick houses. It had tossed a school bus into a farm field, as though some fairy-tale giant had crumpled the bus like a soda can. The tornado had knocked over the town's water tower and dragged it a hundred yards to the east. Amazingly enough, the tornado did not kill anyone. Clearly, the early warning that the weather bureau sent out saved lives.

Most people would find paragraph 16a more interesting and readable because it uses passive voice to put the emphasis on the *damage* caused by the tornado. Paragraph 16b is not as interesting, because the emphasis is repeatedly placed on the *tornado*.

Used properly, passive voice can be a helpful tool in your efforts to write plain sentences and paragraphs. Passive voice is misused only when the readers are left wondering who or what is doing the action in the sentence. In these cases, the sentences should be restructured to put the doers in the subject slots.

Persuasive Style

Link

For more information on reader analysis, go to Chapter 2, page 22.

There are times when you will need to do more than simply present information clearly. You will need to influence your readers to take action or make a decision. In these situations, you should shift to *persuasive style.* When used properly, persuasive style can add emphasis, energy, color, and emotion to your writing.

The following four persuasion techniques will help give your writing more impact. A combination of these, properly used, will make your writing more influential and vivid.

Elevate the Tone

Tone is the resonance or pitch that the readers will "hear" as they read your document. Of course, most people read silently to themselves, but all readers have an inner voice that sounds out the words and sentences. By paying attention to tone, you can influence the readers' inner voice in ways that persuade them to read the document with a specific emotion or attitude.

One easy way to elevate tone in written texts is to first decide what emotion or attitude you want the readers to adopt. Then, use logical mapping to find words and phrases that evoke that emotion or attitude (Figure 17.3).

For example, let's say you want your readers to feel excited as they are looking over your document. You would first put the word "excitement" in the middle of your screen or a sheet of paper. Then, as shown in Figure 17.3, you can map out descriptions and feelings associated with that emotion.

Mapping an Emotional Tone

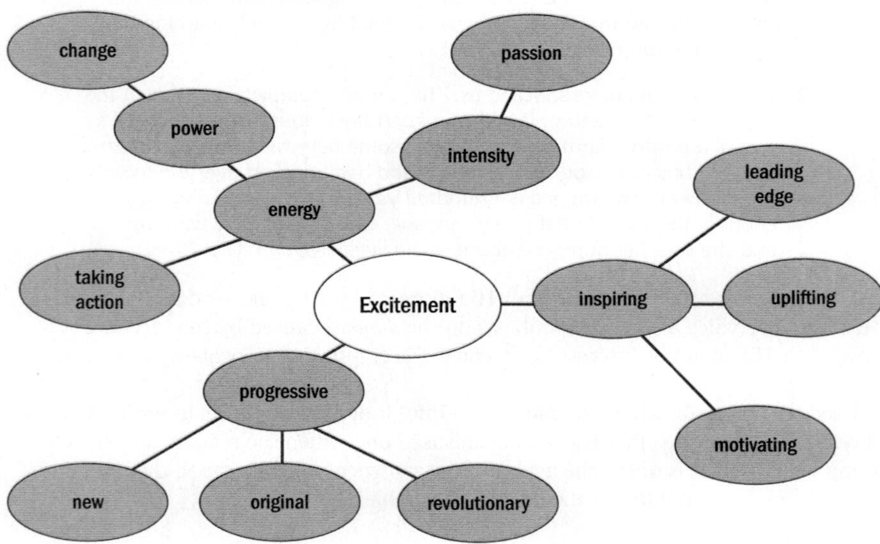

Figure 17.3: You can set an emotional tone by mapping it out and "seeding" the text with words associated with that tone.

GO TO THE NET

For more information on improving tone, go to **www.pearsonhighered.com/johnsonweb4/17.15**

Mapping an Authoritative Tone

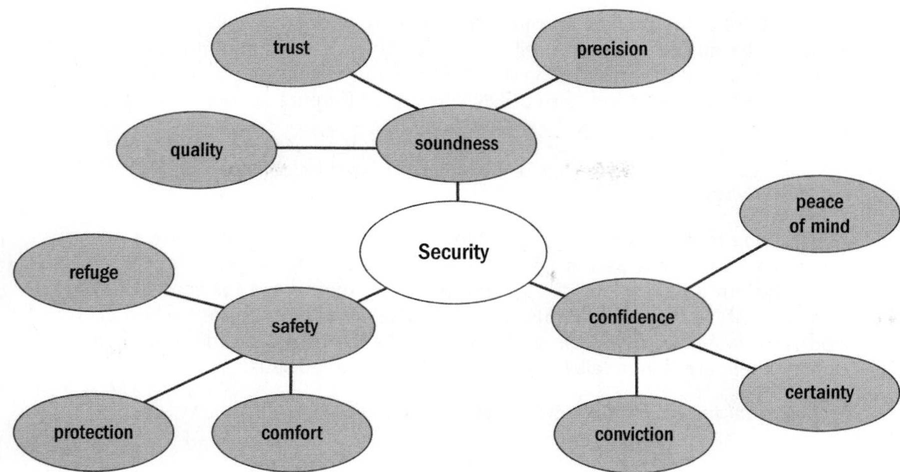

Figure 17.4: If you use these words associated with "security" in your text at strategic points, readers will perceive the sense of security that you are trying to convey.

You can use these words at strategic moments in your document. Subconsciously, the readers will detect this tone in your work. Their inner voice will begin reinforcing the sense of excitement you are trying to convey.

Similarly, if you want to create a specific attitude in your text, map out the words associated with it. For instance, let us say you want your document to convey a feeling of "security." A map around "security" might look like the diagram in Figure 17.4.

If you use these words associated with "security" at strategic points in your text, readers will sense the feeling you are trying to convey.

One warning: Use these words sparingly throughout the text. If you overuse them, the emotion or attitude you are setting will be too obvious. You need to use only a few words at strategic moments to set the tone you are seeking.

Use Similes and Analogies

Similes and analogies are rhetorical devices that help writers define difficult concepts by comparing them to familiar things. For example, let's say you are writing a brochure that describes an integrated circuit to people who know almost nothing about these devices. Using a simile, you might describe it this way:

> An integrated circuit is like a miniature Manhattan Island crammed into a space less than 1 centimeter long.

In this case, the simile ("X is like Y") creates an image that helps readers visualize the complexity of the integrated circuit.

For more examples of similes and analogies, go to **www.pearsonhighered.com/johnsonweb4/17.16**

Here are a couple of other examples of similes:

> A cell is like a tiny cluttered room with a nucleus inside and walls at the edges. The nucleus is like the cell's brain. It's an enclosed compartment that contains all the information that cells need to form an organism. This information comes in the form of DNA. It's the differences in our DNA that make each of us unique.
>
> *(Source: Genetic Science Learning Center, http://gslc.genetics.utah.edu/units/cloning/whatiscloning.)*

> Everything that you see is made of things called atoms. An atom is like a very, very small solar system. As you know, the sun is at the center of the solar system. Just like the solar system, there is something at the center of an atom. It is called the nucleus. So the nucleus of an atom is like the sun in our solar system. Around the sun there are planets. They orbit the sun. In an atom, there are things called electrons that orbit the nucleus.
>
> *(Source: Chickscope, http://chickscope.beckman.uiuc.edu/about/overview/mrims.html.)*

Analogies are also a good way to help readers visualize difficult concepts. An analogy follows the structure "A is to B as X is to Y." For example, a medical analogy might be

> Like police keeping order in a city, white blood cells seek to control viruses and bacteria in the body.

In this case, both parts of the analogy are working in parallel. "Police" is equivalent to "white blood cells," and "keeping order in a city" is equivalent to "control viruses and bacteria in the body."

Use Metaphors

Though comparable to similes and analogies, metaphors work at a deeper level in a document. Specifically, metaphors are used to create or reinforce a particular perspective that you want readers to adopt toward your subject or ideas. For example, a popular metaphor in Western medicine is the "war on cancer."

> **Off Target in the War on Cancer**
>
> We've been fighting the war on cancer for almost four decades now, since President Richard M. Nixon officially launched it in 1971. It's time to admit that our efforts have often targeted the wrong enemies and used the wrong weapons.
>
> *(Source: Devra Davis, Washington Post, November 4, 2007, http://www.washingtonpost.com/wp-dyn/content/article/2007/11/02/AR2007110201648.html.)*

> **Are We Retreating in the War on Cancer?**
>
> They are America's foot-soldiers in the war on cancer—young scientists whose research may someday lead to better treatments and even cures.

Want to learn more about metaphors? Go to **www.pearsonhighered.com/johnsonweb4/17.17**

But experts worry this small elite army is leaving the field in droves because government funding, which once allowed cancer research to flourish, is now drying up.

(Source: CBS News, May 20, 2008, http://www.cbsnews.com/stories/2008/05/20/eveningnews/main4111776.shtml?source=related_story.)

As shown in these examples, by employing this metaphor, the writer can reinforce a particular perspective about cancer research. A metaphor like "war on cancer" would add a sense of urgency, because it suggests that cancer is an enemy that must be defeated, at any cost. Metaphors are powerful tools in technical writing because they tend to work at a subconscious level.

But what if a commonly used metaphor like "war on cancer" is not appropriate for your document? Perhaps you are writing about cancer in a hospice situation, where the patient is no longer trying to "defeat" cancer. In these cases, you might develop a new metaphor and use it to invent a new perspective. For example, perhaps you want your readers to view cancer as something to be managed, not fought.

To manage their cancer, patients become supervisors who set performance goals and lead planning meetings for teams of doctors and nurses. Our doctors become consultants who help patients become better managers of their own care.

In a hospice situation, a "management" metaphor would be much more appropriate than the usual "war" metaphor. It would invite readers to see their cancer from a less confrontational perspective.

If you look for them, you will find that metaphors are commonly used in technical documents. Once you are aware of these metaphors, you can use the ones that already exist in your field, or you might create new metaphors to shift your readers' perspective about your subject. In Figure 17.5, for example, the writer uses metaphors to recast the debate over the use of nuclear energy.

Change the Pace

You can also control the readers' pace as they move through your document. Longer sentences tend to slow down the reading pace, while shorter sentences tend to speed it up. By paying attention to the lengths of sentences, you can increase or decrease the intensity of your text.

Long sentences (low intensity)

These long sentences slow the reading down.

According to behavior problem indices, children who experience food insecurity suffer more psychological and emotional difficulties than other children. They exhibit more aggressive and destructive behaviors, as well as more withdrawn and distressed behaviors. Children who are experiencing a great deal of psychological and emotional distress in response to issues of food insecurity will often react to this distress with a range of negative behavioral responses, including acting out and violence toward others. Moreover, a child's psychological and emotional well-being is also negatively affected by food insecurity, which may also have implications for other child development outcomes. Their higher levels of psychological

A Persuasive Argument

The author uses metaphors to add energy and imagery.

The "argument is war" metaphor is used throughout the article.

Patrick Moore: Going nuclear over global warming
By Patrick Moore

For years the Intergovernment Panel on Climate Change (IPCC) of the United Nations has warned us that greenhouse gas emissions from our fossil fuel consumption threaten the world's climate in ways we will regret. This year IPCC won the Nobel Peace Prize for its efforts.

It is the IPCC that former colleagues in Greenpeace, and most of the mainstream environmental movement, look to for expert advice on climate change. Environmental activists take the rather grim but measured language of the IPCC reports and add words like "catastrophe" and "chaos," along with much speculation concerning famine, pestilence, mass extinction and the end of civilization as we know it.

Until the past couple of years, the activists, with their zero-tolerance policy on nuclear energy, have succeeded in squelching any mention by the IPCC of using nuclear power to replace fossil fuels for electricity production. Burning fossil fuels for electricity accounts for 9.5 billion tons of global carbon dioxide emissions while nuclear emits next to nothing. It has been apparent to many scientists and policymakers for years that this would be a logical path to follow. The IPCC has now joined these growing ranks advocating for nuclear energy as a solution.

In its recently issued final report for 2007, the IPCC makes a number of unambiguous references to the fact that nuclear energy is an important tool to help bring about a reduction in fossil fuel consumption. Greenpeace has already made it clear that it disagrees. How credible is it for activists to use the IPCC scientists' recommendations to fuel apocalyptic fundraising campaigns on climate change and then to dismiss the recommendations from the same scientists on what we should do to solve it?

Greenpeace is deliberately misleading the public into thinking that wind and solar, both of which are inherently intermittent and unreliable, can replace baseload power that is continuous and reliable. Only three technologies can produce large amounts of baseload power: fossil fuels, hydroelectric, and nuclear. Given that we want to reduce fossil fuels and that potential hydroelectric sites are becoming scarce, nuclear is the main option. But Greenpeace and its allies remain in denial despite the fact that many independent environmentalists and now the IPCC see the situation clearly.

I have long realized that in retrospect we made a big mistake in the early years of Greenpeace when we lumped nuclear energy together with nuclear weapons as if they were all part of the same holocaust. We were totally fixated, and rightly so, on the threat of all-out nuclear war between the Soviet Union and the United States, and we thought everything nuclear was evil. We failed to distinguish the beneficial and peaceful uses of nuclear technology from its destructive and even evil uses.

The approach would be akin to including nuclear medicine with nuclear weapons just because nuclear medicine uses radioactive materials, most of which are produced in nuclear reactors. Nuclear medicine successfully diagnoses and treats millions of people every year, and it would be ludicrous to ban its use.

Greenpeace and company are basically stuck in the 1970s when it comes to their energy policy as it relates to climate change. They should accept the wisdom of the scientists at the IPCC and recognize that nuclear energy is a big part of the climate change solution. And they should stop misleading the public into thinking that wind and solar can do the job on their own. I will be the first to commend them for their courage.

Figure 17.5: In this article, the author uses numerous persuasive style techniques to add energy to his writing. Is the author persuasive?

An even pace is created with consistent sentence length.

The author uses an analogy to make an important comparison.

Source: Sacramento Bee, December 12, 2007, http://www.sacbee.com/110/story/560569.html. This article is protected by copyright and should not be printed or distributed for anything except personal use. The Sacramento Bee, 2100 Q St., P. O. Box 15779, Sacramento, CA 95852 Phone: (916) 321-1000.

and emotional distress may cause problems in other areas such as school achievement, and these difficulties may interfere with a number of out-of-school activities in which they are involved.

Short sentences (high intensity)

Shorter sentences raise the "heartbeat" of the text.

The effect of children's food insecurity can be crucial. According to behavior problem indices, food-insecure children have more psychological and emotional difficulties. They exhibit more aggressive and destructive behaviors. They are more withdrawn and distressed. They react to this distress with a range of behavioral responses, including acting out and violence toward others. The child's overall well-being is negatively affected. This insecurity may have implications for other child development outcomes. These children may have more trouble at school. They also tend to struggle at a number of out-of-school activities.

AT A GLANCE

Persuasive Style Techniques

- Elevate the tone.
- Use similes and analogies.
- Use metaphors.
- Change the pace.

If a situation is urgent, using short sentences is the best way to show that something needs to be done right away. Your readers will naturally feel compelled to take action. On the other hand, if you want your readers to be cautious and deliberate, longer sentences will decrease the intensity of the text, giving readers the sense that there is no need to rush.

Balancing Plain and Persuasive Style

When you are drafting and revising a document for style, look for appropriate places to use plain and persuasive style. Minimally, a document should use plain style. Sentences should be clear and easy to read. Your readers should not have to struggle to figure out what a sentence or paragraph is about.

Persuasive style should be used to add energy and color. It should also be used in places in the document where readers are expected to make a decision or take action. The use of tone, similes, analogies, and metaphors in strategic places should encourage readers to do what you want. You can use short or long sentences to adjust the intensity of your prose.

In the end, developing good style takes practice. At first, revising a document to make it plain and persuasive might seem difficult. Before long, though, you will start writing better sentences while drafting. You will have internalized the style guidelines presented in this chapter.

CHAPTER REVIEW

- The style of your document can convey your message as strongly as the content; *how you say something is often what you say.*
- Style can be classified as plain, persuasive, or grand. Technical communication most often uses plain or persuasive style.
- To write plain sentences, follow these eight guidelines: (1) identify your subject, (2) make the "doer" the subject of the sentence, (3) put the doer's action in the verb, (4) move the subject of the sentence close to the beginning of the sentence, (5) reduce nominalizations, (6) remove excess prepositional phrases, (7) eliminate redundancy, and (8) make sure that sentences are "breathing length."
- Writing plain paragraphs involves the use of four types of sentences: transitional and point sentences, which are optional, and topic and support sentences, which are necessary.
- Check sentences in paragraphs for subject alignment, and use the given/new method to weave sentences together.
- Techniques for writing persuasively include elevating the tone; using similes, analogies, and metaphors; and changing the pace.

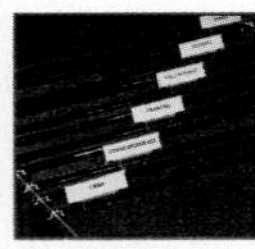

Individual or Team Projects

1. In your workplace or on campus, find three sentences that seem particularly difficult to read. Use the eight "plain style" guidelines discussed in this chapter to revise them to improve their readability. Make a presentation to your class in which you show how using the guidelines helped make the sentences easier to read.

2. Find a document and analyze its style. Underline key words and phrases in the document. Based on the words you underlined, what is the tone used in the document? Does the document use any similes or analogies to explain difficult concepts? Can you locate any metaphors woven into the text? How might you use techniques of persuasive style to improve this document? Write a memo to your instructor in which you analyze the style of the document. Do you believe the style is effective? If so, why? If not, what stylistic strategies might improve the readability and persuasiveness of the document?

3. While revising some of your own writing, try to create a specific tone by mapping out an emotion or attitude that you would like the text to reflect. Weave a few concepts from your map into your text. At what point is the tone too strong? At what point is the tone just right?

For these exercises, sample sentences, paragraphs, and documents are available at **www.pearsonhighered.com/johnsonweb4/17.18**

4. Read the Case Study at the end of this chapter. Imagine that you are Henry and write a memo to the managers of NewGenSport expressing your concerns about the safety of the product and its advertising campaign. Make sure you do some research on ephedra so you can add some technical support to your arguments.

Collaborative Project

Metaphors involving "war" are common in American society. We have had wars on poverty, cancer, drugs, and even inflation. More recently, "war on terrorism" has been a commonly used metaphor. With a group of classmates, choose one of these war metaphors and find examples of how it is used. Identify places where the metaphor seems to fit the situation. Locate examples where the metaphor seems to be misused.

With your group, discuss the ramifications of the war metaphor. For example, if we accept the metaphor "war on drugs," who is the enemy? What weapons can we use? What level of force is necessary to win this war? Where might we violate civil rights if we follow this metaphor to its logical end?

Then, try to create a new metaphor that invites people to see the situation from a different perspective. For example, what happens when we use "managing drug abuse" or "healing drug abuse" as new metaphors? How do these new metaphors, for better or worse, change how we think about illegal drugs and react to them?

In a presentation to your class, use examples to show how the war metaphor is used. Then, show how the metaphor could be changed to consider the situation from a different perspective.

For additional technical writing resources, including interactive sample documents, document design tutorials and guidelines, and more, go to **www.mytechcommlab.com.**

For sites that use or discuss war metaphors, go to **www.pearsonhighered.com/johnsonweb4/17.19**

CASE STUDY

Going Over the Top

Henry Wilkins is a nutritionist who works for NewGenSport, a company that makes sports drinks. His company's best seller, Overthetop Sports Drink, is basically a fruit-flavored drink with added carbohydrates and salts. As a nutritionist, Henry knows that sports drinks like Overthetop do not really do much for people, but they don't harm them either. Overwhelmingly, the primary benefit of Overthetop is the water, which people need when they exercise. The added carbohydrates and salts are somewhat beneficial, because people deplete them as they exercise.

The advertisements for Overthetop paint a different picture, however. They show overly muscular athletes drinking Overthetop and then dominating opposing players. Henry thinks the advertisements are a bit misleading, but harmless. The product won't lead to the high performance promised by the advertising. But, Henry reasons, that's advertising.

One day, Henry was asked to evaluate a new version of Overthetop, called Overthetop Extreme. The new sports drink would include a small amount of an ephedra-like herbal stimulant formulated to enhance performance and cause slight weight loss. Henry also knew that ephedra had been linked to a few deaths, especially when people took too much. Ephedra was banned by the FDA in 2003, but this new "ephedra-like" herbal stimulant was different enough to be legal. Henry suspected this stimulant would have many of the same problems as its sister drug, ephedra.

Henry's preliminary research indicated that 36 ounces (three bottles) a day of Overthetop Extreme would have no negative effect. But, more than a few bottles a day might lead to complications, perhaps even death in rare cases.

But before Henry could report his findings, the company's marketing team pitched its new advertising campaign for the debut of the new product. In the campaign, as usual, overly muscular athletes were shown guzzling large amounts of Overthetop Extreme and then going on to victory. It was clear that the advertisements were overpromising—as usual. But, they were also suggesting the product be used in a way that would put people at risk. The advertisements were technically accurate, but their style gave an inflated sense of the product's capabilities. In some cases, Henry realized, the style of the advertisements might lead to overconsumption, which could lead to problems.

The company's top executives were very excited about the new product. The advertisements only made them more enthusiastic. They began pressuring Henry to finish his evaluation of Overthetop right away so that they could start advertising and putting the product on store shelves.

Henry, unfortunately, was already concerned about the product's safety. But he was especially concerned that the advertisements were clearly misleading consumers in a dangerous way.

If you were Henry, how would you handle this situation? In your report to the company's management, how would you express your concerns? What else might you do?

Want to know more about ephedra? Go to **www.pearsonhighered.com/johnsonweb4/17.20**

CHAPTER

18 Designing Documents and Interfaces

Five Principles of Design *482*
Design Principle 1: Balance *482*
Design Principle 2: Alignment *492*
Design Principle 3: Grouping *493*
Design Principle 4: Consistency *500*
Design Principle 5: Contrast *505*
Cross-Cultural Design *508*
Using the Principles of Design *511*
A Primer on Binding and Paper *513*
Chapter Review *516*
Exercises and Projects *516*
Case Study: Scorpions Invade *518*

In this chapter, you will learn:

- How to design technical documents for electronic and paper-based media.
- To use the five principles of design: balance, alignment, grouping, consistency, and contrast.
- To use balancing techniques that enhance readability in a document.
- To use alignment and grouping strategies to add visual structure to text.
- To use consistency and contrast to balance uniformity and difference in the design of a document or interface.
- To anticipate the design expectations of international readers.
- How to use a process to design documents.
- About options for binding documents, as well as the sizes, colors, and weights of paper available.

Document design has become ever more important with the use and availability of computers. Today, readers expect paper-based documents to be attractive and easy to read, and to include images and color. Meanwhile, screen-based documents like websites and multimedia texts are highly visual, so readers expect the interface to be well designed. Readers don't just *prefer* well-designed documents—they *expect* the design to highlight important ideas and concepts.

Five Principles of Design

People rarely read technical documents word for word, sentence by sentence. Instead, they tend to look over technical documents at various levels, skimming some parts and paying closer attention to others.

Good design creates a sense of order and gives readers obvious "access points" to begin reading and locating the information they need. Good design is not something to be learned in a day. However, you can master some basic principles that will help you make better decisions about how your document should look. Here are five principles to consider as you design documents:

Balance—The document looks balanced from left to right and top to bottom.

Alignment—Images and words on the page are aligned to show the document's structure, or hierarchy.

Grouping—Related images and words are placed near each other on the page.

Consistency—Design features in the document are used consistently, so the document looks uniform.

Contrast—Items in the document that are different look significantly different.

These principles are based on theories of Gestalt psychology, a study of how the mind recognizes patterns (Arnheim, 1969; Koffka, 1935). Designers of all kinds, including architects, fashion designers, and artists, have used Gestalt principles in a variety of ways (Bernhardt, 1986). You will find these five principles helpful as you learn about designing documents.

Design Principle 1: Balance

Balance is perhaps the most prominent feature of design in technical documents. On a balanced page or screen, the design features should offset each other to create a feeling of stability.

To balance a text, pretend your page or screen is balanced on a point. Each time you add something to the left side of the page, you need to add something to the right side to maintain balance. Similarly, when you add something to the top of the page, you need to add something to the bottom. Figure 18.1, for example, shows an example of a balanced page and an unbalanced page.

To learn more about Gestalt psychology and design, go to **www.pearsonhighered.com/johnsonweb4/18.1**

Balanced and Unbalanced Page Layouts

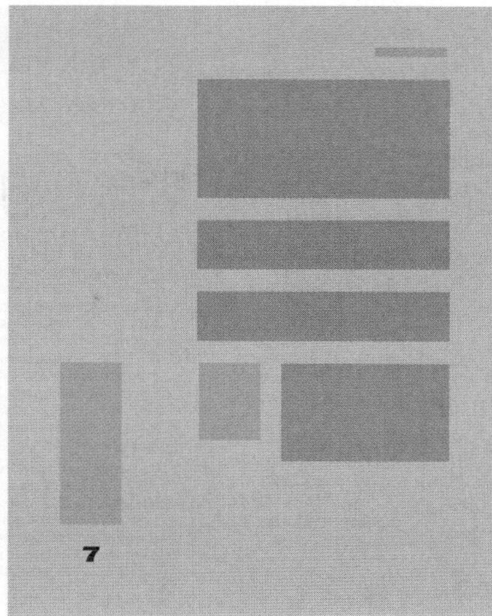

Figure 18.1: The balanced page on the left feels more stable and comfortable. The unbalanced page on the right creates more tension.

In Figure 18.1, the page on the left is balanced because the design features offset each other. The page on the right is unbalanced because the items on the right side of the page are not offset by items on the left. Also, the right page is top-heavy because the design features are bunched at the top of the page.

Balanced page layouts can take on many forms. Figures 18.2 and 18.3 show examples of balanced layouts. The idea is not to create symmetrical pages (i.e., where left and right, top and bottom mirror each other exactly). Instead, you want to balance pages by putting text and images on all sides.

Balance is also important in screen-based documents. In Figure 18.4, the screen interface is balanced because the items on the left offset the items on the right.

Weighting a Page or Screen

When balancing a page or screen, graphic designers will talk about the "weight" of the items on the page. What they mean is that some items on a page or screen attract readers' eyes more than others—these features have more weight. A picture, for example, has more weight than printed words because readers' eyes tend to be drawn toward pictures. Similarly, an animated figure moving on the screen will capture more attention than static items.

Here are some basic weighting guidelines for a page or screen:

- Items on the right side of the page weigh more than items on the left.
- Items on the top of the page weigh more than items on the bottom.
- Big items weigh more than small items.

To see samples of unbalanced pages, go to **www.pearsonhighered.com/johnsonweb4/18.2**

A Balanced Design with Energy

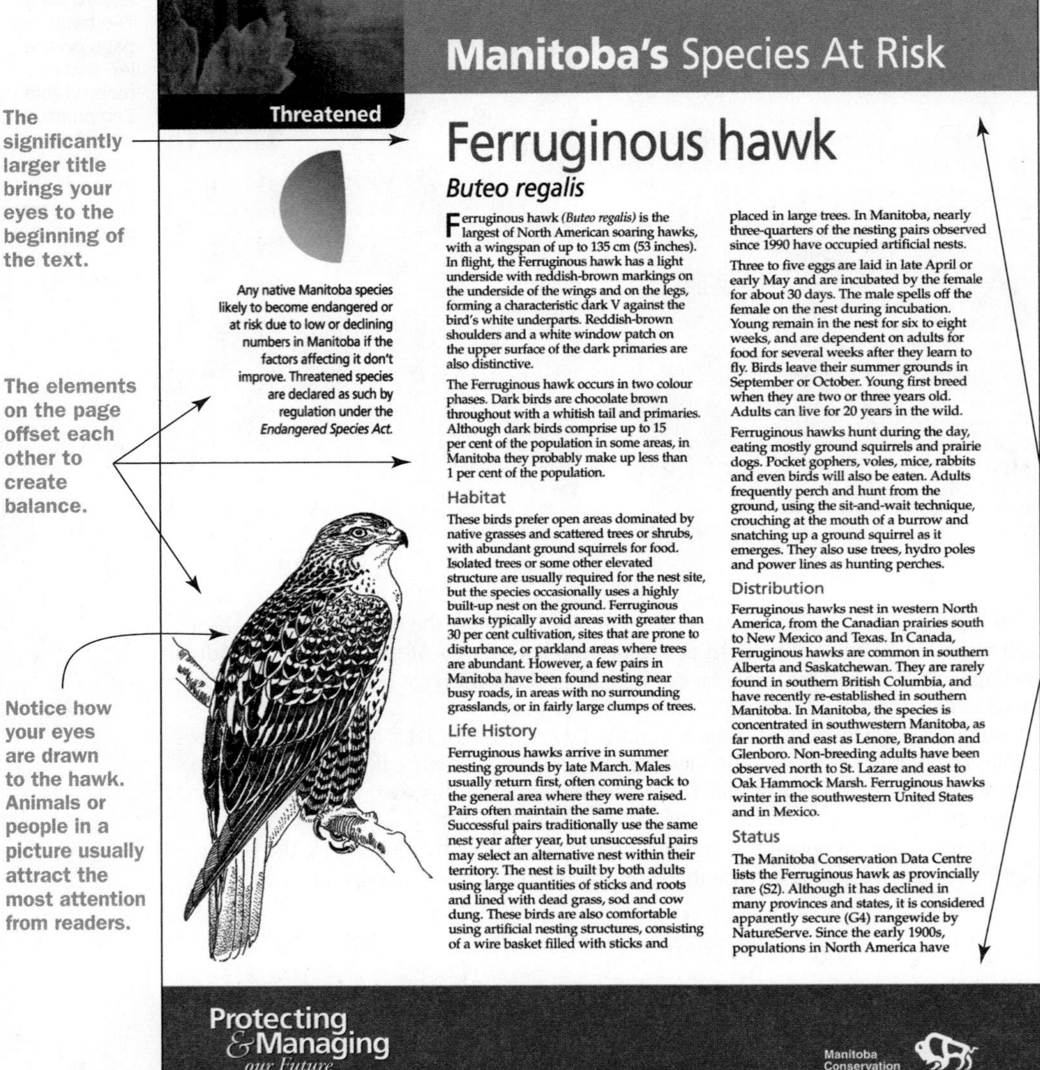

The significantly larger title brings your eyes to the beginning of the text.
The elements on the page offset each other to create balance.
Notice how your eyes are drawn to the hawk. Animals or people in a picture usually attract the most attention from readers.

Manitoba's Species At Risk

Threatened

Ferruginous hawk

Buteo regalis

Any native Manitoba species likely to become endangered or at risk due to low or declining numbers in Manitoba if the factors affecting it don't improve. Threatened species are declared as such by regulation under the *Endangered Species Act*.

Ferruginous hawk *(Buteo regalis)* is the largest of North American soaring hawks, with a wingspan of up to 135 cm (53 inches). In flight, the Ferruginous hawk has a light underside with reddish-brown markings on the underside of the wings and on the legs, forming a characteristic dark V against the bird's white underparts. Reddish-brown shoulders and a white window patch on the upper surface of the dark primaries are also distinctive.

The Ferruginous hawk occurs in two colour phases. Dark birds are chocolate brown throughout with a whitish tail and primaries. Although dark birds comprise up to 15 per cent of the population in some areas, in Manitoba they probably make up less than 1 per cent of the population.

Habitat

These birds prefer open areas dominated by native grasses and scattered trees or shrubs, with abundant ground squirrels for food. Isolated trees or some other elevated structure are usually required for the nest site, but the species occasionally uses a highly built-up nest on the ground. Ferruginous hawks typically avoid areas with greater than 30 per cent cultivation, sites that are prone to disturbance, or parkland areas where trees are abundant. However, a few pairs in Manitoba have been found nesting near busy roads, in areas with no surrounding grasslands, or in fairly large clumps of trees.

Life History

Ferruginous hawks arrive in summer nesting grounds by late March. Males usually return first, often coming back to the general area where they were raised. Pairs often maintain the same mate. Successful pairs traditionally use the same nest year after year, but unsuccessful pairs may select an alternative nest within their territory. The nest is built by both adults using large quantities of sticks and roots and lined with dead grass, sod and cow dung. These birds are also comfortable using artificial nesting structures, consisting of a wire basket filled with sticks and placed in large trees. In Manitoba, nearly three-quarters of the nesting pairs observed since 1990 have occupied artificial nests.

Three to five eggs are laid in late April or early May and are incubated by the female for about 30 days. The male spells off the female on the nest during incubation. Young remain in the nest for six to eight weeks, and are dependent on adults for food for several weeks after they learn to fly. Birds leave their summer grounds in September or October. Young first breed when they are two or three years old. Adults can live for 20 years in the wild.

Ferruginous hawks hunt during the day, eating mostly ground squirrels and prairie dogs. Pocket gophers, voles, mice, rabbits and even birds will also be eaten. Adults frequently perch and hunt from the ground, using the sit-and-wait technique, crouching at the mouth of a burrow and snatching up a ground squirrel as it emerges. They also use trees, hydro poles and power lines as hunting perches.

Distribution

Ferruginous hawks nest in western North America, from the Canadian prairies south to New Mexico and Texas. In Canada, Ferruginous hawks are common in southern Alberta and Saskatchewan. They are rarely found in southern British Columbia, and have recently re-established in southern Manitoba. In Manitoba, the species is concentrated in southwestern Manitoba, as far north and east as Lenore, Brandon and Glenboro. Non-breeding adults have been observed north to St. Lazare and east to Oak Hammock Marsh. Ferruginous hawks winter in the southwestern United States and in Mexico.

Status

The Manitoba Conservation Data Centre lists the Ferruginous hawk as provincially rare (S2). Although it has declined in many provinces and states, it is considered apparently secure (G4) rangewide by NatureServe. Since the early 1900s, populations in North America have

Protecting & Managing *our Future*

Manitoba Conservation

Source: Manitoba Conservation Wildlife and Ecosystem Protection Branch.

Figure 18.2: Graphic designers are especially careful about balance. This page layout uses the image of the hawk to balance two columns of written text. Meanwhile, the bold header and footer anchor the text at the top and bottom.

The green header and footer at the top and bottom of the page make it feel balanced and stable.

A Simpler Design That Is Elegant

Figure 18.3: This page from a magazine is simpler in design than the document shown in Figure 18.2. Notice how the elements on the page offset each other to create a balanced, stable look.

The two-column format balances the text.

Your eyes naturally flow down the page.

Restoring V-Site—Birthplace of the *Gadget*

V-Site is located deep inside the current high explosives (HE) research and development area at Los Alamos National Laboratory. This site is significant because the activities that took place in six wooden sheds and the events leading up to those activities transformed the world and ushered in the Atomic Age. The buildings of V-Site are among the most historically significant buildings of the 20th century.

V-Site buildings were typical of World War II temporary wood structures at military installations. The buildings were wood post-and-frame construction that rested on concrete slab floors. Asbestos shingles covered the exterior. Earthen berms, which served as protection against HE accidents, surrounded the buildings and were secured by heavy wood post-and-beam retaining walls.

The Manhattan Project

The Manhattan Project (1942–1946) consisted of two major efforts: production of fissile material and the research, design, and production of a new class of weapon that could end World War II. Manhattan Project installations at Oak Ridge, Tennessee, and Hanford, Washington, focused on production of enriched uranium and plutonium that could be used with new weapons designed at Los Alamos.

Los Alamos, known as Project Y during the Manhattan Project, was the location of the secret research and development efforts to design and build the first atomic weapons. Project Y brought together physicists, engineers, and the Special Engineering Detachment of the US Army to design and build the weapons.

The initial plans called for a gun-type design employing Oak Ridge's enriched uranium and Hanford's plutonium. The gun design was conceptually simple and involved shooting one subcritical mass of fissile material into another subcritical mass. The two subcritical masses would form a critical mass, thereby releasing a tremendous amount of nuclear energy.

An early alternative to the gun design was the implosion method. The implosion method, a technically efficient approach, was intended to be a backup to the gun design. In 1943, J. Robert Oppenheimer, the Laboratory's first director, allowed a small number of scientists to pursue the implosion method.

In 1944, Los Alamos scientists determined that the gun design was not suitable for use with plutonium. The main reason was that plutonium produced in nuclear reactors, such as the plutonium produced at

Fat Man assembly at V-Site.

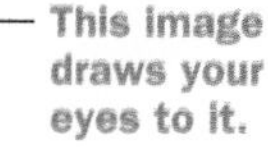
This image draws your eyes to it.

Nuclear Weapons Journal, Issue 1, 2007 13

Source: Nuclear Weapons Journal, p. 13.

A Balanced Interface

The banner at the top of the screen offsets text and images lower on the page.

Left and right images and text are used to balance the screen

Figure 18.4: This screen is balanced, even though it is not symmetrical. The items on the left offset the items on the right.

Source: Fermilab.

- Pictures weigh more than written text.
- Graphics weigh more than written text.
- Colored items weigh more than black-and-white items.
- Items with borders around them weigh more than items without borders.
- Irregular shapes weigh more than regular shapes.
- Items in motion weigh more than static items.

As you are balancing a page, use these weight guidelines to help you offset items. For example, if an image appears on the right side of the page, make sure that there is something on the left to create balance.

Using Grids to Balance a Page Layout

When designing a page or screen, your challenge is to create a layout that is balanced but not boring. A time-tested way to devise a balanced page design is to use a *page grid* to evenly place the written text and graphics on the page. Grids divide the page vertically into two or more columns. Figure 18.5 shows some standard grids and how they might be used.

Figure 18.6 (on page 489) shows the use of a three-column grid in a report. Notice how the graphics and text offset each other in the page layout.

GO TO THE NET

To see samples of unbalanced webpages, go to **www.pearsonhighered.com/johnsonweb4/18.3**

Grids for Designing Page Layouts

One-column grid

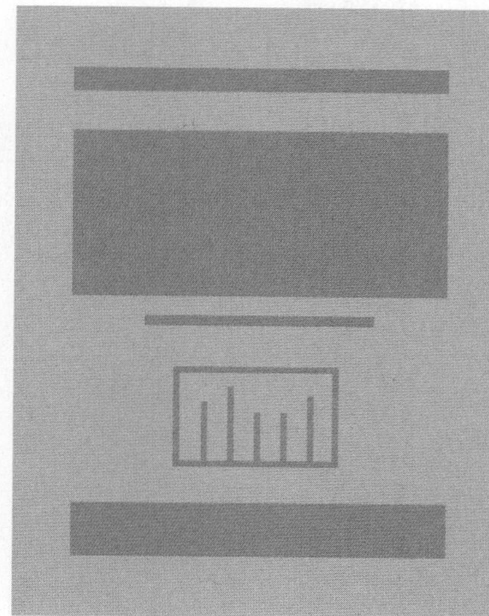

Figure 18.5: Grids can help you place items on a page in a way that makes it look balanced.

One-column grids offer simplicity, but not much flexibility.

Two-column grid

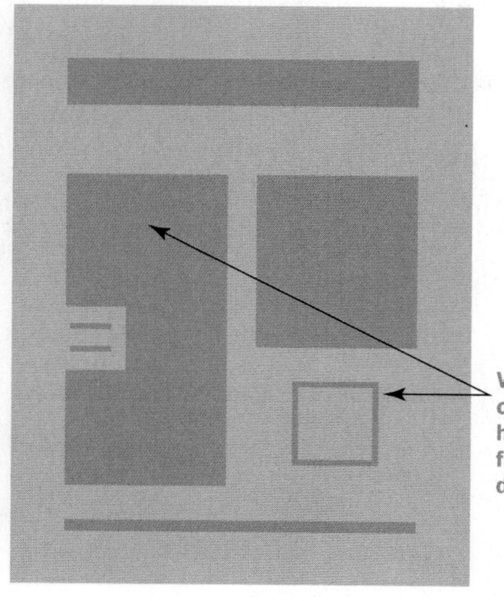

With more columns, you have more flexibility for design.

(continued)

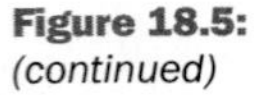

Figure 18.5: *(continued)*

Three-column grid

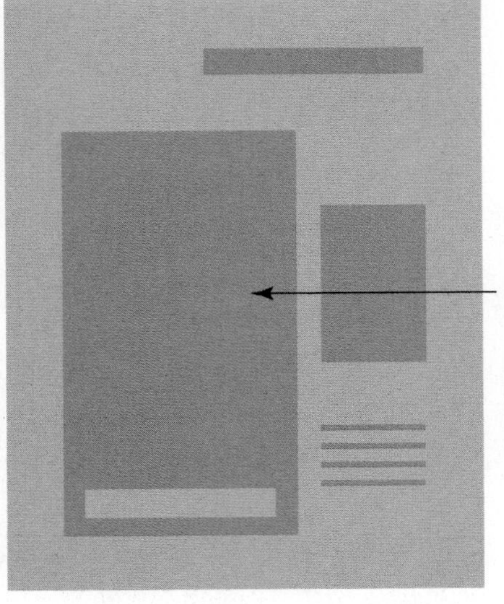

Notice how the text can go over two columns, leaving a large margin on one side.

Four-column grid

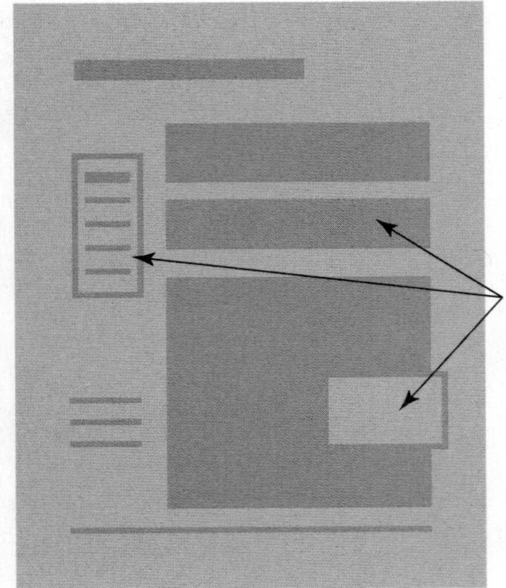

A four-column grid offers plenty of opportunities for creativity.

GO TO THE NET

For downloadable templates, go to **www.pearsonhighered.com/johnsonweb4/18.4**

Using Grids to Lay Out a Page

A three-column layout structures the whole page.

Images on left and bottom create balance with text.

An image can cross two columns, as this one does.

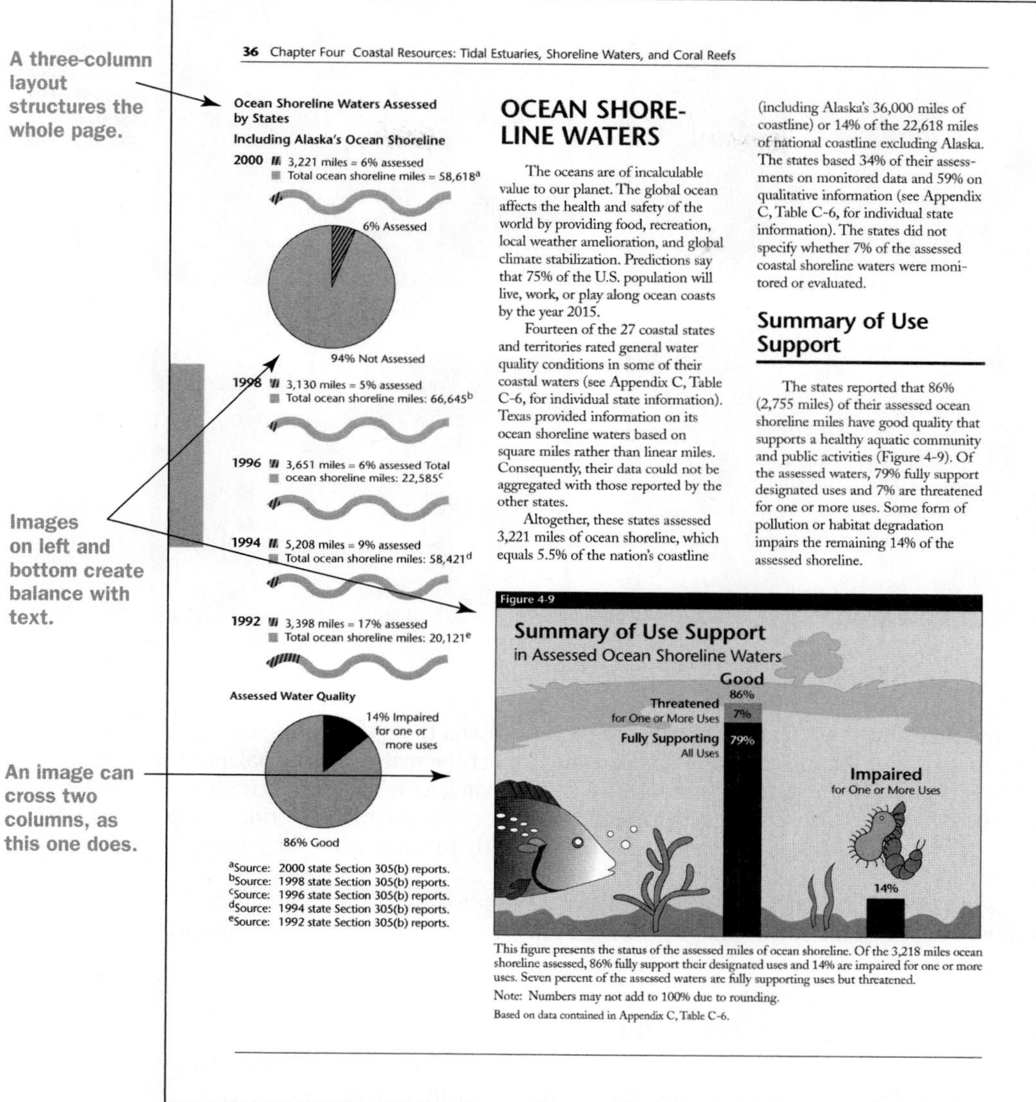

36 Chapter Four Coastal Resources: Tidal Estuaries, Shoreline Waters, and Coral Reefs

Ocean Shoreline Waters Assessed by States

Including Alaska's Ocean Shoreline

2000 3,221 miles = 6% assessed
Total ocean shoreline miles = 58,618[a]

6% Assessed

94% Not Assessed

1998 3,130 miles = 5% assessed
Total ocean shoreline miles: 66,645[b]

1996 3,651 miles = 6% assessed Total ocean shoreline miles: 22,585[c]

1994 5,208 miles = 9% assessed
Total ocean shoreline miles: 58,421[d]

1992 3,398 miles = 17% assessed
Total ocean shoreline miles: 20,121[e]

Assessed Water Quality

14% Impaired for one or more uses

86% Good

[a]Source: 2000 state Section 305(b) reports.
[b]Source: 1998 state Section 305(b) reports.
[c]Source: 1996 state Section 305(b) reports.
[d]Source: 1994 state Section 305(b) reports.
[e]Source: 1992 state Section 305(b) reports.

OCEAN SHORELINE WATERS

The oceans are of incalculable value to our planet. The global ocean affects the health and safety of the world by providing food, recreation, local weather amelioration, and global climate stabilization. Predictions say that 75% of the U.S. population will live, work, or play along ocean coasts by the year 2015.

Fourteen of the 27 coastal states and territories rated general water quality conditions in some of their coastal waters (see Appendix C, Table C-6, for individual state information). Texas provided information on its ocean shoreline waters based on square miles rather than linear miles. Consequently, their data could not be aggregated with those reported by the other states.

Altogether, these states assessed 3,221 miles of ocean shoreline, which equals 5.5% of the nation's coastline (including Alaska's 36,000 miles of coastline) or 14% of the 22,618 miles of national coastline excluding Alaska. The states based 34% of their assessments on monitored data and 59% on qualitative information (see Appendix C, Table C-6, for individual state information). The states did not specify whether 7% of the assessed coastal shoreline waters were monitored or evaluated.

Summary of Use Support

The states reported that 86% (2,755 miles) of their assessed ocean shoreline miles have good quality that supports a healthy aquatic community and public activities (Figure 4-9). Of the assessed waters, 79% fully support designated uses and 7% are threatened for one or more uses. Some form of pollution or habitat degradation impairs the remaining 14% of the assessed shoreline.

Figure 4-9

This figure presents the status of the assessed miles of ocean shoreline. Of the 3,218 miles ocean shoreline assessed, 86% fully support their designated uses and 14% are impaired for one or more uses. Seven percent of the assessed waters are fully supporting uses but threatened.

Note: Numbers may not add to 100% due to rounding.

Based on data contained in Appendix C, Table C-6.

Figure 18.6: A three-column grid was used to lay out this text in a balanced manner.

Source: U.S. Environmental Protection Agency, National Water Quality Inventory, 2000 Report, 2000, p. 36.

Grids for Interfaces

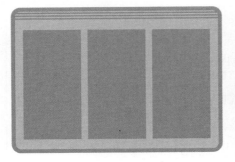

Figure 18.7: Items on screen-based pages should also be evenly placed. This approach creates a sense of stability.

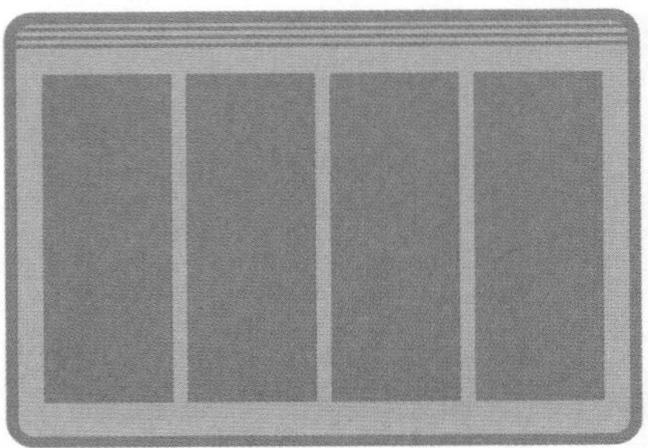

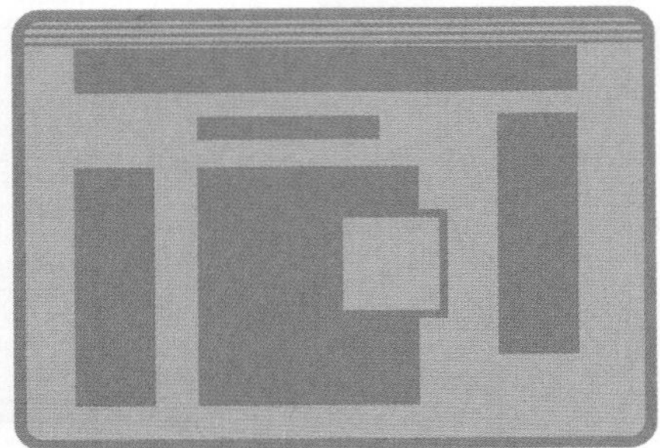

Link

To learn more about designing screen interfaces, go to Chapter 22, page 626.

Grids are also used to lay out screen-based texts. Even though screen-based texts tend to be wider than they are long, readers still expect the material to be balanced on the interface. Figure 18.7 shows possible designs using three- and four-grid templates. Figure 18.9 (on page 492) demonstrates the use of a five-column grid. Notice how the features of the interface are spaced consistently to create an orderly feel to the text.

In many cases, the columns on a grid do not translate directly into columns of written text. Columns of text and pictures can often overlap one or more columns in the grid.

Using Other Balance Techniques

Word-processing programs and desktop publishing software also give you the ability to use professional design features like *margin comments, sidebars,* and *pullouts* (Figure 18.8). In screen-based documents, you can also use *navigation bars* and *banners* to balance the design. These features provide "access points" where people can begin reading the text.

MARGIN COMMENTS Margin comments summarize key points or highlight quotations in the margin of the document. When a grid is used to design the page, one of

For more advice about interface design, go to **www.pearsonhighered.com/johnsonweb4/18.5**

Using Other Balancing Techniques for Print Documents

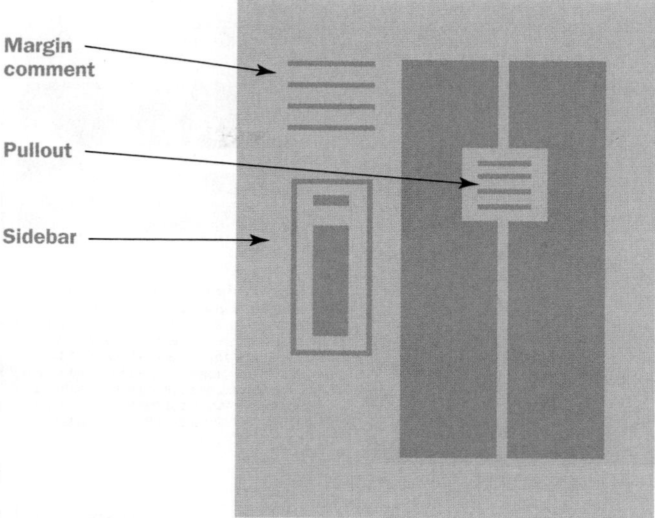

Figure 18.8: Margin comments, pullouts, and sidebars can make a text more accessible.

> **Balancing the Design of a Page or Screen**
>
> AT A GLANCE
>
> - Weight items on the page or screen.
> - Use grids to balance the layouts.
> - Use design features such as margin comments, sidebars, pullouts, navigation bars, and banners.

the margins often leaves enough room to include an additional list, offer a special quote, or provide a simple illustration (Figure 18.8). Also, the page in Figure 18.6 shows how a whole column can be used to offer supplementary information to the main text.

NAVIGATION BARS In screen interfaces, navigation bars are used to highlight links, usually as buttons or text (Figure 18.9). Navigation bars are typically placed along the left side of the screen or the top, but they can appear anywhere on the page.

SIDEBARS Sidebars are used to provide examples and facts that support the main text. Sidebars should never contain essential information that readers must have in order to understand the subject or make a decision. Rather, they should offer supplemental information that enhances the readers' understanding.

PULLOUTS Pullouts are quotes or paraphrases drawn from the body text and placed in a special text box to catch the readers' attention. A pullout should draw its text from the page on which it appears. Often the pullout is framed with rules (lines) or a box, and the text wraps around it (Figure 18.8).

BANNERS Banners run along the top of the screen interface. In websites, they may include the name of the organization or an advertiser. In multimedia documents, they may identify the name of the document. Banners often include links, but they don't need to.

Want to learn about other balancing techniques? Go to **www.pearsonhighered.com/johnsonweb4/18.6**

Other Balancing Techniques for Interfaces

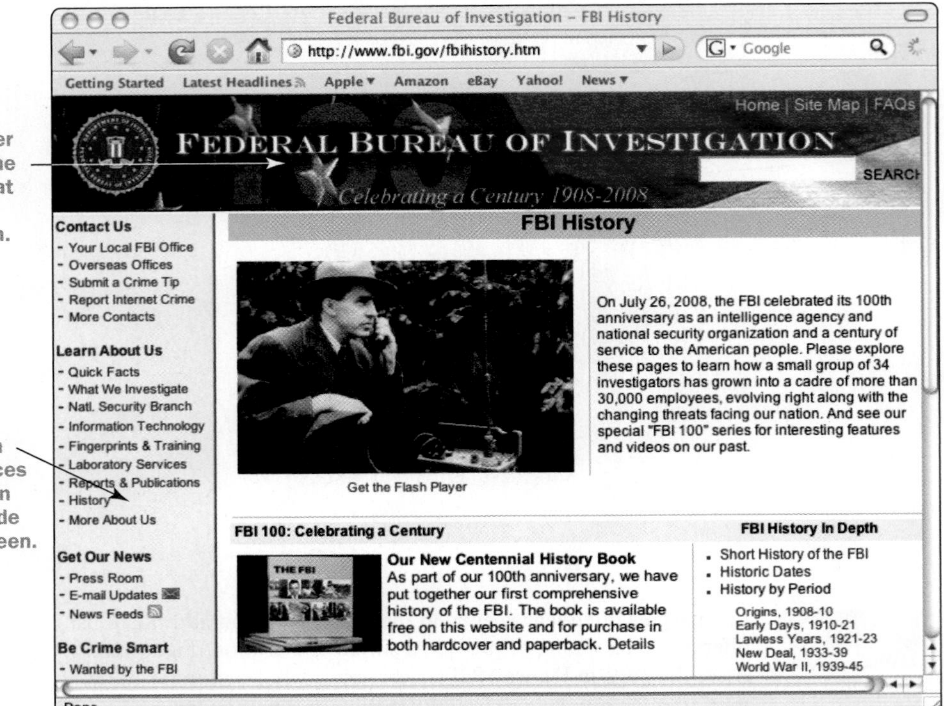

This banner anchors the interface at the top of the screen.

This navigation bar balances the text on the left side of the screen.

Figure 18.9: Usually, what works on paper can work on screen. You can use pull-outs, margin comments, and sidebars on a screen interface. Navigation bars and banners can also be added to screen-based documents for balance.

Source: Federal Bureau of Investigation, http://www.fbi.gov/fbihistory.htm.

Design Principle 2: Alignment

Items on a page or screen can be aligned vertically and horizontally. By aligning items *vertically* on the page, you can help readers identify different levels of information in a document. By aligning items *horizontally,* you can connect them visually so readers view them as a unit.

In Figure 18.10, for example, the page on the left gives no hint about the hierarchy of information, making it difficult for a reader to scan the text. The page on the right, meanwhile, uses alignment to clearly signal the hierarchy of the text.

Alignment takes advantage of readers' natural tendency to search out visual relationships among items on a page. If a picture, for example, is aligned with a block of text on a page, readers will naturally assume that they go together.

In paper-based documents, look for ways you can use margins, indentation, lists, headings, and graphics to create two or three levels in the text. If you use a consistent

GO TO THE NET To see sample texts that use alignment well and poorly, go to **www.pearsonhighered.com/johnsonweb4/18.7**

Using Vertical Alignment

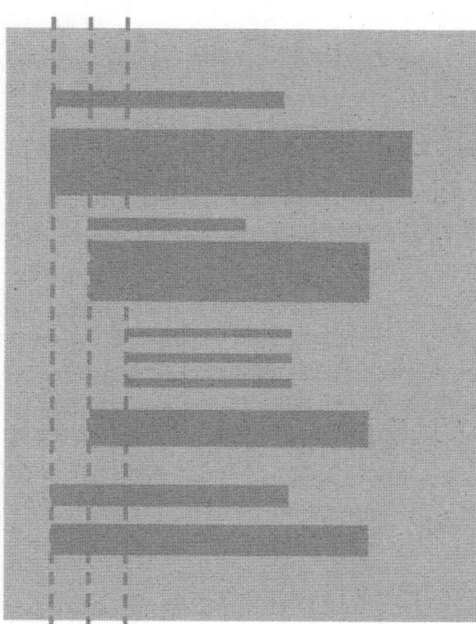

Figure 18.10: Alignment allows readers to see the hierarchy of information in a text.

alignment strategy throughout the text, you will design a highly readable and accessible document.

In technical documents, items are usually aligned on the left side. In rare cases, you might try aligning titles and headings on the right side. But you should use centering only for titles, because it causes alignment problems in the text. Figure 18.11 on page 495 shows how centering can create unpredictable vertical lines in the text.

Alignment is also very important in on-screen documents. To create a sense of stability, pay attention to the horizontal and vertical alignments of features on the interface. For example, the screen in Figure 18.12 on page 496 shows how you can align text and graphics to make an interface look stable.

Design Principle 3: Grouping

The principle of grouping means that items on a page that are near each other will be seen as one unit. Grouping allows you to break up the information on a page by dividing the text into scannable blocks.

Humans naturally see items that are placed near each other as a whole unit. So, if two items are placed near each other, like a picture and a caption, readers will assume that they belong together. In Figure 18.12, notice how pictures are put near paragraphs so that they are seen as units. The banner at the top of the page is supposed to be seen as a block unto itself.

AT WORK

Jennifer Martin

CREATIVE DIRECTOR, IBM CORPORATION, CHICAGO

The Centers for IBM e-business Innovation help companies use design and technology to solve problems and deliver solutions.

How can design make an interface more accessible for people with disabilities?

Accessibility is one of the most important issues when developing a website. The term "accessibility" refers to improving access for everyone regardless of his or her abilities. These improvements can also benefit users of limited-function devices, older technologies, or slower connection speeds.

There are many simple techniques that you can use to make your website more accessible for people with disabilities:

- Make the page layout easily scannable by breaking content into manageable chunks, while including sufficient white space to help users with special needs.
- Verify that there is sufficient contrast between foreground and background elements to improve legibility for users with color vision deficiencies.
- Ensure that links and buttons are large enough so that users with limited fine motor skills are able to easily target them.
- Include heading tags and intra-page links such as "skip to main content." These tags and links are made invisible to those using traditional browsers and greatly improve navigation when using a screen reader.
- Set the Alt attribute to null (" ") for images that provide no value to visually impaired users, so that they will be ignored by screen readers.
- Use an accessibility checker to validate that your website complies with the World Wide Web Consortiums Content Accessibility Guidelines and any applicable laws.

Designing an accessible website should be thought of as something that is integral to the design process, rather than a separate task or skill.

Grouping is also referred to as "using white space" to frame items on the page. White spaces are places where no text or images appear on the page and include:

- the margins of the document.
- the space around a list.
- the area between an image and the body text.
- the space between two paragraphs.

These spaces create frames around the items on the page so readers can view them as groups. For example, the white space around a vertical list (like the one in Figure 18.12) helps readers see that list as one unit.

Alignment Problems with Centered Text

Proposal for an
Improved
Wireless Network on Campus

Written to Sally Johnson, Vice President for Electronic Advancement

Abstract: In this proposal, we argue that Western State Community College needs to improve its wireless network on campus. Currently, the network only covers specific "hot spots" like the Student Union, the Biotechnology Center, and the Computer Science Center. As a result, a majority of students cannot take advantage of it. We recommend creating a wireless cloud over the campus by establishing a grid of Cisco 350 Series Wireless Access Points (WAPs) with a maximum transmission rate of 11 Mbps. The network could be made secure with IPSec and a Cisco 3030 VPN Concentrator. The cost for this network would likely be about $120,000. The benefits, however, would greatly outstrip the costs, because the university could minimize its hardwired computer classrooms.

The Innovation Team

Introduction to Technical Communication, Humanities 381

Figure 18.11: Centering is fine for headings, but too much centered material can make the text look chaotic.

Using Headings

One way to group information is to use headings. When used properly, headings will help your readers quickly understand the structure of your document and how to use it.

Your computer's word-processing program makes it easy for you to use headings by changing fonts and font sizes. Or, you can use your word processor's Style feature to create standard headings for your documents.

Different types of headings should signal the various levels of information in the text.

- **First-level headings** should be sized significantly larger than second-level headings. In some cases, first-level headings might use all capital letters ("all caps") or small capital letters ("small caps") to distinguish them from the font used in the body text.
- **Second-level headings** should be significantly smaller and different from the first-level headings. Whereas the first-level headings might be in all caps, the second-level headings might use bold lettering.
- **Third-level headings** might be italicized and a little larger than the body text.

Want more information on using headings? Go to **www.pearsonhighered.com/johnsonweb4/18.8**

An Interface That Uses Alignment and Grouping Well

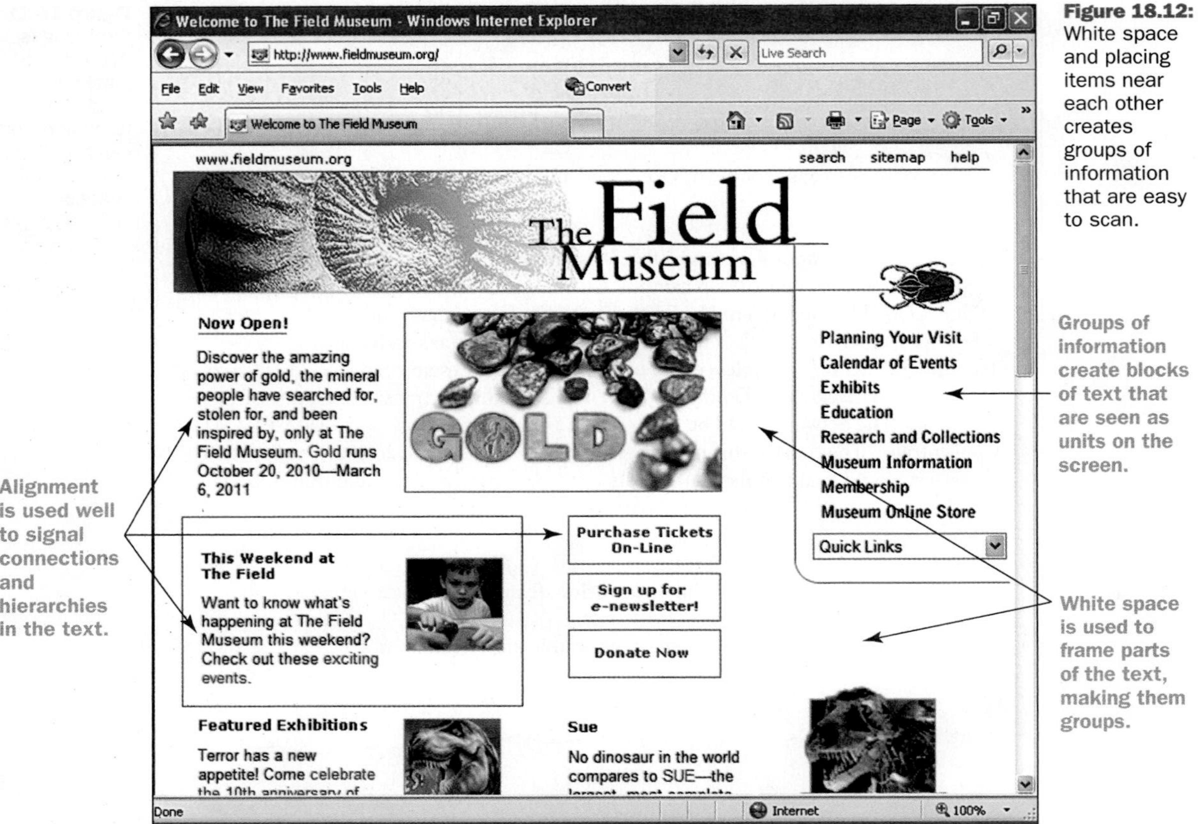

Figure 18.12: White space and placing items near each other creates groups of information that are easy to scan.

Source: The Field Museum, Chicago, Illinois, http://www.fieldmuseum.org.

- **Fourth-level headings** are about as small as you should go. They are usually boldfaced or italicized and placed on the same line as the body text.

Figure 18.13 shows various levels of headings and how they might be used.

In most technical documents, the headings should use the same typeface throughout (e.g., Avante Garde, Times, Helvetica). In other words, the first-level heading should use the same typeface as the second-level and third-level headings. Only the font size and style (bold, italics, small caps) should be changed.

Headings should also follow consistent wording patterns. A consistent wording pattern might use gerunds (*-ing* words) to lead off headings. Or, to be consistent, questions might be used as headings.

Levels of Headings

DOCUMENT TITLE

FIRST-LEVEL HEADING

This first-level heading is 18-point Avant Garde, boldface with small caps. Notice that it is significantly different from the second-level heading, even though both levels are in the same typeface.

Second-Level Heading

This second-level heading is 14-point Avant Garde with boldface.

Third-Level Heading

This third-level heading is 12-point Avant Garde italics. Often, no extra space appears between a third-level heading and the body text, as shown here.

Fourth-Level Heading. This heading appears on the same line as body text. It is designed to signal a new layer of information without standing out too much.

Figure 18.13: The headings you choose for a document should be clearly distinguishable from the body text and from each other so that readers can see the levels in the text.

Inconsistent Headings

Global Warming

What Can We Do about Global Warming?

Alternative Energy Sources

Consistent Headings Using Gerunds

Defining Global Warming

Doing Something about Global Warming

Finding Alternative Energy Sources

Consistent Headings Using Questions

What Is Global Warming?

How Can We Do Something about Global Warming?

Are Alternative Energy Sources Available?

Headings should also be specific, clearly signaling the content of the sections that follow them:

Unspecific Headings

Kinkajous

Habitat

Food

Specific Headings

Kinkajous: Not Rare, But Hard to Find

The Habitat of Kinkajous

The Food and Eating Habits of Kinkajous

Headings serve as *access points* for readers, breaking a large text into smaller groups. If headings are used consistently, readers will be able to quickly access your document to find the information they need. If headings are used inconsistently, readers may have difficulty understanding the structure of the document.

Using Borders and Rules

In document design, *borders* and straight lines called *rules* can be used to carve a page into smaller groups of information. They can also help break the text into more manageable sections for the readers.

Borders completely frame parts of the document (Figure 18.14). Whatever appears within a border should be able to stand alone. For example, a bordered warning statement should include all the information readers need to avoid a dangerous situation. Similarly, a border around several paragraphs (like a sidebar) suggests that they should be read separately from the main text.

Rules are often used to highlight a banner or carve a document into sections. They are helpful for signaling places to pause in the document. But when they are overused, they can make the text look and feel too fragmented.

Borders and rules are usually easy to create with any word processor. To put a border around something, highlight that item and find the Borders command in your word processor (Figure 18.15). In the window that appears, you can specify what kind of border you want to add to the text.

Rules can be a bit more difficult to use, so you might want to use a desktop layout program, like Adobe InDesign or QuarkXPress. However, if your document is small or simple, you can use the Draw function of your word processor to draw horizontal or vertical rules in your text.

More advice about using rules and borders is available at **www.pearsonhighered.com/johnsonweb4/18.9**

Using Rules and Borders to Group Information

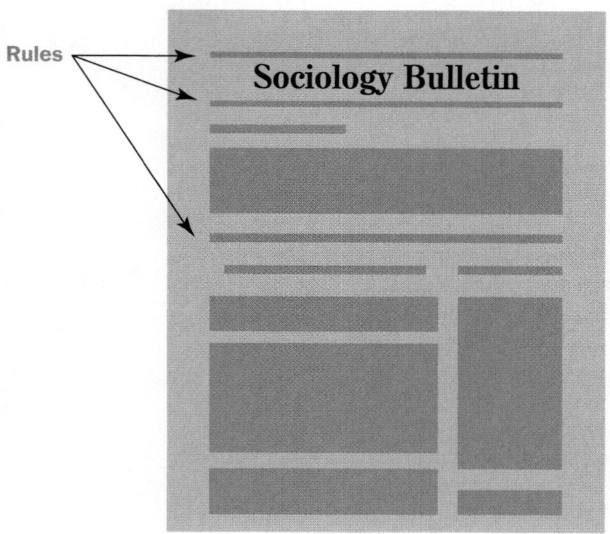

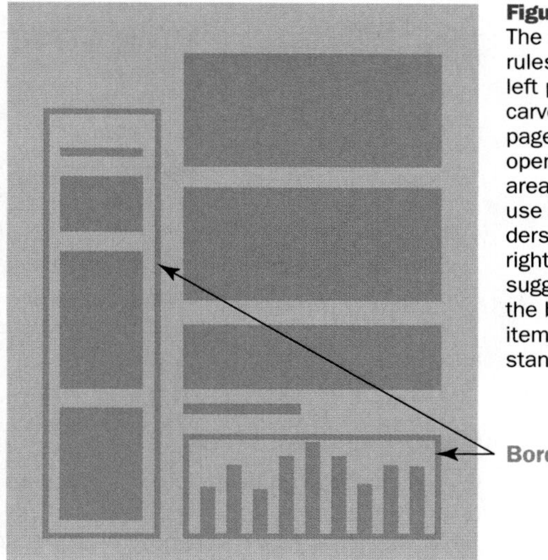

Figure 18.14: The use of rules on the left page carves the page into open-ended areas. The use of borders on the right page suggests that the bordered items can stand alone.

Making Borders

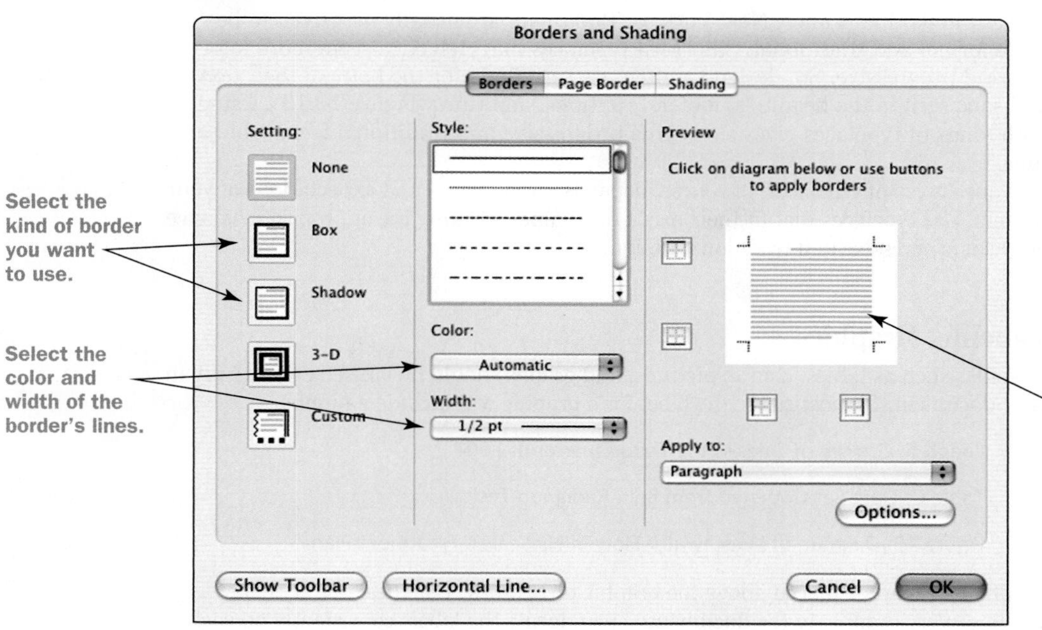

Figure 18.15: The Borders command in your word processor will allow you to put boxes or draw lines around items.

Design Principle 4: Consistency

The principle of consistency suggests that design features should be used consistently throughout a document or website:

- Headings should be predictable.
- Pages should follow the same grid.
- Lists should use consistent bulleting or numbering schemes.
- Page numbers should appear in the same place on each page.

Consistency is important because it creates a sense of order in a document while limiting the amount of clutter. A consistent page design will help your readers access information quickly, because each page is similar and predictable (Figure 18.16). When design features are used inconsistently, readers will find the document erratic and hard to interpret.

Choosing Typefaces

Consistency should be an important consideration when you choose typefaces for your document. As a rule of thumb, a document should not use more than two typefaces. Most page designers will choose two typefaces that are very different from each other, usually a *serif* typeface and a *sans serif* typeface.

A serif typeface like Times Roman, New York, or Bookman has small tips (serifs) at the ends of the main strokes in each letter (Figure 18.17). Sans serif typefaces like Arial or Helvetica do not have these small tips.

Serif fonts, like Times, New York, or Bookman, are usually perceived to be more formal and traditional. Sans serif typefaces like Helvetica seem more informal and progressive. So, designers often use serif fonts for the body of their text and sans serif in the headings, footers, captions, and titles (Figure 18.18). Using both kinds of typefaces gives a design a progressive and traditional feel at the same time.

Your choice of typefaces, of course, depends on the needs and expectations of your readers. Use typefaces that fit their needs and values, striking the appropriate balance between progressive and traditional design.

Labeling Graphics

Graphics such as tables, charts, pictures, and graphs should be labeled consistently in your document. In most cases, the label for a graphic will include a number.

Graph 5: Growth of Sales in the Third Quarter of 2004

Table C: Data Set Gathered from Beta Radiation Tests

Figure 10: Diagram of Wastewater Flow in Hinsdale's Treatment Plant

Link

For more information on labeling graphics, see Chapter 19, page 527.

The label can be placed above the graphic or below it. Use the same style for labeling every graphic in the document. Then, locate the labels consistently on each graphic.

For websites that discuss the history of typefaces, go to **www.pearsonhighered.com/johnsonweb4/18.10**

Consistent Layout

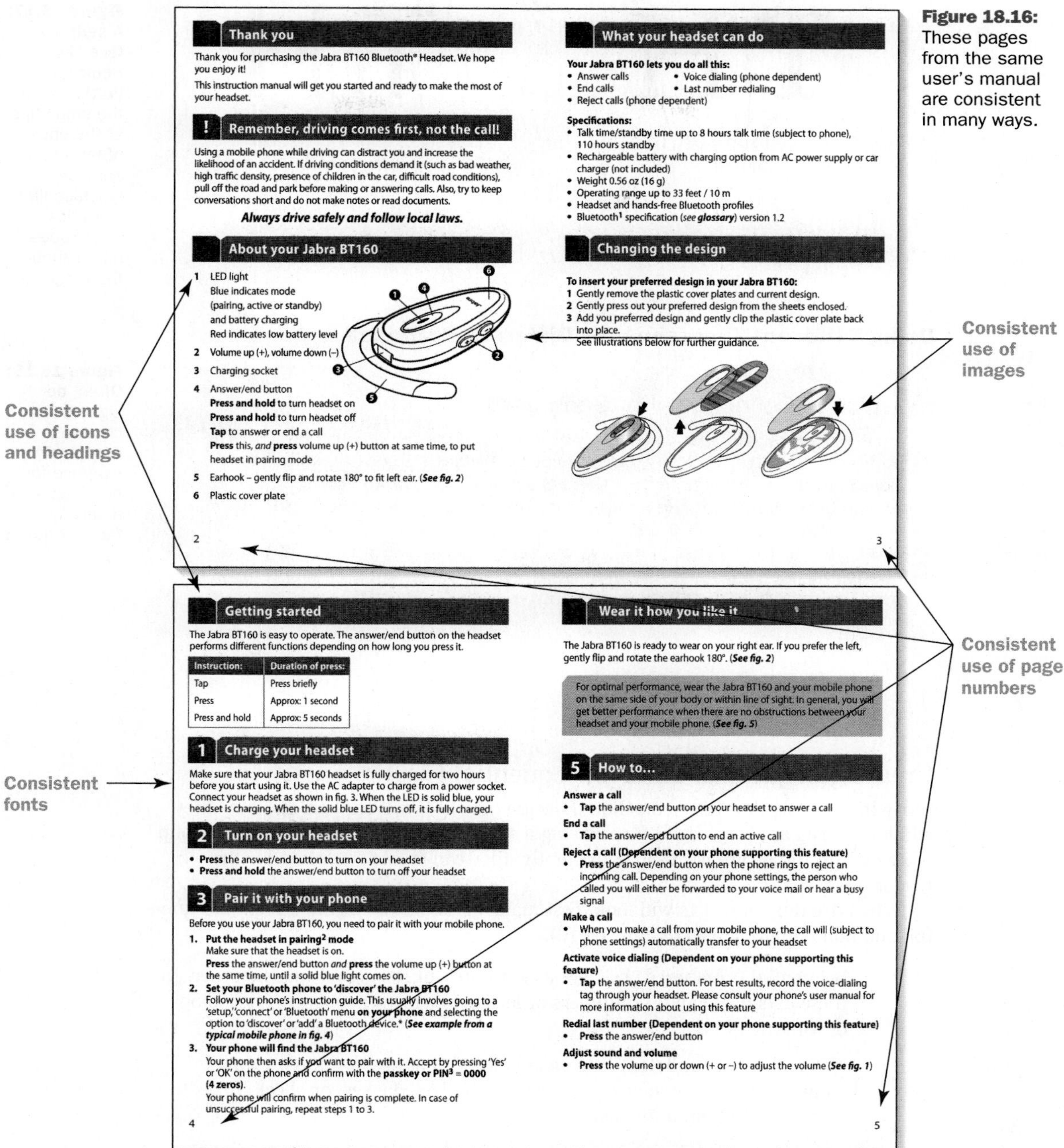

Thank you

Thank you for purchasing the Jabra BT160 Bluetooth® Headset. We hope you enjoy it!

This instruction manual will get you started and ready to make the most of your headset.

! Remember, driving comes first, not the call!

Using a mobile phone while driving can distract you and increase the likelihood of an accident. If driving conditions demand it (such as bad weather, high traffic density, presence of children in the car, difficult road conditions), pull off the road and park before making or answering calls. Also, try to keep conversations short and do not make notes or read documents.

Always drive safely and follow local laws.

About your Jabra BT160

1 LED light
Blue indicates mode
(pairing, active or standby)
and battery charging
Red indicates low battery level

2 Volume up (+), volume down (–)

3 Charging socket

4 Answer/end button
Press and hold to turn headset on
Press and hold to turn headset off
Tap to answer or end a call
Press this, *and* **press** volume up (+) button at same time, to put headset in pairing mode

5 Earhook – gently flip and rotate 180° to fit left ear. (***See fig. 2***)

6 Plastic cover plate

2

What your headset can do

Your Jabra BT160 lets you do all this:

- Answer calls
- End calls
- Reject calls (phone dependent)
- Voice dialing (phone dependent)
- Last number redialing

Specifications:

- Talk time/standby time up to 8 hours talk time (subject to phone), 110 hours standby
- Rechargeable battery with charging option from AC power supply or car charger (not included)
- Weight 0.56 oz (16 g)
- Operating range up to 33 feet / 10 m
- Headset and hands-free Bluetooth profiles
- Bluetooth[1] specification (*see* ***glossary***) version 1.2

Changing the design

To insert your preferred design in your Jabra BT160:

1 Gently remove the plastic cover plates and current design.
2 Gently press out your preferred design from the sheets enclosed.
3 Add you preferred design and gently clip the plastic cover plate back into place.
See illustrations below for further guidance.

3

Getting started

The Jabra BT160 is easy to operate. The answer/end button on the headset performs different functions depending on how long you press it.

Instruction:	Duration of press:
Tap	Press briefly
Press	Approx: 1 second
Press and hold	Approx: 5 seconds

1 Charge your headset

Make sure that your Jabra BT160 headset is fully charged for two hours before you start using it. Use the AC adapter to charge from a power socket. Connect your headset as shown in fig. 3. When the LED is solid blue, your headset is charging. When the solid blue LED turns off, it is fully charged.

2 Turn on your headset

- **Press** the answer/end button to turn on your headset
- **Press and hold** the answer/end button to turn off your headset

3 Pair it with your phone

Before you use your Jabra BT160, you need to pair it with your mobile phone.

1. **Put the headset in pairing[2] mode**
   Make sure that the headset is on.
   Press the answer/end button *and* **press** the volume up (+) button at the same time, until a solid blue light comes on.
2. **Set your Bluetooth phone to 'discover' the Jabra BT160**
   Follow your phone's instruction guide. This usually involves going to a 'setup,' 'connect' or 'Bluetooth' menu **on your phone** and selecting the option to 'discover' or 'add' a Bluetooth device.* (***See example from a typical mobile phone in fig. 4***)
3. **Your phone will find the Jabra BT160**
   Your phone then asks if you want to pair with it. Accept by pressing 'Yes' or 'OK' on the phone and confirm with the **passkey or PIN[3] = 0000 (4 zeros)**.
   Your phone will confirm when pairing is complete. In case of unsuccessful pairing, repeat steps 1 to 3.

4

Wear it how you like it

The Jabra BT160 is ready to wear on your right ear. If you prefer the left, gently flip and rotate the earhook 180°. (***See fig. 2***)

For optimal performance, wear the Jabra BT160 and your mobile phone on the same side of your body or within line of sight. In general, you will get better performance when there are no obstructions between your headset and your mobile phone. (***See fig. 5***)

5 How to...

Answer a call
- **Tap** the answer/end button on your headset to answer a call

End a call
- **Tap** the answer/end button to end an active call

Reject a call (Dependent on your phone supporting this feature)
- **Press** the answer/end button when the phone rings to reject an incoming call. Depending on your phone settings, the person who called you will either be forwarded to your voice mail or hear a busy signal

Make a call
- When you make a call from your mobile phone, the call will (subject to phone settings) automatically transfer to your headset

Activate voice dialing (Dependent on your phone supporting this feature)
- **Tap** the answer/end button. For best results, record the voice-dialing tag through your headset. Please consult your phone's user manual for more information about using this feature

Redial last number (Dependent on your phone supporting this feature)
- **Press** the answer/end button

Adjust sound and volume
- **Press** the volume up or down (+ or –) to adjust the volume (***See fig. 1***)

5

Figure 18.16: These pages from the same user's manual are consistent in many ways.

Source: Jabra BT160 Bluetooth User Manual, pp. 2–5. GN Netcom, Inc.

Serif and Sans Serif Typefaces

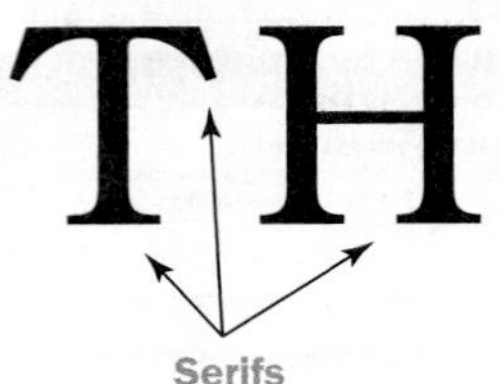

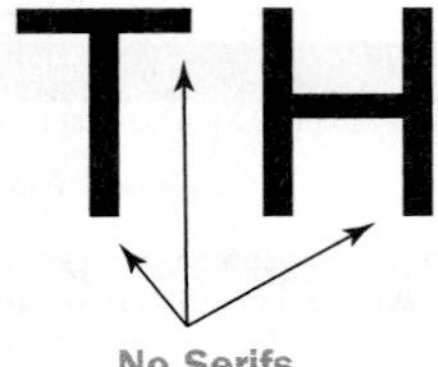

Figure 18.17: A serif typeface like Bookman (left) includes the small tips at the ends of letters. A sans serif typeface like Helvetica (right) does not include these tips.

Using Different Typefaces for Different Purposes

Which Is Better? Serif or Sans Serif?

Serif typefaces are often used in traditional-looking texts, especially for the body text. Studies have suggested inconclusively that serif typefaces like Bookman are more legible, but there is some debate as to why. Some researchers claim that the serifs create horizontal lines in the text that make serif typefaces easier to follow. These studies, however, were mostly conducted in the United States, where serif typefaces are common. In other countries, like Britain, where sans serif typefaces are often used for body text, the results of these studies might be quite different.

Figure 18.18: Often, designers will choose a sans serif typeface for headings and a serif typeface for the main text.

Creating Sequential and Nonsequential Lists

Early in the design process, you should decide how lists will be used and how they will look. Lists are very useful for showing a sequence of tasks or setting off a group of items. But if they are not used consistently, they can create confusion for the readers.

When deciding how lists will look, first make decisions about the design of sequential and nonsequential lists (Figure 18.19).

Sequential (numbered) lists are used to present items in a specific order. In these lists, you can use numbers or letters to show a sequence, chronology, or ranking of items.

Nonsequential (bulleted) lists include items that are essentially equal in value or have no sequence. You can use bullets, dashes, or check boxes to identify each item in the list.

GO TO THE NET

For advice about using lists, go to **www.pearsonhighered.com/johnsonweb4/18.11**

Using Sequential and Nonsequential Lists

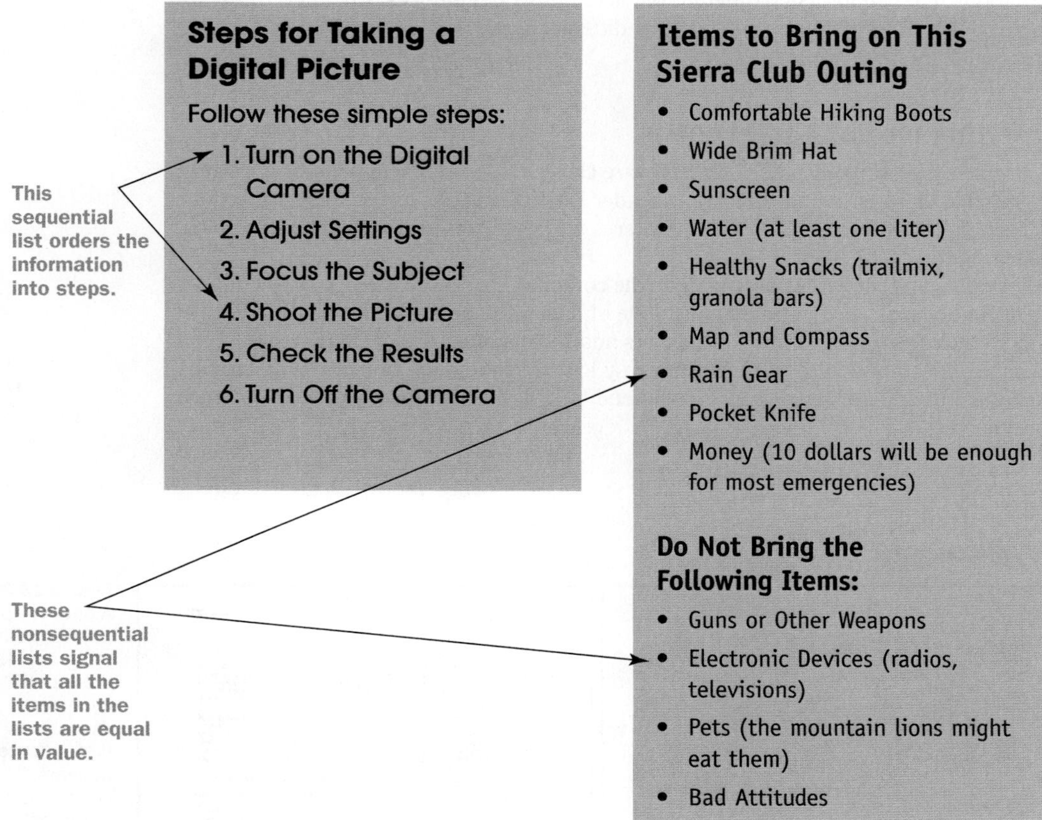

Steps for Taking a Digital Picture
Follow these simple steps:
1. Turn on the Digital Camera
2. Adjust Settings
3. Focus the Subject
4. Shoot the Picture
5. Check the Results
6. Turn Off the Camera
Items to Bring on This Sierra Club Outing
• Comfortable Hiking Boots
• Wide Brim Hat
• Sunscreen
• Water (at least one liter)
• Healthy Snacks (trailmix, granola bars)
• Map and Compass
• Rain Gear
• Pocket Knife
• Money (10 dollars will be enough for most emergencies)
Do Not Bring the Following Items:
• Guns or Other Weapons
• Electronic Devices (radios, televisions)
• Pets (the mountain lions might eat them)
• Bad Attitudes
This sequential list orders the information into steps.
These nonsequential lists signal that all the items in the lists are equal in value.

Figure 18.19: The sequential list on the left shows an ordering of the information, so it requires numbers. The nonsequential list on the right uses bullets because there is no particular ordering of these items.

> **AT A GLANCE**
>
> **Checking for Consistency**
>
> The following items should be used consistently in your document:
>
> - Typefaces (serif and sans serif)
> - Labeling of graphics
> - Lists (sequential and nonsequential)
> - Headers and footers

Lists make information more readable and accessible. So, you should look for opportunities to use them in your documents. If, for example, you are listing steps to describe how to do something, you have an opportunity to use a sequential list. Or, if you are creating a list of four items or more, ask yourself whether a nonsequential list would present the information in a more accessible way.

Just make sure you use lists consistently. Sequential lists should follow the same numbering scheme throughout the document. For example, you might choose a numbering scheme like (1), (2), (3). If so, do not number the next list 1., 2., 3. and others A., B., C., unless you have a good reason for changing the numbering scheme. Similarly, in nonsequential lists, use the same symbols when setting off lists. Do not use bullets (•) with one list, check marks (✓) with another, and boxes

(■) with a third. These inconsistencies only confuse the readers. Of course, there are situations that call for using different kinds of nonsequential lists. If you need lists to serve completely different purposes, then different symbols will work—as long as they are used consistently.

Inserting Headers and Footers

Even the simplest word-processing software can put a header or footer consistently on every page. As their names suggest, a header is text that runs across the top margin of each page in the document, and a footer is text that runs along the bottom of each page (Figure 18.20).

Headers and footers usually include the company's name or the title of the document. In documents of more than a couple of pages, the header or footer (not both) should include the page number. Headers and footers often also include design features like a horizontal rule or a company logo. If these items appear at the top or bottom of each page, the document will tend to look like it is following a consistent design.

Headers and Footers

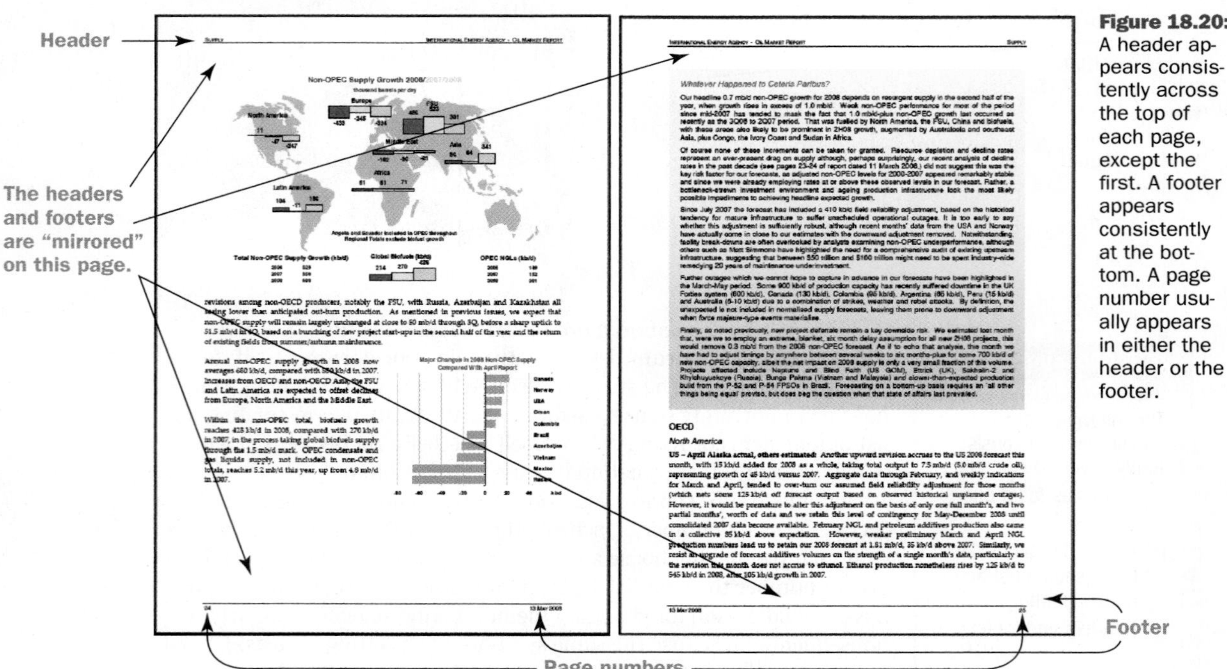

Figure 18.20: A header appears consistently across the top of each page, except the first. A footer appears consistently at the bottom. A page number usually appears in either the header or the footer.

Source: Oil Market Report, May 13, 2008.

Need a template? Go to **www.pearsonhighered.com/johnsonweb4/18.12**

Five Principles of Document Design

AT A GLANCE

- Balance
- Alignment
- Grouping
- Consistency
- Contrast

Design Principle 5: Contrast

Contrast makes items look distinct and different, adding energy and sharpening boundaries among the features on the page or screen.

A good guideline is to "make different things on the page look very different." Contrast, as shown in Figure 18.21, makes design elements lively.

When designing a page, consider your use of contrast carefully. Word processors offer many tools for adding contrast in ways that capture readers' attention. Sometimes, though, you can accidentally create contrast problems with different colors or shading in the background—or just too much clutter on the page. In this section are some helpful tips for using contrast in documents and interfaces.

Adding Shading and Background Color

When used properly, shading and background color can help highlight important text in a document. However, these design features can also make texts hard to read. For example, Figure 18.22 shows how a lack of contrast can make text hard to read.

Contrast in a Webpage

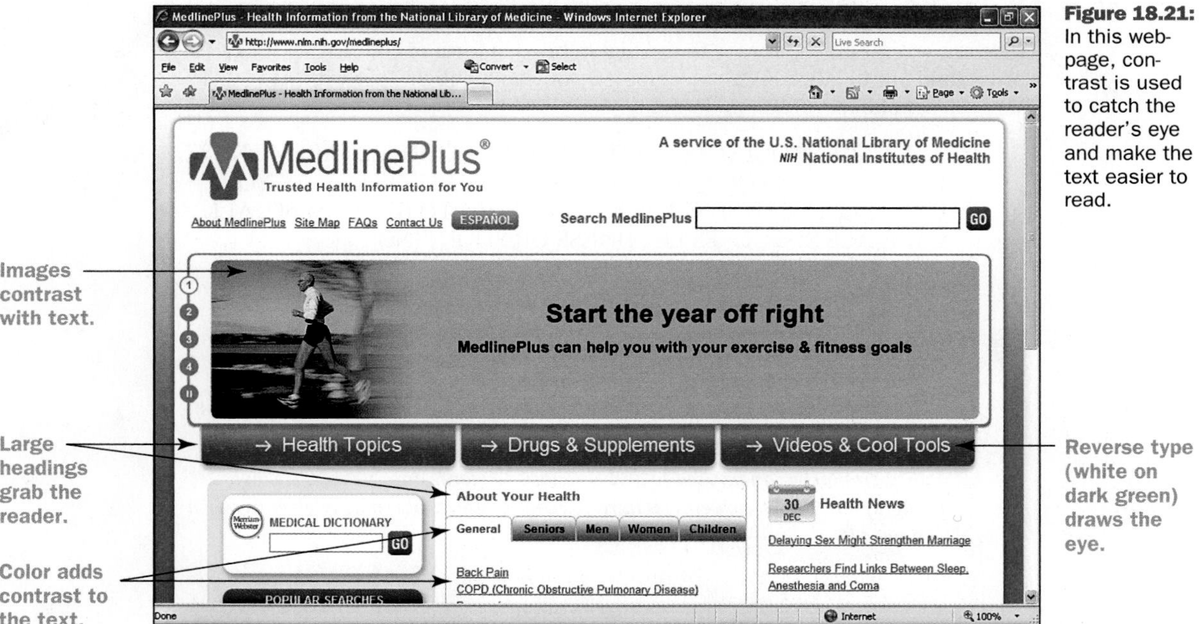

Figure 18.21: In this webpage, contrast is used to catch the reader's eye and make the text easier to read.

Source: Medline Plus, http://www.medlineplus.gov.

To view webpages with contrast problems, go to **www.pearsonhighered.com/johnsonweb4/18.13**

Shading and Contrast

This text is set against a gray background. It is difficult to read because there is not enough contrast between the words and the background against which they are set.	This text is set against a white background. It is easier to read, because the contrast between the words and the white background is sharp and definite.

Figure 18.22: Shaded versus unshaded text: The shaded text on the left is more difficult to read because the words do not contrast significantly with the background shading.

Background color or images can also make text on computer screens difficult to read. So use them carefully and be sure to check how backgrounds work (or don't work) with the text on the screen.

It is fine to use shading, background color, and background images. However, make sure the words on the page contrast significantly with the background.

Highlighting Text

There are several different ways to use highlighting, a form of contrast, to make words stand out in a text. With a computer, you can highlight text with ALL CAPS, SMALL CAPS, **boldface**, *italics*, underlining, and color.

> ALL CAPS SHOULD BE USED IN MODERATION, USUALLY ONLY FOR HEADINGS OR SHORT WARNING STATEMENTS. AS YOU CAN SEE, THESE SENTENCES ARE HARD TO READ. ALSO, ALL CAPITALS SUGGESTS LOUDNESS OR SHOUTING.
>
> SMALL CAPS ARE A LITTLE EASIER TO READ THAN ALL CAPS BUT THEY SHOULD STILL BE USED SPARINGLY. THEY ARE ESPECIALLY USEFUL FOR HEADINGS.
>
> **Boldface is useful for highlighting individual words or whole sentences. When used in moderation, boldface helps words or sentences stand out. When used too often, however, the highlighting effect is reduced because readers grow used to the bold text.**
>
> *Italics, similarly, can be used to add emphasis to words or sentences. A whole paragraph in italics, however, is hard to read, so use italics sparingly—only when you need it.*
>
> Color is perhaps the most prominent way to highlight words or sentences. It is especially helpful for making headings more prominent. However, color can be distracting when it is overused. So use it selectively and consistently.

Underlining is rarely used now that we have other ways to highlight text. Underlining was effective when typewriters were widely used, because it was the only way (except for all caps) to highlight information. Computers, however, now give us better options for highlighting like italics, boldface, and color.

Highlighting is an easy way to create contrast in a document. When used selectively, it can make important words or sentences pop off the page. If it is overused, though, readers become immune to it.

Using Font Size and Line Length

Your decisions about font size, line length, and line spacing should depend on the readers of your document.

APPROPRIATE FONT SIZE The font size you choose for your text depends on how your readers will be using the text.

For most readers, text in 11-point or 12-point font sizes is easy to read.

Fonts smaller than 10 points can be difficult to read, especially when used for a longer stretch of text. When a page length is required for a document, it is often tempting to use a smaller font, so more words can be put on a page. But most people quickly grow tired of straining their eyes to read the text. You are better off using a normal-sized font with fewer words. That way, at least the readers will actually read the text.

Large font sizes (above 14 points) should be used only for special purposes, such as documents written for older readers or situations where the text needs to be read at a distance. When used at length, these larger sizes can make a text look childish. Extra-large font sizes usually suggest that the writer is simply filling a page to hide a lack of content.

PROPER LINE LENGTH Line length is also an important choice in a document. Readers who scan usually prefer a shorter line length, making columns especially useful for documents that will be read quickly.

However, when lines are
short, they force
readers to quickly look

back and forth.
Eventually, these short
lines will frustrate
readers because they
require such rapid
movement of the eyes.
They also seem to suggest
fragmented thinking on
the part of the writer,
because the sentences
look fragmented.

Longer line lengths can have the opposite effect. Readers quickly grow tired of following the same line for several inches across the page. The lines seem endless, giving readers the impression that the information is hard to process. More important, though, readers will find it difficult to locate the next line on the left side. A line should never be wider than 6 inches across.

As you consider font size and line length, you should anticipate how your readers will use your document. If they will be scanning, shorter line lengths will help them read quickly. If they will not be scanning, longer lines are fine. Then, choose the font size or line length that suits their needs.

Cross-Cultural Design

As the global economy grows, designing documents for cross-cultural readers may be one of the greatest challenges facing technical communicators. Today, most international readers are adjusting to Western design practices. But, with the global reach of the Internet and the growth of economies around the world, international readers are beginning to expect documents and interfaces to reflect their own cultural design conventions.

When designing cross-culturally, your first consideration is whether your document or interface needs a "culturally deep" or a "culturally shallow" design.

- **Culturally deep** documents and interfaces use the language, symbols, and conventions of the target culture to reflect readers' design preferences and expectations. To develop a culturally deep design, you probably need help from designers or consultants who are familiar with the target culture and understand its design expectations.
- **Culturally shallow** documents and interfaces usually follow Western design conventions, but they adjust to reflect some of the design preferences of the cultures in which they will be used. They also avoid any cultural taboos of the people who are likely to use the text. Culturally shallow designs tend to be used in documents or interfaces that need to work across a variety of cultures.

Unless your company is targeting its products or services to a specific culture (e.g., a nation like Korea or Zimbabwe), most of your documents or interfaces will need to be culturally shallow so that they can work across a variety of cultures.

Culturally shallow designs usually consider four design issues: use of color, use of people, use of symbols, and direction of reading.

Use of color—Choice of colors in a cross-cultural document can influence how readers interpret the message, because colors can have different meanings across cultures. For instance, the use of red in Japan signals anger, while in China red signals happiness. The use of red in Egypt symbolizes death. Meanwhile, the color green in France symbolizes criminality, while in the United States green symbolizes moving forward or environmental consciousness. Figure 18.23 shows how some common colors are perceived across cultures. When designing your document or interface, you should use colors that reflect the expectations of the likely readers (or at least avoid colors that have negative associations).

Images of people—Cross-cultural texts should use images of people carefully. Avoid big smiles, highly emotional expressions, suggestive behavior, and flashy clothing. In pictures, interactions between women and men should avoid sending mixed signals. In some cultures, especially Islamic cultures, images of people are used only when "needed." The definition of "need" varies among Islamic subcultures, but images tend to be used only for purposes of identification.

Colors in Other Cultures

Color	Japan	France	China	Egypt	United States
Red	Anger, danger	Aristocracy	Happiness	Death	Danger, stop
Blue	Villainy	Freedom, peace	Heavens, clouds	Virtue, faith, truth	Masculine, conservative
Green	Future, youth, energy	Criminality	Ming Dynasty, heavens	Fertility, strength	Safe, go, natural
Yellow	Grace, nobility	Temporary	Birth, wealth, power	Happiness, prosperity	Cowardice, temporary
White	Death	Neutrality	Death, purity	Joy	Purity, peace, marriage

Figure 18.23: Colors can have very different meanings in different cultures. In some cases, the meanings of colors may even be contradictory among cultures.

Source: From Patricia Russo and Stephen Boor, "How Fluent Is Your Interface? Designing for International Users." In S. Ashlund, K. Mullet, A. Henderson, E. Hollnagel, and T. White, eds., Proceedings of the INTERACT '93 and CHI '93 Conference on Human Factors in Computing Systems, *pp. 342–347, Table 1. © 1993 ACM, Inc. Reprinted by permission. http://doi.acm.org/10.1145/169059.169274.*

Link

For more information on international and cross-cultural symbols, go to Chapter 19, page 542.

Link

For more help on working with cross-cultural readers, go to Chapter 2, page 28.

Use of symbols—Common symbols can have very different meanings in different cultures. For example, in many cultures, the "OK" hand signal is highly offensive. Uses of crescent symbols (i.e., moons) or crosses can have a variety of religious meanings. White flowers or a white dress can signify death in many Asian cultures. To avoid offending readers with symbols, a good approach is to use only simple shapes (e.g., circles, squares, triangles) in cross-cultural documents.

Direction of reading—Many cultures in the Middle East and Asia read right to left instead of left to right. As a result, some of the guidelines for balancing a page design discussed earlier in this chapter should be reversed. For example, a document or interface that reads right to left tends to be anchored on the right side. Otherwise, the text will look unbalanced to a right-to-left reader. Figure 18.24, for example, shows a website that is designed right to left for Middle Eastern readers.

Cross-cultural design can be very challenging. The secret is to consult with people from the target culture and/or use consultants to help you design your documents and interfaces. Then, be ready to learn from your mistakes.

A Right-to-Left Interface Design

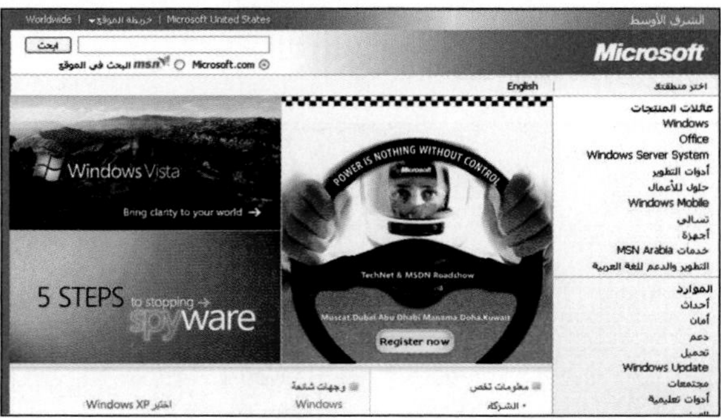

Figure 18.24: These interfaces demonstrate a right-to-left versus left-to-right design of the same webpage. Note how the Middle Eastern webpage (top) is anchored on the right, mirroring the design in the Western webpage (bottom).

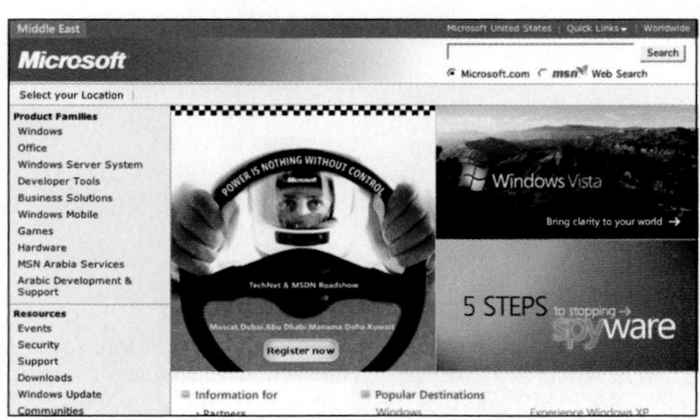

Source: Microsoft, http://www.microsoft.com.

Using the Principles of Design

Balance, alignment, grouping, consistency, contrast: These five basic design principles should help you create easy-to-read page layouts and screen interfaces that highlight important information and attract your readers.

In many ways, designing documents is like drafting the written text in your documents. You should go through a process—in this case, a *design process.*

Analyze your readers and the document's context of use.

Use thumbnails to sketch out the design.

Design the document.

Revise and edit the design.

As with drafting, each stage in the process takes you closer to a finished document.

Analyze Your Readers and the Document's Context of Use

You should begin designing a document by first looking back at your analyses of your readers and the contexts in which they will use your document. Different kinds of readers will have different expectations and preferences for the design of your document. A more traditional reader, for example, might prefer a conservative layout with simple headings and a one-column or two-column layout. A more progressive reader might prefer a bolder layout with more color and contrast.

As you consider your readers, try to match your design to the tone you are setting in the written part of the document. Positive, exciting themes in the written text should be reflected in the color and boldness of the design. Serious or somber themes should be reflected with subdued color schemes and a plainer design.

The context in which your document will be used should also play an important role in helping you decide on the appropriate design. Pay special attention to the *physical context* in which the readers will be using your document. Will they be in a meeting? At a worksite? In their living rooms?

Each of these contexts might shape how you will design the document. For example, if you know your readers will be discussing your document in a meeting, you might use margin notes and lists to highlight information that will be discussed at the meeting. That way, your readers will be able to quickly find "talking points" in your document that they can share with others in the meeting.

Use Thumbnails to Sketch Out the Design

Next, start sketching out how you think the design will look. Professional graphic designers often use *thumbnails* to sketch out possible layouts for their documents or screen interfaces. Thumbnails can be sketched by hand, usually with pencil, or with a computer drawing program (Figure 18.25). They are miniature versions of the pages.

Thumbnailing is especially helpful when you are working with a team. With little effort, you can quickly sketch out sample pages or screen interfaces. When the group agrees, you can then use these sketches to help you lay out the page. The simplest way to thumbnail pages is to fold a regular sheet of paper into four parts (i.e., fold it once lengthwise and once horizontally). The folds will then give you four separate

For more advice on thumbnailing, go to **www.pearsonhighered.com/johnsonweb4/18.14**

Sample Page and Interface Thumbnails

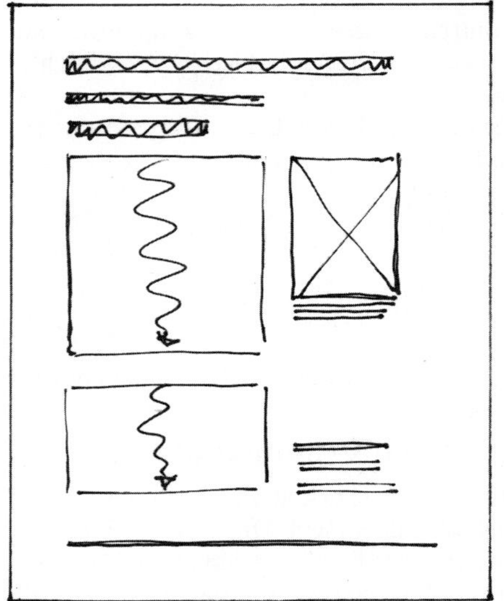

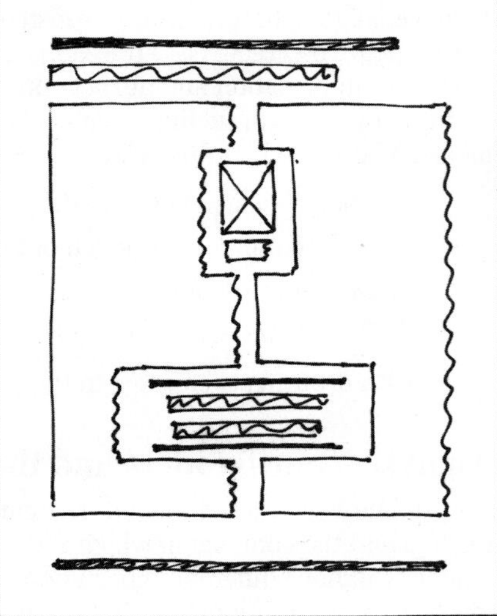

Figure 18.25: Thumbnails are miniature sketches of possible page or screen layouts. They help you find a good design with a minimum amount of effort.

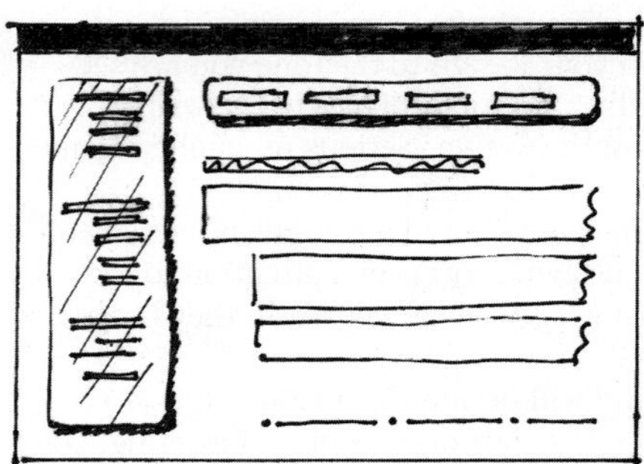

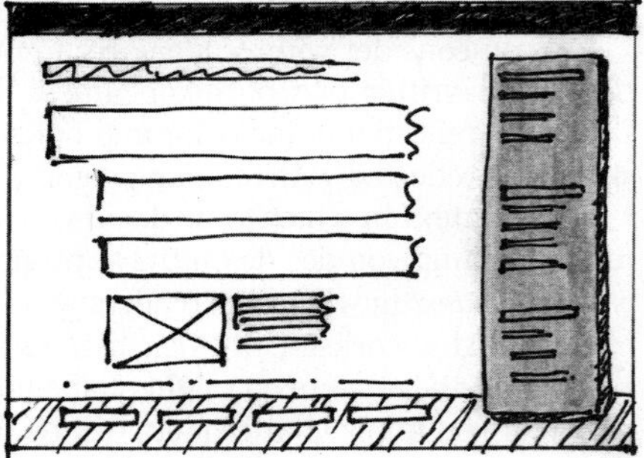

"pages" to practice on. For computer interfaces, simply turn the sheet sideways. You will have four rectangles that are each about the shape of a screen.

One nice thing about designing with thumbnails is how easy they are to use. It takes only seconds to sketch out a few layouts. If you or other people on your team don't like the sketches, you can toss them in the recycling bin and start again. Thumbnails require much less effort than an attempt to develop a full layout of a page or document. As a result, you will have more freedom to be creative and to try out a few different designs for the document.

Design the Document

Usually, the design is developed as the document is drafted. You don't need to wait until the entire document is written before you start laying out some of its pages. In fact, it is better to begin designing *before* the whole document is drafted. That way, the design might help you (1) identify gaps in the content; (2) find places where visuals, like graphs and charts, are needed; and (3) locate places in the document where supplemental text, like sidebars or margin notes, might be added to enhance the readability.

When you or your team have drafted a few pages of written text, you should try turning your thumbnails into page designs. Of course, the text will change as the drafting continues. But the actual written words will help you see how the document is going to look on paper or on the screen. Then, you can start using templates and/or grids to lay out information on the page. Make sure you create a flexible enough design for your document, leaving plenty of room for graphics and other visual elements.

As the drafting of the document nears completion, use your sample pages to guide the design of the rest of the pages.

Revise and Edit the Design

Revising and editing a document's design is similar to revising and editing the written text. Start looking over the document globally, paying attention to issues of balance, alignment, grouping, consistency, and contrast. Then, examine the document page by page to see if the visuals are consistent and appropriate.

Finally, proofread the design. While proofreading, you are going to find places where smaller alterations and corrections are needed. You should take the time to iron out these little wrinkles. After all, small inconsistencies and errors in the design will draw readers' attention away from the written text.

A Primer on Binding and Paper

In the technical workplace, most documents are printed on standard-sized paper—the kind you use in your copier or printer. There will be times, though, when you will need to make decisions about how your document will be bound, as well as decisions about the size, weight, and color of the paper. As copiers become more sophisticated, you may find yourself making these decisions with more frequency.

Here is a quick primer to help you choose binding and paper.

Binding

In most cases, technical documents are bound with a paper clip, a staple, or perhaps a three-ring binder. But if you are interested in a more permanent type of binding, here are a few of your options.

PERFECT BINDING In this method, glue is applied to the back edge of the pages, and the pages are then inserted into a cover (Figure 18.26). Perfect binding is a relatively inexpensive option for binding larger documents. The downside is that, with time, the binding will eventually come loose. When the binding begins to fail, pages will fall out of the document.

To learn more about binding documents, go to **www.pearsonhighered.com/johnsonweb4/18.15**

Different Kinds of Bindings

Perfect Binding

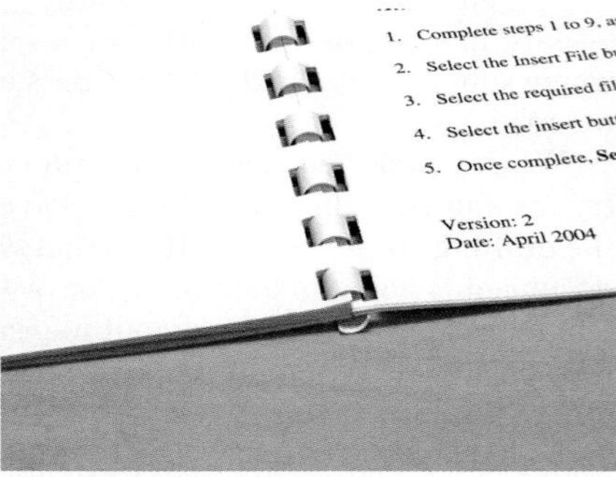

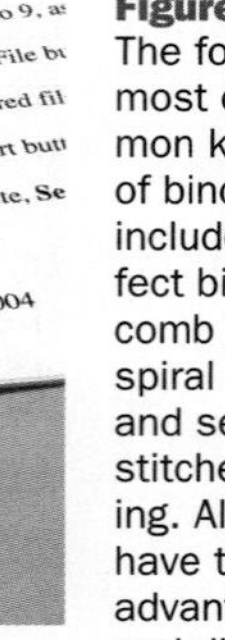

Comb Binding

Figure 18.26: The four most common kinds of bindings include perfect binding, comb binding, spiral binding, and sewn or stitched binding. All four have their advantages and disadvantages.

Spiral Binding

Stitched Binding

COMB BINDING For comb binding, slots are punched along the back edge of the pages. Then, a machine is used to insert a plastic comb binder. This kind of binding is usually the least expensive, especially for smaller documents. However, comb binding often comes loose with heavy usage. Also, over time, the plastic combs become brittle and break.

SPIRAL BINDING This method is similar to comb binding, except holes are drilled or punched into the pages and a wire spiral is threaded through the holes. Compared to comb binding, spiral binding is much more secure but is also more expensive. This binding is great for documents that need to lay flat when open.

SEWN OR STITCHED BINDING Here, sheets of paper are sewn or stapled together down the middle fold. Many hardback books are made with one of these two methods. Expensive hardbacks that are meant to last will be sewn with thread. Less expensive hardbacks will have two or three staples down the middle fold. In large books, sections of the book (called *signatures*) are sewn or stitched together. Then, these signatures are placed side by side and bound into the cover.

Selecting the Paper

Your choice of paper depends on the needs of your readers and the contexts in which the document will be used.

SIZE A standard sheet of paper in North America is 8.5 × 11 inches, while most other countries use a metric sheet of paper called A4, which is slightly larger at 21.0 centimeters × 29.7 centimeters. You can fold standard-sized sheets in half lengthwise or widthwise to make a nice booklet. You can fold a standard-sized sheet into three panels to make a brochure. Meanwhile, a double sheet (11 × 17 inches or 42 centimeters × 29.7 centimeters) is available in many copiers. You can fold one of these sheets in half to make two standard sheets.

Other sizes can also be used. You should talk to a printer or copy shop about what sizes are available. If you want an irregular size of paper, the cost of printing will go up significantly.

WEIGHT Paper comes in a variety of weights. The paper that you use in your printer or a copier is probably *20-pound bond,* which is a relatively lightweight paper. You can choose heavier paper if you want to add firmness to the document. For a cover, you might even choose a 60-pound or heavier card stock to make the document stiffer.

COLOR Paper comes in a variety of colors. When choosing a color, you should keep two things in mind. First, you want the words and images to *contrast* significantly with the page. Black words printed on a gray-colored sheet of paper, for instance, will often be difficult to read. Second, you do not want the color to interfere with any duplicating that the readers might need to do. A mauve-colored document, for example, is not going to duplicate well on a copy machine. To avoid problems with contrast and duplication, in most cases, your best choices for paper colors are white or off-white.

GLOSSY OR MATTE Increasingly, people are able to choose whether the paper is glossy or matte. Glossy paper is smooth and shiny and is often used in magazines or annual reports. Matte paper has a rougher texture and is not shiny, like the paper in your printer or copier. Most professional printers can offer you glossy paper for important documents (but at a higher expense).

Desktop publishing systems are giving us more options for binding and paper choices. As copiers grow more sophisticated, you might find yourself making these kinds of choices about the binding and paper used in your documents.

Interested in learning more about paper? Go to **www.pearsonhighered.com/johnsonweb4/18.16**

CHAPTER REVIEW

- Good design allows readers to (1) easily scan a document, (2) access the document to find the information they need, (3) understand the content more readily, and (4) appreciate the appearance of the document.
- The five principles of document design are balance, alignment, grouping, consistency, and contrast.
- On a balanced page, elements offset each other to create a stable feeling in the text. Unbalanced pages add tension to the reading process.
- Alignment creates relationships among items in a text and helps readers determine the hierarchical levels in the text.
- Grouping divides text into "scannable" blocks by using headings, rules, and borders to make words and images easier to comprehend.
- Consistency makes documents more accessible by making features predictable; inconsistent documents are harder to read and interpret.
- Contrast can cause text features or elements to stand out, but contrast should be used with restraint so that text elements work together rather than compete against each other.
- Documents and interfaces that need to work cross-culturally can use a culturally shallow design or a culturally deep design.
- Use a design process that includes (1) analyzing your readers and your document's context of use, (2) using thumbnails to rough out the design, (3) designing the document, and (4) revising and editing the design.
- You can choose from a variety of bindings and paper sizes, weights, and colors.

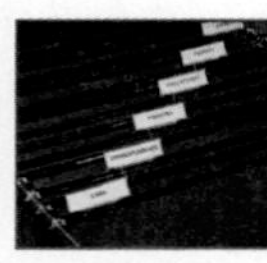

EXERCISES AND PROJECTS

Individual or Team Projects

1. On campus or at your workplace, find a poorly or minimally designed document. If you look on any bulletin board, you will find several good documents you can use. In a memo to your instructor, critique this document using the five design principles discussed in this chapter. Explain how the document fails to follow the principles.

2. Read the Case Study at the end of this chapter. Using the design principles discussed in this chapter, sketch out some thumbnails of a better design for this document. Then, using your word processor, develop an improved design. In a memo to your instructor, compare and contrast the old design with the new one, showing why the new design is superior.

3. Find a document that illustrates good design. Then, change some aspect of its rhetorical situation (purpose, readers, context). Redesign the document to fit that

altered rhetorical situation. For example, you might redesign a user's manual to accommodate the needs of 8-year-old children. You might turn a normal-sized document into a poster that can be read from a distance. In a memo to your instructor, discuss the changes you made to the document. Show how the changes in the rhetorical situation led to alterations in the design of the document. Explain why you think the changes you made were effective.

4. On the Internet, find an international company's website that is intended to work cross-culturally. In a presentation to your class, explain why you believe the site is culturally shallow or culturally deep. How have the designers of the website made adjustments to suit the expectations of people from a different culture? How might they improve the design to make it more effective for the target readers?

Collaborative Project

With other members of your class, choose a provider of a common product or service (for example, a car manufacturer; mobile phone, computer, clothing, or music store; museum; or theater). Then, find the websites of three or four competitors for this product or service.

Using the design principles you learned in this chapter, critique these websites by comparing and contrasting their designs. Considering its target audience, which website design seems the most effective? Which is the least effective? Explain your positive and negative criticisms in some depth.

Then, using thumbnails, redesign the weakest site so that it better appeals to its target audience. How can you use balance, alignment, grouping, consistency, and contrast to improve the design of the site?

In a presentation to your class, discuss why the design of one site is stronger and others are weaker from a design perspective. Then, using an overhead projector, use your thumbnails to show how you might improve the design of the weakest site you found.

For additional technical writing resources, including interactive sample documents, document design tutorials and guidelines, and more, go to **www.mytechcommlab.com.**

GO TO THE NET

Sample documents and websites for critiquing and redesigning are available at **www.pearsonhighered.com/johnsonweb4/18.17**

CASE STUDY

Scorpions Invade

Lois Alder is a U.S. Environmental Protection Agency biologist in western Texas who specializes in integrated pest management (IPM). The aim of IPM is to use the least toxic methods possible for controlling insects, rodents, and other pests. Lois's job is to help companies, schools, and government organizations develop IPM strategies to minimize their use of pesticides, especially where people live, work, and play.

Recently, a rural school district outside El Paso, Texas, was experiencing an infestation of scorpions. Scorpions were finding their way into classrooms, and they were regularly found in the schoolyards around the buildings. Teachers and children were scared, though no one had been stung yet. Parents were demanding that the school district do something immediately.

In a conference call, Lois talked to the principals and custodians of the schools. The custodians said they just wanted to "nuke" the scorpions by spraying large amounts of pesticides in and around the buildings. But this solution would expose the children, teachers, and staff at the schools to high levels of toxins. Lois convinced them that they should first try an IPM approach to the problem. She said she would quickly set up a training program to help them. Lois knew, though, that she needed to send these school officials some information right away.

From the EPA website, Lois downloaded a well-written chapter, "IPM for Scorpions in Schools" (Figure A) (http://www.epa.gov/pesticides/ipm/schoolipm/). The design of the document, however, was a bit bland, and Lois was concerned that school officials, especially the custodians, would not read it carefully. Moreover, she knew that some of the custodians had poor reading skills.

So, she decided she would need to redesign the document to make it more accessible. She wanted to highlight the need-to-know information while stripping out any information that the school officials would not need. If you were Lois, how would you use the principles of design discussed in this chapter to redesign this document? How could you make this document more accessible to the people who need to use it? How could you make it more attractive and interesting to its likely readers?

A Document Design That Could Be Improved

Figure A: This document's design is rather minimal and perhaps less than effective for the readers who might use it. How could the design be improved?

CHAPTER 13
IPM FOR SCORPIONS IN SCHOOLS

INTRODUCTION

Scorpions live in a wide variety of habitats from tropical to temperate climates and from deserts to rain forests. In the United States they are most common in the southern states from the Atlantic to the Pacific. All scorpions are beneficial because they are predators of insects.

The sting of most scorpions is less painful than a bee sting. There is only one scorpion of medical importance in the United States: the sculptured or bark scorpion, Centruroides exilicauda (=sculpturatus). The danger from its sting has been exaggerated, and its venom is probably not life-threatening. This species occurs in Texas, western New Mexico, Arizona, northern Mexico, and sometimes along the west bank of the Colorado River in California.

IDENTIFICATION AND BIOLOGY

Scorpions range from 3/8 to 8 1/2 inches in length, but all scorpions are similar in general appearance.

Scorpions do not lay eggs; they are viviparous, which means they give birth to live young. As embryos, the young receive nourishment through a kind of "placental" connection to the mother's body. When the young are born, they climb onto the mother's back where they remain from two days to two weeks until they molt (shed their skin) for the first time. After the first molt, the young disperse to lead independent lives. Some scorpions mature in as little as six months while others take almost seven years.

All scorpions are predators, feeding on a variety of insects and spiders. Large scorpions also feed on small animals including snakes, lizards, and rodents. Some scorpions sit and wait for their meal to come to them while others actively hunt their prey. Scorpions have a very low metabolism and some can exist for 6 to 12 months without food. Most are active at night. They are shy creatures, aggressive only toward their prey. Scorpions will not sting humans unless handled, stepped on, or otherwise disturbed.

It is rare for scorpions to enter a building since there is little food and temperatures are too cool for their comfort. There are some exceptions to this rule. Buildings in new developments (less than three years old) can experience an influx of scorpions because the construction work has destroyed the animals' habitat. In older neighborhoods, the heavy bark on old trees provide good habitat for scorpions, and they may enter through the more numerous cracks and holes in buildings in search of water, mates, and prey. Also, buildings near washes and arroyos that are normally dry may become refuges for scorpions during summer rains.

Scorpions do not enter buildings in winter because cold weather makes them sluggish or immobile. They are not active until nighttime low temperatures exceed 70°F. Buildings heated to 65½ or 70°F provide enough warmth to allow scorpions to move about. Scorpions found inside buildings in cold weather are probably summer visitors that never left. Although scorpions prefer to live outdoors, they can remain in buildings without food for long periods of time.

STINGS

A scorpion sting produces considerable pain around the site of the sting, but little swelling. For four to six hours, sensations of numbness and tingling develop in the region of the sting, then symptoms start to go away. In the vast majority of cases, the symptoms will subside within a few days without any treatment.

If the sting is from a bark scorpion, symptoms can sometimes travel along nerves, and tingling from a sting on a finger may be felt up to the elbow, or even the shoulder. Severe symptoms can include roving eyes, blurry vision, excessive salivation, tingling around the mouth and nose, and the feeling of having a lump in the throat. Respiratory distress may occur. Tapping the sting can produce extreme pain. Symptoms in children also include extreme restlessness, excessive muscle activity, rubbing at the face, and sometimes vomiting. Most vulnerable to the sting of the bark scorpion are children under five years and elderly persons who have

IPM for Schools | 103 | **Chapter 13 • Scorpions**

Source: U.S. Environmental Protection Agency, http://www.epa.gov/pesticides/ipm/schoolipm.

(continued)

Figure A:
(continued)

Avoiding Scorpion Stings

Schools in areas where encounters with scorpions are likely should teach children and adults how to recognize scorpions and to understand their habits. Focus on scorpion biology, behavior, likely places to find them, and how to avoid disturbing them.

At home, children and parents should be taught to take the following precautions to reduce the likelihood of being stung:

- Wear shoes when walking outside at night. If scorpions are suspected indoors, wear shoes inside at night as well.
- Wear leather gloves when moving rocks, boards, and other debris.
- Shake out shoes or slippers before they are worn, and check beds before they are used.
- Shake out damp towels, washcloths, and dishrags before use.
- When camping, shake out sleeping bags, clothes, and anything else that has been in contact with the ground before use.
- Protect infants and children from scorpions at night by placing the legs of their cribs or beds into clean wide-mouthed glass jars and moving the crib or bed away from the wall. Scorpions cannot climb clean glass.

an underlying heart condition or respiratory illness. The greatest danger to a child is the possibility of choking on saliva and vomit.

Antivenin for the bark scorpion is produced at Arizona State University in Tempe and is available in Arizona but not in other states. The therapeutic use of antivenin is still experimental. People have been treated without antivenin for many years, and in areas where antivenin is unavailable, people are monitored closely by medical professionals until the symptoms subside.

Detection and Monitoring

To determine where scorpions are entering, inspect both the inside and outside of the building at night (when scorpions are active) using a battery-operated camp light fitted with a UV (black) fluorescent bulb. Scorpions glow brightly in black light and can be spotted several yards away.

Always wear leather gloves when hunting scorpions. Places to check inside the building include under towels, washcloths, and sponges in bathrooms and kitchen; under all tables and desks, since the bark scorpion may climb and take refuge on a table leg or under the lip of a table; and inside storage areas. Outside, check piles of rocks and wood, under loose boards, and in piles of debris. After the following treatment strategies have been implemented, monitoring with the black light should continue to verify population reduction.

Management Options

Physical Controls

In most cases, physical controls will be adequate to manage a scorpion problem.

Removal of Scorpions

Any scorpions found during monitoring can be picked up using gloves or a pair of kitchen tongs, and transferred to a clean, wide-mouthed glass jar. Scorpions cannot climb clean glass. You can also invert a jar over a live scorpion and then slide a thin piece of cardboard under the mouth of the jar to trap the scorpion inside. Once a scorpion is captured, drop it into a jar of alcohol or soapy water (water without soap will not work) to kill it.

Habitat Modification

If you discover areas near school buildings that harbor a number of scorpions, you can try to alter the habitat to discourage them. Wood piles, rocks, loose boards, and other debris should be removed from the immediate vicinity of the building.

If there are slopes on the school grounds that are faced with rip rap (large rocks placed on a slope or stream bank to help stabilize it), or other similar areas highly attractive to scorpions, place a barrier of aluminum flashing between the rip rap and the school to prevent scorpion access. The flashing must be bent in an "L" shape away from the building. The other edge of the flashing should be buried a short distance from the rocks, deep enough in the soil so that the L shape will not fall over and lean on the rip rap. Make the height of the barrier before the bend greater than the height of the rip rap to prevent scorpions from standing on the rocks and jumping over the barrier.

IPM for Schools 104 **Chapter 13 • Scorpions**

Figure A: *(continued)*

Carry a caulking gun during nighttime inspections inside and outside the building to seal any cracks and crevices found. If scorpions are entering through weep holes in windows or sliding doors, cover the holes with fine-mesh aluminum screening, available from hardware or lumber stores. The ends of pipes that are designed as gray water drains should be fitted with loose filter bags, or makeshift end-pieces made from window screen. The screened end will prevent scorpion access to the drainpipe, sink, and other parts of the building.

It is important to continue nighttime patrols and caulking until all entryways have been located and sealed, and all the scorpions in the building have been captured and killed. Once the access routes are sealed, and all indoor scorpions have been removed, only doorways provide access, unless the scorpions ride in on logs and other materials. Glazed tiles can be placed around the perimeter of the buildings, and under or around doors and windows as part of the decor and as practical scorpion barriers. Scorpions have difficulty crossing smooth tiles unless the grout line is wide. Wood storage should be elevated above the ground since scorpions like contact with moist soil. Before bringing materials such as logs inside, bang them on a stone to dislodge any scorpions.

Traps

A simple trap made of damp gunny sacks laid down near the building in the evenings may be useful for monitoring and trapping. Scorpions may seek out the moist environment under the sacks where they can be collected in the morning. This trap is most effective when used before summer rains.

Chemical Controls

In general, preventing scorpion problems is better than trying to kill these creatures with pesticide. Spraying the perimeters of buildings is not only unnecessary but also ineffective. Scorpions can tolerate a great deal of pesticide in their environment. The pesticide will only be harmful to humans, especially children, and to other wildlife. Using physical controls along with education to reduce the fear of scorpions will help prevent unnecessary treatments.

First Aid for Scorpion Stings

Most scorpion stings are similar to a bee or wasp sting, and like bee or wasp stings, the majority of scorpion stings can be treated at school or the victim's home.

First aid for a scorpion sting includes the following:

- Calm the victim.
- Do not use a tourniquet.
- Wash the area with soap and water.
- Apply a cool compress (an ice cube wrapped in a wet washcloth), but do not apply ice directly to the skin or submerge the affected limb in ice water.
- To reduce pain, over-the-counter pain relievers such as aspirin or acetaminophen can help.
- Elevate or immobilize the affected limb if that feels more comfortable.
- Do not administer sedatives such as alcohol.

Bibliography

Bio-Integral Resource Center (BIRC). 1996. 1997 directory of least-toxic pest control products. IPM Practitioner 18(11/12):1-39.

Cloudsley-Thompson, J.L. 1968. Spiders, Scorpions, Centipedes and Mites. Pergamon Press, New York. 278 pp.

Keegan, H.L. 1980. Scorpions of Medical Importance. Jackson University Press, Jackson, MS. 140 pp.

Mallis, A. 1982. Handbook of Pest Control. Franzak and Foster, Cleveland, OH. 1101 pp.

Olkowski, W., S. Daar, and H. Olkowski. 1991. Common-Sense Pest Control: Least-toxic solutions for your home, garden, pets and community. Taunton Press, Newtown, CT. 715 pp.

Polis, G.A., ed. 1990. The Biology of Scorpions. Stanford University Press, Stanford, CA. 587 pp.

Smith, R.L. 1982. Venomous animals of Arizona. Cooperative Extension Service, College of Agriculture, University of Arizona, Tucson, AZ, Bulletin 8245. 134 pp.

Smith, R.L. 1992. Personal communication. Associate Professor, Entomology Dept., University of Arizona at Tucson.

This document was produced for USEPA (Document #909-B-97-001) by the Bio-Integral Resource Center, P.O. Box 7414, Berkeley, CA 94707, March 1997.

CHAPTER

19

Creating and Using Graphics

Guidelines for Using Graphics 523

Displaying Data with Graphs, Tables, and Charts 528

Using Pictures, Drawings, and Screen Shots 536

Using Cross-Cultural Symbols 542

Using Video and Audio 545

Chapter Review 547

Exercises and Projects 547

Case Study: Looking Guilty 549

In this chapter, you will learn:

- The importance of visuals in documents and presentations.
- Four guidelines for using visuals effectively.
- How to use tables, charts, and graphs.
- Strategies for taking photographs and using them in documents and presentations.
- How drawings, icons, and clip art can be used effectively to enhance understanding.
- How to use video and audio in multimedia and Internet-based documents.

Graphics are an essential part of any technical document or presentation. Your readers will often pay more attention to the visuals in your document than to the written text. For example, think about how you began reading this chapter. More than likely, you did not begin reading at the top of this page. Instead, you probably took a quick glance at the graphics in the chapter to figure out what it is about. Then, you started reading the written text. Your readers will approach your documents the same way.

Guidelines for Using Graphics

As you draft your document, you should look for places where graphics could be used to support the text. Graphics are especially helpful in places where you want to reinforce important ideas or help your readers understand complex concepts or trends.

Graphics should be used to enhance and clarify your message—to slice through the details and numbers to make the information easier to process. In Figure 19.1, for

Written Text vs. Table

In 2007, the state of Michigan reported 16 new cases of West Nile Virus (WNV) among humans. Four of those cases resulted in death. The most cases were in Wayne County, which saw 7 total cases with 5 males and 2 females contracting the virus. Wayne County had two deaths, half of the state's total for 2007. Kent, Macomb, and Oakland Counties each had two cases apiece. Kent County reported two females (ages 19–65) with WNV, Macomb County had two females (one 19–65 and one over 65), and Oakland County counted two males (19–65). The Macomb County victim over 65 died. Kalamazoo, Lapeer, and St. Clair Counties each had one reported case. Kalamazoo County's victim, which resulted in death, was a male over 65. Lapeer County reported one male (19–65), and St. Clair County reported one female (19–65).

2007 HUMAN WNV CASES

COUNTY	AGE (YEARS)	MALE	FEMALE	TOTAL CASES	DEATHS
Kalamazoo	0–18				
	19–65				
	>65	1		1	1
Kent	0–18				
	19–65		2	2	
	>65				
Lapeer	0–18				
	19–65	1		1	
	>65				
Macomb	0–18				
	19–65		1	1	
	>65		1	1	1
Oakland	0–18				
	19–65	2		2	
	>65				
St. Clair	0–18				
	19–65		1	1	
	>65				
Wayne	0–18	1		1	
	19–65	3	1	4	
	>65	1	1	2	2
TOTALS (includes probable and confirmed cases, all clinical syndromes)		9	7	16	4

Source: State of Michigan, http://www.michigan.gov/emergingdiseases.

Figure 19.1: Putting figures into a table makes them much easier to access. The table cannot completely replace the written text, but it can reinforce it by organizing the information more effectively.

To see more examples of text-to-table conversions, go to **www.pearsonhighered.com/johnsonweb4/19.1**

example, the written text and the table provide essentially the same information. And yet, the information in the table is much easier to access.

To help you create and use graphics effectively and properly, there are four guidelines you should commit to memory.

Guideline One: A Graphic Should Tell a Simple Story

A graphic should tell the "story" about your data in a concise way. In other words, your readers should be able to figure out at a quick glance what the graphic says. If your readers need to pause longer than a moment, there is a good chance they will not understand what the graphic means.

Figure 19.2, for example, shows how a graph can tell a simple story. Almost immediately, a reader will recognize that obesity rates around the world are going up dramatically. It's also obvious that the United States is the most obese nation and growing worse.

A Graph That Tells a Simple Story

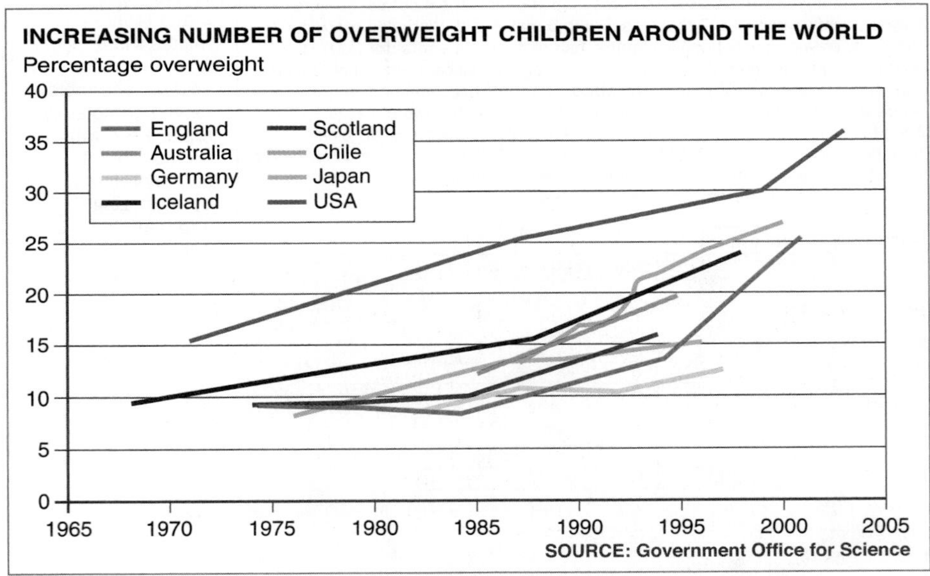

Figure 19.2: This graph tells a simple story about obesity that readers can grasp at a glance.

Source: British Broadcasting Company, http://news.bbc.co.uk/2/hi/health/7151813.stm.

This first guideline—tell a simple story—also applies to photographs in a document (Plotnik, 1982). At a glance, your readers should be able to figure out what story a photograph is telling. The photograph in Figure 19.3, for example, is not complex, but it tells a clear story about the markings on a Desert Checkerspot butterfly.

GO TO THE NET

For websites that offer other guidelines for using graphics, go to **www.pearsonhighered.com/johnsonweb4/19.2**

A Photograph Should Tell a Simple Story, Too

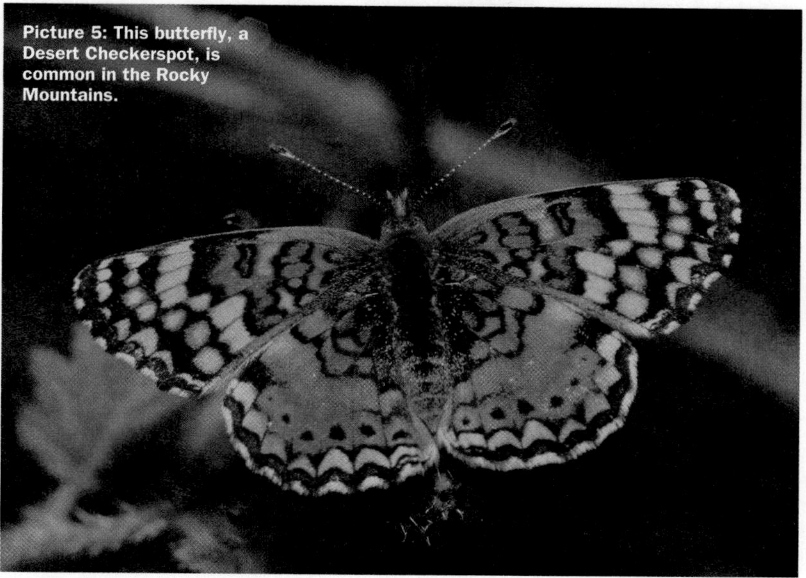

Picture 5: This butterfly, a Desert Checkerspot, is common in the Rocky Mountains.

Figure 19.3: This photograph tells a simple story that reinforces the written text.

Source: Corel.

Guideline Two: A Graphic Should Reinforce the Written Text, Not Replace It

Graphics should be used to support the written text, but they cannot replace it altogether. Since technical documents often discuss complex ideas or relationships, it is tempting to simply refer the readers to a graphic (e.g., "See Chart 9 for an explanation of the data"). Chances are, though, that if you cannot explain something in writing, you won't be able to explain it in a graphic, either.

Instead, your written text and visuals should work with each other. The written text should refer readers to the graphics, and the graphics should support the written information. For example, the written text might say, "As shown in Figure 13.1, the number of high school students who report being in fights has been declining." The graph would then support this written statement by illustrating this trend (Figure 19.4).

The written text should tell readers the story that the graphic is trying to illustrate. That way, readers are almost certain to understand what the graphic is showing them.

Guideline Three: A Graphic Should Be Ethical

Graphs, charts, tables, illustrations, and photographs should not be used to hide information, distort facts, or exaggerate trends. In a bar chart, for example, the scales can be altered to suggest that more growth has occurred than is actually the case (Figure 19.5). In a line graph, it is tempting to leave out data points that won't allow a smooth line to be drawn. Likewise, with computers, photographs can be distorted or doctored.

To see other graphs that tell simple stories, go to **www.pearsonhighered.com/johnsonweb4/19.3**

A Graph That Reinforces the Written Text

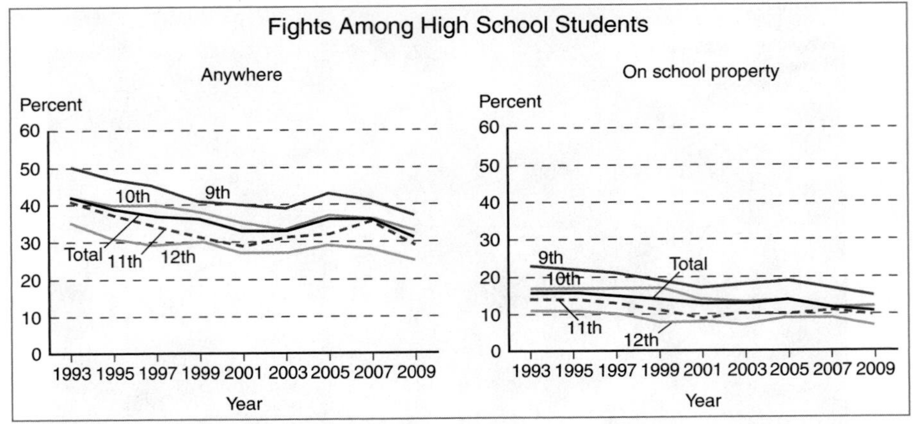

Figure 19.4: A line graph typically shows a trend over time. This graph shows how fights among high school students have been declining.

Source: National Center for Education Statistics, "Indicators of School Crime and Safety, 2010," http://nces.ed.gov/programs/crimeindicators/crimeindicators2010/figures/figure_13_1.asp.

Unethical and Ethical Bar Charts

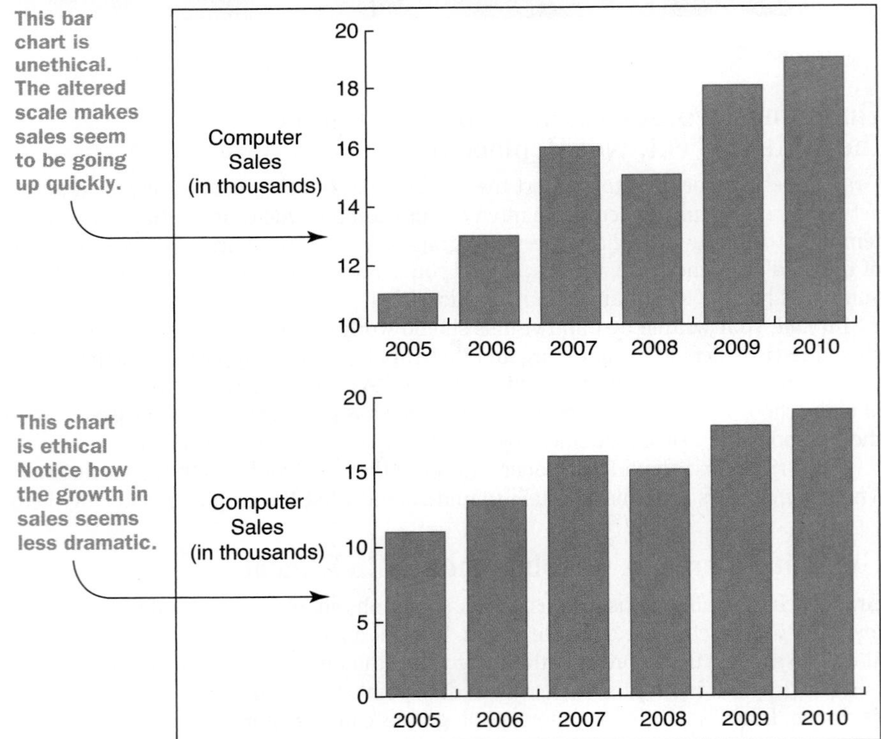

This bar chart is unethical. The altered scale makes sales seem to be going up quickly.

This chart is ethical Notice how the growth in sales seems less dramatic.

Figure 19.5: The top bar chart is unethical because the y-axis has been altered to exaggerate the growth in sales of computers. The second bar chart presents the data ethically.

Want to see other unethical graphics? Go to **www.pearsonhighered.com/johnsonweb4/19.4**

A good rule of thumb with graphics—and a safe principle to follow in technical communication altogether—is to always be absolutely honest with the readers. Your readers are not fools, so attempts to use graphics to distort or stretch the truth will eventually be detected. Once detected, unethical graphics can erode the credibility of an entire document or presentation (Kostelnick & Roberts, 1998). Even if your readers only *suspect* deception in your graphics, they will begin to doubt the honesty of the whole text.

Link

For more information on the ethical use of data, see Chapter 4, page 87.

Guideline Four: A Graphic Should Be Labeled and Placed Properly

Proper labeling and placement of graphics help readers move back and forth between the main text and images. Each graphic should be labeled with an informative title (Figure 19.6). Other parts of the graphic should also be carefully labeled:

- The x- and y-axes of graphs and charts should display standard units of measurement.
- Columns and rows in tables should be labeled so readers can easily locate specific data points.
- Important features of drawings or illustrations should be identified with arrows or lines and some explanatory text.
- The source of the data used to make the graphic should be clearly identified underneath.

Link

For more information on designing page layouts, see Chapter 18, page 482.

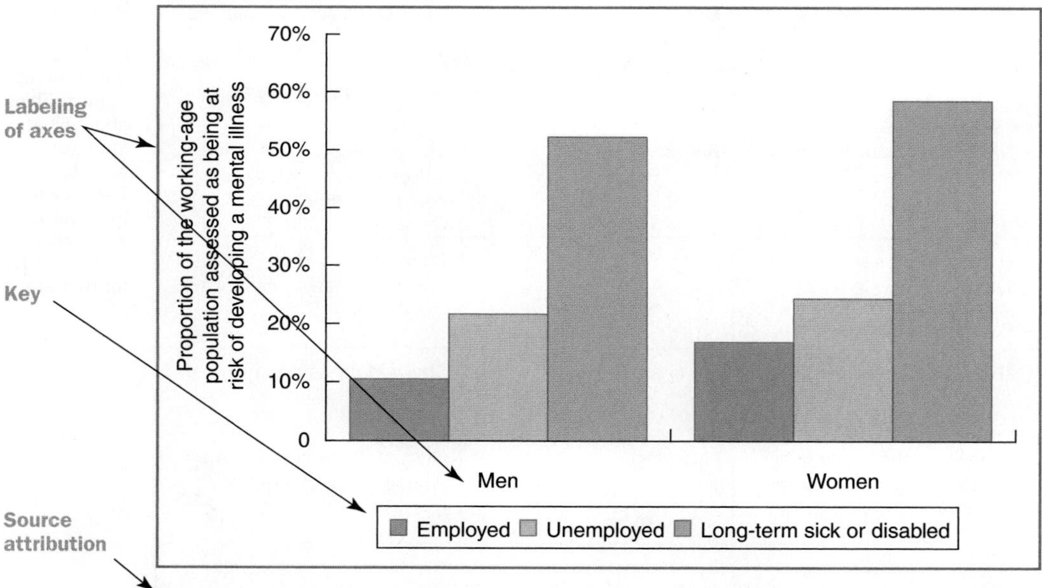

Figure 19.6: Good labeling of a graphic is important so that readers can understand it.

AT A GLANCE

Guidelines for Using Graphics

- A graphic should tell a simple story.
- A graphic should reinforce the written text, not replace it.
- A graphic should be ethical.
- A graphic should be labeled and placed properly.

If you include a title with the graph, an explanatory caption is not needed. Nevertheless, a sentence or two of explanation in a caption can often help reinforce or clarify the story the graphic is trying to tell.

When placing a graphic, put it on the page where it is referenced or, at the farthest, put it on the following page. Readers will rarely flip more than one page to look for a graphic.

Even if they *do* make the effort to hunt down a graphic that is pages away, doing so will take them out of the flow of the document, inviting them to start skimming.

Readers should be able to locate a graphic with a quick glance. Then, they should be able to quickly return to the written text to continue reading. When labeled and placed properly, graphics work seamlessly into the flow of the whole text.

Displaying Data with Graphs, Tables, and Charts

To decide which graphic is best for the data you want to display, first decide what story you want to tell. Then, choose the type of graphic that best fits that story. The chart in Figure 19.7 will help you decide which one works best.

Choosing the Appropriate Graphic

The Story to Be Told	Best Graphic	How Data Are Displayed
"I want to show a trend."	Line graph	Shows how a quantity rises and falls, usually over time
"I want to compare two or more quantities."	Bar chart	Shows comparisons among different items or the same items over time
"I need to present data or facts for analysis and comparison."	Table	Displays data in an organized, easy-access way
"I need to show how a whole is divided into parts."	Pie chart	Shows data as a pie carved into slices
"I need to show how things, people, or steps are linked together."	Flowchart	Illustrates the connections among people, parts, or steps
"I need to show how a project will meet its goals over time."	Gantt chart	Displays a project schedule, highlighting the phases of the work

Figure 19.7: Different kinds of graphics tell different stories. Think about what story you want to tell. Then, locate the appropriate graph, table, or chart for that story.

GO TO THE NET

For more information on labeling and placing graphs, go to **www.pearsonhighered.com/johnsonweb4/19.5**

Line Graphs

Line graphs are perhaps the most familiar way to display data. They are best used to show measurements over time. Some of their more common applications include the following:

Showing trends—Line graphs are especially good at showing how quantities rise and fall over time (Figure 19.8). Whether you are illustrating trends in the stock market or charting the changes in temperature during a chemical reaction, a line graph can show how the quantity gradually increases or decreases. When two or more lines are charted on a line graph, you can show how quantities rise and fall in tandem (or don't).

Showing relationships between variables—Line graphs are also helpful when charting the interaction of two different variables. Figure 19.9, for example, shows a line graph that illustrates how a rise in the temperature of a gas is accompanied by a rise in the volume of gas.

A Line Graph Showing a Trend

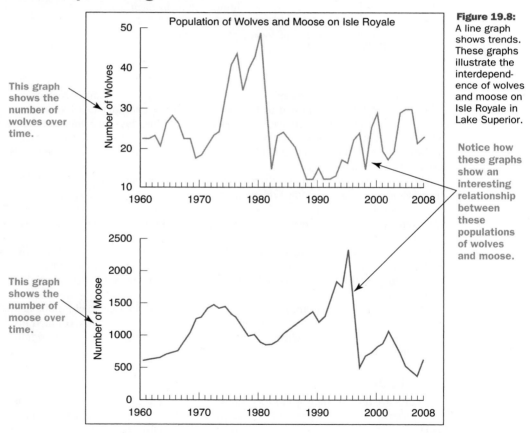

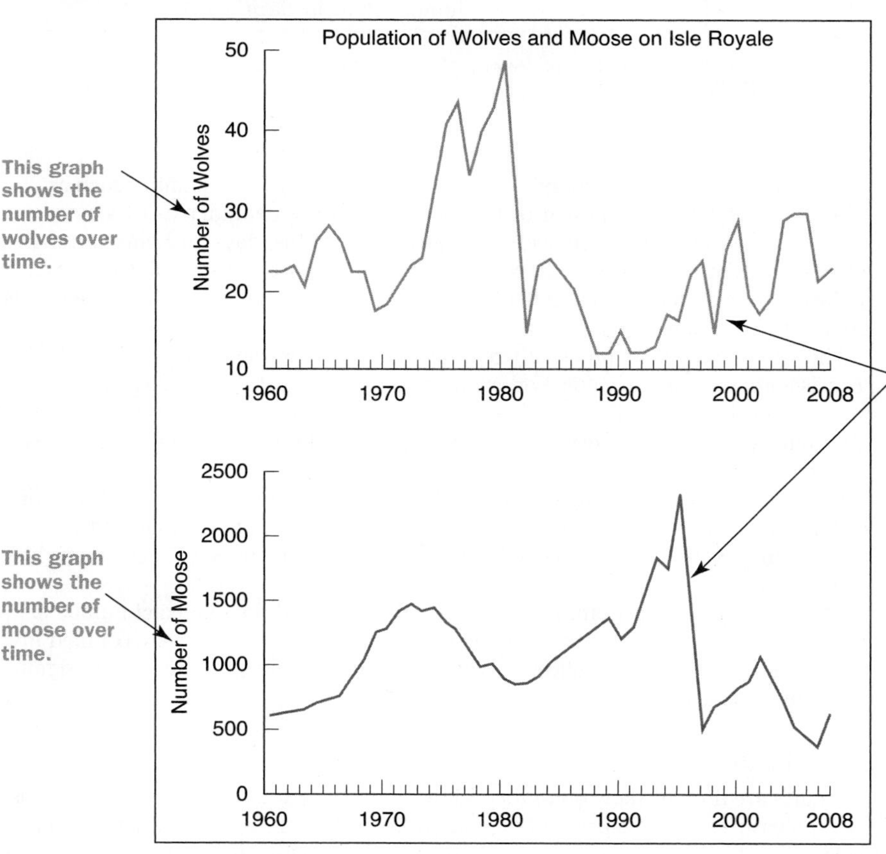

This graph shows the number of wolves over time.

This graph shows the number of moose over time.

Figure 19.8: A line graph shows trends. These graphs illustrate the interdependence of wolves and moose on Isle Royale in Lake Superior.

Notice how these graphs show an interesting relationship between these populations of wolves and moose.

Data Source: http://www.isleroyalewolf.org.

Examples of these documents are available at **www.pearsonhighered.com/johnsonweb4/19.6**

A Line Graph Showing a Relationship Between Variables

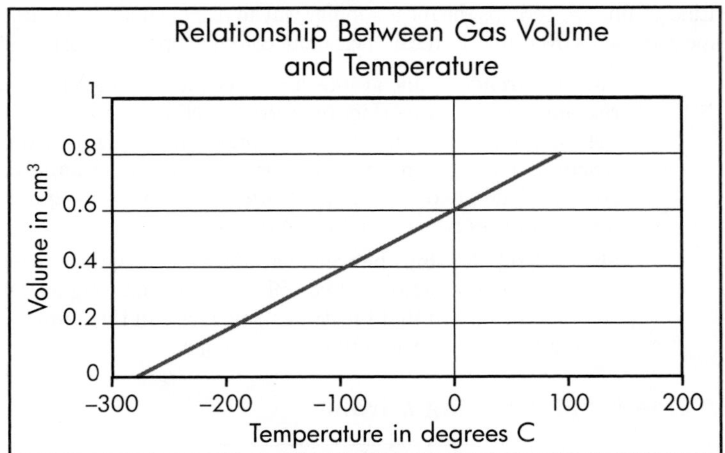

Figure 19.9: Here, the volume of a gas is plotted against the temperature. In this case, an extrapolation of the line allows us to estimate "absolute zero," the temperature at which all molecular activity stops.

Source: The Safetyline Institute, http: //www.safetyline.wa.gov.au/institute/level2/course16 /lecture47/147_02.asp.

In a line graph, the vertical axis (y-axis) displays a measured quantity such as sales, temperature, production, growth, and so on. The horizontal axis (x-axis) is usually divided into time increments such as years, months, days, or hours. As shown in Figure 19.8, in a line graph, the x- and y-axes do not need to start at zero. Often, by starting one or both axes at a nonzero number, you can better illustrate the trends you are trying to show.

The x-axis in a line graph usually represents the "independent variable," which has a consistently measurable value. For example, in most cases, time marches forward steadily, independent of other variables. So, time is often measured on the x-axis. The y-axis often represents the "dependent variable." The value of this variable fluctuates over time.

You can use more than one line to illustrate trends in a line graph. Depending on your printer, computers also give you the ability to use colors to distinguish the lines. Or, you can use dashes, dots, and solid lines to help your readers distinguish one line from the others.

The drawback of line graphs is their inability to present data in exact numbers. For example, in Figure 19.8, can you tell exactly how many wolves were counted in 2001? Line graphs are most effective when the trend you are showing is more significant than the exact figures.

Bar Charts

Bar charts are used to show quantities, allowing readers to make visual comparisons among measurements (Figure 19.10). The width of the bars is kept the same, while the length of the bars varies to represent the quantity measured.

For more examples of line graphs, go to **www.pearsonhighered.com/johnsonweb4/19.7**

A Bar Chart

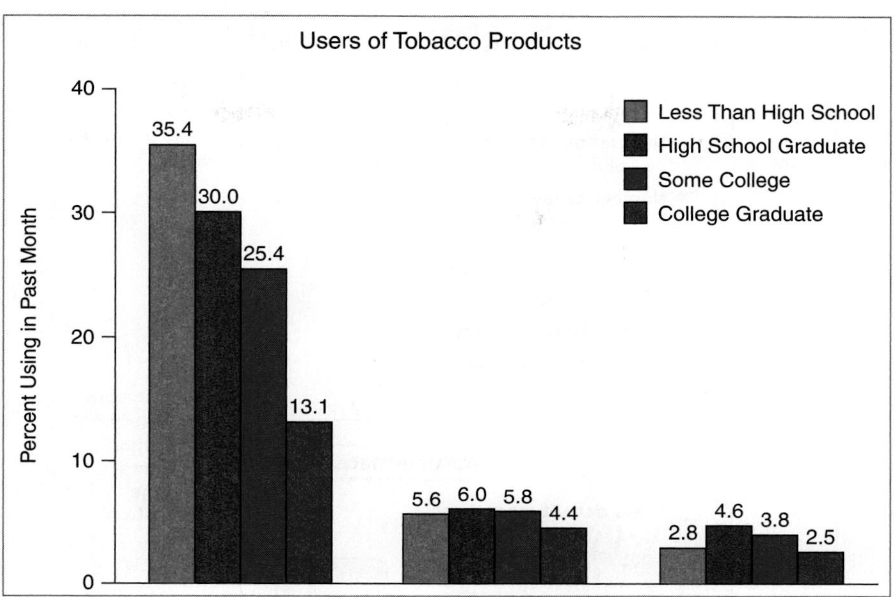

Figure 19.10: A bar chart is especially effective for making comparisons.

Source: Substance Abuse and Mental Health Services Administration, 2009 National Survey on Drug Use and Health.

Computers can be used to enhance bar charts even further. Coloring and shading the bars will enhance readers' ability to interpret the data and identify trends.

Tables

Tables provide the most efficient way to display data or facts in a small amount of space. In a table, information is placed in horizontal rows and vertical columns, allowing readers to quickly find specific numbers or words that address their needs.

Creating a table takes careful planning, but computers can do much of the hard work for you. For simpler tables, you can use the Table function on your word-processing software (Figure 19.11). It will allow you to specify how many rows and columns you need (make sure you include enough columns and rows for headings in the table). Then, you can start typing your data or information into the cells.

If the Table function in your word processor is not sufficient for your needs, spreadsheet programs like Microsoft Excel and Corel Quattro Pro also allow you to make quick tables (see the Help box in this chapter).

After creating the basic table, you should properly label it. In most cases, the table's number and title should appear above it (Figure 19.12). Down the left column, the *row headings* should list the items being measured. Along the top row, the *column headings* should list the qualities of the items being measured. Beneath the table, if needed, a citation should identify the source of the information.

To see more sample bar charts, go to **www.pearsonhighered.com/johnsonweb4/19.8**

Inserting a Table

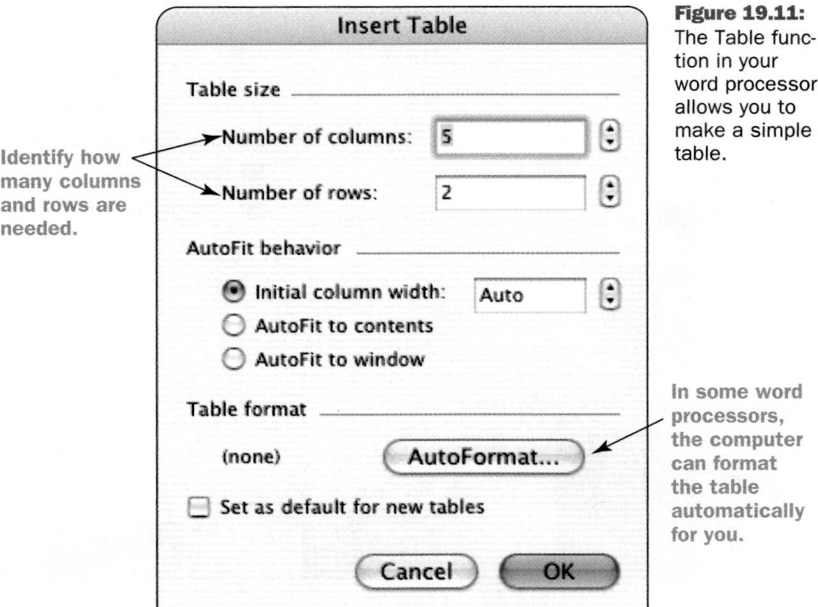

Figure 19.11: The Table function in your word processor allows you to make a simple table.

Parts of a Table

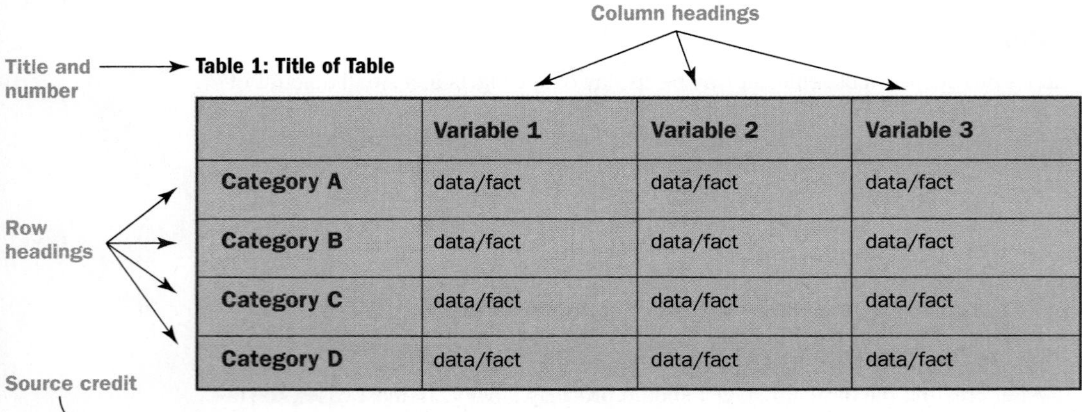

	Variable 1	Variable 2	Variable 3
Category A	data/fact	data/fact	data/fact
Category B	data/fact	data/fact	data/fact
Category C	data/fact	data/fact	data/fact
Category D	data/fact	data/fact	data/fact

Figure 19.12: The parts of a table are rather standard. Rows and columns align in ways that allow readers to locate specific pieces of information.

In some cases, tables can be used to present verbal information rather than numerical data. In Figure 19.13, for example, the table is being used to verbally provide health information. With this table, readers can quickly locate their age and find the cancer screening test they need.

A Table That Presents Verbal Information

CANCER DETECTION			
TEST OR PROCEDURE			
Age	**Frequency**	**Females**	**Males**
18–20	One time	Complete Health Exam	Complete Health Exam
	Yearly	Pap smear	
	Monthly	Skin self-exam, Breast self-exam	Skin self-exam
20–40	Every 3 years	Complete Health Exam, Clinical breast exam, Pelvic exam	Complete Health Exam
	Yearly	Pap smear	
	Monthly	Skin self-exam, Breast self-exam	Skin self-exam, Testis self-exam
40–50	Every 3 years	Complete Health Exam	Complete Health Exam, Prostate specific antigen (PSA) blood test
	Yearly	Cinical breast exam, Mammogram, Endometrial biopsy, Pap smear, Pelvic exam, Digital rectal exam, Stool blood test	Digital rectal exam, Stool blood test
	Monthly	Skin self-exam, Breast self-exam	Skin self-exam, Testis self-exam
50–65	Every 5–10 years	Colonoscopy, Procto, Double-contrast barium enema (DCBE)	Colonoscopy, Procto, Double-contrast barium enema (DCBE)
	Yearly	Complete Health Exam, Cinical breast exam, Endometrial biopsy, Mammogram, Pap smear, Pelvic exam, Digital rectal exam, Stool blood test	Complete Health Exam, Prostate specific antigen (PSA) blood test, Digital rectal exam, Stool blood test
	Monthly	Skin self-exam, Breast self-exam	Skin self-exam, Testis self-exam
65+	Every 5–10 years	Colonoscopy, Procto, Double-contrast barium enema (DCBE)	Colonoscopy, Procto, Double-contrast barium enema (DCBE)
	Yearly	Complete Health Exam, Cinical breast exam, Mammogram, Endometrial biopsy, Pap smear, Pelvic exam, Digital rectal exam, Stool blood test	Complete Health Exam, Prostate specific antigen (PSA) blood test, Digital rectal exam, Stool blood test
	Monthly	Skin self-exam, Breast self-exam	Skin self-exam, Testis self-exam

Source: National Foundation for Cancer Research.

Figure 19.13: Tables can also present verbal information concisely. In this table, a great amount of information is offered in an easy-to-access format.

Need more help making a table? Go to **www.pearsonhighered.com/johnsonweb4/19.9**

When adding a table to your document, think about what your readers need to know. It is often tempting to include large tables that hold all your data. However, these large tables might clog up your document, making it difficult for readers to locate specific information. You are better off creating small tables that focus on the specific information you want to present. Move larger tables to an appendix, especially if they present data not directly referenced in the document.

Pie Charts

Pie charts are useful for showing how a whole divides into parts (Figure 19.14). Pie charts are popular, but they should be used sparingly. They take up a great amount of space in a document while usually presenting only a small amount of data. The pie chart in Figure 19.14, for instance, uses a third of a page to plot a mere eleven data points.

Pie charts are difficult to construct by hand, but your computer's spreadsheet program (Excel or Quattro Pro) can help you create a basic pie chart of your data. When labeling a pie chart, you should try to place titles and specific numbers in or near the graphic. For instance, in Figure 19.14, each slice of the pie chart is labeled and includes a measurement to show how the pie was divided. These labels and measurements help readers compare the data points plotted in the chart.

A Pie Chart

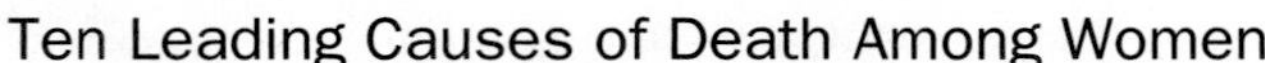

Source: U.S. Department of Health and Human Services, Women's Health 2007.

Figure 19.14: A pie chart is best for showing how a whole can be divided into parts.

Do you want to see other pie charts? Go to **www.pearsonhighered.com/johnsonweb4/19.10**

The key to a good pie chart is a clear story. For example, what story is the pie chart in Figure 19.14 trying to tell? Heart disease and cancer are the most significant causes of death among women.

Flowcharts

Flowcharts are used to visually guide readers through a series of decisions, actions, or steps. They typically illustrate a process described in the written text. Arrows are used to connect parts of the flowchart, showing the direction of the process.

As shown in Figure 19.15, flowcharts are helpful for illustrating instructions, especially when judgment calls need to be made by the user of the instructions. A flowchart

Link

For more information on writing instructions, go to Chapter 7, page 166.

A Flowchart

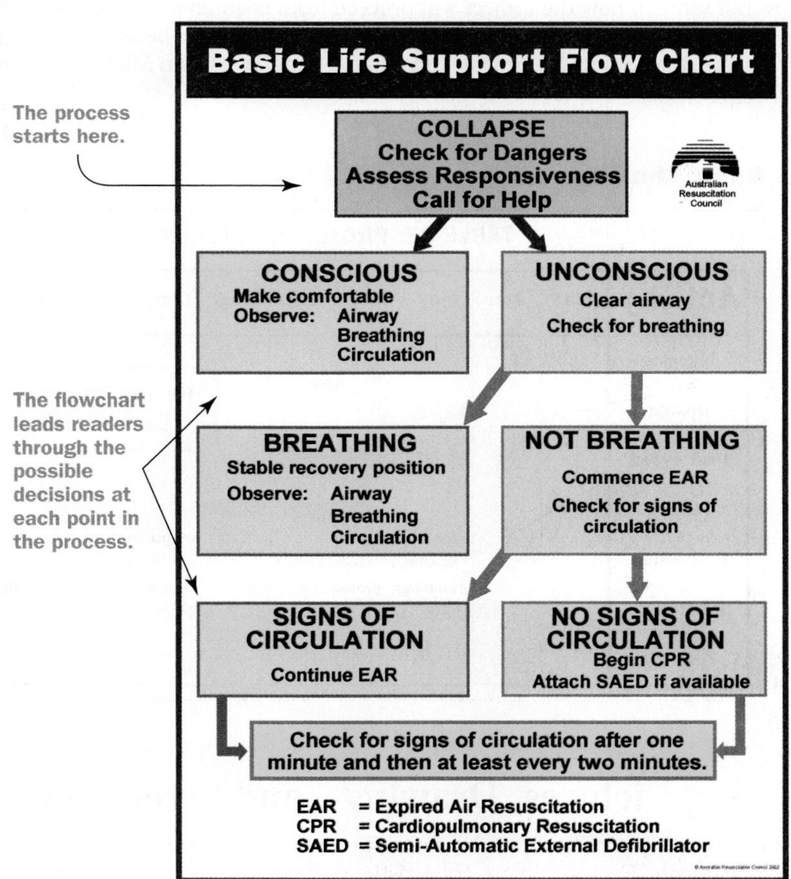

Figure 19.15: A flowchart is often useful for illustrating a process.

Source: Australian Resuscitation Council, http://www.resus.org.au/public/bls_flow_chart.pdf.

GO TO THE NET

Other flowcharts can be viewed at
www.pearsonhighered.com/johnsonweb4/19.11

typically cannot replace written instructions, especially if the steps are complex. But it can illustrate the steps in the process to help readers understand the written text.

Flowcharts can be found in a variety of other forms, such as organization charts or circuit diagrams. An organization chart illustrates the hierarchy of decision making in an organization. In a circuit diagram, a flowchart is used to chart the path of electricity.

Gantt Charts

Gantt charts have become quite popular in technical documents, especially proposals and progress reports. Gantt charts, like the one in Figure 19.16, are used to illustrate a project schedule, showing when the phases of the project should begin and end.

Visually, these charts illustrate the interrelations among different aspects of a large project. That way, people who are working on one part of the project will know what other teams are doing. Another benefit to a Gantt chart is that it gives readers an overall sense of how the project will proceed from beginning to end.

Gantt charts are becoming increasingly simple to create, because project-planning software like ArrantSoft, Signals' Basecamp, Omnifocus, and Microsoft Project can easily generate them for use in technical documents.

A Gantt Chart

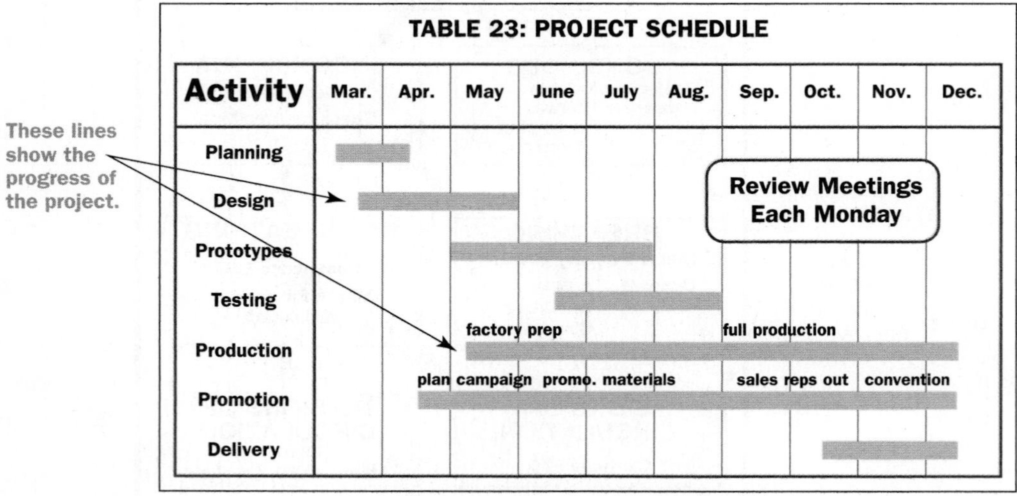

Figure 19.16: A Gannt chart is often used to show how the stages in a project interrelate. In this chart, you can easily see how the promotion of the product is started, even before the product is fully tested.

Using Pictures, Drawings, and Screen Shots

Increasingly, computers give you the ability to include pictures, drawings, and video in documents. Even if you are not artistic, you can quickly use a digital camera, a drawing program, a scanner, or a video camera to add life to your documents.

The purpose of a picture, drawing, or video is to show what something looks like. These kinds of visuals are especially helpful when your readers may not be familiar with something, like an animal or a piece of equipment. They are also helpful for showing the condition of something, like a building under construction or damage to a car.

Photographs

Digital cameras and scanners are making the placement of photographs in technical documents easier than ever. A good first step is to ask what *story* you want the photograph to tell. Then, set up a shot that tells that story.

PHOTOGRAPHING PEOPLE If you need to include a picture of a person or a group of people standing still, take them outside and photograph them against a simple but scenic background. Photographs taken in the office tend to look dark, depressing, and dreary. Photographs taken outdoors, on the other hand, imply a sense of openness and freethinking. When photographing people working, a good strategy is to show people doing what they *actually* do (Figure 19.17).

If you need to photograph people inside, put as much light as possible on the subjects. If your subjects will allow it, use facial powder to reduce the glare off their cheeks, noses, and foreheads. Then, take their picture against a simple backdrop to reduce background clutter.

If you are photographing an individual, take a picture of his or her head and shoulders. People tend to look uncomfortable in full-body pictures.

One general photography guideline that works well in most situations, especially when photographing people, is the "Rule of Thirds." The Rule of Thirds means the focal point of a picture (e.g., a subject's eyes, the key feature of an object) will appear where the top third of the picture begins. For example, in Figure 19.17, the welder's

A Photograph of a Person in Action

Figure 19.17: Try to capture people in action, close up.

Source: Corel.

To learn more tips about photography, go to **www.pearsonhighered.com/johnsonweb4/19.12**

A Photograph of an Object

Note the plain background behind the subject of the photograph.

Figure 19.18: When photographing objects, try to reduce the amount of clutter around your subject.

Source: The Internet Public Library, http://www.ipl.si.umich.edu/div/pottery/image15.htm.

goggles, which are the focal point of this picture, appear where the top third of the picture meets the middle third. Similarly, in Figure 19.18, notice how the focal point of the pottery (the design and bulge) is where the top third of the picture starts.

PHOTOGRAPHING OBJECTS When taking pictures of objects, try to capture a close-up shot while minimizing any clutter in the background (Figure 19.18). It is often a good idea to put a white drop cloth behind the object to block out the other items and people in the background. Make sure you put as much lighting as possible on the object so it will show up clearly in your document.

When photographing machines or equipment, try to capture them close-up and in action. After all, a picture of equipment sitting idle on the factory floor is rather boring. But if you show the machine being used or focus on the moving parts, you will have a much more dynamic picture.

PHOTOGRAPHING PLACES Places are especially difficult to photograph. When you are at the place itself, snapping a picture seems simple enough. But the pictures often come out flat and uninteresting. Moreover, unless people are in the picture, it is often difficult to tell the scale of the place being photographed.

When photographing places, focus on people doing something in that place. For example, if you need to photograph a factory floor, you should show people doing their jobs. If you are photographing an archaeological site, include someone working on the site. The addition of people will add a sense of action and scale to your photograph.

Inserting Photographs and Other Images

A digital camera will usually allow you to save your photographs in a variety of memory sizes. High-resolution photographs (lots of pixels) require a lot of memory in the camera and in your computer. They are usually saved in a format called a .tiff file.

To learn more about using electronic images, go to **www.pearsonhighered.com/johnsonweb4/19.13**

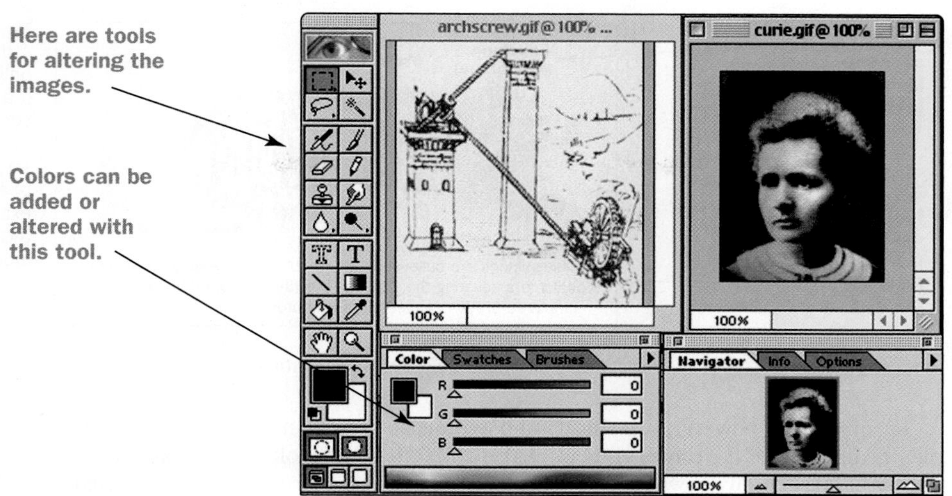

Figure 19.19: Software such as Adobe Photoshop allows you to work with photographs and other kinds of images.

Lower-resolution photographs (fewer pixels) are saved as .gif or .jpg files. Usually, .gif and .jpg files are fine for print and online documents. However, if the photograph needs to be of high quality, a .tiff file might be the best choice.

Once you have downloaded an image to your computer, you can work with it using software programs like Microsoft Paint or Adobe Photoshop (Figure 19.19). These programs will allow you to touch up the photographs or, if you want, completely alter them.

When you have finished touching up or altering the image, you can then insert it into your document or presentation. Most word-processing programs have an Insert Picture command. To insert the picture, put your cursor where you want the image to appear in the document. Then, select "Insert Picture." A box will open that allows you to locate the image on your computer's hard drive. Find and select the image you want to insert.

At this point, your computer will insert the image into your document. Usually, you can then do a few simple alterations to the file, like cropping, with the Picture toolbar in your word processor.

Illustrations

Illustrations are often better than photographs at depicting buildings, equipment, maps, and schematic designs. Whereas photographs usually include more detail than needed, a good illustration highlights only the most important features of the subject.

LINE DRAWINGS AND DIAGRAMS A line drawing or diagram is a semirealistic illustration of the subject being described. You can create simple drawings and diagrams with the Draw function of most word-processing programs. As the drawings grow more complex, however, most writers will hire professional artists to transform rough sketches into finished artwork.

Line drawings offer several advantages. They can provide a close-up view of important features or parts. They can also be easily labeled, allowing you to point out important features to the readers.

To see examples of line drawings, go to **www.pearsonhighered.com/johnsonweb4/19.14**

A Diagram

Labels are added to identify features.

Explanatory text can be added to clarify the meaning of the diagram.

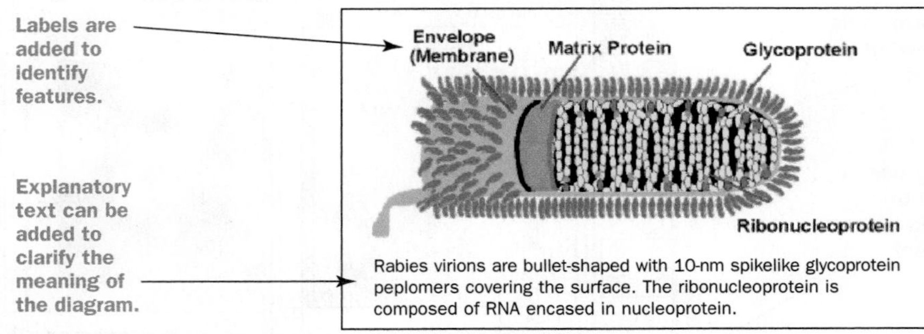

Figure 19.20: A drawing is only partially realistic. It concentrates on relationships instead of showing exactly what the subject looks like.

Source: Centers for Disease Control, http://www.cdc.gov/rabies/virus.htm.

In some ways, however, drawings and diagrams are less than realistic. For example, the diagram of the rabies virus in Figure 19.20 does not look exactly like the actual virus. Instead, it shows only how the larger parts of the virus are interconnected and work together.

MAPS Maps offer a view of the subject from above. You can use them to show geographic features like streets, buildings, or rivers. They can be used to portray the rooms in a building or illustrate a research site (Figure 19.21). In some cases, they can be used to show readers where a particular event occurred, allowing them to see

A Map

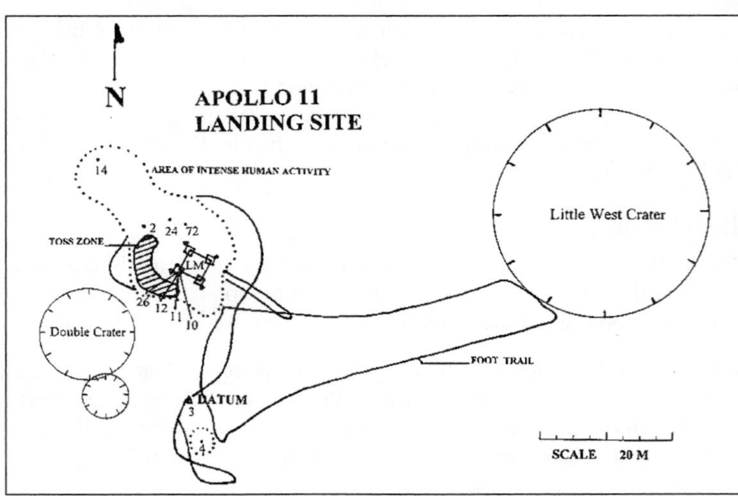

Figure 19.21: A map is useful for showing a place from above.

Source: New Mexico State University, http://www.spacegrant.nmsu.edu/lunarlegacies/images/scan2.gif.

For links to Internet mapping sites, go to **www.pearsonhighered.com/johnsonweb4/19.15**

Common Icons

Sources: Centers for Disease Control, http://www.cdc.gov/diabetes/pubs/images/balance.gif, and International Association for Food Protection, http://www.foodprotection.org.

Figure 19.22: Icons are widely available on the Internet. The person in the middle is supposed to be sneezing, but, as with many icons, it could convey an unintentional meaning.

where a particular place fits into an overall geographic area. When including a map, zoom in on the area that is being discussed in the text.

Maps may be easier to generate than you think. Today, Internet sites like Mapquest (www.mapquest.com), Google Maps (maps.google.com), and Topozone (www.topozone.com) allow you to create very detailed maps. Instead of trying to draw your own map, you might let these Internet sites do it for you.

ICONS AND CLIP ART Icons play an important role in technical documentation. In some documents, they are used as warning symbols. They can also serve as signposts in a text to help readers quickly locate important information (Figure 19.22). If you need to use an icon, standard sets of symbols are available on the Internet for purchase or for free.

Link

For more information on copyright law, go to Chapter 4, page 85.

Clip art drawings are commercially produced illustrations that can be purchased or used for free. Usually, when you purchase a collection of clip art, you are also purchasing the rights to use that clip art in your own documents.

It is tempting to advise you not to use clip art at all. When desktop publishing first came into the workplace, clip art was an original way to enhance the message and tone of a document. But now, most readers are tired of those little pictures of people shaking hands, pointing at whiteboards, and climbing ladders. In some cases, clip art becomes decorative fluff that takes readers' attention away from the document's message. Use it sparingly and only when it *truly* contributes to your message.

Screen Shots

With a combination of keystrokes, most computers will allow you to make a *screen shot* of your computer screen. Essentially, a screen shot is a picture of whatever is on the screen (Figure 19.23). You can use screen shots to insert a variety of images into your document or presentation.

You can make screen shots quickly with a PC or a Mac. With a PC, press the Print Screen button on your keyboard. The computer will put an image of the screen on your "clipboard." Look for it there. With a Mac, press the three keys Apple-Shift-3 at the same time. You should hear a camera clicking sound. The picture will then appear on your screen's desktop.

GO TO THE NET

For websites that offer free icons and clip art, go to **www.pearsonhighered.com/johnsonweb4/19.16**

A Screen Shot

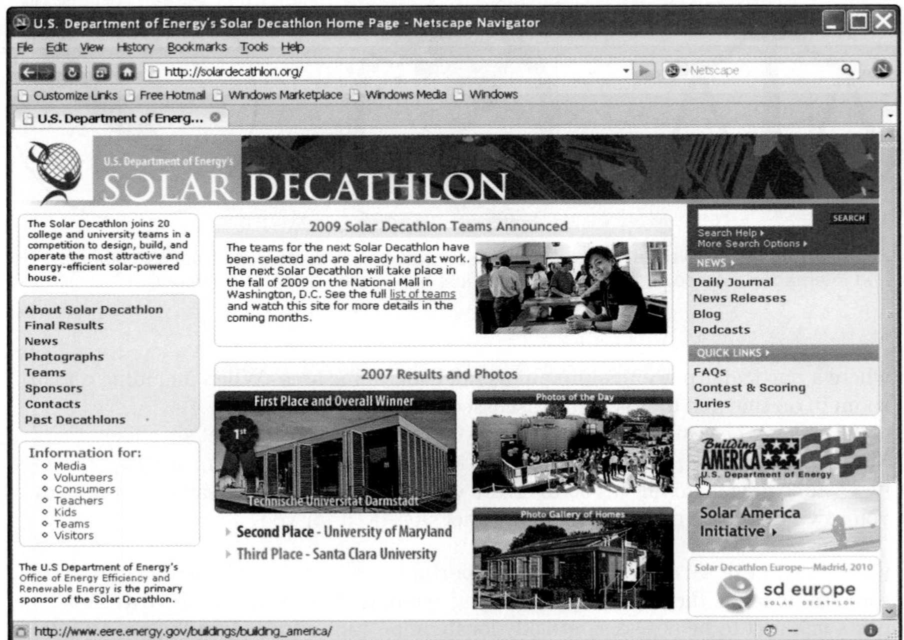

Figure 19.23: Screen shots are a simple way to add images to your document or presentation.

Source: Office of Energy Efficiency and Renewable Energy, http://www.eere.energy.gov/solar_decathlon.

You can insert the screen shot into your document with the Insert Picture command in your word processor. Using your Drawing tools, you can then "crop" (trim) the image to remove any extra details.

Screen shots are useful in a variety of situations. For example, you can use a screen shot to put a picture of a webpage in a printed document. Or, after drawing an illustration, you can make a screen shot of it and add it to a webpage.

Using Cross-Cultural Symbols

Symbols often translate among cultures better than words. They can also be more memorable and enhance comprehension for second-language readers (Horton, 1993, p. 683).

Symbols, however, don't always translate exactly across cultures, so you need to check your use of symbols in documents and websites with readers from other cultures. Otherwise, your symbols might lead to unintended consequences. For example, international dockworkers have been known to roughly toss boxes labeled with the broken wine glass symbol (meaning "fragile"), because they assumed the boxes contained broken glass.

Mr. Yuk Versus the Skull-and-Crossbones Symbol

Figure 19.24: The poison symbol, Mr. Yuk, was intended to avoid problems with the traditional skull-and-crossbones. But the symbol has its own problems when it crosses cultures. The skull-and-crossbones shown here is from the European Union's standardized set of symbols.

In another case, the "Mr. Yuk" poison symbol has had mixed results in its bid to replace the traditional skull-and-crossbones symbol (Figure 19.24). One problem is that "Yuk" is a common name in Asia, especially Korea. Meanwhile, the use of "Mr." suggests elder status to many Asian children, implying the face deserves added respect.

In a research study, a majority of international children did not understand the Mr. Yuk image or see it as negative, and a few thought the symbol meant the product was good to eat (Jackson-Smith & Essuman-Johnson, 2002). In North America, some health organizations now prohibit the use of Mr. Yuk because of these kinds of problems. The symbol was so heavily promoted that it now has a friendly undertone for children, attracting them to dangerous products.

To avoid misunderstandings, designers have developed symbols that are intended to cross cultures. The American Institute of Graphic Arts (AIGA) created the symbol system that is familiar to North Americans and is used globally (Figure 19.25). The European Union and International Standards Organization (ISO) have also created sets of international symbols that are widely used.

International Symbols

Figure 19.25: The AIGA, European Union, and International Standards Organization (ISO) have created a set of symbols that work internationally.

People in Symbols

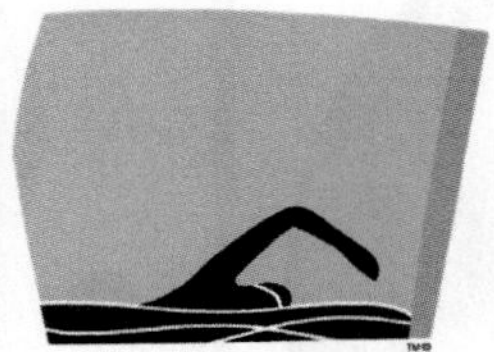

Figure 19.26: Simple pictographs are often used for human icons that need to cross cultures. These are examples of icons used at the Olympics.

Here are a few helpful guidelines for using symbols cross-culturally:

Keep human icons simple—Icons of humans should not be more than simple pictographs (Figure 19.26). Distinctive clothing or facial features could lead to unintended interpretations or confusion. Smiles, frowns, winks, or smirks can have very different meanings across cultures, so symbols that use faces are particularly problematic.

Use hand signals carefully—Just about any hand signal is considered offensive in some culture, including the thumbs-up signal, "OK" sign, V-symbol, a pointing finger, and even the palm out "halt" signal. If you can imagine an entire user's manual that uses an extended middle finger to point to things, you will get the idea about why hand signals can be problematic.

Avoid culture-specific icons—Mailboxes, phonebooths, and eating utensils, among other items, can look very different in other cultures, so symbols representing them might not translate. The typical North American mailbox on a street corner, for example, looks nothing like the canister mailboxes in England, while some cultures don't have public mailboxes at all. In another case, much of the world uses chopsticks for eating, so a fork would not properly symbolize "eat" or "food" to many readers.

Avoid religious symbols—Using crosses, crescents, stars, wings, candles, yin and yang, and other religious symbols can be interpreted very differently in other cultures. The symbol for the Red Cross, for example, is the Red Crescent in Islamic cultures, and the Red Crystal is used in Israel.

Avoid animal symbols and mascots—Animals can mean very different things in other cultures. In Western societies, the owl symbolizes wisdom, but in Southeast Asia, owls are considered unintelligent and vicious. Rats are considered clever and intelligent in many Asian countries, while in Western countries they are thought to be diseased and threatening. In some Islamic cultures, dogs are considered "unclean," making them particularly bad cartoon mascots for products. Meanwhile, the word *mouse* is not associated with computers in some cultures, so using a mouse symbol to represent a computer's pointing device would be confusing.

Link

For more information on cross-cultural readers, go to Chapter 2, page 28.

Symbols can be very helpful in technical documents because they enhance translation and comprehension. Your best approach is to use internationally accepted symbols whenever they are available and to always check your use of symbols with likely cross-cultural readers.

Using Video and Audio

One of the options available with computers is the use of video and audio in multimedia or Internet-based documents. With a digital video or audio recorder, you can add movies, music, and sound to any on-screen document.

Video

The ability to add video to a text is a mixed blessing. On the one hand, nothing will grab readers' attention like a video presentation. A video, for example, might show an important experiment or a current event. It might show an expert talking about a product or service (Figure 19.27). On the other hand, poorly made videos seem amateurish and quickly become tedious to watch. Another problem is that videos can take a long time to download.

If you want to add a video to your document, first decide exactly what you want the video to do. Then, write a script that achieves that goal in the least amount of time, because long videos take up lots of memory (and they can get a little boring).

As you create a video, think of it as a moving photograph in your document. Many of the same guidelines for shooting photographs will work with videos. For example, try to reduce background clutter so viewers can focus on the subject. Also, put as much light on the subject as possible. And, if you are filming people, you might want to put some powder on their foreheads, noses, and cheeks to reduce glare.

Using a Video

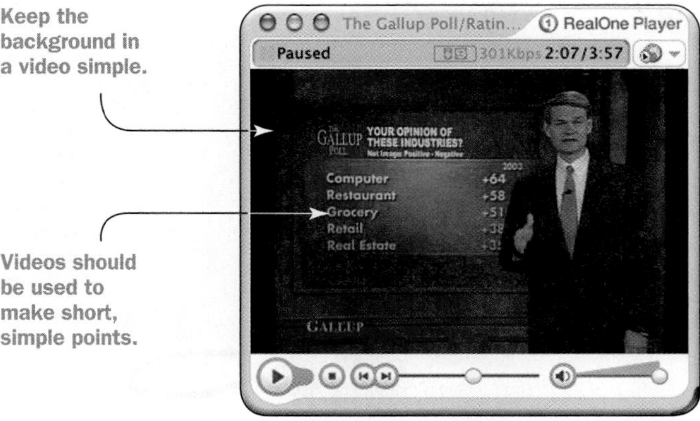

Figure 19.27: Video editing software can help you make videos to be inserted into multimedia documents, websites, or presentations.

Source: The Gallup Organization, http://www.gallup.com.

To learn more about video editing software, go to **www.pearsonhighered.com/johnsonweb4/19.17**

After you shoot the video, you can use video editing software like Apple iMovie, Adobe Premiere, and Microsoft Movie Maker to edit the final piece into a concise format.

When you are finished filming and editing the video, you can place it into your document, usually with the Insert Movie command. Even simple word processors like Microsoft Word and Corel WordPerfect will allow you to put a video into the text. Videos can also be inserted into presentations made with Microsoft PowerPoint, Corel Presentation, and other presentation software packages.

Audio, Podcasting, and Music

You can also add audio clips or background music to your electronic documents, including websites. Most word-processing programs, presentation software, and website design software will allow you to "insert" music or audio. The best way to insert music is from a compact disc (CD) or by downloading a song in MP3 format off the Internet.

Several software applications, like Apple iTunes or Windows Media Player, will allow you to download music and podcasts legally off the Internet, usually at minimal cost. Music and podcasts can be downloaded legally at apple.com, emusic.com, pandora.com, and rhapsody.com.

When you select Insert, the computer will usually ask you to select a file. Choose the audio track or music file to be inserted. Then, if you want the music to play continuously, tell the computer to "Loop" the music when you are in the "Audio" text box.

Podcasting is becoming a popular way to insert audio files into documents, especially websites. Podcasts are radio-like audio files that can be downloaded to computers and MP3 players (not only Apple iPods). If you have a microphone and a computer with Internet access, you can start creating podcasts right away. Software packages for making podcasts include Podcast Station, Propaganda, Audacity, and Adobe Audition. Many companies offer a "podcasting kit" that includes a microphone and software (Figure 19.28).

Podcasting

Figure 19.28: A podcasting mic, like this one from Alesis, will allow you to add professional-quality audio to websites or multimedia documents.

Source: Alesis, http://www.alesis.com.

GO TO THE NET For links to legal music download sites, go to **www.pearsonhighered.com/johnsonweb4/19.18**

CHAPTER REVIEW

- Computers have made including graphics in technical documents easier, so readers have come to expect them.
- Graphics should: (1) tell a simple story; (2) reinforce the text, not replace it; (3) be ethical; and (4) be properly labeled and placed on the page.
- Various kinds of graphs, tables, and charts allow you to tell different stories with data or facts.
- Digital cameras and scanners are making the placement of photographs in documents easier than ever.
- Use icons and clip art only when they enhance the readability and comprehension of the document. Clip art, especially, can simply clutter a document.
- Graphics need to be carefully considered when documents need to work cross-culturally. Images and symbols can have very different meanings in other cultures.
- Increasingly, computers allow you to insert video and audio into documents. Treat these items as you would treat photographs.

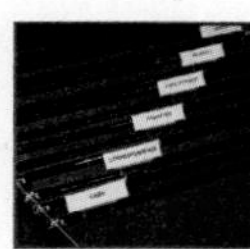

EXERCISES AND PROJECTS

Individual or Team Projects

1. On the Internet, find a chart or graph that you can analyze. Using the four guidelines for graphics discussed in this chapter, critique the chart or graph by discussing its strengths and places where it might be improved. Present your findings to your class.

2. Find a set of data. Then, use different kinds of charts and graphs to illustrate trends in the data. For example, you might use a bar chart, line graph, and pie chart to illustrate the same data set. How does each type of graphic allow you to tell a different "story" with the data? What are the strengths and limitations of each kind of graphic? Which kind of chart or graph would probably be most effective for illustrating your data set?

3. Using a digital camera, practice taking pictures and inserting those pictures into documents. Take pictures of people, objects, and places. When taking pictures of people, compare pictures taken inside and outside. Take full-body pictures and head shots. When taking pictures of objects, first leave the background behind the object cluttered. Then, use a backdrop to unclutter the picture. When photographing places, try to make images that tell a story about the place.

   When you are finished, compare and contrast your photographs. Which types of photographs seem to work best in a document? What kinds of photographs tend not to work?

Sample charts, graphs, and data sets for these exercises are available at **www.pearsonhighered.com/johnsonweb4/19.19**

Collaborative Project

With a group of classmates, locate a large document that has few or no visuals. Then, do a "design makeover" in which you find ways to use visuals to support and clarify the written text. Try to include at least one visual for every two pages in the document. Use graphs, photographs, and drawings to illustrate important points in the document. Then, add icons and clip art to reinforce important points or themes in the document.

When you are finished, write a brief report to your instructor in which your group discusses how you made over the document. Critique the original draft of the document, showing how the lack of adequate visuals made the information in the document hard to access. Then, discuss the ways in which your revised version improves on the original. Finally, discuss some of the following issues about the amount and types of visuals used in this kind of document:

- At what point are there too many graphics?
- Do some graphics work better than others?
- How can you balance the written text with visuals to avoid making the document too text-heavy or visual-heavy?
- How do the needs and characteristics of the expected readers of the document shape the kinds of visuals that are used?

Your report might offer some additional guidelines, beyond the ones discussed in this chapter, for using visuals more effectively.

For additional technical writing resources, including interactive sample documents, document design tutorials and guidelines, and more, go to **www.mytechcommlab.com.**

GO TO THE NET

You can find sample documents for the collaborative project assignment at **www.pearsonhighered.com/johnsonweb4/19.20**

CASE STUDY

Looking Guilty

Thomas Helmann was recently promoted to sergeant with the campus police at Southwest Vermont University. One of his added responsibilities was mentoring Officer Sharon Brand who had been hired a couple of weeks ago. Newly hired officers were usually given the "paperwork jobs" that the other officers didn't want to do. One of those jobs was putting together the "Annual Campus Crime Statistics Report" for the university's Executive Committee.

The report was a bore to write, and generally the officers assumed it was never looked at by the Executive Committee. Criminal activities at SWVU were rare, except for the usual problems with under-aged drinking, bicycle thefts, and graffiti that happened on every college campus.

So, it wasn't a surprise when Helen Young, the captain of the campus police, gave Sharon the job of collecting all the statistics and writing the report. At their weekly mentoring session, Thomas assured Sharon it was a good opportunity to get to know the campus and the kinds of problems that she would be dealing with. He advised Sharon, "Just look at last year's report and include the same kinds of facts and figures. The secretary can give you all our activity reports and statistics from last year."

A couple of weeks later, Sharon submitted the report to Thomas. At first glance, Thomas thought Sharon did a good job. He didn't have time to read the report closely, but the figures all looked accurate, and it covered the same issues as last year. Same old boring report.

Sharon also inserted several photographs to add some life and color to the report. Most of the photos showed places where petty crimes had happened, including dorms, bike racks, and parking lots. A couple of pictures showed students drinking beer at a football tailgate party.

There were also pictures of a few students, who Sharon obviously recruited, pretending they were stealing bikes, tagging walls, and taking computers out of dorm rooms. Thomas thought the staged photos were a bit silly, but he didn't think they were a problem. Nobody would read this report anyway.

He sent a dozen copies to the Executive Committee.

A few days later, Thomas was called into Captain Young's office. She was really angry about the report. "Sergeant, we have a big problem with the Crime Statistics Report."

A bit surprised, Thomas asked her what the problem was.

"Well, a member of the University's Executive Committee pointed out that all the staged pictures of criminal activities used African-Americans and Hispanics as models."

She handed the report to Thomas. Sure enough, all the people pretending to wheel off bicycles, steal computers, and spraypaint walls were minorities. Meanwhile, the students drinking at the football tailgater were white. He was shocked he hadn't noticed the racist tone of the photographs, but now the problem was glaringly obvious.

Stunned, he tried to offer an apology. The captain snapped back, "I'm not the person you should be apologizing to, though my butt is in the fire, too. This report makes the campus police look like a bunch of racists."

Thomas knew the offensive photographs could be easily removed from the report, but he wasn't sure what to do about properly reprimanding Sharon, his mentee. Thomas also didn't know who he should apologize to and how. Even more, though, Thomas found it troubling that he hadn't noticed the problem in the first place.

How do you think Thomas should respond to this issue? What should he tell Sharon? Who should he and/or Sharon apologize to? What else do you think should happen at this point?

For examples of unethical graphs, go to **www.pearsonhighered.com/johnsonweb4/19.21**

CHAPTER

20 Revising and Editing for Usability

Levels of Edit 552
Revising: Level 1 Editing 552
Substantive Editing: Level 2 Editing 554
Copyediting: Level 3 Editing 556
Proofreading: Level 4 Editing 558
Using Copyediting Symbols 562
Lost in Translation: Cross-Cultural Editing 563
Document Cycling and Usability Testing 566
Chapter Review 570
Exercises and Projects 571
Case Study: Wrong Version 572

In this chapter, you will learn:

- How to use revising and editing as a form of quality control.
- How to apply the four levels of editing to your text.
- How to use copyediting symbols.
- Strategies for revising your draft by reviewing its rhetorical situation (subject, purpose, readers, and context of use).
- Strategies for reviewing the content, organization, and design of your document.
- How to revise your sentences, paragraphs, headings, and graphics.
- How to proofread for grammar, punctuation, and word usage.
- Strategies for document cycling and usability testing.

Link

For more information on quality control, turn to Chapter 3, page 63.

Revising and editing are important because they are forms of *quality control* in a document. Your documents should reflect the quality standards that are held by your company (or maintain an even higher quality standard). The revising and editing phase is where your documents will go from "adequate" to "excellent."

Levels of Edit

Revising and editing requires more than checking a document for errors and perhaps changing a few sentences. In most cases, this kind of "revising" is not sufficient for technical documents.

Instead, when you are revising and editing, you need to approach the document from several different points of view. Professional editors use a tool called the "levels of edit" to assess how much editing a document needs before the deadline:

> **Level 1: Revision**—revises the document as a whole, which is why this level of edit is often called "global editing." Revision pays attention to the document's subject, purpose, readers, and context of use.
>
> **Level 2: Substantive editing**—pays special attention to the content, organization, and design of the document.
>
> **Level 3: Copyediting**—concentrates on revising the style for clarity, persuasion, and consistency, especially at the sentence and paragraph levels.
>
> **Level 4: Proofreading**—catches only the grammar mistakes, misspellings, and usage problems.

Which level of edit is appropriate for your document? The answer to this question depends on how much time you have and on the quality needed in the document. Given enough time, a writer will ideally go through all four levels, beginning with revision and ending with proofreading (Figure 20.1). But, sometimes the deadline is looming, leaving you time to only copyedit and proofread.

So, as you begin the revising and editing phase, start out by determining what level of editing is possible and/or needed to produce the desired quality of document. Then, begin reworking the document at that level.

Revising: Level 1 Editing

While you were drafting the document, you probably revised it as you wrote, sharpening your ideas on the subject, reconsidering your purpose, and adjusting the design. Now that the document is completely drafted, you can start revising it as a whole.

Revision is a process of "re-visioning" the document. In other words, you are trying to gain a new perspective on your text so you can ensure that its subject and purpose are appropriate for your intended readers. You also want to revise the document so it will work in its context of use.

To see other variations of the levels of edit, go to **www.pearsonhighered.com/johnsonweb4/20.1**

The Levels of Editing

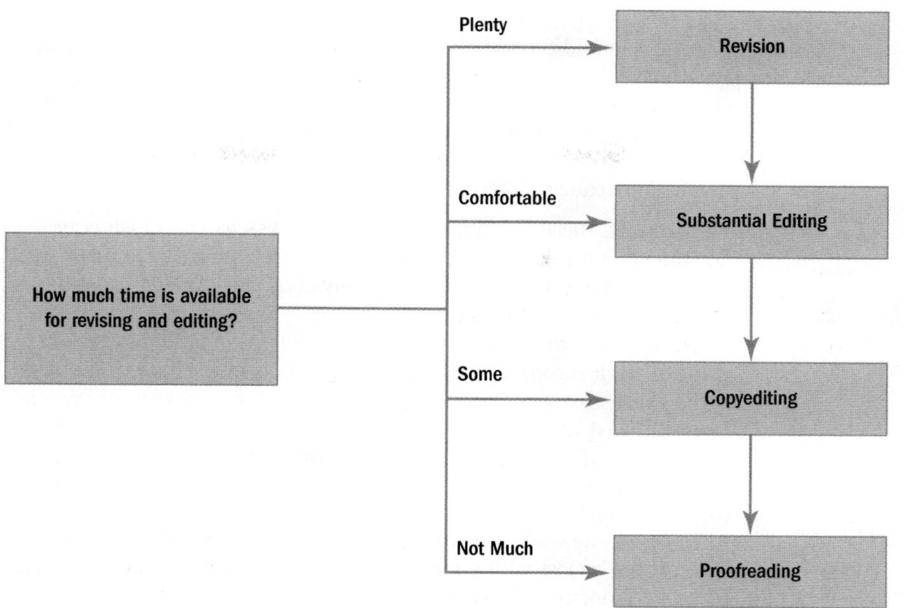

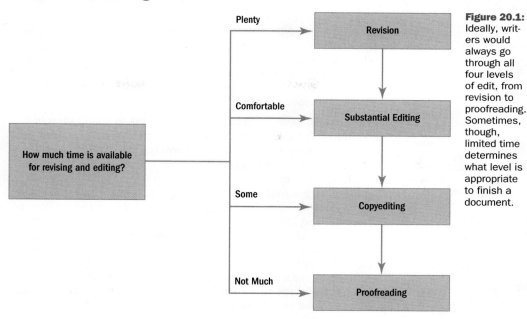

Figure 20.1: Ideally, writers would always go through all four levels of edit, from revision to proofreading. Sometimes, though, limited time determines what level is appropriate to finish a document.

To revise (re-vision) your document, look back at your initial decisions about the rhetorical situation at the beginning of your writing process:

SUBJECT Check whether your subject needs to be narrowed or broadened.

- Has your subject changed or evolved?
- Did you limit or expand the scope of your subject?
- Has your document strayed from the subject anywhere?

PURPOSE Make sure the document is achieving its purpose.

- What do you want the document to achieve?
- Is your document's purpose still the same?
- Has your purpose become more specific or has it broadened?

READERS Looking back at your original profile of your readers, think about the characteristics of the primary readers and other possible readers.

- Do you now know more about your primary readers' needs?
- Have you fully anticipated your readers' values and attitudes?
- Have you thought about the secondary, tertiary, and gatekeeper readers?

Link

For more information on defining the rhetorical situation, go to Chapter 1, page 5.

CONTEXT OF USE Consider the contexts in which your document might be read or used.

- Do you better understand the physical places where your readers will read or use the document?
- Do you better understand the economic, political, and ethical issues that will influence how your readers interpret your document?
- Have you anticipated the personal, corporate, and industry-related issues that will also shape your readers' interpretation?

AT A GLANCE

Guidelines for Revising (Level 1)

- Subject—Is the subject too narrow or too broad?
- Purpose—Does the document achieve its stated purpose?
- Readers—Is the document appropriate for the readers?
- Context of use—Is the document appropriate for its context of use?

Certainly, these are difficult questions to ask as you finish your document. But they are worth asking. You need to make adjustments to the document if its purpose has evolved or you have gained a better understanding of your readers.

Once you have reconsidered the document's rhetorical situation, you can work through the text to see if it stays focused on the subject and achieves its purpose (Figure 20.2). Look for places where you can revise the text to better suit your readers and the contexts in which they will use the information you are providing.

Revision at this global level requires some courage. You may discover that parts of your document need to be completely rewritten. In some cases, the whole document may need to be reconceived. But it is better to be honest with yourself at this point and make those changes. After all, a document that fails to achieve its purpose is a waste of all the time you've spent.

Substantive Editing: Level 2 Editing

While doing substantive editing, you should concentrate on the content, organization, and design of the document (Figure 20.3). A good approach to substantive editing is to consider the document from three different perspectives:

CONTENT Look for any gaps or digressions in the content.

- Are there any places (gaps) where you are lacking proof or support for your claims?
- Do you need to do more research to support your points?
- Are there any places where you have included information that the readers do not need to know to make a decision or take action?

ORGANIZATION A document should conform to a recognizable genre, and it should have an identifiable introduction, body, and conclusion.

- Are there any places where you have deviated from the organizational pattern of the genre you are following? Are these deviations helpful toward achieving your purpose? Or, should you reorganize the document to suit the genre?

Need more help with revision? Go to **www.pearsonhighered.com/johnsonweb4/20.2**

Revising the Document

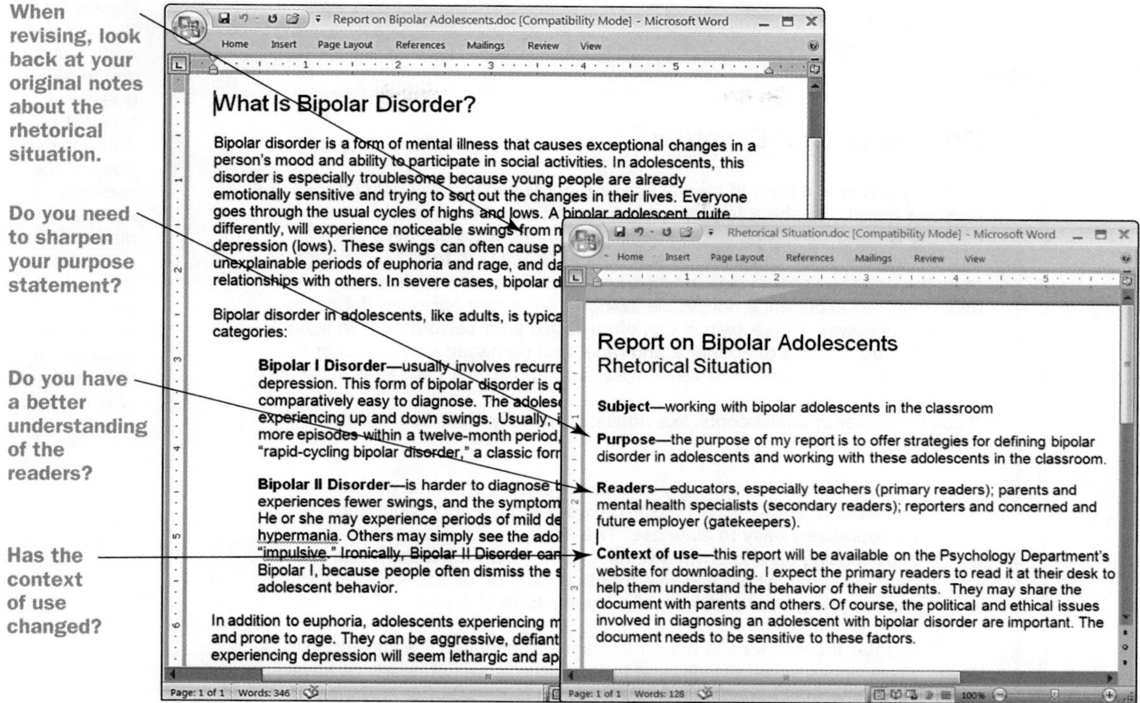

Figure 20.2: Look back at your original notes about the rhetorical situation. Have any of these elements changed or evolved? Does your document reflect these original decisions? If not, does the document need to change or does your understanding of the rhetorical situation need to change?

- Does the introduction clearly identify the subject, while stating your purpose and your main point? Should the introduction include more background information or stress the importance of the subject?
- Does the conclusion restate your main point, re-emphasize the importance of the subject, and look to the future?

DESIGN The document should be designed for its readers and the contexts in which it will be used.

- Is the text readable (scannable) in the situations and places where people will use it?
- Does the design reflect your readers' values and attitudes? Is it straightforward for conservative readers, or is it more innovative for progressive readers?

Link

For more information about organization, see Chapter 16, page 430.

AT A GLANCE

Guidelines for Substantive Editing (Level 2)

- Content—Are there any digressions or gaps in content?
- Organization—Does the document conform to a recognizable genre or pattern?
- Design—Do the page layout and graphics enhance the readability of the document?

GO TO THE NET

To learn more about editing design, go to **www.pearsonhighered.com/johnsonweb4/20.3**

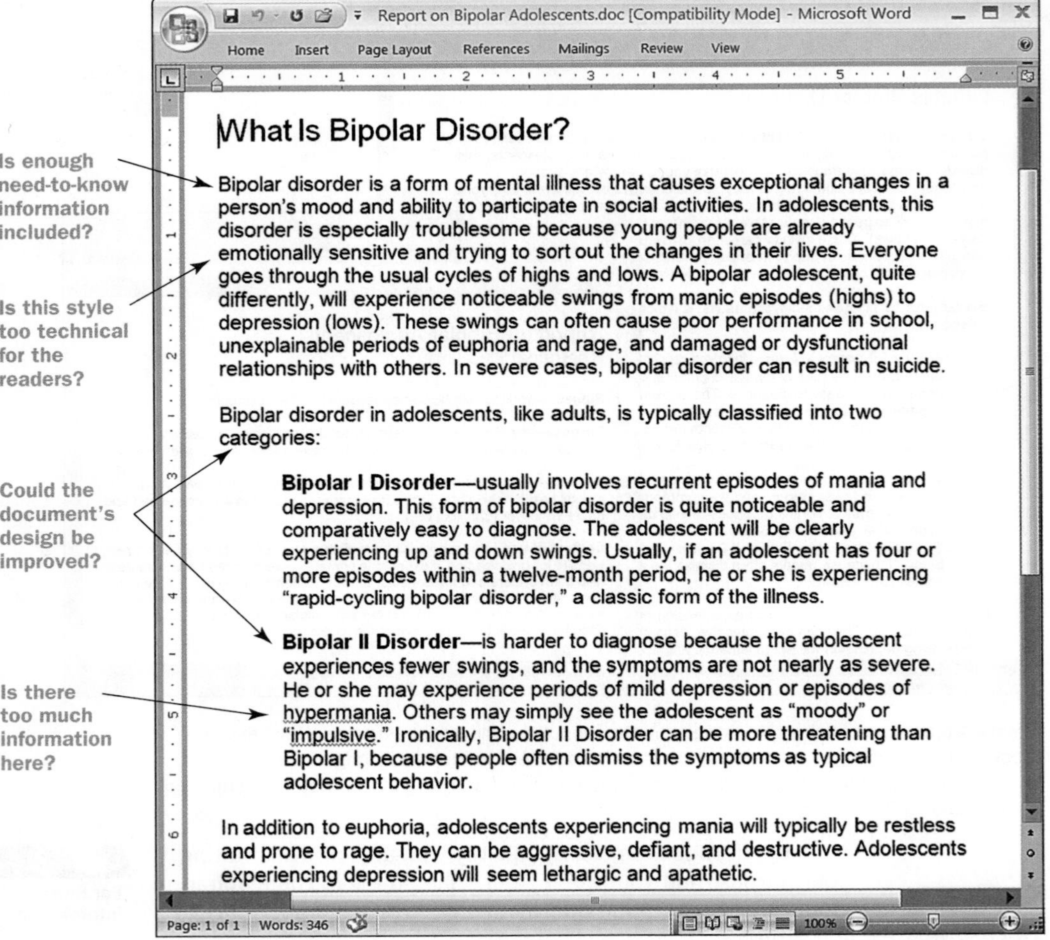

Figure 20.3: Substantive editing urges you to ask questions about the appropriateness of the content, style, and design of your document. When answering these questions, think about the needs of your readers.

Link

For more document design strategies, see Chapter 18, page 482.

- Does the design properly use principles of balance, alignment, grouping, consistency, and contrast?
- Does the design clarify the structure of the text with titles and subheads?
- Do the graphics support the text, and do they clarify difficult points?

Copyediting: Level 3 Editing

When you are copyediting, you should concentrate on improving style and consistency, especially at the sentence and paragraph levels. You should also look over the headings and graphics to make sure they are appropriate and accurate.

GO TO THE NET

To learn more about copyediting, go to **www.pearsonhighered.com/johnsonweb4/20.4**

Tracking Changes While Copyediting and Proofreading

Report on Bipolar Adolescents.2.doc.bin [Compatibility Mode] - Microsoft Wo...

Home Insert Page Layout References Mailings Review View

3

What Is Bipolar Disorder?

Bipolar disorder is a form of mental illness that causes noticeable ~~exceptional~~ changes in a person's mood and ability to participate in social activities. ~~In adolescents,~~ ~~t~~This disorder is especially troublesome in adolescents because young people are already emotionally sensitive. They are ~~and~~ trying to sort out all the changes in their lives.

Adolescents, like adults, ~~Everyone~~ goes through the usual cycles of highs and lows. Unlike most adults, though, ~~A~~ a bipolar adolescent~~, quite differently,~~ will experience very noticeable swings from manic episodes (highs) to depression (lows). These swings can often cause poor performance in school, unexplainable periods of euphoria and rage, and damaged or dysfunctional relationships with others. In severe cases, bipolar disorder can result in siucide. Bipolar disorder in adolescents, like adults, is typically classified into two categories:

> **Bipolar I Disorder**—~~u~~Usually involves recurrent episodes of mania and depression. This form of bipolar disorder is quite noticeable and comparatively easy to diagnose. The adolescent will be clearly experiencing up and down swings. Usually, if an adolescent has four or more episodes within a twelve-month period, he or she is experiencing "rapid-cycling bipolar disorder," a classic form of the illness.

Page: 1 of 1 Words: 353 100%

Color text highlights changes made to text.

These lines in the margin help locate changes.

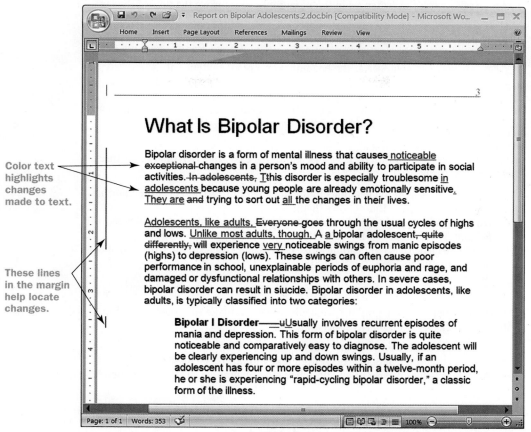

Figure 20.4: With the Track Changes function, your word processor can keep track of the changes you (or others) make to the document.

While copyediting, you might find the Track Changes feature of your word processor especially helpful (Figure 20.4). It will show places where changes were made to the text. Later, you can decide if you want to keep those changes or not. This feature is especially helpful if you are working with a team on a document. That way, any changes are recorded for the others in the group to approve.

When copyediting, pay close attention to the sentences, paragraphs, headings, and graphics.

SENTENCES Look over the sentences to make sure they are clear and concise.

- Are the subjects of the sentences easy to locate?
- Do the verbs of the sentences express the actions of the sentences?
- Can you eliminate any unnecessary prepositional phrases?
- Are the sentences breathing length?

Link

For help on improving style at the sentence level, go to Chapter 17, page 456.

Want to improve you grammar? Go to **www.pearsonhighered.com/johnsonweb4/20.5**

AT A GLANCE

Guidelines for Copyediting (Level 3)

- Sentences—Are the sentences clear and concise?
- Paragraphs—Do the paragraphs have a clear topic sentence and support?
- Headings—Do the headings help the readers scan for important information?
- Graphics—Do the graphics support the written text?

PARAGRAPHS Make sure the paragraphs support specific claims. Rework the sentences in the paragraphs to improve the flow of the text.

- Does each paragraph have a clear topic sentence (a claim) and enough support to back it up?
- Would any paragraphs be stronger if you included a transition sentence at the beginning or a point sentence at the end?
- Are the subjects in the paragraph aligned, or could you use given/new strategies to smooth out the text?
- Would transitions or transitional phrases help bridge any gaps between sentences?

HEADINGS The headings should be easy to understand and consistently used.

- Do the headings in the document properly reflect the information that follows them?
- Do the headings make the document scannable, highlighting places where important information can be found?
- Are there clear levels of headings that help readers identify the structure of the document and the importance of each part of the document?

Link

For more information on using headings, see Chapter 18, page 495.

GRAPHICS Look over the graphics in the document to make sure they support the written text. Check the graphics for accuracy.

- Does each graphic tell a simple story?
- Does each graphic support the written text without replacing it?
- Are the graphics clearly titled and referred to by number in the written text?

Link

For more help using graphics, turn to Chapter 19, page 523.

Proofreading: Level 4 Editing

Proofreading assumes that the document is complete in almost every way. Now you need to focus only on the mechanical details of the document, like the grammar, spelling, punctuation, and word usage. While proofreading, you should focus on marking "errors" and making only minor stylistic changes to the text.

Grammar

In technical documents, correct grammar and punctuation are expected. You should always remember that your readers *will* notice the grammatical mistakes. So, these mistakes can often sabotage an otherwise solid document.

Most word-processing programs have a grammar checker, but these checkers are notoriously unreliable. So follow any advice from the grammar checker cautiously, because your computer will often flag grammatically correct sentences. There is no substitute for mastering grammar rules.

GO TO THE NET

For quick answers to grammar issues, go to **www.pearsonhighered.com/johnsonweb4/20.6**

Common Grammatical Errors

Error	Explanation
comma splice	Two or more distinct sentences are joined only by a comma.
run-on sentence	The sentence is composed of two or more distinct sentences.
fragment	The sentence is incomplete, usually missing a subject or verb.
dangling modifier	A modifier (usually an introductory phrase) implies a different subject than the one in the sentence's subject slot.
subject-verb disagreement	A singular or plural subject does not agree with the verb form.
misused apostrophe	An apostrophe is used where it doesn't belong (usually confusing *it's* and *its*).
misused comma	A comma signals an unnecessary pause in a sentence.
pronoun-antecedent disagreement	A pronoun does not agree with a noun earlier in the sentence.
faulty parallelism	A list of items in a sentence is not parallel in structure.
pronoun case error	The case of a pronoun is incorrect (usually due to confusion about when to use *I* or *me*).
shifted tense	Sentences inconsistently use past, present, and future tenses.
vague pronoun	It is unclear what the pronoun refers to.

Figure 20.5: Here are the usual grammatical culprits. If you avoid these simple errors, your document will have almost no grammatical problems.

As you proofread the document, pay attention to your own reactions. When you stumble or pause, chances are good that you have found a grammatical mistake. Mark these places, and then identify what grammatical mistake has been made. Figure 20.5 describes some of the more common grammatical errors.

In the Grammar and Punctuation Guide (Appendix A), you will find examples of and remedies for these common errors. If you are not familiar with one or more of these errors, you should spend a little time learning how to avoid them.

Link

For more on grammar rules, go to Appendix A, page A-2.

In the end, most readers can usually figure out the meaning of a document even when it has errors in grammar. But these errors make the document look sloppy or seem unprofessional. After stumbling over a few errors, readers will begin to doubt the quality or soundness of the document and the information it contains. Even worse, they may doubt your abilities or commitment to quality. Readers expect correct grammar in technical documents.

Punctuation

Punctuation reflects the way we speak. For example, a *period* is supposed to reflect the amount of time (a period) that it takes to say one sentence. If you were to read a document out loud, the periods would signal places to breathe. Similarly, commas are used to signal pauses. When you come across a comma in a sentence, you pause slightly.

By understanding the physical characteristics of punctuation, you can learn how to use the marks properly (Figure 20.6).

Link

For a more detailed discussion of punctuation, go to Appendix A, page A-9.

For short essays on grammar issues, go to **www.pearsonhighered.com/johnsonweb4/20.7**

Physical Characteristics of Punctuation

Punctuation Mark	Physical Characteristic
capitalization	signals a raised voice to indicate the beginning of a sentence or a proper name
period [.]	signals a complete stop after a statement
question mark [?]	signals a complete stop after a question
exclamation mark [!]	signals a complete stop after an outcry or objection
comma [,]	signals a pause in a sentence
semicolon [;]	signals a longer pause in a sentence and connects two related, complete statements
colon [:]	signals a complete stop but joins two equal statements; or, it indicates the beginning of a list
hyphen [-]	connects two or more words into a compound word
dash [—]	sets off a comment by the author that is an aside or interjection
apostrophe [']	signals possession, or the contraction of two words
quotation marks [" "]	signal a quotation, or when a word or phrase is being used in a "unique way"
parentheses [( )]	enclose supplemental information like an example or definition

Figure 20.6: Punctuation mirrors the physical characteristics of speech.

Punctuation is intended to help readers understand your text. But when the marks are misused, they can create confusion. The Grammar and Punctuation Guide (Appendix A) includes a more detailed explanation of punctuation usage.

Spelling and Typos

Spelling errors and typos can be jarring for readers. One or two spelling errors or typos in a document may be forgiven, but several errors will cause your readers to seriously question your commitment to quality. Here are some ways to avoid those errors.

Use the spell check feature on your computer—Most word-processing programs come with a spelling checker that is rather reliable. Even if you are a good speller, the spelling checker will often catch those annoying typos that inevitably find their way into texts (Figure 20.7).

However, a spelling checker is not perfect, so you will still need to pay careful attention to spelling in your documents. After all, words may be spelled correctly, but they may not be the words you intended. Here is a sentence, for example, that has no errors according to a spelling checker:

> Eye sad, they're our many places four us to sea friends and by good she's stakes in Philadelphia.

To avoid these embarrassing errors, don't rely exclusively on the spelling checker.

For websites that discuss the history of punctuation marks, go to **www.pearsonhighered.com/johnsonweb4/20.8**

Running the Spelling Checker

Report on Bipolar Adolescents.3.doc [Compatibility Mode] - Microsoft ...

Home Insert Page Layout References Mailings Review View

Adolescents, like adults, go through the usual cycles of highs and lows. Unlike most adults, though, a bipolar adolescent will experience very noticeable swings from manic episodes (highs) to depression (lows). These swings can often cause poor performance in school, unexplainable periods of euphoria and rage, and damaged or dysfunctional relationships with others. In severe cases, bipolar disorder can result in siucide. Bipolar disorder in adolescents, like adults, is typically classified into two categories:

Spelling and Grammar: English (U.S.)

Not in Dictionary:

In severe cases, bipolar disorder can result in siucide.

Ignore Once

Ignore All

Add to Dictionary

Suggestions:

suicide

Change

Change All

AutoCorrect

Check grammar

Options...

Undo

Cancel

In addition to euphoria, adolescents experiencing mania will typically be restless and prone to rage. They can be aggressive, defiant, and destructive. Adolescents experiencing depression will seem lethargic and apathetic.

Word is checking the spelling and grammar in the document...

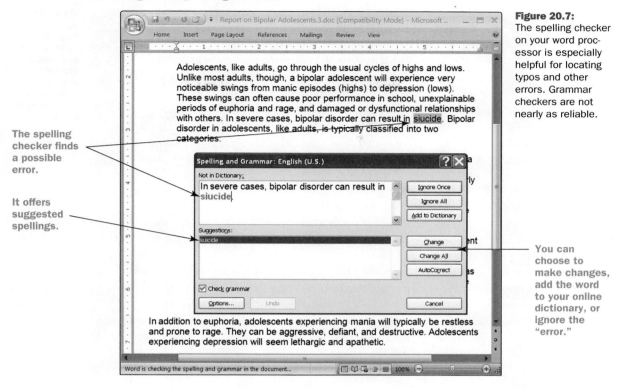

The spelling checker finds a possible error.

It offers suggested spellings.

You can choose to make changes, add the word to your online dictionary, or ignore the "error."

Figure 20.7: The spelling checker on your word processor is especially helpful for locating typos and other errors. Grammar checkers are not nearly as reliable.

Guidelines for Proofreading (Level 4)

AT A GLANCE

- Grammar—Are all the sentences grammatically correct?
- Punctuation—Are the sentences properly punctuated?
- Spelling and typos—Are there any spelling errors or typos?
- Word usage—Are all the words used properly?

Keep a dictionary close by—Technical documents often use terms and jargon that are not in your computer's spelling checker. A dictionary is helpful in these cases for checking spelling and usage.

On the Internet, you may be able to find dictionaries that define specialized words in your field. There are, for example, numerous online dictionaries for engineering or medicine.

Overall, though, you should be quick to grab the dictionary or look up a word online when there are any doubts about the spelling of a word. The best editors are people who do not hesitate to look up a word, even when they are almost certain that they know how the word is spelled.

Word Usage

In English, many words seem the same, but they have subtle differences in usage that you should know (Figure 20.8). There are several

For help using your spelling checker, go to **www.pearsonhighered.com/johnsonweb4/20.9**

Want to try out some great online dictionaries? Go to **www.pearsonhighered.com/johnsonweb4/20.10**

Common Usage Problems in Technical Documents

Confused Words	Explanation of Usage
accept, except	*accept* means to receive or agree to; *except* means to leave out
affect, effect	*affect* is usually used as a verb; *effect* is usually used as a noun
anyone, any one	*anyone* is a pronoun that refers to a person; *any one* means any one of a set
between, among	*between* is used for two entities; *among* is used for more than two entities
capitol, capital	*capitol* is the seat of a government; *capital* is money or goods
criterion, criteria	*criterion* is singular; *criteria* is plural
ensure, insure	*ensure* means to make certain; *insure* means to protect with insurance
complement, compliment	to *complement* is to complete something else or make it whole; a *compliment* is a kind word or encouragement
discreet, discrete	*discreet* means showing good judgment; *discrete* means separate
farther, further	*farther* refers to physical distance; *further* refers to time or degree
imply, infer	*imply* means to suggest indirectly; *infer* means to interpret or draw a conclusion
its, it's	*its* is always possessive; *it's* is always a contraction that means "it is"; *its'* is not a word
less, fewer	*less* refers to quantity; *fewer* refers to number
personal, personnel	*personal* refers to an individual characteristic; *personnel* refers to employees
phenomenon, phenomena	*phenomenon* is singular; *phenomena* is plural
precede, proceed	*precede* means to come before; *proceed* means to move forward
principle, principal	a *principle* is a firmly held belief or law; a *principal* is someone who runs a school
their, there, they're	*their* is a possessive pronoun; *there* is a place; *they're* is a contraction that means "they are"
whose, who's	*whose* is a possessive pronoun; *who's* is the contraction of "who is"
your, you're	*your* is a possessive pronoun; *you're* is a contraction that means "you are"
who, whom	*who* is a subject of a sentence; *whom* is used as the object of the sentence

Figure 20.8: You can avoid some of the most common usage problems in technical documents by consulting this list.

good *usage guides* available in print that explain the subtle differences among words. More usage guides are becoming available over the Internet.

If you are unsure about how a word should be used, look it up. Dictionaries are usually sufficient for answering usage questions. When a dictionary cannot answer your question, though, look up the word in an online or print usage guide.

Using Copyediting Symbols

While editing, you might find it helpful to use the same editing symbols that professional editors use. To mark any stylistic changes and inconsistencies, editors have developed a somewhat universal set of copyediting marks (Figure 20.9). They are easy to use and widely understood. See Figure 20.10 for examples of how they are used.

GO TO THE NET

Need to check the usage of a word? Go to **www.pearsonhighered.com/johnsonweb4/20.11**

Editing Symbols and Their Uses

Symbol	Use	Symbol	Use
∧	insert	⊙	add period
⟶	delete	∧	add comma
⌢	close up space	∧	add colon
# ∧	insert space	∧	add semicolon
∿	transpose	∨ ∨	add quotation marks
≡	capital letters	∨	add apostrophe
/	lowercase	¶	begin new paragraph
/—	lowercase, several letters	⊂	remove paragraph break
——	italics	]	indent text
~~	boldface	[	move text left
////	delete italics or boldface	⌐	block text
(rom)	normal type (roman)	(sp)	spell out (abbreviations or numbers)

Figure 20.9: Copyediting symbols offer a standardized method for editors and writers to work together on a document.

Lost in Translation: Cross-Cultural Editing

Today, translation software from Language Weaver, free services from World Lingo (www.worldlingo.com), and Altavista's Babel Fish (babelfish.altavista.com) make it possible to translate documents from one language to another with up to 80 percent accuracy. But that means 20 percent of any "machine-translated" document might be flawed or confusing. The limited success of these programs is often due to cultural subtleties that can cause confusing results.

When working with cross-cultural documents, you need to put extra effort into the revising and editing process. Several companies have experienced some classic gaffes when they have not taken the extra time to edit their texts from their readers' cultural point of view. For example, the baby food brand name Gerber means "vomit" in French. Also, when Gerber began introducing their jars of baby food in Africa, they kept its familiar design with a plump, happy baby on the front of the jar. In parts of Africa, though, where many are illiterate, the labels of jars and cans typically show what kind of food is inside.

Meanwhile, the Germanic word for travel is "Fahrt," which causes some classic problems with travel-related products that are sold in both Central Europe and English-speaking countries. For example, the Swedish furniture maker IKEA once sold a mobile workbench called the "Fartfull" in the United States and Europe.

Moreover, slogans often don't translate well. In 1996, Japanese designers at Panasonic created a web browser that used Woody Woodpecker as a mascot. Japanese executives were horrified to learn that the English meaning of their slogan "Touch Woody—The Internet Pecker" was not exactly what they had intended (Yoshida, 1996). Fortunately, most cross-cultural gaffes cause only confusion, not embarrassment. Nevertheless, you need to check your documents carefully when editing.

GO TO THE NET

Want to learn more about using copyediting symbols? Go to **www.pearsonhighered.com/johnsonweb4/20.12**

Copyediting a Text

Figure 20.10: Copyediting marks can be used to identify changes in the text.

Eruption History of Kilauea

Can you be more specific here? Are more accurate estimates available?

When Kilauea began to form is not known, but various estimates are 300,000–600000 years ago. The volcano had been active ever since, with no prolonged periods of quiescence known. geologic studies of surface exposures and examination of drillhole samples showthat Kilauea is made mostly of lava flows locally interbedded with deposites of explosive eruptions. Probably what we have seen hapen in the past 200 years is a good guide to what has happened ever since Kilauea emerged from the sea as an island perhaps 50,000–100,000 ago.

Lava Erupts from Kilauea's Summit and Rift Zones

Define the jargon in this paragraph.

Thrughout its history Kilauea has erupted from three main areas: its summit and two rift zones. Geologists debate whether Kilauea has Always had a caldera at the summit or whether it is a relatively recent feature of past few thousand years. It seems most likely that the caldera has come and gnoe throughout the life of Kilauea.

The summit of the volcano is high because eruptions are more frequent there than at any other single location on the volcano.

It's hard to figure out what you're describing here. A diagram would help.

However, more eruptions actually occur on the long rift zones than in the summit area, but they are not localized instead construct ridges of lower elevation than the summit. Eruptions along the east and southwest rift zones have build ridges reaching outward from the summit some 125 KM and 35 KM respectively

Most eruption are relatively gentle, sending lava flows downslope from fountains a few meters to a few hundred meters high. OVER and over again these eruptions occur, gradually building up the volcano and giving it a gentle, shield-like form. Every few decades to centuries, however powerful explosions spread ejecta across the landscape. Such explosions can be lethal, as the one in 1790 that killed scores of people in a war party near the summitting of Kilauea. Such explosions can take place from either the bridging summit or the upper rift zones.

Source: U.S. Geological Survey, http://hvo.wr.usgs.gov/kilauea/history/main.html. Errors in the text were added and did not appear in the original.

International business specialists Carol Leininger and Rue Yuan (1998) offer the following advice for creating and editing cross-cultural documents:

Use short, direct sentences that follow subject, verb, object order—Second-language readers and translation software will be more successful if they can easily locate the subjects and verbs of sentences. Longer sentences should be cut into shorter sentences.

Use positive sentences and minimize negative sentences—Negative sentences sometimes translate more harshly than originally intended. A negative sentence that offers a simple caution to the reader can translate into one that makes dire predictions of harm or death.

Use a limited set of words—Most international companies, such as Caterpillar and IBM, have developed standard language guides of English words to be used in international documents. Documents that use these words are easier for people and translation software to translate.

Avoid humor or jokes—Jokes are highly culture-specific and situational, so they rarely translate well into other cultures and languages. Usually, they are just confusing, but sometimes they can be embarrassing for the writer or insulting to the reader.

Minimize jargon and slang—Jargon words and slang phrases are also culturally dependent and difficult to translate. These terms should be translated into their common meanings even though they might lose some of their original flair.

Check any sayings, clichés, or idioms—These turns of phrase often do not translate well. For example, in North America people "cross their fingers" for luck, but in Germany, people "hold their thumbs." Machines "run" in English, but they "walk" in Spanish.

Avoid obvious metaphors—Metaphors cannot be completely avoided, but obvious ones should be removed. In German, calling a company a "shooting star" would suggest that it will be successful long into the future. In North America, "shooting star" means the company will rise quickly and fail spectacularly. Meanwhile, sports metaphors like, "She hit a home run" or "He just punted" will be confusing to most international readers. Metaphors that use body parts (e.g., "I'll keep an eye on the project") or animals (e.g., "He's a workhorse") can have very different and disturbing meanings when translated.

Check slogans—Slogans usually rely on a cultural twist of words, so they are particularly risky when translated. In Taiwan, Pepsi's slogan "Come alive with the Pepsi Generation" translated into, "Pepsi will bring your ancestors back from the dead" (Pendergrast, 1994).

Check product names—Names of products can also translate in embarrassing ways. Products like the Pinto, Puffs, Waterpik, and latte, among others, have sexually suggestive meanings in other languages. The Chevy Nova didn't sell well in Mexico and Latin America because "no va" means "It doesn't go" in Spanish.

To ensure that your documents will work cross-culturally, your best strategy is to user-test your documents with readers from likely target cultures because, unfortunately, translation software will rarely catch the subtleties of languages. So, you need to do it. Also keep in mind that your translation dictionary probably won't include insulting phrases or sexually suggestive slang that a test reader would catch.

Document Cycling and Usability Testing

When you are completing your document, it is important that you gain an outside perspective. Often, while drafting, we become too close to our documents. Consequently, we can no longer edit or assess our own work objectively. Two ways to gain that outside perspective are *document cycling* and *usability testing.*

Document Cycling

Document cycling is a method for letting others at your company look over your draft. When you *cycle* a document, you pass it among your co-workers and supervisors to obtain feedback (Figure 20.11).

Computers give you the ability to quickly send your document around to others for their suggestions for improvement. You can send it as an attachment or make multiple paper copies. Then, your supervisors, colleagues, and even your primary readers can look it over and offer suggestions for improvement.

When revising and editing your documents, it is important to let others look over your work. Document cycling is an important part of a *quality feedback loop,* a central principle of quality management. If you rely on yourself alone to edit your work, the quality of your document might suffer.

Usability Testing

Usability testing means trying out your document on real readers. This kind of testing on real readers can be informal or formal, depending on the importance of your document and the time you have to test it.

> **Informal usability testing**—Minimally, you should ask other people to look over the document and mark places where they stumble or find it difficult to understand your meaning.

Document Cycling in the Workplace

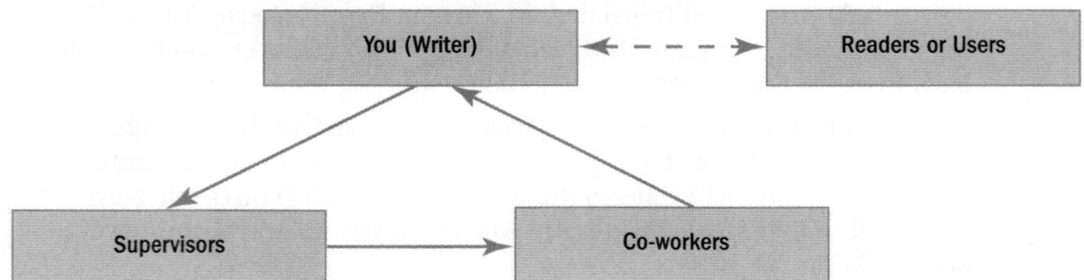

Figure 20.11: Document cycling gathers feedback by letting others look over the document.

Types of Usability Testing

	Usability Test	How It Is Conducted
informal testing	document markup	Readers are asked to read through a document, marking places where they stumble or fail to understand.
↑	read and locate test	Readers are asked to locate specific kinds of information in a document. They are timed and videotaped.
	summary test	Readers are asked to summarize the important information in a document.
	protocols	Readers are asked to talk out loud as they are using the text. Their comments are taped and transcribed.
	journal or tape recording	Readers are asked to keep a written or taped journal at their workplace to record their experiences with the document.
	surveying	Readers are given a questionnaire after they use the document, asking them about their experience.
	interviewing	Readers are interviewed about their experiences using a document.
↓	focus groups	Groups of readers look over a document and discuss their reactions to the work.
formal testing	laboratory testing	Through cameras and a one-way mirror, readers are carefully observed using a text.

Figure 20.12: A variety of methods are available to user-test a document.

Formal usability testing—More formally, you can run experiments with sample readers to measure how well they can understand and use your document. Instructions and user's manuals are commonly user-tested in this way. But, increasingly, focus groups of readers are being used to test persuasive documents like proposals. In some cases, companies will create *usability laboratories* to observe the reactions of readers as they use documents.

The more similar your test subjects are to your target readers, the better the results of your testing will be. In other words, the folks around the office are fine for an informal test. But if you want a more accurate assessment of how your document will be used, you should try to locate a group of people who are most like your readers. Figure 20.12 lists some of the more common methods for testing usability, ranking them from informal methods to formal methods.

Most usability testing is designed to answer four questions (Figure 20.13):

Can they find it?—*Read-and-locate tests* are used to determine whether users can locate important parts of the document and how quickly they can do so. Often, the users are videotaped and timed while using the document.

Can they understand it?—*Understandability tests* are used to determine if the users retain important concepts and remember key terms. Users are often asked to summarize parts of the document or define concepts.

To learn more about usability testing, go to **www.pearsonhighered.com/johnsonweb4/20.13**

Usability Questions

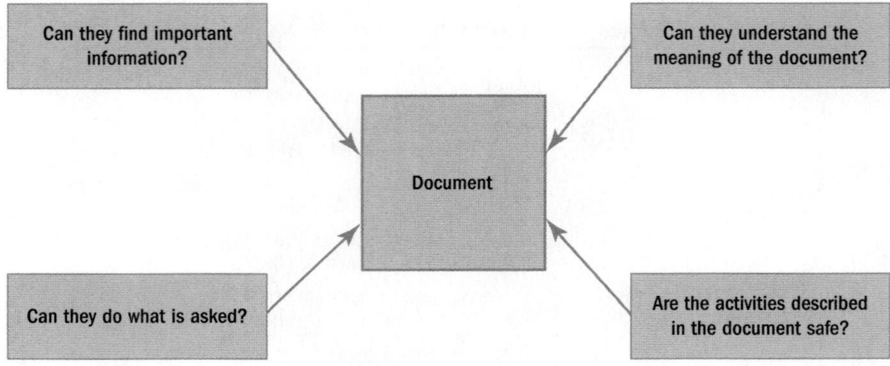

Figure 20.13: Usability testing is designed to answer four simple questions.

Can they do it?—*Performance tests* are used to determine whether users can perform the actions the document describes. These tests are often used with instructions and procedures.

Is it safe?—*Safety tests* are used to study whether the activities described in the document, especially in instructions or user's manuals, are safe. These tests carefully watch for possible safety problems by having sample readers use the product documentation.

As you devise a usability test or series of usability tests, you should set quantifiable objectives that will allow you to measure *normal* and *minimal* user performance with the document. In other words, first define what you would expect *typical* users to be able to accomplish with your document. Then, define what you would *minimally* expect them to be able to do.

READ-AND-LOCATE TESTS: CAN THEY FIND IT? To run a read-and-locate test, list five to seven important pieces of information that you want readers to locate in the document. Then, while timing and/or videotaping them, see how long it takes them to find that information.

Videotaping your subjects is especially helpful, because you can observe how readers go about accessing the information in your document. Do they go right to the beginning or the middle? Do they flip through the text looking at the headings or graphics? Do they look at the table of contents or index (if these features exist)?

After your subjects locate the major pieces of information you asked them to look for, have them tell you about these major points orally or in writing. Then, check their answers against your original list. If they successfully found four or five items from your list of most important pieces of information, your document is likely well written and well designed. If, however, they struggled to find even a few of your major points, you probably need to revise your document to ensure that the important information is easy to locate.

UNDERSTANDABILITY TESTS: CAN THEY UNDERSTAND IT? When running an understandability test, you want to determine how well the users of your document grasped its meaning. Before running the test, write down your document's purpose and main point. Then, write down three important concepts or points that anyone should retain after reading the document.

Give your readers a limited amount of time to read through the document or use it to perform a task. Then have them put the document away so they cannot use it. Verbally or in writing, ask them:

- What is the purpose of this document?
- What is the document's main point?
- Can you tell me three major points that are made in the document?

If their answers to these questions are similar to the ones you wrote down, your document is likely understandable. If, however, your readers struggle to answer these questions, or get them wrong, you should think seriously about revising the document to highlight the information you intended your readers to retain.

PERFORMANCE TESTS: CAN THEY DO IT? Almost all technical documents are written to help readers take some kind of action. A set of instructions, obviously, asks readers to follow a procedure. A report might make some recommendations for change.

To do a performance test, have the users perform the procedure the document describes. Or, ask them to react to your recommendations. Here again, videotaping the users is a good way to keep a record of what happened. Did they seem to find the document easy to use? Where did they stumble or show frustration? When did they react positively or negatively to the tasks or ideas described in the document?

Usability Testing a Document

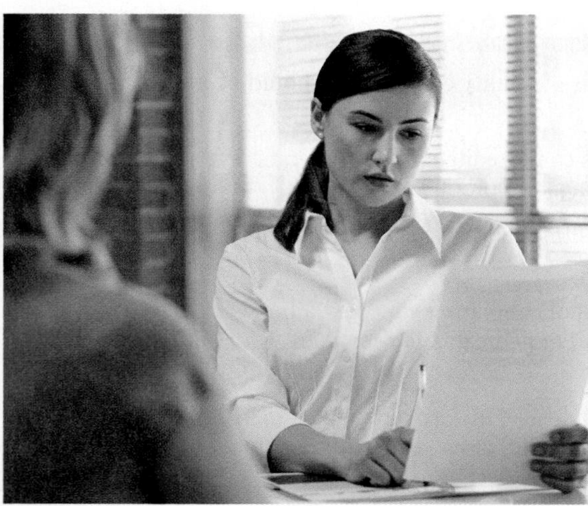

To test the usability of a document, you can run experiments with people who represent real readers. Videotaping is an especially good way to see how people actually use a document.

Ultimately, performance tests are designed to find out whether the users can do what the document asks of them. But it is also important to determine their attitude toward performing these tasks. You want to ensure not only that they *can* do it, but also that they *will* do it.

SAFETY TESTS: IS IT SAFE? Above all, you want your documentation and products to be safe. It is impossible to reduce all risk of injury, but you should try to reduce the risk as much as possible. Today, it is common for companies to be sued when their documentation or products are shown to be inadequate. Often, in product liability lawsuits, documents like instructions and user's manuals are used to prove or deny a company's negligence for an injury.

Without putting test subjects at risk themselves, safety tests are usually designed to locate places where users may make potentially injurious mistakes. They also ask readers about the warnings and cautions in the document to determine whether the reader noticed and understood these features.

SETTING OBJECTIVES AND MEASURING RESULTS The challenge to effective usability testing is to first identify some objectives for the document. These objectives could refer to (1) how well the users can find information, (2) how well they understand important ideas, and (3) how well they perform tasks described in the document. Then, measure the results of your usability testing against these objectives.

It's often quite sobering to watch people fumble around with your document, misunderstand its meaning, and not follow its directions. But, the results of your tests should help you revise the document to improve its usability.

No form of usability testing will ensure that your document is a success. However, feedback from users is usually the best way to gain new insights into your document and to solicit suggestions for improvement.

CHAPTER REVIEW

- Revising and editing are forms of quality control that should be a regular part of your writing process.
- Documents and presentations can be edited at four different levels: revision, substantive editing, copyediting, and proofreading.
- Editorial tools such as copyediting marks are helpful even for nonprofessional editors.
- Document cycling is a process of circulating your text among colleagues and your supervisor. You can use e-mail attachments to send your work out for review by others.
- Usability testing can involve informal or formal methods to test the effectiveness of your document. You might ask sample readers to test documents that will be used by a broad readership.

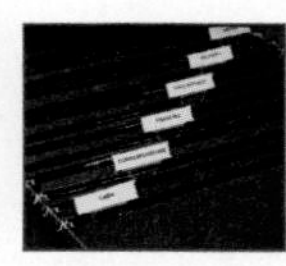

EXERCISES AND PROJECTS

Individual or Team Projects

1. Find a document on campus or at your workplace that needs editing. Edit the document by working backward from a level 4 edit (proofreading) to a level 1 edit (revision). As you apply each level of edit to the document, pay attention to the different kinds of actions and decisions you make at each level. How is the document evolving as you edit it?

2. Exchange a text with a member of your class. Then, do a level 2 edit (substantive editing) of the draft, using copyediting symbols to reflect the changes you think should be made to the document. Write a cover letter to the author in which you explain the changes you want made. Hand in a copy of this documentation to your instructor.

3. Find a text on the Internet that needs to be edited. Do a level 3 edit (copyediting) of the text, using the copyediting marks shown in this chapter. Write a memo to your instructor in which you discuss how you edited the text. Discuss some of the places where you struggled to mark the changes you wanted to make.

4. Find a text that needs editing and use the online editing strategies discussed in this chapter to edit it. In an e-mail to your instructor (with the edited version attached), discuss some of the differences and difficulties between paper-based and online editing. Discuss which form of editing you prefer and why.

Collaborative Project

With your group, locate a longer document on the Internet that needs editing. The document might be a report or proposal, or perhaps even a website. Then, do a level 2 edit (substantive editing) on the document, including the two levels of editing below it (copyediting and proofreading).

As you are editing, set up a document cycling routine that allows you to distribute your document among your group members. You can cycle the document in a paper version or use online editorial strategies to keep track of versions and changes in the document.

When your group has finished editing the document, revise it into a final form. Then, conduct usability tests on the document, using other members of your class as subjects.

In a memo to your instructor, discuss the evolution of the document at each stage of the editorial process. Tell your instructor (1) how you used substantive editing to improve the document, (2) how you cycled the document among your team members, and (3) how you used usability testing to identify places where the document might be improved.

For additional technical writing resources, including interactive sample documents, document design tutorials and guidelines, and more, go to **www.mytechcommlab.com.**

Longer documents that you can use for these exercises are available at **www.pearsonhighered.com/johnsonweb4/20.14**

CASE STUDY

Wrong Version

Wayne Hamilton was a design engineer at NextGen Manufacturers, a tool and die manufacturer in Los Angeles, California. He and his boss, Jim VanMeter, had been working on a proposal for a new account with the Jet Propulsion Laboratory. They knew only a few companies would be able to supply the specialized dies needed for manufacturing a new type of satellite.

The proposal was a truly collaborative effort. Wayne and Jim were working in different office buildings, so the document went back and forth as an attachment. As they put the proposal together, they would insert sections written by people in other parts of the company. They used Microsoft Word's Track Changes feature and Comment feature to keep track of versions and make suggestions for improvements.

On the day it was due, Wayne was responsible for sending in the final draft. He printed the final version out and read it through one last time. He made some small proofreading changes. Then, he called his boss to make sure there was nothing more to add. Jim said, "Nope. It's as good as it's going to get. Send it off."

Wayne attached the Word file to an e-mail. He also copied (cc'd) the e-mail to himself to make sure the file came through all right. Then, he hit send.

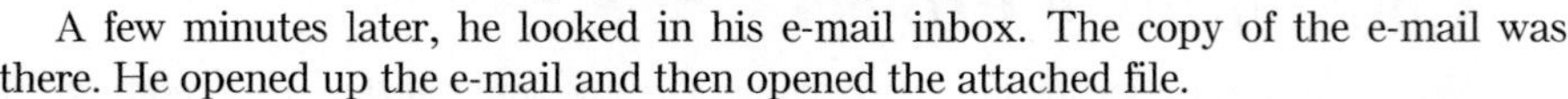

A few minutes later, he looked in his e-mail inbox. The copy of the e-mail was there. He opened up the e-mail and then opened the attached file.

In shock, he saw that he had forgotten to delete the comments in the margins of the file. There, in big red text bubbles, were all the comments and revisions from his team. Most of the margin comments just showed the editorial changes that had been made. Some of them, though, talked about how much they should charge for the service and what they should offer. A lower price than the final price was mentioned in the comments.

Even worse, one of the writers had written some unflattering comments about the JPL readers in the margin. He wrote that the folks at JPL might not be smart enough to understand some of the technical specifics, so those items would need to be "dumbed down for people who aren't that bright."

There was a good chance the readers at JPL would never see those comments. More than likely, the file would be printed out by a staff member and distributed with the other proposals. On the other hand, there was a good chance the file would just be forwarded electronically, and those comments would pop up for the readers, just as they had for Wayne.

Fortunately, a few hours remained before the final deadline. Wayne had a chance to accept all the changes, delete all the comments, and resubmit the proposal with some kind of excuse. But would resubmitting signal a problem with the original proposal? Would it hint to the readers that they were careless in some way? What if they looked at the original file anyway and saw those comments? What do you think Wayne should do? And do you think he should tell his boss and others about his mistake? If so, how should he tell them?

CHAPTER

21

Preparing and Giving Presentations

Planning and Researching Your Presentation *574*

Choosing the Right Presentation Technology *579*

Organizing the Content of Your Presentation *582*

Help: Giving Presentations with Your iPod, MP3, or Mobile Phone *586*

Choosing Your Presentation Style *594*

Creating Visuals *596*

Delivering the Presentation *599*

Rehearsing *602*

Working Cross-Culturally with Translators *604*

Chapter Review *608*

Exercises and Projects *608*

Case Study: The Coward *610*

In this chapter, you will learn:

- The importance of being able to prepare and deliver public presentations.
- How to define the rhetorical situation (subject, purpose, audience, and context of use) for your presentation.
- Strategies for organizing the content of presentations.
- How to create an effective presenting style.
- How to create and use visuals in presentations.
- The importance of practice and rehearsal.
- How to work effectively with translators in cross-cultural situations.

If you don't like giving public presentations, you are not alone. Each year, surveys show that people fear speaking in public more than anything else—even more than death.

Yet giving public presentations is an essential part of most technical careers. More than likely, you will find yourself regularly giving presentations to clients, supervisors, and colleagues. Presenting information in public is a crucial skill in today's technical workplace.

Public Speaking Is More Important Than Ever

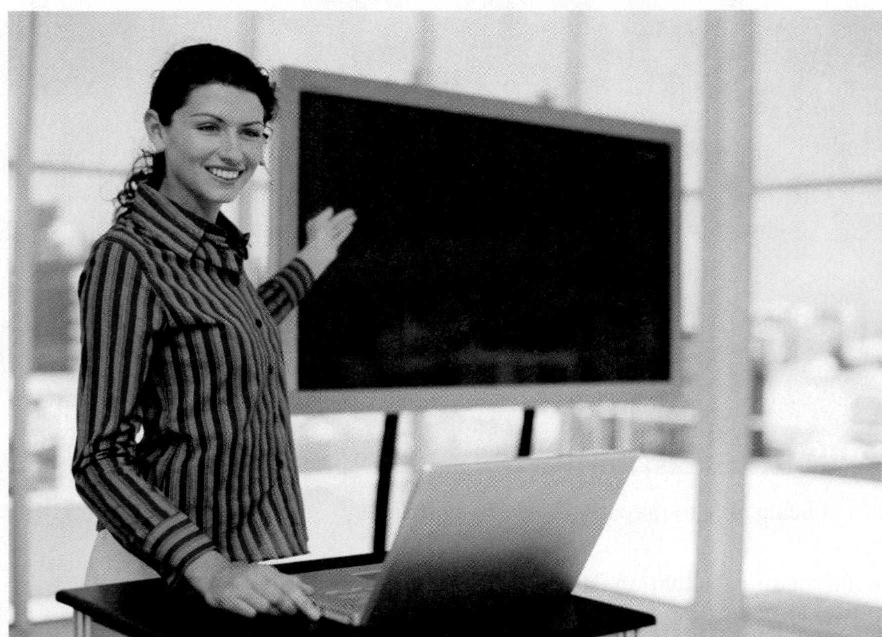

Public presentations are easier than ever with computers. You will find, though, that audiences now expect polished, professional presentations with plenty of visuals and visual appeal.

Planning and Researching Your Presentation

The preparation of a public presentation should follow a process, just like preparing written documents (Figure 21.1). Before getting up in front of an audience, you will need to

- plan and research your subject.
- organize your ideas.
- choose an appropriate presentation style.
- create graphics and slides.
- practice and rehearse your presentation.

The Preparation Process for a Presentation

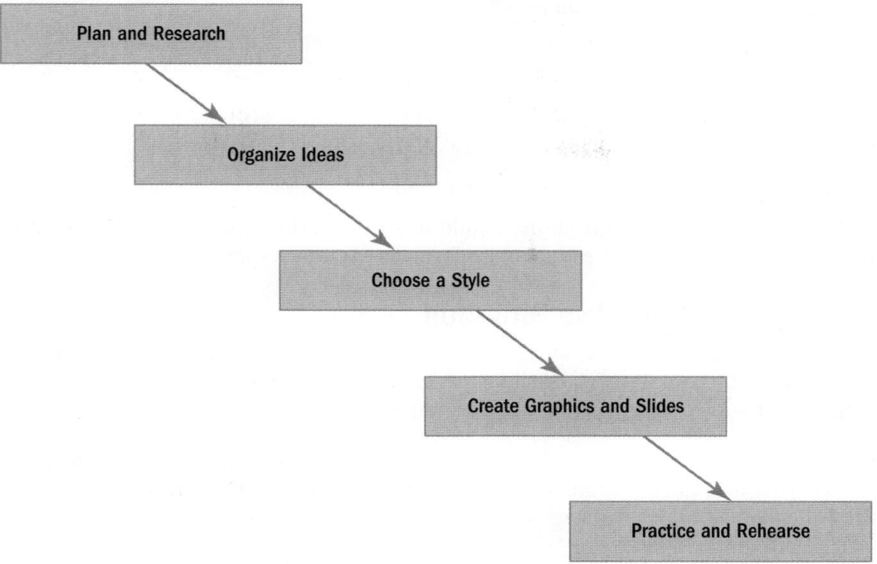

Figure 21.1: Just as with written documents, you should follow a process as you prepare a presentation.

Even if you are already an accomplished public speaker, it is important that you spend significant time on each stage of the process. After all, a presentation cannot succeed on content alone. You need to pay attention to issues of organization, style, design, and delivery.

As you begin planning your talk, remember that presentations can be either formal or informal:

> **Formal presentations**—These presentations often include the use of a podium, speaking notes, and slides made with presentation software. Formal presentations include speeches, workshops, trainings, briefings, demonstrations, and panel discussions. They are made to clients, management, and colleagues.
>
> **Informal presentations**—Most presentations at work are informal. At monthly meetings, you will be asked to report on your team's progress. If you have a new idea, you will need to pitch it to your boss. And, if your supervisor stops by and says, "Hey, in 10 minutes, could you come by my office to tell the vice president how the project is going?" you are about to make an informal public presentation.

In both formal and informal presentations, solid planning is the key to success. A good way to start the planning phase is to analyze the rhetorical situation of your presentation. Begin by asking some strategic questions about your subject, purpose, and audience and the context in which you will be giving your talk.

GO TO THE NET For websites that discuss giving presentations, go to **www.pearsonhighered.com/johnsonweb4/21.1**

The Five-W and How Questions give you a good place to start.

Who will be in my audience?

What kind of information do the audience members need or want?

Why am I presenting this information to this audience?

Where will I be presenting?

When will I need to give my talk?

How should I present this information?

Your answers to these questions should help you start crafting your materials into something that will be interesting and informative to the audience.

Defining the Rhetorical Situation

Now that you have considered the Five-W and How Questions, you can think a little deeper about the situation in which you will be speaking (Figure 21.2). Consider closely your subject, purpose, audience, and the context of use.

SUBJECT Identify what the audience needs to know, putting emphasis on information they require to take action or make a decision. You might also ask yourself what the audience does *not* need to know. That way, you can keep your presentation concise and to the point.

Asking Questions About the Rhetorical Situation

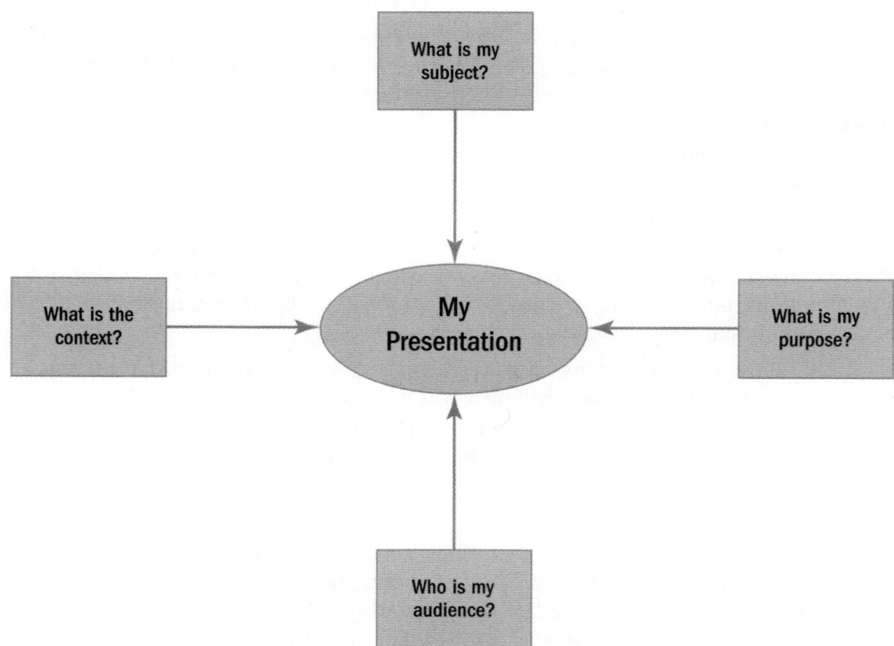

Figure 21.2: Asking questions about the subject, purpose, audience, and context will help you anticipate the information you need and how that information should be presented.

GO TO THE NET
To learn more about analyzing audiences, go to **www.pearsonhighered.com/johnsonweb4/21.2**

PURPOSE You need to know exactly what you want to achieve in your presentation. Try to state your purpose in one sentence.

> My goal is to persuade elected officials that global warming is a looming problem for our state.
>
> We need to demonstrate the G290 Robot Workstation to the CEO of Geocom Industries, showing her how it can be used to clean up toxic spills.
>
> I need to motivate our technical staff to improve quality so we can meet the standards demanded by our new client.

If you need more than one sentence to express your purpose, you are probably trying to do too much in your presentation.

Link

For more help defining your purpose, go to Chapter 1, page 6.

AUDIENCE Members of your audience will come to your presentation with various needs, values, and attitudes. You should anticipate these characteristics and shape your presentation to their specific requirements and interests.

Primary audience (action takers)—For most presentations, the primary audience is the most important, because it is made up of people who will be making a decision or taking action.

Secondary audience (advisors)—These members of the audience might advise the primary audience on what actions to take. They might be experts in your field or people who have information or opinions on your subject.

Tertiary audience (evaluators)—Others in the audience may have an interest in what you are saying. They might be journalists, lawyers, activists, or concerned citizens.

Gatekeepers (supervisors)—Your supervisors and others at your company will often need to see your presentation before you give it to an audience. They will be looking for accuracy and checking whether you are achieving your purpose and fulfilling the mission of the company.

Link

For more audience analysis techniques, go to Chapter 2, page 22.

CONTEXT OF USE Context is always important in technical communication, but it is especially important in public presentations. You need to be fully aware of the physical, economic, ethical, and political factors that will shape your presentation and how your audience will react to it.

Physical context—Take time to familiarize yourself with the room in which you will be speaking and the equipment you will be using. You will need to adjust your presentation, including your visuals, to the size, shape, and arrangement of the room. Also, find out what kind of furniture and equipment you will have available.

Will you be using a podium or lectern?

Will you be sitting behind a table or standing out in the open?

Will there be a microphone?

Will a projector be available, and will you be able to use it?

For worksheets to help you analyze your audience and the context of the presentation's use, go to **www.pearsonhighered.com/johnsonweb4/21.3**

Will you need to bring your own projector and computer?

Will other visual aids like whiteboards, flip charts, and large notepads be available?

Will the audience be eating, and will drinks be available?

When will the audience need breaks?

It is astonishing how many presentations fail because speakers are not prepared for the physical characteristics of the room. They show up with slides that can't be read because the room is too large, or they try to talk to a large audience with no public address system.

These sorts of problems might be someone else's fault, but it's your presentation. You will be the person who looks unprepared. Proper preparation for the physical context will help you avoid these problems.

Link

For more information on ethics, go to Chapter 4, page 69.

Economic context—As always, money will be a central concern for your audience members. So, consider the microeconomic and macroeconomic factors that will influence how they will receive your presentation. Microeconomic issues might include budgetary concerns or constraints. Macroeconomic issues might include your audience's economic status, economic trends in the industry, or the state of the local or national economy.

Ethical context—Presentations almost always touch on ethical issues in one way or another. As you prepare, identify and consider any rights, laws, or issues of common concern that might shape your presentation.

Link

To learn more about analyzing context of use, go to Chapter 2, page 26.

Political context—Politics also play a role in presentations. In some cases, politics might simply involve the usual office politics that shape how people react to you and your subject. In other situations, larger political issues may come into play. National issues (energy policies, conservation, gun rights, privacy, security, etc.) will evoke different political responses among members of your audience.

Allotting Your Time

As the speaker, you have an unstated "contract" with the audience. According to this contract, your audience is allowing you a specific number of minutes. It's your responsibility to fill that time productively—*and not go over that amount of time.* Nothing annoys an audience more than a speaker who runs past the time allotted.

So, as you are planning your presentation, first determine how much total time you have to speak. Then, *scale* your presentation to fit the allotted time. Figure 21.3 shows how a few common time periods might be properly budgeted.

Of course, if you have fewer or more than four topics, you should make adjustments to the times allowed for each. Also, you might need to spend more time on one topic than another. If so, adjust your times accordingly.

There are two things you should notice about the times listed in Figure 21.3.

- Longer presentations do not necessarily allow you to include substantially longer introductions and conclusions. No matter how long your presentation is scheduled to run, keep your introductions and conclusions concise.

GO TO THE NET

For more time management strategies, see **www.pearsonhighered.com/johnsonweb4/21.4**

Allotting Your Presentation Time

	15-Minute Presentation	30-Minute Presentation	45-Minute Presentation	One-Hour Presentation
Introduction	1 minute	1–2 minutes	2 minutes	2–3 minutes
Topic 1	2 minutes	5 minutes	8 minutes	10 minutes
Topic 2	2 minutes	5 minutes	8 minutes	10 minutes
Topic 3	2 minutes	5 minutes	8 minutes	10 minutes
Topic 4	2 minutes	5 minutes	8 minutes	10 minutes
Conclusion	1 minute	1–2 minutes	2 minutes	2 minutes
Questions	2 minutes	5 minutes	5 minutes	10 minutes

Figure 21.3: When planning your presentation, carve up your time carefully to avoid going over the total allotted time.

- You should not budget all the time available. Always leave yourself some extra time in case something happens during your talk or you are interrupted.

In the end, the unstated contract with your audience is that you will finish within the time scheduled. If you finish a few minutes early, you won't hear any complaints. However, if you run late, your audience will not be pleased.

Choosing the Right Presentation Technology

As you plan your presentation, it is a good idea to think about what presentation technology you will use for your talk. Will you use presentation software with a digital projector? Are you going to use a whiteboard? Are you going to make transparencies for an overhead projector? The kind of presentation technology you need depends on the type of presentation you will be making (Figure 21.4).

Fortunately, presentation software like PowerPoint, Presentations, and Keynote makes it easy to create slides and use graphics in your presentations. These programs will help you create visually interesting presentations for a variety of situations.

Each kind of visual aid offers specific advantages and disadvantages. Here are some pros and cons of the more common types of visuals:

Digital projector with a computer—Most companies have a digital projector available for your use. The projector can display the slides from your computer screen onto a large screen. The advantages of digital projectors are their ease of use and their ability to create highly attractive, colorful presentations. The disadvantage is that the projected slides often dominate the room because the lights need to be turned down. As a result, the audience can become fixated on the slides and stop listening to what you are saying.

Overhead projector with transparencies—The overhead projector is the tried-and-true method for giving presentations. You can use an overhead projector

GO TO THE NET

To find a selection of presentation software packages, go to **www.pearsonhighered.com/johnsonweb4/21.5**

Presentation Technologies

Type of presentation	Visuals
Presentation to a group of more than 10 people	Digital projector with computer
	Overhead projector with transparencies
	35-mm slide projector
	Whiteboard or chalkboard
Presentation to a group of fewer than 10 people	Digital projector with computer
	Overhead projector with transparencies
	35-mm slide projector
	Flip charts
	Large notepads
	Digital video on TV monitor (DVD or CD-ROM)
	Posters
	Handouts
	Computer screen
	Whiteboard or chalkboard

Figure 21.4: There are many different ways to present materials. You should choose the one that best fits your subject and audience.

to project slides onto a large screen. Transparencies for the projector can be made on a paper copier. The advantages of overhead projectors are that they are commonly available in workplaces and they are more reliable than digital projectors. Plus, you can use a marker to write on transparencies as you interact with the audience. The disadvantage is that presentations using overheads often seem more static and lifeless than ones made with digital projectors. The colors are not as sharp, and the pictures can be blurry.

Whiteboard, chalkboard, or large notepad—People often forget about the possibility of using a whiteboard, chalkboard, or large notepad in a room. But if you are giving a presentation that requires interaction with the audience, you can use these items to make visuals on the fly. The advantage is that you can create your visuals in front of the audience. Your listeners won't feel that they are receiving a canned presentation in which they have little input. The disadvantage is that you need to think on your feet. You need to find ways to translate your and the audience's comments into visuals on the board.

Flip charts—For small, more personal presentations, a flip chart is a helpful tool. As the speaker talks, he or she flips a new page forward or behind with each new topic. The advantage of flip charts is their closeness to the audience. You can give a flip chart presentation to a small group. The disadvantage of flip charts is that they are too small to be seen from a distance. If you have more than a handful of people in the audience, the flip chart won't work.

GO TO THE NET

Want to know more about presentation technologies? Go to **www.pearsonhighered.com/johnsonweb4/21.6**

Using a Digital Projector

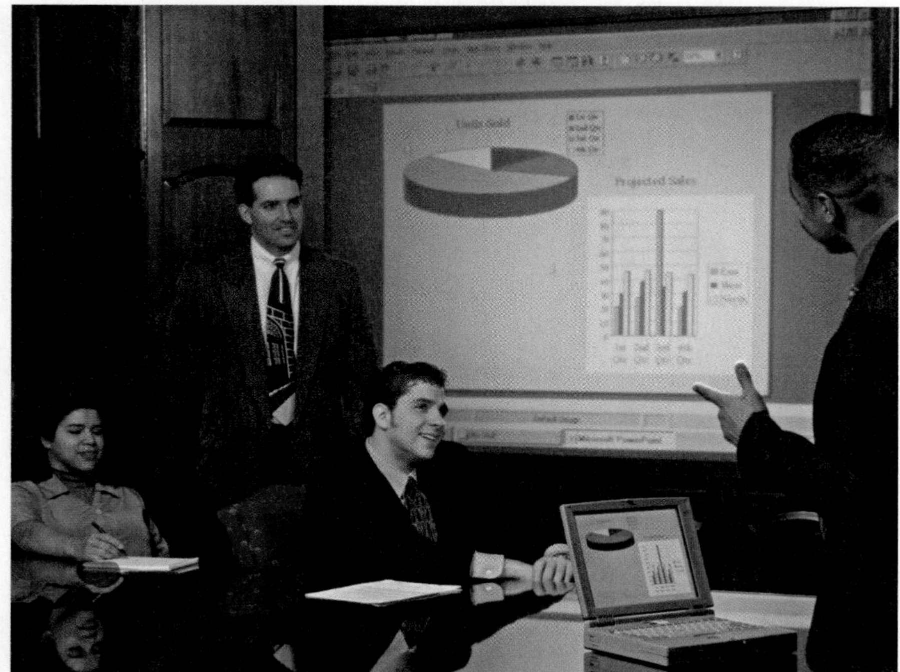

Digital projectors are increasingly common. They project your computer screen onto a large screen.

Posters—In some cases, you might be asked to present a poster. A poster usually includes about five to seven slides that describe a product or show a procedure. The slides are often used to summarize an experiment. The advantage of a poster is that everything covered in the talk is visually available to the audience. In some cases, the poster might be left alone on a display or a wall so readers can inspect it on their own time. Posters, however, have the same disadvantage as flip charts. They can be used only in presentations to a handful of people. They are too hard to see from a distance.

Handouts—Handouts can be helpful in some cases, but they are not always appropriate. When used properly, they can reinforce points made in your presentation or provide data that won't be visible with a projector. Also, handouts made with presentation software can be formatted to leave room for note taking. Handouts, though, can also be very distracting. In a large room, handouts can take a few minutes to be passed around, causing the speaker to lose momentum.

These kinds of technology decisions are best made up front—while you are planning—because your choice of technology will often shape your decisions about the content, organization, style, and design/delivery of your information.

GO TO THE NET

To see sample poster presentations, go to **www.pearsonhighered.com/johnsonweb4/21.7**

Using a Flip Chart

Flip charts are low tech, but they are especially effective in presentations to a few people.

Organizing the Content of Your Presentation

One problem with organizing public presentations is that you usually end up collecting more information than you can talk about in the time allowed. Of course, you cannot tell the audience everything, so you need to make some hard decisions about what they need to know and how you should organize that information.

Link

To help you collect content for your presentation, see Chapter 15, page 408.

Keep your purpose and audience foremost in your mind as you make decisions about what kind of content you will put in the presentation. You want to include only need-to-know information and cut out any want-to-tell information that is not relevant to your purpose or audience. As you make decisions about what to include or cut, you should keep the following in mind: *The more you say, the more they will forget.*

Building the Presentation

There is an old adage about public presentations that has been used successfully for years: *Tell them what you're going to tell them. Tell them. Tell them what you told them.*

Remember Your Audience

Always keep your audience in mind. Most audiences prefer a concise, to-the-point presentation.

In other words, like any document, your presentation should have a beginning (introduction), a middle (body), and an end (conclusion) (Figure 21.5).

The Introduction: Tell Them What You're Going to Tell Them

The beginning of your presentation is absolutely crucial. If you don't catch the audience's attention in the first few minutes, there is a good chance they will tune out and not listen to the rest of your presentation. Your introduction slide(s) for the presentation should present at least the subject, purpose, and main point of your presentation (one slide). As shown in Figure 21.6, you might also use a second slide to forecast the structure of the talk.

Like the introduction to a document, your presentation should begin by making up to six moves:

MOVE 1: DEFINE THE SUBJECT Make sure the audience clearly understands the subject of your presentation. You might want to use a *grabber* to introduce your

GO TO THE NET

Want more ideas on how to start a presentation? Go to **www.pearsonhighered.com/johnsonweb4/21.8**

Basic Pattern for a Presentation

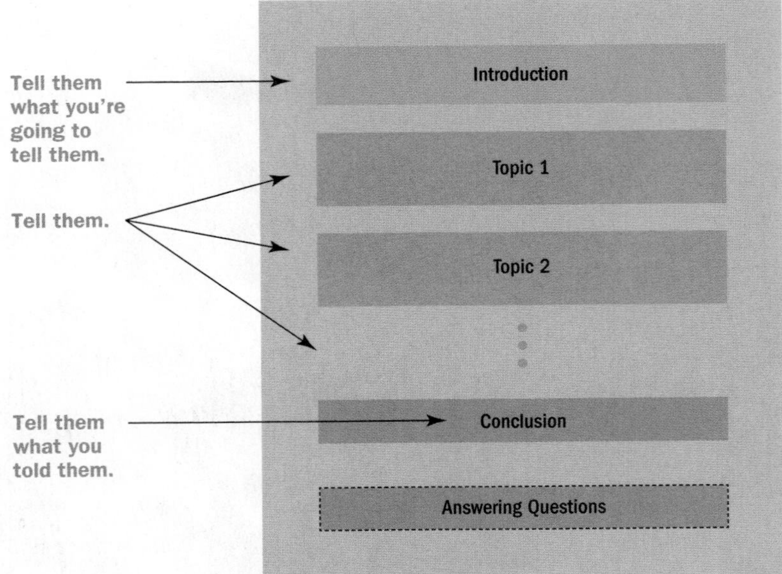

Figure 21.5: A presentation has a beginning, a middle, and an end. Usually, time is left for questions at the end.

subject. A grabber states something interesting or challenging to capture the audience's attention. Some effective grabbers include the following:

A rhetorical question—"Have you ever thought about changing your career and becoming a professional chef?"

A startling statistic—"A recent survey shows that 73 percent of children aged 15 to 18 in the Braynard area have tried marijuana. Almost a third of Braynard teens are regular users of drugs."

A compelling statement—"Unless we begin to do something about global warming soon, we will see dramatic changes in our ecology within a couple of decades."

An anecdote—"A few years ago, I walked into a computer store, only to find the place empty. I looked around and didn't find a salesperson anywhere. Now, I've always been an honest person, but it did occur to me that I could pocket thousands of dollars of merchandise without being caught."

A quotation—"William James, the famous American philosopher, once said, 'Many people believe they are thinking, but they are merely rearranging their prejudices.'"

GO TO THE NET

Need a good grabber? Go to **www.pearsonhighered.com/johnsonweb4/21.9**

Introduction Slides for a Presentation

Introduction
• What Is Climate Change?
• When Did Our Climate Begin to Change?
• Why Is Climate Change Especially Important to People Living in the West?
• Answer: Our Ecosystem Is Too Fragile to Adjust to These Changes
Define the subject.
Offer background information.
Stress the importance of the subject to the audience.
State the purpose and main point of the presentation.

Today's Presentation
• Causes of Climate Change
• Effects of Climate Change in the West
• Solutions to This Problem
• Recommendations
Forecast the structure of the presentation.

Figure 21.6: An introduction builds a context for the body of the presentation.

A show of hands—"How many of you think children watch too much violent television? Raise your hands." Follow this question with an interesting, startling statistic or compelling statement.

But where do you find grabbers? The Internet is a source of endless material for creating grabbers. There are plenty of reference websites like Bartleby.com or Infoplease.com that will help you find quotations, statistics, and anecdotes.

Otherwise, you can use search engines like Google.com, Ask.com, or Yahoo.com to find interesting information for grabbers. Type in your subject and a keyword or

How do you set the right tone at the start of your presentation? For answers, go to **www.pearsonhighered.com/johnsonweb4/21.10**

HELP

Giving Presentations with Your iPod, MP3, or Mobile Phone

One major hassle about giving presentations with digital projectors is lugging along the laptop that holds your presentation. Even if a laptop is provided with the projector, it's always risky to show up with only a flash drive, DVD, or CD that holds your presentation. After all, you never know whether the computer's hardware and/or software is compatible with your presentation until you arrive and plug it in. If something doesn't work, you're in trouble.

The solution? Why not use your iPod, MP3 player, or mobile phone to store and give your presentation? Your player or phone is portable (you were taking it anyway, right?) and you can hold it in your hand while you're talking. Also, depending on your player, you can add background music and video to your presentation and play it right through the projector or television (Figure A).

Using Your iPod or MP3 Player to Give a Presentation

Figure A: Your iPod or MP3 player is a lightweight way to transport your presentation. It also eliminates some of the problems of connecting laptops to projectors or televisions.

An added advantage is that you can go over your slides and practice your presentation any time without firing up that laptop or even bringing it along to campus or on your trip.

How can you turn your music player or phone into a presentation tool? You will need an iPod, MP3 player, or mobile phone that has color photo or video capability and an AV port (not just a headphone port). You will also need an AV cable that can be plugged into a projector or television. This kind of cable usually plugs into the place where your headphones are plugged in, and it has three connectors at the other end.

Then, create your presentation in PowerPoint, Keynote, or any other presentation software. When you're done, select "Save As" and choose jpeg (photographs) or PDF (files) format. (For iPods, there is also a free software package called iPresent It by ZappTek that makes the process even easier and allows more options for music and video. This software works on Windows and Mac OS.)

Then, download your files into your iPod, MP3 player, or phone as photographs or PDF files. You can then use "Settings" to add transitions and background music if you want.

When you are ready to give your presentation, plug your AV cord into the projector or television, matching the colors of the three connectors to the colors of the ports. Your presentation should appear on the screen. (If not, turn off the projector and turn it back on.)

Then, while you do your presentation, use the forward and back buttons to move through your slides as you talk.

phrase that sets a specific tone that you want your grabber to establish. The search engine will locate stories, quotes, statistics, and other information that you can use to create an interesting grabber.

MOVE 2: STATE THE PURPOSE OF YOUR PRESENTATION In public presentations, you can be as blunt as you like about what you are trying to achieve. Simply tell the audience your purpose up front.

> The purpose of this presentation is to prove to you that global warming is real, and it is having a serious impact on Nevada's ecology.

> In this demonstration, our aim is to show you that the G290 Robot is ideal for cleaning up toxic spills.

MOVE 3: STATE YOUR MAIN POINT Before moving into the body of your presentation, you should also state your main point. Your main point holds the presentation together, because it is the one major idea that you want the audience to take away from your talk.

> Global warming is a serious problem for our state, and it is growing worse quickly. By switching to nonpolluting forms of energy, we can do our part to minimize the damage to our ecosystem.

> The G290 Robot gives you a way to clean up toxic spills without exposing your hazmat team to dangerous chemical or nuclear materials.

MOVE 4: STRESS THE IMPORTANCE OF THE SUBJECT TO THE AUDIENCE At the beginning of any presentation, each of the audience members wants to know, "Why is this important to me?" So, tell them up front why your subject is important.

> Scientists predict that the earth's overall temperature will minimally rise a few degrees in the next 30 years. It might even rise 10 degrees. If they are correct, we are likely to see major ecological change on this planet. Oceans will likely rise a foot as the polar ice caps melt. We will also see an increase in the severity of storms like hurricanes and tornadoes. Here in Nevada, we will watch many of our deserts simply die and blow away.

> OSHA regulations require minimal human contact with hazardous materials. This is especially true with toxic spills. As you know, OSHA has aggressively gone after companies that expose their employees to toxic spills. To avoid these lawsuits and penalties, many companies are letting robot workstations do the dirty work.

AT A GLANCE

Opening Moves in a Presentation

- Define the subject.
- State the purpose of your presentation.
- State your main point.
- Stress the importance of the subject to the audience.
- Offer background information on the subject.
- Forecast the structure of the presentation.

MOVE 5: OFFER BACKGROUND INFORMATION ON THE SUBJECT Providing background information on your subject is a good way to build a framework for the audience to understand what you are going to say. You can give the audience a little history on the subject or perhaps tell about your relationship with it.

MOVE 6: FORECAST THE STRUCTURE OF THE PRESENTATION In your introduction, tell the audience how you have organized the rest of the presentation. If you are going to cover four topics, say something like,

> In this presentation, I will be going over four issues. First, I will discuss . . . . Second, I will take a look at . . . . Third, I will identify some key objectives that . . . . And finally, I will offer some recommendations for . . . .

Forecasting gives your audience a mental framework to follow your presentation. A major advantage to forecasting is that it helps the audience pay attention. If you say up front that you will be discussing four topics, the audience will always know where you are in the presentation.

The Body: Tell Them

The body of your presentation is where you are going to do the heavy lifting. Start out by dividing your subject into two to five major topics that you want to discuss.

Experience and research show that people can usually remember only five to seven items comfortably. So, if a presentation goes beyond five topics, the audience will start feeling overwhelmed or restless. If you have more than five topics, try to consolidate smaller topics into larger ones.

If you have already written a document, you might follow its organizational structure. If you are starting from scratch, you can follow some of the basic organizational patterns listed below.

PROBLEM, NEED, SOLUTION This pattern is most effective for proposing new ideas. After your introduction, offer a clear definition of the problem or opportunity you are discussing. Then, specify what is needed to solve the problem or take advantage of the opportunity. Finally, offer a solution/plan that achieves the objective.

CHRONOLOGICAL When organizing material chronologically, divide the subject into two to five major time periods. Then, lead the audience through these time periods, discussing the relevant issues involved in each. In some cases, a three-part *past-present-future* pattern is a good way to organize a presentation.

SPATIAL You might be asked to explain or demonstrate visual spaces, like building plans, organizational structures, or diagrams. In these cases, divide the subject into

Karen Paone

OWNER AND LEAD CONSULTANT, PAONE & ASSOCIATES, ALBUQUERQUE, NEW MEXICO

Paone & Associates is a communication consulting firm that trains employees to make effective presentations.

How can I overcome my fear of speaking in public?

The fear of public speaking routinely ranks ahead of death among the top three fears of most adults. Indeed, there's nothing quite as horrifying as the rush of adrenaline, the rising sense of panic in the pit of your stomach, increased heart rate, dry mouth, and complete loss of memory that assault you the very moment you realize your turn at the podium is imminent.

So what's the difference between these terrifying sensations and the energy, enthusiasm, and engaging style of a polished public speaker? In fact, there is very little difference between the physical reaction of an inexperienced individual presenting at a conference for the first time and an engaged, seasoned presenter. Both experience a heightened sense of awareness and both feel a strong sensation in the solar plexus. Why is one likely to succeed and the other to fail at presenting the information at hand?

- **Rehearsal**—The seasoned speaker creates a realistic rehearsal situation in which she either enlists a practice audience or visualizes her anticipated audience. She knows that simply thinking about the presentation is insufficient to prepare for a real audience. She stands in front of a practice audience or a mirror and runs the entire presentation at least three times. She reworks the rough spots until the entire presentation flows smoothly, anticipates questions, and practices responding with the answers.
- **Heightened awareness**—The seasoned presenter recognizes that the rush of adrenaline is very normal, and that a heightened sense of awareness is essential for an interesting presentation. She uses deep breathing, stretching, and smiling to focus that energy into a confident and enthusiastic stage presence.
- **Communication and connection**—Finally, the seasoned presenter knows that the secret to managing stage fright is focusing on communicating the content at hand in a sincere and personal way to each and every member of the audience. By focusing on connecting with each listener, the presenter's concern is with the audience rather than the way she or he looks, sounds, or feels.

The prescription for stage fright is to rehearse until you can deliver your presentation smoothly from start to finish. Recognize that the physical response to fear and excitement is essentially the same. What differs is how you use those sensations. And finally, when all of your energy is focused on connecting with your audience rather than observing yourself, you too will be perceived as a polished and successful public speaker.

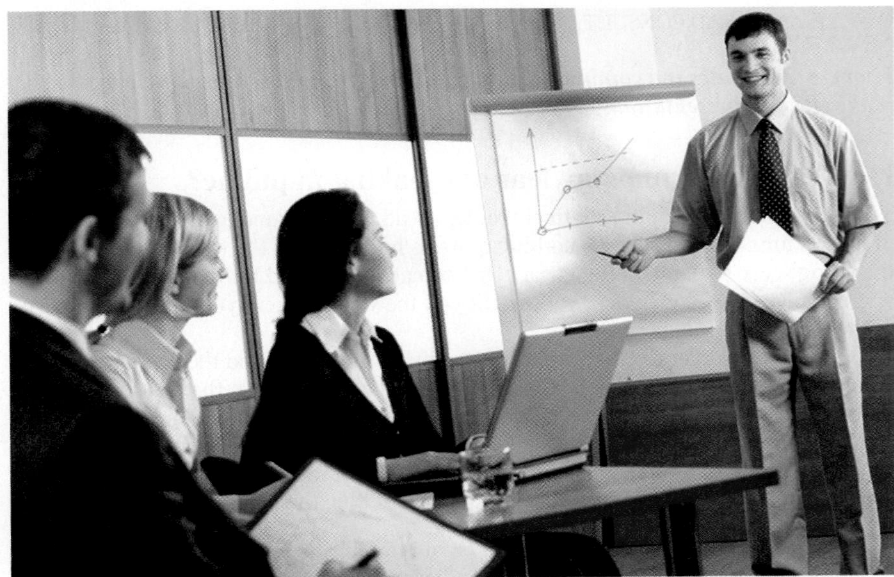

Keep your presentation to two to five major points. You risk losing the audience if you try to cover more than five points.

two to five zones. Then, walk the audience through these zones, showing each zone individually and discussing how it relates to the zones around it.

NARRATIVE Audiences always like stories, so you might organize your presentation around the narrative pattern. Narratives typically (1) set a scene, (2) introduce a complication, (3) evaluate the complication, (4) resolve the complication, and (5) explain what was learned from the experience.

Link

For more information on presenting research, go to Chapter 10, page 303.

METHODS, RESULTS, DISCUSSION This pattern is commonly used to present the results of research. This pattern (1) describes the research plan or methodology, (2) presents the results of the study, (3) discusses and interprets the results, and (4) makes recommendations.

CAUSES AND EFFECTS This pattern is common for problem solving. Begin the body of the presentation by discussing the causes for the current situation. Then, later in the body, discuss the effects of these causes and their likely outcomes. You can also alternate between causes and effects. In other words, discuss a cause and its effect together. Then discuss another cause and its effect, and so on.

Link

To learn more about describing products or processes, turn to Chapter 6, page 146.

DESCRIPTION BY FEATURES OR FUNCTIONS If you are demonstrating a product or process, divide your subject into its two to five major features or functions. Then, as you discuss each of these major features/functions, you can discuss the minor features/functions that are related to them.

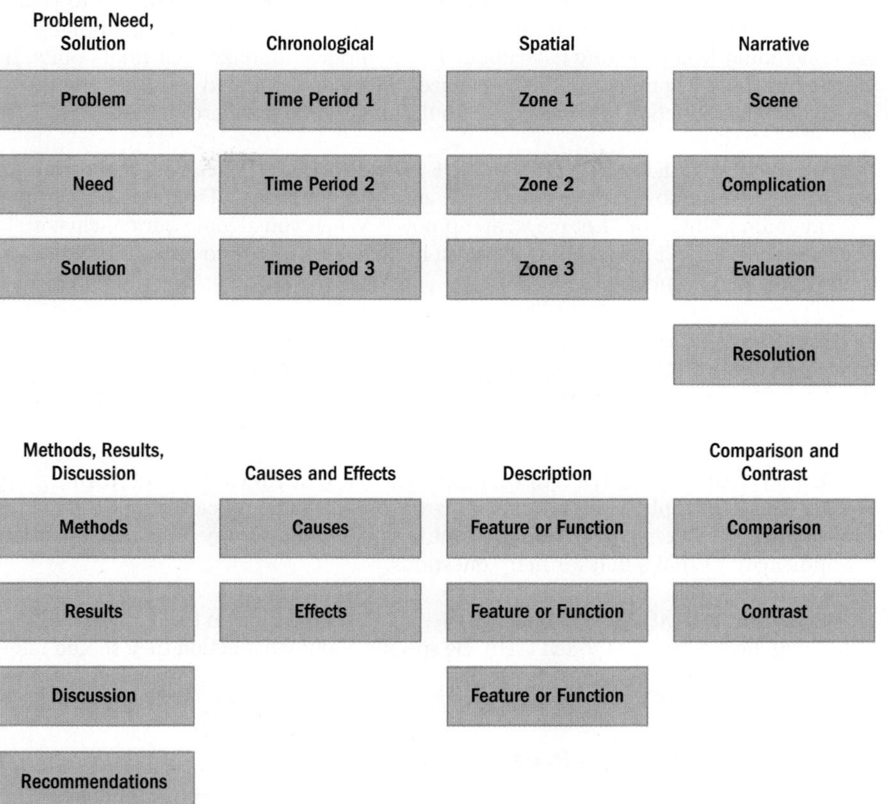

Figure 21.7: These common patterns will help you organize the body of your presentation.

COMPARISON AND CONTRAST Usually this pattern is followed when the speaker is comparing something new or unfamiliar with something that the audience knows well. Choose two to five major points on which these two things can be compared and contrasted. Then, compare and contrast them point by point.

There are countless patterns available for organizing the body of your presentation. The ones shown in Figure 21.7 are some of the most common in technical communication. These patterns are not formulas to be followed in lockstep. Rather, they can be manipulated to fit a variety of speaking situations.

Link

For more ideas about organizing body information, go to Chapter 16, page 433.

The Conclusion: Tell Them What You Told Them

The conclusion is often the most important part of any presentation, and yet speakers consistently make mistakes at the ends of their talks. They often end the presentation by shrugging their shoulders and saying, "Well, that's all I have to say. Any questions?"

GO TO THE NET

Not sure how to end your presentation? For ideas, go to **www.pearsonhighered.com/johnsonweb4/21.11**

Your conclusion needs to do much more. Specifically, you want to summarize your key points, while leaving the people in your audience in a position to say *yes* to your ideas. But you don't have much time to do all these things. Once you signal that you are concluding, you probably have about 1 to 3 minutes to make your final points. If you go beyond a few minutes, your audience will become agitated and frustrated.

Like the introduction, a conclusion should make some standard moves.

MOVE 1: SIGNAL CLEARLY THAT YOU ARE CONCLUDING When you begin your conclusion, use an obvious transition such as, "In conclusion," "Finally," "To summarize my main points," or "Let me wrap up now." When you signal your conclusion, your audience will sit up and pay attention because they know you are going to tell them your main points.

MOVE 2: RESTATE YOUR KEY POINTS Summarize your key points for the audience, including your overall main point (Figure 21.8). Minimally, you can simply list them and go over them one last time. That way, if your audience remembers anything about your presentation, it will be these most important items.

MOVE 3: RE-EMPHASIZE THE IMPORTANCE OF YOUR SUBJECT TO THE AUDIENCE Tell the people in your audience again why they should care about this subject. Don't tell them why it is important to you—that's assumed. Instead, answer the audience's "What's in it for me?" questions.

MOVE 4: CALL THE AUDIENCE TO ACTION If you want people in the audience to do something, here is the time to tell them. Be specific about what action they should take.

Conclusion Slide for a Presentation

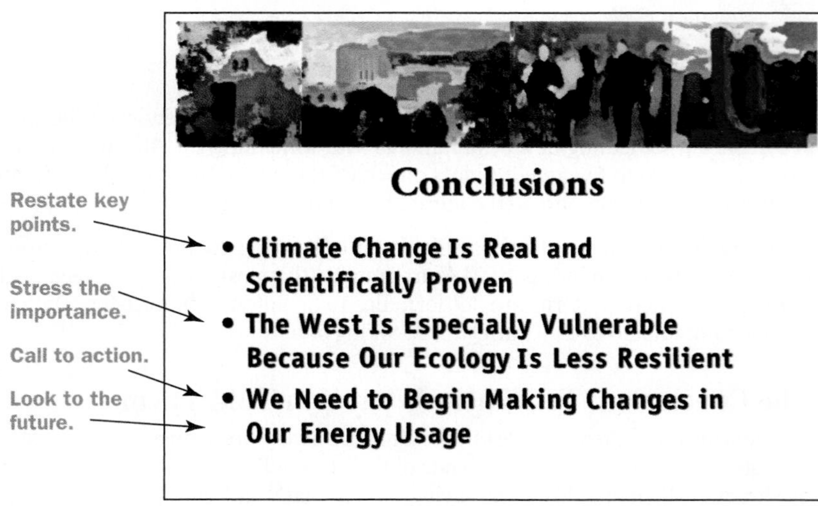

Figure 21.8: The conclusion slide should drive home your main points and look to the future.

AT A GLANCE

Concluding Moves in a Presentation

- Signal that you are concluding.
- Restate your key points.
- Re-emphasize the importance of the subject.
- Call the audience to action.
- Look to the future.
- Say thank you.
- Ask for questions.

MOVE 5: LOOK TO THE FUTURE Briefly, offer a vision of the future, usually a positive one, that will result if they agree with your ideas.

MOVE 6: SAY THANK YOU At the end of your presentation, don't forget to thank the audience. Saying, "Thank you" signals that you are really finished. Often, it will also signal the audience to applaud, which is always a nice way for a presentation to end.

MOVE 7: ASK FOR QUESTIONS Once the audience has stopped applauding, you can ask for questions.

Preparing to Answer Questions

While preparing and researching your presentation, you should spend some time anticipating the kinds of questions you might be asked. Questions are an opportunity to interact with the audience and clarify your ideas. You will generally be asked three types of questions: elaboration, hostile, and heckling.

THE ELABORATION OR CLARIFICATION QUESTION Members of the audience might ask you to expand on your ideas or explain some of your concepts. These questions offer you an opportunity to reinforce some of your key points. You should not feel defensive or threatened by these questions. The person asking the question is really giving you a chance to restate some of your main points or views.

When you receive one of these kinds of questions, start out by rephrasing the question for the audience. For example, "The question is whether global warming will have an impact that we can actually observe with our own eyes." Your rephrasing of the question will allow you to shift the question into your own words, making it easier to answer.

Then, offer more information or reinforce a main point. For instance, "The answer is 'yes.' Long-time desert residents are already reporting that desert plants and animals are beginning to die off. One example is the desert willow. . . ."

THE HOSTILE QUESTION Occasionally, an audience member will ask you a question that calls your ideas into doubt. For example, "I don't trust your results. Do you really expect us to believe that you achieved a precision of .0012 millimeters?"

Here is a good three-step method for deflecting these kinds of questions:

1. **Rephrase the question**—"The questioner is asking whether it is possible that we achieved a precision of .0012 millimeters."

2. **Validate the question**—"That's a good question, and I must admit that we were initially surprised that our experiment gave us this level of precision."

3. **Elaborate and move forward**—"We achieved this level of precision because . . . ."

You should allow a hostile questioner only one follow-up remark or question. After giving the hostile questioner this second opportunity, do not look at that person. If you look elsewhere in the room, someone else in the audience will usually raise his or her hand and bail you out.

For helpful question-and-answer strategies, go to **www.pearsonhighered.com/johnsonweb4/21.12**

Answering Questions

Prepare in advance for the kinds of questions you might be asked after your presentation.

THE HECKLING QUESTION In rare cases, a member of the audience will be there only to heckle you. He or she will ask rude questions or make blunt statements like, "I think this is the stupidest idea we've heard in 20 years."

In these situations, you need to recognize that the heckler is *trying* to sabotage your presentation and cause you to lose your cool. Don't let the heckler do that to you. You simply need to say something like, "I'm sorry you feel that way. Perhaps we can meet after the presentation to talk about your concerns." Usually, at this point, others in the audience will step forward to ask more constructive questions.

It is rare that a heckler will actually come to talk to you later. That's not why he or she was there in the first place. If a heckler does manage to dominate the question-and-answer period, simply end your presentation. Say something like,

> Well, we are out of time. Thank you for your time and attention. I will stick around in the room for more questions.

Then, step back from the podium or microphone and walk away. Find someone to shake hands with. Others who have questions can approach you one-on-one.

Choosing Your Presentation Style

Your speaking style is very important. In a presentation, you can use style to add flair to your information while gaining the audience's trust. Poor style, on the other hand, can bore the audience, annoy them, and even turn them against you.

GO TO THE NET Need to improve your style? Go to **www.pearsonhighered.com/johnsonweb4/21.13**

There are many ways to create an appropriate style for your presentation, but four techniques seem to work best for technical presentations. Each of these techniques will help you project a particular tone in your speaking.

DEVELOP A PERSONA In ancient Greek, the word *persona* meant "mask." So, as you consider your presentation style, think about the mask you want to wear in front of the audience. Then, step into this character. Put on the mask. You will find that choosing a persona will make you feel more comfortable, because you are playing a role, much like an actor plays a role. The people in the audience aren't seeing and judging *you.* They are seeing the mask you have chosen to wear for the presentation.

SET A THEME A theme is a consistent tone you want to establish in your presentation. The best way to set a theme is to decide which one word best characterizes how you want the audience to *feel* about your subject. Then, use logical mapping to find words and phrases associated with that feeling (Figure 21.9).

As you prepare your talk, use these words regularly. When you give your presentation, the audience will naturally pick up on your theme. For example, let's say you want your audience to be "concerned." Map out words and phrases related to this feeling. Then, as you prepare your presentation, weave these words and phrases into your speech. If you do it properly, your audience will feel your concern.

A Theme in a Presentation

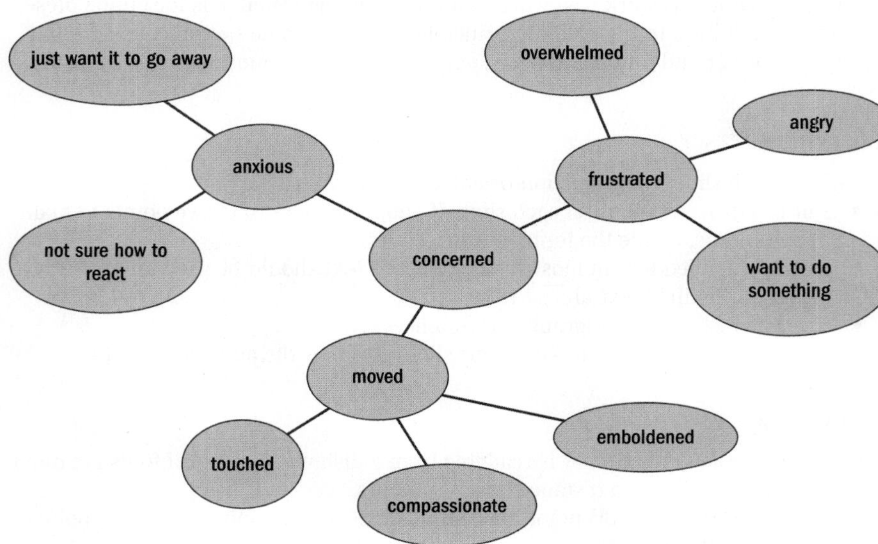

Figure 21.9: By mapping around a key term and seeding your speech with related words, you can set a theme that creates a specific feeling in your audience.

To see websites that offer advice for presenting well, go to **www.pearsonhighered.com/johnsonweb4/21.14**

AT A GLANCE

Creating Your Presentation Style

- Develop a persona.
- Set a theme.
- Show enthusiasm.
- KISS: Keep It Simple (Stupid).

SHOW ENTHUSIASM If you're not enthusiastic about your subject, your audience won't be either. So, get excited about what you have to say. Be intense. Get pumped up. Show that you are enthusiastic about this subject (even if you aren't) and the audience members will be too.

KISS: KEEP IT SIMPLE (STUPID) The KISS principle is always something to keep in mind when you are presenting, especially when it comes to style. Speak in plain, simple terms. Don't get bogged down in complex details and concepts.

In the end, good style is a choice. You may have come to believe that some people just have good style and others don't. That's not true. In reality, people with good style are just more conscious of the style they want to project.

Creating Visuals

In this visual age, you are really taking a risk if you try to present information without visuals. People not only want visuals, they *need* them to fully understand your ideas (Munter & Russell, 2011).

Designing Visual Aids

One of the better ways to design visual aids is to use the presentation software (PowerPoint, Keynote, or Presentations) that probably came bundled with your word-processing software. These programs are rather simple to use, and they can help you quickly create the visuals for a presentation. They also generally ensure that your slides will be well designed and readable from a distance.

The design principles discussed in Chapter 18 (balance, alignment, grouping, consistency, and contrast) work well when you are designing visual aids for public presentations. In addition to these design principles, here are some special considerations concerning format and font choices that you should keep in mind as you are creating your visuals:

FORMAT CHOICES

- Title each slide with an action-oriented heading.
- Put five or fewer items on each slide. If you have more than five points to make about a topic, divide the topic into two slides.
- Use left-justified text in most cases. Centered text should be used infrequently and right-justified text almost never.
- Use lists instead of paragraphs or sentences.
- Use icons and graphics to keep your slides fresh for the audience.

FONT CHOICES

- Use a simple typeface that is readable from a distance. Sans serif fonts are often more readable from a distance than serif fonts.
- Use a minimum of a 36-point font for headings and a minimum of a 24-point font for body text.

For advice about creating visuals, go to **www.pearsonhighered.com/johnsonweb4/21.15**

Sample Slides from a Presentation

Cover Slide

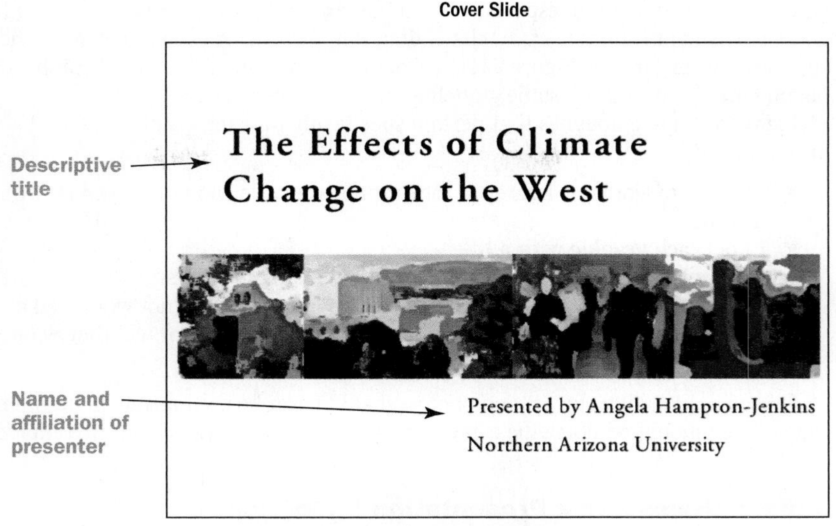

Descriptive title
The Effects of Climate Change on the West
Name and affiliation of presenter
Presented by Angela Hampton-Jenkins
Northern Arizona University

Slide from Body of Presentation

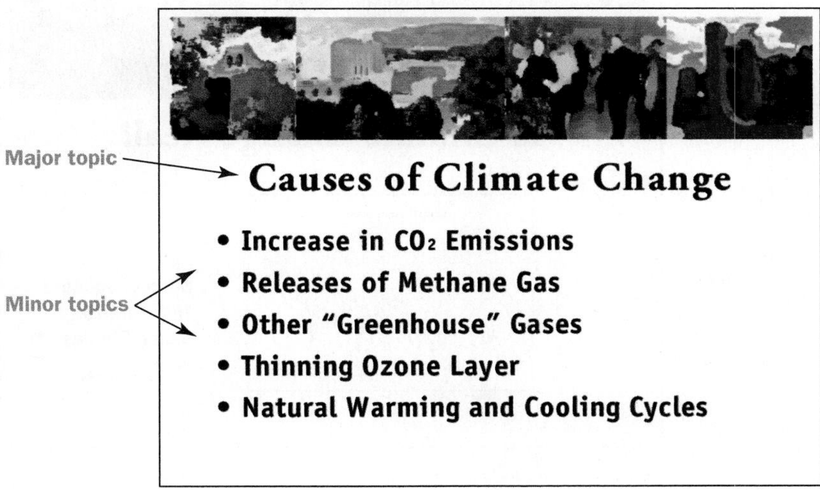

Major topic
Causes of Climate Change
Minor topics
• Increase in CO_2 Emissions
• Releases of Methane Gas
• Other "Greenhouse" Gases
• Thinning Ozone Layer
• Natural Warming and Cooling Cycles

Figure 21.10: These two slides show good balance, simplicity, and consistency. Keep your slides simple so that the audience doesn't have to work too hard to read them.

- Use color to keep slides interesting and to improve retention.
- Do not use ALL UPPERCASE letters because they are hard to read from a distance.

Overall, it is best to keep your slides as simple as possible (Figure 21.10). After all, if your audience needs to puzzle through your complex slides, they probably won't be listening to you.

Using Graphics

Graphics are also helpful, especially when you are trying to describe something to the audience. An appropriate graph, chart, diagram, picture, or even a movie will help support your argument (Figure 21.11). Chapter 19 discusses the use of graphics in documents. Most of those same guidelines apply to presentations.

Link

For more information on creating and using graphics, go to Chapter 19, page 523.

Here are some guidelines that pertain specifically to using graphics in a presentation:

- Make sure words or figures in the graphic are large enough to be read from a distance.
- Label each graphic with a title.
- Keep graphics uncomplicated and limited to simple points.
- Keep tables small and simple. Large tables full of data do not work well as visuals because the audience will not be able to read them—nor will they want to.
- Use clip art or photos to add life to your slides.

Graphics, including clip art and photos, should never be used merely to decorate your slides. They should reinforce the content, organization, and style of your presentation.

Using a Graphic on a Presentation Slide

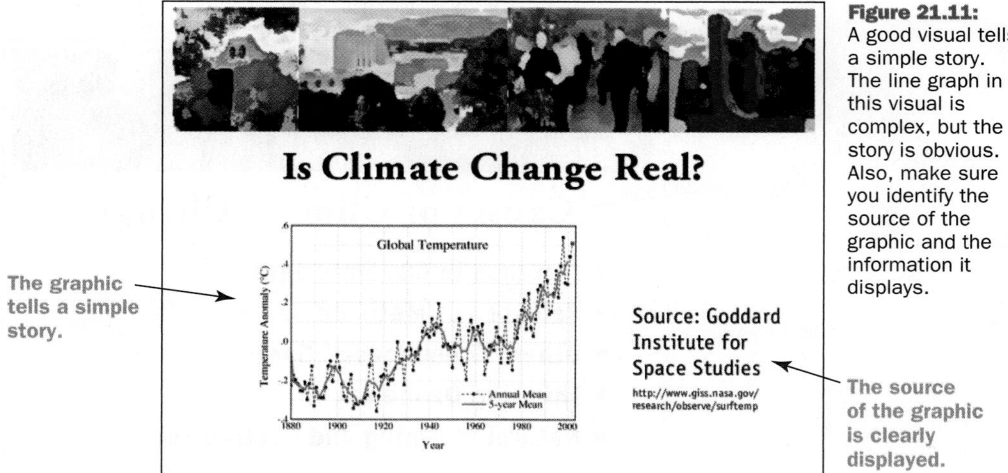

Figure 21.11: A good visual tells a simple story. The line graph in this visual is complex, but the story is obvious. Also, make sure you identify the source of the graphic and the information it displays.

Slides to Avoid

All of us have been to a presentation in which the speaker used ineffective slides. He or she put up a transparency with a 12-point type font and minimal design (Figure 21.12). Or, the speaker put up a table or graph that was completely indecipherable because the font was too small or the graphic was too complex.

These kinds of slides are nothing short of painful. The only thing the audience wants from such a slide is for the speaker to remove it—as soon as possible. Always

An Ineffective Slide

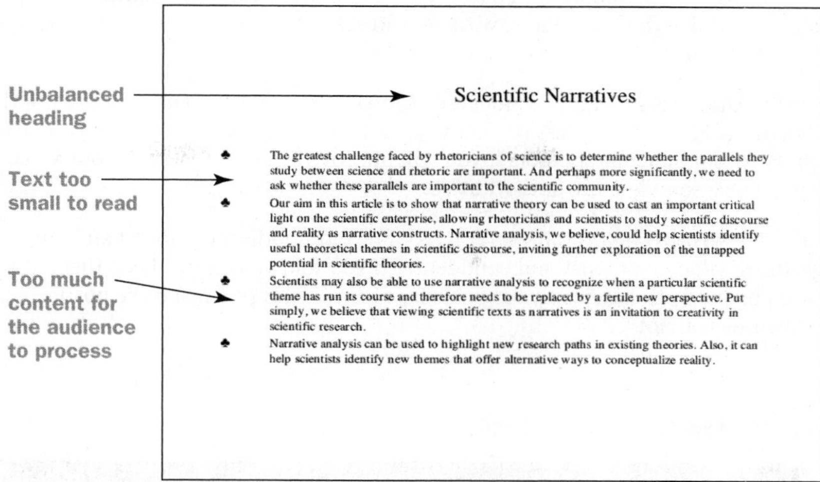

Figure 21.12: An ineffective slide often says a great amount—if the audience can read it.

remember that we live in a visual culture. People are sensitive to bad design. So take the time to properly create slides that enhance your presentation, not detract from it.

Delivering the Presentation

Why do people go to presentations, especially if a printed or screen version of the talk is available?

People attend presentations because they want to see you perform the material. They want to see how you act and interact with them. They want you to put a human face on the material. For this reason, you should pay close attention to your delivery so the audience receives a satisfying performance.

The usual advice is to "be yourself" when you are presenting. Of course, that's good advice if you are comfortable talking to an audience. Better advice is to "be the person the audience expects." In other words, like an actor, play the role that seems to fit your material and your audience.

Body Language

The audience will pay close attention to your body language. So use your body to reflect and highlight the content of your talk.

DRESS APPROPRIATELY How you dress should reflect the content and importance of your presentation. A good rule of thumb is to dress a level better than how you expect your audience to dress. For example, if the audience will be in casual attire, dress a little more formally. A female speaker might wear a blouse and dress pants or a nice skirt. Men might wear a shirt, tie, and dress pants. If the audience will be wearing suits and "power" dresses, you will need to wear an even *nicer* suit or an even *better* power dress.

GO TO THE NET Want more advice on delivering presentations? Go to **www.pearsonhighered.com/johnsonweb4/21.16**

STAND UP STRAIGHT When people are nervous, they have a tendency to slouch, lean, or rock back and forth. To avoid these problems when you speak, keep your feet squarely under your shoulders, with your knees slightly bent. Keep your shoulders back and your head up.

DROP YOUR SHOULDERS Under stress, people also have a tendency to raise their shoulders. Raised shoulders restrict your airflow and make the pitch of your voice go up. By dropping your shoulders, you will improve airflow and lower your voice. A lower voice sounds more authoritative.

USE OPEN HAND AND ARM GESTURES For most audiences, open hand and arm gestures will convey trust and confidence. If you fold your arms, keep them at your sides, or put both hands in your pockets, you will convey a defensive posture that audiences will not trust.

Delivering the Presentation

You can use your body and hands to highlight important parts of your presentation.

MAKE EYE CONTACT Everyone in the audience should believe that you made eye contact with him or her at least once during your presentation. As you are presenting, make it a point to look at all parts of the room at least once. If you are nervous about making eye contact, look at audience members' foreheads instead. They will think you are looking them directly in the eye.

There are exceptions to these generally accepted guidelines about gestures and eye contact. In some cultures, like some Native American cultures, open gestures and eye contact might be considered rude and even threatening. If you are speaking to an unfamiliar audience, find out which gestures and forms of eye contact are appropriate for that audience.

MOVE AROUND THE STAGE If possible, when you make important points, step toward the audience. When you make transitions in your presentation from one topic to the next, move to the left or right on the stage. Your movement across the stage will highlight the transition.

POSITION YOUR HANDS APPROPRIATELY Nervous speakers often strike a defensive pose by using their hands to cover specific parts of their bodies (perhaps you can guess which parts). Keep your hands away from these areas.

Voice, Rhythm, and Tone

A good rule of thumb about voice, rhythm, and tone is to *speak lower and slower than you think you should.*

Why lower and slower? When you are presenting, you need to speak louder than normal. As your volume goes up, the pitch of your voice will go up. So your voice will seem unnaturally high (even shrill) to the audience. By consciously lowering your voice, you should sound just about right to the audience.

Meanwhile, nervousness usually causes you to speak faster than you normally would. By consciously slowing down, you will sound more comfortable and more like yourself.

USE PAUSES TO HIGHLIGHT MAIN POINTS When you make an important point, pause for a moment. Your pause will signal to audience members that you just made an important point that you want them to consider and retain.

USE PAUSES TO ELIMINATE VERBAL TICS Verbal tics like "um," "ah," "like," "you know," "OK?" and "See what I mean?" are simply nervous habits that are intended to fill gaps between thoughts. If you have problems with a verbal tic (who doesn't?), train yourself to pause when you feel like using one of these sounds or phrases. Before long, you will find them disappearing from your speech altogether.

Using Your Notes

The best presentations are the ones that don't require notes. Notes on paper or index cards are fine, but you need to be careful not to keep looking at them. Nervousness will often lead you to keep glancing at them instead of looking at the audience. Some speakers even get stuck looking at their notes, glancing up only rarely at the audience. Looking down at your notes makes it difficult for the audience to hear you. You want to keep your head up at all times.

The following are some guidelines for making and using notes.

USE YOUR SLIDES AS MEMORY TOOLS You should know your subject inside out. So you likely don't need notes at all. Practice rehearsing your presentation with your slides alone. Eventually, you will be able to completely dispense with your written notes and work solely off your visual aids while you are speaking.

Need help improving your voice, rhythm, and tone? Go to **www.pearsonhighered.com/johnsonweb4/21.17**

TALK TO THE AUDIENCE, NOT TO YOUR NOTES OR THE SCREEN Make sure you are always talking to the audience. It is sometimes tempting to begin looking at your notes while talking. Or, in some cases, presenters end up talking to the screen. You should only steal quick glances at your notes or the screen. Look at the audience instead.

PUT WRITTEN NOTES IN A LARGE FONT If you need to use notes, print them in a large font on the top half of a piece of paper. That way, you can quickly find needed information without effort. Putting the notes on the top half of the paper means you won't need to glance at the bottom of the piece of paper, restricting airflow and taking your eyes away from the audience.

USE THE NOTES VIEW FEATURE IN YOUR PRESENTATION SOFTWARE Presentation software usually includes a Notes View feature that allows you to type notes under or to the side of slides (Figure 21.13). These notes can be helpful, but be wary of using them too much. First, they force you to look at the bottom of a sheet of paper, restricting your airflow. Second, if you are nervous, you may be distracted by these notes or start reading them to the audience.

IF SOMETHING BREAKS, KEEP GOING If your overhead projector dies, your computer freezes, or you drop your notes all over the front row, don't stop your presentation to fumble around trying to fix the situation. Keep going. You need only acknowledge the problem: "Well, it looks like some gremlins have decided to sabotage my projector. So, I will move forward without it." Then do so.

Rehearsing

For some presentations, you might want to rehearse your presentation in front of a test audience. At your workplace, you can draft people into listening to your talk. These practice audiences might be able to give you some advice about improving your presentation. If nothing else, though, they will help you anticipate how the audience will react to your materials. Groups like Toastmasters also can provide a test audience (Figure 21.14). At Toastmasters meetings, people give presentations and work on improving their public speaking skills.

The more you rehearse, the more comfortable you are going to feel with your materials and the audience. Practicing will help you find the errors in your presentation, but rehearsal will help you put the whole package together into an effective presentation.

Evaluating Your Performance

Evaluation is an important way to receive feedback on your presentation. In some speaking situations, the audience will expect to evaluate your presentation. If an evaluation form is not provided, audience members will expect you to hand one out so they can give you and supervisors feedback on your performance. Figure 21.15 shows a typical evaluation form that could be used for a variety of presentations.

For more practice and rehearsing tips, go to **www.pearsonhighered.com/johnsonweb4/21.18**

The "Notes View" Feature

The copy of the slide is up here.

Causes of Climate Change

- **Increase in CO_2 Emissions**
- **Releases of Methane Gas**
- **Other "Greenhouse" Gases**
- **Thinning Ozone Layer**
- **Natural Warming and Cooling Cycles**

The notes appear down here.

Causes of Climate Change

Increase in CO_2 Emissions—scientists at the Hanson Laboratories report that carbon dioxide emissions, especially from cars, is on the rise. We have seen an increase of 15 percent in emissions.

Releases of Methane Gas—methane is naturally occurring. Cows for example, belch methane gas. The problem is with larger and larger releases of methane gas due to industrialization.

Other Greenhouse Gases—a variety of other greenhouse gases, like nitrous oxide and halocarbons, need to be controlled.

Thinning Ozone Layer—the thinning has slowed, but it still continues. As developing countries become more industrialized, they are using more chemicals that thin the ozone layer.

Natural Warming and Cooling Cycles—the earth does naturally warm and cool—creating "ice ages" and "hot periods." These changes are very gradual, though, and could only minimally account for the climate change we are experiencing now.

5

Figure 21.13: The Notes View feature in your presentation software allows you to put notes below or to the side of a slide. Your audience won't be able to see them, but you can print out each slide with your notes so that you don't need to keep looking up at your slides.

Places to Improve Your Presentation Skills

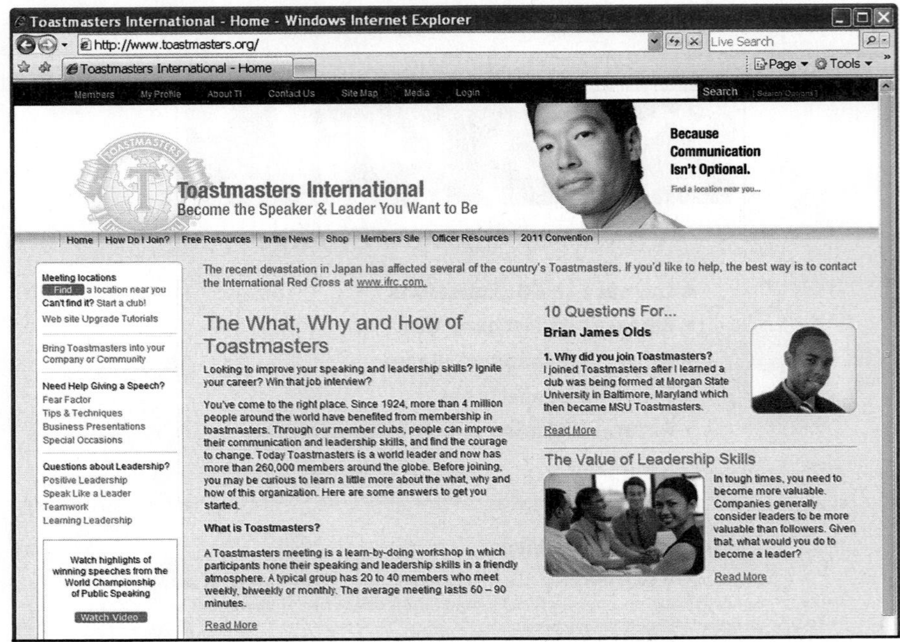

Figure 21.14: Toastmasters is a group that helps people practice and improve their public speaking skills. Local groups can be found in almost any large town.

Working Cross-Culturally with Translators

When speaking to an international audience, you will likely need the services of a translator. A translator does more than simply convert one language into another. He or she will also modify your words to better capture your intent and adjust them to the cultural expectations of your audience. An effective translator will help you better express any subtle points while avoiding cultural taboos and gaffes.

Here are some strategies to help you work more effectively with a translator. These strategies can also be helpful when speaking in any cross-cultural situation.

KEEP YOUR SPEECH SIMPLE The words and sentences in your speech should be as plain and simple as possible. Figures of speech, clichés, or complex sentences will be difficult to translate, especially when the audience is right in front of you.

AVOID JOKES Translators cringe when speakers decide to tell jokes, because jokes rarely translate well into another culture. Jokes that are funny in one culture are often not funny in another. Meanwhile, jokes often rely on turns of phrase or puns that are impossible to translate. Translators have been known to tell the audience, "The speaker is now telling a joke that doesn't translate into our language. I will tell you when to laugh." Then, as the speaker finishes the joke, the translator signals that the audience should laugh.

GO TO THE NET

To locate groups like Toastmasters, go to **www.pearsonhighered.com/johnsonweb4/21.19**

An Evaluation Form for Presentations

Presentation Evaluation Form

Please answer these questions with one or more sentences.

Content

Did the speaker include more information than you needed to know? If so, where?

Where did the speaker not include enough need-to-know information?

What kinds of other facts, figures, or examples would you like to see included in the presentation?

Organization

In the introduction, was the speaker clear about the subject, purpose, and main point of the presentation?

In the introduction, did the speaker grab your attention effectively?

In the body, was the presentation divided into obvious topics? Were the transitions among these topics obvious to you?

In the conclusion, did the speaker restate the presentation's main point clearly?

Did the speaker leave enough time for questions?

Style and Delivery

Did the speaker speak clearly and loudly enough?

Figure 21.15: An evaluation form is a good way to receive feedback on your presentation. Have your audience comment on your presentation's content, organization, style, and delivery.

(continued)

SPEAK SLOWLY A translator will struggle to keep up with someone who is speaking at a faster-than-normal pace, leading to errors in translation. Meanwhile, the structure of some languages (e.g., German) can cause translation to take a little longer.

GO TO THE NET To download this evaluation form, go to **www.pearsonhighered.com/johnsonweb4/21.20**

Did the speaker move effectively, using hands and body language to highlight important points?

Did the speaker have any verbal tics (uh, um, ah, you know)?

Did the speaker look around the room and make eye contact?

Was the speaker relaxed and positive during the presentation?

Visuals

Did the visuals effectively highlight the speaker's points in the presentation?

Did the speaker use the visuals effectively, or did they seem to be a distraction?

Did the speaker use notes effectively? Was the speaker distracted by his or her notes?

Concluding Remarks

List five things you learned during this presentation.

What did you like about this presentation?

What did you not like about this presentation?

What suggestions can you offer to make this presentation better?

Figure 21.15: *(continued)*

MINIMIZE SLANG, JARGON, AND SAYINGS These words and phrases rarely translate easily into other languages, because they are culturally dependent. For example, if the speaker says, "Instead of doing another kickoff meeting, we just need to sit down and hammer out an agreement," the translator would struggle to translate three concepts in this sentence: "doing" a meeting, "kicking off" that meeting, and "hammering out" an agreement. The meanings of these words are dependent on the culture of the speaker and might have little meaning to the audience.

Working with a Translator

Translators are becoming increasingly important as technology and manufacturing become more international.

AVOID RELIGIOUS REFERENCES In most cases, it is risky to include religious themes or terms in cross-cultural speeches. Even seemingly harmless phrases like, "God help us" or "Let's pray that doesn't happen" can translate in unexpected ways. Meanwhile, attempts to incorporate the sayings of a religious figure or scripture can be potentially insulting and even sacrilegious.

KNOW YOUR TRANSLATOR Whenever possible, check your translator's level of fluency and understanding of your subject matter. One of your bilingual colleagues may be able to help you determine your translator's abilities. Also, you should hire a translator from your audience's specific culture. Just because a translator knows Spanish doesn't mean he or she can handle the dialects and colloquialisms in all Spanish-speaking cultures.

PROVIDE YOUR SPEECH, VISUALS, AND HANDOUTS IN ADVANCE Giving your translator your speech ahead of time will greatly improve the accuracy of the translation, because your translator will have time to become familiar with the topic and anticipate ideas that are difficult to translate.

STAND WHERE YOUR TRANSLATOR CAN SEE YOUR FACE A translator may have trouble hearing you correctly if you are turned away from him or her. Also, translators sometimes read lips or facial expressions to help them figure out difficult words or concepts.

For now, English speakers are fortunate that their language has become an international language of business and technology. Consequently, many people in your audience will be able to understand your speech without the help of a translator. Before too long, though, people from other cultures will expect business to be conducted in their languages, too. At that point, translators will become even more critical in technical fields.

CHAPTER REVIEW

- Public presentations are an essential part of communicating effectively in technical workplaces.
- Planning is the key to successful public presentations. You need a firm understanding of the subject, purpose, audience, and context of your presentation.
- A well-organized presentation "tells them what you're going to tell them, tells them, and tells them what you told them."
- Visual aids are essential in presentations. Software programs are available to help you make these aids.
- People come to presentations because they want to see you *perform* the material.
- Good delivery means paying attention to your body language, appearance, voice, rhythm, tone, and effective use of notes.
- Practice is important, but so is rehearsal. Use practice to iron out the rough spots in the presentation, and then rehearse, rehearse, rehearse.
- Translators can help you appropriately present your message to a cross-cultural audience. Keeping your words simple and speaking at a normal pace will help translators more accurately reflect your intended meaning.

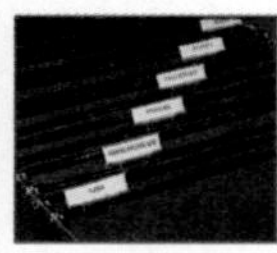

EXERCISES AND PROJECTS

Individual or Team Projects

1. Attend a public presentation at your campus or workplace. Instead of listening to the content of the presentation, pay close attention to the speaker's use of organization, style, and delivery. In a memo to your instructor, discuss some of the speaker's strengths and suggest places where he or she might have improved the presentation.

2. Using presentation software, turn a document you have written for this class into a short presentation with slides. Your document might be a set of instructions, a report, a proposal, or any other document. Your task is to make the presentation interesting and informative to the audience.

3. Pick a campus or workplace problem that you think should be solved. Then, write a 3-minute oral presentation in which you (a) identify the problem, (b) discuss what is needed to solve the problem, and (c) offer a solution. As you develop your presentation, think about the strategies you might use to persuade the audience to accept your point of view and solution.

4. Interview one of your classmates for 10 minutes about his or her hometown. Then, from your notes alone, make an impromptu presentation to your class in which you introduce your classmate's hometown to the rest of the class. In your presentation, you should try to persuade the audience members that they should visit this town.

Collaborative Projects

Group presentations are an important part of the technical workplace because projects are often team efforts. Therefore, a whole team of people often needs to make the presentation, with each person speaking about his or her part of the project.

Turn one of the collaborative projects you have completed in this class into a group presentation. Some of the issues you should keep in mind are:

- Who will introduce the project and who will conclude it?
- How will each speaker "hand off" to the next speaker?
- Where will each presenter stand while presenting?
- Where will the other team members stand or sit while someone else is presenting?
- How can you create a consistent set of visuals that will hold the presentation together?
- How will questions and answers be handled during and after the presentation?
- How can the group create a coherent, seamless presentation that doesn't seem to change style when a new speaker steps forward?

Something to keep in mind with team presentations is that members of the audience are often less interested in the content of the presentation than they are in meeting your team. So, think about some ways in which you can convey that personal touch in your team presentation while still talking effectively about your subject.

For additional technical writing resources, including interactive sample documents, document design tutorials and guidelines, and more, go to **www.mytechcommlab.com.**

CASE STUDY

The Coward

Jennifer Sandman is an architect who works for a small firm in Kansas City. The firm is completing its largest project ever, designing a building for a GrandMart superstore. Jennifer and her boss, a lead architect, Bill Voss, are in charge of the project.

At an upcoming city council meeting, they will need to give a public presentation on their plans for the new GrandMart. At the meeting, the mayor and city council will look over the plans and listen to comments from people in the audience. Failure to persuade the audience might mean the loss of the project as well as future business with GrandMart, an important client.

Everyone knows that the meeting will be contentious, with many angry residents arguing against the plans. And yet, GrandMart's surveys have shown that a silent majority would welcome the new store in the neighborhood. Meanwhile, GrandMart's marketing specialists are certain that the store would be profitable, because there are no competitors within 10 miles.

Jennifer has one problem—her boss is a coward. He scheduled a business trip to coincide with the meeting, and he promptly delegated the presentation to her. As her boss walked out the door, he said, "It will be a good growth opportunity for you. Just don't make the firm look bad."

For Jennifer, giving presentations isn't a problem. She is a whiz with PowerPoint, and she has given enough presentations to feel comfortable in front of people. Her concern, though, is the hostile people in the audience. There might even be some hecklers. These hostile people will never accept the idea that a GrandMart should be built in their neighborhood. Others in the audience, though, will be undecided. So she wants to devise ways in her presentation to soften and deflect some of the hostile criticisms, while gaining the support of people in the community who want the store built.

If you were Jennifer, how would you craft your presentation to fit these difficult conditions? How would you prepare for the hostile and heckling questions? How would you design and deliver your presentation to avoid being seen as defensive?

CHAPTER

22

Designing Websites

Basic Features of a Website *612*

Planning and Researching a Website *615*

Organizing and Drafting a Website *620*

Using Style in a Website *626*

Designing the Website *626*

Uploading, Testing, and Maintaining Your Website *630*

Chapter Review *632*

Exercises and Projects *632*

Case Study: Slowed to a Crawl *634*

In this chapter, you will learn:

- How websites are both similar to and different from print documents.
- The basic features of websites.
- How to plan and research a website.
- How to organize and draft a website.
- How to use web-authoring software.
- How to use style to make a website more readable.
- How design principles are used to create an effective website interface.
- The importance of testing and maintaining websites to keep them current.

In a short time, websites have become an essential form of communication in technical workplaces. Today, websites are regularly written and designed by engineers, scientists, and other technical personnel. In some high-tech fields it is expected that new employees will know how to write for the web.

The first thing you should remember about websites is that *they are documents.* They may look different from paper-based documents and they may be used differently, but they are still *written texts* with words and images. As a result, you can apply many of the communication strategies in this book to developing these screen-based documents.

An important difference, though, is that people tend to read websites visually and spatially, scanning from one block of information to another block of information. They *navigate* within the website among blocks of information. Because websites are visual-spatial texts, they are composed and designed differently than paper-based documents. This chapter will show you how to take advantage of the visual-spatial qualities of on-screen documents.

Basic Features of a Website

A website is a group of related webpages that are linked together into a whole document. Websites have a few different kinds of pages.

> **Home page**—The home page is usually the first page of the site. It identifies the subject and purpose of the website, while forecasting its overall structure. In many ways, the home page is like an introduction and a table of contents for the website (Figure 22.1).

Similarities Between Websites and Other Documents

Website		Paper-Based Document
Home page	→	Introduction, or table of contents
Node page	→	Chapter or section
Page	→	Page
Linking	→	Turning a page
gifs, jpegs, tiffs	→	Graphics (pictures, drawings)
Navigating	→	Reading and scanning
Surfing the web	→	Reading and scanning
Site map	→	Index
Search engine	→	Index
Navigation bar	→	Table of contents or tabs

GO TO THE NET

Want to learn more about how the Internet works? Go to **www.pearsonhighered.com/johnsonweb4/22.1**

A Home Page

Figure 22.1: A website uses visual-spatial strategies to organize information. In this website from the U.S. Forestry Service, notice how the text and images are presented visually for easy access to the information on the page and site.
Source: U.S. Forestry Service, http://www.fs.fed.us.

Node pages—These pages tend to be linked to the home page. They subdivide the website's content into smaller topics. For example, a university's home page will have links that go to node pages like *Colleges and Departments, Libraries, Students,* and *Faculty and Staff.* These are all nodes in the website.

Pages—Individual pages contain the facts, details, images, and other information that readers are seeking.

Navigation pages—The navigation pages help readers find the information they are seeking. These pages include site maps that show the contents of the site, much like an index in a book. They also include search engines that allow readers to search the site with keywords.

Splash page—This optional page comes before the home page and acts like a cover to a book. Splash pages are mostly decorative, using animation or images to set a specific tone before readers enter the website through the home page.

GO TO THE NET
To see some interesting splash pages, go to
www.pearsonhighered.com/johnsonweb4/22.2

AT WORK

Mitch Curren

PUBLIC RELATIONS INFORMATION QUEEN, BEN & JERRY'S ICE CREAM

Mitch Curren writes copy for both printed materials and the web for Ben & Jerry's Ice Cream in South Burlington, Vermont.

How is writing for a website different from other kinds of writing?

In many ways, writing for the web is similar to other kinds of writing. As Ben & Jerry's Chief S'creamwriter & LexiConeHead, I'm responsible for writing and editing copy that communicates the company's *brand* voice, the voice that conveys Ben & Jerry's unique brand personality in ways our customers find recognizable and engaging.

Whether I'm creating copy for the back of an ice cream pint, for a printed brochure, or for our website and e-newsletter audience, ensuring brand voice and style are constant objectives. I also work closely with various project managers to meet project-specific goals and objectives.

Quite often that means I put on my "Composition 101" instructor's hat to make sure folks' copy requests contain basic copy-directional essentials:

- What's the message?
- Who's your audience?
- What do you want the message to achieve?

While the same writing basics and project dynamics apply to copy written for our website, the dynamics of the web itself are different. Most people don't "read" web content the same way they'd read a printed document; hypertext and hyperlinks enable all kinds of browsing, surfing, and scanning opportunities no linear print materials can match.

Those and other opportunities are what I keep in mind when I write stuff for our website. Whatever topic I'm covering or message I'm conveying needs to be organized into informational "chunks"—often with links to additional chunks—that offer a quick gist for folks who like to scan as well as links for folks who want further information.

In the end, though, don't be too concerned about all the website jargon. If you think about it, all these features of websites are very similar to features in paper-based documents. For example, the home page is just like the introduction to any other document. Selecting a link is just like turning a page in a book or magazine.

One significant difference, however, involves the organization of the information. As shown in Figure 22.2, websites usually use spatial (or nonlinear) organizational patterns. In a spatial organizational pattern, readers can follow a variety of paths after reading the home page. The website lets readers move around freely in the document to find the information they need. Keep in mind that websites come in many shapes and sizes, and they are usually not as symmetrical as the diagram shown in Figure 22.2.

Basic Pattern for a Website

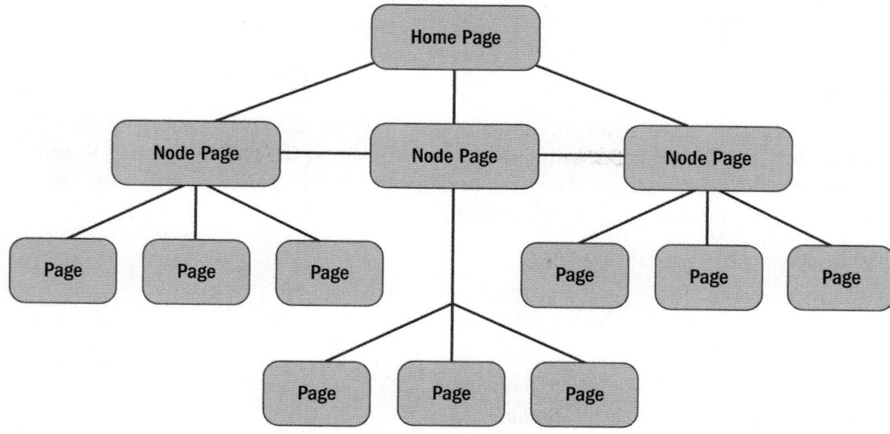

Figure 22.2: Websites are spatial rather than linear. They allow readers to move in a variety of different directions. In this example, readers can go directly to any of three pages from a node page without going through other pages.

Planning and Researching a Website

When planning a new website, you should start by making some strategic guesses about the kinds of people who might be interested in using it. Begin the planning phase by answering the Five-W and How Questions about your website:

Who will be visiting this website?

Why are they using this website?

What information or kind of information are they looking for?

Where will the information in the site be useful to them?

When will they use this information?

How might readers use this website?

With these questions answered, you can think more deeply about how the website should be developed. Consider the website's subject, purpose, readers, and context of use.

Subject

Clearly define the boundaries of your website by determining what is "inside" the subject area and what is "outside." Websites can hold almost limitless amounts of information. So, you should include only information that your readers might need. For example, the boundaries of your subject might be defined in the following way:

> My website will be about Nobel physicist Richard Feynman's career as a scientist and professor. It won't be about his personal life.

Link

For more information on defining the boundaries of a subject, go to Chapter 1, page 7.

GO TO THE NET: To see other organizational structures for websites, go to **www.pearsonhighered.com/johnsonweb4/22.3**

We want to create a website that explains the paleontology research being pursued at Dinosaur National Monument and allows paleontologists to network with each other.

Purpose

Link

For more tips on defining a document's purpose, see Chapter 1, page 6.

Like any other document, you should define the purpose of your website. For the most part, websites are meant to be informative, meaning they are developed to provide information to readers, rather than to persuade them. Consequently, some keywords in a statement of purpose might include the following:

to inform	*to report*
to explain	*to summarize*
to train	*to collect*
to acquaint	*to familiarize*
to teach	*to show*
to instruct	*to exhibit*
to present	
to display	
to demonstrate	
to illustrate	

A purpose statement for a website might read something like the following sentences:

This site is designed to present the life and works of Richard Feynman.

Our intent is to familiarize paleontologists with prior research at Dinosaur National Monument so they can plan future explorations and digs.

Make sure you clearly define your purpose *before* you start creating the website. You should be able to state your website's purpose in one sentence. If you cannot boil it down to one sentence, your website will probably not be focused enough.

Readers

Even though just about anyone can use your website, you should target specific readers by paying attention to their needs. As discussed in Chapter 2, you should begin analyzing your readers with a Writer-Centered Chart (Figure 22.3).

PRIMARY READERS (ACTION TAKERS) Your primary readers will use the information on your website to make a decision or take action. What are their needs, values, and attitudes? Are they researching your company's services or products? Are they collecting information for a report? Are they looking for help to solve a problem?

GO TO THE NET For worksheets that will help you analyze your website's rhetorical situation, go to **www.pearsonhighered.com/johnsonweb4/22.4**

A Writer-Centered Chart for a Website

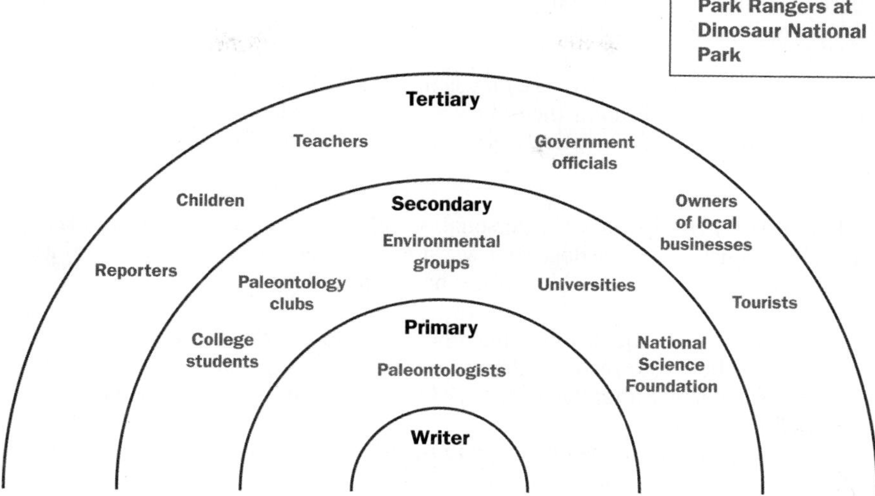

Figure 22.3: The writer-centered chart is a helpful way to identify and sort out the readers who would most likely be interested in using your website.

SECONDARY READERS (ADVISORS) These readers will likely be advisors or supervisors to the primary readers. Are they reviewing your information for accuracy, or are they confirming facts? Are they advising the primary readers about a decision? What is their level of expertise with your website's subject?

AT A GLANCE

Defining a Website's Rhetorical Situation

- Subject: What information is inside the scope of the website, and what isn't?
- Purpose: In one sentence, what is the purpose of the website?
- Readers: Who will be reading the site, and what kind of information are they looking for?
- Context: What physical, economic, ethical, and political factors will shape how the website is written and read?

TERTIARY READERS (EVALUATORS) Other readers may also have various reasons for visiting your site. How might your company's competitors use the information on your website? How might a journalist or lawyer use the information? What kinds of proprietary or personal information would not be appropriate for public access?

GATEKEEPER READERS (SUPERVISORS) Your gatekeepers are mostly interested in the accuracy of the site and whether it reflects the image and values the company wants to promote. Specifically, your supervisors will want to check your site to make sure it reflects the mission of the company.

Link

For more information on analyzing readers, see Chapter 2, page 23.

Context of Use

You should also consider the physical, economic, ethical, and political issues that will shape how your readers use your website.

GO TO THE NET

For worksheets that will help you analyze your website's audience, go to **www.pearsonhighered.com/johnsonweb4/22.5**

PHYSICAL CONTEXT Involves the physical places where people might use your website. Will readers be accessing the site from home, in their office, in a meeting room, or perhaps at an Internet cafe? Are they at a desk, on the factory floor, or in an airport? Are they accessing through a wireless network? If so, is the network Wi-Fi, WiMAX, or Bluetooth? How fast is their connection to the Internet?

ECONOMIC CONTEXT Involves the financial issues that shape how readers will interpret your website. What is their likely economic situation or status? What are the trends in the industry or national economy? What kinds of economic decisions are you asking readers to make?

Link

For more information on ethics, see Chapter 4, page 69.

ETHICAL CONTEXT Involves the personal, social, and environmental issues that your website touches. Where does your website involve issues of rights, laws, utility, or caring? How should these ethical issues be handled?

Link

For more information on defining a document's context of use, go to Chapter 2, page 26.

POLITICAL CONTEXT Involves the micropolitical and macropolitical issues that might influence how your readers interpret the website. For example, does your website step on some toes within the company? How does it fit in with your company's overall political stance?

As you think about the rhetorical situation for your website, put yourself in your readers' place. Try to anticipate the kinds of information they are looking for and why they need that information. Putting yourself in your readers' place will give you valuable insights as you are writing and designing the website.

Websites for International and Cross-Cultural Readers

It is becoming increasingly important to design websites for international and cross-cultural readers. Up to this point, English has been the unofficial language of the web, but increasingly, non-English-speaking users around the world are accessing the Internet.

Of course, it would be impossible to anticipate the needs of all potential readers around the world, but you can make your website more usable in a few important ways.

Translate the website—The global marketplace allows your company to attract potential clients and customers around the world. So, if your company regularly does business with people from another country or culture, it might be a good idea to make your website, or at least parts of your website, available in the readers' language. You can use Google Translate (translate.google.com) to translate your website for you, but it isn't as reliable as a human translator (Figure 22.4).

Use common words—Try to use words that are commonly defined in English. The meanings of slang and jargon words change quickly, sometimes leaving international readers confused.

Avoid clichés and colloquialisms—Informal American English includes phrases like "piece of cake" or "miss the boat" that might be meaningless to people from other cultures. Also, sports metaphors like "kickoff meeting"

Want to know more about ethical issues involving websites? Go to **www.pearsonhighered.com/johnsonweb4/22.6**

Translating a Website with Google Translate

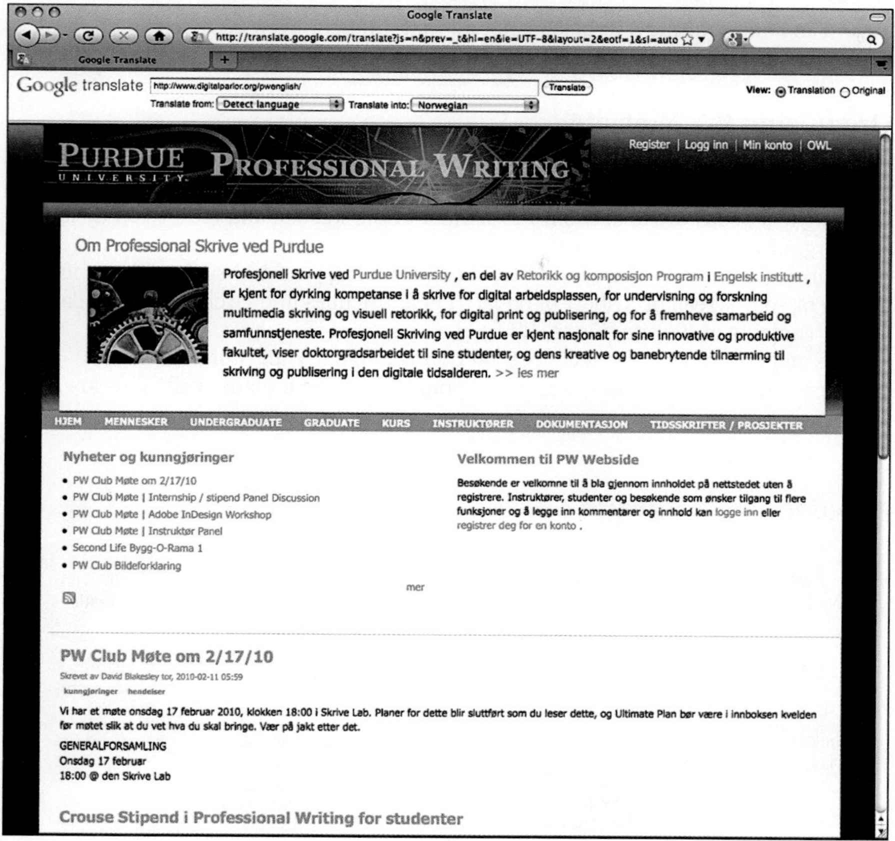

Figure 22.4: Here is a Norwegian translation of the Professional Writing Program's website at Purdue University done with Google Translate (translate.google.com).

Source: Purdue University, Professional Writing Program, http://www.digitalparlor.org/pwenglish.

or "hit a home run" sound very odd to people who are not familiar with American football and baseball.

Avoid cultural icons—Symbols, especially religious symbols, should be avoided where possible and carefully used where necessary.

Minimize humor—American humor does not translate well into other cultures, so attempts to be funny on your website might be offensive or just confusing.

Chapter 2 discusses writing for international and cross-cultural audiences in more depth. In most cases, these guidelines for writing for cross-cultural readers are applicable to websites also.

Link

For more information on writing for international and cross-cultural readers, go to Chapter 2, page 28.

GO TO THE NET

For more advice about writing websites for international and cross-cultural readers, go to **www.pearsonhighered.com/johnsonweb4/22.7**

Organizing and Drafting a Website

Organizing and drafting a website is not all that different from organizing and drafting a print document.

Organizing the Website

Professional website developers often prefer to start creating a new site by mapping out its contents on a whiteboard, a large piece of paper, or a computer screen. Logical mapping is a good way to sort out the content of the website and to develop an efficient organizational scheme for it.

To map out the contents of the site, start by writing your website's purpose in the middle of your screen or a sheet of paper. Then, use lines and circles to begin identifying the contents of the site (Figure 22.5).

You might also try a low-tech method that is popular with professional web designers—using sticky notes to map out your website on a blank wall. This low-tech

Mapping Out a Website

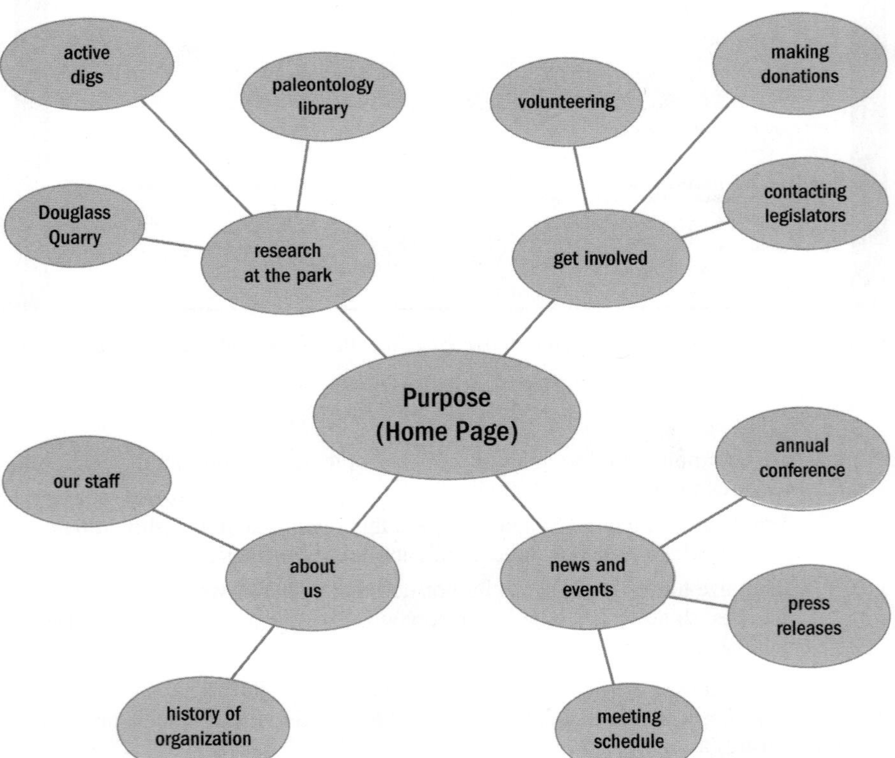

Figure 22.5: Mapping out a website is a good way to see the visual relationships among the site's pieces of information.

method allows you and your team to move the notes around to look at different ways to organize the site. Another advantage to sticky notes is that they can be added or crumpled up with ease.

Creating Levels in the Website

Your logical map and research will likely give you a sense of how many levels are needed in the website. The map shown in Figure 22.5, for example, might translate into a structure like the one shown in Figure 22.6.

Levels in a Website

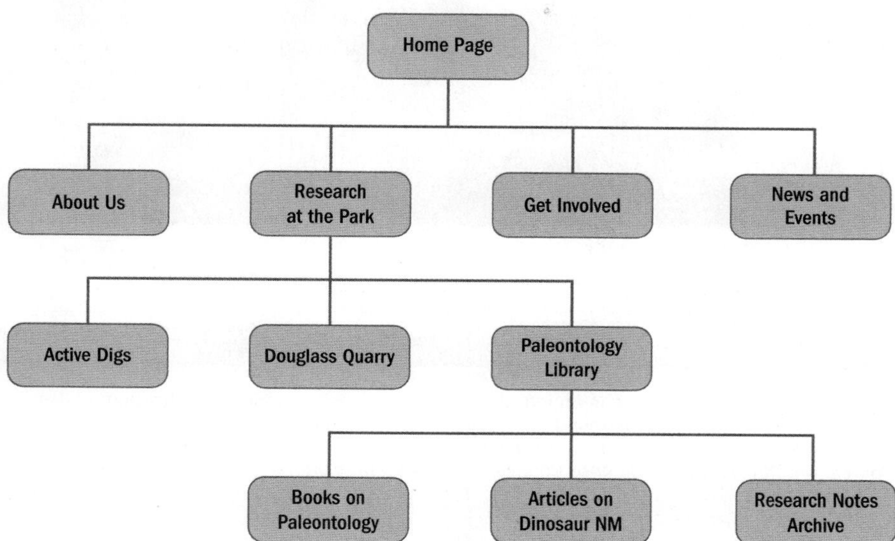

Figure 22.6: A logical map can be turned into levels of information. Each level offers information that is increasingly specific.

How many levels should you create in a website? Professional website designers use the following guidelines to determine the number of levels needed in a website:

- a maximum of three links for the most important information.
- a maximum of five links for 80 percent of all information.
- a maximum of seven links for all information.

These guidelines are helpful but they aren't firm rules, so you can ignore them if needed. However, if you force your readers, especially customers, to wade through too many pages, you risk losing them. If you make them work too hard, they will grow frustrated and give up.

Drafting the Home Page

Your home page should be similar to the introduction in a paper-based document. This page should set a context by clearly signaling the subject, purpose, and main point of the site. It might also stress the importance of the site's information to your readers.

GO TO THE NET

To see examples of good and bad home pages, go to **www.pearsonhighered.com/johnsonweb4/22.8**

A Home Page

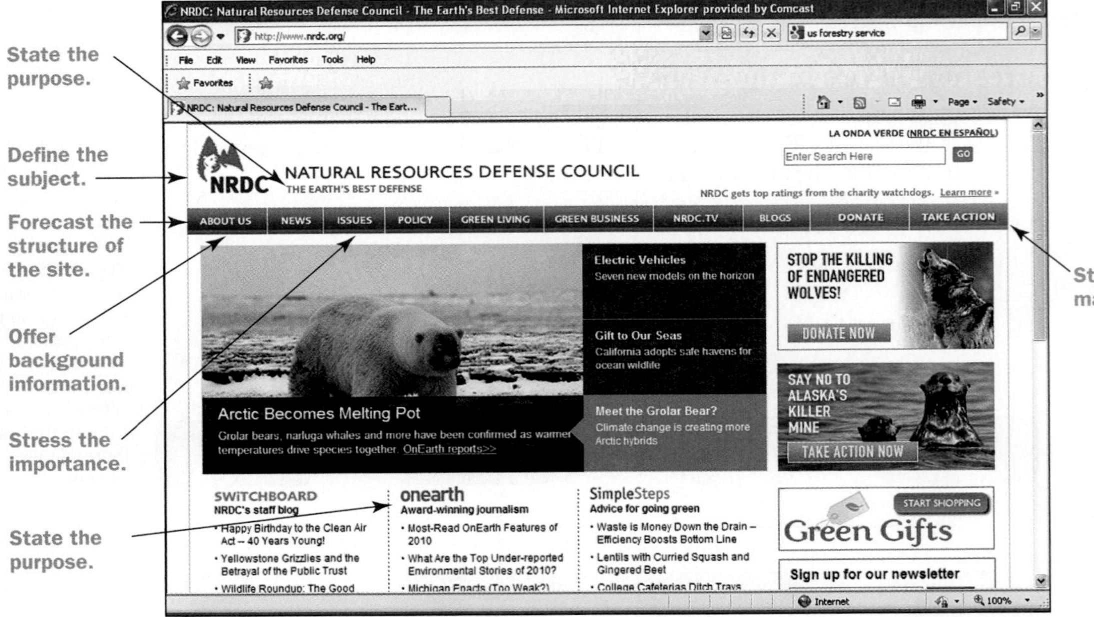

Figure 22.7: The home page is an introduction to the website. Notice how this home page makes all the common moves found in the introduction of a paper-based document. These introductory moves can be made in a few different places on the screen.
Source: Natural Resources Defense Council, http://www.nrdc.org.

SUBJECT Your site's subject should be clear as soon as readers access the home page. You don't need to say something like, "The subject of this site is . . .," but you can use a title or a graphic to quickly indicate what the site is about (Figure 22.7).

PURPOSE Likewise, the purpose of your site should also be clear on the home page.

> In this website, you will find helpful information about paleontology research and resources at Dinosaur National Monument. This information will support the planning of your own research at the site.

Link

For more information on writing introductions, see Chapter 16, page 430.

Often, websites do not directly state their purpose on the home page. In these situations, the purpose of the website should be obvious to the readers.

MAIN POINT Since websites tend to be more informative than persuasive, they often do not state a main point on the home page. Nevertheless, you should think about the overall point you want to stress.

> Being informed about prior and current research at Dinosaur National Monument is essential to designing your own research project, especially if you are proposing a dig.

STATEMENT OF IMPORTANCE The importance of your subject may be immediately obvious to you, but it might not be obvious to your website's readers. You should stress the importance of the subject on the home page to encourage readers to use the site.

> Resources at the monument are very limited, so projects are approved only if they are likely to succeed and do not duplicate prior projects. The information in this website should help you avoid proposing research projects that will not be approved. This website might also help you locate other paleontologists with whom you can collaborate on large projects.

In some cases, a picture or graphic can also signal the importance of the site.

BACKGROUND INFORMATION On most home pages, background information is limited, because only a limited amount of space is available on the screen. You might, however, give the readers some background information on your subject to help them gain a quick overview.

> The Dinosaur National Monument Paleontologist Network was created in 1986 to foster communication among paleontologists who work in the park. Since then, our network has grown to include over 200 scientists.

FORECASTING Usually, home pages include a navigation bar that forecasts the structure of the site. The navigation bar can appear just about anywhere, but it typically appears along the top of the page or on the left-hand side (Figure 22.7).

Drafting Node Pages

Node pages direct traffic on your website. These pages typically follow links from the home page, further dividing the subject of the site. Each node page introduces readers to one of the site's major topics. For example, the node page shown in Figure 22.8 was a link displayed on the home page in Figure 22.7.

A well-designed node page allows readers to select among other further topics. In essence, it serves as a miniature home page to this part of the site. In large websites, node pages are very much like home pages because they tell readers the *subject, purpose,* and *main point* of this part. They include mostly contextual information that helps direct readers toward more specific details further into the website.

Drafting Basic Pages

There really is no sharp distinction between node pages and basic pages. As readers move further into the site, the pages should become more detail oriented.

Basic pages increasingly provide the content (e.g., facts, data, examples, details, descriptions) that readers are looking for. For example, the basic page shown in Figure 22.9 includes mostly facts, examples, and reasoning. More than likely, this information is why readers came to the website in the first place. But they needed to navigate through the home page and node pages to arrive here.

To see other sample node pages, go to **www.pearsonhighered.com/johnsonweb4/22.9**

A Node Page

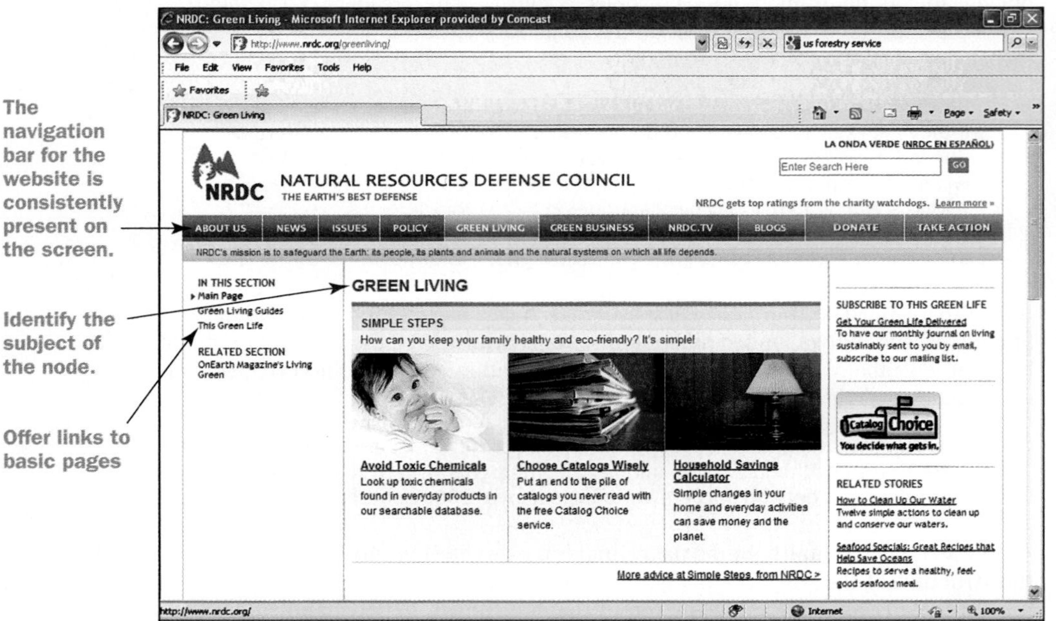

Figure 22.8: A node page introduces a specific topic in the website. Here is the "Green Living" node page for this website. It allows readers to find more information on research about whales.

Source: Natural Resources Defense Council, http://www.nrdc.org/greenliving.

In a basic page, you should keep the discussion limited to a topic that can be handled in about a screen and a half. Readers of websites rarely have enough patience to scroll down further. So, if you need more than two screens to cover a particular issue, you might need two or more basic pages.

AT A GLANCE

Types of Website Pages

- Home page—introduction page for the website
- Node pages—opening pages for sections of the website
- Basic pages—pages containing the content readers are looking for
- Navigational pages—site maps, Help area, search engine

Drafting Navigational Pages

Websites also include other kinds of pages that can help readers navigate the site.

SITE MAPS A site map is like an index in a book. The site map lists all the topics covered in a website, large or small. That way, if readers are having trouble finding what they want, they can go straight to the site map to find a link to the information they need.

HELP Software programs like RoboHelp, DocToHelp, and Help & Manual will allow you to create Help areas for your website. Usually, only the largest websites include a Help area. These Help areas offer additional information and usually provide definitions or tutorials that readers may need.

GO TO THE NET For more information about Help software packages, go to **www.pearsonhighered.com/johnsonweb4/22.10**

A Basic Page

Notice that the navigation bar is the same as before.

Specific information about an issue is included.

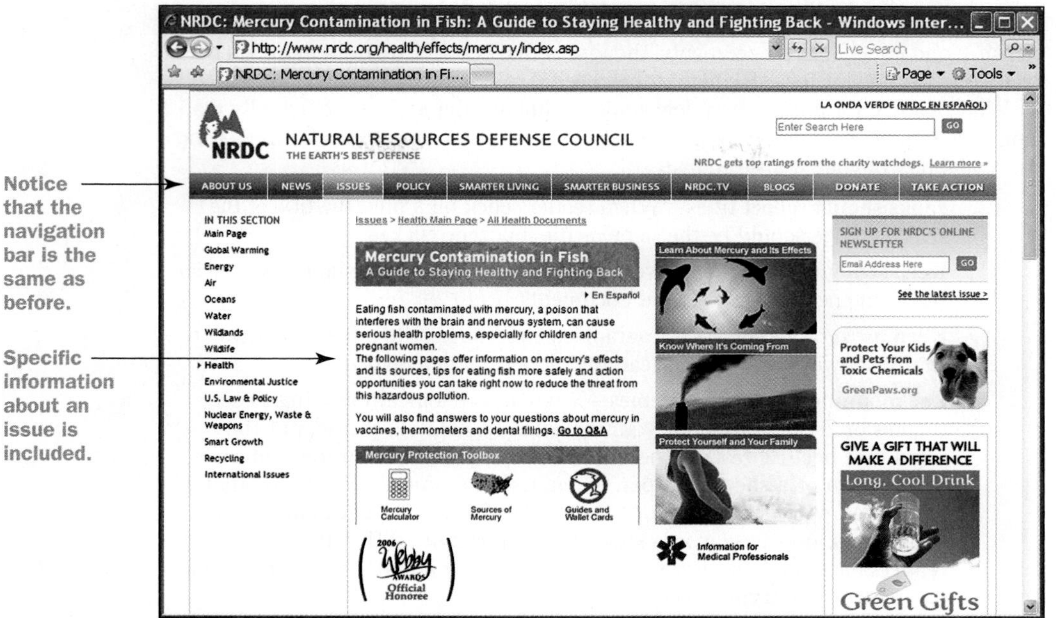

Figure 22.9: A basic page contains the specific information that readers are looking for. This basic page (linked from the "Green Living" node page) discusses a specific issue: mercury contamination.

Source: Natural Resources Defense Council, http://www.nrdc.org/health/effects/mercury/index.asp.

SEARCH ENGINES Larger websites may also include an internal search engine that helps readers locate information by typing in keywords. Search engines are available, usually free, from the larger search engine providers like Excite (excite.com), Lycos (lycos.com), and Google (google.com). However, putting a search engine into your website takes some programming skill.

A Warning About Copyright and Plagiarism

Inventing and collecting the content for websites is not much different from creating the content for print documents. You still need to thoroughly research the subject in electronic, print, and empirical sources. The information you invent or collect will be the substance of the site.

To avoid any problems with copyright or plagiarism, remember that the same rules apply to websites as apply to print documents. If you want to use an image off someone else's website, you need to ask permission. Also, you cannot take passages of text from other sites and use them in your own unless you properly cite them.

Link

To learn more about copyright and plagiarism, go to Chapter 4, page 88.

For more information on copyright issues and the Internet, go to **www.pearsonhighered.com/johnsonweb4/22.11**

Using Style in a Website

As in any other document, the style of a website is important. The difference between websites and paper-based documents, though, is that readers are even more likely to "raid" the website for the information they need.

When raiding for information, few readers will have the patience to actually read long paragraphs or sentences on a screen. So, here are some strategies for improving the readability of your website.

Links should reflect titles—When readers click on a link, the title of the selected page should be the same as the link they clicked.

Keep sentences short—On average, sentences in websites should be shorter than sentences in paper-based documents.

Keep paragraphs short—Paragraphs should be kept to a few sentences or less. That way, readers can scan the paragraph in a glance.

Link

For more information about using logical mapping to improve style, turn to Chapter 17, page 472.

Use mapping to develop themes—You can use logical mapping to set themes or create a tone for your website. To create a theme, put the word that best represents the theme or tone you desire in the center of your screen or a sheet of paper. Then, use mapping to find words that are associated with that theme or tone (Figure 22.10). As with paper-based documents, use these words strategically throughout the site. When your readers encounter these words, they will subconsciously sense the theme or tone you are trying to create.

Designing the Website

The design of your website is important for several reasons.

Readers prefer attractive, well-designed webpages—They are more likely to spend time looking over webpages that please their eyes.

Well-designed websites are more navigable and more effective—If readers can quickly locate paths to the information or products they are seeking, they will be much more willing to do business with your company.

Well-designed websites give readers confidence—A professional-looking website will make your company and its products seem more trustworthy.

Here, we will go over a few basic principles of web design. As you improve, you will want to learn how to use Cascading Style Sheets (CSS) to design your webpages and sites.

Designing the Interface

The *interface* of your website concerns how the pages look on the screen. The design of your website's interface determines in great part whether your website will be attractive to readers and whether it will be easy to use.

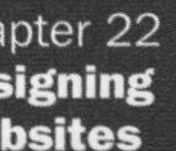

For websites that discuss Internet style, go to **www.pearsonhighered.com/johnsonweb4/22.12**

Mapping a Theme for a Website

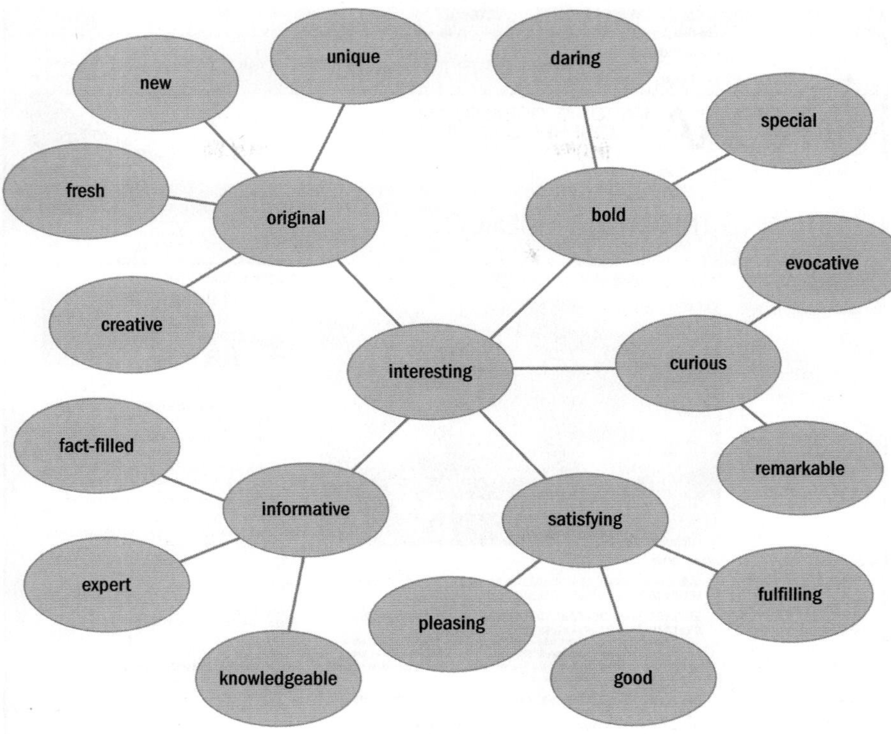

Figure 22.10: Logical mapping can help you create themes or tones that hold the website together. A website that uses this map would set the theme "interesting."

As discussed in Chapter 18, when designing an interface for a website, you should use the *Five Principles of Design:*

Link
For more information on designing interfaces, see Chapter 18, page 492.

> **Balance**—Items placed on the screen offset each other to create a feeling of stability on the page or screen.
>
> **Alignment**—Aligned items on the page help readers identify different levels of information in the interface.
>
> **Grouping**—Items that are near each other on the screen will be seen as "grouped."
>
> **Consistency**—Each page in the website should look similar to the others, minimizing chaos in the interface.
>
> **Contrast**—Contrast sharpens the relationships between images and text, headings and text, and text and the background.

The interface shown in Figure 22.11, for example, demonstrates all of these principles successfully. Notice how the text is balanced from side to side. Information is also aligned in clear vertical lines, and you can see where specific information has

A Well-Designed Interface

Figure 22.11: This home page uses all the design principles successfully. Notice how the interface uses balance, alignment, grouping, consistency, and contrast to create an attractive and functional page.
Source: National Oceanic and Atmospheric Administration (NOAA), http://www.noaa.gov.

been grouped into larger blocks. Meanwhile, consistency and contrast are used to make the interface interesting but consistent.

These five principles should be used to develop a consistent interface, called a template, for your webpages. The template will give your website a standardized, predictable design, allowing your readers to quickly locate the information they are seeking.

Adding Images

Images for your website can be saved as separate files on your computer and on the server. These files can include any pictures, graphs, or drawings you might want to include on your webpage. They also include elements of the interface, such as ban-

GO TO THE NET Want to know more about designing interfaces? Go to **www.pearsonhighered.com/johnsonweb4/22.13**

Two Major Types of Screen-Based Images

Figure 22.12: Most images on websites use either the jpeg or the gif format. The image of Hurricane Katrina on the left is a jpeg. The Purdue logo on the right is a gif.

Sources: NASA, http://rapidfire.sci.gsfc.nasa.gov/gallery/?search=katrina, and Purdue University.

ners on the screens, corporate logos, and icons. Images can be saved in a variety of *file formats.* The two formats most commonly used for websites are *jpeg* and *gif* formats (Figure 22.12).

> **jpeg (joint photographic experts group)**—The jpeg file format is widely used for photographs and illustrations with many colors. Images in jpeg format can use millions of different colors, allowing them to better capture the subtleties of photographs. The main limitation of jpeg images is their higher memory requirements, causing them to download more slowly, especially on computers with slow connections to the Internet.
>
> **gif (graphic interchange format)**—The gif format is primarily used for illustrations, logos, and simple graphics. Gif images can use a maximum of only 256 colors, making them less useful for photographs. Their advantage is that they use less memory than jpeg files, making them quicker to download.

You can tell the difference between jpeg and gif files by the extensions on their file names. A jpeg file will have the extension ".jpg" added to it (e.g., cougarpic.jpg), and a gif file will have the extension ".gif" added (e.g., cougarpic.gif).

To put images in your webpages, you need to do three things.

1. **Create or capture the image.** You can use a digital camera, scanner, or drawing software to create an image. Then, you can use a program like Adobe Photoshop to turn the file into a jpeg or gif. Or, you might find an image on a website or CD-ROM and download it to your hard drive.

Want to learn more about using images? Go to **www.pearsonhighered.com/johnsonweb4/22.14**

To capture an image off a website, put your cursor on it and hold down the button on your mouse. The browser will bring up a window that gives you a few choices. Choose "Save Image As." The browser will let you save the image to your computer's hard drive.

If the image is large, you can "compress" it with Adobe Photoshop, iPhoto, or MS Office Picture Manager. A compressed image will load faster for your readers.

2. **Place the image.** Put the image file in the folder on your computer that holds the other files for your website. When possible, keep images with the pages that reference them. Otherwise, there is a chance the image will not be transferred with the page, causing the image to not be shown on the page (see "Uploading, Testing, and Maintaining Your Website" below).

3. **Insert the image.** Using your web-authoring software, open the page in which the image should appear. Then, use the Insert Image command (or equivalent) to add the image to the page.

When you are finished inserting the image, it should appear on the webpage.

Uploading, Testing, and Maintaining Your Website

Once you are finished revising, editing, and testing the usability of your website, it is time to upload it to the Internet. Once it is uploaded, you will need to test the site yet again to see if it works the way you want it to. After correcting any problems, you will need to maintain the website to keep it up to date.

Uploading the Website

To make your website available to the public, you need to upload it to a server—a large computer that connects your computer to the Internet. Before uploading to the server, most writers of websites prefer to complete the website on their computer's hard drive. Then, they use a file transfer protocol (FTP) software program to copy the whole website to the server at once (Figure 22.13).

Increasingly, universities and corporations are making this process easier by creating a special WWW folder for each person or division on a server. To upload your website, you only need to move the files from your hard drive to the WWW folder.

When you copy your webpage files, don't forget to also copy your image file at the same time. If the image files are not included with the webpage files, the pictures, illustrations, graphs, and other images you want placed on your website will not appear.

Testing the Website

Once your website is on the server, it is time to test it. Use at least a couple of different browsers (e.g., Firefox, Opera, Explorer, Safari) to look over the pages on the site.

For more editing strategies for websites, go to **www.pearsonhighered.com/johnsonweb4/22.15**

Uploading the Website

Your Hard Drive

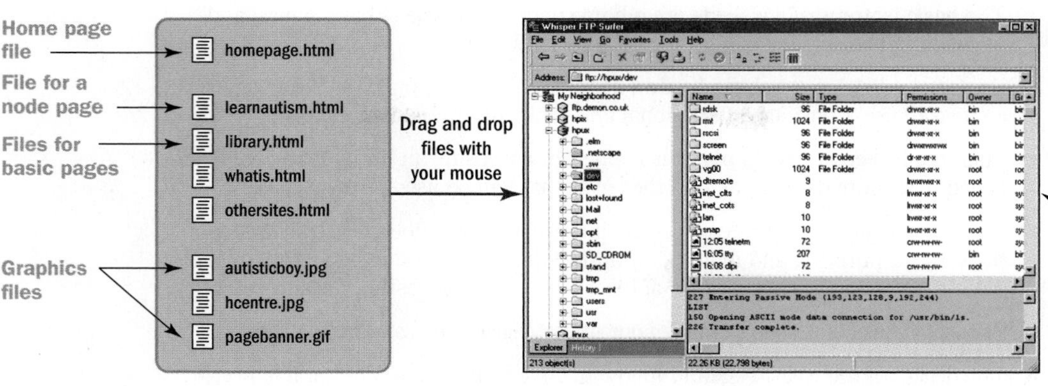

Figure 22.13: An FTP program copies files from your computer's hard drive to a server. The server is connected to the Internet.

Files and folders in server computer are shown here.

Source: Whisper Technology, http://www.whispertech.com/surfer/index.htm.

Try out the links and resize the browser window a few different ways. Usually, you will find that the website has some formatting problems that did not show up in your web-authoring program's version of the site. For example, perhaps an image failed to appear or the spaces between lines are uneven. Another common problem is that links to other pages don't work, which means that readers will see error messages when they choose links to these pages.

To fix these problems, reopen the files with your web-authoring program. Make the corrections and changes needed. Then, re-upload your revised pages to the website.

The testing phase can be a little frustrating at times, but it is important. You want to make certain that your website looks the way you intend it to. Those little problems won't fix themselves, and they will be annoying to your readers. Once you have fixed these small problems, though, your website will convey the sense of quality you want to project.

Maintaining the Website

Most websites cannot be left alone for long. You will find yourself wanting to add or change information on your website regularly. You will find new links to other websites, or you will discover that information on your site is out of date. You should plan to regularly update your website, usually weekly or monthly.

To update a page, go back to the file saved on your hard drive and open it in your web-authoring program. Make the changes you want and then use FTP software to copy the updated file onto the server. The server will replace the old file with your new file.

Most companies hire or contract a webmaster to maintain their site. The webmaster regularly updates the site, incorporating any changes and eliminating any dated materials. When you are your own site's webmaster, though, you should revisit your site regularly, looking for items to update.

GO TO THE NET Want to learn more about FTP? Go to **www.pearsonhighered.com/johnsonweb4/22.16**

CHAPTER REVIEW

- Websites are important for both corporate and personal communication.
- The basic features of a website are a home page, node pages, basic pages, and navigational pages.
- Prepare to develop a website by first defining its subject, purpose, readers, and context of use. Consider international and multicultural audiences as well.
- Mapping is a good way to plan the content of a website; an initial map can be followed by a more detailed plan of the levels that will be used in the website.
- The home page of a website is like the introduction to a document. Tell readers the subject, purpose, and main point of the site, while offering background information, stressing the importance of the subject, and forecasting the site's structure.
- Choose a consistent style to make your website easier to scan and interpret.
- The design of the website should follow the five principles of design: balance, alignment, grouping, consistency, and contrast.
- Revising and editing are especially important in websites, because you will change the text regularly.
- When your website is finished, you need to upload, test, and maintain it.

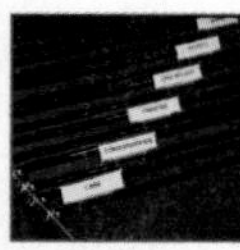

EXERCISES AND PROJECTS

Individual or Team Projects

1. On paper or whiteboard, map out the contents of a website you find on the Internet. Identify its home page, node pages, basic pages, and navigational pages. Then, in a presentation to your class, discuss the organization of the website. Show its strengths and identify any places where the website's organization might be improved.

2. Imagine a website you would like to build. Identify its subject, purpose, readers, and context of use. Then, diagram the site on paper or a whiteboard, showing how it would be organized. Thumbnail some sample pages on paper, sketching out how a few pages would look on the site (home page, node pages, basic pages). Attach these drawings to a memo of transmittal to your instructor in which you discuss the content and organization of your website.

3. Find a website at your college or workplace that is ineffective. Write a report to a specific reader in which you discuss the shortcomings of the site and make recommendations for improvement (you don't need to send the report). In your report, consider the content, organization, style, and design of the site. In your recommendations, show your sample reader pages and organizational schemes that would improve the site.

Samples websites for you to critique are available at **www.pearsonhighered.com/johnsonweb4/22.17**

Collaborative Project

You and your group have been asked to develop a "virtual tour" of your university or workplace. The tour will allow visitors to your website to look around and familiarize themselves with important places on your campus or in your office. The tour should allow visitors to quickly navigate the site, finding the information they need to locate important places and people.

Start out by discussing and describing in depth the subject, purpose, readers, and context of use for the site. Then, using paper or a whiteboard, describe how the site would be organized. Make decisions about the style of the site and how the various pages should be designed.

Write a proposal to your university administrators or company managers, showing how this virtual tour would be a nice addition to the university's or company's website. Your proposal should offer a work plan for making the website a reality and describe its costs and benefits, especially the benefits to visitors.

Then, if the software and hardware are available, create part of the website. Of course, you probably cannot create the whole virtual tour, but you can make the home page, a couple of node pages, and a few sample basic pages that show important places on campus or your workplace.

For additional technical writing resources, including interactive sample documents, document design tutorials and guidelines, and more, go to **www.mytechcommlab.com.**

Slowed to a Crawl

Alice Franklin worked for Kent Medical's Outreach Division, which was part of a large medical services company. She worked her way up through the company as a biomedical engineer, but she jumped at the chance to transfer to the Outreach Division because it allowed her to work with people in other parts of the world.

Her job was to help Kent Medical deliver free medical services and supplies to people in poor countries that did not have advanced medical services. She went home every night knowing she had made a difference. Plus, her trips to South America and Africa were exhilarating and fulfilling.

A few months ago, Alice's boss, Kathryn Young, had hired Zigda Communications, a high-priced web design firm, to redesign the Outreach Division's website. The website was a critical part of the division's public relations and fundraising efforts. It was also the portal through which clients requested help and communicated with Kent Medical.

Zigda agreed to do the work for a fraction of their usual fee, and they did an incredible job redesigning the site. The Flash animation was amazing, and the embedded videos were very effective at telling the story of Kent Medical's efforts to help people in other parts of the world. A week after the site went online, it was featured as a "Best Website" on a major news website. That week, the website got over 700,000 hits, more than it ever had before. Plus, the Outreach Division had already raised over $1.2 million in donations through online contributions.

However, Alice noticed a problem. The Outreach Division's clients in South America and Africa were not accessing the site as much as they had before. Requests for services and supplies had dropped off dramatically.

Trying to figure out what was happening, Alice called up one of the missing clients in Brazil. A representative explained to her that they were not able to access the website because their computers were using dial-up connections. So all those fancy animations, video, and forms were overwhelming their computers.

Even when they *could* get through to a request form for medical services and supplies, the new forms were far too advanced for their browsers. Upgrading the browsers, however, would clog up their computers' memory and cause their computers to crash.

Alice wasn't sure what to do. Hiring Zigda had been the decision of her boss. Bringing up this problem was going to be very embarrassing because the "successful" new website was undermining their core mission of providing medical services and supplies to their clients. And yet, the website was also generating millions of dollars in new donations. How might you handle this situation if you were in Alice's place?

CHAPTER

23 Using Social Networking Tools (Web 2.0)

Social Networking in the Workplace *636*

Blogs and Microblogs *638*

Internet Videos and Podcasts *640*

Wikis *642*

Help: Using Skype for Collaboration *642*

Chapter Review *644*

Exercises and Projects *644*

Case Study: My Boss Might Not "Like" This *646*

In this chapter, you will learn:

- How to start a social networking site.
- Strategies for using social networking to enhance your career.
- Tips for contributing to corporate social networking sites.
- How to create your own blog.
- Techniques for using blogs and microblogs to collaborate with co-workers.
- How to post a video or podcast.
- How to add an article to a wiki.

Social networking sites, like Facebook and Twitter, are becoming common forms of communication in technical workplaces. Web 2.0 tools, such as social networking sites, blogs, microblogs, and wikis, are regularly used to provide information to clients and customers, while helping co-workers and colleagues communicate with each other.

Companies, especially ones in technical fields, are using these same social networking sites to interact with their clients and customers, while keeping the public informed about new products, corporate decisions, and services. Similarly, career-related social networking sites, like LinkedIn and Spoke, can help you network with potential employers, colleagues, and business associates.

Social Networking in the Workplace

Until recently, there were still regular debates about whether companies should allow employees to access social networking sites at work. Managers were concerned that their employees would be distracted by their Facebook accounts or Twitter feeds. Today, companies have mostly begun to realize that these social networking tools are useful for developing new clients and building customer loyalty.

Do employees occasionally waste time and procrastinate with Facebook and Twitter? Sure they do. But companies are discovering that a net savvy, interactive workforce is more important than monitoring whether somebody is catching up with friends on Facebook. The benefits of social networking outweigh the costs. Moreover, the "social" aspects of social networking can create stronger bonds among co-workers, leading to higher morale and productivity.

You should probably have two social networking sites. Your personal site, like Facebook, should be for staying in touch with your friends. Your professional site, like LinkedIn, should be used exclusively for your professional life.

Your company or team may also keep one or more corporate Facebook or Twitter sites that you can use to collaborate with co-workers or work with clients.

Using a Personal Social Networking Site

Starting a social networking site is easy. When you have decided which ones would work best for you, go to their websites and select "Sign up for an Account" or similar. For example, Figure 23.1 shows the homepage for LinkedIn, which is a social networking site for professionals.

To start an account, you will need to enter some basic information. The site will then lead you through the set-up process.

CHOOSE YOUR "FRIENDS" (WISELY) If you already have a social networking site, you probably have a long list of friends, including people you may or may not know personally. As you start your professional life, you should be more selective about who has access to your site. You want to remove any people who you don't know or who might cause a future or present employer to question your judgment.

MAINTAIN YOUR SOCIAL NETWORKING SITES You should check your social networking site regularly to keep it clean and up to date. You should never assume

Want to start a social networking site? Go to **www.pearsonhighered.com/johnsonweb4/23.1**

A Social Networking Site

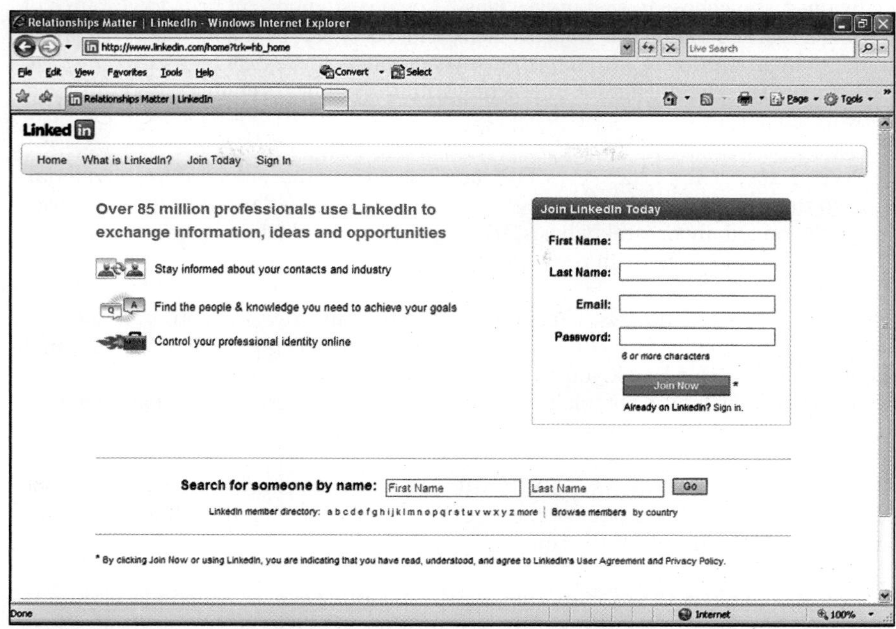

Figure 23.1: LinkedIn is a popular social networking site. LinkedIn is similar to Facebook, except it's for professional networking.

Source: LinkedIn, http://www.LinkedIn.com.

that your personal website is a "safe place" from recruiters and supervisors. Today, recruiters are finding ways to access personal sites as they do background checks on possible employees. Meanwhile, it's easy for someone to forward something from your social networking site to one of your supervisors. So you don't want to say anything or post something that might cause them to question your judgment.

Link

To learn more about collaborating with co-workers, turn to Chapter 4, page 45.

COLLABORATE, BUT CAREFULLY You can collaborate with your coworkers through Facebook and other social networking sites, but you need to be extra careful about posting anything that might be proprietary information. It's probably fine to set up meeting times or make announcements on a social networking site, but any proprietary information should be handled through a more secure medium, like print or e-mail.

Using a Corporate Social Networking Site

Your company may also have an official presence on sites like Facebook or Twitter. If you are asked or allowed to contribute to the corporate site, you need to first understand why the company has created the site. Many companies see social networking as a way to enhance their brand and build customer loyalty. So they are generally looking for ways to improve their image and perhaps counter negative images from other social networking accounts.

Or, your company may be looking for ways to interact with its customers, especially ones who choose to sign up as "fans." Loyal customers can be a good source of feedback and suggestions for improvement. Plus, they can be among the first indicators if there is a problem with the company's products or services.

Here are some strategies for interacting with the public through a corporate social networking site:

ACTIVELY MAINTAIN THE SITE The "fans" on a corporate site need fresh material to keep them coming back. If a site goes cold for even a few days, they will stop visiting it. So, on a daily or weekly basis, try to give them product news, chances to do polls, and questions they can respond to.

ANSWER POSTS FROM FANS People who take the time to visit a corporate social networking site and leave a comment want to be acknowledged. If they offer a compliment, respond with a thank you or some additional information about the product or service. If they have a complaint, try to explain how the company is handling the situation and perhaps improving it.

SHOW FANS HOW THEIR COMMENTS ARE BEING USED Sometimes fans have good ideas for improving the product or service. Where possible, tell them how their ideas are being used by people at the company.

You should never view a corporate social networking site as a good place to vent your frustrations with clients or customers, nor should you express your displeasure with the company itself. Again, even seemingly "safe" social networking sites are not good places to go negative on your employer, because it is too easy for someone to forward what you post to others.

Plus, your company's public relations team is probably going to find a way to be a friend or fan on any website that refers to the company in some way. One of their main responsibilities is to head off any negative or unflattering comments about the company. If those kinds of comments are coming from one of their employees, there is a good chance that person will be found out and fired.

Blogs and Microblogs

Blogs are websites that contain a series of posts written by a person or a team of people. Usually, they include written entries, but there are also an increasing number of photo blogs, video blogs, and audio blogs on the Internet.

Blogs can be more secure than social networking sites, so supervisors and coworkers often use them for making announcements, soliciting new ideas, updating colleagues on projects, and gathering feedback about new ideas and proposed changes in policies.

Microblogs, like Twitter, are similar to regular blogs, except they limit posts to a specific character amount, like 140 characters. In technical workplaces, microblogs are especially useful because they allow colleagues, clients, and customers to "follow" you, your team, or your company, receiving updates when things happen. You can also follow others. Managers are increasingly using microblogs to send out workplace announcements, set up meetings, or alter schedules.

Both blogs and microblogs can be used for collaborating with your team on a project. Here are some tips for setting up and using blogging and microblogging sites.

CHOOSE YOUR BLOG OR MICROBLOG SITE You shouldn't need to pay for a blogging site. Some popular free blogging host sites include Blogger, Wordpress, Blogsome, and Moveable Type (Figure 23.2). Each one has its strengths and weaknesses, so you might look at them all to determine which one will fit your needs and reach the people you want to speak to.

Twitter is still the dominant microblogging site, but there are others like Plurk, Jaiku, and Poodz that you might consider trying out. These sites do basically the same things, though some offer more video or audio capability than others.

START YOUR BLOG Once you choose your blog or microblog host, you should sign up for an account. The home page for the blogging site will usually ask for some basic information, such as your name and e-mail address. It will also let you choose your blog's name and pick a template. You should pick a template that fits your interests and perhaps your company.

Creating Your Blog

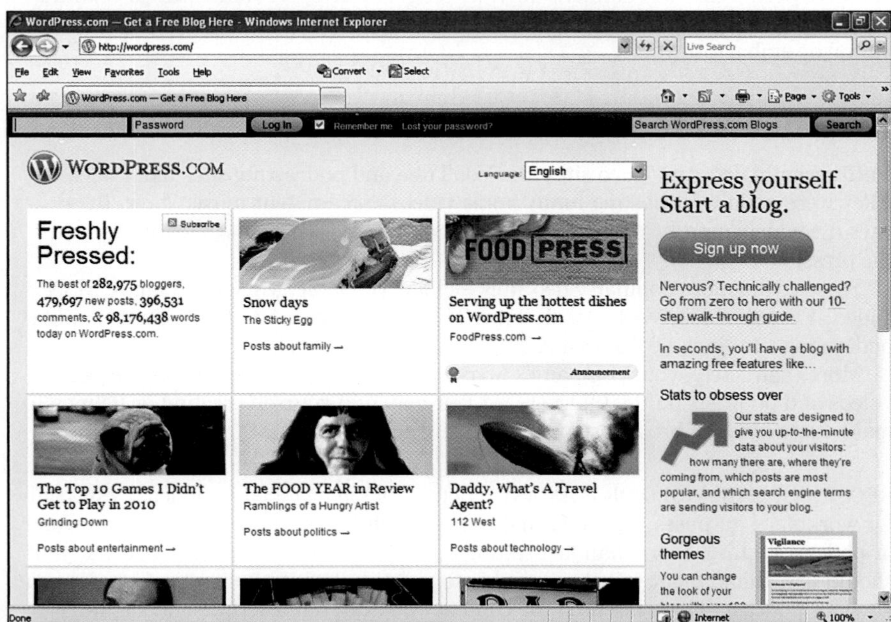

Figure 23.2: Starting a blog is simple. Choose the blog host site that best fits you. Then, sign up for an account. Here is the sign-up page from WordPress, one of the more common blog host sites.

Source: WordPress, http://www.wordpress.com.

WRITE AND MAINTAIN YOUR BLOG On your blogging site, look for the buttons on the screen that allow you to "compose," "edit," and "publish" what you write. You can also personalize your blog by adding photographs, profiles, polls, newsreels, icons, and other gadgets.

Your blog or microblog should be regularly maintained, just like your personal networking site. You can use it to share your ideas and comment on what is happening around you. Keep in mind, though, that a blog can be a public site. If you want to blog about things happening at work, be careful about any information you share or opinions you express. You don't want to say or reveal anything that will cause problems among you, your co-workers, and your supervisors. Also, you don't want to share information that might give your company's competitors proprietary information or some other kind of advantage.

Your blog and microblog are not places to share gossip or complain about people at work. Your unkind comments can be forwarded to people you didn't want to offend.

LET OTHERS JOIN THE CONVERSATION The initial settings for your blogging site will give you strict control over the content of your blog. You and you alone will be able to post comments on your blog. As you grow more comfortable, though, you might want to loosen up your settings to allow others to add comments. If so, you should first decide what kinds of people should be able to make comments on your blog. Then, in your settings, identify the "registered users" who you give permission to comment on your blog. Never open your settings to allow "anyone" to comment, because strangers and spammers will contribute posts that annoy or embarrass you.

Internet Videos and Podcasts

Until recently, Internet video sites like YouTube and podcasting sites like Podcast Alley were mainly for posting funny home videos and amateur music. Now, these sites are widely used by corporations to communicate with clients, customers, and the press.

Some of the more popular video sites include YouTube, MySpace Videos, MSN Video, Yahoo! Video, Veoh, Joost, iFilm, Hulu, Metacafe, and blip.tv. Some popular podcasting sites include Podcast Alley, iTunes, Digg, and Podcast Pickle.

More than likely, your company's marketing team will be responsible for putting videos and podcasts on the Internet, but you may find yourself creating or helping to make these kinds of broadcasts yourself. Here's how to do it.

Link

For more information on creating a presentation, turn to Chapter 21, page 576.

RECORDING THE VIDEO OR PODCAST Before making a video or podcast, especially for workplace purposes, you need to do some careful preparation. Recording a video or podcast requires more than just setting up a camera or microphone. Instead, you should first think about your topic, purpose, readers, and the contexts of use for your video or podcast. Then, research the content and draft out a script. You should also make some strategic decisions about where your video will be filmed and what kinds of background scenery you will need.

GO TO THE NET

Ready to begin blogging? Some good places to get started can be found at **www.pearsonhighered.com/johnsonweb4/23.2**

As you draft your script, keep it as concise as possible. Long, rambling videos and audios tend to bore the audience. Keep the message brief and to the point. Avoid writing or doing anything that will put you or your company in a bad light. Trying to be funny can be risky, especially in corporate videos and podcasts. You also want to avoid broadcasting something that might help your company's competitors or give them information to use against you or your company.

EDITING YOUR VIDEO OR PODCAST One major difference between an amateurish effort and a professional product is editing of the video and sound. Some good video editing software packages include Corel VideoStudio, MS Movie Maker, Adobe Premiere, Final Cut, and iMovie. The most common sound editing software packages for podcasts include Adobe Audition, Audacity, GarageBand, and Cubase (Figure 23.3). One or more of these editing tools may have already been preloaded onto your computer, so look for them in your applications before you buy something new.

Editing software will allow you to cut and paste segments of your broadcast to eliminate parts that you don't want to include in the final product. You can also add titles and transitions, while eliminating any background hiss.

UPLOADING YOUR VIDEOS OR PODCASTS When you have finished editing your video or podcast, go to the host website, like YouTube or Podcast Alley, where you want it to appear. The site will ask you to create an account, and it will ask for some basic information.

Once your account is created, click the "upload" button on the screen. The site will lead you through the process. More than likely, it will ask you for a title and description of your video or podcast. You will also have an opportunity to include some keywords or "tags" that will help people find your video.

Editing Your Podcast

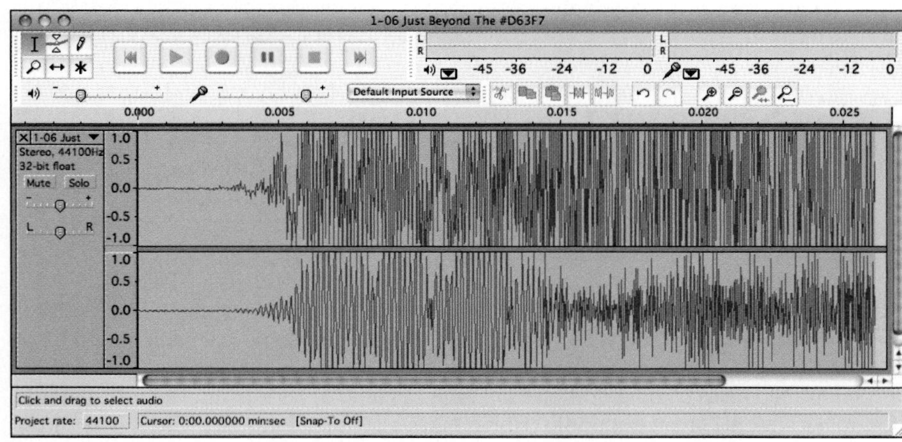

Figure 23.3: You might already have video and editing software installed on your computer. Here is the editing screen from Audacity, which is a good podcast editing tool.

Source: Audacity, http://audacity.sourceforge.net.

GO TO THE NET Want to make your own podcast? Go to **www.pearsonhighered.com/johnsonweb4/23.3**

Wikis

Wikis are websites that let users add and modify the content. You probably know some of the popular wikis, like Wikipedia, WikiHow, and Wikicars. In the technical workplace, wikis are becoming important tools for keeping documentation and specifications up to date and doing customer service. In fact, user-generated wikis are often better than corporate-run websites for troubleshooting.

COMPOSE THE TEXT There isn't anything special about writing a wiki article. You should begin by identifying your subject, purpose, readers, and the contexts in which your article would be used. Research your subject thoroughly and draft the article. Add any graphics or videos. Below your article, you should also create a reference list of your sources and identify any "external links" that readers might want to explore on your topic. Then, edit and proofread your work carefully.

In most cases, you should compose and edit your article completely in a word processor. It is possible to compose in the wiki itself, but the interface is often not as flexible as your word processor, making the work much harder.

POSTING YOUR ARTICLE TO THE WIKI On most wiki homepages, you can find a button that says something like "Create an Article" or "Start the X Article." When you are ready to upload your contribution, click on that button. Then, cut and paste your article into the window provided. Before saving your article in the wiki, edit and proofread one more time. It's easier to catch and correct problems at this point, rather than trying to fix them after the article is posted.

Other people will have the ability to rewrite and edit your wiki article. That's what wikis are all about. So you should regularly return to your articles, especially ones on contentious issues, to make sure no one has added anything inaccurate. The nice thing about wikis is that other people will add to and refine what you wrote. You might be pleasantly surprised by what they have to offer.

HELP

Using Skype for Collaboration

Skype is a web-based application that allows people to communicate in a variety of ways through the Internet. Most people know Skype as a good way to make free long-distance calls through their computer. But Skype has many other features that can be useful for communicating and collaborating with others, especially in university settings.

Corporations have access to web conferencing services, such as GoToMeeting, Adobe Acrobat Connect, IBM Lotus Sametime, and MS Live Meeting, but these services are expensive and universities rarely support them. Skype is a good option for students.

To use Skype as a collaboration tool, each member of your team needs to sign up for an account at Skype.com. Subscribing takes a couple of minutes. Skype can then be used as an application on your computer or even on your mobile phone.

Skype has several features that your team will find useful.

Audio Conference Calling You and your team can all connect at once for a real-time conference call. The audio quality is usually almost as good as a typical conference phone call.

Video Conference Calling Your team can also see each other if you have video cameras installed on your computers (Figure A). Skype can send video feeds to your group members and display their images on your screen. While you discuss the project, they can see you and you can see them.

Instant Messaging While using audio and video with your team, you can also use instant messaging, text messaging, and file sharing. These features are helpful because they allow team members to communicate in writing while others are talking.

Skype can also be used in conjunction with file-sharing programs like Google Docs, allowing you and your team to work on a document as you are talking. If you have a fast enough Internet connection, collaborating through Skype is as good as working in the same room with each other—perhaps even better. After all, Skype allows everyone on your team to be at a computer, working as you talk.

Depending on your computer, you may need a headset to use Skype. Some computers are equipped with adequate microphones, speakers, and video cameras to carry on a real-time conversation. But others struggle with audio interference, especially when more than two people are part of the conference call. So you might want to have a headset ready for use.

Give Skype a try. More than likely, collaboration will increasingly be done through the Internet, so practice using these web-based collaboration tools now.

Skype as a Conference Calling Tool

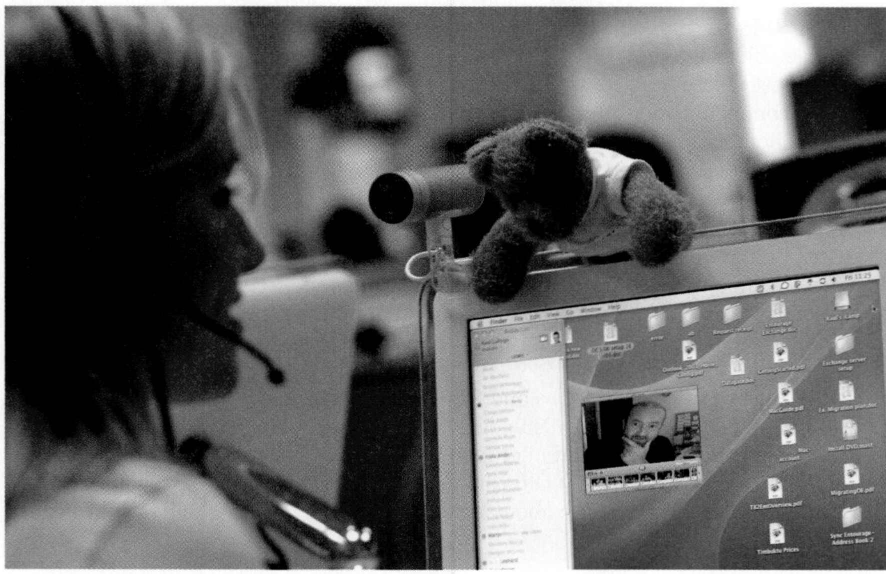

Figure A: Skype is an inexpensive way to collaborate on a project with others.

GO TO THE NET

If you're interested in using Skype, go to **www.pearsonhighered.com/johnsonweb4/23.4**

CHAPTER REVIEW

- Social networking sites are important for both corporate and personal communication.
- Social networking and other Web 2.0 tools are becoming more common in today's workplaces.
- Create both a personal social networking site and a professional site.
- Corporate social networking sites need to stay fresh and interactive with their "fans."
- Blogs and microblogs are websites that allow you to make regular commentaries about issues that interest you.
- You can create videos and podcasts with a variety of software products, including ones that are probably already loaded on your computer.
- A wiki is a website that lets users add articles and modify the existing content.

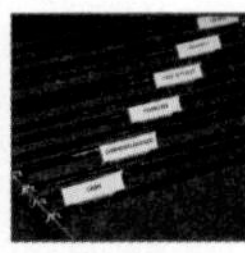

EXERCISES AND PROJECTS

Individual or Team Projects

1. Write a white paper in which you explore the use of social networking, blogs, microblogs, and wikis in your future career. Do some research on the Internet and conduct at least one e-mail or text interview with someone who is working in your field. How are these Web 2.0 tools being used now? How will they likely be used in the future? What are some other ways you can imagine these tools being used in the workplace?

2. Set up a blog or microblog of your own. Go to one of the free blogging sites, like Blogger, Blogspot, or Wordpress. Sign up for an account. Then, each day for a week, post at least one comment related to your major or future career. Your posts can be comments about things that are happening in your field, including comments about stories or companies you have found on the Internet. Send the blog address to your instructor.

3. Find a Web 2.0 application that is not mentioned in this chapter. Write a technical description of this application, introducing it to the other members of your class. What are its major features? How are people using this Web 2.0 application? What are some possible uses for this application in the future, especially in the technical workplace? Send your technical description to your class and your instructor.

4. Create a list of the "Top 10 Dos and Don'ts" for social networking. Start out by listing the items for a personal site, like Facebook or MySpace. Then, convert your list into something that would offer helpful tips for using social networking sites in the technical workplace. As you create and convert your Top 10 list, you might do an Internet search for articles about how social networking can help or harm your job search.

Collaborative Project

Imagine that your group has been hired to create or renovate the Facebook site for your college or university. Start out by first exploring the rhetorical situation that shapes how this site works and how people use it (subject, purpose, readers, context of use). What are its current strengths? What are its weaknesses?

Now, rethink the rhetorical situation to sharpen the focus of the site. What kind of content do you think would be appropriate for the readers of this site? What do you think the site should try to achieve? How can the site be improved to attract the kinds of readers it seeks? How could the site get these readers involved and keep them coming back?

Write a two-page proposal to your university's public relations department in which you describe how the Facebook site could be improved. Describe the problems with the current site, show how the site could be improved, and explain why making these changes would make the site more effective.

To make this collaborative project more challenging, try to accomplish all this work through Skype and/or Google Docs.

For additional technical writing resources, including interactive sample documents, document design tutorials and guidelines, and more, go to **www.mytechcommlab.com.**

CASE STUDY

My Boss Might Not "Like" This

Henry Blackburg was a biomedical engineer for a manufacturer of CAT scan machines in Stamford, Connecticut. For the most part, he was good at keeping his professional and personal lives separate. People at work didn't even know he was gay, and he liked it that way. When they asked about whether he was seeing anyone or whether they could fix him up with a female date, he would tell them he was dating someone in New York, which was true. He went to New York regularly, so his cover worked well.

Henry's boss, John Hamilton, occasionally made negative comments about gays, and he was not particularly open-minded. Henry didn't think he would be fired or held back if his boss found out about his sexual orientation, but he didn't want to take any chances.

Facebook was Henry's lifeline to the gay community in New York. It was how he stayed in touch and found out where social events were going to happen. His friends would also post photos from weekend parties. They were tame by most standards, but it was obvious that these weren't heterosexual events.

Until last week, all was well. Henry's boss had probably heard of Facebook, but he didn't have an account. The company, however, encouraged everyone to get on Facebook to see the new corporate fansite. So Henry's boss signed up. Facebook began prompting him to friend people from his high school, university, and, yes, his workplace. So, Henry's boss sent him a friend request.

Henry didn't know what to do. He wasn't embarrassed about being gay, but he wanted to keep that aspect of his personal life separate from his work life. Plus, he didn't want to "straighten up" his Facebook site to keep his boss from figuring out he was gay. There was no way he could be a friend with his boss and continue to stay connected with his real friends.

Henry ignored his boss's friend request, hoping it was just a passing thing. A few days later in a meeting, though, his boss announced that he would be sending work-related updates over Facebook, so if anyone hadn't responded to his friend request, they should do so right away.

Other people in the room looked uneasy. Henry realized he wasn't the only person who didn't want to be a "friend" with the boss on Facebook.

How do you think Henry should handle this situation?

GO TO THE NET

For more advice about Facebook issues and the workplace, go to **www.pearsonhighered.com/johnsonweb4/23.5**

APPENDIX

A Grammar and Punctuation Guide

This guide is a reference tool to help you handle mechanical issues in writing. The guide has two parts: (1) the Top Ten Grammar Mistakes, and (2) a Punctuation Refresher. The Top Ten Grammar Mistakes will show you how to avoid the ten most common grammar mistakes. The Punctuation Refresher will update you on punctuation rules, showing you how to properly use punctuation marks.

Consistent grammatical correctness and punctuation are essential if you are going to clearly express yourself, especially in technical documents and presentations. Any grammar and punctuation mistakes in your writing will undermine even your best ideas. Moreover, readers will make judgments about you and your work based on your grammar and punctuation. If you send a document littered with grammatical errors to your supervisor or your company's clients, they will question your attention to quality, your commitment to the project, and even your intelligence.

Mastering basic grammar and punctuation does not take long. If you haven't worked on improving your grammar since high school, take some time to refresh these rules in your mind.

The Top Ten Grammar Mistakes

Don't be one of those people who is always apologizing with statements like, "I'm not good at grammar." These kinds of statements do not excuse your grammatical problems; they only indicate that you have not taken the time to learn basic grammar rules. If you are one of these grammar apologists, commit to spending the little time necessary to master the rules. An hour or two of study will make those grammatical problems disappear.

Comma Splice

A significant percentage of grammar errors are comma splices. A comma splice occurs when two complete sentences are joined together with a comma.

Incorrect

The machine kept running, we pulled the plug.

We moved the telescope just a little to the left, the new nova immediately came into view.

In these examples, notice how the parts before and after the commas could stand alone as sentences. The comma is *splicing* these sentences together.

How can you fix these comma splices? There are a few options.

Correct

The machine kept running, so we pulled the plug.

We moved the telescope just a little to the left, and the new nova immediately came into focus.

(Add a conjunction [*so, and, but, yet*] after the comma.)

Correct The machine kept running; we pulled the plug.

We moved the telescope just a little to the left; the new nova immediately came into focus.

(Replace the comma with a semicolon.)

The machine kept running. We pulled the plug.

We moved the telescope just a little to the left. The new nova immediately came into focus.

(Replace the comma with a period and create two grammatically correct sentences.)

The machine kept running; therefore, we pulled the plug.

We moved the telescope just a little to the left; consequently, the new nova immediately came into view.

(Replace the comma with a semicolon and add a conjunctive adverb [*therefore, consequently, however, thus*].)

Since the machine kept running, we pulled the plug.

Because we moved the telescope just a little to the left, the new nova immediately came into view.

(Insert a subordinating conjunction [*since, because*] at the beginning of the sentence.)

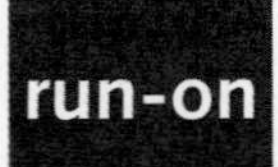

Run-On Sentence

The run-on sentence error is a close cousin of the comma splice. In a run-on sentence, two or more sentences have been crammed into one.

Incorrect The computer suddenly crashed it had a virus.

The Orion nebula lies about 1500 light-years from the sun the nebula is a blister on the side of the Orion molecular cloud that is closest to us.

Run-on sentences are corrected the same way comma splices are corrected. You can use conjunctions (*and, but, or, nor, for, because, yet, however, furthermore, hence, moreover, therefore*) to fix them. Or, you can divide the sentences with a semicolon or period.

Correct The computer suddenly crashed because it had a virus.

The computer suddenly crashed; we guessed it had a virus.

The computer suddenly crashed. It had a virus.

The Orion nebula lies about 1500 light-years from the sun. The nebula is a blister on the side of the Orion molecular cloud that is closest to us.

Want more information on comma splices? Go to **www.pearsonhighered.com/johnsonweb4/A.1**

The Orion nebula lies about 1500 light-years from the sun; moreover, the nebula is a blister on the side of the Orion molecular cloud that is closest to us.

In most cases, run-on sentences are best fixed by adding a period, thus separating the two sentences completely.

frag

Fragment

A fragment, as the name suggests, is an incomplete sentence. A fragment typically occurs when the sentence is missing a subject or a verb, or it lacks a complete thought:

Incorrect

Because the new motherboard was not working.

(This fragment contains a subject and verb but does not express a complete thought.)

The report missing important data.

(This fragment contains a subject but no verb.)

The first fragment can be corrected in the following ways:

Correct

The new motherboard was not working.

(Remove the conjunction [*because*].)

Because the new motherboard was not working, we returned it to the manufacturer.

(Join the fragment to a complete sentence [an independent clause].)

The second fragment can be corrected in the following ways:

Correct

The report was missing important data.

(Insert a verb [*was*].)

The report, missing important data, was corrected immediately.

(Insert a verb [*was corrected*] and an adverb [*immediately*].)

Sometimes writers, especially creative writers, will use fragments for the purpose of jarring their readers. In technical writing, you don't want to jar your readers. Leave the creative uses of fragments to creative writers.

dm

Dangling Modifier

A dangling modifier occurs when a phrase does not properly explain the subject.

Incorrect

While eating lunch, the acid boiled over and destroyed the testing apparatus.

(The acid is apparently eating lunch while it does damage to the testing apparatus. That's some acid!)

Worksheets on run-ons, fragments, and comma splices are available at **www.pearsonhighered.com/johnsonweb4/A.2**

After driving to Cleveland, our faithful cat was a welcome sight.

(The cat apparently drove the car to Cleveland. Bad kitty!)

These kinds of errors are common—and often funny. To avoid them (and your readers' grins), make sure that the introductory phrase is modifying the subject of the sentence. Here are corrections of these sentences:

Correct

While we were eating lunch, the acid boiled over and destroyed the testing apparatus.

After driving to Cleveland, we were glad to see our faithful cat.

Notice how the information before the comma modifies the subject of the sentence, which immediately follows the comma.

Subject-Verb Disagreement

Subject-verb disagreements occur when the subject of the sentence does not match the verb. Singular subjects should go with singular verbs, while plural subjects should have plural verbs. Here are a few sentences with subject-verb disagreement:

Incorrect

The windows, we discovered after some investigation, was the reason for heat loss in the house.

(The singular verb *was* does not match the plural subject *windows*.)

The robin, unlike sparrows and cardinals, do not like sunflower seeds.

(The subject *robin* is singular, while the verb phrase *do not like* is plural.)

Either my DVD player or my stereo were blowing the fuse.

(The subject *DVD player or my stereo* is singular, while the verb *were* is plural.)

Here are the correct sentences:

Correct

The windows, we discovered after some investigation, were the reason for heat loss in the house.

Robins, unlike sparrows and cardinals, do not like sunflower seeds.

Either my DVD player or my stereo was blowing the fuse.

When *or* is used in the subject, as it is in the third example, the verb needs to agree with the noun that follows *or*. In this sentence, the use of *or* means we are treating the DVD player and the stereo separately. So a singular verb is needed.

This third example brings up an interesting question: What if one or both of the words flanking the *or* is plural? Again, the answer is that the verb will agree with the noun that comes after the *or*.

The speeding cars or the reckless motorcyclist was responsible for the accident.

Either the falling branches or the high winds were responsible for the damage.

Some great grammar handbooks are online. Go to **www.pearsonhighered.com/johnsonweb4/A.3**

Collective nouns (*crowd, network, group*) that name a group as a whole take a singular verb:

The group uses a detailed questionnaire to obtain useful feedback.

Pronoun-Antecedent Disagreement

Pronoun-antecedent disagreement usually occurs when a writer forgets whether the subject is plural or singular.

Incorrect

Anyone who thinks Wii is better than Sony Playstation should have their head examined.

Like the scientists, we were sure the rocket was going to blast off, but it wasn't long before you knew it was a dud.

In these cases, words later in the sentence do not agree with pronouns earlier in the sentence. In the first sentence, *anyone* is singular while *their* is plural. In the second sentence, *we* is a first-person noun while *you* is a second-person pronoun. Here are a couple of ways to correct these sentences:

Correct

People who think Wii is better than Sony Playstation should have their heads examined.

(The subject is made plural [*people*] and an *s* is added to *head* to make it plural.)

Anyone who thinks Wii is better than Sony Playstation should have his or her head examined.

(The subject [*anyone*] is kept singular, and the plural *their* is changed to the singular *his or her*.)

Like the scientists, we were sure the rocket was going to blast off, but it wasn't long before I knew it was a dud.

(The *you* was changed to *I*, keeping the whole sentence in first person.)

Like the scientists, we were sure the rocket was going to blast off, but it wasn't long before we knew it was a dud.

(The *you* was changed to *we*, again keeping the whole sentence in first person.)

In most cases, the secret to avoiding these subtle errors is to check whether the pronouns later in the sentence match the subject earlier in the sentence.

Faulty Parallelism

Lists can be difficult to manage in sentences. A good rule of thumb is to remember that each part of the list needs to be parallel in structure to every other part of the list.

Incorrect After the interview, we went out for dinner, had a few drinks, and a few jokes were told.

(In this sentence, the third part of the list, *jokes were told,* is not parallel to the first two parts, *went out for dinner* and *had a few drinks*.)

Our survey shows that people want peace, want to own a home, and that many are worried about their jobs.

(In this sentence, the second item in the list, *want to own a home,* is not parallel to the first and third items.)

To correct these sentences, make the items in the list parallel.

Correct After the interview, we went out for dinner, had a few drinks, and told a few jokes.

Our survey shows that people want peace, most want to own a home, and many are worried about their jobs.

To avoid faulty parallelism, pay special attention to any lists you write. Check whether the items are parallel in phrasing.

pc

Pronoun Case Error (*I* and *Me, We* and *Us*)

Often, people are confused about when to use *I* or *me* and *we* or *us.* Here is a simple way to make the right decision about which one to use: If you are using the word as the subject of the sentence or phrase, use *I* or *we.* Anywhere else, use *me* or *us.*

Incorrect Jones's team and me went down to the factory floor to see how things were going.

(The word *me* is misused here because it is part of the subject of the sentence. In this case, *I* should have been used.)

When the roof fell in, the manager asked Fred and I to start developing a plan for cleaning up the mess.

(The word *I* is misused in this sentence because the phrase *Fred and I* is not the subject. In this case, the phrase *Fred and me* should have been used.)

Things were getting pretty ugly in there, so us unimportant people slipped out the back door.

(The phrase *us unimportant people* is the subject of the phrase that follows the comma. Therefore, the phrase *we unimportant people* should have been used.)

Remember, if the word is being used as the subject of a phrase or sentence, use *I* or *we.* Anywhere else, use *me* or *us.* Here are the correct versions of these sentences:

Correct Jones's team and I went down to the factory floor to see how things were going.

For practice avoiding parallelism problems, go to **www.pearsonhighered.com/johnsonweb4/A.4**

When the roof fell in, the manager asked Fred and me to start developing a plan for cleaning up the mess.

Things were getting pretty ugly in there, so we unimportant people slipped out the back door.

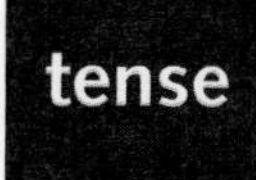

Shifted Tense

Sentences can be written in past, present, or future tense. In most cases, neighboring sentences should reflect the same tense. Shifting tenses can make readers feel like they are hopping back and forth in time.

Incorrect

Few countries possess nuclear weapons, but many countries tried to build them.

(Here, *possess* is in present tense, while *tried* is in past tense.)

Parts flew everywhere on the factory floor as the robot finally breaks down.

(Here, *flew* is in past tense, while *breaks* is in present tense.)

The advances in microchip technology allowed electronics to become much smaller, which leads to today's tiny electronic devices.

(The word *allowed* is in past tense, while *leads* is in present tense.)

We found ourselves staring in disbelief. The excavation site is vandalized by people who wanted to have a campfire.

(Here, *found* is in past tense, while *is vandalized* is in present tense.)

To revise these sentences, make the tenses consistent.

Correct

Few countries possess nuclear weapons, but many countries are trying to build them.

Parts flew everywhere on the factory floor as the robot finally broke down.

The advances in microchip technology allowed electronics to become much smaller, which led to today's tiny electronic devices.

We found ourselves staring in disbelief. The excavation site had been vandalized by people who wanted to have a campfire.

Tense shifts are not always wrong. Sometimes a sentence or phrase needs to be in a different tense than those around it. When checking sentences for unnecessary tense shifts, look for places where tense shifts cause more confusion than clarity.

vp

Vague Pronoun

Occasionally, a writer uses a pronoun, seeming to know exactly who or what the pronoun refers to, while readers are left scratching their heads, trying to figure out what the writer means.

Incorrect Fred and Javier went to the store, and then he went home.

(Who does *he* refer to, Fred or Javier?)

They realized that the inspection of the building was not going well. It was fundamentally unsound.

(In this sentence, *It* could refer to the inspection or the building.)

We really had a great week. Our program review went well, and we made huge strides toward finishing the project. This is why we are taking all of you to lunch.

(What does *This* refer to? The great week? The program review? The huge strides? All of them?)

The camera captured the explosion as it ripped apart the car. It was an amazing experience.

(In these sentences, the multiple uses of *it* are confusing. In the first sentence, *it* might refer to the camera or the explosion. In the second sentence, *It* might refer to the taking of the picture or the explosion. Or, the final *It* might just be a weak subject for the second sentence and doesn't refer directly to anything in the previous sentence.)

Correcting these sentences mostly involves rewording them to avoid the vague pronoun.

Correct Fred and Javier went to the store, and then Javier went home.

They realized that the inspection of the building was not going well. The inspection process was fundamentally unsound.

We really had a great week. Our program review went well, and we made huge strides toward finishing the project. For these reasons, we are taking all of you to lunch.

The camera captured the explosion as it ripped apart the car. Seeing the explosion was an amazing experience.

A common cause of vague pronoun use is the overuse of "It is . . . " and "This is . . . " to begin sentences. These kinds of sentences force readers to look back at the previous sentence to figure out what *It* or *This* refers to. In some cases, two or three possibilities might exist.

To avoid these problems, train yourself to avoid using "It is . . . " and "This is . . . " sentences. Occasionally, these sentences are fine, but some writers rely on them too much. You are better off minimizing their use in your writing.

Punctuation Refresher

More than likely, you already have a good sense of the punctuation rules. However, there are quirks and exceptions that you need to learn as your writing skills advance to a new level. So spend a little time here refreshing your memory on punctuation rules.

Period, Exclamation Point, Question Mark

Let's start with the most basic marks—the period, exclamation point, and question mark. These punctuation marks signal the end of a sentence, or a full stop.

We need to test the T6 Robot on the assembly line.

The acid leaked and burned through the metal plate beneath it!

Where can we cut costs to bring this project under budget?

The period signals the end of a standard sentence. The exclamation point signals surprise or strong feelings. The question mark signals a query.

Periods can also be used with abbreviations and numbers.

C.E.

Feb.

Fig. 8

Dr. Valerie Hanks

56.21 cm

Question marks can also be used in a series.

How will global warming affect this country? Will it flood coastal cities? Create drought in the southwest? Damage fragile ecosystems?

Using periods, exclamation points, and question marks with quotes

A common mistake with periods, exclamation points, and question marks is their misuse with quotation marks. In almost all cases, these punctuation marks are placed inside the quotation marks.

Incorrect He said, "The audit team reported that we are in compliance".

We asked him, "Do you really think we are going to finish the project on time"?

Correct He said, "The audit team reported that we are in compliance."

We asked him, "Do you really think we are going to finish the project on time?"

The one exception to this rule is when a quoted statement is placed within a question.

Did he really say, "The company will be closing the Chicago office"?

Commas

In the English language, commas are the most flexible, useful, and therefore problematic punctuation mark. In most cases, a comma signals a pause in the flow of a sentence.

For websites on punctuation, go to **www.pearsonhighered.com/johnsonweb4/A.5**

When he hiked in the mountains east of Ft. Collins, he always took along his compass and map to avoid getting lost.

(This comma signals that an introductory phrase is finished.)

My PDA is a helpful organizing tool, and it makes a great paperweight too.

(This comma signals that two independent clauses are joined with a conjunction [*and, yet, but, so, because*].)

Our company's CEO, the engineering genius, is scheduled to meet with us tomorrow.

(These commas set off information that is not essential to understanding the sentence.)

We are, however, having some luck locating new sources of silicon on the open market.

(These commas set off a conjunctive adverb [*therefore, however, furthermore, nevertheless, moreover*].)

The archaeological dig yielded pottery shards, scraping tools, and a corn meal grinding stone.

(These commas separate items in a series.)

Using commas with quotation marks

Using commas with quotation marks can be problematic. Just remember that commas almost always go before, not after, the quotation mark.

Albert Einstein said, "God does not play dice."

(Here the comma sets off a speaker tag. Notice that the comma is placed before the quotation mark.)

"I'm having trouble hearing you," she said, "so I'm switching over to a new phone."

(Again, the commas in this sentence offset the speaker tag. Note that both commas come before the quotation marks.)

Using commas in numbers, dates, and place names

Commas are also used with numbers, dates, and place names.

Reporters estimated that the rally drew nearly 10,000 people.

We first noticed the problem on January 10, 2009.

For great pizza you need to go to Uncle Pete's in Naperville, Illinois.

Need more help with commas? Go to **www.pearsonhighered.com/johnsonweb4/A.6**

Removing excess commas

There is some flexibility in the use of commas. Many editors recommend an "open" punctuation style that eliminates commas where they do not help comprehension. For example, the following sentences are both correct:

> In minutes, she whipped up the most amazing apple pie.
>
> In minutes she whipped up the most amazing apple pie.

Here, the comma after *minutes* is not aiding understanding, so it can be removed. Be careful, though. Sometimes the lack of a comma can cause some confusion.

Confusing Soon after leaving the airplane needed to turn back for mechanical reasons.

Not Confusing Soon after leaving, the airplane needed to turn back for mechanical reasons.

Semicolon and Colon

Semicolons and colons are less common than periods and commas, but they can be helpful in some situations. In a sentence, semicolons and colons signal a partial stop.

The trick to properly using these marks is to remember this simple rule: In most cases, the phrase on either side of a semicolon or colon should be able to stand alone as a separate sentence (an independent clause).

> We were not pleased with the results of the study; however, we did find some interesting results that gave us ideas for new research.
>
> > (The semicolon joins two independent clauses. The second clause [the part starting with *however*] supports the first clause.)
>
> Commuting to work by car requires nerves of steel: Each mile brings you into contact with people who have no respect for the rules of the road.
>
> > (Here, the colon also divides two independent clauses. In this example, though, the colon signals that the two parts of the sentence are equivalent to each other.)

How do you know when to use a semicolon or a colon? It depends on whether the part of the sentence following the mark is *lesser than* or *equal to* the first part. If lesser, use the semicolon. If equal, use the colon.

If you want to avoid problems, use semicolons and colons only when necessary. When considering these punctuation marks in a sentence, you should ask yourself whether a period or a conjunction (*and, but, so, yet*) would make the sentence easier to understand. After all, joining sentences together with semicolons and colons can often create long, difficult-to-read sentences.

As with most punctuation rules, there are exceptions to the rule that semicolons and colons are used with independent clauses.

GO TO THE NET Websites to help you master the semicolon and colon are available at **www.pearsonhighered.com/johnsonweb4/A.7**

Using semicolons in a series

Semicolons can be used to punctuate complicated lists in a sentence.

We have offices in Boston, Massachusetts; Freetown, New York; and Sedona, Arizona.

Using colons to lead off lists

Colons can be used to signal a list.

Four steps are required to complete the process: (1) preparing the workspace, (2) assembling the model, (3) painting the model, and (4) checking quality.

Keep in mind the following issues when searching for a job:

- You can't get the job if you don't apply.
- Jobs don't always go to the people with the most experience.
- What makes you different makes you interesting.

Using colons in titles, numbers, and greetings

Colons are commonly used in titles of books and articles, in numbers, and in greetings in letters and memos.

The Awakening: The Irish Renaissance in Nineteenth-Century Boston

Genesis 2:18

11:45 A.M.

They won by a 3:1 ratio.

Dear Mr. Franklin:

Using semicolons and colons with quotation marks

Unlike commas and periods, semicolons and colons should appear after the quotation mark in a sentence.

Land Commissioner George Hampton claimed, "The frogs will survive the draining of the lake"; but he was clearly wrong.

One of my favorite chapters in Leopold's *A Sand County Almanac* is "Thinking Like a Mountain": This essay is his best work.

Whenever possible, though, you should avoid these kinds of situations. In both of these examples, the sentences could be rearranged or repunctuated to avoid these awkward, though correct, uses of the semicolon or colon.

Misusing the colon

A common misuse of the colon is using it with an incomplete sentence.

Incorrect The reasons for our dissatisfaction are: low quality, late work, and slow response.

(The colon is misused, because the phrase before the colon cannot stand alone as a separate sentence.)

For example:

(Again, the phrase before the colon cannot stand alone as a sentence. In this case, a dash or comma should be used. Or, turn the phrase *For example* into a complete sentence.)

In his report, Bill Trimble claims: "We have a golden opportunity to enter the Japanese market."

(Yet again, the information before the colon cannot exist as a separate sentence. In this case, a comma should have been used instead of the colon.)

As a rule, the information before the colon should *always* be able to stand alone as a complete sentence. Here are the correct versions of these sentences:

Correct The reasons for our dissatisfaction are the following: low quality, late work, and slow response.

For example, consider these interesting situations:

In his report, Bill Trimble makes this important statement: "We have a golden opportunity to enter the Japanese market."

Notice how all three of these examples have independent clauses (full sentences) before the colon.

Apostrophe

The apostrophe has two important jobs in the English language: (1) to signal contractions and (2) to signal possession.

Using an apostrophe to signal a contraction

An apostrophe that signals a contraction identifies the place where two words have been fused and letters removed.

They're going to the store today.

He really isn't interested in the project.

They shouldn't have taken that road.

Contractions should be used only in informal writing. They signal a familiarity with the readers that could seem too informal in some situations. Some other common contractions include *won't, it's, I'm, you've, wouldn't,* and *couldn't.*

Using an apostrophe to signal possession

An apostrophe is also used to signal possession. With a singular noun, an *'s* is added to signal possession. Joint possession is usually signaled with an *s'*.

We have decided to take Anna's car to the convention.

The players' bats were missing before the game.

When plural nouns do not end in an *s,* you should use an *'s* to create the plural.

We rode the children's bikes.

The men's briefcases were left near the door.

When singular nouns end in an *s,* you should add an *'s* to show possession.

They met in Mary Jones's office.

Charles's computer was shorting out.

Using apostrophes to show possession with two or more nouns

When you are showing possession with multiple nouns, your use of the apostrophe depends on your meaning. If two nouns are acting as one unit, only the last noun needs an apostrophe to signal possession.

We decided to accept Grim and Nether's proposal.

But if you are signaling possession for several separate nouns, each needs an apostrophe.

I found it difficult to buy meaningful gifts for Jane's, Valerie's, and Charles's birthdays.

Using apostrophes to signal plurals of numbers, acronyms, and symbols

You can use apostrophes to signal plurals of numbers, acronyms, and symbols, but do so sparingly. Here are a couple of situations where apostrophes would be appropriate:

The *a*'s just kept appearing when I typed *x*'s.

Is it necessary to put ©'s on all copyrighted documents?

In most cases, though, do not include apostrophes to show a plural if they do not aid the meaning of the text.

The police discovered a warehouse full of stolen TVs.

The 1870s were a tough time for immigrants.

In the basement, a crate of dead CPUs sat unnoticed.

Quotation Marks

Quotation marks are used to signal when you are using someone else's words. Quotation marks should not be used to highlight words. If you need to highlight words, use italics.

Using quotation marks to signal a quote

Quotation marks are used to frame an exact quotation from another person.

In *The Panda's Thumb,* Gould states, "The world, unfortunately, rarely matches our hopes and consistently refuses to behave in a reasonable manner."

More help with quotation marks is available at
www.pearsonhighered.com/johnsonweb4/A.8

"Not true" was her only response to my comment.

He asked me, "Are you really working on that project?"

Use quotation marks only when you are copying someone else's exact words. If you are only paraphrasing what someone else said, do not use quotation marks.

In *The Panda's Thumb,* Gould argues that nature often does not meet our expectations, nor does it operate in predictable ways.

She rejected my comment as untrue.

He asked me whether I was working on the project.

Also, when paraphrasing, avoid the temptation to highlight words with quotation marks.

Using quotation marks to signal titles

Titles of works that are part of larger works, like articles, songs, or documents, should be set off with quotation marks.

Time published an article called "The Silicon Valley Reborn."

A classic blues tune covered by the Yardbirds was "I'm a Man."

The report, "Locating Evidence of Ancient Nomads in Egypt," is available online.

Titles of books and other full works should not be set in quotation marks. They should be italicized.

Taking the Quantum Leap, by Fred Wolf, is a very helpful book.

Revolver is one of the Beatles' best albums.

Using single quotation marks to signal a quote or title within another quote

When quoting something within another quote, you should use single quotation marks to set it off.

Tim Berra shows the weakness of the creationist argument by quoting one of its strongest advocates: "Morris wrote 'the only way we can determine the true age of the earth is for God to tell us what it is.'"

One of the physicists at the conference remarked, "I cannot believe that Einstein's 1905 paper on special relativity, 'On the Electrodynamics of Moving Bodies,' is already a century old."

Using quotation marks to signal irony

Quotation marks are often used incorrectly to highlight words and slang terms.

Incorrect One problem with "free-trade policies" is that the laborers who work for Third World countries work almost for free.

I found working with her to be "wonderful," because she is so "attentive" and "understanding."

The sentences above do not need quotation marks to set off these words. If you do need to highlight words that are not direct quotes, use italics.

You can, however, use quotation marks to signal irony by quoting another person's misuse of a term or phrase.

Conservation is more than Vice President Cheney's notion of a "personal virtue."

The Matrix is an entertaining film, but it's hard to accept the "biblical significance" that Clarke and others claim for this highly violent movie.

Using quotation marks with in-text citations

One of the exceptions to placing periods inside quotation marks is when quotation marks are used with in-text citations.

In his article on ancient dams, Abbas points out that "water-driven power systems have been around for thousands of years" (p. 67).

Here, note that the period comes after the in-text citation, not within the closing quotation mark.

Dashes and Hyphens

The uses of dashes and hyphens follow some rather specific rules. There are actually two types of dashes, the "em dash" and the "en dash." The em dash is the longer of the two (the width of an *m*), and it is the more widely used dash. The en dash is a bit shorter (the width of an *n*), and it is less widely used. Hyphens are shorter than the two dashes.

Using em dashes to highlight asides from the author or the continuation of a thought

An em dash is typically used to insert comments from the author that are asides to the readers.

At the meeting, Hammons and Jenkins—this is the ironic part—ended up yelling at each other, even though they both intended to be peacemakers.

We must recognize the continuing influence of Lamarckism in order to understand much social theory of the recent past—ideas that become incomprehensible if forced into the Darwinian framework we often assume for them.

An em dash can be made with two hyphens (--). Most word processors will automatically change two dashes into an em dash. Otherwise, a series of keystrokes (usually, shift-command-hyphen) will create this longer dash (—).

For more help with dashes, hyphens, parentheses, and brackets, go to **www.pearsonhighered.com/johnsonweb4/A.9**

Using en dashes in numbers and dates

It might seem trivial, but there is a difference between en dashes and em dashes. An en dash is almost always used with numbers and dates.

> Copernicus (1473–1543) was the first European to make a cogent argument that the earth goes around the sun rather than the sun going around the earth.

> Young and Chavez argue conclusively that Valles Bonita is really a dormant sunken volcano, called a *caldera* (pp. 543–567).

As you can see in these examples, the en dash is slightly shorter than the em dash.

Using the hyphen to connect prefixes and make compound words

The hyphen is mainly used to connect prefixes with words or to connect two or more words to form compound words.

> neo-Platonists
>
> one-to-one relationship
>
> trisomy-21
>
> four-volume set of books

One thing you should notice is how hyphens are used to create compound adjectives but not compound nouns. You can write "four-volume set of books," where *four-volume* is an adjective. But you would need to write "the four volumes of books," because the word *volumes* is being used as a noun. Hyphens are generally used to make compound adjectives, but not compound nouns.

Parentheses and Brackets

Parentheses and brackets are handy for setting off additional information, like examples, definitions, references, lists, and asides to the readers.

Using parentheses to include additional information

Parentheses are often used to include additional information or refer readers to a graphic.

> When hiking through the Blanca Mountains, you will be surprised by the wide range of animals you will see (e.g., elk, deer, hawks, eagles, and the occasional coyote).

> The data we collected show a sharp decline in alcohol use when teens become involved in constructive, nontelevision activities (see Figure 3).

> These unicellular organisms show some plant-like features (many are photosynthetic) and others show more animal-like features.

Using parentheses to clarify a list

Parentheses can be used to clarify the elements of a long list.

> When meeting up with a bear in the wild, (1) do not run, (2) raise your arms to make yourself look bigger, (3) make loud noises, and (4) do not approach the animal.

> Only three things could explain the mechanical failure: (1) the piston cracked, (2) one of the pushrods came loose, or (3) the head gasket blew.

Using brackets to include editorial comments or to replace a pronoun

Brackets are less common than parentheses, but they can be helpful for inserting editorial comments or replacing a pronoun in a quote.

> Though pictures of the moon are often spectacular, *any view of the moon from earth is slightly blurred* [emphasis mine].

> Shea points out, "Whether he intended it or not, [Planck] was the originator of the quantum theory."

In this second example, the second *he* was replaced with *Planck* to make the meaning of the quote clearer.

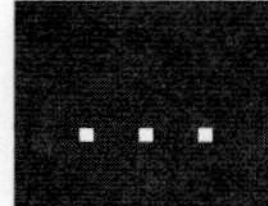

Ellipses

Ellipses are used to show that information in a quote was removed or to indicate the trailing off of a thought. Ellipses are made with three dots, with spaces between each dot (. . .), not (...).

Using ellipses to signal that information in a quote has been removed

Sometimes a passage, especially a longer one, includes more information than you want to quote. In these cases, ellipses can be used to trim out the excess.

> As historian Holton writes, "What Bohr had done in 1927. . . was to develop a point of view that allowed him to accept the wave-particle duality as an irreducible fact" (117).

Using ellipses to show that a thought is trailing off

At the end of a sentence, you might use ellipses to urge the reader to continue the thought.

> For those who don't want to attend the orientation, we can find much less pleasant ways for you to spend your day. . . .

When ellipses end the sentence, use an additional dot to make four (. . . .). The extra dot, after the last word, is a period that signals the end of the sentence.

APPENDIX B

English as a Second Language Guide

The English language is a composite of many languages. As a result, English has many maddening exceptions and inconsistencies in spelling, syntax, and usage. If English is not your native language, you should pay special attention to these irregularities so that your writing and speaking will be consistent with the writing and speaking of fluent speakers.

This English as a Second Language (ESL) guide will not help you learn English. Instead, this guide concentrates on three major sources of irregularities in technical English—use of articles, word order of adjectives and adverbs, and verb tenses. As you master the English language, you should first concentrate on understanding these three sources of irregularities. Then, you can turn to other ESL resources to help you refine your use of English.

Using Articles Properly

Perhaps the most significant source of ESL problems is the use of articles (*a, an, the*) in English. For example, the sentences "The computer broke down" and "A computer broke down" have significantly different meanings in English. The use of *the* suggests that a specific computer broke down. The use of *a* suggests that one computer—among many computers—broke down.

Using *the* to refer to specific items

If you are referring to a specific item, use *the* to signal the noun.

> The planet Saturn has been bright during the last week.
>
> The car stalled, so we started walking to the nearest service station.
>
> The professor asked us to work harder on the next assignment.

Using *a* to refer to nonspecific items

If you are referring to a nonspecific item, use *a* to signal the noun.

> A planet was found circling a star in the Orion system.
>
> A car stalled in the road, so we needed to drive around it.
>
> A professor asked us to attend the party.

Misusing an article with uncountable things

Articles should be used only with things that can be counted, like *the eight cars, the five bikes,* or *an orange.* When items are not counted or cannot be counted, do not use an article.

Links to ESL sources on the Internet are available at **www.pearsonhighered.com/johnsonweb4/B.1**

He surfaced for air.

They decided to have tea with dinner.

Rice grown in Asia tastes better than rice grown in North America.

Putting Adjectives and Adverbs in the Correct Order

Compared to some languages, English is flexible in its syntax. Nevertheless, word order in sentences is important for expressing the meaning you intend. In this section, we will go over the two major sources of word-order problems for ESL writers—adjectives and adverbs.

Using adjectives in the proper order

In English, adjectives should be placed in the proper order.

Improper The red beautiful sailboat came into the bay.

Proper The beautiful red sailboat came into the bay.

Fluent speakers of English will still understand the improper sentence, but it will sound odd to them. To properly order adjectives, you can use the following hierarchy of adjectives, which has been modified from *The New Century Handbook*, 5e by Hult and Huckin (© 2011, Pearson). Hult and Huckin offer a more comprehensive approach.

1. article, determiner, or possessive (*a, an, the, this, that, those, my, our, their, Lisa's*)
2. ordinal (*first, second, third, final, next*)
3. quantity (*one, two, three, more, some, many*)
4. size and shape (*big, tiny, large, circular, square, round*)
5. appearance (*beautiful, filthy, clean, damaged, old, young, ancient*)
6. color (*red, yellow, black, green*)
7. substance (*wool, copper, wood, plastic*).

When properly ordered, adjectives can be strung together indefinitely in a sentence.

A third, large, beautiful, ancient, red, wooden sailboat came into the bay.

Keep in mind, though, that you should not string together too many adjectives. More than three or four adjectives strung together can be difficult to understand.

Using adverbs in proper places

Adverbs usually modify verbs in clauses. Adverbs can be used in a variety of places in the clause. Three guidelines are especially helpful for placing adverbs.

GUIDELINE 1: ***Adverbs involving time and place usually go after the verb.***

The cat went *outside* when the children came over.

Want to practice adjective order? Go to **www.pearsonhighered.com/johnsonweb4/B.2**

She arrived *late* to the lecture.

He went *promptly* to his professor's office for help.

GUIDELINE 2: ***Adverbs that show frequency usually go before the verb.***

The cat *typically* runs outside when the children come over.

She *usually* arrives late to the lecture.

He *always* goes to his professor's office for help.

GUIDELINE 3: ***Do not put an adverb between the verb and the object of a clause.***

Improper Maria drives *recklessly* her car on the interstate.

Kim eats *quickly* his breakfast before he goes to class.

Proper Maria drives her car *recklessly* on the interstate.

Kim eats his breakfast *quickly* before he goes to class.

Using Verb Tenses Appropriately

Proper use of verbs and verb phrases can be difficult in English, even for fluent speakers. So, learn them as best you can, and be patient while you are mastering the numerous English tenses.

To use tenses appropriately, remember that English, like most languages, has *past, present,* and *future* tenses. Each of these tenses also has four verbal aspects, which are called *simple, progressive, perfect,* and *perfect progressive.*

Simple—indicates whether the event happened, is happening, or will happen.

Progressive—indicates that the event was, is, or will be in progress at a specific time.

Perfect—indicates that the event was, is, or will be completed by a specific point in time.

Perfect progressive—indicates whether the event was, is, or will be progressing until a specific point in time.

Altogether, English has twelve tenses that need to be learned. Let's consider them separately.

Past Tense

Past tense refers to events that have already happened.

Simple Past Tense

Victor walked to the store yesterday.

More information on adverbs is available at
www.pearsonhighered.com/johnsonweb4/B.3

Past Progressive Tense

Victor was walking to the store yesterday, when he was nearly hit by a car.

Past Perfect Tense

Before going to class, Victor had walked to the store.

Past Perfect Progressive Tense

Victor had been walking to the store before class.

Present Tense

Present tense refers to events that are happening at the moment.

Simple Present Tense

I enjoy talking with my friends.

Present Progressive Tense

I am enjoying talking with my friends this evening.

Present Perfect Tense

I have enjoyed talking with my friends this evening.

Present Perfect Progressive Tense

I have been enjoying talking with my friends this evening.

Future Tense

Future tense refers to events that will happen.

Simple Future Tense

The movie will start at 6:00 this evening.

Future Progressive Tense

The movie will be starting at 6:00 this evening.

Future Perfect Tense

By 6:10, the movie will have started already.

Future Perfect Progressive Tense

By 6:10, the movie will have been playing for nearly 10 minutes.

Still not sure about verb tenses? Go to **www.pearsonhighered.com/johnsonweb4/B.4**

When learning English, you should begin by mastering the simple past, present, and future forms of verbs. Then, as you grow more comfortable with the language, you can begin using the progressive, perfect, and perfect progressive aspects.

When you are unsure about whether to use progressive, perfect, or perfect progressive, just revert to the simple form. In most cases, these simple sentences will be correct, though perhaps a bit awkward sounding to fluent speakers of English.

APPENDIX C

Documentation Guide

Documenting sources is an important part of doing research. As you collect information on your subject, you should keep track of the sources from which you drew quotes and ideas. Then, cite these sources in your text and use them to create a list of references at the end of your document.

When should you cite and document a source? The answer to this question depends on the kind of document you are writing. A scientific report, for example, requires more citation than a technical description or a set of instructions. The best way to determine the necessary level of documentation is to consider your readers' needs. How much citing and documenting will they need to feel confident in your work?

Some commonly documented materials include the following:

Quotes or ideas taken from someone else's work—If others wrote it or thought it before you did, you should cite them as the owners of their words and ideas. Otherwise, you might be accused of lifting their work. In important cases, you or your company might be sued for using someone else's ideas.

Materials that support your ideas—You can build the credibility of your work by showing that others have discussed the topic before. Readers are going to be highly skeptical of your work if they think you are pulling your ideas out of thin air.

Sources of any data or facts—Any numbers or facts that you did not generate yourself need to be carefully cited and documented. That way, readers can check your sources for accuracy.

Materials that refer to your subject—You want to demonstrate that you are aware of the broader conversation on your subject. By citing sources, including those with which you disagree, you show that you have a complete understanding of the issues involved.

Historical sources on your subject—To build a background for readers to understand your subject, you should include any sources that might help them understand its history.

Graphics taken from online or print sources—On the Internet, it is easy to download graphs, tables, images, and photographs. Sometimes you will need permission to use these items. Minimally, though, you should cite the sources from which you obtained them.

Let's review the three most common documentation styles in technical communication:

- The **APA documentation style** from the American Psychological Association is widely used in engineering and the sciences.
- The **CSE documentation style** from the Council of Science Editors is used primarily in biological and medical sciences, though it is gaining popularity in other scientific fields.

Links to sources on APA, CSE, and MLA styles can be found at **www.pearsonhighered.com/johnsonweb4/C.1**

- The **MLA documentation style** from the Modern Language Association is used in the humanities. Although this style is not commonly used in engineering and science, it is sometimes used when scholars approach technical issues from cultural, historical, rhetorical, or philosophical perspectives.

Literally hundreds of other documentation styles are available. So find out which documentation style is used in the organization or company for which you work.

Citing and documenting a source requires two items, an *in-text citation* and a *full entry* in the References or Works Cited list at the end of the document.

Footnotes and endnotes are not common in APA or CSE style, and they are increasingly rare in MLA style. If you want to use footnotes or endnotes, consult the style guide you are using. Footnotes and endnotes will not be covered here because they are rarely used in technical documents.

We will discuss the most common in-text citations and full-entry patterns. If one of the following models does not fit your needs, you should consult the style guide (APA, CSE, or MLA) that you are following.

APA Documentation Style

APA documentation style is most common in the natural and human sciences, except in fields related to biology and medicine. The official source for this style is the *Publication Manual of the American Psychological Association,* Sixth Edition (2010).

When using APA style, you will need to include in-text citations and a list of alphabetically arranged references at the end of your document.

APA In-Text Citations

APA style follows an author-year system for in-text citations, meaning the author and year are usually cited within the text.

Individual Authors

Individual authors are cited using their last name and the date of the article.

> One study reports a significant rise in HIV cases in South Africa in 1 year (Brindle, 2000).
>
> One study reported a 12.2% rise in HIV cases in only 1 year (Brindle, 2000, p. 843).
>
> Brindle (2000) reports a significant rise in HIV cases in South Africa in 1 year.
>
> Brindle (2000) reports a 12.2% rise in HIV cases in South Africa in 1 year (p. 843).

In most cases, only the author and year need to be noted, as shown in the first example. If you are reporting a specific fact or number, however, you should cite the page from which it was taken.

To find online sources that discuss APA style, go to **www.pearsonhighered.com/johnsonweb4/C.2**

Multiple Authors

If an article has two or more authors, you should use the ampersand symbol (&) to replace the word *and* in the in-text citation. The word *and* should be used in the sentence itself, however.

(Thomas & Linter, 2001)

According to Thomas and Linter (2001) . . .

Technical documents often have more than two authors. If the work has less than six authors, cite all the names the first time the work is referenced. After that, use the last name of the first author followed by "et al."

First Citation of Work

(Wu, Gyno, Young, & Reims, 2003)

As reported by Wu, Gyno, Young, and Reims (2003) . . .

Subsequent Citations of Work

(Wu et al., 2003)

As reported by Wu et al. (2003) . . .

If the work has six or more authors, only the first author's last name should be included, followed by "et al." This approach should be used with all citations of the work, including the first in-text citation.

Corporate or Unknown Authors

When the author of the document is a corporation or is unknown, the in-text citation uses the name of the corporation or the first prominent word in the title of the document.

First Citation of Work

(National Science Foundation [NSF], 2004)

("Results," 2002)

(*Silent,* 2002)

Subsequent Citations of Work

(NSF, 2004)

("Results," 2002)

(*Silent,* 2002)

Notice in these examples that the first word of a journal article title should be put in quotation marks, while the first word of a book title should be put in italics.

Paraphrased Materials

When citing paraphrased materials, usually only the year and page number are needed because the authors' names are typically mentioned in the sentence. In many cases, only the year is needed.

> Franks and Roberts report that aptitude for visual thinking runs in families (2003, p. 76).
>
> The instinct for survival, according to Ramos (2004), is strong in the Mexican wolf.
>
> Jones (2001) argues that finding a stand of dead trees near an industrial plant is a good indicator that something is seriously wrong (pp. 87–88).

Two or More Works in Same Parentheses

In some cases, several documents will state similar information. If so, you should cite them all and separate the works with semicolons.

> Studies have shown remarkable progress toward reviving the penguin population on Vostov Island (Hinson & Kim, 2004; Johnson & Smith, 2001; Tamili, 2002).

Personal Communication and Correspondence

APA style discourages putting any forms of personal communication in the References list. Personal communication includes conversations, e-mails, letters, and even interviews. So, in-text citations are the only citations for these sources in a document.

> Bathers (personal communication, December 5, 2003) pointed out to me that . . .

These sources are not listed in the References list because they do not provide information that is retrievable by readers.

The References List for APA Style

When using APA style, your references should be listed in alphabetical order at the back of the document. In your references, you should list only the items that you actually cited in your document. Leave any documents that you consulted but did not cite out of the list of references. After all, if a document is important enough to list in your references, it should be important enough to cite in the text itself.

In most cases, the reference list is identified by the centered heading "References." Entries should be double-spaced. Also, each reference should use a hanging indent style (i.e., the second line and subsequent lines should be indented). The first line should be flush with the left margin.

The following list includes examples of APA style. This list is not comprehensive. If you do not find a model here for a document you are adding to your references list, you should check the *Publication Manual of the American Psychological Association,* Sixth Edition (2010). Or check one of many websites that offer examples of APA style.

1. **Website or Webpage, Author Known**

   Jaspers, F. (2001). *Einstein online.* Retrieved from http://www.einsteinonlinetoo.com

2. **Website or Webpage, Corporate Author**

   National Wildlife Service. (2002). *Managing forest on your land.* Retrieved from http://www.nws.gov/manageyourforest.htm

3. **Webpage, Author Unknown**

   Skin cancer treatments debated. (2004, January 1). *CNN.com.* Retrieved from http://www.cnn.com/2004/HEALTH/conditions/01/19/skincancer.treatment.ap/index.html

4. **Book, One Author**

   Jones, S. (2001). *Darwin's ghost: The origin of species updated.* New York, NY: Balantine Books.

5. **Book, More Than One Author**

   Pauling, L., & Wilson, E. B. (1935). *Introduction to quantum mechanics.* New York, NY: Dover Publications.

6. **Book, Corporate or Organization Author**

   American Psychiatric Association. (1994). *Diagnostic and statistical manual of mental disorders* (4th ed.). Washington, DC: Author.

7. **Book, Edited Collection**

   Mueller-Vollmer, K. (Ed.). (1990). *The hermeneutics reader.* New York, NY: Continuum.

8. **Book, Translated**

   Habermas, J. (1979). *Communication and the evolution of society.* (T. McCarthy, Trans.). Boston, MA: Beacon Press.

9. **Book, Author Unknown**

   Handbook for the WorkPad c3 PC Companion. (2000). Thornwood, NY: IBM.

10. **Book, Second Edition or Beyond**

    Williams, R., & Tollet, J. (2000). *The non-designer's web book* (2nd ed.). Berkeley, CA: Peachpit.

11. **Book, Dissertation or Thesis**

    Simms, L. (2002). *The hampton effect in fringe desert environments: An ecosystem under stress.* (Unpublished doctoral dissertation). University of New Mexico, Albuquerque, NM.

12. **Book, Electronic**

    Darwin, C. (1862). *On the various contrivances by which British and foreign orchids are fertilised by insects.* Retrieved from http://pages.britishlibrary.net/charles.darwin3/orchids/orchids_fm.htm

Having trouble finding an APA model for a source? Go to **www.pearsonhighered.com/johnsonweb4/C.3**

13. **Document, Government Publication**

Greene, L. W. (1985). *Exile in paradise: The isolation of Hawaii's leprosy victims and development of Kalaupapa settlement, 1865 to present.* Washington, DC: U.S. Department of the Interior, National Park Service.

14. **Document, Pamphlet**

The Colorado Health Network. (2002). *Exploring high altitude areas.* Denver, CO: Author.

15. **Film or Video Recording**

Jackson, P. (Director), & Osborne, B., Walsh, F., & Sanders, T. (Producers). (2002). *The lord of the rings: The fellowship of the ring* [Motion picture]. Hollywood, CA: New Line.

16. **Article, Journal with Continuous Pagination**

Boren, M. T., & Ramey, J. (1996). Thinking aloud: Reconciling theory and practice. *IEEE Transactions on Professional Communication, 39,* 49–57. doi:1109/147.867942

17. **Article, Journal without Continuous Pagination**

Kadlecek, M. (1991). Global climate change could threaten U.S. wildlife. *Conservationist, 46*(1), 54–55.

18. **Article, Journal with Digital Object Identifier (DOI)**

Tomlin, R. (2008). Online FDA regulations: Implications for medical writers. *Technical Communication Quarterly, 17*(3), 289–310. doi:10.1080/10572250802100410

19. **Article, Edited Book**

Katz, S. B., & Miller, C. R. (1996). The low-level radioactive waste siting controversy in North Carolina: Toward a rhetorical model of risk communication. In G. Herndl & S. C. Brown (Eds.), *Green culture: Environmental rhetoric in contemporary America* (pp. 111–140). Madison: University of Wisconsin Press.

20. **Article, Magazine**

Appenzeller, T. (2004, February). The case of the missing carbon. *National Geographic,* 88–118.

21. **Article, Online Magazine**

Grinspoon, D. (2004, January 7). Is Mars ours? *Slate Magazine.* Retrieved from http://slate.msn.com/id/2093579

22. **Article, Newspaper**

Hall, C. (2002, November 18). Shortage of human capital envisioned, Monster's Taylor sees worker need. *The Chicago Tribune,* p. E7.

23. **Article, Author Unknown**

The big chill leaves bruises. (2004, January 17). *Albuquerque Tribune,* p. A4.

Need more help with online sources? Go to **www.pearsonhighered.com/johnsonweb4/C.4**

24. **Article, CD-ROM**

Hanford, P. (2001). Locating the right job for you. *The electronic job finder* [CD-ROM]. San Francisco, CA: Career Masters.

25. **Blog Posting**

Katie. (2007, 17 September). 30 days and tech writing. [Web log post]. Retrieved from http://techwriterscrum.blogspot.com

26. **Podcast**

DMN Communications. (Producer). (2008, May 18). Talking wikis with Stewart Mader [Audiopodcast]. *Communications from DMN.* Retrieved from http://dmn.podbean.com/2008/05

27. **Song or Recording**

Myer, L. (1993). Sometimes alone. *Flatlands* [CD]. Ames, IA: People's Productions.

28. **Television or Radio Program**

Harris, R. (2003, January 6). *Destination: The south pole.* Washington, DC: National Public Radio. Retrieved from http://discover.npr.org/features /feature.jhtml?wfld=904848

29. **Personal Correspondence, E-Mail, or Interview**

This result was confirmed by J. Baca (personal communication, March 4, 2004). (In APA style, a personal correspondence is not included in the References list. Instead, the information from the correspondence should be written in the in-text citation.)

Creating the APA References List

In APA style, the References list is placed at the end of the document on a separate page or in an appendix. The sources cited in the document should be listed alphabetically by author's last name.

References

Assel, R., Cronk, K., & Norton, D. (2003). Recent trends in Laurentian Great Lakes ice cover. *Climatic Change, 57,* 185–204.

Hoffmann, A., & Blows, M. (1993). Evolutionary genetics and climate change: Will animals adapt to global warming? In P. M. Kareiva, J. G. Kingsolver, & R. B. Huey (Eds.), *Biotic interactions and global change* (pp. 13–29). Sunderland, MA: Sinauer.

Houghton, J. (1997). *Global warming: The complete briefing* (2nd ed). Cambridge, MA: Cambridge University Press.

Kadlecek, M. (1991). Global climate change could threaten U.S. wildlife. *Conservationist, 46*(1), 54–55.

Sherwood, K., & Idso, C. (2003). *Is the global warming bubble about to burst?* Retrieved from http://www.co2science.org/edit/v6_edit/v6n37edit.htm

CSE Documentation Style (Citation-Sequence)

The CSE documentation style is most commonly used in biological and medical fields, though it is gaining popularity in other scientific fields. The official source for this style is *Scientific Style and Format: The CSE Manual for Authors, Editors, and Publishers,* Seventh Edition (2006).

The *CSE Manual* describes two citation methods. The first method, called the *author-year* system, is very similar to APA style, so it will not be discussed here. The second method, called the *citation-sequence* system, will be discussed here because it offers a good alternative to APA style. If you need to use the CSE author-year system, you can consult the *CSE Manual* or websites that offer examples of this system.

In the citation-sequence system, sources are referred to by number within the text, usually with a superscript number similar to a footnote.

> This bacteria has been shown [1] to grow at a significant rate when exposed to black light.

When referring to multiple sources, a dash is used to signal the range of sources.

> Several studies [3–8, 10] have illustrated this relationship.

In some situations, editors will ask for the citations to use numbers in parentheses or brackets instead of superscript numbers:

> This relationship between the virus and various illnesses has been demonstrated in numerous studies (3, 12–15).

> Franklin and Chou argued this point in their influential research on HIV mutation [3], in which they explained its tendency to seek out new paths for replication.

In the References list at the end of the document, the sources are numbered and listed in the order in which they were cited in the text. Then, other references to that source in the document will use the same number.

The advantage of the citation-sequence system is that readers feel less disruption than with the author-year system, because the superscript numbers are less intrusive. However, a disadvantage is that readers need to flip back to the list of references to see author names for any sources of information.

The References List for CSE Citation-Sequence Style

The following list includes examples of CSE citation-sequence style. This list is not comprehensive. If you do not find a model here for a document you are adding to your references, you should check *Scientific Style and Format: The CSE Manual for Authors, Editors, and Publishers,* Seventh Edition.

Need to use the CSE author-year system? Go to **www.pearsonhighered.com/johnsonweb4/C.5**
For more information on CSE style updates, go to **www.pearsonhighered.com/johnsonweb4/C.6**

The format of the list of references for CSE citation-sequence style is somewhat different from that of reference lists following APA or MLA style:

- Sources are numbered (1, 2, 3, and so on) to reflect the order in which they were cited.
- The items in the References list are all flush left against the margin (no hanging indent).
- When a citation refers to a specific page or set of pages in a stand-alone document, the full text reference includes the page number(s) after a *p* (e.g., *p 23* or *p 123–36*). If the citation is referring to the whole work, the page numbers are not needed.

Items in the References list should be single-spaced.

1. **Website or Webpage, Author Known**

   12. Jaspers F. Einstein online [Internet]. Downers Grove (IL): Einstein Inc.; c2001 [cited 2003 Mar 9]. Available from: http://www.einsteinonlinetoo.com

2. **Website or Webpage, Corporate Author**

   34. National Wildlife Service. Managing forest on your land [Internet]. Washington: NWS; c2003 [cited 2004 Sep 8]. Available from: http://www.nws.gov/manageyourforest.htm

3. **Webpage, Author Unknown**

   3. Skin cancer treatments debated [Internet]. Atlanta (GA): CNN.com; c2004 [cited 2004 Jan 1]. Available from: http://www.cnn.com/2004/HEALTH/conditions/01/19/skincancer.treatment.ap/index.html

4. **Webpage, Online Periodical**

   7. Grinspoon D. Is Mars ours? Slate Magazine [Internet]. 2004 Jan 7 [cited 2004 Jan 19]. Available from: http://slate.msn.com/id/2093579

5. **Book, One Author**

   23. Jones S. Darwin's ghost: the origin of species updated. New York: Balantine Books; 2001. p 86–92.

6. **Book, More Than One Author**

   2. Pauling L, Wilson EB. Introduction to quantum mechanics. New York: Dover Publications; 1935. p 38.

7. **Book, Corporate or Organization Author**

   11. American Psychiatric Association. Diagnostic and statistical manual of mental disorders. 4th ed. Washington: American Psychiatric Association; 1994.

8. **Book, Edited Collection**

   22. Mueller-Vollmer K, editor. The hermeneutics reader. New York: Continuum; 1990. p 203–12.

9. **Book, Translated**

   14. Habermas J. Communication and the evolution of society. McCarthy T, translator. Boston (MA): Beacon Press; 1979. p 156.

10. Book, Author Unknown

13. Handbook for the WorkPad c3 PC Companion. Thornwood (NY): IBM; 2000.

11. Book, Second Edition or Beyond

21. Williams R, Tollet J. The non-designer's web book. 2nd ed. Berkeley (CA): Peachpit; 2000. p 123–27.

12. Book, Dissertation or Thesis

18. Simms L. The hampton effect in fringe desert environments: an ecosystem under stress [dissertation]. [Albuquerque (NM)]: Univ of New Mexico; 2002.

13. Book, Electronic

13. Darwin C. On the various contrivances by which British and foreign orchids are fertilised by insects [Internet]. London: John Murray; c1862 [cited 2002 Sep 5]. Available from: http://pages.britishlibrary.net/charles.darwin3/orchids/orchids_fm.htm

14. Document, Government Publication

6. Greene LW. Exile in paradise: the isolation of Hawaii's leprosy victims and development of Kalaupapa settlement, 1865 to present. Washington: Department of Interior (US); 1985. Available from: U.S. Department of the Interior, National Park Service, Washington, DC.

15. Document, Pamphlet

23. The Colorado Health Network. Exploring high altitude areas. Denver (CO); 2002. Available from: TCHN, Denver, CO.

16. Film or Video Recording

16. The lord of the rings: the fellowship of the ring [DVD]. Jackson P, director. Osborne B, Walsh F, Sanders T, producers. Hollywood (CA): New Line Productions; 2002.

17. CD-ROM

7. Geritch T. Masters of renaissance art [CD-ROM]. Chicago: Revival Productions; 2000. 2 CD-ROMs: sound, color, 4 ¾ in.

18. Article, Journal with Continuous Pagination

34. Boren MT, Ramey J. Thinking aloud: reconciling theory and practice. IEEE Trans on Prof Comm 1996; 39:49–57.

19. Article, Journal Without Continuous Pagination

32. Lenhoff R, Huber L. Young children make maps! Young Children 2000; 55(5):6–12.

20. Article, Edited Book

1. Katz SB, Miller CR. The low-level radioactive waste siting controversy in North Carolina: toward a rhetorical model of risk communication. In: Herndl G, Brown SC, editors. Green culture: environmental rhetoric in contemporary America. Madison (WI): Univ of Wisconsin Pr; 1996. p 111–40.

21. Article, Magazine

12. Appenzeller T. The case of the missing carbon. National Geographic 2004 Feb: 88–118.

22. Article, Newspaper

6. Hall C. Shortage of human capital envisioned, Monster's Taylor sees worker need. Chicago Tribune 2002 Nov 18; Sect E:7(col 2).

23. Article, Author Unknown

3. The big chill leaves bruises. Albuquerque Tribune 2004 Jan 17; Sect A:4(col 1).

24. Article, CD-ROM

21. Hanford P. Locating the right job for you. The electronic job finder [CD-ROM]. San Francisco: Career Masters; 2001. CD-ROM: sound, color, 4¾ in.

25. Song or Recording

12. Myer L. Sometimes alone. Flatlands [CD]. Ames (IA): People's Productions; 1993.

26. Television or Radio Program

4. Harris R. Destination: the south pole [recording]. Washington: National Public Radio; 2003 Jan 6 [cited 2004 Jan 19]. Available from: http://discover.npr.org/features/feature.jhtml?wfld=904848

27. Personal Correspondence, E-Mail, or Interview

These complications seem to have been resolved (2006 e-mail from FH Smith to me) while others seem to have emerged.

(References that refer to personal correspondences or personal interviews should be placed within the text and not in the References list.)

Creating the CSE References List (Citation-Sequence Style)

In CSE style, the References list is placed at the end of the document or in an appendix. The sources are listed by number in the order in which they were first referenced in the text.

References

1. Hoffmann A, Blows M. Evolutionary genetics and climate change: will animals adapt to global warming? In: Kareiva P, Kingsolver J, Huey R, editors. Biotic interactions and global change. Sunderland (MA): Sinauerl; 1993. p 13–29.
2. Sherwood K, Idso C. Is the global warming bubble about to burst? [Internet]. Tempe (AZ): Center for the Study of Carbon Dioxide and Global Change; c2003 [cited 2004 Mar 4]. Available from: http://www.co2science.org/edit/v6 edit/v6n37edit.htm

Can't find a CSE model for your source? Go to **www.pearsonhighered.com/johnsonweb4/C.7**

3. Assel R, Cronk K, Norton D. Recent trends in Laurentian Great Lakes ice cover. Climatic Change 2003; 57:185–204.

4. Kadlecek M. Global climate change could threaten U.S. wildlife. Conservationist 1991; 46(1):54–55.

5. Houghton J. Global warming: the complete briefing. 2nd ed. Cambridge (MA): Cambridge Univ Pr; 1997. p 12.

MLA Documentation Style

The MLA documentation style is not commonly used in technical or scientific fields; it is most commonly used in the arts and humanities. Nevertheless, there are occasions where MLA style is requested, because it is a widely used documentation style. The official source for this style is the *MLA Style Manual and Guide to Scholarly Publishing,* Third Edition (2008).

When using MLA style, you will need to use in-text citations and a list of alphabetically arranged references, called "Works Cited," at the end of your document.

MLA In-Text Citations

MLA style follows an *author-page number* system for in-text citations, meaning the author and page number are usually cited within the text.

Individual Authors

Individual authors are cited using their last name and the page number(s) from which the information was drawn. If the year is significant, put it after the author's name in parentheses.

> One study reports a significant rise in HIV cases in South Africa in one year (Brindle 834).
>
> One study reported a 12.2 percent rise in HIV cases in only one year (Brindle 834).
>
> Brindle (2000) reports a significant rise in HIV cases in South Africa in one year.
>
> Brindle (2000) reports a 12.2 percent rise in HIV cases in South Africa in one year (834).

In most cases, only the author and page number need to be noted, as shown in the first example above. In MLA style, the year of publication is not usually a large concern, so include the year only if it is necessary.

Multiple Authors

If an article has two or more authors, you should use the word *and* to connect the authors' last names.

More information on MLA style is available at **www.pearsonhighered.com/johnsonweb4/C.8**

(Thomas and Linter 130)

According to Thomas and Linter (2001) the number of mammals in this area was dramatically reduced during the Ice Age (130).

Technical documents often have more than two authors. In these cases, cite all the names the first time the work is referenced. Afterward, you can repeat all the names or use the last name of the first author followed by "et al."

First Citation of Work

(Wu, Gyno, Young, and Reims 924)

Subsequent Citations of Work

(Wu et al. 924)

As reported by Wu et al., the Permian Age . . .

Corporate or Unknown Authors

When the author of the document is a corporation or unknown, the in-text citation uses the name of the corporation or the first prominent word in the title of the document.

(National Science Foundation 76)

("Results" 91)

(*Silent* 239)

As shown here, if the source is an article, put the first prominent word in quotes. If it is a book, put it in italics.

Paraphrased Materials

Because the authors' names are typically mentioned in the sentence, citing paraphrased materials usually requires only a mention of the page number.

Franks and Roberts report that aptitude for visual thinking runs in families (76).

The instinct for survival, according to Ramos, is strong in the Mexican wolf (198–201).

Jones argues that finding a stand of dead trees near an industrial plant is a good indicator that something is seriously wrong (87–88).

Two or More Works in Same Parentheses

In some cases, several documents will state similar information. In these cases, you should cite them all and separate them with semicolons.

Studies have shown remarkable progress toward reviving the penguin population on Vostov Island (Hinson and Kim 330; Johnson and Smith 87; Tamili 102).

The Works Cited List for MLA Style

When using MLA style, your Works Cited list should be in alphabetical order at the back of the document. In your list, you should include only the items that you actually cited in your document. Leave any documents that you consulted but did not cite out of the list of works cited. After all, if a document is important enough to include in your Works Cited, it should be important enough to cite in the text itself.

In most cases, the list is identified by the centered heading "Works Cited." Entries should be double-spaced. Also, each reference should use a hanging indent style (i.e., the second line and subsequent lines should be indented). The first line should be flush with the left margin.

The following list includes examples of MLA style. This list is not comprehensive. If you do not find a model here for a source you are adding to your Works Cited, you should check the *MLA Style Manual and Guide to Scholarly Publishing,* Third Edition. Or check the several websites available that offer examples of MLA style.

1. **Website or Webpage, Author Known**

   Einstein Online. Ed. Fred Jaspers. 9 Mar. 2003. Web. 13 Dec. 2008.

2. **Website or Webpage, Corporate Author**

   Managing Forest on Your Land. National Wildlife Service. 8 Sept. 2002. Web. 10 Oct. 2008.

3. **Webpage, Author Unknown**

   "Skin Cancer Treatments Debated." CNN.com. 1 Jan. 2004. Web. 9 Aug. 2008.

4. **Webpage, Online Periodical**

   Grinspoon, David. "Is Mars Ours?" *Slate Magazine* 7 Jan. 2004. Web. 19 Jan. 2008.

5. **Book, One Author**

   Jones, Steve. *Darwin's Ghost: The Origin of Species Updated.* New York: Balantine Books, 2001. Print.

6. **Book, More Than One Author**

   Pauling, Linus, and E. Bright Wilson. *Introduction to Quantum Mechanics.* New York: Dover Publications, 1935. Print.

7. **Book, Corporate or Organization Author**

   American Psychiatric Association. *Diagnostic and Statistical Manual of Mental Disorders.* 4th ed. Washington: American Psychiatric Association, 1994. Print.

8. **Book, Edited Collection**

   Mueller-Vollmer, Kurt, ed. *The Hermeneutics Reader.* New York: Continuum, 1990. Print.

9. **Book, Translated**

   Habermas, Jurgen. *Communication and the Evolution of Society.* Trans. Thomas McCarthy. Boston: Beacon Press, 1979. Print.

10. Book, Author Unknown

Handbook for the WorkPad c3 PC Companion. Thornwood: IBM, 2000. Print.

11. Book, Second Edition or Beyond

Williams, Robin, and John Tollet. *The Non-Designer's Web Book.* 2nd ed. Berkeley: Peachpit, 2000. Print.

12. Book, Dissertation or Thesis

Simms, Laura. "The Hampton Effect in Fringe Desert Environments: An Ecosystem Under Stress." Diss. U of New Mexico, 2002. Print.

13. Book, Electronic

Darwin, Charles. *On the Various Contrivances by which British and Foreign Orchids Are Fertilised by Insects.* London: John Murray, 1862. Web. 1 Jan. 2008.

14. Document, Government Publication

Greene, Linda W. *Exile in Paradise: The Isolation of Hawaii's Leprosy Victims and Development of Kalaupapa Settlement, 1865 to Present.* Washington: US Department of the Interior, National Park Service, 1985. Print.

15. Document, Pamphlet

Exploring High Altitude Areas. Denver: TCHN, 2002. Print.

16. Film or Video Recording

The Lord of the Rings: The Fellowship of the Ring. Dir. Peter Jackson. Prod. Barrie Osborne, Peter Jackson, Fran Walsh, and Tim Sanders. New Line Productions, 2002. Film.

17. CD-ROM

Geritch, Thomas. *Masters of Renaissance Art.* CD-ROM. Chicago: Revival Productions, 2000.

18. Article, Journal with Continuous Pagination

Boren, M. Ted, and Judith Ramey. "Thinking Aloud: Reconciling Theory and Practice." *IEEE Transactions on Professional Communication* 39 (1996): 49–57. Print.

19. Article, Journal without Continuous Pagination

Lenhoff, Rosalyn, and Lynn Huber. "Young Children Make Maps!" *Young Children* 55.5 (2000): 6–12. Print.

20. Article, Edited Book

Katz, Steven B., and Carolyn R. Miller. "The Low-Level Radioactive Waste Siting Controversy in North Carolina: Toward a Rhetorical Model of Risk Communication." *Green Culture: Environmental Rhetoric in Contemporary America.* Ed. Carl G. Herndl and Stuart C. Brown. Madison: U of Wisconsin P, 1996. 111–40. Print.

21. Article, Magazine

Appenzeller, Tim. "The Case of the Missing Carbon." *National Geographic* Feb. 2004: 88–118. Print.

Can't find an MLA model for your source? Go to **www.pearsonhighered.com/johnsonweb4/C.9**

22. **Article, Newspaper**

Hall, Cheryl. "Shortage of Human Capital Envisioned, Monster's Taylor Sees Worker Need." *Chicago Tribune* 18 Nov. 2002: E7. Print.

23. **Article, Author Unknown**

"The Big Chill Leaves Bruises." *Albuquerque Tribune* 17 Jan. 2004: A4. Print.

24. **Article, CD-ROM**

Hanford, Peter. "Locating the Right Job for You." *The Electronic Job Finder.* CD-ROM. San Francisco: Career Masters, 2001.

25. **Song or Recording**

Myer, Larry. "Sometimes Alone." *Flatlands.* Ames: People's Productions, 1993. CD.

26. **Television or Radio Program**

"Destination: The South Pole." Narr. Richard Harris. *All Things Considered.* National Public Radio. 6 Jan. 2003. Web. 4 Feb. 2004.

27. **Personal Correspondence, E-Mail, or Interview**

Baca, James. Personal interview. 4 Mar. 2004.

Creating the MLA Works Cited List

In MLA style, the Works Cited list is placed at the end of the document on a separate page or in an appendix. The sources referenced in the document should be listed alphabetically, and each entry should be double-spaced.

Works Cited

Assel, Robert, Kevin Cronk, and David Norton. "Recent Trends in Laurentian Great Lakes Ice Cover." *Climatic Change* 57 (2003): 185–204. Print.

Hoffmann, Amber, and Marlin Blows. "Evolutionary Genetics and Climate Change: Will Animals Adapt to Global Warming?" *Biotic Interactions and Global Change.* Ed. Paul M. Kareiva, John G. Kingsolver, and Renee B. Huey. Sunderland: Sinauer, 1993. 13–29. Print.

Houghton, James. *Global Warming: The Complete Briefing.* 2nd ed. Cambridge, MA: Cambridge UP, 1997. Print.

Kadlecek, Mary. "Global Climate Change Could Threaten U.S. Wildlife." *Conservationist* 46.1 (1991): 54–55. Print.

Sherwood, Kevin, and Craig Idso. "Is the Global Warming Bubble About to Burst?" 4 Mar. 2004. Web. 8 Nov. 2008.

References

Alcatel-Lucent. (1998). *What is a laser?* Retrieved from http://www.bell-labs.com/history/laser/laser_def.html

Alesis. (2008). *Podcasting microphone* [Photo]. Retrieved from http://www.alesis.com

Allen, L., & Voss, D. (1997). *Ethics in technical communication: Shades of gray.* New York, NY: Wiley.

American Institute of Aeronautics and Astronautics. (2003). *How design engineers evaluate their education* [Bar chart]. Retrieved from http://www.aiaa.org/tc/de

American Psychological Association. (2010). *Publication manual of the American Psychological Association* (6th ed.). Washington, DC: Author.

American Red Cross. (2008). Home page. Retrieved from http://www.redcross.org

American Society of Civil Engineers. (2008). *Welcome to the ASCE online research library.* Retrieved from http://ascelibrary.org

Aristotle. (1991). *On rhetoric.* (G. Kennedy, Trans.). New York: Oxford University Press.

Arnheim, R. (1969). *Visual thinking.* Berkeley: University of California Press.

Ask.com. (2008). *Tsunami definition.* Retrieved from http://www.ask.com

Australian Resuscitation Council. (2002). *Basic life support flowchart.* Retrieved from http://www.resus.org.au/public/bls_flow_chart.pdf

Aversa, A. (2003). *Galaxy simulations.* Retrieved from http://www.u.arizona.edu/~aversa/galaxysims.pdf

Barret, M., & Levy, D. (2008). *A practical approach to managing phishing* [White paper]. Retrieved from http://files.shareholder.com/downloads/PAY

Barrow, L., & Rouse, C. E. (2005). Does college still pay? *The Economist's Voice, 2*(4), 1–8.

Belbin, M. (1981). *Management teams.* New York, NY: Wiley.

Benson, P. (1985). Writing visually: Design considerations in technical publications. *Technical Communication, 32,* 35–39.

Berg, A. (2003). *Counseling to prevent tobacco use and tobacco-caused disease* [Brochure]. Washington, DC: U.S. Preventive Services Task Force.

Bernhardt, S. (1986). Seeing the text [Survey]. *College Composition and Communication, 30,* 66–78.

Bledsoe, L., & Sar, B. K. (2001). *Campus survey report: Safety perception and experiences of violence.* Retrieved from http://www.louisville.edu

Blizzard Entertainment. (2008). *World of Warcraft anti-harassment policies.* Retrieved from http://us.blizzard com/support

Blum, D., & Knudson, M. (1997). *Field guide for science writers.* Oxford, UK: Oxford University Press.

BMW of North America. (2006). *Cooper Mini unauthorized owner's manual* [Manual]. Retrieved from http://www.motoringfile.com/files/unauth_manual.pdf

Boisjoly, R. (1985). SRM o-ring erosion/potential failure criticality. In *Report of the Presidential Commission on the Space Shuttle* Challenger *Accident.* Retrieved from http://science.ksc.nasa.gov/shuttle/missions/51-l/docs/rogers-commission/table-of-contents.html

Boor, S., & Russo, P. (1993). How fluent is your interface? Designing for international users. *Proceedings of INTERCHI '93,* 342–347.

Booth, W., Columb, G. C., & Williams, J. (1995). *The craft of research.* Chicago, IL: University of Chicago Press.

Bosley, D. (2001). *Global contexts.* Boston, MA: Allyn & Bacon.

Brody, M. (1998). *Speaking your way to the top.* Boston, MA: Allyn & Bacon.

Brusaw, C., Alred, G., & Oliu, W. (1993). *Handbook of technical writing* (4th ed.). New York, NY: St. Martin's Press.

Buckinghamshire County Council. (2003). *Flooding—cause and effect.* Retrieved from http://www.southbucks.gov.uk/documents/Flooding%20FACTS3.doc

Burke, K. (1969). *A rhetoric of motives.* Berkeley, CA: University of California Press.

Buzan, T. (1994). *The mind map book: How to use radiant thinking to maximize your brain's untapped potential.* New York, NY: Dutton.

Campbell, G. M. (2002). *Bulletproof presentations.* Franklin Lake, NJ: Career Press.

Carey, S. S. (1998). *A beginner's guide to the scientific method* (2nd ed.). Belmont, CA: Wadsworth.

Carson, R. (2002). *Silent spring.* New York, NY: Mariner.

CBS News (2008). *Are we retreating in the war on cancer?* Retrieved from www.cbsnews.com/stories/2008/05/20/eveningnews

Chaney, L., & Martin, J. (2004). *Intercultural business communication* (3rd ed.). Upper Saddle River, NJ: Pearson Prentice Hall.

Chohan, R. (2005). *Scientists confirm Earth's energy is out of balance.* Washington, DC: National Aeronautics and Space Administration. Retrieved from www.nasa/vision/earth/environment/earth_energy.html

Clark, H., & Haviland, S. (1977). Comprehension and the given-new contract. In R. Freedle (Ed.), *Discourse production and comprehension* (pp. 1–40). Stamford, CT: Ablex.

Clark, R. (1971). *Einstein: His life and times.* New York, NY: World Publishing.

Clement, D. E. (1987). Human factors, instructions, and warning, and product liability. *IEEE Transactions on Professional Communication, 30,* 149–156.

Consumers Against Supermarket Privacy Invasion and Numbering [CASPIAN]. (2008). *What Is RFID?* Retrieved from http://www.spychips.com/what-is-rfid.html

Council of Science Editors. (2006). *The CSE manual for authors, editors, and publishers* (6th ed.). Reston, VA: Author.

Curlew Communications. (2003). *Hemyock Castle.* Retrieved from http://www.hemyockcastle.co.uk/bodiam.htm

Davis, D. (2007, Nov. 4). Off target in the war on cancer. *Washington Post.* Retrieved from http://www.washingtonpost.com/wp-dyn/content/article/2007/11/02/AR2007110201648.html

Deming, W. E. (2000). *Out of crisis.* Cambridge, MA: MIT Press.

De Vries, M. A. (1994). *Internationally yours: Writing and communicating successfully in today's global marketplace.* New York, NY: Houghton Mifflin.

Dombrowski, P. (2000). *Ethics in technical communication.* Boston, MA: Allyn & Bacon.

Dragga, S. (1996). A question of ethics: Lessons from technical communicators on the job. *Technical Communication Quarterly, 6,* 161–178.

Dumas, J. S., & Redish, J. C. (1993). *A practical guide to usability testing.* Norwood, NY: Ablex.

Einstein, A. (1939). *August 2, 1939, letter to Franklin Roosevelt* [Letter]. Retrieved from http://www.anl.gov/OPA/frontiers96arch/aetofdr.html

Fagen, W. T., & Coish, D. (1999). *Validating the IN and FOR distinctions of a workplace literacy program.* Retrieved from http://www.mun.ca/educ/faculty/mwatch/fall99/fagancoish.html

Farkas, D., & Farkas, J. (2002). *Principles of web design.* New York, NY: Longman.

Fermi National Accelarator Laboratory. (2008). *Fermilab.* Retrieved from http://www.fermilab.gov

Field Museum. (2008). Home page. Retrieved from http://www.fieldmuseum.org

Geothermal Energy Association. (2011). Home page. Retrieved from www.geo-energy.org/gea/gea.asp

Gibaldi, J. (2009). *MLA handbook for writers of research papers* (7th ed.). New York, NY: Modern Language Association.

GN Netcom (2005). *User manual for Jabra BT160 Bluetooth headset* [Manual].

Goddard Institute for Space Studies. (2004). *Global temperature trends: 2002 summation.* Retrieved from http://www.giss.nasa.gov/research/observe/surftemp

Google. (2008). *Google.* Retrieved from http://www.google.com

Google Chrome. (2008). *Google Chrome Internet browser* [Manual]. Retrieved from http://www.google.com/chrome

Google Docs. (2008). *Google Docs interface.* Retrieved from http://docs.google.com

Google Translate. (2008). *Translation of Purdue University professional writing program home page.* Retrieved from http://translate.google.com

Guilford, C. (1996). *Paradigm online writing assistant.* Retrieved October 5, 2003, from http://www.powa.org

Gurak, L. (2000). *Oral presentations for technical communication.* Boston, MA: Allyn & Bacon.

Hallberg, B. A. (2003). *Networking: A beginner's guide* (3rd ed.). New York, NY: McGraw-Hill.

Haneda, S., & Shima, H. (1983). Japanese communication behavior as reflected in letter writing. *Journal of Business Communication, 19,* 19–32.

Haviland, S., & Clark, H. (1974). What's new: Acquiring new information as a process in comprehension. *Journal of Verbal Learning and Verbal Behavior, 13,* 512–521.

Helfman, E. S. (1967). *Signs and symbols around the world.* New York, NY: Lothrop, Lee, and Shepard.

Hemyock Castle. (2008). *Comparison between Hemyock and Bodiam castles.* Retrieved from http://www.hemyockcastle.co.uk/bodiam.htm

Herrington, T. (2003). *A legal primer for the digital age.* New York, NY: Pearson Longman.

Hoft, N. (1995). *International technical communication.* New York, NY: Wiley.

Horton, W. (1993). The almost universal language: Graphics for international documents. *Technical Communication, 40,* 682–683.

Hult, C., & Huckin, T. (2011). *The new century handbook* (5th ed.). New York, NY: Longman.

Husqvarna. (2002). *Working with a chainsaw* [Manual]. Åsbro, Sweden: Electrolux.

IBM. (2008). *IBM Lotus Notes* [Software]. Retrieved from http://demos.dfw.ibm.com/on_demand/streamed/IBM_Demo_Lotus_Notes_7-1-Nov05.html?S=index

Indiana Department of Transportation. (2007). *Dry flow testing of flowable backfill materials* [Fact sheet]. Retrieved from http://www.in.gov/indot

Institute of Electrical and Electronics Engineers. (2006). *IEEE code of ethics.* Retrieved from http://www.ieee.org/portal/pages/iportals/aboutus/ethics/code.html

Institute for Social and Economic Research. (2007). *British household panel survey* [Survey]. New Policy Institute. Retrieved from http://www.poverty.org.uk

International Association for Food Protection. (2008). *Sneezing icon* [Graphic]. Retrieved from http://www.foodprotection.org/aboutIAFP/iconmania.asp

International Energy Agency. (2008, May 13). *Oil market report.* Retrieved from http://omrpublic.iea.org/omrarchive

International Olympic Committee. (2004). Symbols for basketball, swimming, and cycling [Graphics].

International Organization for Standardization. (2008). *ISO 9000 and ISO 14000.* Retrieved from http://www.iso.org

Israelson, A. (2007, May 30). Transcript of interview with Bill Gates and Steve Jobs [Interview transcript]. *Ubiqus Reporting.* Retrieved from http://d5.allthingsd.com/d5gates-job-transcript

Jackson, M. (2007). *Memorandum for all DHS employees.* Retrieved from http://www.slate.com

Johnson-Sheehan, R. (2002). *Writing proposals: A rhetoric for managing change.* New York, NY: Longman.

Johnson-Sheehan, R., & Baehr, C. (2001). Visual-spatial thinking: Thinking differently about hypertexts. *Technical Communication, 48,* 37–57.

Jones, D. (1998). *Technical writing style.* Boston, MA: Allyn & Bacon.

Karjane, H., Fisher, B., & Cullen, F. (2005). *Sexual assault on campus: What colleges and universities are doing about it.* Washington, DC: U.S. Department of Justice, National Institute of Justice.

Koffka, K. (1935). *Principles of gestalt psychology.* New York, NY: Harcourt.

Kostelnick, C., & Roberts, D. (1998). *Designing visual language.* Boston, MA: Allyn & Bacon.

Kotler, P. (2002). *Marketing management* (11th ed.). Englewood Cliffs, NJ: Prentice Hall.

Laib, N. (1993). *Rhetoric and style.* Englewood Cliffs, NJ: Prentice Hall.

Lakoff, G. (2004). *Don't think of an elephant.* White River Junction, VT: Chelsea Green.

Lanham, R. (1999). *Revising business prose* (4th ed.). New York, NY: Pearson Longman.

Leininger, C., & Yuan, R. (1998). Aligning international editing efforts with global business strategies. *IEEE Transactions on Professional Communication, 41,* 16–23.

Leopold, A. (1966). *A sand county almanac.* New York, NY: Ballantine.

Lincoln, A. (2001). Lost in translation. *CFO.com.* Retrieved from http://www.cfo.com/article.cfm/3000717

Lipnack, J., & Stamps, J. (2000). *Virtual teams: People working across boundaries with technology.* New York, NY: Wiley.

Llamagraphics. (2008). *Life balance screenshot* [Software]. Llamagraphics, Inc.

Manitoba Conservation Wildlife and Ecosystem Protection Branch. (2004). *Manitoba's species at risk: Ferruginous hawk* [Brochure].

Markel, M. M. (1991). A basic unit on ethics for technical communicators. *Journal of Technical Writing and Communication, 21,* 327–350.

Mathes, J., & Stevenson, D. (1976). *Designing technical reports.* Indianapolis, IN: Bobbs-Merrill.

Medline Plus. (2011). Home page. Retrieved from http://www.medlineplus.gov

Merriam-Webster. (2008). Definition of "gene." *Merriam-Webster Collegiate® Dictionary* (11th ed.) Retrieved from www.merriam-webster.com

Metacrawler. (2011). Home page. Retrieved from http://www.metacrawler.com

Microsoft. (2006). Middle Eastern web page. Retrieved from www.microsoft.com/middleeast

Microsoft. (2007). *Project Standard 2007 overview* [Software]. Retrieved from http://www.office.microsoft.com/en-us/project/HA101656381033.aspx

MLA. (2008). *Style manual and guide to scholarly publishing.* (3rd ed.). New York, NY: Modern Language Association of America.

Moore, P., & Fitz, C. (1993). Using gestalt theory to teach document design and graphics. *Technical Communication Quarterly, 2,* 389–410.

Mozilla Firefox. (2006). *Help.* Retrieved from www.mozilla.com

Munter, M. & Russell, L. (2011) *Guide to presentations* (3rd ed). Upper Saddle River, NJ: Prentice Hall.

National Aeronautics and Space Administration. (2004). *Mars Exploration Rover* [Press release]. Retrieved from http://www.jpl.nasa.gov/news/presskits/merlandings.pdf

National Aeronautics and Space Administration. (2005). *Hurricane Katrina approaching the gulf coast* [Graphic]. Washington, DC. Retrieved from http://rapidfiresco.gsfc.nasa.gov/gallery/?search=Katrina

National Aeronautics and Space Administration. (2008). *Why the moon?* [Poster]. Retrieved from http://www.nasa.gov/pdf/163561main_why_moon2.pdf

National Commission on Writing. (2004). *A ticket to work . . . or a ticket out: A survey of business leaders.* New York: College Board.

National Foundation for Cancer Research. (2006). *Cancer detection guide.* Bethesda, MD: Author.

National Human Genome Research Institute. (2008). Home page. Retrieved from http://www.genome.gov

National Library of Medicine. (2001). *National Library of Medicine recommended formats for bibliographic citation: Supplement: Internet formats.* Bethesda, MD: Author.

National Library of Medicine. (2003). *Medline Plus.* Retrieved from http://www.nlm.nih.gov/medlineplus/medlineplus.html

National Marine Sanctuaries. (2004). *Project summary.* Retrieved from http://sanctuaries.noaa.gov/news/bwet

National Oceanic and Atmospheric Administration. (2011). Home page. Retrieved from http://www.noaa.gov

National Ocean Service. (2005). *National Ocean Service Accomplishments.* Washington, DC: National Ocean Service.

National Science Foundation. (2011). Home page. Retrieved from http://www.nsf.gov

National Survey on Drug Use and Health. (2006). *Alcohol use and risks among young adults by college enrollment status* [Survey]. Washington, DC: Office of Applied Studies, Substance Abuse and Mental Health Services Administration.

Netscape Communicator Browser Window. (2008). Netscape browsers.

Nikon. (2002). *The Nikon guide to digital photography with the Coolpix 885 digital camera* [Manual]. Tokyo, Japan: Nikon.

Occupational Safety and Health Administration. (2007). *Fact sheet, flood cleanup* [Fact sheet]. Retrieved from http://www.osha.gov/OshDoc/data_Hurricane_Facts/floodcleanup.pdf

Office of Applied Studies, Substance Abuse and Mental Health Services Administration. (2009). *National survey on drug use and health* [Survey]. Washington, DC: Author.

Ong, W. (1982). *Orality and literacy.* London, UK: Routledge.

Online Ethics Center for Engineering and Science. (2008). Home page. Retrieved from http://www.onlineethics.org

Pacific Northwest Search and Rescue. (2008). *Welcome.* Retrieved from http://www.pnwsar.org

Pakiser, L., & Shedlock, K. (1997). *Earthquakes* [Fact sheet]. Retrieved from http://pubs.usgs.gov/gip/earthq1/earthqkgip.html

Pearsall, T. (2001). *Elements of technical writing.* Boston, MA: Allyn & Bacon.

Peckham, G. (2003). Safety symbols. *Compliance engineering* [Graphic]. Retrieved from http://www.ce-mag.com/archive/02/03/peckham.html

Pendergrast, M. (1994). *For god, country, and Coca-Cola.* New York: Collier.

Pew Research Center. (2008). *Reports.* Retrieved from http://www.pewinternet.org/reports.asp

Piven, J., & Borgenicht, D. (1999). How to jump from a moving car. *The worst-case scenario survival handbook.* San Francisco, CA: Chronicle.

Piven, J., & Borgenicht, D. (2003). How to use a defribrillator to restore a heartbeat. *Worst-case scenarios online.* Retrieved from http://www.worstcasescenario.com

Plotnik, A. (1982). *The elements of editing.* New York, NY: Macmillan.

Polk Audio. (2005). *XRt12 tuner owner's manual* [Manual]. Baltimore, MD: Author.

Price, J. (1999). *Outlining goes electronic.* Stamford, CT: Ablex.

Reynolds, S., & Valentine, D. (2004). *Guide to cross-cultural communication.* Upper Saddle River, NJ: Pearson Prentice Hall.

Rooksby, E. (2002). *E-mail and ethics: Style and ethical relations in computer-mediated communication.* New York, NY: Routledge.

Rubin, J. (1994). *Handbook of usability testing.* New York, NY: Wiley.

Rude, C. (1998). *Technical editing* (2nd ed.). Boston, MA: Allyn & Bacon.

Ryobi. (2000). *510r 4-cycle garden cultivator operator's manual* [Manual]. Chandler, AZ: Ryobi.

Safetyline Institute. (1998). *Gas laws* [Fact sheet]. Retrieved from http://www.safetyline.wa.gov.au/institute/level2/course16/lecture47/l47_02.asp

Schriver, K. (1997). *Dynamics in document design.* New York, NY: Wiley.

Schultz, H. (2000). *The elements of electronic communication.* Boston, MA: Allyn & Bacon.

Shapiro, G. (1998, March). The ABCs of asthma. *Discover, 35,* 30–33.

Shenk, D. (1998). *Data smog: Surviving the information glut.* San Francisco, CA: HarperCollins.

Silyn-Roberts, H. (1998). Using engineers' characteristics to improve report writing instruction. *Journal of Professional Issues in Engineering Education and Practice, 124,* 12–16.

Silyn-Roberts, H. (2000). *Writing for science and engineering: Papers, presentations, and reports.* Boston, MA: Butterworth-Heinemann.

Smith-Jackson, T., Essuman-Johnson, A., & Leonard, S. D. (2002) Symbol printes: Cross-cultural comparison of symbol representation. *Proceedings of the 15th Triennial Congress of the International Ergonomics Association,* Seoul, Korea.

State of Michigan. (2007). *Emerging diseases: 2007 human WNV cases* [Fact sheet]. Retrieved from http://www.michigan.gov/emergingdiseases

Stauber, J., & Rampton, S. (1995). *Toxic sludge is good for you! Lies, damn lies and the public relations industry.* Monroe, ME: Common Courage Press.

Stevens, B. (2005). What communication skills do employers want? Silicon valley recruiters respond. *Journal of Employment Counseling, 42*(1), 2–9.

Stewart-MacDonald. (2003). *Violin kit assembly instructions.* Retrieved from http://www.stewmac.com

Terminello, V., & Reed, M. (2003). *E-mail: Communicate effectively.* Upper Saddle River, NJ: Prentice Hall.

TiVo. (2002). *Start here* [Manual]. Tokyo: TiVo and Pioneer Corporation.

Toastmasters International. (2010). Home page. Retrieved from http://www.toastmasters.org

Tuckman, B. W. (1965). Development sequence in small groups. *Psychological Bulletin, 63,* 384–399.

Tufte, E. (1983). *The visual display of quantitative information.* Cheshire, MA: Graphics Press.

University of Chicago Press. (2010). *Chicago manual of style* (16th ed.). Chicago, IL: Author.

University of Minnesota Libraries. (2008). Home page. Retrieved from http://www.lib.umn.edu

U.S. Census Bureau. (2010). Home page. Retrieved from http://www.census.gov

U.S. Centers for Disease Control. (2003). *Fight the bite* [Fact sheet]. Retrieved from http://www.cdc.gov/ncidod/dvbid/westnile/index.htm

U.S. Centers for Disease Control. (2003). *The rabies virus* [Fact sheet]. Retrieved from http://www.cdc.gov/rabies/virus.htm

U.S. Centers for Disease Control. (2003). *West Nile Virus (WNV) infection: Information for clinicians* [Fact sheet]. Retrieved from http://www.cdc.gov/ncidod/dvbid/westnile/index.htm

U.S. Centers for Disease Control. (2007). *Balance scale.* Retrieved from http://www.cdc.gov/diabetes/pubs/images/balance.gif

U.S. Copyright Office. (2011). Home page. Retrieved from http://www.loc.gov/copyright

U.S. Department of Energy. (2003). *Fuel cell technology: How it works* [Fact sheet]. Retrieved from http://www.fe.doe.gov/coal_power/fuelcells/fuelcells_howitworks.shtml

U.S. Department of Health and Human Services. (2007). *Women's health* [Fact sheet]. Retrieved from http:///www.cdc.gov/lcod.htm

U.S. Environmental Protection Agency. (2000). *National water quality inventory, 2000 report.* Washington, DC: Author.

U.S. Environmental Protection Agency. (2002). *Global warming impacts: Forests.* Retrieved from http://yosemite.epa.gov/oar/globalwarming.nsf/content/ImpactsForests.html

U.S. Environmental Protection Agency. (2004). *IPM for scorpions in schools: A how-to manual* [Manual]. Washington, DC: Author.

U.S. Federal Bureau of Investigation. (2008). *FBI history.* Retrieved from http://www.fbi.gov/fbihistory.htm

U.S. Geological Survey. (1997). *Predicting earthquakes* [Fact sheet]. Retrieved from http://pubs/usgs/gov/gip/earthq1/predict.html

U.S. Geological Survey. (2002). *Eruption history of Kilauea* [Fact sheet]. Retrieved from http://hvo.wr.usgs.gov/kilauea history/main.html

U.S. Geological Survey. (2003). *A proposal for upgrading the national-scale soil geochemical database for the United States.* Washington, DC: Author.

U.S. Office of Energy Efficiency and Renewable Energy. (2008). *Solar decathlon, 2075.* Retrieved from http://www.solardecathlon.org

Velasquez, M. G. (2002). *Business ethics: Concepts and cases* (5th ed.). Upper Saddle River, NJ: Prentice Hall.

Velotta, C. (1987). Safety labels: What to put in them, how to write them, and where to place them. *IEEE Transactions on Professional Communication, 30,* 121–126.

VeriMed Corporation. (2006). *VeriMed* [Brochure]. Retrieved from http://www.verimedinfo.com/files/Patient_VM_003R2(web).pdf

Vucetich, J. (2003). *Population data from the wolves and moose of Isle Royale* [Data file]. Retrieved from http://www.isleroyalewolf.org

Washington State Department of Ecology. (2003). *North Creek water cleanup plan* [Fact sheet]. Retrieved from http://www.ecy.wa.gov/programs/wq/tmdl/watershed/north-creek/solution.html

West Virginia Office of Emergency Medical Services. (2007). *EMT-paramedic treatment protocol 4202* [Manual]. Morgantown, WV: Trauma and Emergency Care System, NOROP Center.

Whisper Technology. (2007). Home page [Software]. Retrieved from http://www.whisper.com

White, J. (1988). *Graphic design for the electronic age.* New York, NY: Watson-Guptill.

Williams, J. (1990). *Style.* Chicago, IL: University of Chicago Press.

Williams, R. (2004). *The non-designer's design book* (2nd ed.). Berkeley, CA: Peachpit.

Williams, R., & Tollett, J. (2000). *The non-designer's web book* (2nd ed.). Berkeley, CA: Peachpit.

Writing Center at Rensselaer Polytechnical Institute. (2003). *Revising prose.* Retrieved from http://www.rpi.edu/dept/llc/writecenter/web/revise.html

Yahoo! Babel Fish. (2008). Home page. Retrieved from http://babelfish.yahoo.com

Yoshida, J. (1996, October 7). A suggestive Woody has Japanese touchy. *Electronic Engineering Times.*

Credits

Text Credits

Page 8 © 2003 by TiVo Inc. and Pioneer Corporation. Reprinted by permission of Pioneer Corporation (Tokyo). TiVo and the TiVo logo are registered trademarks of TiVo Inc. TiVo, Season Pass, Wishlist, TiVo Basic, TiVo Plus, and the Series2 logo are trademarks of TiVo Inc. VCR Plus+ is a registered trademark of Gemstar Development Corp. The DVD logos are trademarks of DVD FLLC. Other trademarks are the properties of their respective owners. Page 10 © Copyright 2008 The American National Red Cross. All Rights Reserved. Reprinted with permission. Page 12 Courtesy: National Human Genome Research Institute. Illustration credit (center of screen shot): Jane Ades, NHGRI. Page 49 Microsoft product screen shot reprinted with permission from Microsoft Corporation. Page 62 Reprint Courtesy of International Business Machines Corporation, copyright 2011 © International Business Machines Corporation. Page 78 World of Warcraft, © 2005–2008 Blizzard Entertainment. All rights reserved. Reprinted with permission. Page 83 Institute of Electrical and Electronics Engineers. © 2006 IEEE. Reprinted with permission of the IEEE. Page 84 Online Ethics Center at the National Academy of Engineering. Reprinted with permission of the National Academy of Sciences, Courtesy of the National Academies Press, Washington, D.C. Copyright © 2003–2007 National Academy of Sciences. Page 137 Line Drawing Credit: MDA. NASA/courtesy of nasaimages.org. Page 139 Reprinted by permission of International Organization for Standardization. Pages 152–153 Reprinted by permission of Honeywell and WindTronics. Page 154 From Google Chrome: Behind the Open Source Browser Project, words by the Google Chrome team, comics adaptation by Scott McCloud, http://www.scottmccloud.com. Pages 157–158 Reprinted with permission of Alcatel-Lucent USA Inc. Page 169 From *The Worst-Case Scenario Survival Handbook*™ by Joshua Piven and David Borgenicht. Copyright © 1999 by Quirk Productions, Inc. Used with permission of Chronicle Books LLC, San Francisco. Visit ChronicleBooks.com. Pages 170–172 Reprinted by permission of the West Virginia Trauma & Emergency Care System, NOROP Center, Morgantown, WV. Page 178 © Copyright 1999–2007 by Quirk Productions, Inc. Worst-Case Scenario® and *The Worst-Case Scenario Survival Handbook*™ are trademarks of Quirk Productions, Inc. www.worstcasescenarios.com. Used with permission. All rights reserved. Page 183 Reprinted with permission of Stewart-MacDonald. Page 185 Reprinted with permission from Nikon. Pages 187–188 From "Working with a Chainsaw." Husqvarna, 2002, pp. 6–7. Reprinted by permission of Husqvarna. Page 191 From *Ryobi 510r 4-Cycle Garden Cultivator: Operator's Manual.* Reprinted by permission of Ryobi. Page 193 Reprinted by permission of Polk Audio. Page 200 © Metropolitan Transportation Authority. Used with permission. Pages 233–234 Reprinted by permission of graFighters LLC. Pages 250–252 From *A Practical Approach to Managing Phishing* by Michael Barrett, Chief Information Security Officer and Dan Levy, Director of Risk Management—Europe. PayPal, April 2008. These materials have been reproduced with the permission of PayPal, Inc. COPYRIGHT © 2008 PAYPAL, INC. ALL RIGHTS RESERVED. Page 258 Life Balance. Reprinted by permission of Llamagraphics, Inc. Page 281 © 2000–2008 Pew Internet & American Life Project. Reprinted with permission. Page 298 GOOGLE is a trademark of Google Inc. Page 310 Used with permission of Nathan Stweart. Page 317 Monster.com screenshot reprinted by permission. Page 348 © 2010 Welch Architecture. Reprinted with permission. Page 359 Diagram created in Inspiration® by Inspiration Software®, Inc. Page 375 © 2006 VeriChip Corporation. Reprinted with permission. Page 395 MetaCrawler, © 2008 InfoSpace, Inc. All rights reserved. Page 396 American Society of Civil Engineers Research Library, © 2008 American Society of Civil Engineers. Reprinted with permission. Page 397 © 2008 Regents of the University of Minnesota. All rights reserved. Page 400 Excerpt from transcript of the interview by Kara Swisher and Walt Mossberg conducted with Microsoft Chairman Bill Gates and Apple CEO Steve Jobs at The Wall Street Journal's D: All Things Digital conference on May 30, 2007. © 2007 Dow Jones & Company Inc. Reprinted with permission. Pages 401–402 Reprinted with permission of Dr. Linda K. Bledsoe and Dr. Bibhuti K. Sar. This work was supported under award 1999-WA-VX-0012 from the U.S. Department of Justice "Grants to Combat Crimes Against Women on Campuses." The opinions, findings, and conclusions expressed in this document are those of the authors and do not necessarily represent the official position or policies of the U.S. Department of Justice. Page 416 Zotero screenshot reprinted by permission of the Center for History and New Media, George Mason University. Page 416 Reproduced by permission of EBSCO Publishing, Inc. Page 416 Reprinted by permission from Macmillan Publishers Ltd. Page 434 Pacific Northwest Search & Rescue. Reprinted by permission of Pacific Northwest Search & Rescue, Inc. Page 438 Reprinted with permission of the Buckinghamshire County Council. Page 439 Richard Sheppard, Curlew Communications Ltd., UK. Reprinted with permission. Page 440 The economist's voice. Copyright 2005 by Berkeley Electronic Press. Reproduced with permission of BERKELEY ELECTRONIC PRESS in the format Textbook and Other book via Copyright Clearance Center. Page 442 Reprinted with permission of Alan G. Aversa. Page 460 Reprinted by permission of Geothermal Energy Association. Weather Channel toolbar Courtesy of weather.com. Page 463 Reproduced with permission of Yahoo! Inc.® 2008 by Yahoo! Inc. YAHOO! and the YAHOO! logo are trademarks of Yahoo! Inc. Page 476 Patrick Moore, "Patrick Moore: Going Nuclear Over Global Warming," Special to The *Sacramento Bee*, December 12, 2007. Reprinted by permission of Dr. Patrick Moore, Greenspirit Strategies, Ltd. www.greenspiritstrategies.com. Page 484 Reproduced by permission of the Queen's Printer for Manitoba. The Queen's Printer for Manitoba does not warrant the accuracy or currency of this material. Page 486 Reprinted with permission of Fermilab. Page 496 Field Museum home page, http://www.fieldmuseum.org, reprinted by permission. Page 499 Microsoft product screen shot reprinted with permission from Microsoft Corporation. Page 501 © 2005 Jabra, GN Mobile A/S (GN Netcom A/S). Reprinted by permission of GN Netcom, Inc. Page 504 Oil Market Report © OECD/IEA, 2008, pages 24–25. Page 509 From Patricia Russo and Stephen Boor, "How Fluent Is Your Interface? Designing for International Users." In S. Ashlund, K. Mullet, A. Henderson, E. Hollnagel, and T. White, eds., *Proceedings of the INTERACT '93 and CHI '93 Conference on Human Factors in Computing Systems,* pp. 342–347, Table 1. © 1993 ACM, Inc. Reprinted by permission. http://doi.acm.org/10.1145/169059.169274. Page 510 Microsoft product screen shot reprinted with permission from Microsoft Corporation. Page 523 2007 Human WNV Cases. Reprinted by permission of Mary Grace Stobierski,

DVM/Michigan Department of Community Health. Page 524 Global Obesity Forecast. World Health Organization 2005. Reprinted by permission. Page 527 Reprinted by permission of New Policy Institute. Page 529 Data from Figure 1 from John A. Vucetich and Rolf O. Petersen, *Ecological Studies of Wolves on Isle Royale: Annual Report 2007–2008.* Houghton, MI: School of Forest Resources and Environmental Science, Michigan Technological University, March 31, 2008, p. 3. Reprinted by permission of John A. Vucetich/Michigan Technological University. Page 533 Reprinted by permission of the National Foundation for Cancer Research. Page 535 Reprinted with permission of the Australian Resuscitation Council. Page 539 Adobe product screen shot reprinted with permission from Adobe Systems Incorporated. Page 540 Reprinted with permission of the New Mexico Space Grant Consortium. Page 544 Reprinted by permission of the International Olympic Committee. Page 545 Reprinted with permission of The Gallup Organization. Page 555 Microsoft product screen shot reprinted with permission from Microsoft Corporation. Page 556 Microsoft product screen shot reprinted with permission from Microsoft Corporation. Page 557 Microsoft product screen shot reprinted with permission from Microsoft Corporation. Page 561 Microsoft product screen shot reprinted with permission from Microsoft Corporation. Page 604 Reprinted by permission of Toastmasters International. Page 613 Screen shot: http://www.fs.fed.us/. Reprinted by permission of US Forestry Service. Page 619 Original English web page © 2008 Purdue University. Reprinted with permission. Translated with Google Translate. Page 622 Reprinted with permission from the Natural Resources Defense Council. Page 624 Reprinted with permission from the Natural Resources Defense Council. Page 625 Reprinted with permission from the Natural Resources Defense Council. Page 629 Purdue logo, reprinted by permission of Purdue University. Page 637 Reprinted by permission of LinkedIn. Page 639 Screenshot reprinted by permission of WordPress.

Photo Credits

Page 1 moodboard/Alamy. Page 2 Fancy Collection/SuperStock. Page 13 Ed Murray/The Image Works. Page 15 Paul Conklin/PhotoEdit. Page 19 Michael Newman/PhotoEdit. Page 20 Jon Feingersh/Corbis. Page 28 Anton Vengo/SuperStock. Page 34 Mary Knox Merrill/The Image Works. Page 40 Michael Newman/PhotoEdit. Pages 44, 60 Dynamic Graphics/Alamy. Page 67 Pete Leonard/Corbis. Page 68 Exactostock/SuperStock. Page 72 Time/Life/Getty Images. Page 77 Robert McCabe Family Collection. Page 87 handengbers/Alamy. Page 92 Jose Luis Pelaex Inc./drr.net. Page 93 Alamy. Page 94 colorblend/Getty Images. Page 130 Royalty Free/Corbis. Page 132 Prisma/The Image Works. Page 163 Helen King/Corbis. Page 165 Peter Chen/The Image Works. Page 168 Bob Daemmrich/PhotoEdit. Page 204 Richard Megna/Fundamental Photographs. Page 205 Roger Bamber/Alamy. Page 245 iStockPhoto. Page 246 Norbert Schwerin/The Image Works. Page 268 iStockPhoto. Page 269 age fotstock/SuperStock. Page 313 Veer. Page 314 Exactostock/SuperStock. Page 340 Dorling Kindersley Media Library. For further information, see www.dkimages.com. Page 343 David Young Wolff/PhotoEdit. Page 351 Veer. Page 352 Rachel Epstein/PhotoEdit. Page 365 Blend/Corbis. Page 366 Fancy Collection/SuperStock. Page 380 David Buffington/Corbis. Page 384 Royalty Free/Corbis. Pages 385, 399 The Image Works. Page 386 Powerstock/age footstock. Page 406 Digital Vision/Getty. Page 407 Ruth Von Briel/PhotoEdit. Page 423 Jeff Greenberg/The Image Works. Page 424 Kate Mitchell/Corbis. Page 453 Patrick Giardino/Corbis. Page 454 Kimberly White/Landov. Page 480 Maximillian Stock/Phototake. Page 481 David Young Wolff/PhotoEdit. Page 514 ALL: Dorling Kindersley Media Library. For further information, see www.dkimages.com. Page 518 Will Sanders/Getty. Page 522 Dan Tardiff/Corbis. Page 525 Corel. Page 537 Corel. Page 538 The Newark Museum/Art Resource. Page 546 Courtesy of Alesis. Page 549 Bob Daemmrich/PhotoEdit. Page 551 Michael Newman/PhotoEdit. Page 569 moodboard/Corbis. Page 572 Helen King/Corbis. Page 573 Warren Morgan/Corbis. Page 574 Royalty Free/Corbis. Page 581 Doug Martin/Photo Researchers. Page 582 Doris Kinderling. Page 583 Exactostock/SuperStock. Page 590 Dmitriy Shironosov/Alamy. Page 594 Cultura RM/Alamy. Page 600 Bob Daemmrich/PhotoEdit. Page 607 Aierdale Brothers/Getty. Page 610 Ralf Schultheiss/Corbis. Page 611 Ann Hermes/The Image Works. Page 634 Royalty Free/Corbis. Page 635 Jeff Greenberg/PhotoEdit. Page 643 David Adamson/Alamy. Page 646 Pixellover/Alamy.

Index

A, use of article, A-19
Abstract, in analytical report, 300–301
Access points, headings as, 498
Acronyms
  apostrophes with, A-14
  in cross-cultural documents, 464
Action verbs. *See also* Verbs
  in analytical report, 276
  in describing purpose, 6–7, 138, 169, 211–212, 276, 327, 616
  in instructions and documentation, 169
  in résumé, 327
  in websites, 616
Active voice
  in application letters, 338
  passive voice and, 338, 470–471
Activity journal, 254, 258 (fig.)
Activity reports, 246–266
  briefings as, 247–248
  conclusion of, 261–263
  context of use, 259–260
  features of, 248
  incident reports, 250, 253–254 (fig.)
  introduction in, 260, 261 (fig.)
  laboratory reports, 250, 254, 255–257 (fig.)
  organizing and drafting of, 260–263
  planning and researching of, 254–260
  progress, 247, 249 (fig.), 262 (fig.)
  purpose of, 247, 259
  readers of, 259
  status reports, 264–265
  style and design of, 263–264
  subject of, 259
  types of, 247–254
  white papers as, 247–248, 250–252 (fig.)
Address, on envelope, 120, 120 (fig.)
Adjectives, order of, A-20
Adjustment letters and memos, 111–114, 113 (fig.)
Adobe Audition, 641
Adobe Dreamweaver, 195
Adobe Photoshop, 539
Adobe Premiere, 641
Adverbs
  order of, A-20–A-21
  proper placement of, A-20–A-21
Advertisements, classified, 319
Agenda, for meeting, 54–56, 55 (fig.)
Agreement
  pronoun-antecedent, A-5
  subject-verb, A-4–A-5
Alignment
  as design element, 482, 492–493, 627
  problems with centered text, 495 (fig.)
  of sentence subjects in paragraph, 467–469
American Institute of Aeronautics and Astronautics (AIAA), survey by, 16
American Institute of Graphic Arts (AIGA), 543
Analogies, 147, 473–474
Analysis charts
  Context Analysis Chart, 26, 27 (fig.)
  Reader Analysis Chart, 25 (fig.), 25–26, 27
  Team Performance Review, 64 (fig.)
  Writer-Centered Chart for identifying readers, 22 (fig.), 22–23
Analytical reports, 269–311
  conclusions and recommendations in, 285
  designing, 304–310, 306–308 (fig.)
  drafting front and back matter for, 299–303
  efficient writing of, 283
  feasibility reports, 272, 286–297 (fig.)
  features of, 270, 271
  as genre, 3
  graphics in, 305
  IMRaD pattern for, 270, 271
  jargon in, 304
  methodology section of, 284
  nature of, 3 (fig.)
  organizing and drafting of, 282–303, 426–428
  planning of, 272–277
  purpose of, 270, 272–276
  research reports as, 270, 273–275 (fig.)
  researching, 277–282
  results section of, 285
  specialized terms defined in, 304
  style of, 303–304
  types of, 270–272
  virtual teaming on, 298 (fig.), 298–299
Anecdotes, in introduction, 584
Animal symbols and mascots, 544
Antecedent, pronoun-antecedent disagreement and, A-5
APA documentation style, 414–415, A-24, A-25–A-31
  reference list for, 415, 415 (fig.), A-27–A-30
Apostrophes, 560 (fig.)
  with plurals of numbers, acronyms, and symbols, A-14
  for possession, A-13–A-14
Appendixes, in analytical report, 303
Application letters, 331–339
  content and organization of, 331–335
  emphasizing education, 336 (fig.)
  emphasizing work experience, 337 (fig.)
  features of, 316
  revising and proofreading, 339
  style of, 335–339
ArrantSoft, 48
Artemis Project Management, 48
Articles (publication)
  in APA documentation style, A-29–A-30
  in CSE documentation style, A-33–A-34
  in journals and newspapers, 397–398
  in MLA documentation style, A-38–A-39
Articles (grammar), for ESL, A-19–A-20
Asia, direction of reading in, 35 (fig.), 510, 510 (fig.)
Asides, en dashes with, A-16
Attachments, to e-mail, 96, 99, 123
Attire
  for interview, 344
  for presentation, 599
Attitudes
  in application letters, 335–338
  nature of, 25
  persuasion through, 377
  of readers, 23–26
Audacity, 641
Audience. *See also* Readers
  international presentations for, 604–606
  for presentation, 577, 583 (fig.), 592, 602
Audio
  in documents, 546, 546 (fig.)
  presentations and, 546, 546 (fig.), 586–587
Audio conference calling, 642–643
Authoritative tone, mapping of, 472–473, 473 (fig.)
Authors
  in APA documentation style, A-25–A-30
  in CSE documentation style, A-31, A-32–A-34
  in MLA documentation style, A-35–A-36
Autodesk Inventor, 141
Awards and activities, in portfolio, 341

Babel Fish, 462, 463 (fig.), 563
Back matter, drafting, 303

Back translation, 463
Background color, in design, 505 (fig.), 505–506
Background information
  for e-mail, 104
  on home page, 623
  for instructions and documentation, 177
  in introduction, 431–432
  for letter or memo, 103 (fig.), 104, 331
  for presentation, 588
  for proposals, 212
  for technical description, 140
Backward-planning project calendar, 48, 50 (fig.)
Bacon, Francis, 399
Balance, as design element, 482–492, 492 (fig.), 627
Banners, in interface, 491, 492 (fig.)
Bar charts, 530–531, 531 (fig.)
"bcc" line, 96
Because, in persuasion process, 370, 373
Better and worse statements, 371, 438–439
Bias
  personal, 418–419
  of sources, 417–418
Binder, portfolio in, 341–342
Binding, 513–514, 514 (fig.)
Bio, 347–348
Blogger, 639
Blogs, 89, 396, 638–640, 639 (fig.)
  in APA documentation style, A-30
Blogsome, 639
Body
  of activity reports, 260–261
  of application letter, 335
  of current situation section (proposals), 216, 217 (fig.)
  of document, 425, 434–435, 435 (fig.)
  of e-mail, 95, 98, 105–106
  of letter or memo, 95, 103–104 (fig.), 105–106
  of methodology section of analytical report, 284
  organizing and drafting, 433–446
  of presentation, 588–591
  of project plan (proposals), 218
  sections of, 433–436
Body language, during presentation, 599–601
Boisjoly, Roger, 425, 427 (fig.)
Books
  in APA documentation style, A-28, A-29
  in CSE documentation style, A-32–A-33
  in MLA documentation style, A-37–A-38
  as research sources, 397
Borders, in design, 498, 499 (fig.)
Boxing, in drafting process, 360–361, 362 (fig.)
Brackets, A-18
Brainstorming, idea generation through, 359
Briefings, 247–248
Browsers, 394
Bureaucratic phrasing, 117–118, 338
Burridge, Brian, 436

Calculations, in analytical report, 303
Calendars
  groupware for, 62, 62 (fig.), 356–357
  planning with online, 356–357
  project, 48, 49 (fig.), 50 (fig.), 356–357, 357 (fig.)
Care, in social ethics, 74–76, 75 (fig.)
Careers, 314–349
  application letters, 316, 331–339
  goals of, 315
  interviewing strategies, 343–347
  networking for, 318
  planning for, 315–319
  professional bio, 347–348
  professional portfolio, 339–343
  résumé for, 316, 320–331, 333–334
  sources for job search, 315–319
Case studies, in research, 399
Catalog of U.S. Government Publications, 398
Cause and effect
  as document pattern of arrangement, 437, 438 (fig.)
  as presentation pattern, 590, 591 (fig.)
  in proposals, 213
  statements of, 370
Caution statements, in instructions, 190
"cc" line, 96, 121
CD-ROMs
  in APA documentation style, A-30
  in CSE documentation style, A-33, A-34
  in MLA documentation style, A-38, A-39
  online documentation and, 195
  for research, 394
Centered text, alignment problems with, 495 (fig.)
Chalkboard, for presentations, 580
Change
  creativity and, 358
  in instructions and documentation, 179
  technical descriptions and, 140
Charts
  bar charts, 530–531, 531 (fig.)
  flowcharts, 535 (fig.), 535–536
  Gantt charts, 536, 536 (fig.)
  pie charts, 534 (fig.), 534–535
Chronological order
  as document pattern of arrangement, 443
  in presentation, 588, 591 (fig.)
Chronological résumé, 320–329, 322–325 (fig.)
Citation manager, 416 (fig.), 416–417
Citation-sequence system, in CSE documentation style, A-31–A-35
Claim letters and memos, 111, 112 (fig.)
Clarification questions, 593
Clark, Herbert, 469
Classified advertisements, job search in, 319
Clichés
  in cross-cultural documents, 565
  in e-mail, 123
  on websites, 618–619
Clip art, 541
Closing of document, 435, 435 (fig.)
  of current situation section (proposals), 216
  letter, 119 (fig.), 120
  memo, 121, 122 (fig.)
  methodology section of analytical report, 284
  project plan (proposals), 218
Clothing
  for interview, 344
  for presentation, 599
Codes of ethics
  of IEEE, 83 (fig.)
  resolving ethical dilemma and, 81–82
Collaboration. *See* Teaming
College placement office, 319
Colloquialisms, on websites, 618–619
Colons, 560 (fig.), A-11–A-12
  for leading off lists, A-12
  misuse of, A-12–A-13
  with quotation marks, A-12
  in titles, numbers, and greetings, A-12
Color
  in design, 505–506
  in highlighting of text, 506–507
  in other cultures, 36, 509, 509 (fig.)
  of paper for document, 515
  researching, 36
Comb binding, 514, 514 (fig.)
Comma, 560 (fig.)
Comma splice, A-1–A-2
Command voice, for instructions and documentation, 184, 201

Commas, A-9–A-11
  in numbers, dates, and place names, A-10
  with quotation marks, A-10
  removing excess, A-11
Commentary, in research process, 412–413
Comments
  in social networking, 638
  with steps in instructions, 189
  about subject of sentence, 456
Common knowledge, citing sources for, 421
Communication. *See* Cross-cultural communication; *specific type of communication*
Comparison and contrast
  as document pattern of arrangement, 437, 439 (fig.)
  as presentation pattern, 591, 591 (fig.)
Compelling statement, in introduction, 584
Complaint letters and memos, 111, 112 (fig.)
Completer/finisher (team role), 61
Completion reports, 270
Compound words, hyphens with, A-17
Computer-aided drafting (CAD), technical description writing and, 141
Computers. *See also* Internet; Software; Word processors
  clip art, 541
  in collaboration process, 11, 12–13
  cyberbullying, 78–81
  electronic networks, 59, 62, 87
  ethics of using electronic files, 85–90
  international and cross-cultural communication and, 28–37
  keeping notes on, 409, 410 (fig.)
  on-screen documentation, 194–195
  plain sentences created with, 460–464
  in research process, 7, 36
  screen shots, 541–542, 542 (fig.)
  spelling and grammar checker on, 463–464, 558, 560–561, 561 (fig.)
  typefaces, 500
  virtual teaming with, 59–60, 62–63, 298 (fig.), 298–299
Conciseness
  in instructions and documentation, 186
  reader desire for, 22
Conclusion
  of activity reports, 261–263
  of analytical reports, 285
  of application letter, 335
  closing moves in, 446–447
  of document, 425, 448 (fig.)
  of e-mail, 95, 98, 106
  indirect approach in, 450
  of instructions and documentation, 192
  of letter or memo, 95, 104 (fig.), 106
  organizing and drafting, 446–448
  of presentation, 591–594
  of proposals, 226, 227 (fig.)
  of technical descriptions, 148 (fig.), 149
Conclusion slides, for presentation, 592 (fig.)
Conditions required, in instructions, 183
Conflict
  management of, in storming phase of teaming, 54–58
  mediation in meetings, 56–58, 57 (fig.)
  resolving by team, 49–50
Conservation ethics, 72, 73 (fig.), 76–77
  confronting issues in, 81 (fig.)
  ethical dilemma and, 80
Consistency
  as design element, 482, 497, 500–504, 501 (fig.), 627
  in language, 462
  in layouts, 501 (fig.)
Constitution (U.S.), 86, 88
Contact information
  in analytical reports, 285
  in conclusion, 447
  in proposals, 226, 227 (fig.)
Content
  of application letter, 331–335
  cultural differences in, 32–33
  editing for, 554
  organizing, 425, 426
  of presentations, 582–594
Context Analysis Chart, 26, 27 (fig.)
Context of use
  for activity reports, 259–260
  for analytical reports, 276–277
  analyzing, 511
  for e-mail, 102
  in editing process, 554
  as element of rhetorical situation, 5 (fig.), 6
  identifying, 26–27
  for instructions and documentation, 177
  for letters and memos, 102
  for presentations, 577–578
  profiling, 26–27, 27 (fig.)
  for proposals, 212
  for technical descriptions, 139
  of websites, 617–618
Contextual cues, 378–381
Continuous Quality Improvement (CQI), 63
Contractions, apostrophes with, A-13
Contrast
  as design element, 482, 505–508, 627
  in webpage, 505 (fig.)
Cookies, computer, 87
Coordinator (team role), 48, 61–62
Copyediting, 10, 552, 556–558
  symbols for, 562, 563 (fig.), 564 (fig.)
  Track Changes, 557, 557 (fig.)
Copyright law, 85, 88 (fig.), 88–90
  avoiding problems with, 89–90, 281
  copyrighting one's work, 90
  plagiarism and, 90
  websites and, 625
Corel VideoStudio, 641
Corporate authors
  in APA documentation style, A-26, A-28
  in CSE documentation style, A-32
  in MLA documentation style, A-36, A-37
Corporate codes, resolving ethical dilemma and, 81–82
Correspondence. *See also* E-mail; Letters
  in APA documentation style, A-27, A-30
  in CSE documentation style, A-34
  in MLA documentation style, A-39
Costs and benefits
  as document pattern of arrangement, 440
  as proposal section, 226, 227 (fig.)
  statements of, 371
Cover letters and memos, 96, 107–111
Cover sheet, for portfolio, 341
Creativity
  generating new ideas, 357–362
  in strategic planning, 357–362
Cropping, digital photography, 150
Cross-cultural communication, 15, 449 (fig.)
  computers and, 28–37
  content and, 32–33
  customs in, 28–37, 121–125
  design and, 35 (fig.), 35–36, 508–510
  e-mail and, 121–125
  editing in, 563–566
  graphics in, 35–36, 508–510, 542–543
  in high-context cultures, 378–381, 448–450
  instructional writing and, 179–180
  listening in, 36
  organization and, 33–34, 448–450
  presentations for international audience, 604–606
  scanning page and, 35, 35 (fig.), 510, 510 (fig.)
  strategies for, 37

Cross-cultural communication (*continued*)
  style and, 34–35
  symbols and, 542–543
  translators and, 604–606, 607
  websites in, 618–619
CSE (citation-sequence) documentation style, A-24, A-31–A-35
Cubase, 641
Cues, contextual, 378–381
Culturally deep documents, 508–510
Culturally shallow documents, 508–510
Culture. *See also* Cross-cultural communication
  design and, 35 (fig.), 35–36, 508–510
  high-context, 378–381
  low-context, 448
Curren, Mitch, 614
Cyberbullying, 78–81
CyberMatrix Office, 59

Danger statements, in instructions, 189
Dangling modifier, A-3–A-4
Dashes, 560 (fig.), A-16–A-17
  em dashes, A-16
  en dashes in numbers and dates, A-17
Data
  in graphs, tables, and charts, 528–536
  persuasion with, 372
Database programs, for note organizing, 409
Dates
  commas in, A-10
  dashes in, A-17
Deliverables, in forming stage of teaming, 47, 47 (fig.)
Delivery, of presentation, 599–602
Deming, W. Edwards, 63
Description, by features/functions, in presentation, 590, 591 (fig.)
Design, 481–516. *See also* Design elements
  of activity reports, 263–264
  of analytical reports, 304–310, 306–308 (fig.)
  cultural differences in, 35 (fig.), 35–36, 508–510
  document, 512–513
  editing for, 555–556
  importance of, 10 (fig.), 14 (fig.), 22
  of instructions and documentation, 180, 192–201
  of letters and memos, 118–120
  of presentation visuals, 596–597
  of proposals, 229–234
  of résumé, 329–331, 333, 334 (fig.)
  revising and editing, 513
  of technical descriptions, 151–158
  thumbnails in, 511–512, 512 (fig.)
  using principles of, 511–513
  of websites, 505 (fig.), 626–630
  in writing process, 4, 4 (fig.), 9, 10 (fig.), 118–120
Design elements
  alignment as, 482, 492–493, 627
  balance as, 482–492, 492 (fig.), 627
  binding and, 513–514, 514 (fig.)
  borders and rules as, 498, 499 (fig.)
  consistency as, 482, 497, 500–504, 501 (fig.), 627
  contrast as, 482, 505–508, 627
  grouping as, 482, 493–499, 627
  paper selection and, 515
  principles of, 627–628
  typefaces as, 500, 502 (fig.)
Designer (team role), 48
Diagrams, 539–540, 540 (fig.)
Dictionary
  online, 561
  for spelling, 561
Digital photographs, in technical description, 149–150
Digital projector, with computer, 581 (fig.)
Direction of reading, 35, 35 (fig.), 510, 510 (fig.)
Discussion lists, 62
Discussion section, in analytical report, 285
Document cycling, 566, 566 (fig.)
Document design. *See* Design
Documentation. *See* Documentation of sources; Instructions and documentation
Documentation of sources, 413–417, A-24–A-39
  APA documentation style, 414–415, A-24, A-25–A-31
  to avoid plagiarism or copyright problems, 89–90, 281, 420–421
  CSE documentation style (citation-sequence), A-24, A-31–A-35
  MLA documentation style, A-25, A-35–A-39
Documents
  audio and, 546, 546 (fig.)
  binding of, 513–514, 514 (fig.)
  comparison with website, 612 (fig.), 614
  cross-cultural, 449 (fig.), 563–566
  culturally deep/culturally shallow, 508–510
  designing. *See* Design
  for general public versus experts, 29–32 (fig.)
  groupware for posting, 63
  outlining of, 360–361, 361 (fig.), 428–430
  paper for, 515
  patterns of arrangement for, 270, 271, 437–446
  reason-based/values-based statements in, 367 (fig.)
  reasoning used in, 368–369 (fig.)
  revising, 552–554
  substantive editing of, 9, 554–556, 556 (fig.)
  usage problems in, 561–562, 562 (fig.)
  video in, 545 (fig.), 545–546
Drafting, 430–448
  of activity reports, 260–263
  of analytical reports, 282–303
  of basic webpages, 623–624, 625 (fig.)
  of body, 433–446
  of conclusion, 446–448
  of home page, 621–623
  of instructions, 181–192
  of introduction, 430–433, 432 (fig.), 433 (fig.)
  of navigational pages, 613, 624–625
  of node page, 623, 624 (fig.)
  outlining and, 360–361, 361 (fig.), 428–430
  of proposals, 213–226
  of technical descriptions, 145–150
  of websites, 621–625
  in writing process, 4, 4 (fig.), 7
Dress
  for interviews, 344
  for presentation, 599
DVDs, for research, 395

Economic context of use, 26
  for analytical reports, 276
  for presentation, 578
  for proposals, 212
  for websites, 618
Ecosystem, as ethical issue, 76–77
Editing
  copyediting. *See* Copyediting
  in cross-cultural communication, 563–566
  for design, 513
  levels of, 552–562, 553 (fig.)
  of podcasts, 641, 641 (fig.)
  proofreading and. *See* Proofreading
  revising. *See* Revising
  spelling and grammar checkers and, 463–464, 558, 560–561, 561 (fig.)
  substantive, 9, 552, 554–556, 556 (fig.)
  of videos, 641
  in writing process, 4, 4 (fig.), 9–10

Editor (team role), 48
Editorial comments, brackets with, A-18
Educational background
in application letter, 336 (fig.)
in portfolio, 341
Einstein, Albert, 69–72, 70–72 (fig.)
Either. . . or, as document pattern of arrangement, 370, 440–442, 442 (fig.)
Elaboration questions, 593
Electronic networks
as intranets, 59
local area networks (LANs), 62
surveillance in, 87
Electronic portfolio, 342 (fig.), 342–343
Electronic sources, in research, 393, 394–396
Elevator pitch, as proposal, 233–234
Ellipses
to show thought is trailing off, A-18
to signal information removal from quote, A-18
Em dashes, A-16
E-mail
in APA documentation style, A-30
attachments, 96, 99, 123
conclusion of, 106
context of use, 102
in CSE documentation style, A-34
features of, 95, 96, 98, 99 (fig.), 102–106
formatting, 96–100, 99 (fig.)
graphics in, 123
header for, 96
of inquiry, 106–107
as interview thank you, 347
introduction of, 95, 96, 102–105
message area, 96–98, 99 (fig.)
in MLA documentation style, A-39
netiquette for, 123–125
organizing and drafting, 102–106
response, 107
signature file for, 99–100, 124
status report, 264–265
style of, 114–121
subject line in, 96
texting at work, 124–125
thank you, 347
transmittal, 96
types of, 106–114
Emergency instructions, 199–201
Emotional tone, mapping of, 117 (fig.), 196 (fig.), 230 (fig.), 472 (fig.), 473 (fig.)
Empirical research, 386, 387, 387 (fig.), 398–404
sources in, 393
Empirical research reports, 270
En dashes, A-17
English as a Second Language, grammar guide for, A-19–A-23
English language, in instructions for cross-cultural markets, 180
Enthusiasm, in presentation, 596
Envelopes, 120 (fig.), 120–121
Environment, conservation ethics and, 72, 73 (fig.), 76–77
Ethical codes
of IEEE, 83 (fig.)
resolving ethical dilemma and, 81–82
Ethical context of use, 27
for analytical reports, 277
for presentation, 578
for websites, 618
Ethical dilemmas
categories of, 74–76
confronting, 80–81
nature of, 69
resolving, 69, 77–84
Ethics, 68–90
balancing issues in, 81 (fig.)
conflicts between individual and company, 82–84
confronting issues in, 80–81
conservation, 72, 73 (fig.), 76–77
copyright law and, 85
defined, 69
dilemmas in, 69–72, 77–84
of graphics, 525–527, 526 (fig.)
personal, 72, 73, 73 (fig.), 80, 81 (fig.)
in proposals, 212
in research, 419
safety information and, 180
social, 72, 73 (fig.), 73–76, 80, 81 (fig.)
sources of, 72–77, 73 (fig.)
in technical communication, 13–15
in technical workplace, 85–90
Ethnographies, in research, 399
Etiquette, for e-mail, 123–125
European Union, symbols created by, 543
Evaluation of presentation
form for, 605–606 (fig.)
of performance, 602, 605–606 (fig.)
Evidence, reasoning with, 367, 371–372
Examples
as document pattern of arrangement, 443
reasoning with, 367, 371
with steps in instructions, 189
Exclamation point, 560 (fig.), A-9
Executive summary, in analytical report, 300–302
Experiences, persuasion with, 371–372
Experiments, in research, 280–281, 398
Experts. *See also* Subject matter experts (SMEs)
documents prepared for general public versus, 29–32 (fig.)
quoting in persuasive argument, 372
Extended definitions, 156
External proposals, 206, 213
Eye contact
cultural differences, 36, 344
during presentations, 600–601
with translator of presentation, 607

Facebook, 79, 636, 637
Facilitator, meeting, 54
Facts, persuasion with, 372, 380
Fair use, of copyrighted material, 86, 89–90
Faulty parallelism, A-5–A-6
Feasibility reports, 272, 286–297 (fig.)
Feedback
in instructions, 189
for presentation, 602, 605–606 (fig.)
quality feedback loop, 63, 566
Field notes and observations
purpose of, 133
in research, 398
Films
in APA documentation style, A-29
in CSE documentation style, A-33
in MLA documentation style, A-38
Final Cut, 641
First Amendment, on trademarked items, 86
First-level headings, 495, 497 (fig.)
Five-W and How Questions
for activity reports, 258–259
for analytical reports, 272
for career goals, 315
in defining rhetorical situation, 5
for drafting introductions, 430
idea generation through, 361
for instructions and documentation, 168
for letters and memos, 100
for presentations, 576, 576 (fig.)
for profiling readers, 20–21, 21 (fig.)
for proposals, 206–207
for technical descriptions, 133–134
for website, 615
Flip charts, for presentations, 580, 582 (fig.)
Flowcharts, 535 (fig.), 535–536
Fonts
in presentation visuals, 596–597
serif and sans serif, 500, 502 (fig.)
size of, 507
Footers, inserting, 504, 504 (fig.)

Forecasting
analytical report's content in introduction, 285
in conclusions, 106, 285, 447, 593
of content in introduction, 432
on home page, 623
presentation's structure, 588
Formal presentations, 575
Formal usability testing, 567, 567 (fig.)
Formatting
of e-mail, 96–100, 99 (fig.)
of envelopes, 120 (fig.), 120–121
of letters, 96, 97 (fig.), 118–120, 119 (fig.)
of memos, 96, 98 (fig.), 121, 122 (fig.)
presentation visuals, 596
Forming stage of teaming, 45 (fig.), 45–53
Fourth-level headings, 496, 497 (fig.)
Fragments, A-3
Frames, in persuasion, 372
Framing
logical mapping for, 378 (fig.)
from reader's perspective, 377
reframing and, 377, 379 (fig.)
Fraud, ethics and, 88
Freewriting, idea generation through, 359–360, 360 (fig.)
Front matter, drafting, 299–302
Functional résumé, 320, 329, 330 (fig.)
Functions, partitioning subject by, 141
Future, looking to, in conclusion, 106, 285, 447, 593
Future perfect progressive tense, A-22
Future perfect tense, A-22
Future progressive tense, A-22
Future tense, A-22

Gantt charts, 536, 536 (fig.)
GarageBand, 641
Gatekeepers (supervisors), 22 (fig.), 23
of analytical reports, 276
of e-mail, 101
of instructions and documentation, 177
of letters and memos, 101
for presentations, 577
of proposals, 212
of technical descriptions, 138
of websites, 617, 617 (fig.)
Gender
cross-cultural communication and, 35
in personal titles, 120, 123
Genres, 2–10
choosing, 428 (fig.)
defined, 426
nature of genres, 2–3
organization by, 426–428, 428 (fig.)
Gif (graphic interchange format) files, 629–630
Given/new method of writing, for paragraphs, 469–470
Global market, 15
Glossary
in analytical report, 303
for technical definitions, 155–158
Glossy paper, for document, 515
Goals
for career, 315
and needs compared, 374
in persuasion, 372
Google, 356–357
Google Docs, 298 (fig.), 298–299
Google Translate, 618, 619 (fig.)
Government publications
in APA documentation style, A-29
in CSE documentation style, A-33
in MLA documentation style, A-38
as research sources, 398
Grabbers, for presentation, 585–587
Grammar
common errors, 559 (fig.)
in e-mails, 124
proofreading for, 558–559, 559 (fig.)
Grammar and punctuation guide, A-1–A-8
adjective order, A-20
adverb order, A-20–A-21
apostrophes, A-13–A-14
articles, A-19–A-20
comma splice, A-1–A-2
dangling modifier, A-3–A-4
for English as a Second Language, A-19–A-23
faulty parallelism, A-5–A-6
fragments, A-3
misused comma, A-11
pronoun-antecedent disagreement, A-5
run-on sentences, A-2–A-3
shifted tense and, A-7
subject-verb agreement, A-4–A-5
vague pronoun, A-7–A-8
verb tenses, A-21–A-23
Grammar checkers, 558, 561 (fig.)
Grand style, 455
Grant proposals, 376
Grants, 206
Graphics, 522–547. *See also* Visuals
in analytical reports, 305
bar charts as, 530–531, 531 (fig.)
choosing, 528 (fig.)
clip art, 541
copyediting, 558
in cross-cultural communication, 35–36, 508–510, 542–543
data display with, 528–536
in e-mail, 123
ethics of, 525–527, 526 (fig.)
flowcharts as, 535 (fig.), 535–536
Gantt charts as, 536, 536 (fig.)
guidelines for, 523–528
on home page, 622 (fig.)
illustrations as, 539–541
in instructions and documentation, 189, 190, 191 (fig.), 197, 198–199 (fig.), 200
labeling and placement of, 500, 527 (fig.), 527–528
line drawings and diagrams, 539–540, 540 (fig.)
line graphs as, 529 (fig.), 529–530, 530 (fig.)
maps, 540 (fig.), 540–541
from online or print sources, A-24
photographs as. *See* Photographs
pie charts as, 534 (fig.), 534–535
in presentations, 598
in proposals, 229, 231 (fig.)
screen shots, 541–542, 542 (fig.)
storytelling with, 524, 524 (fig.), 525, 525 (fig.)
tables as, 531–534, 532 (fig.), 533 (fig.)
in technical communication, 13, 14 (fig.)
in technical descriptions, 140, 151–156
written text reinforced with, 525, 526 (fig.), 533 (fig.)
written text versus table, 523 (fig.)
Graphs
data display with, 529 (fig.), 529–530, 530 (fig.)
storytelling with, 524, 524 (fig.)
Greetings
colons in, A-12
interview, 344
in letter, 118, 119 (fig.)
Grids
for interfaces, 490 (fig.)
for page layout, 486–490, 487–488 (fig.), 489 (fig.)
Grouping, as design element, 482, 493–499, 627
Groupware, 59, 62 (fig.), 62–63, 356–357
for activity reports, 258
for virtual teams, 298 (fig.), 298–299

Hand gestures
cross-cultural communication and, 35, 36, 180, 544
in presentations, 600, 601

Handouts, for presentations, 581, 607
Harassment
computer, 78–81
through questions, 593–594
Haviland, Susan, 469
Hazard statements, in instructions, 190, 191 (fig.)
Headers
e-mail, 96
inserting, 504, 504 (fig.)
of memo, 121, 122 (fig.)
Headings
as access points, 498
consistency in, 497
copyediting, 558
levels of, 495–497, 497 (fig.)
Heckling questions, 594
Heightened awareness, public speaking and, 589
Help programs, for drafting of website, 624
High-context cultures, 378–381, 448–450
Highlighting of text, 506–507
Historical records, resolving ethical dilemma and, 82
Home page, 612, 613 (fig.), 622 (fig.)
drafting of, 621–623
as introduction, 434 (fig.)
Horizontal alignment, 492
Hostile questions, 593
Human icons, 544, 544 (fig.)
Humor
in cross-cultural communication, 565, 604
in e-mail, 123
on websites, 619
Hyphens, 560 (fig.), A-16–A-17
in compound words, A-17
to connect prefixes, A-17
Hypothesis
in analytical reports, 278, 282
in research, 282, 387 (fig.), 389–390

I and *me*, pronoun case error with, A-6–A-7
Icons
common, 541 (fig.)
culture-specific, 544
in emergency instructions, 200
as graphics, 541, 541 (fig.)
human, 544, 544 (fig.)
in instructions for cross-cultural markets, 180, 190
on websites, 619
Ideals, in persuasion, 372
Idioms, in cross-cultural documents, 565
If . . . then, as document pattern of arrangement, 370, 440, 441 (fig.)
IHMC Concept Mapping Software, 358
Illustrations, 539–541
Images
animal, 544
human icons, 544, 544 (fig.)
inserting, 538–539, 539 (fig.), 630
in other cultures, 509
in photography, 537 (fig.), 537–538
on website, 628–630
iMovie, 641
Implementor (team role), 61
Importance of task, in instructions, 181
IMRaD pattern, for analytical reports, 270, 271
In-text citations, 410, 413
in APA documentation style, 415, A-25–A-27
in MLA documentation style, A-35–A-36
quotation marks with, A-16
Incident reports, 250, 253–254 (fig.)
Indexes, to print research sources, 397–398
Indirect approach
in conclusion, 450
in introduction, 450
of letter, 449 (fig.)
Informal presentations, 575
Informal usability testing, 566, 567 (fig.)
Information
for analytical report, 282
analyzing, 282
appraising sources of, 417–420
documenting sources of, A-24–A-39
organization for website, 626
Information glut, 386
Information management, 386, 408–413, 425
Information sharing, 87. *See also* Groupware
Inquiry letters and memos, 106–107, 108 (fig.)
Insert Picture command, 539
Inside address, 118, 119 (fig.)
Inspiration (software), 358, 359 (fig.)
Instant messaging (IM), 62, 642–643
Instructions and documentation
categories of, 166
context of use, 179–180
for cross-cultural readers, 179–180
danger statements in, 189
design of, 180, 192–201
drafting, 181–192
emergency instructions, 199–201
examples, 169 (fig.), 170–176, 178 (fig.)
features of, 165, 167
as genre, 7, 8 (fig.)
graphics in, 189, 190, 191 (fig.), 197, 198–199 (fig.), 200
online, 194–195
organizing, 181–192
parts, tools, and conditions required in, 182–183
planning, 168–177, 179–180
researching, 177–179
safety information and symbols in, 189–190, 191 (fig.), 199–201
sequentially ordered steps in, 183–189, 184 (fig.), 185 (fig.)
style of, 192–201
troubleshooting in, 192, 193 (fig.)
Interactivity, computers for, 11, 12 (fig.)
Interfaces
accessible, 494
alignment of, 496 (fig.)
balanced, 486 (fig.), 492 (fig.)
design of, 626–628, 628 (fig.)
grids for, 490 (fig.)
grouping, 496 (fig.)
Internal proposals, 206, 208–210 (fig.)
International communication. *See* Cross-cultural communication
International Organization for Standardization (ISO)
symbols approved by, 190, 543
website, 139, 139 (fig.)
International symbols, 190, 543
Internet. *See also* E-mail; Search engines; Social networking; Webpages; Websites
fraud on, 88
job search process on, 315–318, 326
obtaining information from, 280, 280 (fig.), 281 (fig.)
online documentation on, 195
online persuasion, 373–374
sources for technical definitions on, 561
Interviews
in APA documentation style, A-30
in CSE documentation style, A-34
in job search process, 343–347
in MLA documentation style, A-39
in research process, 281, 398, 400 (fig.)
strategies for, 343–347
Intranets, 59
Introduction
in activity reports, 260, 261 (fig.)
in analytical reports, 282–284
in application letter, 332–335
to document, 425, 430–433
drafting, 430–433

Introduction (*continued*)
in e-mail, 95, 96, 102–105
indirect approach in, 450
of instructions and documentation, 181–182
in letter or memo, 95, 102–105, 103 (fig.)
opening moves in, 431–432
of presentation, 583–588, 585 (fig.)
of proposals, 214, 215 (fig.)
of technical descriptions, 145–146
versions of, 432 (fig.), 433 (fig.)
Irony, quotation marks with, A-15–A-16
Isaacs, Kim, 326
ISO 9000/ISO 14000 standards, 139, 139 (fig.), 149, 190, 543

Jargon, 304
in cross-cultural communication, 464, 565, 605–606
Job search process
application letter in, 316, 331–339
Internet in, 315–318, 326
interviews in, 343–347
job-searching cycle, 319 (fig.)
planning, 315–319
professional bio in, 347–348
professional portfolio in, 339–343
résumé in, 316, 320–331, 339
Jokes. *See* Humor
Journalist's questions. *See* Five-W and How Questions
Journals
in APA documentation style, A-29–A-30
in CSE documentation style, A-33
in MLA documentation style, A-38
as research sources, 397
Jpeg (joint photographic experts group) files, 629–630
Justice, in social ethics, 74–76, 75 (fig.)

Keywords, in website purpose, 616
KISS (Keep It Simple Stupid) principle, using in presentations, 596

Labeling, of graphics, 500, 527 (fig.), 527–528
Laboratory reports, 250, 254, 255–257 (fig.)
Land ethic, 76, 77 (fig.)
Language Weaver, 563
Laws
copyright, 85, 88 (fig.), 88–90
resolving ethical dilemma and, 81–82
Layouts. *See* Page layouts
Legal issues
ethical, 81, 82
in social ethics, 74
in technical communication, 13–15
Leininger, Carol, 37, 565
Leopold, Aldo, 76, 77 (fig.)
Letterhead, 118, 119 (fig.)
Letters
adjustment, 111–114, 113 (fig.)
claim, 111, 112 (fig.)
complaint, 111, 112 (fig.)
conclusion of, 106
context of use, 102
features of, 94, 95, 97 (fig.), 118–120, 119 (fig.)
formatting, 96, 97 (fig.), 118–120, 119 (fig.)
inquiry, 106–107, 108 (fig.)
introduction of, 102–105
organizing and drafting, 102–106
planning and researching, 100–102
of reference, in portfolio, 341
refusal, 114, 115 (fig.)
with reports, 299
response, 107, 109 (fig.)
style of, 114–121
thank you, 345, 346 (fig.)
tone in, 116–117
transmittal, 96, 107–111, 299
types of, 106–114
Libel, ethics and, 88
Library research, 280, 390–391
Library search engine, 397 (fig.)
Life Balance software (Llamagraphics), 258
Line drawings, 539–540
Line graphs, 529 (fig.), 529–530, 530 (fig.)
Line length, 507–508
LinkedIn, 637 (fig.)
Links
in e-mail, 98
to webpages, 98, 626
Listening, in cross-cultural communication, 36
Lists
colons in, A-12
idea generation through, 359
parentheses with, A-18
sequential and nonsequential, 502–504, 503 (fig.)
Listservs, 394
Literature review, of source, 420
Local area networks (LANs), 62
Logic, reasoning with, 367, 367 (fig.), 370–371
Logical mapping
of authoritative tone, 472–473, 473 (fig.)
in creating tone, 116–117, 117 (fig.)
of current situation (proposals), 214, 216 (fig.)
to develop a research methodology, 278–279, 279 (fig.)
of emotional tone, 117 (fig.), 196 (fig.), 230 (fig.), 472 (fig.), 472–473, 473 (fig.)
for framing, 378 (fig.)
for generating ideas, 358, 359 (fig.), 388, 388 (fig.)
to identify steps in instructions, 183, 184 (fig.)
in instructions and documentation, 177, 179 (fig.)
for letter or memo, 105, 105 (fig.), 116–117
partitioning of subject with, 141–142, 142 (fig.)
of plan, 219 (fig.), 354–355, 355 (fig.)
in proposals, 212, 213 (fig.), 214, 216 (fig.), 230 (fig.)
of research methodology, 391, 392 (fig.)
software for, 358, 359 (fig.)
use by researchers, 388, 388 (fig.), 391
of websites, 620 (fig.), 620–621, 626, 627 (fig.)
Lotus Notes (IBM), 59, 62, 258
Low-context cultures, 448

Magazines
in APA documentation style, A-29
in CSE documentation style, A-34
in MLA documentation style, A-38
as research sources, 397–398
Main point
of analytical report, 284, 285
of e-mail, 102–104, 106
of home page, 622
in introduction, 431
of letter or memo, 102–104, 106
of presentation, 587
restating in conclusion, 106, 447, 592
of technical description, 146
Maps
as graphics, 540 (fig.), 540–541
logical. *See* Logical mapping
site, 624
Margin comments, as balance technique, 490–491
Martin, Jennifer, 494
Mascots, graphic uses of, 544
Matte paper, for document, 515
McCloud, Scott, 155
Me and *I*, pronoun case error with, A-6–A-7
Measurements, technical description and, 140
Measuring results, in usability testing, 570

Mediation
  of conflict in meetings, 56–58, 57 (fig.)
  in ethical conflict with company, 82
Meetings
  agenda for, 54–56, 55 (fig.)
  avoiding texting during, 124
  recording decisions in, 56
  running, 54–56
  scheduling in task list, 356
Memos
  adjustment, 111–114
  body, 103–104 (fig.)
  claim, 111
  complaint, 111
  conclusion of, 106
  context of use, 102
  in ethical conflict with company, 82
  features of, 94–96, 98 (fig.), 121, 122 (fig.)
  formatting, 96, 98 (fig.), 121, 122 (fig.)
  of inquiry, 106–107
  introduction of, 102–105, 103 (fig.)
  organizing and drafting, 102–106
  planning and researching, 100–102
  refusal, 114
  with reports, 299
  response, 107
  style of, 114–121
  tone in, 116–117
  transmittal, 96, 107–111, 110 (fig.), 299
  types of, 106–114
Message
  in letter, 119 (fig.), 120
  of memo, 121, 122 (fig.)
Message area, in e-mail, 99 (fig.)
MetaCrawler, 395 (fig.)
Metaphors, 474–475
  in cross-cultural documents, 565
  in technical descriptions, 147
Methodology, research, 278–279, 279 (fig.), 387 (fig.), 391–393
  describing, 391–392
  logical mapping of, 391, 392 (fig.)
  outlining, 393 (fig.)
  revising, 392
  using, 392
Methodology section, in analytical report, 284
Methods/results/discussion pattern, in presentation, 590, 591 (fig.)
Microblogs, 638–640
Microform/microfiche, as research sources, 398
Microsoft Expression Web, 195
Microsoft Movie Maker, 641
Microsoft Outlook, 59, 62, 258
Microsoft Paint, 539
Microsoft Project, 48, 49 (fig.)
Middle East, direction of reading in, 35 (fig.), 510, 510 (fig.)
MindManager, 358
Mission statement, in forming stage of teaming, 46
MLA documentation style, A-25, A-35–A-39
  for in-text citations, A-35–A-36
  for works cited list, A-37–A-39
Modifier, dangling, A-3–A-4
Monitor/evaluator (team role), 61
Motion
  in instruction and documentation, 179
  technical description and, 140
Motivation, in instructions, 182
Moveable Type, 639
MP3 players, 395, 546, 586–587
Music, 546, 546 (fig.)
  in APA documentation style, A-30
  in CSE documentation style, A-34
  in MLA documentation style, A-39
MySpace, 79

Names, in introductions for cross-cultural markets, 180
Narrative pattern, in presentation, 590, 591 (fig.)
NASA, technical description of Mars Explorer, 142, 143–144 (fig.)
NASA Godard Institute, reasoning in document by, 367, 368–369 (fig.)
National Commission on Writing, 16
National Science Foundation (NSF) website, 206, 211 (fig.)
Nauman, Mary, 376
Navigation bar, for balance, 491, 492 (fig.)
Navigation pages, 613, 624–625
Need-to-know information, 22, 105–106, 408 (fig.), 409
Needs
  and goals compared, 374
  nature of, 25
  of readers, 23–26
Negative information, avoiding "you" style with, 116
Netiquette, for e-mail, 123–125
Networking. *See also* Social networking
  electronic networks, 59, 62, 87
  personal, 318
  professional, 318
Newspapers
  in APA documentation style, A-29
  in CSE documentation style, A-34
  in MLA documentation style, A-39
  as research sources, 397–398
Node pages, 613, 623, 624 (fig.)
Nominalizations, eliminating, 304, 458
Nonsequential lists, designing, 502–504, 503 (fig.)
Norming stage of teaming, 45 (fig.), 58–63
Note taking
  paraphrasing as, 411–412
  for quotations, 409–410
  in research process, 408–413
Notepad, for presentations, 580
Notes
  with steps in instructions, 189
  using during presentation, 601–602, 607
Notes View in presentation software, 602, 603 (fig.)
Nouns, apostrophes for possession with, A-14
Numbered steps
  in instructions and documentation, 186–189
  paragraph style instructions, 186, 187–188 (fig.)
Numbers
  apostrophes with, A-14
  colons and, A-12
  commas and, A-10
  dashes in, A-17

Object, in cross-cultural documents, 565
Objectives
  in forming stage of teaming, 46, 47 (fig.)
  of meeting, 54, 55 (fig.)
  in norming stage of teaming, 58
  revising, 58
  in strategic planning process, 353–354
  for usability testing, 570
Objects, photographing, 538, 538 (fig.)
Observations
  field notes and, 133, 398
  for instruction research, 177
  persuasion with, 371–372
  reasoning with, 371–372
  in research, 280–281
Online calendars, 356–357, 357 (fig.)
Online documentation. *See* Instructions and documentation
Online Ethics Center for Engineering and Science, National Academy of Engineering, 84, 84 (fig.), 86
Online Help features, 194–195
Online sources, for technical definitions, 561
Opening, 434, 435 (fig.)
  of current situation section (proposals), 216
  of methodology section of analytical report, 284
  of project plan (proposals), 218

Oppenheimer, Robert, 72 (fig.)
Organization, 424–451
of activity reports, 260–263
of analytical reports, 282–303
for any document, 425, 426 (fig.)
of application letter, 331–335, 332 (fig.), 336 (fig.)
of conclusion, 446–448
cultural differences in, 33–34, 448–450
and drafting of body, 433–446
and drafting of introduction, 430–433
of e-mail, 102–106
editing for, 554–555
by genre, 426–428, 428 (fig.)
of instructions, 181–192
of letters and memos, 102–106, 331–335, 332 (fig.)
outline and, 360–361, 361 (fig.), 428–430
of portfolio, 341
of presentation, 582–594
presentation software for, 601–602
of proposals, 213–226
role of genres in, 3, 3 (fig.), 426–428
of technical descriptions, 145–150
of websites, 620–621
in writing process, 4, 4 (fig.), 7
Outcomes
in forming stage of teaming, 47, 47 (fig.)
in norming stage of teaming, 58
Outline View, in word processor, 360, 361 (fig.), 429–430, 430 (fig.)
Outlining
of document, 360–361, 361 (fig.), 428–430
in drafting process, 360–361
of research methodology, 393 (fig.)
Outlook (Microsoft), 59, 62, 258
Overhead projector, with transparencies, 579–580

Page layouts
for analytical report, 305 (fig.)
balance in, 483 (fig.), 484 (fig.), 485 (fig.), 486–490
consistency in, 501 (fig.)
grids for, 486–490
for instructions and documentation, 196–197
for proposals, 230, 231 (fig.)
for technical description, 151
weighting, 483–486
Pamphlets
in APA documentation style, A-29
in CSE documentation style, A-33
in MLA documentation style, A-38
Paone, Karen, 589
Paper, for document, 515
Paragraphs
in active and passive voice, 470–471
aligning sentence subjects in, 467–469
copyediting, 558
elements of, 464–466
given/new method of writing, 469–470
paragraph style instructions, 186, 187–188 (fig.)
in plain writing style, 263, 464–470
using sentence types in, 466–467
on websites, 626
Parallelism, faulty, A-5–A-6
Paraphrased materials
APA documentation style for, A-27
executive summary, 300–302
MLA documentation style for, A-36
recording, 411–412
Parentheses, 560 (fig.), A-17–A-18
Partitioning of subject
with logical mapping, 141–142, 142 (fig.)
in technical description, 140–144, 142 (fig.)
Parts list, in instructions, 182, 183 (fig.)
Parts of a sentence, 456
Passion, creativity and, 358
Passive voice
in application letter, 338
decision to use, 470–471
Past perfect progressive tense, A-22
Past perfect tense, A-22
Past progressive tense, A-22
Past tense, A-21–A-22
Patchwriting, 90, 421
Patents
ethics and, 85, 86
purpose of, 133
Patterns of arrangement
for documents, 270, 271, 437–446
for presentations, 588–591, 591 (fig.)
Pauses, in presentations, 601
Pdf (portable document format), 195
People, photographing, 537 (fig.), 537–538
Perfect binding, 513, 514 (fig.)
Perfect progressive tenses, A-21
Perfect tenses, A-21
Performance tests, 569–570
Performing stage of teaming, 45 (fig.), 63
Period, 560 (fig.), A-9
Periodical index, 397–398
Perkins, Jane, 228
Permission
copyright and, 89–90
for using source material, 421
Persona, and presentation style, 595
Personal ethics, 72, 73, 73 (fig.)
ethical dilemma and, 80, 81 (fig.)
Personal networking, 318
Persuasion, 366–382
in analytical reports, 303–304
in ethical conflict with company, 82
in high-context cultures, 378–381
with reason, 367–372
reasoning-based, 367–372
types of, 367 (fig.)
values-based, 367 (fig.), 372–378, 375 (fig.), 378 (fig.)
Persuasive style, 9
balance between plain style and, 477
for proposals, 229
sample persuasive argument, 476 (fig.)
techniques for, 367 (fig.), 472–477
types of, 455
uses of, 472–477
Peterson, Chris, 141
Pew Research Center, archive of, 281 (fig.)
Photographs, 537 (fig.), 537–539, 538 (fig.)
digital, 149–150
in e-mail, 123
inserting, 538–539, 539 (fig.)
storytelling with, 525 (fig.)
in technical descriptions, 149–150
Physical context of use, 26
for analytical reports, 276
for presentation, 577–578
for proposals, 212
for websites, 618
Pie charts, 534 (fig.), 534–535
Places
commas in place names, A-10
photographing, 538
Plagiarism
avoiding, 89–90, 281, 420–421
defined, 420
on websites, 625
Plain style, 9
for activity reports, 263
for analytical reports, 303–304
balance between persuasive style and, 477
for instructions and documentation, 195–196
paragraphs in, 464–470
for proposals, 229
sentences in, 455–464
technical description, 151
writing plain sentences, 455–464
Planning. *See also* Strategic planning
for activity reports, 254–260

for analytical reports, 272–277
for career, 315–319
for e-mail, 100–102
for instructions and documentation, 168–177, 179–180
for letters and memos, 100–102
with online calendars, 356–357, 357 (fig.)
for presentations, 574–579
for proposals, 206–212
for technical descriptions, 133–139
timeline in, 355–356
of website, 615–619
in writing process, 4, 4 (fig.), 5–7
Plant (team role), 61
Plurals, apostrophes with, A-14
Podcasts/podcasting, 395, 546, 546 (fig.), 640–641
in APA documentation style, A-30
editing, 641, 641 (fig.)
in presentations, 586–587
recording, 640–641
uploading, 641
Point sentences, in paragraph, 466, 467
Political context of use, 26
for analytical reports, 276–277
for presentation, 578
for proposals, 212
for websites, 618
Political dimensions, in technical communication, 13–15
Portfolio
assembling in binder, 341–342
electronic, 342 (fig.), 342–343
organizing, 341
professional, 339–343, 345
uses of, 345
Possession, apostrophes for, A-13–A-14
Posters
poster presentation, 309–310, 310 (fig.), 581
in reframing process, 377, 379 (fig.)
Posture, for presentation, 600
Power Translator, 462
Prefixes, hyphens with, A-17
Prepositional phrases, 459
Present perfect progressive tense, A-22
Present perfect tense, A-22
Present progressive tense, A-22
Present tense, A-22
Presentation software, 601–602, 603 (fig.)
Presentation style, 594–596
enthusiasm and, 596
persona and, 595
theme and, 595, 595 (fig.)
Presentations, 573–608
answering questions after, 593–594
audience for, 577, 583 (fig.), 592, 602, 604–606
audio and, 546, 546 (fig.), 586–587
body of, 588–591
conclusion, 591–594
delivering, 599–602
grabbers for, 585–587
for international audience, translators and, 604–606
introduction, 583–588
organizing, 582–594
overcoming fear of public speaking and, 589
patterns of arrangement, 584 (fig.), 588–591, 591 (fig.)
planning and researching, 574–579
practice and rehearsal for, 602–604
preparation for, 575 (fig.)
slides for, 585 (fig.), 592 (fig.), 597 (fig.), 598–599, 599 (fig.), 601
subject of, 576, 583–587, 592
technology for, 579–582, 580 (fig.)
time management, 578–579, 579 (fig.)
visuals for, 596–599, 607
Primary audience (action takers), for presentations, 577
Primary readers (action takers), 22 (fig.), 23
of analytical reports, 276
of e-mail, 101
of instructions and documentation, 176
of letters and memos, 101
of proposals, 212
of technical descriptions, 138
of websites, 616, 617 (fig.)
Primary research, 386, 387, 387 (fig.), 398–404
Print documents, balancing techniques for, 491 (fig.)
Print sources
in research, 393
uses of, 396–398
Prioritization, of issues in meeting, 57
Privacy issues, 87
Problem/need/solution pattern
as document pattern of arrangement, 443, 444–446 (fig.)
for presentations, 588, 591 (fig.)
Procedures/protocols. *See also* Instructions and documentation
example, 170–172 (fig.)
features of, 167
nature of, 166
organizing, 426–428
Product names, in cross-cultural documents, 565
Professional codes, resolving ethical dilemma and, 81–82
Professional networking, 318
Professional portfolio, 339–343, 345
Profiling
of contexts of use, 26–27
of readers, 20–26, 21 (fig.)
using to own advantage, 27–28
Progress reports, 247, 249 (fig.), 262 (fig.)
Progressive tenses, A-21–A-22
Project calendar, 48, 49 (fig.), 50 (fig.), 356–357, 357 (fig.)
Project planning
logical mapping for, 219 (fig.), 354–355, 355 (fig.)
software for, 48, 49 (fig.)
Pronouns
brackets to replace, A-18
case error (*I* and *me, we* and *us*), A-6–A-7
pronoun-antecedent disagreement, A-5
vague, A-7–A-8
Proofreading, 10, 552, 558–562
of application letter, 339
Track Changes, 557, 557 (fig.)
Proposals, 205–234
categories of, 206
conclusion of, 226, 227 (fig.)
context of use, 212
costs and benefits section of, 226, 227 (fig.)
current situation section of, 214–217
design, 229–234
elements of successful, 228
elevator pitch, 233–234
features of, 206, 207
grant, 206, 376
graphics in, 229, 231 (fig.)
legal and ethical issues in, 212
logical mapping for, 212, 213 (fig.), 214, 216 (fig.), 230 (fig.)
organizing and drafting of, 213–226
planning of, 206–212
project plan for, 218–219, 220–223 (fig.)
purpose of, 206
qualification sections of, 219–225, 224–225 (fig.)
researching, 212–213
solicited, 206, 208–210 (fig.)
style of, 229
tone in, 229
unsolicited, 206
writing persuasive, 376
Proprietary information, ethics and, 87
Protocols. *See* Procedures/protocols

Public speaking. *See also* Presentations
overcoming fear of, 589
Publication Manual of the American Psychological Association, 414
Pullouts, for design balance, 491
Punctuation
copyediting, 559–560
physical characteristics, 560 (fig.)
Punctuation guide, A-8–A-18
apostrophe, A-13–A-14
brackets, A-18
colons, A-11–A-13
commas, A-9–A-11
dashes, A-16–A-17
ellipses, A-18
exclamation point, A-9
hyphens, A-16–A-17
parentheses, A-17–A-18
period, A-9
question marks, A-9
quotation marks, A-9
quotes, A-9
semicolons, A-11–A-12
Puns, in cross-cultural documents, 464
Purpose
of activity reports, 259
of analytical reports, 270, 272–276
defining, 5, 6 (fig.), 6–7
of e-mail, 100–101, 102
in editing process, 553
as element of rhetorical situation, 5 (fig.), 6, 100–101
of home page, 622
of instructions and documentation, 169, 181
in introduction, 431
of letters and memos, 100–101, 102
of presentations, 577, 587
of proposals, 211–212
of technical descriptions, 138, 145
of website, 616

Qualifications sections, of proposals, 219–225
Qualitative research, 398
Quality, improvement of, 63, 566
Quality control, editing as, 552
Quality feedback loop, 63, 566
Quantitative research, 398
Question mark, 560 (fig.), A-9
Questionnaires, for research, 398–399, 401–402 (fig.)
Questions. *See also* Five-W and How Questions
answering after presentation, 593–594
closed-ended/open-ended, 398–399
in interview, 344–345
research, 389–390
survey, 281, 398–399, 401–402 (fig.)
in usability testing, 567–568, 568 (fig.)
Quotation marks, 560 (fig.), A-14–A-16
colons and semicolons with, A-12
commas with, A-10
with in-text citations, A-16
for irony, A-15–A-16
in quotations, 409–410
quotes and, A-14–A-15
for title within a quote, A-15
Quotations
documenting, A-24
ellipses with, A-18
from experts, 372
in introduction, 584
periods, exclamation points, and question marks with, A-9
quotation marks for, A-14–A-15
recording, 409–410

Radio programs
in APA documentation style, A-30
in CSE documentation style, A-34
in MLA documentation style, A-39
for research, 395
Raine, Kristy, 390–391
Read-and-locate tests, 568
Reader Analysis Chart, 25 (fig.), 25–26, 27
Readers. *See also* Audience; Primary readers (action takers); Secondary readers (advisors); Tertiary readers (evaluators)
of activity reports, 259
of analytical reports, 276
analyzing, 20–26, 511
categories of, 22 (fig.), 22–23
cross-cultural, for instructions, 179–180
of e-mail, 101
in editing process, 553
as element of rhetorical situation, 5 (fig.), 6
general public versus experts as, 29–32 (fig.)
identifying, 22–23
of instructions and documentation, 176–177, 182
of letters and memos, 101
nature of, 20–26
needs, values, and attitudes of, 23–26
profiling, 20–26, 22 (fig.)
of proposals, 212
as raiders for information, 9, 21
of technical communication, 12, 20–26, 101
of technical descriptions, 138
of website, 616–617
Writer-Centered Chart for identifying, 22 (fig.), 22–23
Reader's Guide to Periodical Literature, 397–398
Reading, direction of, 35, 35 (fig.), 510, 510 (fig.)
Reasoning
document using, 368–369 (fig.)
with examples and evidence, 367, 371–372
with logic, 367, 367 (fig.), 370–371
persuading with, 367–372, 380
Recipient address, on envelope, 120, 120 (fig.)
Recommendation reports, 272
Recordings
in APA documentation style, A-30
in CSE documentation style, A-34
in MLA documentation style, A-39
Redundancy, 459
Reference librarians, in research process, 390–391
Reference list, 413
in APA documentation style, 415, 415 (fig.), A-27–A-30
in CSE documentation style, A-31–A-35
MLA works cited list and, A-37–A-39
Reference materials, as research sources, 398
Reframing, 377, 379 (fig.)
Refusal letters and memos, 114, 115 (fig.)
Rehearsal
to overcome fear of public speaking, 589
of presentation, 589, 602–604
Reliability, of sources, 387, 387 (fig.), 393–394, 396, 417–418, 420
Religion
religious references, 606–607
religious symbols, 544
Reply text, e-mail, 98
Reports. *See* Activity reports; Analytical reports
Research, 385–404
for activity reports, 254–260
for analytical reports, 277–282
beginning, 388–389
bias in, 417–419
in cross-cultural communication, 36
into culture, 37
defining subject, 388–389
electronic sources for, 394–396, 395 (fig.)

SAMPLE DOCUMENTS

Memos

Agenda *55*

Memos *98, 103–104, 122, 128–129, 131*

Memo That Does Not Consider Tertiary Readers *24*

Memo of Transmittal *110*

"Smoking Gun" Memo *427*

Work Plan *51–53*

Letters

Adjustment Letter *113*

Claim Letter *112*

Envelope Formatting *120*

Indirect Approach Letter *449*

Letter *97*

Letter of Application Emphasizing Education *336*

Letter of Application Emphasizing Work Experience *337*

Letter to F. D. Roosevelt About the Atom Bomb *70–71*

Letter Formatting *119*

Letter of Inquiry *108*

Letter That Uses Reasoning *368–369*

Professional Biography *348*

Refusal Letter *115*

Response Letter *109*

Thank You Letter *346*

Brochures and Booklets

Booklet That Appeals to Values *375*

Brochure with a Balanced Design *484*

Brochure with Good Design *14*

Technical Definitions

Extended Definition: Print *157–158*

Specification *152–153*

Webpages

Basic Page *625*

Blog Host Site *639*

Electronic Portfolio: Welcome Page *342*

Google Docs Interface *298*

Home Pages *211, 434, 613, 622*

Home Page with a Balanced Interface *486*

Home Page Containing Multiple Balancing Techniques *492*

Home Page with a Good Design *10*

Home Page That Demonstrates the Six Moves of an Introduction *434*

Home Page That Uses Alignment and Grouping Well *496*

Home Page That Uses Contrast *505*

Internet Search Engine *395*

Library Search Engine *397*

Model Webpage *12*

Node Page *624*

Online Translation Tool *463*

Research Database *396*

Social Networking Home Page *637*

Translating a Website with Google Translate *619*

Webpage That Uses Plain Language *460*

Webpage That Uses Right-to-Left Interface Design *510*

Well-Designed Interface *628*

Evaluations

Evaluation Form for Presentations *605–606*

Team Performance Review *64*

Résumés

Chronological Résumés *322, 323*

Chronological Résumé: Student Who Returned to College *324–325*

Functional Résumé *330*

Professional Biography *348*

Scannable Résumé *334*

improving the style, 4, 4 (fig.), 9
organizing and drafting, 4, 4 (fig.), 7
overview of stages, 4, 4 (fig.)
planning and researching, 4, 4 (fig.), 5–7
revising and editing, 4, 4 (fig.), 9–10
team member responsibilities for, 48
for work plan, 48–49, 51–53 (fig.)
Written text
reinforced with graphics, 525, 526 (fig.), 533 (fig.)
tables versus, 523 (fig.)
Yahoo!, 356–357
"You" style
application letter, 335–338
command voice in instructions, 184, 201
in letters and memos, 116
in online persuasion, 373
Yuan, Rue, 565